Lecture Notes in Artificial Intelligence 1611

Subseries of Lecture Notes in Computer Science
Edited by J. G. Carbonell and J. Siekmann

Lecture Notes in Computer Science

Edited by G. Goos, J. Hartmanis and J. van Leeuwen

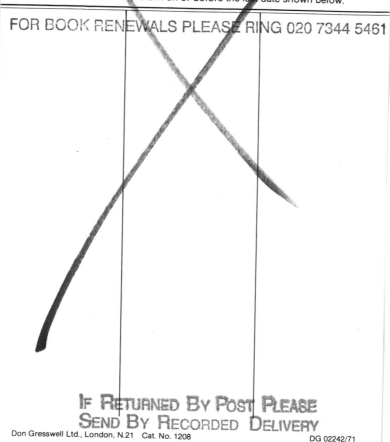

Springer
Berlin
Heidelberg
New York
Barcelona
Hong Kong
London
Milan
Paris
Singapore
Tokyo

Ibrahim Imam   Yves Kodratoff
Ayman El-Dessouki   Moonis Ali (Eds.)

Multiple Approaches to Intelligent Systems

12th International Conference
on Industrial and Engineering Applications
of Artificial Intelligence and Expert Systems
IEA/AIE-99
Cairo, Egypt, May 31 - June 3, 1999
Proceedings

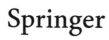 Springer

Series Editors

Jaime G. Carbonell, Carnegie Mellon University, Pittsburgh, PA, USA
Jörg Siekmann, University of Saarland, Saarbrücken, Germany

Volume Editors

Ibrahim F. Imam
Thinking Machines Corporation
16 New England Executive Park, Burlington, MA 01803, USA
E-mail: ifi@think.com

Yves Kodratoff
Université Paris-Sud, LRI, Bâtiment 490
F-91405 Orsay, Paris, France
E-mail: yk@lri.fr

Ayman El-Dessouki
Electronics Research Institute, National Research Center Building
El-Tahrir Street, Dokki, Giza, Egypt
E-mail: ayman@eri.sci.eg

Moonis Ali
Southwest Texas State University, Department of Computer Science
601 University Drive, San Marcos, TX 78666-4616, USA
E-mail: ma04@swt.edu

Cataloging-in-Publication data applied for

Die Deutsche Bibliothek - CIP-Einheitsaufnahme

Multiple approaches to intelligent systems : proceedings / 12th International
Conference on Industrial and Engineering Applications of Artificial
Intelligence and Expert Systems, IEA/AIE-99, Cairo, Egypt, May 31 - June 3,
1999. Ibrahim Imam ... (ed.). - Berlin ; Heidelberg ; New York ; Barcelona ;
Hong Kong ; London ; Milan ; Paris ; Singapore ; Tokyo : Springer, 1999
(Lecture notes in computer science ; Vol. 1611 : Lecture notes in artificial
intelligence)
ISBN 3-540-66076-3

CR Subject Classification (1998): I.2, J.2, J.6

ISBN 3-540-66076-3 Springer-Verlag Berlin Heidelberg New York

© Springer-Verlag Berlin Heidelberg 1999
Printed in Germany

Typesetting: Camera-ready by author
SPIN 10705131   06/3142 – 5 4 3 2 1 0   Printed on acid-free paper

Preface

We never create anything,
We discover and reproduce.

The Twelfth International Conference on Industrial and Engineering Applications of Artificial Intelligence and Expert Systems has a distinguished theme. It is concerned with bridging the gap between the academic and the industrial worlds of Artificial Intelligence (AI) and Expert Systems. The academic world is mainly concerned with discovering new algorithms, approaches, and methodologies; however, the industrial world is mainly driven by profits, and concerned with producing new products or solving customers' problems. Ten years ago, the artificial intelligence research gap between academia and industry was very broad. Recently, this gap has been narrowed by the emergence of new fields and new joint research strategies in academia. Among the new fields which contributed to the academic-industrial convergence are knowledge representation, machine learning, searching, reasoning, distributed AI, neural networks, data mining, intelligent agents, robotics, pattern recognition, vision, applications of expert systems, and others. It is worth noting that the end results of research in these fields are usually products rather than empirical analyses and theoretical proofs. Applications of such technologies have found great success in many domains including fraud detection, internet service, banking, credit risk and assessment, telecommunication, etc. Progress in these areas has encouraged the leading corporations to institute research funding programs for academic institutes. Others have their own research laboratories, some of which produce state of the art research.

As this conference marks the end of the 20th century and the beginning of a new century, we have to think very seriously about this problem, which exists in fields that are less demanding on our daily life. It is true that the most important factor of the academic-industrial convergence is the individual demand for technology. For example, medical research, automobile research, and food research are closely associated with the industrial world. Moreover, the success of such research depends highly on the products produced. For this reason, it has been more difficult to achieve academic-industrial convergence in mathematical research and other pure science fields.

The industrial world of artificial intelligence is growing rapidly. A very high percentage of today's corporations utilize AI in their products. We expect by early next century, AI will be embraced in some way in all machinery products. It is our view that as this becomes true, academic-industrial research will increasingly converge. Most current attempts to converge academic-industrial research cover only one type of convergence. The other possible alternative is to utilize AI for enhancing demanded products rather than building AI products to be demanded. To achieve such approach of convergence, competent management is necessarily needed. There are many advantages and disadvantages on both sides. We list some of them here to be considered, enhanced, or ignored.

Advantages:
- Upgrading technology in industry is usually driven by customer needs.
- Research in academia is mature and competition is very comprehensive.
- The technology installment in industrial corporations is usually based on academic research.

Disadvantages:
- Competitive marketing forces corporations to hide their technology.
- Publications have no effect on carrier advancement in industry, and in many cases, it is prohibited.
- Fulfilling a graduate degree or earning a promotion often drives the advance of technology in academia.
- Even though the technology installment in industry is usually based on academic research, industrial technology rarely follows the state of the art in academic research.
- The requirements of the Doctoral of Philosophy degree do not include any industrial standards.

Recommendations:
Finally, here are our recommendations for such convergence:
- Increasing the industrial support to academic research.
- Introducing a new Doctoral degree (not in philosophy) that is oriented toward solving real world-problems.
- Educating conference organizers to allow research that hides industrial technology, which may influence the market (an on-site demo may be requested in return).
- Computer Science Departments should require industrial involvement in all graduate advisory committees and prepare an industrial oriented program for corporate employees.
- Industrial corporations should encourage their employees to complete their graduate study.

The 12th International Conference on Industrial and Engineering Applications of Artificial Intelligence and Expert Systems (IEA/AIE-99) attracted many researchers and engineers from all over the world. Over 140 papers were submitted to the conference. These proceedings contain 91 papers from 32 countries. Most of the papers in this volume present applications in domains that are mentioned above. It is worth mentioning that one third of the papers in this volume are concerned with Intelligent Agents, Expert Systems, and Pattern Recognition. One can notice the strong presence of academia. This may also reflect the growing rate of industrial applications in academia.

On behalf of the organizing committee, we would like to thank all those who contributed to the 12th International Conference on Industrial and Engineering Applications of Artificial Intelligence and Expert Systems. We would like to thank the program committee members for their valuable assistance in reviewing the submitted papers. We would like to extend our thanks to the auxiliary reviewers. We thank Nihal Abosaif for her assistance in editing these proceedings, and reviewing their contents. We would like to thank Cheryl Morriss for handling the administration aspects of the conference. We thank Vladan Devedzic, Hans W. Guesgen, Howard Hamilton, Gholamreza Nakhaeizadeh, Michael Schroeder, and Jan Treur for their professional organizational effort.

We especially thank all the sponsor organizations including the International Society of Applied Intelligence (ISAI), Association for Computing Machinery (ACM/SIGART), the American Association for Artificial Intelligence (AAAI), Canadian Society for Computational Studies of Intelligence (CSCSI); Institution of Electrical Engineers (IEE); International Neural Network Society (INNS); Japanese Society of Artificial Intelligence (JSAI); and Southwest Texas State University (SWT). We also thank Thinking Machines Corporation, the Electronic Research Institute of Egypt, and all supportive organizations. We would also like to thank the Harvey travel agency for local and international arrangement of the conference. A special thanks to Ahmed Kadry El-Zoheiry and Hesham El-Badry.

Finally, we thank all authors who submitted papers to the IEA/AIE-99.

Ibrahim F. Imam
Yves Kodratoff
Ayman El-Dessouki
Moonis Ali

May 1999

The 12th International Conference on Industrial and Engineering Applications of Artificial Intelligence and Expert Systems IEA/AIE-99

Le Meridien Cairo, Cairo, Egypt, May 31- June 3, 1999

Sponsored by:
International Society of Applied Intelligence
Organized in Cooperation with:
AAAI, ACM/SIGART, CSCSI, IEE, INNS, JSAI, SWT

General Chair
Moonis Ali, Southwest Texas State University, USA

Program Chairs
Ayman El-Dessouki, Electronics Research Institute, Egypt
Ibrahim F. Imam, Thinking Machines Corporation, USA
Yves Kodratoff, Université Paris Sud, France

Organizing Committee
Ashraf H. Abdel-Wahab, Electronics Research Institute, Egypt
Nihal Y. Abosaif, Harvard University, USA & Cairo University, Egypt
Moonis Ali, Southwest Texas State University, USA
Vladan Devedzic, University of Belgrade, Yugoslavia
Ayman El-Dessouki, Electronics Research Institute, Egypt
Howard Hamilton, University of Regina, Canada
Ibrahim F. Imam, Thinking Machines Corporation, USA
Maged F. Imam, Cairo University, Egypt
Yves Kodratoff, Université Paris Sud, France
Cheryl Morriss, Southwest Texas State University, USA
Gholamreza Nakhaeizadeh, Daimler-Benz AG, Germany
Michael Schroeder, City University, UK
Alaa F. Sheta, Electronics Research Institute, Egypt
Jan Treur, Vrije Universiteit, The Netherlands

Local Committee
Local Chair: Ashraf H. abdel Wahab, Electronics Research Institute, Egypt
Publicity Chair: Alaa F. Sheta, Electronics Research Institute, Egypt
Maged F. Imam, Cairo University, Egypt
Registration Chair: Cheryl Morriss, Southwest Texas State University, USA

Program Committee
Frank D. Anger, National Science Foundation, USA
Grigoris Antoniou, Griffith University, Australia
Osman Badr, Ain Shams University, Egypt
Senen Barro Ameneiro, University Santiago de Comp, Spain
Kai H. Chang, Auburn University, USA
Luca Chittaro, University di Udine, Italy
Roger Debreceny, Nanyang Technological University, Singapore
Robert Engels, University of Karlsruhe, Germany
Ibrahim Farag, Cairo University, Egypt
Klaus Fischer, DFKI GmbH, Germany
Hans W. Guesgen, University of Auckland, New Zealand
Howard Hamilton, University of Regina, Canada
Gerard Lacey, Trin. College Dublin, Ireland
David Leake, Indiana University, USA
Manton Matthews, University of South Carolina, USA
Richard B. Modjeski, Florida Institute of Technology, USA
L. Monostori, Hungarian Academy of Sciences, Hungary
Angel P. del Pobil, Jaume-I University, Spain
Don Potter, University of Georgia, USA
Gokul C. Prabhakar, Bell Laboratories, USA
Claude Sammut, University of New South Wales, Australia
Samir I. Shaheen, Cairo University, Egypt
Ron Sun, University of Alabama, USA
Jan Treur, Vrije Universiteit Amsterdam, The Netherlands
Yasushi Umeda, University of Tokyo, Japan
Gerhard Widmer, Austrian Research Institute, Austria
Mike Wooldridge, Queen Mary & Westfield College, UK

Auxiliary Reviewers
Alberto Bugarin, Universidade de Santiago de Compostela, Spain
Purificacion Carinena, Universidade de Santiago de Compostela, Spain
Maria J. Carreira, Universidade de Santiago de Compostela, Spain
Mohamed S. El Sherif, Electronics Research Institute, Egypt
Paulo Felix, Universidade de Santiago de Compostela, Spain
Roberto Iglesias, Universidade de Santiago de Compostela, Spain
Samia Mashali, Electronics Research Institute, Egypt
Rajkumar Thirumalainambi, University of Auckland, New Zealand

Table of Contents

Genetic Algorithms

Search

Reasoning

Expert Systems and Applications

Case-Base Reasoning

Intelligent Agents

Distributed AI

Pattern Recognition and Vision

Machine Learning

Knowledge Representation

Planning and Scheduling

Tutoring and Manufacturing Systems

Intelligent Software Engineering

Formalisations of Uncertain Reasoning

Flávio Soares Corrêa da Silva

Instituto de Matemática e Estatística da Universidade de São Paulo
Cidade Universitária "ASO" – São Paulo (SP) – Brazil
Email: fcs@ime.usp.br

*Extended abstract for the invited talk at the International Conference on Industrial &
Engineering Applications of Artificial Intelligence & Expert Systems, IEA/AIE-99:*

> *"Something has existed from eternity"*
> *Samuel Clarke (1675-1729)*
> *As quoted by George Boole (1854)*

Introduction

The proper construction of a knowledge-based system has as a pre-requisite the (in most cases logic-based) formalisaton of the problem domain knowledge and inference procedures related to the system. This is a necessary step if one desires to convey scientific accreditation to the results eventually obtained from the system and used for solving problems.

As proposed by Curry [2], "the first stage in formalisation is the formulation of the discipline (a.k.a. problem domain, in our terminology) as a deductive theory, with such exactness that the correctness of its basic inferences can be judged objectively by examining the language in which they are expressed."

We have some possible interpretations for "formulation" and for "objective judgement":

- formulation: we may be interested in a formal description of the problem domain itself, thus providing our language with the appropriate machinery to solve problems in that domain, or we may be interested in a formal description of the problem-solving procedures, and in this latter case we must provide the language with expressive power to provide explicit representations of those procedures. These two interpretations were taken into account in [2], and they were also the subject matter of [1], in which issues related specifically to uncertain reasoning were discussed.

- objective judgement: at least two criteria can be taken into account for judgement of the appropriateness of results drawn through inference procedures from a deductive language. The results may be deemed appropriate given the mathematical well-formedness of the language (that ensure, for example, closure and non-ambiguity of the inference procedures, as well as consistent mapping from objects of the language to semantic objects of its interpreting models) or – quite independently – given the high correlation between them and the observable phenomena that stimulated the construction of the language.

The latter is harder to ensure, since it is domain-specific and so will also be the methods to ascertain it. Within the realm of uncertain reasoning, nevertheless, many authors have proposed general classifications of the sorts of uncertainty that one can face when building a knowledge-based system, in order to obtain proto-methodologies to serve as guidance for the design of systems for real applications, or to serve as basis for development of general theories of uncertainty [3, 4, 5, 6, 7].

In [5] we find a tree-based classification "aimed at classifying uncertainty in terms of those aspects which are most relevant to the development of AI applications", which we reproduce in figure 1.

Confidence Propensity is related to subjective degrees of belief; Equivocation and Ambiguity are respectively related to overdeterminacy and underdeterminacy of truth-values; anomalous systems are those systems that can contain errors in their construction.

To each class of uncertainty can be associated a standard approach to revise or to extend formal systems, and in this presentation we intend to bring for discussion a personal view of how the approaches for formal treatment of uncertainty can be characterised.

Our working definition of formal system shall be:

- Two formal languages, with their corresponding alphabets and grammars to build well-formed sentences, heretofore called the syntactical language and the semantical language;
- A collection of rules to produce new sentences from given ones for the syntactical language, called the inference system; and
- A class of homomorphic mappings from the syntactical to the semantical language, called the interpretations of the syntactical language in terms of the semantical language.

In order to make the discussion more concrete, we propose well-known representatives for each class of uncertainty. These representatives are shown in *italic* in figure 1.

- **Unary**
- **Ignorance**
- **Vagueness** *(fuzzy logics)*
- **Confidence Propensity** *(bayesian and Dempster-Shafer theories)*
- **Conflict**
- **Equivocation** *(lattice-based multiple-valued logics)*
- **Ambiguity** *(default logics)*
- **Anomaly (error)**
- **Set-theoretic**
- **Ignorance**
- **Incompleteness** *(paracomplete logics)*
- **Irrelevance** *(revision systems)*
- **Conflict**
- **Inconsistency** *(paraconsistent logics)*

Anomaly (error)

Figure 1: Uncertainty Classification for AI Systems (adapted from [5])

Starting from classical propositional and first-order logics, we propose systematic transformations of formal systems to accommodate uncertainties of each class.

References

[1] F. S. Corrêa da Silva. On Reasoning With and Reasoning About Uncertainty in Artificial Intelligence. In European Summer Meeting of the Association of Symbolic Logic (abstract published in The Bulletin of Symbolic Logic, v 3 (2), pp. 255-256). Spain. 1996.

[2] M. B. Curry. Foundations of Mathematical Logic. McGraw-Hill. 1977.

[3] T. L. Fine. Theories of Probability. Academic Press. 1973.

[4] I. Hacking. The Emergence of Probability. Cambridge University Press. 1975.

[5] P. Krause and D. Clark. Representing Uncertain Knowledge – An Artificial Intelligence Approach. Kluwer. 1993.

[6] R. Kruse, E. Schwecke and J. Heinsohn. Uncertainty and Vagueness in Knowledge-based Systems – Numerical Methods. Springer-Verlag. 1991.

[7] A. Motro and Ph. Smets (editors). Uncertainty Management in Information Systems – From Needs to Solutions. Kluwer. 1997.

Agent-Oriented Software Engineering
Nicholas R. Jennings

Dept. Electronic Engineering, Queen Mary & Westfield College,
University of London, London E1 4NS, UK.
n.r.jennings@qmw.ac.uk

*Extended abstract for the invited talk at the International Conference on Industrial &
Engineering Applications of Artificial Intelligence & Expert Systems, IEA/AIE-99:*

1. Introduction

Increasingly many computer systems are being viewed in terms of autonomous agents. Agents are being espoused as a new theoretical model of computation that more closely reflects current computing reality than Turing Machines. Agents are being advocated as the next generation model for engineering complex, distributed systems. Agents are also being used as an overarching framework for bringing together the component AI sub-disciplines that are necessary to design and build intelligent entities. Despite this intense interest, however, a number of fundamental questions about the nature and the use of agents remain unanswered. In particular:

- what is the essence of agent-based computing?
- what makes agents an appealing and powerful conceptual model?
- what are the drawbacks of adopting an agent-oriented approach?
- what are the wider implications for AI of agent-based computing?

These questions can be tackled from many different perspectives; ranging from the philosophical to the pragmatic. This paper proceeds from the standpoint of using agent-based software to solve complex, real-world problems. Building high quality software for complex real-world applications is difficult. Indeed, it has been argued that such developments are one of the most complex construction tasks humans undertake (both in terms of the number and the flexibility of the constituent components and in the complex way in which they are interconnected). Moreover, this statement is true no matter what models and techniques are applied: it is a consequence of the "essential complexity of software" [2]. Such complexity manifests itself in the fact that software has a large number of parts that have many interactions [8]. Given this state of affairs, the role of software engineering is to provide models and techniques that make it easier to handle this complexity. To this end, a wide range of software engineering paradigms have been devised (e.g. object-orientation [1] [7], component-ware [9], design patterns [4] and software architectures [3]). Each successive development either claims to make the engineering process easier or to extend the complexity of applications that can feasibly be built. Although evidence is emerging to support these claims, researchers continue to strive for more efficient and powerful techniques, especially as solutions for ever more demanding applications are sought.

In this article, it is argued that although current methods are a step in the right direction, when it comes to developing complex, distributed systems they fall short in one of three main ways: (i) the basic building blocks are too fine grained; (ii) the interactions are too rigidly defined; or (iii) they posses insufficient mechanisms for dealing with organisational structure. Furthermore, it will be argued that: *agent-oriented approaches can significantly enhance our ability to model, design and build complex (distributed) software systems.*

2. The Essence of Agent-Based Computing

The first step in arguing for an agent-oriented approach to software engineering is to precisely identify and define the key concepts of agent-oriented computing. Here the key definitional problem relates to the term *"agent"*. At present, there is much debate, and little consensus, about exactly what constitutes agenthood. However, an increasing number of researchers find the following characterisation useful [10]:

an agent is an encapsulated computer system that is situated in some environment, and that is capable of flexible, autonomous action in that environment in order to meet its design objectives

There are a number of points about this definition that require further explanation. Agents are: (i) clearly identifiable problem solving entities with well-defined boundaries and interfaces; (ii) situated (embedded) in a particular environment—they receive inputs related to the state of their environment through sensors and they act on the environment through effectors; (iii) designed to fulfil a specific purpose—they have particular objectives (goals) to achieve; (iv) autonomous—they have control both over their internal state and over their own behaviour; (v) capable of exhibiting flexible problem solving behaviour in pursuit of their design objectives—they need to be both reactive (able to respond in a timely fashion to changes that occur in their environment) and proactive (able to opportunistically adopt new goals) [11].

When adopting an agent-oriented view of the world, it soon becomes apparent that most problems require or involve multiple agents: to represent the decentralised nature of the problem, the multiple loci of control, the multiple perspectives, or the competing interests. Moreover, the agents will need to interact with one another: either to achieve their individual objectives or to manage the dependencies that ensue from being situated in a common environment. These interactions can vary from simple information interchanges, to requests for particular actions to be performed and on to cooperation, coordination and negotiation in order to arrange inter-dependent activities. Whatever the nature of the social process, however, there are two points that qualitatively differentiate agent interactions from those that occur in other software engineering paradigms. Firstly, agent-oriented interactions occur through a high-level (declarative) agent communication language. Consequently, interactions are conducted at the knowledge level [6]: in terms of which goals should be followed, at what time, and by whom (cf. method invocation or function calls that operate at a purely syntactic level). Secondly, as agents are flexible problem solvers, operating in an environment over which they have only partial control and observability, interactions need to be handled in a similarly flexible manner. Thus, agents need the computational apparatus to make context-dependent decisions about the nature and scope of their interactions and to initiate (and respond to) interactions that were not foreseen at design time.

In most cases, agents act to achieve objectives either on behalf of individuals/companies or as part of some wider problem solving initiative. Thus, when agents interact there is typically some underpinning organisational context. This context defines the nature of the relationship between the agents. For example, they may be peers working together in a team or one may be the boss of the others. In any case, this context influences an agent's behaviour. Thus it is important to explicitly represent the relationship. In many cases, relationships are subject to ongoing change: social interaction means existing relationships evolve and new relations are created. The temporal extent of relationships can also vary significantly: from just long enough to deliver a particular service once, to a permanent bond. To cope with this variety and dynamic, agent researchers have: devised protocols that enable organisational groupings to be formed and disbanded, specified mechanisms to ensure groupings act together in a coherent fashion, and developed structures to characterise the macro behaviour of collectives [5] [11].

3. Agent-Oriented Software Engineering

The most compelling argument that could be made for adopting an agent-oriented approach to software development would be to have a range of quantitative data that showed, on a standard set of software metrics, the superiority of the agent-based approach over a range of other techniques. However such data simply does not exist. Hence arguments must be qualitative in nature.

The structure of the argument that will be used here is as follows. On one hand, there are a number of well-known techniques for tackling complexity in software. Also the nature of complex software systems is (reasonably) well understood. On the other hand, the key characteristics of the agent-based paradigm have been elucidated. Thus an argument can be made by examining the degree of match between these two perspectives.

Before this argument can be made, however, the techniques for tackling complexity in software need to be introduced. Booch identifies three such tools [1]:

- *Decomposition:* The most basic technique for tackling any large problem is to divide it into smaller, more manageable chunks each of which can then be dealt with in relative isolation.
- *Abstraction:* The process of defining a simplified model of the system that emphasises some of the details or properties, while suppressing others.
- *Organisation*[1]: The process of identifying and managing interrelationships between various problem-solving components.

---

1 Booch actually uses the term "hierarchy" for this final point [1]. However, the more neutral term "organisation" is used here.

Next, the characteristics of complex systems need to be enumerated [8]:

- Complexity frequently takes the form of a hierarchy. That is, a system that is composed of inter-related sub-systems, each of which is in turn hierarchic in structure, until the lowest level of elementary sub-system is reached. The precise nature of these organisational relationships varies between sub-systems, however some generic forms (such as client-server, peer, team, etc.) can be identified. These relationships are not static: they often vary over time.
- The choice of which components in the system are primitive is relatively arbitrary and is defined by the observer's aims and objectives.
- Hierarchic systems evolve more quickly than non-hierarchic ones of comparable size. In other words, complex systems will evolve from simple systems more rapidly if there are *stable intermediate forms*, than if there are not.
- It is possible to distinguish between the interactions *among* sub-systems and the interactions *within* sub-systems. The latter are both more frequent (typically at least an order of magnitude more) and more predictable than the former. This gives rise to the view that complex systems are *nearly decomposable*: sub-systems can be treated almost as if they are independent of one another, but not quite since there are some interactions between them. Moreover, although many of these interactions can be predicted at design time, some cannot.

With these two characterisations in place, the form of the argument can be expressed: (i) show agent-oriented decompositions are an effective way of partitioning the problem space of a complex system; (ii) show that the key abstractions of the agent-oriented mindset are a natural means of modelling complex systems; and (iii) show the agent-oriented philosophy for dealing with organisational relationships is appropriate for complex systems.

Merits of Agent-Oriented Decomposition

Complex systems consist of a number of related sub-systems organised in a hierarchical fashion. At any given level, sub-systems work together to achieve the functionality of their parent system. Moreover, within a sub-system, the constituent components work together to deliver the overall functionality. Thus, the same basic model of interacting components, working together to achieve particular objectives occurs throughout the system.

Given this fact, it is entirely natural to modularise the components in terms of the objectives they achieve[2]. In other words, each component can be thought of as achieving one or more objectives. A second important observation is that complex systems have multiple loci of control: "real systems have no top" [7] pg. 47. Applying this philosophy to objective-achieving decompositions means that the individual components should localise and encapsulate their own control. Thus, entities should

2 Indeed the view that decompositions based upon functions/actions/processes are more intuitive and easier to produce than those based upon data/objects is even acknowledged within the object-oriented community [7] pg. 44.

have their own thread of control (i.e. they should be active) and they should have control over their own choices and actions (i.e. they should be autonomous).

For the active and autonomous components to fulfil both their individual and collective objectives, they need to interact with one another (recall complex systems are only nearly decomposable). However the system's inherent complexity means it is impossible to know *a priori* about all potential links: interactions will occur at unpredictable times, for unpredictable reasons, between unpredictable components. For this reason, it is futile to try and predict or analyse all the possibilities at design-time. It is more realistic to endow the components with the ability to make decisions about the nature and scope of their interactions at run-time. From this, it follows that components need the ability to initiate (and respond to) interactions in a flexible manner.

The policy of deferring to run-time decisions about component interactions facilitates the engineering of complex systems in two ways. Firstly, problems associated with the coupling of components are significantly reduced (by dealing with them in a flexible and declarative manner). Components are specifically designed to deal with unanticipated requests and they can spontaneously generate requests for assistance if they find themselves in difficulty. Moreover because these interactions are enacted through a high-level agent communication language, coupling becomes a knowledge-level issue. This, in turn, removes syntactic concerns from the types of errors caused by unexpected interactions. Secondly, the problem of managing control relationships between the software components (a task that bedevils traditional functional decompositions) is significantly reduced. All agents are continuously active and any coordination or synchronisation that is required is handled bottom-up through inter- agent interaction. Thus, the ordering of the system's top-level goals is no longer something that has to be rigidly prescribed at design time. Rather, it becomes something that is handled in a context-sensitive manner at run-time.

From this discussion, it is apparent that a natural way to modularise a complex system is in terms of multiple, interacting, autonomous components that have particular objectives to achieve. In short, agent-oriented decompositions aid the process of developing complex systems.

Appropriateness of Agent-Oriented Abstractions

A significant part of the design process is finding the right models for viewing the problem. In general, there will be multiple candidates and the difficult task is picking the most appropriate one. When designing software, the most powerful abstractions are those that minimise the semantic gap between the units of analysis that are intuitively used to conceptualise the problem and the constructs present in the solution paradigm. In the case of complex systems, the problem to be characterised consists of sub-systems, sub-system components, interactions and organisational relationships. Taking each in turn:

- Sub-systems naturally correspond to agent organisations. They involve a number of constituent components that act and interact according to their role within the larger enterprise.

- The appropriateness of viewing sub-system components as agents has been made above.
- The interplay between the sub-systems and between their constituent components is most naturally viewed in terms of high-level social interactions: "at any given level of abstraction, we find meaningful collections of entities that collaborate to achieve some higher level view" [1] pg. 34. This view accords precisely with the knowledge-level treatment of interaction afforded by the agent-oriented approach. Agent systems are invariably described in terms of "cooperating to achieve common objectives", "coordinating their actions" or "negotiating to resolve conflicts".
- Complex systems involve changing webs of relationships between their various components. They also require collections of components to be treated as a single conceptual unit when viewed from a different level of abstraction. Here again the agent-oriented mindset provides suitable abstractions. A rich set of structures are typically available for explicitly representing and managing organisational relationships. Interaction protocols exist for forming new groupings and disbanding unwanted ones. Finally, structures are available for modelling collectives. The latter point is especially useful in relation to representing sub-systems since they are nothing more than a team of components working to achieve a collective goal.

Need for Flexible Management of Changing Organisational Structures

Organisational constructs are first-class entities in agent systems. Thus explicit representations are made of organisational relationships and structures. Moreover, agent-based systems have the concomitant computational mechanisms for flexibly forming, maintaining and disbanding organisations. This representational power enables agent-oriented systems to exploit two facets of the nature of complex systems. Firstly, the notion of a primitive component can be varied according to the needs of the observer. Thus at one level, entire sub-systems can be viewed as a singleton, alternatively teams or collections of agents can be viewed as primitive components, and so on until the system eventually bottoms out. Secondly, such structures provide a variety of stable intermediate forms, that, as already indicated, are essential for rapid development of complex systems. Their availability means that individual agents or organisational groupings can be developed in relative isolation and then added into the system in an incremental manner. This, in turn, ensures there is a smooth growth in functionality.

4. Conclusions and Future Work

This paper has sought to justify the claim that agent-based computing has the potential to provide a powerful suite of metaphors, concepts and techniques for conceptualising, designing and implementing complex (distributed) systems. However if this potential is to be fulfilled and agent-based systems are to reach the mainstream of software engineering, then the following limitations in the current state of the art need to be overcome. Firstly, a systematic methodology that enables developers to clearly analyse and design their applications as multi-agent systems needs to be devised. Secondly, there needs to be an increase in the number and sophistication of industrial-strength tools for building multi-agent systems. Finally, more flexible and scalable techniques need to be devised for enabling heterogeneous agents to inter-operate in open environments;

References

[1]   G. Booch (1994) "Object-oriented analysis and design with applications" Addison Wesley.

[2]   F. P. Brooks (1995) "The mythical man-month" Addison Wesley.

[3]   F. Buschmann, R. Meunier, H. Rohnert, P. Sommerlad, and M. Stahl (1998) "A System of Patterns" Wiley.

[4]   E. Gamma, R. Helm, R. Johnson and J. Vlissides (1995) "Design Patterns" Addison Wesley.

[5]   N. R. Jennings and M. Wooldridge (eds.) (1998) "Agent technology: foundations, applications and markets" Springer Verlag.

[6]   A. Newell, (1982) "The Knowledge Level" Artificial Intelligence 18 87-127.

[7]   B. Meyer (1988) "Object-oriented software construction" Prentice Hall.

[8]   H. A. Simon (1996) "The sciences of the artificial" MIT Press.

[9]   C. Szyperski (1998) "Component Software" Addison Wesley.

[10] M. Wooldridge (1997) "Agent-based software engineering" IEE Proc Software Engineering 144 26-37.

[11] M. Wooldridge and N. R. Jennings (1995) "Intelligent agents: theory and practice" The Knowledge Engineering Review 10 (2) 115-152.

[12] M. J. Wooldridge and N. R. Jennings (1998) "Pitfalls of Agent-Oriented Development" Proc 2nd Int. Conf. on Autonomous Agents (Agents-98), Minneapolis, USA, 385-391.

A Unified-Metaheuristic Framework

Ibrahim H Osman

Institute of Mathematics & Statistics
University of Kent,
Canterbury, Kent CT2 7NF
E-mail: I.H.Osman@ukc.ac.uk
And
Faculty of Administrative Sciences
Department of QM and IS
P.O. Box 5486 Safat
13055 Kuwait
E-mail: osmanih@kucOl.kuniv.edu.kw

*Extended abstract for the invited talk at the International Conference on Industrial &
Engineering Applications of Artificial Intelligence & Expert Systems, IEA/AIE-99:*

Introduction

In recent years, there have been significant advances in the theory and application of metaheuristics to approximate solutions of complex optimization problems. A metaheuristic is an iterative master process that guides and modifies the operations of subordinate heuristics to efficiently produce high quality solutions, [6] [8]. It may manipulate a complete (or incomplete) single solution or a collection of solutions at each iteration. The subordinate heuristics may be high (or low) level procedures, or a simple local search, or just a construction method. The family of metaheuristics includes, but is not limited to, Adaptive memory programming, Ants systems, Evolutionary methods, Genetic algorithms, Greedy randomised adaptive search procedures, Neural networks, Simulated annealing, Scatter search, Tabu search and their hybrids.

Metaheuristics provide decision-makers with robust tools that obtain high quality solutions, in a reasonable computational effort, to important applications in business, engineering, economics and the sciences. Finding exact solutions to these applications still poses a real challenge despite the impact of recent advances in computer technology and the great interaction between computer science, management science/operations research and mathematics. The widespread successes of metaheuristics have been demonstrated by the establishment of the Metaheuristics International Conference Series (MIC-95, MIC-97 and MIC-99), the modern-heuristics digest mailing list (1994), the existence of many sessions at the most important conferences in operational research and computer science, and the devotion of entire journals and special issues to study the advances in the theory and applications of Metaheuristics. For more details, we refer to most recent books [1] [5] [8] [9] [10] [11] [12] and the comprehensive bibliography in [7].

The primary components of these metaheuristics are the space of solutions, the neighbourhood (local search) structures, the memory learning structures, and the search engines. In this paper, we briefly review recent and efficient metaheuristics and describe for each metaheuristic its associated components. We then present a unified-metaheuristic framework initially proposed in [6]. It will be shown how the existing metaheuristics can fit into the presented framework and how it can allow for more scope and innovations. It also invites extra research into designing new and better approaches.

Moreover, we shall discuss our research experience [2] [3] [4] [13] on metaheuristics for solving successfully some hard combinatorial optimization problems, namely, the capacitated clustering problem, the weighted maximal planar graph problem and the vehicle routing problem and its variants. Finally, we conclude by highlighting current trends and future research directions in this active area of the science of heuristics.

References

[1] E.H.L. Aarts and J.K. Lenstra, *Local Search in Combinatorial Optimization*, *Wiley*, Chichester, 1997.

[2] S. Ahmadi, *Metaheuristics for the Capacitated Clustering Problem*, Ph.D. thesis, Canterbury Business School, University of Kent, U.K., 1998.

[3] M.A. Barakeh, *Approximate Algorithms for the Weighted Maximal Planar Graph*, M.Sc. thesis, Institute of Mathematics and Statistics, University of Kent, U.K., 1997.

[4] N. Kalafatis, *Metaheuristic Techniques for the Weighted Maximal Planar Graph*, M.Sc. thesis, Canterbury Business School, University of Kent, U.K. 1998.

[5] Z. Michalewicz, *Genetic Algorithms + Data Structures = Evolutionary Programs*, *Spinger*Verlag, Berlin, 1992.
Modern-heuristics http address: http://www.mailbase.ac.uk/lists/modem-heuristics/

[6] I.H. Osman, An Introduction to Metaheuristics, in: *Operational Research Tutorial Papers*, Editors: M. Lawrence and C. Wilsdon (Operational Research Society Press, Birmingham, 1995) 92-122.

[7] I.H. Osman and G. Laporte, Metaheuristics: A bibliography, in: *Metaheuristics in Combinatorial Optimization*, Annals of Operational Research, Vol. 63 (1996) 513-628.

[8] I.H. Osman and J.P. Kelly, *Metaheuristics. Theory and Applications*, Kluwer, Boston, 1996.

[9] V.J., Rayward-Smith, I.H. Osman, C.R. Reeves and G.D. Smith, *Modem Heuristic Search Methods*, Wiley, Chichester, 1996.

[10] C.R. Reeves, *Modern Heuristic Techniques for Combinatorial Problems*, Blackwell, Oxford, 1993.

[11] G. Glover, and M. Laguna, *Tabu Search*, Kluwer, Boston, 1997.

[12] S. Voss, S. Martello, I.H. Osman, and C. Roucairol, *Metaheuristics: Advances and Trends in Local Search Paradigms for Optimization*, Kluwer, Boston, 1998.

[13] N. Wassan, *Tabu Search Metaheuristic for a Class of Routing Problem*, Ph.D. thesis, Canterbury Business School, University of Kent, U.K., 1998.

A Fuzzy Knowledge Representation and Acquisition Scheme for Diagnostic Systems

Shyue-Liang Wang[1], Yi-Huey Wu[2]

[1]Department of Information Management, I-Shou University
1, Section 1, Hsueh-Cheng RD, Ta-Hsu Hsiang,
Kaohsiung, Taiwan, R.O.C.
slwang@csa500.isu.edu.tw
[2]Institute of Information Engineering, I-Shou University
1, Section 1, Hsueh-Cheng RD, Ta-Hsu Hsiang,
Kaohsiung, Taiwan, R. O.C.
m863202m@csa500.isu.edu.tw

Abstract. This paper presents a fuzzy knowledge representation, acquisition and reasoning scheme suitable for diagnostic systems. In addition to fuzzy sets and fuzzy production rules, we propose to using proximity relations for representing the interrelationship between symptoms in the antecedence of fuzzy production rules. A systematic generation method for acquiring proximity relations is proposed. An approximate reasoning algorithm based on such representation is also shown. Application to vibration cause identification in rotating machines is illustrated. Our scheme subsumes other fuzzy set based knowledge representation and reasoning approaches when proximity relation is reduced to identity relation.

1 Introduction

Production rule [2] has been one of the most successful knowledge representation methods developed. Incorporating fuzzy logic introduced by Zadch [13], fuzzy production rules (FPRs) have been widely used to represent and capture uncertain, vague and fuzzy domain expert knowledge. While FPRs have been applied to solve a wide range of problems, there have been many different forms of FPRs in order to fully represent the specific problem at hand. Certainty factor, threshold value, local and global weights are some of the parameters embedded into the fuzzy production rules. However, these representations have always assumed that the propositions in the antecedence of FPRs are independent from each other. Nevertheless, in some diagnostic problems, these propositions or symptoms may be dependent. In addition, the relationship between the propositions in antecedence may depend on the consequence in the rules.

In this paper, we present a fuzzy knowledge acquisition, representation and reasoning scheme suitable for diagnostic problems, where proximity relations, fuzzy

---

[1] To whom all correspondences should be addressed.

sets and fuzzy production rules are represented for field experience and textbook knowledge. A proximity relation is typically used to represent the degree of "closeness" or "similarity" between elements of a scalar domain. We use it to represent the interrelationship between symptoms in the antecedence. A systematic generation method for acquiring the proximity relations is also presented. In addition, our method can perform fuzzy matching using similarity measure between observed symptom manifestations and the antecedent propositions of FPR to determine the presence of consequence or causes of the symptoms.

This paper is organized as followings. Section 2 introduces concepts of fuzzy set theory, fuzzy production rules, and proximity relation. In section 3, a generation method for proximity relations is described. In section 4, the definitions of similarity function are described. In section 5, a fuzzy reasoning technique is presented. The final section describes the conclusion and future work.

2 Knowledge Representation

In this section, we review concepts of fuzzy sets, fuzzy production rules, and proximity relation.

2.1 Fuzzy Sets

The theory of fuzzy sets was proposed by Zadeh [13] in 1965. Roughly speaking, a fuzzy set is a class with fuzzy boundaries. It can be defined as follows [3]:

Definition 2.1.1: Let U be the universe of discourse and let u designate the generic element of U. A fuzzy set F of U is characterized by a membership function $\chi_F: U \rightarrow [0,1]$, which associates with each element $u \in U$ a number $\chi_F(u)$ representing the grade of membership of u in F, and is designated as: $F = \{(u , \chi_F(u))\mid u \in U \}$.
Let A and B be two fuzzy sets in the universe of discourse U, i.e.,
$U = \{u_1, u_2, \cdots, u_p\}$,
$A = \{(u_i , \chi_A(u_i))\mid u_i \in U \}$,
$B = \{(u_i , \chi_B(u_i))\mid u_i \in U \}$,
where χ_A and χ_B are the membership functions of the fuzzy sets A and B respectively.

Definition 2.1.2: The union operation between fuzzy sets A and B is designated as:
$A \cup B = \{(u_i , \chi_{A \cup B}(u_i))\mid \chi_{A \cup B}(u_i) = \text{Max} (\chi_A(u_i), \chi_B(u_i)), u_i \in U\}$.

2.2 Fuzzy Production Rule

The FPR allows the rule to contain some fuzzy quantifiers [7] (i.e., always, strong, medium, weak, etc.), and certainty factor. The types of fuzzy quantifiers and certainty factor are numerical. The definitions of fuzzy quantifiers and their corresponding numerical intervals are given in Table 1. Furthermore, the level of certainty factor and their corresponding numerical intervals are given in Table 2. The general formulation of the rule R_i adopted here, $1 \le i \le n$, is as follows:

Let R be a set of FPRs, $R = \{R_1, R_2, R_3, \cdots, R_n\}$.

$$R_i: \text{IF } S_i \text{ THEN } C_i \ (CF = \mu_i), (Th = \lambda),$$

where

(1) S_i represents the antecedent portion of R_i which may contain some fuzzy quantifiers. For machinery diagnosis applications, it may represent symptoms.

(2) C_i represents the consequent portion of R_i. For machinery diagnosis applications, it may represent causes.

(3) μ_i is the certainty factor of the rule and it represents the strength of belief of the rule.

(4) λ is threshold value which triggered for R_i.

For example, let U be a set of symptoms, $U = \{40\text{-}50\%$ oil whirl frequency, lower multiple, very high frequency$\}$, and seal rub be a concluded cause, then knowledge may be represented by the rule R_1 as follows:

R_1: IF {strong 40-50% oil whirl frequency $\wedge$ medium lower multiple $\wedge$ medium very high frequency}

THEN seal rub (CF= quite certain), λ=0.70

which can be represented as following rule when fuzzy quantifiers and certainty level are replaced by numbers, using Tables 1 and 2.

R_1: IF {(40-50% oil whirl frequency, 0.93), (lower multiple, 0.55), (very high frequency, 0.55)}

THEN seal rub (CF=0.70), λ=0.70

Table 1. Fuzzy quantifiers and their corresponding numerical intervals

Fuzzy quantifiers	Numerical intervals
always	[1.00, 1.00]
strong	[0.86, 0.99]
more or less strong	[0.63, 0.85]
medium	[0.47, 0.62]
more or less weak	[0.21, 0.46]
weak	[0.01, 0.20]
no	[0.00, 0.00]

Table 2. Certainty levels and their corresponding numerical intervals

Certainty levels	Numerical intervals
absolutely certain	[1.00, 1.00]
extremely certain	[0.94, 0.99]
very certain	[0.78, 0.93]
quite certain	[0.62, 0.77]
fairly certain	[0.46, 0.61]
more or less certain	[0.31, 0.45]
little certain	[0.16, 0.30]
hardly certain	[0.01, 0.15]
absolutely uncertain	[0.00, 0.00]

2.3 Proximity Relation

Shenoi and Melton [11] proposed proximity relation instead of similarity relation to represent analogical data defined on discrete domains. Both proximity and similarity relations have been used to represent fuzzy data in fuzzy database systems. A proximity relation is defined in the following manner.

Definition 2.3.1: A proximity relation is a mapping, p_j: $D_j \times D_j \rightarrow [0,1]$, such that for x, $y \in D_j$,

(i) $p_j(x, x) = 1$ (reflexivity),
(ii) $p_j(x, y) = p_j(y, x)$ (symmetry).

The property of reflexivity and symmetry are appropriate for expressing the degree of 'closeness' or 'proximity' between elements of a scalar domain. For machinery diagnosis applications, proximity relation of related symptoms can be observed during a period of time or is given by the field expert. A fuzzy proximity relation usually can be represented by a matrix form in a given domain. For example, Fig.1 shows the relative relations between three symptoms, rotor and stator resonant frequency, 1×running frequency, and higher multiple.

	rotor and stator resonant frequency	1×running frequency	higher multiple
rotor and stator resonant frequency	1.00	0.40	0.57
1×running frequency	0.40	1.00	0.24
higher multiple	0.57	0.24	1.00

Fig. 1. An example of proximity relation

3   Generation Method for Proximity Relations

In general, most of the fuzzy theory researchers assume that proximity relations are already given. However, knowledge acquisition such as these relations or production rules has long been recognized as a difficult task. Only few of researchers have considered to systematically generate proximity relations up to now [5, 6]. In this section, we describe a systematic generation method based on [5] for the acquisition of proximity relations for machinery diagnosis.

A proximity relation, such as Fig.1, is typically used to represent the degree of " closeness" or " similarity" between elements of a scalar domain. The determination of the similarity between domain elements is usually based on the elements' common features or characteristics. For example, the three symptoms appeared in Fig. 1 all have the characteristics of catching the causes permanent low or lost, casing distortion, and misalignment with different degrees of certainty (Table 3). Since these data are much easier to be collected or acquired, they can be used to generate the proximity relation for symptoms.

Table 3. Feature values for symptoms

Causes Symptoms	Feature		values
	permanent low or lost	casing distortion	misalignment
rotor and stator resonant frequency	0.30	0.10	0.05
1×running frequency	0.90	0.60	0.30
higher multiple	0.05	0.10	0.10

Let θ_i represent the domain element of proximity relation. For machinery diagnosis application, it may represent symptoms. Let x_j represent the feature or characteristic of symptom. For machinery diagnosis application, it represents the possible causes. Let $\mu_j(x)$ represent the feature value for x_j, $\mu_j(x) \in [0,1]$. We can then represent feature values for each symptom as following:

θ_1 (rotor and stator resonant frequency) = $\{\mu_1=0.30, \mu_2=0.10, \mu_3=0.05\}$,

θ_2 (1×running frequency) = $\{\mu_1=0.90, \mu_2=0.60, \mu_3=0.30\}$,

θ_3 (higher multiple) = $\{\mu_1=0.05, \mu_2=0.10, \mu_3=0.10\}$.

To calculate the relationship between each pair of symptoms, a geometric distance based similarity measure is adopted [9].

$$S_{\theta_i,\theta_j} = 1 - \frac{\sum\limits_{x \in X} \left| \mu_{\theta_i}(x) - \mu_{\theta_j}(x) \right|}{\sum\limits_{x \in X} \left(\mu_{\theta_i} + \mu_{\theta_j} \right)} \qquad 1 \le i, j \le n, \tag{1}$$

According to (1), the proximity relation can be calculated as following and its result is shown in Fig. 1.

$$S_{\theta_1,\theta_2} = 1 - \frac{|0.3 - 0.9| + |0.1 - 0.6| + |0.05 - 0.3|}{(0.3 + 0.9) + (0.1 + 0.6) + (0.05 + 0.3)}$$
$$= 1 - 0.6 = 0.4$$

$$S_{\theta_1,\theta_3} = 1 - \frac{|0.3 - 0.05| + |0.1 - 0.1| + |0.05 - 0.1|}{(0.3 + 0.05) + (0.1 + 0.1) + (0.05 + 0.1)}$$
$$= 1 - 0.429 = 0.571 \approx 0.57$$

$$S_{\theta_2,\theta_3} = 1 - \frac{|0.9 - 0.05| + |0.6 - 0.1| + |0.3 - 0.1|}{(0.9 + 0.05) + (0.6 + 0.1) + (0.3 + 0.1)}$$
$$= 1 - 0.756 = 0.243 \approx 0.24$$

4 Similarity Measure

According to section 2.1, let U be the universe of discourse and let A and B be two fuzzy sets of U. i.e.

$U=\{u_1, u_2, \cdots, u_p\}$,

$A=\{(u_1, a_1), (u_2, a_2), \cdots, (u_p, a_p)\}$,

$B=\{(u_1, b_1), (u_2, b_2), \cdots, (u_p, b_p)\}$.

where $a_i, b_i \in [0,1]$, and $1 \le i \le p$. A and B can be represented by the vectors $\overline{A}$ and $\overline{B}$, where

$\overline{A} = \, < a_1, a_2, a_3, \cdots, a_p >$,

$\overline{B} = \, < b_1, b_2, b_3, \cdots, b_p >$.

In this section, we propose using a similarity measure function T [9]. We use it to calculate similarity value between A and B. Assume that each u_i has the same degree of importance. The similarity measure idefined as follows:

$$S_{\overline{B},\overline{A}} = 1 - \frac{\sum_{i=1}^{n} |a_i - b_i|}{\sum_{i=1}^{n} (a_i + b_i)} \qquad (2)$$

The larger the value of $S_{\overline{B},\overline{A}}$, the higher the similarity between A and B.

Example 4-1: Assume that

$U=\{$40-50% oil whirl frequency, lower multiple, very high frequency$\}$,
$A=\{$(40-50% oil whirl frequency, 0.70), (lower multiple, 0.92), (very high frequency, 0.64)$\}$,
$B=\{$(40-50% oil whirl frequency, 0.50), (lower multiple, 0.80), (very high frequency, 0.45)$\}$.

$\overline{A} = \, < 0.70, 0.92, 0.64 >$,

$\overline{B} = \, < 0.50, 0.80, 0.45 >$.

By equation (2),

$$S_{\overline{B},\overline{A}} = 1 - \frac{|0.50 - 0.70| + |0.80 - 0.92| + |0.45 - 0.64|}{(0.50 + 0.70) + (0.80 + 0.92) + (0.45 + 0.64)}$$

$$= 1 - 0.127 = 0.872 \approx 0.87$$

Therefore, the similarity value between A and B is about 0.87.

5 A Fuzzy Reasoning Method

We propose a fuzzy reasoning method with proximity relation to infer causes of diagnostic problems. We use fuzzy matching based on similarity measure between appearance of machinery's symptom manifestations and the antecedent portion of the rule. If the value of similarity measure of fuzzy matching is larger than threshold value (λ), then the rule can be fired. Moreover, the degree of certainty for consequence C_i of a rule is equal to the value of similarity measure times certainty factor (u_i) of the rule.

According to section 2.2, let

$$\overline{S_i} = <q_1, q_2, q_3, \cdots, q_p>,$$

represent the vector of antecedent portion of R_i, $1 \le i \le n$, where q_l represents symptom's fuzzy quantifier in R_i, $1 \le l \le p$. If we assume that some symptom manifestation are given and others are unknown. Since proximity relation is used to represent the relationships between symptoms in a rule, we can then use proximity relation, which is acquired from the field experts, to assume values of unknown symptom manifestations. For each rule R_i, let $\overline{RSM_j}$ represent relationship between symptom manifestations with respect to known symptom j which can be obtained from row j of a proximity relation.

Let r_j denote the degree of manifestation regarding to a symptom, i.e., the degree of an observed symptom with respect to symptom in the rule. Then the total effect of all symptom manifestations is

$$\overline{DRSM} = <m_1, m_2, m_3, \cdots, m_p>$$

$$= \bigcup_{j=1}^{p} \overline{DRSM_j} \tag{3}$$

where $\overline{DRSM_j} = r_j * \overline{RSM_j}$.

By equation (2), the similarity between the machinery's symptom manifestations and the antecedence of the rule can be calculated as

$$S_{\overline{DRSM},\overline{S_i}} = 1 - \frac{\sum_{k=1}^{p}|m_k - q_k|}{\sum_{k=1}^{p}(m_k + q_k)} \tag{4}$$

where $S_{\overline{DRSM},\overline{S_i}} \in [0,1]$. The larger the value of $S_{\overline{DRSM},\overline{S_i}}$, the higher the similarity between RSM and S_i.

If $S_{\overline{DRSM},\overline{S_i}} \ge \lambda$, then the rule R_i can be fired; it indicates that the machinery might have the cause with the degree of certainty of about c_i, where $c_i = S_{\overline{DRSM},\overline{S_i}} * u_i$ and $c_i \in [0,1]$. If $S_{\overline{DRSM},\overline{S_i}} < \lambda$, then the rule R_i can not be fired.

Example 5-1: Assume that U is a set of symptoms and knowledge base contains the following fuzzy production rule R_1, where

$U = \{$rotor and stator resonant frequency, $1\times$running frequency, higher multiple$\}$,

R_1: IF $\{$(rotor and stator resonant frequency, 0.34), ($1\times$running frequency, 0.93), (higher multiple, 0.55)$\}$

THEN misalignment (μ_i=0.70), λ=0.60

By vector representation,

$$\overline{S}_1 = <0.34, 0.93, 0.55>$$

(I) Assuming that has only one symptom manifestation with possible degree 0.30 has been observed, e.g.,

SM={(rotor and stator resonant frequency, 0.30)}.

According to Fig.1,

RSM_i={(rotor and stator resonant frequency, 1.00), (1×running frequency, 0.40), (higher multiple, 0.57)}.

Because the symptom has the degree of 0.30, we get

$DRSM_i$=0.30*{(rotor and stator resonant frequency, 1.00), (1×running frequency, 0.40), (higher multiple, 0.57)}.

By equation (3) and (4), we can get

$$\overline{DRSM} = <0.30, 0.12, 0.17>,$$

$$S_{\overline{DRSM} \cdot \overline{S}_1} = 1 - \frac{|0.3 - 0.34| + |0.12 - 0.93| + |0.17 - 0.55|}{(0.3 + 0.34) + (0.12 + 0.93) + (0.17 + 0.55)}$$

$$= 1 - 0.480 = 0.52$$

Because 0.52 < λ=0.60, so the rule can't be fired.

(II) Assuming that two symptom manifestations have been observed with possible degree: rotor and stator resonant frequency has possible degree of 0.30 and 1×running frequency has possible degree of 0.50.

SM={(rotor and stator resonant frequency, 0.30), (1×running frequency, 0.50)}.

According to Fig.1,

RSM_i={(rotor and stator resonant frequency, 1.00), (1×running frequency, 0.40), (higher multiple, 0.57)}.

RSM_2={(rotor and stator resonant frequency, 0.40), (1×running frequency, 1.00), (higher multiple, 0.24)}.

Then,

$DRSM_i$=0.30*{(rotor and stator resonant frequency, 1.00), (1×running frequency, 0.40), (higher multiple, 0.57)}.

$DRSM_2$=0.50*{(rotor and stator resonant frequency, 0.40), (1×running frequency, 1.00), (higher multiple, 0.24)}.

By the representation of vector,

$$\overline{DRSM}_1 = <0.30, 0.12, 0.17>,$$

$$\overline{DRSM}_2 = <0.20, 0.50, 0.12>.$$

$$\overline{DRSM} = \text{Max}\{DRSM_1, DRSM_2\}$$
$$= \text{Max}\{\overline{DRSM}_1, \overline{DRSM}_2\}$$
$$= <0.30, 0.50, 0.17>,$$

and by equation (4), we can get

$$S_{\overline{DRSM}.\overline{s_1}} = 1 - \frac{|0.3 - 0.34| + |0.50 - 0.93| + |0.17 - 0.55|}{(0.3 + 0.34) + (0.50 + 0.93) + (0.17 + 0.55)}$$

$$= 1 - 0.305 = 0.695 \approx 0.70$$

Because $0.70 \geq \lambda = 0.60$, so the rule can be fired. We can obtain $c_i = 0.70*0.70 \approx 0.49$. From Table 2, we can see that the corresponding certainty level of c_i is "fairly certain" in linguistic term.

(III) Assuming that the machinery has three symptom manifestations observed with possible degrees as follows: rotor and stator resonant frequency and 1×running frequency and higher multiple have possible degrees, 0.30, 0.50, 0.32, respectively.

  SM={(rotor and stator resonant frequency, 0.30), (1×running frequency, 0.50), (higher multiple, 0.32)}.

According to Fig.1,

  RSM_1={(rotor and stator resonant frequency, 1.00), (1×running frequency, 0.40), (higher multiple, 0.57)}.

  RSM_2={(rotor and stator resonant frequency, 0.40), (1×running frequency, 1.00), (higher multiple, 0.24)}.

  RSM_3={(rotor and stator resonant frequency, 0.57), (1×running frequency, 0.24), (higher multiple, 1.00)}.

Then,

  $DRSM_1$=0.30*{(rotor and stator resonant frequency, 1.00), (1×running frequency, 0.40), (higher multiple, 0.57)}.

  $DRSM_2$=0.50*{(rotor and stator resonant frequency, 0.40), (1×running frequency, 1.00), (higher multiple, 0.24)}.

  $DRSM_3$=0.32*{(rotor and stator resonant frequency, 0.57), (1×running frequency, 0.24), (higher multiple, 1.00)}.

By the representation of vector,

$$\overline{DRSM_1} = <0.30, 0.12, 0.17>,$$

$$\overline{DRSM_2} = <0.20, 0.50, 0.12>,$$

$$\overline{DRSM_3} = <0.18, 0.08, 0.32>.$$

$$\overline{DRSM} = \text{Max}\{DRSM_1, DRSM_2, DRSM_3\}$$
$$= \text{Max}\{\overline{DRSM_1}, \overline{DRSM_2}, \overline{DRSM_3}\}$$
$$= <0.30, 0.50, 0.32>,$$

and by equation (4), we can get

$$S_{\overline{DRSM}.\overline{s_1}} = 1 - \frac{|0.3 - 0.34| + |0.50 - 0.93| + |0.32 - 0.55|}{(0.3 + 0.34) + (0.50 + 0.93) + (0.32 + 0.55)}$$

$$= 1 - 0.238 = 0.762 \approx 0.76$$

Because $0.76 > \lambda = 0.60$, so the rule can be fired. We can obtain $c_i = 0.76*0.70 \approx 0.53$. From Table 2, we can see that the corresponding certainty level of c_i is "fairly certain" in linguistic term.

6 Conclusion

In this paper, we present a fuzzy knowledge representation, fuzzy knowledge acquisition and approximate reasoning scheme suitable for diagnostic problems. The antecedent propositions in fuzzy production rules have always been assumed to be independent from each other in the literature. However, symptoms appearing in the antecedence of fuzzy production rules may be dependent in diagnostic applications. In our work, we propose using a proximity relation to represent the interdependency between symptoms. In addition, we propose a systematic generation method for acquiring these proximity relations based on observations from the filed experts. Furthermore, a fuzzy reasoning method that performs fuzzy matching between symptom manifestations and the antecedence of fuzzy production rule as well as the determination of causes is also given with application to machinery diagnosis.

Our proposed fuzzy knowledge representation and fuzzy reasoning algorithm subsumes the traditional approach when proximity relation is reduced to identity relation. For future work, we are now concentrating on extending our scheme to multiple input, multiple output production rule systems and multiple rules inferences.

References

1. K. P. Adlassnig, "Fuzzy set theory in medical diagnosis," IEEE Trans. Syst. Man Cybern., 16, 2, 1986, pp. 260-265.
2. B. G. Buchanan and E. H. Shortliffe, "Rule-based expert system: The MYCIN experiments of the Stanford heuristic programming projects," Readings, MA: Addison-Wesley, 1984.
3. S. M. Chen, "A weighted fuzzy reasoning algorithm for medical diagnosis," Decision Support Systems, 11, 1994, pp. 37-43.
4. D. L. Hudson and M. E. Cohen, "Fuzzy logic in medical expert systems," IEEE Engineering in Medicine and Biology, 1994, pp. 693-698.
5. C. S. Kim, S. C, Park, and S. J. Lee, "Systematic generation method and efficient representation of proximity relations for fuzzy relational database systems," IEEE Proceedings of Twentieth Euromicro Conference on System Architecture and Integration, 1994, pp. 549-555.
6. D. H. Lee and M. H. Kim, "Elicitation of semantic knowledge for fuzzy database systems," Conference on Korea Information and Science Society, 1993, pp. 113-116.
7. K. S. Leung and W. Lam, "Fuzzy concepts in expert systems," IEEE Computer, 21, 9, 1988, pp. 43-56.
8. K. S. Leung, W. S. Felix Wong, and W. Lam, "Applications of a novel fuzzy expert system shell," Expert Systems, 6, 1, 1989, pp. 2-10.
9. T. W. Liao and Z. Zhang, "A review of similarity measures for fuzzy systems," Proceedings of IEEE 5th International Conference on Fuzzy System, 1996, pp. 930-935.
10. C. Siu, Q. Shen, and R. Milne, "A fuzzy expert system for vibration cause identification in rotating machines," Proceedings of the 6th IEEE International Conference on Fuzzy Systems, 1997, pp. 555-560.
11. S. Shenoi and A. Melton, "Proximity relations in the fuzzy relational database model," Fuzzy Sets and Systems, 7, 1989, pp. 285-296.
12. L. A. Zadeh, "The role of fuzzy logic in the management of uncertainty in expert systems," Fuzzy sets and Systems, 11, 1983, pp. 199-227.
13. L. A. Zadeh, "Fuzzy sets," Information and Control, 8, 1965, pp. 338-353.
14. H. J. Zimmermann(eds.), Fuzzy Set Theory and Its Application, 1996.

Modelling Fuzzy Sets Using Object-Oriented Techniques

Gary Yat Chung Wong, Hon Wai Chun

City University of Hong Kong
Department of Electronic Engineering
Tat Chee Avenue
Kowloon, Hong Kong
e-mail: eehwchun@cityu.edu.hk

Abstract. This paper describes a new approach to model fuzzy sets using object-oriented programming techniques. Currently, the most frequently used method to model fuzzy sets is by using a pair of arrays to represent the set of ordered pairs, elements and their grades of membership. For continuous fuzzy sets, it is impossible to use infinite number of elements to model, therefore a limited number of array elements are used. Because the grade of membership can be calculated by a membership function, we introduced an approach that models fuzzy set using membership functions directly. Our new approach reduces memory required to model fuzzy sets. Furthermore, grades of membership are calculated dynamically only when needed. Compare with the approach mentioned before, our approach seems to offer advantages in memory space and computation time when modelling systems with complex continuous fuzzy sets.

1   Introduction

In traditional approach, software model to fuzzy set [8, 9] use a pair of array to store the element and grade of membership to represent order pairs [8, 9]:

$$A = \{(x, \mu_A(x)) | x \in X\} , \tag{1}$$

where $\mu_A(x)$ is the membership function [8, 9] representing the grade of membership of x in fuzzy set A. In other words, it maps the fuzzy set domain X to membership space 0 to 1 for a normalised fuzzy set [9]. It is straightforward to model discrete fuzzy set by use of this method. For continuous fuzzy set, limited number sample of membership function pre-processed and stored as discrete fuzzy set [2, 5]. By using sampled ordered pairs, grades of membership between samples can be found by interpolation. In our approach, we would like to store membership function directly instead of storing discrete ordered pairs by means of object-oriented programming techniques.

In the following Sections, the traditional approach will be described briefly. Then, our new object-oriented approach and its implementation will be shown.

2   Traditional Approach

For discrete fuzzy set, using a pair of array to store ordered pairs [2, 5] is the simplest model. Also, operations on discrete fuzzy set using this model are straightforward. For example, the ordered pairs of fuzzy set *A* and *B* are *{(1,0.2), (2,0.5), (3,0.8), (4,1), (5,0.7), (6,0.3)}* and *{(3,0.2), (4,0.4), (5,0.6), (6,0.8), (7,1), (8,1)}* respectively, then *A∩B={(3,0.2), (4,0.4), (5,0.6), (6,0.3)}*. This intersection operation on discrete fuzzy set in software can be done by the following procedure.

---

1. Sort fuzzy set *A* by element
2. Sort fuzzy set *B* by element
3. Expand fuzzy set *A* by adding element with zero grade of membership, which exist in fuzzy set *B* but not *A*.
4. Expand fuzzy set *B* by adding element with zero grade of membership, which exist in fuzzy set *A* but not *B*.
5. Start intersection operation
   For $k = 0$ to $size(A) - 1$
   $GradeOfMembership = min(\mu_A(x_k), \mu_B(x_k))$
   If $GradeOfMembership \neq 0$
   Add $(x_k, GradeOfMembership)$ to resultant fuzzy set

Other fuzzy operations [2, 6, 8, 9] on discrete fuzzy sets can be done by replacing the $min(\mu_A(x_k), \mu_B(x_k))$ function with other functions, such as maximum for union.

For continuous fuzzy set, sampling of membership function is required to initialize fuzzy set. The procedure is shown below:

---

1. *SamplingPeriod = (DomainMax-DomainMin)/(MaxNumberOf Sample-1)*
2. *Sample = DomainMin*
3. *While Sample<DomainMax*
   Add (Sample, μ(Sample))
   Sample = Sample + SamplingPeriod

According to the domain and number of sampling points, a similar process can be used find the resultant fuzzy set after fuzzy operation. Instead of adding the ordered pair, ordered pair of sample and *operation (μ_A(Sample), μ_B(Sample))* is used, where *operation* is a function which process fuzzy set operation. For example, if the operation is intersection, *min (μ_A(Sample), μ_B(Sample))* is used instead of *μ(Sample)*. As shown in the procedure below:

---

1. *SamplingPeriod = (DomainMax-DomainMin)/(MaxNumberOf Sample-1)*
2. *Sample = DomainMin*
3. *While Sample<DomainMax*
   Add (Sample, min (μ_A(Sample), μ_B(Sample)))
   Sample = Sample + SamplingPeriod

To implement continuous fuzzy set operations using the traditional approach, an "adequate" number of samples should be taken. First, if the sampling period is too large, the membership function itself may have distortion. Second, if the resultant fuzzy set sampling period is not small enough, distortion may occur, even the operand fuzzy sets do not have any distortion. As see the example below.

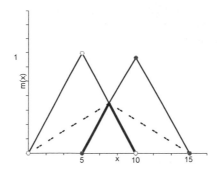

Fig. 2.1. Intersection of two fuzzy sets with triangular membership function which number of sampling is three and domain are *0* to *10* and *5* to *15* respectively. The resultant fuzzy set domain is *0* to *15* and number of sample is three, which is not large enough.

As see from Fig. 2.1, the bold line is the expected result and the dotted line is the result, which do not have enough samples. This problem can be solved by increase the number of sample, however, it is a difficult task to find an "adequate" number of sample because it depends on the membership function. If the number of sampling is too large, it wastes memory. If it is too small, the membership function distorts. We therefore propose an alternative modelling method using object-oriented techniques.

3 Object-Oriented Approach

In this paper, we would like to introduce another approach to model fuzzy sets. According to the first paper in fuzzy set [8] by Professor Zadeh:

A fuzzy set (class) A in X is characterized by a membership (characteristic) function $f_A(x)$ which associate with each point in X a real number in the interval [0, 1], with the value of $f_A(x)$ at x representing the "grade of membership" of x in A.

It is quite natural to describe the above concept using object-oriented programming concepts. We consider fuzzy set as a class that has an attribute to describe the domain space and associated with a "membership function class". This *functor* class contains the actual membership function method, *func*, which maps each point in the domain to the interval *[0, 1]*. As shown in the UML [3, 4] class model in Fig. 3.1.

Fig. 3.1. Class diagram of fuzzy set and fuzzy membership function

FuzzyMembershipFunc is an abstract class; all fuzzy membership function classes should inherit from this class and override the *func* abstract method with its own membership function. The class diagram below describes this structure.

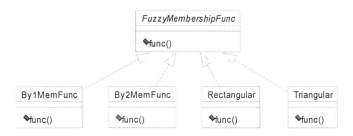

Fig. 3.2. Class diagram of fuzzy membership functions (*By1MemFunc* and *By2MemFunc* will be explained later)

By use of this implementation model, grades of membership are calculated when required. No sampling is needed to initialize fuzzy set because the *func* method defined in the fuzzy membership function classes already captures this relationship. To create a new fuzzy set, just need to assign a suitable fuzzy membership function object to that fuzzy set object.

The UML sequence diagram below shows the process of an application program requesting the grade of membership of a fuzzy set object.

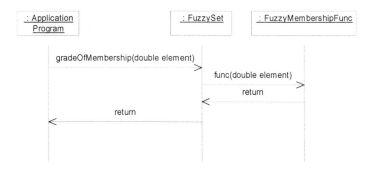

Fig. 3.3. Application program request grade of membership of an element in the fuzzy set

When an application program request the grade of membership of an element, the element is passed to the fuzzy membership function object via the fuzzy set object by calling *gradeOfMembership(element)*, then *func* method in *FuzzyMembershipFunc* invoked and compute the grade of membership in runtime and finally to the application program.

For fuzzy set operations, beside the classes stated above, operator objects, such as complement, alpha cut, maximum, minimum … etc are required. For each operator class, a method is used to describe fuzzy set operation. For example, the method in complement operator class is $f_{A'}(x) = 1 - f_A(x)$ where A' is the resultant fuzzy set. Fig. 3.4 shows the class hierarchy of fuzzy operators.

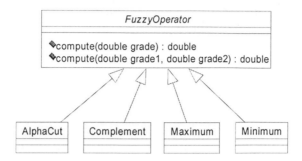

Fig. 3.4. Fuzzy operators class hierarchy

There are two *compute* methods in *FuzzyOperator* class, for unary operations and binary operations. *FuzzyOperator* is an abstract class; all operator classes should inherit from this class. Operators for unary operation such as *AlphaCut, Complement* … etc need to override the *compute(double grade)* method. Operators for binary operation such as *Maximum, Minimum* … etc need to override the *compute(double grade1, double grade2)* method. To create new fuzzy set object after fuzzy operation, beside the operand fuzzy set(s), the membership function associate with it and fuzzy operator object, we also need two special fuzzy membership functions, *By1MemFunc* and *By2MemFunc* for unary and binary operation respectively. In the following paragraphs, we will describe how these two special function work with fuzzy operation. The structure of *By1MemFunc* and *By2MemFunc* are shown in the UML class diagram below:

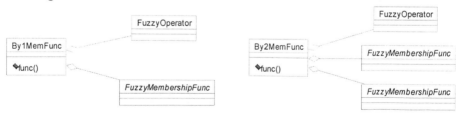

Fig. 3.5. Structure of *By1MemFunc* and *By2MemFunc*

The *FuzzyMembershipFunc* object associate with *By1MemFunc* and *By2MemFunc* object is copied from the membership function object of the operand fuzzy set(s). Fig. 3.6 shows the sequence diagram of unary operation.

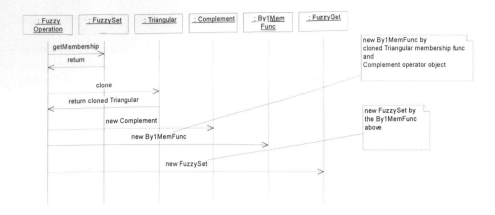

Fig. 3.6. Example: Complement of fuzzy set with triangular membership function

The fuzzy set generated after fuzzy operation in the figure above is different from the fuzzy set in Fig. 3.3. The membership function which associate with this fuzzy set is a *By1MemFunc* which link a cloned triangular object and a fuzzy operator, but the membership function described in Fig. 3.3 do not link any other object and able to return the grade of membership directly. The sequence diagram below shows how the grade of membership of the resultant fuzzy set returned. (All the objects below are generate by the process describe in Fig. 3.6)

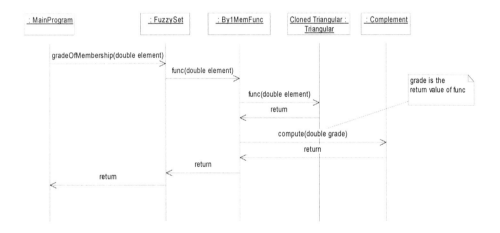

Fig. 3.7. Grade of membership return from the fuzzy set generated by unary operation

When the grade of membership of an element is required from this resultant fuzzy set, the element is passed to the unary fuzzy membership function (*By1MemFunc*) object from fuzzy set object. Then, *By1MemFunc* object obtains the grade of membership of the element from the cloned operand fuzzy membership function object. After that, the grade of membership passed to the operator object. After calculation, the resultant grade of membership obtained and returned to the fuzzy set and to the main program that request grade of membership in runtime.

Binary operation is similar to unary operation (see Fig. 3.6), but it needs to clone one more membership function and use *By2MemFunc* instead of *By1MemFunc*. Fig. 3.8 below shows an example of binary operation.

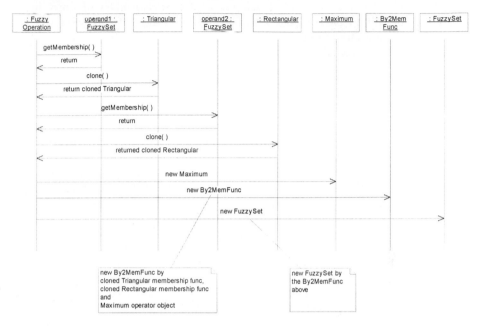

Fig. 3.8. Example: Maximum of fuzzy sets with triangular and rectangular membership function

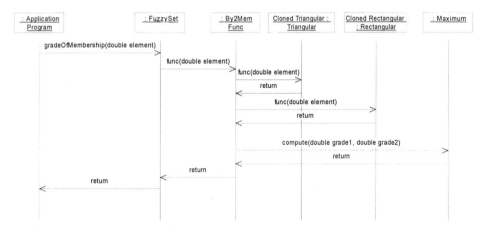

Fig. 3.9. Grade of membership return from the fuzzy set generated by binary operation

When grade of membership is required from the resultant fuzzy set of binary operation. The two-parameter *compute* method in *Operator* object is called instead of the one-parameter *compute* method. As shown in the sequence diagram in Fig. 3.9.

Using the architecture described above, we have implemented a Java class library called **FuzzySys** that facilitates the development of a general class of fuzzy logic systems. In the following Sections, the Java implementation of **FuzzySys** will be described in detail. The concepts we have described can easily be implemented in other object-oriented languages as well.

4   Implementation

In the following, some Java code attached are used to explain the concept described in above Section more clearly.

One of the important techniques to make this approach works, is the concept of copying object. To implement both *FuzzyMembershipFunc* and *FuzzyOperator* class, we need to make it clonable [1], so fuzzy operations can copy the operand fuzzy set's membership function class to create resultant fuzzy set.

For some special membership functions which contain complex data type as attribute, such as *By1MemFunc* and *By2MemFunc*, beside override the *func* method, they also need to override the *clone* method, so the cloned object will not share the same attribute (object) with the original object [1]. Following is part of the *By2MemFunc* code which show the *clone* method and also the *func* method which responsible for a part of operation on Fig. 3.9.

```
public final class By2MemFunc extends FuzzyMembershipFunc {

    protected FuzzyMembershipFunc memFunc1, memFunc2;
    protected FuzzyOperator        operator;

    // Constructor, accessor and modifier define here ...

    public Object clone() {
        By2MemFunc fmem = (By2MemFunc)super.clone();
        fmem.setMembershipFunc1((FuzzyMembershipFunc)memFunc1.clone());
        fmem.setMembershipFunc2((FuzzyMembershipFunc)memFunc2.clone());
        fmem.setOperator( (FuzzyOperator)operator.clone() );
        return fmem;
    }

    public double func(double input) {
        double gradeOfMembership1 = memFunc1.func(input);
        double gradeOfMembership2 = memFunc2.func(input);
        return operator.compute(gradeOfMembership1, gradeOfMembership2);
    }
}
```

For other membership function class without complex data type, it only needs to extends (inherit) from this class and override the *func* method. Following is part of *Rectangular* membership function class Java code.

```
public final class Rectangular extends FuzzyMembershipFunc {

    protected double start, end;

    // Constructor, accessor and modifier define here...

    public double func(double input) {
        if ( input >= start && input <= end )
                return 1;
        else
                return 0;
    }
}
```

Following is part of the Alpha Cut, fuzzy operator code:

```
public final class AlphaCut extends FuzzyOperator {

    protected double alpha;

    // Constructor, accessor and modifier define here ...

    public double compute(double grade) {
        if ( grade < alpha )
                return 0;
        else
                return grade;
    }
}
```

As mention before, fuzzy operations create the resultant fuzzy set by copy the operand(s) fuzzy set's membership function object and new an operator object(s), such as alpha cut, complement ...etc. Following is an example showing how fuzzy operation is done by static methods (See Fig. 3.6 and Fig. 3.8) in *FuzzyOperation* class.

```
public class FuzzyOperation {

    public static FuzzySet complement(FuzzySet op) {
        FuzzyMembershipFunc fmc =
                        (FuzzyMembershipFunc)op.getMembership().clone();
        FuzzyOperator fo = new Complement();
        FuzzyMembershipFunc fmn = new By1MemFunc(fmc, fo);
        return new FuzzySet(op.getDomainMin(), op.getDomainMax(), fmn);
    }

    // Other operations defined here...

}
```

By the implementation above, we do not use array to store sampled order pair to represent fuzzy set; instead we use a function directly and use some attribute to store the characteristics of that function. For example: *start* and *end* in *Rectangular* membership function. Therefore, for system with many fuzzy sets, our approach minimizes memory required. Furthermore, in our implementation, we do not have any loop and the growth of function [7] in our approach is constant. For traditional approach, the growth of functions is in terms of the number of samples. So, our new approach seems to offer advantages in memory space and computation time.

5   Conclusion

The major advantage to our new object-oriented modelling of fuzzy sets is that we avoid problems of over- or under-sampling and the distortion problem in general. Furthermore, because the resultant fuzzy set contains all the fuzzy membership functions and operators involved, the fuzzy set operations are traceable and make application debugging easier. In addition, due to our object-oriented structure, the fuzzy membership functions and operators can be reused and managed in a flexible way. Although increasing cascaded operations will increase the number of computation, this overhead is not large. On the other hand, grades of membership will only be calculated when needed and avoid unnecessary computation of traditional approaches. Therefore in general the performance of our approach will be better.

The traditional method of modelling fuzzy sets with a pair of arrays to store elements and their grades of membership may cause distortion if sampling is not enough. Our object-oriented model solves this problem. We have successfully implemented a Java class library, called **FuzzySys,** which encodes this concept. Our class library not only solves the distortion problem, but also provides user a flexible and efficient way to develop full-scaled fuzzy logic systems.

References

1. Arnold Gosling: The Java Programming Language. Addison Wesley (1997) 77–82
2. Cox, E.: Fuzzy System Handbook. AP Professional (1998) 81–216
3. Flowler Martin. UML Distilled, Applying the Standard Object Modeling Language. Addison Wesley (1997)
4. Grady Booch, James Rumbaugh, Ivar Jacobson: The Unified Modeling Language User Guide. Addison Wesley (1999)
5. Granino A. Korn: Neural Networks and Fuzzy-Logic Control on Personal Computers and Workstation. MIT Press (1991) 315–322
6. R. Lowen: Fuzzy Set Theory: Basic Concepts, Techniques, and Bibliography. Kluwer Academic Publishers (1996) 49–132
7. T.H. Cormen, C.E. Leiserson, R.L. Rivest: Introduction to Algorithms. MIT Press/McGraw-Hill (1990) 23–31
8. Zadeh, Lotfi A.: Fuzzy Set. Information and Control **8** (1965) 338–353
9. Zimmermann, H.J.: Fuzzy Set Theory and Its Applications. Kluwer Academic Publishers (1996)

A Fuzzy Approach to Map Building

H. Zreak, M. Alwan, and M. Khaddour

Higher Institute of Applied Sciences and Technology
Automation Department
P. O. Box 31983, Damascus, Syria.
Fax: +963-11-223 7710, Phone: +963-11-512 0547 (Ext. 2554)
E-mail: m.alwan@ic.ac.uk

Abstract. In this paper, we present test results for a fuzzy map building method (introduced in [2]). We also introduce modifications to the sensor's model and map-update scheme and compare results. Both sensor models and map updates gave encouraging results manifested by recognising cluttered objects as separate entities, detecting small objects and discovering relatively small gaps between objects. The proposed methods can update the map in real-time. The fuzzy map update fusion functions provided a means of tweaking the map adaptability in dynamic environments. Thus, the proposed methods are suited to real-world applications. It has been shown that the direct update scheme suits real-time obstacle avoidance applications, while the fuzzy fusion functions are more appropriate for environment mapping and path-planning applications.

1. Introduction

Building a map of the robot's environment is crucial in autonomous mobile robots for path planning, particularly in dynamic environments. Consequently, map building became the focus of interest for many researchers in the field of mobile robotics.

Various types of perception and sensing devices can be employed to achieve this goal, such as: cameras, laser range finders, infrared sensors, and ultrasonic transducers. Notably, ultrasonic sensors are widely used in autonomous mobile robots for many reasons: they are light, reliable, low cost, simple to operate, and enable the robot to operate in dark environments. However, ultrasonic range finders have their problems. Their first drawback is the width of the sensor's beam, which translates into a low angular resolution and hence implies appreciable angular uncertainty in the position of the obstacle with respect to the sensor's axis. The second is the limited accuracy in measuring the distance of objects, which results in distance uncertainty. The third shortcoming is due to the nature of the ultrasonic waves as well as the reflecting object (such as multiple reflections, absorption, cross talk between sensors, and the possibility of losing the echo in a series of reflections). Hence, these uncertainties need to be included in the world's model, such that they can be taken into consideration when the map is to be used for path planning to yield paths that are likely to execute successfully in the real world [1].

In this paper, we will shed some light on related works, outline the methods we are proposing to model the uncertainties inherent in ultrasonic perception systems, present our map update schemes, discuss the results and draw conclusions upon them.

2. Literature Review

The literature on perception and map building techniques is abundant. We will review in what follows some of the map building strategies that rely on ultrasonic transducers identical to the ones we are using. We will mainly concentrate on methods that model the measurement uncertainties and include such uncertainties in the map. Moreover, we have focused on grid representation of the map, as this representation is of particular interest since it suits the path planning method we introduced in [1] and [2].

The map building methods encountered in the literature differ primarily in the proposed sensor model, map representation, and the fusion methods that combine the information from the sensors with the previous map.

Elfes et al. [3], Poloni et al. [4], and Pagac et al. [5] build two layered maps, one holds occupancy information, and the other is for emptiness information. The three researchers and their co-workers used sensor models based on probabilistic distribution functions along the distance and the angle. However, the sensor models proposed by Poloni and Pagac explicitly reduce the occupancy probability as the range increases, to express less credibility in farther distance measurements that may be caused by multiple reflections. Elfes and his colleagues employed a Bayesian update approach, whereas Poloni and his co-researchers used fuzzy functions for fusing the newly acquired map with the previous one. While Pagac's sensor model diverged from the Bayesian world to evidence theory, since the summation of occupancy probabilities on the arc, where the reflecting object lies, is not equal to one; consequently, Pagac made use of the Dempster's rule of combination from the evidence theory to update the map.

Poloni's method is not suited to dynamic environments, where objects in the environment may move. The method produced grid cells containing contradictory emptiness and occupancy information. This is due to updating the occupancy layer without reducing the emptiness possibility for the corresponding grid cells on the emptiness layer, and vice versa. Moreover, the complex sensor's model makes the method require demanding and time-consuming computations. Elfes' method can cater for dynamic changes in the perceived work space, but the speed of convergence is quite slow (i.e. the environment has to be scanned many times for objects to clearly appear/disappear); the Bayesian update formulae used by Elfes and his colleagues are also computationally expensive. In contrast, Pagac's method gave very good results, but has proven to be slower than the fuzzy methods we are proposing. Finally, none of the above methods made use of the readings of contiguous sensors.

In the rest of this paper, we will present two fuzzy based methods for modelling the sensor's uncertainty. The sensor's model is dependent on the readings coming from contiguous sensors (i.e. whether there is supporting evidence coming from the adjacent sensors on either side).

3. The Proposed Methods

The map building methods we propose rely, in principle, on 24 ultrasonic range finders from Polaroid, having range accuracy equal to ±1% of the measured distance and a beam angle of 30 degrees [6]. The sensors are evenly distributed around the test-bed vehicle's body.

The use of 24 sensors, instead of 12, results in a 50% overlap between the adjacent visibility sectors. This greatly improves the angular resolution of the perception system, and eliminates the chance of loosing echoes from objects on the boundary of the visibility sectors of adjacent sensors, particularly if the sensor's model takes the adjacent sensors' readings into account.

To simplify the sensor's model and to make the distance uncertainty measurement independent, we assume a worst-case and consider the error to be equal to its maximum value always. This renders the width of uncertainty region equal to 2*0.01*10.5m=21cm. The uncertainty region is divided into three rows of cells each is 7 cm wide, as shown in the distance uncertainty profile below. Therefore, the map grid cells can be 7cm x 7cm. However, and since the graphical representation of the map on the screen may introduce an error equivalent to one cell in either of the horizontal or vertical directions, we have chosen a higher map resolution and set the grid size to 3cm x 3cm.

The distance uncertainty profile adopted in both of the proposed sensor models was originally introduced in [2]. Figure 1 illustrates the occupancy possibility as a function of distance when the sensor detects an obstacle at distance R.

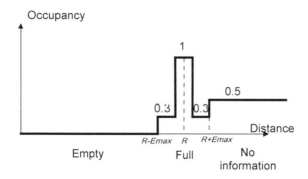

Fig. 1. Distance Uncertainty Profile.

The areas just before and after the distance measurement R could possibly be occupied, while the sonar beam conveys no information about the area beyond the object that reflects the beam. However, we are proposing two different angular uncertainty profiles and two different map update schemes. The distance uncertainty is used as it is with the first model, and used as a modifier multiplied with the angular possibility occupancy profile to give the final occupancy possibility to be filled into the 15 possibly occupied grid cells.

3.1 The First Model

In the first model, we divide the visibility sector of every sensor into three regions, just as we did with distance uncertainty profile. The model illustrated in figure 2 first appeared in [2].

The grid cells in the centre of the beam have the highest occupancy possibility, while grid cells inside the visibility sector on either side are possibly occupied; the areas outside the visibility sector are treated as unknown for this particular sensor.

This method fills grid cells falling in the areas covered by overlapping visibility sectors of adjacent sensors with the maximum occupancy possibility if it is different from 0.5 (which corresponds to lack of information).

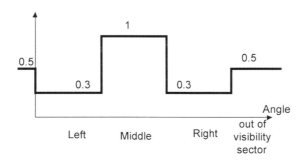

Fig. 2. The Angular Uncertainty Profile.

The map-update scheme, in this method, fills the grid cells with the newly acquired occupancy possibility if the new occupancy possibility is different from 0.5; thus throwing away old information. However, the old occupancy possibility information is retained in the grid cells for which the latest scan did not yield any useful information.

3.2 The Second Model

The visibility sector of every sensor is divided into five sub-sectors. In this model, nonetheless, we distinguish between three different cases:

1. If the distance measurement from adjacent sensors does not match that of the current sensor, this would mean that a narrow obstacle is most likely situated in the middle of the visibility sector. In this case, the proposed occupancy possibility profile is given in figure 3.

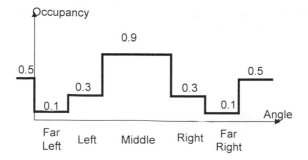

Fig. 3. The Angular Occupancy Possibility Profile for a single sensor's reading.

1. If a sensor's distance measurement is close to the distance measurement of only either of the contiguous sensors, this will increase the occupancy possibility of the sub-sectors situated nearest to that sensor. For example, when the sensor to the right of the current sensor yields a distance reading close to that of the current sensor, the highest occupancy possibility is shifted to the right as illustrated in figure 4.

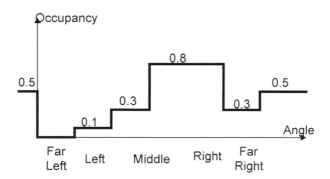

Fig. 4. The Angular Occupancy Possibility Profile for close measurements of two adjacent sensors.

2. If both of the contiguous sensors give distance measurements close to that of the middle one, then we have either of the two possible scenarios shown in figure 5.

38

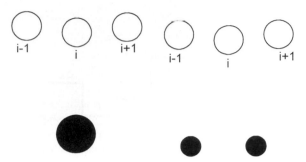

Fig. 5. The Two Possible Scenarios when three contiguous sensors give similar distance readings.

We decided to adopt the right-hand scenario in order to minimize the chance of missing a gap, which can accommodate the robot, between two objects. The corresponding occupancy possibility profile is thus given in figure 6.

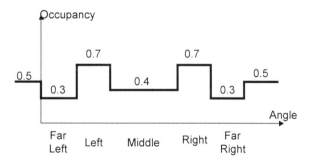

Fig. 6. The Occupancy Possibility Profile adopted when three contiguous sensors give close distance measurements.

We employed parameterised fuzzy combination functions, such as Dombi and Yager to fuse the newly acquired map with the previous one. Dombi's function is given by:

$$Dombi(a,b) = \frac{1}{1+\left[\left(\frac{1}{a}-1\right)^{-\lambda}+\left(\frac{1}{b}-1\right)^{-\lambda}\right]^{-\frac{1}{\lambda}}} \qquad \text{for } \lambda \in [0,\infty] \qquad (1)$$

Where a and b are the old and the new occupancy possibilities of a grid cell, and λ is a tuning parameter. We chose λ to be 2.

Whereas Yager's function is expressed as follows:

$$Yager(a,b) = \min\left[1, \sqrt[p]{a^p + b^p}\right] \qquad \text{with} \qquad p > 0 \qquad (2)$$

Where a and b are the old and the new occupancy possibilities of a grid cell, and p is a tuning parameter. We chose p to be 2.

3.3 Inclusion of Prohibited Areas

In some environments, prohibited areas, where the robot should not go may exist. These areas may be dangerous although they may not appear to the robot's sensors (e.g. a staircase). To protect the robot and to make it avoid such areas, we can store a separate map containing the prohibited areas as obstacles. The map of prohibited areas can be fused with the acquired map using the max operator.

4. Implementation and Test Scenario

In the current implementation, we used only 15 sensors out of 24, covering the front and the sides of our test-bed vehicle. To increase the rate of scanning the environment and keep the possibility of sensors' cross talk to a minimum, we chose to fire sensors in pairs having 180 degrees between them.

The two proposed methods were tested in an office-like environment shown in figure 7.

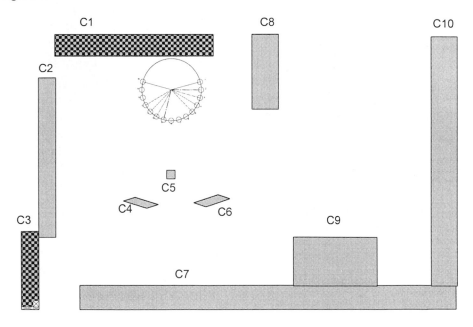

Fig. 7. The Distribution of Objects in the Test Environment.

Here is a brief description of each object in the figure: C1 and C9 are tables. C2, C7 and C10 are walls. C3 is the door. C4 and C6 are two chairs positioned at distance

1.3 m and perpendicular to sensors 7 and 9 respectively. This arrangement has been chosen such that the distance between C4 and C6 is 1 meter. C5 is an object of a small reflecting surface (9cm wide), and positioned 1m away from the opposite sensor. C8 is a fixed object at distance 2.75 meters (from the central sensor No. 8).

5. Results and Discussion

Figure 8 presents the results of scanning the above-described environment according to the first fuzzy model and update method. The map shown in figure 9 has been built using the second fuzzy model and Dombi's update formula, while the map in figure 10 employs the same fuzzy model together with Yager's update scheme.

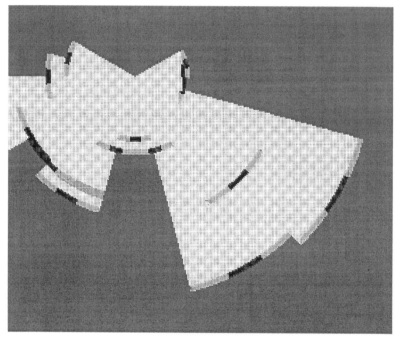

Fig. 8. The Map obtained using the first sensor model and direct update.

It is clear from the figures that both models and update methods were able to detect the small object C5 and represent it on the map. Moreover, both sensor models and map updates managed to discover the gap between the two objects C4 and C6, which can accommodate the test-bed vehicle, as they are supposed to. Both methods and update schemes were also able to detect the three objects C4, C5 and C6, as discern entities despite the small distance between C4, C5 and C6.

We can clearly notice that the wall C7, C9 and C10 have been detected as two arcs in all maps. This is due to the properties of ultrasonic waves, where the echo is received from the nearest point situated in the sector.

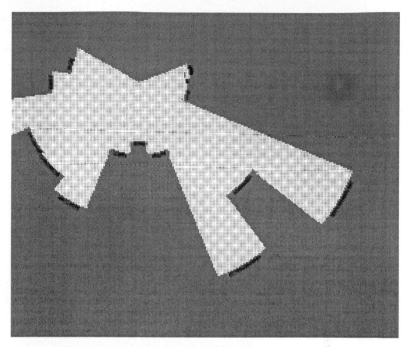

Fig. 9. The Map obtained using the second sensor model and Dombi's fusion function.

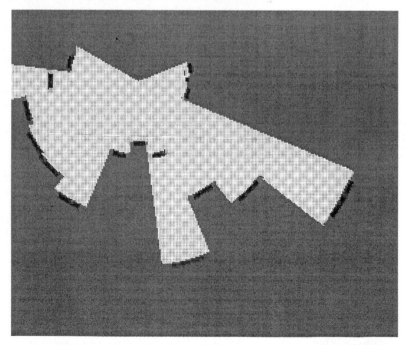

Fig. 10. The Map obtained using the second sensor model and Yager's fusion function.

The main difference between Dombi and Yager's fusion functions was the speed at which an object clearly appeared on the map or disappeared from it. The speed of adaptation of both functions, however, is dependent on the function's parameters; the adaptability is thus tunable.

Moreover, the first fuzzy model has a rather wide uncertain region on both sides of the obstacle, whereas the second fuzzy model gives a better definition of objects.

All of the three proposed update techniques can scan the work space and update the occupancy possibility of the map grid cells in real-time. However, the first fuzzy model together with the first direct update mode was the fastest. Nevertheless, Dombi's and Yager's update functions retain some of the old information. Consequently, the direct update approach is suited to real-time obstacle avoidance on the move, while Dombi's and Yager's are appropriate for building maps for path planning purposes.

6. Conclusions

The map building methods presented in this paper provide an effective tool for modelling uncertainties of sonar range finders and capturing the uncertainties of the world. Both of the proposed methods could detect small objects and could discern relatively small gaps between object.

All of the proposed models and update methods could scan the environment and update memory cells corresponding to the map grid cells in real-time. The speed of adaptation in dynamic environments is tunable, when fuzzy fusion functions are used. Hence, the proposed methods are suitable for real-time real-world applications. The direct update scheme, however, is more appropriate for real-time navigation and obstacle avoidance, whereas the proposed fuzzy update functions are better suited to mapping and path planning applications.

The technique, nevertheless, needs to be investigated further on the mobile platform to get a better view of the environment from different positions and orientations.

References

1- Alwan M, and Cheung P Y K, 1994: "Handling Uncertainty, Ambiguity, and Kinematics in Path Planning", in *Proc. of The Third Conference on Automation, Robotics and Computer Vision (ICARCV '94)*, Singapore, November 9-11, 1994, pp. 257-261.
2- Alwan M, and Cheung P Y K, 1997: "Modelling and Handling Uncertainties in Mobile Robotics", *Journal of Intelligent and Fuzzy Systems*, Vol. 5, No. 3, 1997, pp. 205-217.
3- Elfes A, 1987: "Sonar Based Real-World Mapping and Navigation", *IEEE Journal of Robotics and Automation*, Vol. 3, No. 3, June 1987, pp. 249-265.
4- Poloni M, Ulivi G, and Venditelli M, 1994: "Fuzzy Logic and Autonomous Vehicles: Experiments in Ultrasound Vision", *Journal of Fuzzy Sets and Systems*, 1994.
5- Pagac D, Nebot E, and Durrant-Whyte H, 1998: "an Evidential Approach to Map-Building for Autonomous Vehicles", *IEEE Transaction on Robotics and Automation*, Vol. 14, No. 4, August 1998, pp. 623-629.
6- Data Sheets of Polaroid Ultrasonic Range Finders, Polaroid Corporation.

Towards an Affirmative Interpretation of Linguistically Denied Fuzzy Properties in Knowledge-Based Systems

Daniel Pacholczyk

LERIA, University of Angers,
2 Boulevard Lavoisier, 49045 Angers Cedex 01, France
pacho@univ-angers.fr

Abstract. In this paper, we present a new model dealing with affirmative or negative information. We propose a formalization of negation in the context of fuzzy set theory grounded on a compatibility level and tolerance threshold to render nuanced strength values of modified properties in the scope of a linguistic negation. Their combination allows us to choose the reference frame from which the possible values of a linguistic negation of A appearing in the statement "x is not A" will be extracted. The plausible interpretations of denied properties are obtained as the result of the different scoping of the negation operator. Moreover, a choice strategy computes, if needed, an intended meaning of each linguistic negation.

1 Introduction

In this paper, we present a general model dealing with nuanced information expressed in affirmative or negative forms as they may appear in knowledge bases including, rules like "If Jack *is not small* then he *is visible* in the crowd" or "if the wage *is not high* then the summer holidays *are not very long*" and facts like, "Jack *is really very tall*" or "the wage *is really low*", for example. This problem is threefold and implies *(1)* the representation of *nuanced properties*, *(2)* the representation of strengthening or softening effects resulting from *adverbial modifiers* bearing on properties, and *(3)* the association of different interpretations due to *different scopes* of a *negation operator* on a compound linguistic element expressing a *nuanced property*. Our main goal has been to define a *general model* dealing with affirmative or negative information *(1) grounded on the models* proposed in ([11-14]), *(2) improving* theirs abilities in the management of knowledge bases, and *(3) including* the main results of linguistic analysis of negation ([9], [3], [4], [1]). In Section 2, we present the initial representation of nuanced information based on an automatic process defining the L-R functions associated with *nuances of properties* ([2]). Section 3 points out the fact that *a modelisation implying denied properties must be viewed as a one to many correspondence,* called here *multiset function.* Section 4 is devoted to the linguistic analysis developed in ([14]) which improve the formalization of linguistic negation proposed in ([11], [12]) by an explicit taking into account the different *scopes* of the linguistic negation in accordance with linguistic theories ([9], [3], [4], [1]). In Section 5 we propose a formalization of linguistic negation based upon a *compatibility level* ρ and a *tolerance threshold* ε to rend nuanced strength values of modified properties in the *scope* of a linguistic negation of A appearing in "x is not A". A combination of ρ and ε allows to define a *reference frame*, denoted as $Neg_{\rho,\varepsilon}(A)$, from which the possible values of linguistic negation of A will be extracted. In Section 6, we specify the links between the different definitions of linguistic negation proposed in ([11-14]). In Section 7, each *scope of a negation operator* is associated with a reference frame subset, denoted as $neg_{\rho,\varepsilon}(x, A)$, which constitutes the *intended meanings of the linguistic negation ρ–compatible with A for x with a tolerance threshold ε*. We propose in Section 8 *a choice strategy leading to an intended meaning of the linguistic negation*. In Section 9 we state new properties of the linguistic negation which improve the abilities of previous models.

2   The Initial Frame of Information Representation

In the following, we suppose that our discourse universe is characterised by a finite number of *concepts* C_i. A set of properties P_{ik} is associated with each C_i, whose description *domain* is denoted as D_i. The properties P_{ik} are said to be the *basic properties* connected with concept C_i. For example as regards to the previous knowledge, the concepts of "height", "wage" and "appearance" should be understood as qualifying individuals of the human type. The concept "wage" can be characterized by the basic fuzzy properties ([16]) "low", "medium" and "high". *Linguistic modifiers* bearing on these basic properties permit to express *nuanced knowledge*. This work uses the methodology proposed in ([2]) to cope with affirmative information expressed in utterances like « x *is* $f_\alpha m_\beta P_{ik}$ » or « x *is not* $f_\alpha m_\beta P_{ik}$ » in the case of negation. In this context, expressing a property like "$f_\alpha m_\beta P_{ik}$" called here *nuanced property*, requires a list of linguistic terms.

Two ordered sets of modifiers are selected depending on their modifying effects:
- The first one groups *translation modifiers* resulting somehow in both a *translation and an eventual precision variation* of the basic property: For example, the set of translation modifiers could be M_7={extremely little, very little, rather little, moderately (∅), rather, very, extremely} totally ordered by the relation: $m_\alpha < m_\beta \Leftrightarrow \alpha < \beta$ (*Cf.* Figure 1).

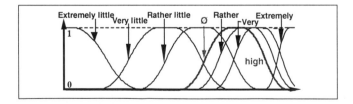

Fig. 1. Translation Modifiers applied to the basic property "high"

- The second one consists of *precision modifiers* which make possible to increase (or decrease) the precision of the previous properties (*Cf.* Figure 2). For example, F_6={vaguely, neighbouring, more or less, moderately, really, exactly} totally ordered by $f_\alpha < f_\beta \Leftrightarrow \alpha < \beta$.

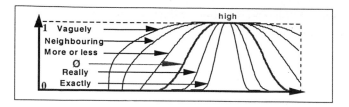

Fig. 2. Precision Modifiers applied to the basic property "high"

3 The Linguistic Negation as a Multiset Function

Let us sketch a main idea underlying our formal model. Utterances like "John is tall" or more generally "x is A" are ways, *natural languages* (henceforth NL) permits to confer some property A to an x. As simple as they could appear at a first glance, their interpretations raise, nevertheless, a lot of difficult problems as soon as the goal is to give them a translation in a formal model entity ([17]). It is well known that first order logic ([8], [5]) appears to be a poor candidate for that, even for these simple utterances, and a formula like A(x) resulting from their translation appears to be far from encompassing linguistics properties of what could inferred from original NL utterances ([6]). Properties of A(x) greatly depends on what A denotes in NL, and modelisations should take all of them into account. This is particularly the case when A is a vague property and fuzzy set theory is generally considered as well suited to cope with this kind of characteristics ([7]). Furthermore, some properties which are generally considered as precise such as *to be closed* gain some vagueness when they are denied. It's the case in particular when the negation of the property is lexicalized through an antonym ([5]). So, for example, *the door in not closed* is not equivalent to *the door is open*. The negation introduces some nuances on the property and in using *not open* instead of *closed* a speaker may exclude *fully open* to mean *half-open*. Moreover, negating some property *introduces a reference frame* from which possible values of the negated property can be extracted. In a sentence like *my hat is not red*, *not red* may refer to some other color for the hat, but also to some nuance of red. It appears clearly that *a modelisation implying denied properties cannot be viewed as a one to one correspondence but as a one to many one*, called here *multiset function*.

Some classical approaches to negation fail to embed all desirable negation properties illustrated in our examples. In the fuzzy context ([16]), "x is not A" receives a unique interpretation "x is $\neg A$" with $\mu_{\neg A}(x)=1-\mu_A(x)$ which corresponds to univocal function. In a similar way, within a *qualitative context*, one can refer to a set of linguistic labels denoted as $L=\{u_0, ..., u_n\}$ totally ordered: $u_i<u_j \Leftrightarrow i<j$ ([10]). But, the linguistic negation (a decreasing involution) verifies $Neg(u_i)= u_{n-i}$ where here also u_{n-i} denotes a sole value in L.

In order to remedy to this deficiency, Torra ([15]) has defined a new *concept of linguistic negation in a totally ordered set L of linguistic labels*. He has proposed a multiset function Neg of L into $\mathcal{P}(L)$ which among the conditions to be satisfied by Neg(A) implies for it to be convex and not void. Unfortunately, these conditions are not totally suited to the representation of linguistic negation (*Cf.* [11, 12]). It is clear that the definitions of linguistic negation proposed in ([11-13]) are based upon multiset functions which alleviate the difficulties of Torra's approach.

4 The Standard Forms Resulting from the Scope of the Linguistic Negation

The development of the model proposed in ([13] [14]) improves the previous ones ([11] [12]) by an explicit taking into account the different *scopes of the linguistic negation*. This permits to make clearer the different interpretations of utterances of the form "x is not A" resulting from scoping effects in accordance with linguistic theories ([9], [3], [4], [1]). In this context, the intended meaning of "x is not A" is conceived as a compositional function applied to the compounds of the denied statement.

Let us now describe a formal model leading to the intended meaning of a linguistic negation. Within the discourse universe, let us denote as: $\mathcal{C}$ the set of distinct concepts C_i, $\mathcal{D}_i$ the domain associated with the concept C_i, $\mathcal{M}$ the set of modifier combinations, $\mathcal{B}_i$ the set of associated basic properties P_{ik} defined on $\mathcal{D}_i$, $\mathcal{N}_{ik}$ the set of all nuances of the basic property P_{ik}, $\mathcal{N}_i$ the set of all nuanced properties associated with C_i. Then let: $\mathcal{D}=\cup_i\mathcal{D}_i$, $\mathcal{B}=\cup_i\mathcal{B}_i$, $\mathcal{N}=\cup_i\mathcal{N}_i$, $\mathcal{E}=\mathcal{M}\cup\mathcal{D}\cup\mathcal{B}\cup\mathcal{N}$.

We define the *reference frame of linguistic negation* as follows.

Definition 4.1. Let Neg a multiset function Neg : $\mathcal{E}\to \mathcal{P}(\mathcal{E})$ verifying the conditions:

L1: $\forall n_\gamma \in \mathcal{M}$, Neg($n_\gamma$)=$\mathcal{M}\backslash\{n_\gamma\}$,

L2: $\forall P_{ik} \in \mathcal{B}_i$, Neg($P_{ik}$)=$\mathcal{B}_i\backslash\{P_{ik}\}$,

L3: $\forall x \in \mathcal{D}_i$, Neg(x)=$\mathcal{D}_i\backslash\{x\}$,

L4: $\forall n_\gamma \in \mathcal{M}$, $\forall P_{ik} \in \mathcal{B}_i$, Neg($n_\gamma P_{ik}$)=$\mathcal{N}_i\backslash\{n_\gamma P_{ik}\}$,

L5: $\forall n_\gamma \in \mathcal{M}$, $\forall P_{ik} \in \mathcal{B}_i$, Neg($n_\gamma(P_{ik})$)=$\mathcal{N}_{ik}\backslash\{n_\gamma P_{ik}\}$, and

L6: $\forall n_\gamma \in \mathcal{M}$, $\forall$Neg(P_{ik}), n_γ(Neg(P_{ik}))=$\mathcal{N}_i\backslash\mathcal{N}_{ik}$.

Note that all of these conditions on the multiset function Neg use the set difference operator \ to built sets of possible values. So, in this view, Negation is characterized as an operator which refers to a subset of values from a set of possible ones. Each condition will be associated with a possible scope of the negation operator which characterizes the reference frame of its intended meaning and the set of values that the choice of this particular scope excludes. From a linguistic point of view "x is not $n_\gamma P_{ik}$" can generally express something corresponding to a nuanced utterance like "y is $n_\delta P_{ij}$".

Following are the different *standard forms* of the nuanced property $n_\delta P_{ij}$ resulting from each possible scope. Depending of the scope of the negation operator, saying "x is not $n_\gamma P_{ik}$", for this x, the speaker (or the user) may refer to:
- *Another object instead of x belonging to the same domain and satisfying the nuanced property.* **[F1]**
So, "x is not $n_\gamma P_{ik}$" means "not(x) is $n_\gamma P_{ik}$" or in other words, not(x)=y$\in$ Neg(x) (cond. **L3**). As an example, "Jack is not guilty" since it is John that is guilty.

- Another nuance of the same property. [F2]

So, "x is not $n_\gamma P_{ik}$"⇔"x is not($n_\gamma (P_{ik})$)"⇔"x is $n_\delta P_{ik}$". So, not($n_\gamma (P_{ik})$)=$n_\delta P_{ik}$∈ Neg($n_\gamma (P_{ik})$) (cond. **L5**). As an example, "Jack is not small" since "Jack is extremely small".

- A nuanced property except $n_\gamma P_{ik}$, which is nonetheless associated with the same concept. [F3]

So, "x is not $n_\gamma P_{ik}$" means "x is not($n_\gamma P_{ik}$)" which is equivalent to "x is $n_\delta P_{ij}$". In other words, not($n_\gamma P_{ik}$)=$n_\delta P_{ij}$ with $n_\delta P_{ij}$∈ Neg($n_\gamma P_{ik}$) (condition **L4**). For example, "the wage being not very high" can be "really low", "medium" or "rather little high".

- A nuance of another basic property associated with the same concept. [F4]

So, "x is not $n_\gamma P_{ik}$" means "x is n_γ(not(P_{ik}))" that is to say "x is $n_\delta P_{ij}$". Here, n_γ(not(P_{ik}))=$n_\delta P_{ij}$, with $n_\delta P_{ij}$∈ n_γ(Neg(P_{ik})) (cond. **L6**). For example, "John is not small" since he is at least "medium".

- A new basic property of the same concept [F5]

In this case "x is not A" means "x is not-A": a new basic property denoted as "not-A" is associated with the same concept. As an example, "this wine is not bad" can induce the new basic property "not-bad".

- Remark. In the following, we do not refer any longer to the first standard form **F1** which seldom occurs in knowledge-bases and suppose that the new property introduced in the last form **F5** appears among other basic properties.

Definition 4.2. We denote as Neg*(u) the subset of Neg(u) leading to one of the previous standard forms defining intended meanings of a linguistic negation.

5 A Negation ρ–compatible with A according to a Tolerance Threshold ε

Linguists ([9], [3], [4], [1]) have pointed out that, asserting that "x is not A", a speaker characterises as negation *i*) the *judgement of rejection* and *2*) the *pragmatic means* exclusively used to notify this rejection. Within the fuzzy context ([11-13]), we have to make explicit the adequacy of the affirmative statement "x is A" with the universe. Intuitively, the speaker considers that this assertion possesses a significant degree of truth. In other words, the membership degree to the fuzzy set associated with A must be greater than a compatibility level with the discourse universe. We can also consider that this value defines the strength with which the assertion is denied. So, *a first parameter* ρ, with 1≥ρ≥0, has been introduced *to take into account this compatibility level and this negation strength*. Then, the judgement of rejection receives as an interpretation: rejection of "x is A"⇒$\mu_A(x)$<ρ.

Moreover, asserting that "x is not A", *if necessary the user refers* to "x is P" as the intended meaning of his negation. The previous analysis only defines the standard forms of the linguistic negation. But, it is obvious that any element of Neg*(u) cannot lead to the intended meaning of "x is not A". Intuitively the speaker understands a real difference between the membership degrees belonging to A and P for their significant values, that is to say:

$\mu_A(x)$ *(resp. $\mu_P(x)$) is greater than* ρ⇒$\mu_P(x)$ (resp. $\mu_A(x)$) *is rather close to 0.*

It is obvious that the expression "rather close to" can receive different translations, each of which is defining the threshold tolerance to which the speaker can accept "x is P" as the intended meaning. So, *a second parameter* ε, *with* $1 \geq \rho \geq \varepsilon \geq 0$, *has been introduced to define the negation strength* ρ *according to a tolerance threshold* ε.

We can now put down a formal context making it possible to define a precise linguistic negation with the aid of a *restriction* to $\mathcal{N}$ of previous multiset function Neg* (*Cf.* [13]).

Definition 5.1 Let ρ, ε such that: $0 \leq \varepsilon \leq \rho \leq 1$. Let us define the multiset function $\text{Neg}_{\rho,\varepsilon}$: $\mathcal{N} \to \mathcal{P}(\mathcal{N})$ as follows:

CD0: $\forall A \in \mathcal{N}$, $\text{Neg}_{\rho,\varepsilon}(A) \subset \text{Neg*}(A)$,

CD1: $\forall P \in \text{Neg}_{\rho,\varepsilon}(A)$, $\forall x$, $\{\mu_A(x) \geq \rho \Rightarrow \mu_P(x) \leq \varepsilon\}$, and

CD2: $\forall P \in \text{Neg}_{\rho,\varepsilon}(A)$, $\forall x$, $\{\mu_P(x) \geq \rho \Rightarrow \mu_A(x) \leq \varepsilon\}$.

Then, $\text{Neg}_{\rho,\varepsilon}(A)$ is said to *be the linguistic negation ρ-compatible with* A *with the tolerance threshold* ε. Moreover, any $P \in \text{Neg}_{\rho,\varepsilon}(A)$ is said to be a linguistic negation ρ–compatible with A with the tolerance threshold ε.

Example. We have collected in Figure 3 $\text{Neg}_{0.75,\,0.35}$ ("low") a set of negations 0.75–compatible with "low" with tolerance threshold 0.35, knowing that Neg*("low") refers to the standard form **F3**.

Remark. As noted before, in some cases a linguistic negation is restricted to a *simple rejection*. So, we add the new standard form defined as follows:
For this x, the user does not refer to an affirmative translation of the negation **[F0]**
We can put Neg*(A)=∅. So, we have $\text{Neg}_{\rho,\varepsilon}(A)=∅$. As an example, "Smith is not guilty" since his alibi is confirmed. Then, this simple rejection of "Smith is guilty" only gives us $\mu_{\text{guilty}}(\text{Smith})<\rho$.

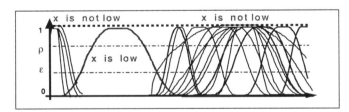

Fig. 3. Negations 0.75–compatible with "low" with tolerance threshold 0.35.

The strength of a linguistic negation can be interpreted as a neighbourhood degree between A and P. For any F-implication $\to$ ([10]), $\mu_A(x) \geq \rho$ gives us $\mu_A(x) \to \mu_P(x) \leq \mu_A(x) \to \varepsilon \leq \rho \to \varepsilon$. Then, $(\text{Min}_x\{\text{Min}\{\mu_P(x) \to \mu_A(x), \mu_A(x) \to \mu_P(x)\}\}) \leq \rho \to \varepsilon$. As pointed out in ([11], [12]), the value $\text{Min}_x\{\text{Min}\{\mu_P(x) \to \mu_A(x), \mu_A(x) \to \mu_P(x)\}\}$ can be considered as the neighborhood degree between A and P.

Proposition 5.1. $P \in \text{Neg}_{\rho,\varepsilon}(A)$ and A are less than $(\rho \to \varepsilon)$ neighbouring.

6 Comparison with Previous Models

We can recall the definition (denoted as **DefI**) introduced in the initial model ([11]). Let ε be a real such that $0.33 \geq \varepsilon \geq 0$ and P a property defined in the same domain as A. If P satisfies the conditions: **C0:** P and A are θ_i similar with $\theta_i <$ moderately, **C1:** $\forall x$, $\mu_A(x) \geq \rho = 0.67 + \varepsilon \Rightarrow \mu_P(x) \leq \mu_A(x) - 0.67$, **C2:** $\forall x$, $\mu_P(x) \geq 0.67 + \varepsilon \Rightarrow \mu_A(x) \leq \mu_P(x) - 0.67$, then "x is P" is said to be a ε-plausible Linguistic Negation of "x is A". The definition (denoted as **Def II**) proposed in ([12]) is completely founded upon a neighborhood relation V. So, conditions **CN1** and **CN2** stand for previous ones : **CN1:** $\forall x$, $(\mu_A(x) \geq \rho) \Rightarrow V(\mu_P(x), \mu_A(x)) \leq 1 - \rho + \varepsilon$, **CN2:** $\forall x$, $(\mu_P(x) \geq \rho) \Rightarrow V(\mu_A(x), \mu_P(x)) \leq 1 - \rho + \varepsilon$. Then "x is P" is said to be a linguistic negation ρ-compatible with "x is A" with tolerance threshold ε.

It is possible to establish the links between solutions P satisfying the conditions 1 and 2 in these definitions (denoted as $\textbf{Cond}_1 = \textbf{C1} + \textbf{C2}$ and $\textbf{Cond}_2 = \textbf{CN1} + \textbf{CN2}$) and the solutions P satisfying the conditions 1 and 2 (denoted as $\textbf{Cond}_3$) appearing in the definition 5.1. Let us choose the neighborhood relation V_L defined as follows (Cf. [11], [12]): $V_L(u, v) = \text{Min}\{u \rightarrow_L v, v \rightarrow_L u\}$ where: $u \rightarrow_L v = 1$ if $u \leq v$ else $1 - u + v$. In this case, we have proved the following results :

1 : if P satisfies $\textbf{Cond}_1$ then $V_L(\mu_A(x), \mu_P(x)) \leq 1 - \rho + \varepsilon$ for $\rho \geq 0.67$, $0 \leq \varepsilon \leq 0.33$ and $\varepsilon \leq \rho$. In other words, there exist ρ and ε such that : $\textbf{Cond}_1 \Rightarrow \textbf{Cond}_2$.

2 : if P satisfies $\textbf{Cond}_2$ then $\mu_A(x) \geq \rho$ (resp. $\mu_P(x) \geq \rho) \Rightarrow \mu_P(x) \leq \varepsilon$ (resp. $\mu_A(x) \leq \varepsilon$) for any $\rho \geq 0.5$ and $\varepsilon \leq \text{Min}\{\rho, 2\rho - 1\}$. So, there exist ρ and ε such that: $\textbf{Cond}_2 \Rightarrow \textbf{Cond}_3$.

3 : if P satisfies $\textbf{Cond}_3$ then $V_L(\mu_A(x), \mu_P(x)) \leq 1 - \rho + \varepsilon$ where $0 \leq \varepsilon \leq \rho \leq 1$. So, given ρ and ε such that $0 \leq \varepsilon \leq \rho \leq 1$, we have: $\textbf{Cond}_3 \Rightarrow \textbf{Cond}_2$.

4: if P satisfies $\textbf{Cond}_3$ then $\mu_A(x) \geq \rho = 0.67 + \varepsilon \Rightarrow \mu_P(x) \leq \mu_A(x) - 0.67$ for any $0 \leq \varepsilon \leq 0.33$. So, given ρ and ε such that $0 \leq \varepsilon \leq 0.33$ and $\rho = 0.67 + \varepsilon$ we have: $\textbf{Cond}_3 \Rightarrow \textbf{Cond}_1$.

It appears clearly that, given the neighborhood relation V_L, the conditions are not equivalent. Generally, the definition 5.1 leads to more restrictive conditions $\textbf{Cond}_3$ defining the linguistic negation. But, for any $\rho \geq 0.5$ and $\varepsilon \leq \text{Min}\{\rho, 2\rho - 1\}$ $\textbf{Cond}_2$ and $\textbf{Cond}_3$ are equivalent. Let us now examine the first condition in the definitions **Def I** and **Def II**. C0 accepts P when P and A are globally less than moderately similar. We can note that, given V_L this condition is satisfied if: $0 \leq \varepsilon \leq \rho - 0.67$. In other words, the definition 5.1, which does not implicitly refer to neighborhood and similarity relations, fulfills this condition in this case. Moreover, the condition **CD0** in the definition 5.1 defines the reference frame Neg*(A) connected with a precise standard form **Fi** of the linguistic negation of "x is A". This condition is in accordance with the linguistic analysis of negation (Cf § 4). But, it is not the case with **Def I** and **Def II**. In other words, the set $\text{Neg}_{\rho,\varepsilon}(A)$ can be viewed as the reference frame of the linguistic negation corresponding to all standard forms **Fi**. It can be noted that the definition 5.1 leads to this set if Neg*(A) refers to the standard form **F3**. As a result, this comparison points out the fact that definition 5.1 can induce previous definitions and realize a better accordance with the linguistic analysis of the negation.

7 The set of Significant Meanings of "x is not A"

The set $Neg_{\rho,\varepsilon}(A)$ defines the reference frame from which we have to extract the intended meanings of the linguistic negation. This comes down to defining explicitly a subset of $Neg_{\rho,\varepsilon}(A)$, denoted as $neg_{\rho,\varepsilon}(x, A)$, which consists of the intended meaning of "x is not A". Since ρ is the compatibility level from which the membership degrees are significant, we accept as intended meanings only the solutions P satisfying this condition at x.

Definition 7.1 Put $neg_{\rho,\varepsilon}(x, A) = \{P \in Neg_{\rho,\varepsilon}(A) \mid \mu_P(x) \geq \rho\}$. Any $P \in neg_{\rho,\varepsilon}(x, A)$ is called an *intended meaning for x of the linguistic negation ρ–compatible with A with tolerance threshold ε*. We say also that "x is P" is an intended meaning with the tolerance threshold ε of the linguistic negation ρ–compatible of "x is A". If no confusion is possible, we simply say that P is an intended meaning of the linguistic negation of A for x.

Example. By using previous solutions (*Cf.* Figure 3), we have collected in the Figure 4 the intended meanings of "x is not low" for the values a (3 solutions based upon "low") and b (4 solutions based upon "medium" and "high").

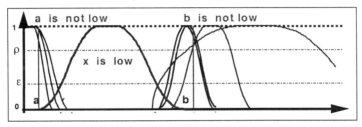

Fig. 4. Intended meanings of negations 0.75–compatible for a or b with "low" with tolerance threshold 0.35.

Remark. This definition proposed in ([13]) and applied in ([14]) is more useful that the previous ones ([11, [12]]) since a set of precise intended meaning among the candidates is computed, and this, without asking the user to define this set of intended meanings.

8 A Choice Strategy of an Intended Meaning

If the user wishes only one interpretation of the linguistic negation, it is possible to use the *choice strategy* proposed in ([13]). A particular choice can be done among the plausible solutions leading to the most significant membership degree and having the weakest complexity.

Definition 8.1 The *complexity* of the nuanced property A, denoted as comp(A), is equal to the number of nuances (different from $\varnothing$) required in its definition.

Definition 8.2 A choice of a nuanced property P satisfying the following conditions: **I1** : P$\in$ neg$_{\rho, \varepsilon}$(x, A), **I2** : μ_p(x)= Max{μ_P(x)|P$\in$ neg$_{\rho, \varepsilon}$(x, A)}., and **I3:** $\forall$Q$\in$ neg$_{\rho, \varepsilon}$(x, A), {μ_Q(x)=ξ(x, A) $\Rightarrow$ comp(P) $\leq$ comp(Q), defines "x is P" as the intended meaning of "x is not A".

Example. By using the solutions collected in Figure 4, "b is not low" receives as internded meaning "b is really medium".

9 Basic Properties of The Linguistic Negation

We can recall the initial properties of linguistic negation presented in ([11-14]).

Property 9.1 The knowledge about "x is A" doesn't automatically define the knowledge about "x is not A".

Property 9.2 The double negation of A does not generally lead to "A".

Property 9.3 {$\rho \leq \rho$', $\varepsilon' \leq \varepsilon \leq \rho$}$\Rightarrow$Neg$_{\rho', \varepsilon'}$ (A)$\subseteq$Neg$_{\rho, \varepsilon}$(A).

Property 9.4 There exists ρ and ε such that the negation ρ-compatible with the tolerance threshold ε takes into account all previous interpretations of "x is not A".

We can now *enrich* the previous set of properties of the linguistic negation. Indeed, it is easy to prove the following new properties.

Property 9.5 P$\in$ Neg$_{\rho, \varepsilon}$(A) $\Rightarrow$ A$\in$ Neg$_{\rho, \varepsilon}$(P).

Property 9.6 $\neg$(P$\in$ neg$_{\rho, \varepsilon}$(x, A) $\Rightarrow$ A$\in$ neg$_{\rho, \varepsilon}$(x, P))

Property 9.7 neg$_{\rho, \varepsilon}$(x, A) can be an empty set

Property 9.8 Even if $\mathcal{N}_i$, the set of all nuanced properties associated with C$_i$, can be totally ordered, for any A$\in \mathcal{N}_i$ the set neg$_{\rho, \varepsilon}$(x, A) can be a not convex set.

Property 9.9 If P$\in$ Neg$_{\rho, \varepsilon}$(A) and Q$\in$ Neg$_{\rho, \varepsilon}$(B), then :
- if Neg*(A$\vee$B)=Neg*(A)$\vee$Neg*(B), then P$\wedge$Q$\in$ Neg$_{\rho, \varepsilon}$(A$\vee$B), and
- if Neg*(A$\wedge$B)=Neg*(A)$\wedge$Neg*(B), then P$\vee$Q$\in$ Neg$_{\rho, \varepsilon}$(A$\wedge$B).

Property 9.10 The model can deal with boolean basic properties without nuances by choosing ρ=1 and ε=0. In this case, the standard forms defining the set Neg*(A) can correspond with the linguistic notion of marked (or not) property (*Cf.* [9], [3]).

Property 9.11 The previous strategy defines P$\in$ neg$_{\rho, \varepsilon}$(a, A) and Q$\in$ neg$_{\rho, \varepsilon}$(b, B) such that the rule "if a is not A, then b is not B" receives as translation the rule "if a is P, then b is Q".

Remark. By using classical systems, deductive process cannot be apply to facts and rules which include linguistic negations. But, it is not the case with our approach to linguistic negation, since the same deductive process can be apply to equivalent rules taking into account the intended meaning of negative information.

10   Conclusion

We have defined a general model of linguistic negation of utterances like "x is not A" in the fuzzy context. This approach to negation, in accordance with linguistic analysis pursues Pacholczyk's preceding works, where the possible interpretations of a denied property, eventually nuanced, are extracted from a reference frame depending on a compatibility level and a tolerance threshold. The model emphasizes the role of the different scopes of the negation operator and negation appears in it as a multiset function bearing on the different elements of the original NL utterance. Moreover, this new system improves the abilities of the previous ones in that the linguistic negation possesses a more powerful set of properties. In particular, classical deductive processes can be maintained since the intended meanings of negative information can be explicitly expressed in affirmative form.

References

1.   Culioli A., Pour une linguistique de l'énonciation: Opérations et Représentations. Tome 1 Ophrys Eds. Paris, 1991.
2.   Desmontils E., and Pacholczyk D., "Towards a linguistic processing of properties in declarative modelling". *Int. Jour. of CADCAM and Comp. Graphics* **12:4**, 351-371, 1997.
3.   Ducrot O., and Schaeffer J.-M., Nouveau dictionnaire encyclopédique des sciences du langage. Seuil Paris, 1995.
4.   Horn L.R., *A Natural History of Negation.* The Univ. of Chicago Press, 1989.
5.   McCawley J. D., Everything That linguists Have always Wanted to Know about Logic (2nd ed.), Chicago University Press, 1993.
6.   Dermott D., "Tarskian Semantics, or no Notation Without denotation", Cognitive Science **2(3)**, 277-282, 1978.
7.   Mel'cuk I. A., *Dependency Syntax: Theory and Practice*, State University of New York Press, 1988.
8.   Moore R. C., "Problems in logical Form" Proceedings of the 19th Annual Meeting of Association for Computational Linguistics, Standford, California, 117-124, 1981.
9.   Muller C., *La négation en français.* Publications romanes et françaises Genève, 1991.
10.   Pacholczyk D., Contribution au traitement logico-symbolique de la connaissance. Thèse d'Etat, 1992.
11.   Pacholczyk D., "An Intelligent System Dealing with Negative Information". Proc. of Int. Conf. ISMIS'97, Charlotte, U. S. A, Lecture Notes in A. I., **1325**, 467-476, 1997.
12.   Pacholczyk D., "A New Approach to the Intended Meaning of Negative Information", Proc. of 13th European Conf. on Art. Int., ECAI98, Brighton, UK, 23-28 August 1998, Pub. by J. Wiley & Sons, 114-118, 1998.

13. Pacholczyk D., "A New Model dealing with Linguistic Negation within the Fuzzy Set Theory Context", Proc. of 6th European Congress on Intelligent Techniques and Soft Computing, EUFIT'98, Aachen, Germany, 573 – 577, 1998.

14. Pacholczyk D, Levrat B., Towards a General Model for the Representation of Negated Nuanced Properties in a Fuzzy Context, Proc. of 5th Workshop of Logic, Language, Inform. and Computation, WOLLIC'98, Sao Paulo, Brazil, 28-31 July 1998, 169-177, 1998

15. Torra V., "Negation Functions Based Semantics for Ordered Linguistic Labels". *Int. Jour. of Intelligent Systems* **11**, 975-988, 1996.

16. Zadeh L.A., "Fuzzy Sets". *Information and Control*, **8**, 338-353, 1965.

17. Zubert R., Implications sémantiques dans les langues naturelles, Editions du CNRS, Paris, France, 1989.

Representational Hierarchy of Fuzzy Logic Concepts in the OBOA Model

Ramo Sendelj[1], Vladan Devedzic[2]

Ramo Sendelj[1], Yugoslav Navy, Bokeskih brigada 44, 85340,Herceg Novi, Montenegro, Yugoslavia, Tel: +381-88-55237, Email: ramo@cg.yu
Vladan Devedzic[2], University of Belgrade, FON - School of Business Administration, Jove Ilica 154, 11000 Belgrade, Yugoslavia, Fax: +381-11-461221 Tel: +381-11-2371440, Email: devedzic@galeb.etf.bg.ac.yu

Abstract. This paper describes hierarchical modeling of fuzzy logic concepts that has been used within the recently developed model of intelligent systems, called OBOA. The model is based on a multilevel, hierarchical, general object-oriented approach. Current methods and software design and development tools for intelligent systems are usually difficult extend, and it is not easy to reuse their components in developing intelligent systems. The OBOA model tries to reduce these deficiencies. The model starts with a well-founded software engineering principle, making clear distinction between generic, low-level intelligent software components, and domain-dependent, high-level components of an intelligent system. This paper concentrates on modeling and implementation of fuzzy logic concepts within the hierarchical levels of the OBOA model. The fuzzy components described are extensible and adjustable. As an illustration of how these components are used in practice, a practical design example is shown. The paper also suggests some steps towards future design of fuzzy components and tools for intelligent systems.

1. Introduction

In the general domain of object-oriented software engineering, hierarchical modeling refers to *layered software architectures* [Batory and O'Malley, 1992], in which:

- each component in a system belongs at a certain conceptual *layer* (layers are sets of classes on the same level of abstraction);
- more complex components are designed starting from simpler components from the same layer or from the lower layers;
- A hierarchically organized tree of components that spans across multiple layers can be drawn to represent the architecture of the system.

One particularly important extension of the concept of layered software architecture is the *orthogonal architecture* [Rajlich and Silva, 1996]. In the orthogonal architecture, classes (objects) are organized into layers and threads. *Threads* consist of classes implementing the same functionality, related to each other by the using relationship [Booch, 1994]. Threads are "vertical", in the sense that their classes belong to different layers. Layers are "horizontal", and there is no using relationship among the classes in the same layer. Hence modifications within a thread do not affect other threads. Layers and threads together form a grid. By the position of a class in the architecture, it is easy to understand what level of abstraction and what functionality it implements. The architecture itself is highly reusable, since it is shared by all programs in a certain domain which have the same layers, but may have different threads.

These general concepts have been recently applied to modeling intelligent software systems in the object-oriented way. As a result, a hierarchical model of intelligent systems, called *OBOA (OBject-Oriented Abstraction)* has been developed [Devedzic and Radovic, 1999]. The model encompasses a wide range of knowledge representation methods and inference techniques commonly used today in designing intelligent systems. The purpose of this paper is to describe how the main concepts of fuzzy logic and fuzzy systems, being important modeling techniques and tools in intelligent systems, are supported in the OBOA model.

The paper is organized as follows. Section 2 is an explicit problem statement. In Section 3, the essence of the OBOA model is described. Section 4 is the central section of the paper. It shows how fuzzy concepts fit into the OBOA model, and presents some design examples. Section 5 shows examples of current implementation of software components for designing fuzzy systems based on the OBOA model. In Section 6, some informal performance analysis is presented. Finally, Section 7 shows the benefits of this kind of modeling fuzzy systems and directions for future research.

2. Problem statement

The purpose of this paper is threefold:

- it shows how the concepts of fuzzy logic and fuzzy systems fit into a more general, object-oriented, hierarchical model of intelligent systems (the OBOA model);
- it explains how design of fuzzy intelligent systems can be facilitated by imposing some hierarchical structure onto the concepts and tools used in the design process;
- It presents an example of how development of practical fuzzy systems can be alleviated using this approach.

3. Previous work

This section illustrates how hierarchical modeling has been included into the OBOA model in order to facilitate design and development of intelligent systems. It also briefly shows how some well-known concepts from the domain of intelligent systems the model supports.

3.1. Levels of abstraction and dimensions in the OBOA model

The OBOA model defines five *levels of abstraction* for designing intelligent systems, Figure 1a. If necessary, it is also possible to define fine-grained sublevels at each level of abstraction. Each level has associated concepts, operations, knowledge representation techniques, inference methods, knowledge acquisition tools and techniques, and development tools. They are all considered as *dimensions* along which the levels can be analyzed, Figure 1b. The concepts of the levels of abstraction and dimensions have been derived starting from the orthogonal architecture.

Level of abstraction	Objective	Semantics	Level of abstraction	Dimension			
				D1	D2	...	D n
Level 1	Integration	Multiple agents or systems	Level 1				
Level 2	System	Single agent or system	Level 2				
Level 3	Blocks	System building blocks	Level 3				
Level 4	Units	Units of blocks	Level 4				
Level 5	Primitives	Parts of units	Level 5				

(a)   (b)

Figure 1. The OBOA model: (a) the levels of abstraction (b) Dimensions

Semantics of the levels of abstractions is easy to understand. In designing intelligent systems, there are *primitives*, which are used to compose *units*, which in turn are parts of *blocks*. Blocks themselves are used to build self-contained *agents* or *systems*, which can be further integrated into more complex systems. For getting a feeling for how the OBOA's levels of abstraction correspond to some well-known concepts from the domain of intelligent systems, consider the following examples. Primitives like plain text, logical expressions, attributes and numerical values are used to compose units like rules, frames, and different utility functions. These are then used as parts of certain building blocks that exist in every intelligent system, e.g. classifiers, controllers, and planners. At the system level, we have self-contained *systems* or *agents* like learning systems, scheduling agents, and knowledge-based diagnostic systems, all composed using different building blocks. Finally, at the integration level there are multiagent systems, distributed intelligent systems, and Web-based intelligent systems.

It should be also noted that the borders between any two adjacent levels are not strict; they are rather approximate and "fuzzy". Several concepts related to intelligent systems can be also treated at different levels of abstraction.

3.2.   Well-known paradigms and the OBOA model

As examples of how some well-known paradigms and techniques are encompassed by the OBOA model, Table 1 shows how neural networks and genetic algorithms fit in the levels of abstraction from Figure 1. Note that several entries in the table are left empty. That is because Table 1 shows only well-known and widely applicable concepts from these two types of intelligent systems.

Table 1. Some examples of modeling neural networks and genetic algorithms in the OBOA model

Level	Objective	Knowledge representation	Operations	Inference methods	Knowledge acquisition
1	Integration	Hybrid intelligent system			
2	System				Monitoring and acquisition of data,
3	Blocks	Neural networks (NN) Genetic algorithms (GA)		RunNM EvolutionGA	NN training GA Reproduction
4	Units	Slab, Layer, DataSet (Training,Test) Population	Get, Put, Propagate, Evaluate, Fitness, MakeChromosome	Propagate, Evaluate Selection	Training, Creating training set and test set
5	Primitives	Neuron, Link, PaternOfData Gen, Hromosome	Get, Put, Activation function Initialisation, Mutation, Crossover, Fitness ...		

4. Fuzzy logic and fuzzy systems in the OBOA model

Table 2 shows how fuzzy logic and fuzzy systems fit in the levels of abstraction from Figure 1.

Table 2. Some examples of modeling fuzzy logic concepts in the OBOA model

Level	Objective	Knowledge representation	Operations	Inference methods	Knowledge acquisition
1	Integration				
2	System	Fuzzy Logic Expert System			Interviews,case studies, learning reasoning strategies
3	Blocks	FuzzyRule	AddBlock, EditBlock, DeleteBlock, GetBlock	Forward chaining, backward chaining, inference explanation	Fuzzy rule training
4	Units	FuzzyVariable, FuzzyNumber, FAM, Premise, Proposition	GetUnit, AddUnit, EditUnit, Initialization, Create	Max-Min inference, Max-produce inference	
5	Primitives	Fuzzy, FuzzyFunction, FuzzySet, FuzzyValue, Operation, Hedge,	Get, Set, Add, Delete, Union, Intersection, Complementation, Concentration, Delation, Indeed, Power	Inheritance	Manual input, measurement

The concepts, operations, methods, etc. at each level of abstraction can be directly mapped onto sets of corresponding components and tools used in designing intelligent systems.

The complexity and the number of these components and tools grow from the lower levels to the higher ones. Consequently, it is quite reasonable to expect further horizontal and vertical subdivisions at higher levels of abstraction in practical applications of the OBOA model for design and development of intelligent systems. Appropriate identification of such subdivisions for some particular types of intelligent systems, such as intelligent tutoring systems and intelligent manufacturing systems, is the topic of our current research.

From the software design point of view, components and tools in Table 2 can be considered as classes of objects. It is easy to derive more specific classes from them in order to tune them to a particular application. The classes are designed in such a way that their semantics is defined horizontally by the corresponding level of abstraction and its sublevels (if any), and vertically by the appropriate key abstractions specified mostly along the concepts.

Class interfaces (method procedures) are defined mostly from the operations and inference methods dimensions at each level. The knowledge acquisition and development tools dimensions are used to specify additional classes and methods at each level used for important development tasks of knowledge elicitation, learning, and knowledge management. At each level of abstraction, any class is defined using only the classes from that level and the lower ones.

Figure 2 shows the *FuzzyElement* class hierarchy, represented using the UML notation [Rumbaugh et al., 1997].

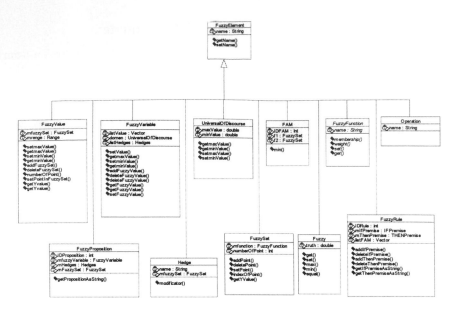

Figure 2. The FuzzyElement class hierarchy

Fuzzy Logic. Fuzzy logic provides the means to both represent and reason with imprecise and common sense knowledge in computer. This ability is extremely valuable to the knowledge engineer responsible for building an intelligent system, which is confronted with an expert that explains the problem-solving tasks in imprecise common-sense terms. Vague terms or rules can be represented and manipulated numerically to provide results that are consistent with the expert's knowledge.

Fuzzy variables. Fuzzy logic is primary concerned with quantifying and reasoning about vague or fuzzy terms that appear in our natural language. In fuzzy logic, these fuzzy terms are referred to as fuzzy variables. In fuzzy intelligent systems, we use fuzzy variables in fuzzy rules. For example:

Fuzzy rule: **IF** Speed is slow   **THEN** Make the acceleration high

We call the range of possible values of a fuzzy variable the variable's universe *of discourse.* In our example, we might give the variable "Speed" the range between 0 and 100 mph. The phrase "Speed is slow" occupies a section of the variable's universe of discourse - it is a fuzzy set. In the OBOA model, the *FuzzyVariable* class is designed as in Figure 3. Names of the interface functions have obvious meanings.

Fuzzy value. Fuzzy values (adjectives) of fuzzy variables are represented using fuzzy sets, which map set elements to degree of belief that the element belongs to the fuzzy set.

59

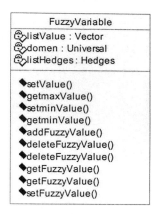

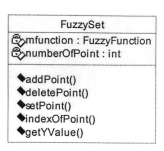

Figure 3. The FuzzyVariable class   **Figure 4.** The *FuzzySet* class

Fuzzy Sets. Let X be universe of discourse, with elements of X denoted as x. Fuzzy set A of X is characterized by a membership function $\mu_A(x)$ that associates each element x with a degree of membership value in A. Membership function defined as:
$\mu_A(x): X \rightarrow [0, 1]$

In the OBOA model, fuzzy sets are represented using the *FuzzySet* class represented in Figure 4. Note that the class interface includes the possibilities to compute hedges (adverbs such as *Very, Somewhat, Indeed, Very Very,* commonly used in practice), as well as functions to perform fuzzy set operations (such as *Intersection, Union, Complement,* etc.).

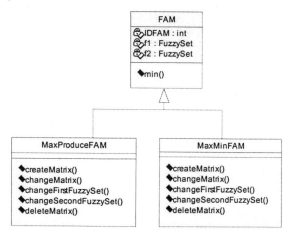

Figure 5. The *FAM* class

Fuzzy Associative Memory (FAM). Fuzzy systems store fuzzy rules as associations. That is, for the rule **IF A THE** B, where A and B are fuzzy sets, a fuzzy system stores the association (A,B) in matrix M. The fuzzy associative matrix M maps fuzzy set A to fuzzy set B. This fuzzy association or fuzzy rule is called a *Fuzzy Associative Memory (FAM)*. A FAM maps a fuzzy set to fuzzy set – the fuzzy inference process. The *FAM* class in the OBOA model is an abstract class and represents Fuzzy

Associative Memory. The two most popular fuzzy inference techniques used in practice are *max-min inference* and *max-product inference*. The corresponding *MaxMinFAM* and *MaxProduceFAM* classes extend the *FAM* class.

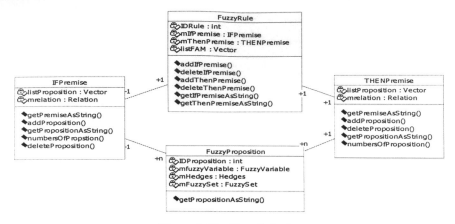

Figure 6. The *FuzzyRule* class

Fuzzy rule. Fuzzy rule establishes a relationship or association between two fuzzy propositions. Fuzzy rule has two premises, IF premise and THEN premise:

IF X is A THEN Y is B

The *Premise* class in the OBOA model represents premises of rules, and they have one or more *fuzzy propositions*. Fuzzy proposition is a statement that asserts a value for some given fuzzy variable:

Fuzzy proposition: X is A

Where A is a fuzzy set on the universe of discourse X. The *FuzzyRule* class represents fuzzy rules. Methods of *FuzzyRule* define interface for this class, Figure 6.

5. Implementation and application

To illustrate design process of a fuzzy system using the OBOA model, we can consider a problem of navigating a golf cart around a golf course [Durkin, 1994]. We can define the following fuzzy variables:

Table 3. Fuzzy Variables in the example system

Fuzzy variable	Range		
	Min Value	Max Value	
Error angle	-180	180	degrees
Tree angle	-180	180	degrees
Speed	0	5	yd/s
Acceleration	-2	1	yd/s/s
Ball distance	0	600	yards

Figure 7 shows how these new fuzzy variables are added to the OBOA-based system designed and implemented using the classes described above.

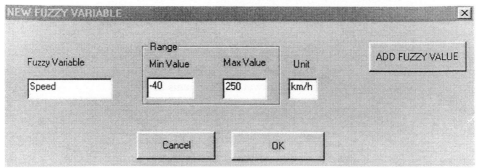

Figure 7. Adding a new fuzzy variable to a fuzzy system

Table 4 shows some fuzzy values in the system, and the corresponding Figure 8 illustrates how they are added to our application.

Table 4. Fuzzy Variables with Fuzzy Values

Error angle	Tree angle	Speed	Acceleration	Ball distance
large Negative	large Negative	Zero	Brake Hard	Zero
Small Negative	Small Negative	Real Slow	Brake Light	Real Close
Zero	Zero	Slow	Coast	Close
Small Positive	Small Positive	Medium	Zero	Medium
Large Positive	Large Positive	Fast	Slight	Far

Figure 8. For example, the new rule might be:

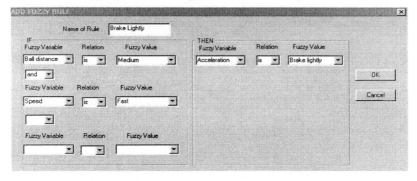

Figure 8. Adding a new fuzzy rule to a fuzzy system

Rule Brake lightly
IF Ball distance is Close **AND** Speed is Fast **THEN** Acceleration Brakelightly

6. Discussion

Component software is an object-based software movement that subsumes compound document as one example of application interoperability. Component software addresses the general problem of designing system from application elements that were constructed independently by different developers using different languages, tools and computing platforms [Szyperski, 1998]. The OBOA model also supports design and development of component-based applications. From the component-based software perspective, it should be noted that all of the classes described in Section 5 are actually developed as software components as well.

The OBOA model is supported by a number of design patterns [Gamma ET al., 1994] and class libraries developed in order to support building of intelligent systems. In fact, designing and developing an intelligent system based on the OBOA model is a matter of first developing a shell and then using it for development of the system itself. In spite of the fact that this means starting the project *without* a shell, it is a relatively easy design and development process, because of the precisely defined hierarchy among the tools and components, as well as the strong software engineering support of the design patterns and class libraries.

Along with the high modularity and reusability provided by the fuzzy class libraries, potential design flexibility is another important advantage of using the fuzzy logic concept. Development of an OBOA-based shell for building fuzzy intelligent applications means putting together only those pieces of software from the relevant class libraries that are really needed for a given application. If any additional class for representing fuzzy concepts is needed, it must be designed and developed by the shell developer. Fortunately, the class hierarchies and design patterns of OBOA provide a firm ground to start from in such an additional development. Most additional subclasses for representing fuzzy concepts can be derived directly from some of the already existing classes. The classes representing fuzzy concepts in the OBOA model are designed in such a way to specify "concept families" using the least commitment principle: each class specifies only the minimum of attributes and inheritance links. That assures the minimum of constraints for designers of new classes.

As an example, consider the job of adding a new fuzzy element when needed. This task doesn't require significant changes in the corresponding module of the fuzzy system (or the fuzzy shell). It is rather a matter of finding out an appropriate place for the new class along the levels of abstraction and in the class hierarchies representing fuzzy concepts, and specifying a few additional attributes and links.

Finally, when developing a fuzzy logic shell, and then using it for development of an intelligent fuzzy system itself, the shell's options are always only the necessary options. Modifications and extensions are made easily and only in accordance with the application's needs.

7. Conclusions

Hierarchical design of fuzzy logic concepts of intelligent systems, presented in the paper, allows for easy and natural conceptualization and design of a wide range of intelligent applications, due to its object-oriented approach. It suggests only general guidelines for developing fuzzy intelligent systems, and is open for fine-tuning and adaptation to particular applications. Fuzzy intelligent systems developed using this model are easy to maintain and extend, and are much more reusable than other similar systems and tools.

The model is particularly suitable for use by developers of software environments (shells) for building fuzzy systems. Starting from a library of classes for fuzzy logic concepts and control needed in the majority of fuzzy systems, it is a straightforward task to design additional fuzzy logic classes needed for a particular fuzzy system shell. Moreover, the model also supports development of component-based intelligent systems, which have started to attract increasing attention among the researchers in the field.

Further development of support for fuzzy logic concepts in the OBOA model is concentrated on development of appropriate classes in order to support a number of different fuzzy systems. The idea is that the system developer can have the possibility to select fuzzy tools from a predefined palette, thus adapting the shell to his/her own design preferences. Such a possibility would enable experimentation with different fuzzy tools and their empirical evaluation.

8. References

1. K. Arnold, J. Gosling: The Java Programming Language. Addison-Wesley, Reading, MA, 1996.
2. D. Batory and S. O'Malley: The Design and Implementation of Hierarchical Software Systems with Reusable Components. ACM Transactions on Software Engineering and Methodology, Vol.1, No.4, 1992, pp. 355-398.
3. G. Booch: Object-Oriented Analysis and Design with Applications. 2nd Edition, Benjamin/Cummings Publishing Company, Inc., Redwood City, CA, 1994.
4. V. Devedzic, D. Radovic: A Framework for Building Intelligent Manufacturing Systems. Accepted for publication in the IEEE Transactions on Systems, Man, and Cybernetics (to appear in 1999).
5. Durkin, J.: Expert Systems - Design and Development. Macmillan Publishing Company, New York, 1994.
6. E. Gamma, R. Helm, R. Johnson and J. Vlissides: Design Patterns: Elements of Reusable Object-Oriented Software. Addison-Wesley, Reading, Massacuhetts, 1994.
7. V. Rajlich, J.H. Silva: Evolution and Reuse of Orthogonal Architecture. IEEE Transactions on Software Engineering, Vol.22, No.2, February 1996, pp. 153-157.
8. Rumbaugh, J., Jacobson, I., Booch, G.: Unified Modelling Language Reference Manual. Addison-Wesley, Reading, MA, 1997.
9. Szyperski, "Component Software: Beyond Object-Oriented Programming", ACM Press/Addison-Wesley, NY/Reading, MA, 1998.
10. R.Sendelj, D.Radovic, V.Devedzic: Java Class for Knowladge Represent. SYMOPIS' 98
    Herceg Novi, September 1998, pp. 337-340.
11. S. Vinoski: CORBA: Integrating Diverse Applications Within Distributed Heterogeneous Environments. IEEE Communications Magazine, Vol. 14, No. 2, February 1997, pp. 28-40.

Design of Fuzzy Sliding Controller Based on Cerebellar Learning Model

Heng-Kang Fan[1], Chih-Ming Chen[2], and Chin-Ming Hong[3]

[1] Department of Electrical Engineering
Kuang-Ku Institute of Technology and Commerce

[2] Department of Electronic Engineering
National Taiwan University of Science and Technology

[3] Department of Industrial Education
National Taiwan Normal University

Abstract. As to the control of fuzzy sliding mode, this paper proposes a cerebellar learning model for on-line learning of the controller. Fuzzy sliding mode has excellent robustness to the system uncertainty and immunity to the noise of the external noise. As for the cerebellar learning mode, it possesses the advantages of easy and fast correction. The combination of the two leads to the design of a fuzzy sliding mode controller with self-learning capability to improve the short-comings of difficulties in setting up the regulations of fuzzy control. It also improves the system stability and enhances the effectiveness of the controller. This paper describes the implementation of a fuzzy sliding controller with cerebellar learning mode. After system simulation and capability test, it is applied to the control of slew-up, stand-on and positioning of a 360 ° inverted pendulum.

Keywords: Variable structure system, Sliding mode, Fuzzy control, Cerebellar Model Articulation Controller, Inverted pendulum.

1. Introduction

A variable structure system[8] can be applied in control mainly through the structure of switching control system and the introduction of the concept of sliding mode. It drives the representative point of the state space to a pre-defined sliding plane, and slides along this plane to the equivalent point of the state space so that the system response can converge steadily in a gradual manner[6].

Fuzzy sliding mode controller (FSMC) combines the merits of variable structure system and fuzzy control. It achieves high degree of control level for the non-linear system, yet maintains the features of robustness and ease of design[15].

To make a controller work efficiently when facing the problems of time varying, the vigorous parameter change of the non-linear system and the inability of accommodating control regulation in time, this paper proposes a cerebellar model articulation controller (CMAC) as the on-line learning system[3][4]. It enables the system parameters to follow the controlled object through continuous correction and to achieve the desired standard .

2. Theory of CMAC Fuzzy Sliding Control
2.1 Control Theory of Fuzzy Sliding Mode

By combining the fuzzy control and the sliding mode control[7] approaches together, and fuzzify the non-ideal sliding plane, this system is driven to slide forward according to the pre-set sliding route.

Suppose the dynamic equation is $\dot{X} = AX + BU$, a sliding function can be designed as:

$$s(x) = C^T X = \sum_i^n c_i x_i \qquad (1)$$

where $C^T = \begin{bmatrix} c_1 & c_2 & .... & c_n \end{bmatrix}^T$ is the coefficient vector of the pre-defined sliding plane.

To make the representative point slide to the sliding plane, it is necessary to satisfy the inequality equation of

$$\lim_{S \to 0} s \cdot \dot{s} \leq 0 \qquad (2)$$

The equation of fuzzy sliding model thus becomes

$$u = u_{eq} + u_d = -\left(K_{eq}^T + K_d^T\right)x \qquad (3)$$

where $K_{eq}^T = \left(C^T B\right)^{-1} C^T A$

$$u_d = -k(x)\,\text{sgn}(s)$$

and sgn(s) is the sign of representative point s.

However, under many circumstances, it is impossible to know AX and B, and also difficult to get a precise information of u_{eq}. Under such uncertain conditions, the control equation will be represented as:

$$u = u_f \qquad (4)$$

where u_f is obtained from the fuzzy sliding mode equation of the following fuzzy rule base:

$$IF \ s \ is \ S_j\left(m_j, \sigma_j\right) THEN \ u_f \ is \ U\left(\phi_j\right) \qquad (5)$$

where s is the system state, S_j is the system fuzzify input, m_j is the mean value of bell-shaped membership function, σ_j is the standard deviation and U(ϕ_j) is the de-fuzzify output.

Using fuzzy control and fuzzify the sliding plane (as shown in Figure 1), the system state can be driven into the sliding plane and proceed to the origin through the sliding route to achieve the result of sliding mode control.

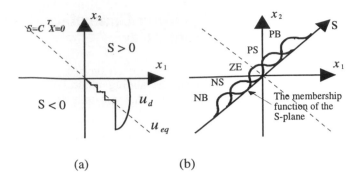

(a)   (b)

Figure 1. (a) The state space of sliding model
(b) Fuzzify sliding plane

2.2 Basic Structure of Cerebellar Model

Cerebellar model articulation controller (CMAC)[1][2] uses a structure which mimics the cerebella cortex to store the information. Under this model, it is necessary to map out the memory structure for the learning of cerebella model. During the learning process, the memory content is modified according to the expected output to achieve the feature of enhancing memory similar to human learning process. Under such model, every learning reference state is quantified and partitioned to many discrete values. These values are then mapped to different memory space. CMAC uses a set of indices to generate the corresponding addresses to fetch memory content, and then generates a set of output signal accordingly. The difference between this output and expected output is feedback to the memory for the update and correction (as shown in Figure 2). This learning process repeats itself until the output converges to a tolerable range. Basically, CMAC uses the technique of table look-up and converges very rapidly.

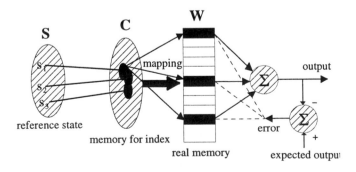

Figure 2. The basic structure of cerebella model

3. Design of Fuzzy Sliding Controller Based on Cerebellar Learning Model

The design of fuzzy sliding-mode controller based on cerebellar learning model (CFSMC) consists of two parts: the design of FSMC as the main part with a learning part the design of CMAC. The system structure is shown in Figure 3.

The FSMC output signal, u_f is the main input of controlled object. And the output of CMAC learning system is the corrected signal source u_c. These two signals are synthesized as the control signal of the controlled object. The learning system

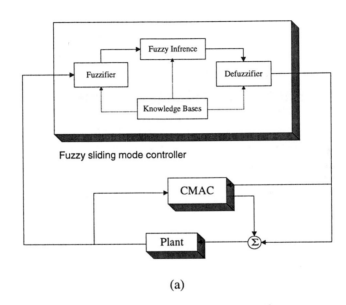

(a)

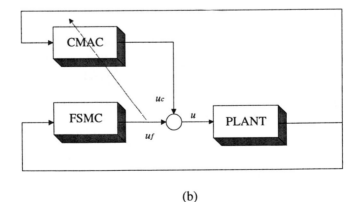

(b)

Figure 3(a). Fussy sliding mode controller with
cerebellar mode learning system
3(b). The structure of CFSMC system

adopts a one-dimensional cerebella model, which uses the output error state as the guiding index to modify the FSMC output.

The initial value of CMAC state memory unit is set to zero. Therefore, the output of FSMC control the system while CMAC is idle in the beginning. However, as time moves on, CMAC continues learning through the system operation and participates the control procedure. Eventually it makes the output of FSMC become zero. This also guarantees the output error reduce to zero to achieve the control goal.

While the system becomes stable, if there occurs any non-identified condition which renders the system parameter change; or a disturbance develops from external to result in errors, FSMC will take control immediately. CMAC only needs to participate part of the learning and it will enable the corrected output value to be adjusted to a better condition already, lest the system diverge. Therefore, it not only reduces the steady state error but also achieves excellent robustness.

4. The experiment of inverted pendulum

The inverted pendulum is often used to verify the effectiveness of control method on the real system because of its non-linear and uncertain features. In this section, the CFSMC controller is used to verify its effectiveness on eliminating the track error, achieving the actions of slew-up, stand-on and positioning actions of the 360 ° inverted pendulum. It is also used to examine the robustness of the controller when the system variables change.

4.1 Hardware Structure

The experimental set-up consists of a straight platform with an inverted pendulum of one-degree freedom, which can swing from 0 to 360 degrees. Pendulum rod is an aluminum rod of 33 cm in length, 6 mm in diameter and 50g in weight. It is driven by an a.c. brushless server motor and moves along a track of 29 cm long in X-direction. The 52g aluminum pendulum tests the robustness of the controller when system changes through the shift of the center of gravity of pendulum rod by pendulum moving.

The angle of inverted pendulum is input to computer interface through a set of encoding counter by an optical encoder. The position of the straight platform is input to computer interface through another set of encoding counter by the optical encoder in the motor.

After the computer receives the information such as the angle and its variation, the positioning and its variation, of the inverted pendulum, CFSMC will calculate the current command for the next moment. It then sends the command to the driver of an a.c. brushless server motor through the D/A converter. The whole experimental system structure is shown in Figure 4.

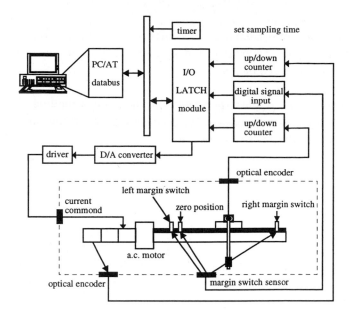

Figure 4. Experimental Set-up Structure

4.2 Software Structure

The main software structure is based on C language, which is coded for I/O interface and the control program of CFSMC. MATLAB is used to present the output of system response. The information propagates between these two languages using ASCII files as interface. After system executes, the input/output states and the capability values are stored in ASCII format and then plotted with MATLAB.

4.3 Experimental Results

In the experiment of inverted pendulum, the complete action sequence includes:
1. Reset
    At the beginning of the program execution, the inverted pendulum platform will be reset and moved to the left. The test positioning is set to zero. The pendulum rod is static as reset state.
2. Positioning
    As the pendulum begins to slew, it must be positioned at the center to gain the maximum degree of freedoms. Positioning is achieved through the use of fuzzy sliding mode control method.
3. Slew-up
    The fuzzy sliding mode will control the platform to swing from left to right and uses the angle of pendulum rod as the control target to achieve the slew-up action.

70

4.   stand-on
When the angle of pendulum rod becomes ready for the control stage of stand-on action, the angle of pendulum rod is maintained at certain control range and positioned toward central point.

It is demonstrated in the experiment that the control method of fuzzy sliding mode can achieve the requirements of positioning and stand-on of the inverted pendulum. The results are shown below:

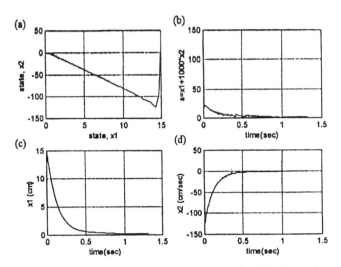

Figure 5. The system response of Positioning of
inverted pendulum.

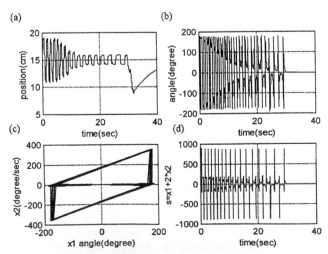

Figure 6. The system response of the slew-up of
inverted pendulum

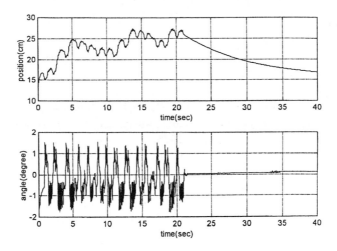

Figure 7. The system response of the stand-on and
positioning of inverted pendulum

5. Conclusions

In the traditional control, due to the change of the mathematical parameters of the controlled device, the operator must respond and adjust the system with empirical rule repeatedly. This results in great inconvenience in practical application. Therefore, the modern control development favors the trend of self-learning, i.e. the system is capable of learning and accommodating the variation of the controlled device parameters to enhance the system adaptability.

In summary, this paper describes the performance of a controller which is designed with the concepts of variable structure, fuzzy control and CMAC. The experimental results reveal two unique features:

1. Through the sliding mode, the multi-input variables are mapped to the sliding plane so that the rule base of the fuzzy control and the design of membership function are related only to the rule and independent of the input variables. This greatly reduces the difficulties of designing the fuzzy controller.
2. Since all the variables are mapped to the sliding plane, the design of CMAC is essentially one-dimensional. It relieves the computation burden of the controller. Fuzzify the sliding plane makes the requirements of CMAC state input the same as fussy theory. This reduces the required memory size and greatly helps the whole system timing.

Acknowledgement

In acknowledgement of being funded by the National Science Council (title : Design of Cerebellar Model Articulation Controller based on Fuzzy Reasoning project # 88-2213-E-003-005)

Reference

[1] Albus,J.S. , "A New Approach to Manipulator Control: The Cerebellar Model Articulation Controller (CMAC)," Journal of Dynamic System, and Control, Transaction of ASME , pp.220-227, September 1975.

[2] Albus, J. S. , "Data Storage in the Cerebellar Model Articulation Controller (CMAC)," Journal of Dynamic Systems, Measurement, and Control, Transactions of ASME , pp. 228-233, September 1975.

[3] Chun-Shin Lin and Hyongsuk Kim, "CMAC-Based Adaptive Critic Self-Learning Control , "IEEE Transactions on Neural Network, Vol. 2, No.5 , pp.530-535, 1991.

[4] Chun-Shin Lin and Hyongsuk Kim, "Selection of Learning Parameters for CMAC-Based Adaptive Critic Learning," IEEE Transactions on Neural Networks, Vol. 6, No.3, pp.642-647, 1995

[5] Ching-Tsan Chiang and Chun-Shin Lin, "CMAC with General Basis Functions," Neural Networks, Vol.9 , No.7 , pp1199-1211, 1996

[6] Giuseppe S. Buja, Roberto Menis and Maria I. Valla, "Variable Structure Control of An SRM Drive," IEEE Transactions on Industrial Electronics, Vol. 40, No 1, February 1993.

[7] Jacob S. Glower and Jeffrey Munighan, "Designing Fuzzy Controller From a Variable Structures Standpoint," IEEE Transactions on Fuzzy Systems. Vol. 5, No. 1, pp. 138-144, February 1997.

[8] John Y. Hring, Weibing Gao and James C. Hung, "Variable Structure Control: A Survey," IEEE Transactions on Industrial Electronics, Vol. 40, No. 1, February 1993.

[9] Jong-Lick Lin and Wen-Tay, "Application of Variable-Structure System in DC-motor velocity-control", Journal of Institute of Engineers. Vol. 10, No. 2, pp201-208, 1987.

[10] Seungrohk Oh and Hassan K. Khalil, "Nonlinear Output-Feedback Tracking Using High-Gain Observer and Variable Structure Control," Automatica, Vol. 33, No. 10, pp. 1845-1856, 1997.

[11] Shinn-cheng Lin and Yung-Yaw Chen, "A GA-Based Fuzzy Controller with Sliding Mode," IEEE INT. Conference on fuzzy system, Vol. 71, No. 3, pp. 1103-1110 1995.

[12] S.K. Bag, S.K. Spurgeon, C., "Output feedback Sliding mode design for linear uncertain systems," IEEE proc.-Control Theory Appl., Vol. 144, No. 3, May 1997.

[13] Susy Thomas and B. Bandyopadhyay, "Variable Structure model following control for industrial robotos," International Journal of Systems Science, Vol. 28, No. 4, pp. 325-334, 1997.

[14] Tian-Ping Zhang and Chun-Bo Feng, "Fuzzy variable structure control via output feedback," International Journal of System Science, Vol. 28, No. 3, pp. 309-319, 1997.

[15] Tong Shaocheng, Chai Tianyou, Shao Cheng, "Adaptive Fuzzy Sliding Mode Control for Nonlinear Systems," IEEE INT.conference on fuzzy system, Vol. 1, pp. 49-54, 1996.

[16] Vadim I. Utkin, "Sliding Mode Control Design Tong Shaocheng, Chai Tianyou, Shao Cheng, Principles and Applications to Electric Drives," IEEE Transactions on Industrial Electronics, Vol. 40, No. 1, February 1993.

[17] Weibing Gao and James C. Hung, " Variable Structure Control of Nonlinear Systems A New Approach," IEEE Transactions on Industrial Electronic, Vol. 40, No. 1, February 1993.

A New Fuzzy Flexible Flow-Shop Algorithm for Continuous Fuzzy Domain

Tzung-Pei Hong[1] and Tzu-Ting Wang[2]

Department of Information Management, I-Shou University
Kaohsiung, 84008, Taiwan, R.O.C.
[1] e-mail: tphong@csa500.isu.edu.tw
[2] e-mail: m873204m@csa500.isu.edu.tw

Abstract. In simple flow shop problems, each machine operation center includes just one machine. If at least one machine center includes more than one machine, the scheduling problem becomes a flexible flow-shop problem. In this paper, we apply triangular fuzzy membership functions to represent uncertainty in processing times for flexible flow shops with two machine centers, and then propose a fuzzy heuristic algorithm for scheduling their jobs. We first use triangular fuzzy LPT algorithm to allocate jobs, and then use triangular fuzzy Johnson algorithm to deal with sequencing the tasks. The proposed method thus provides a more flexible way of scheduling jobs than conventional scheduling methods.

1 Introduction

Scheduling jobs in flexible flow shops is considered an NP-complete problem [5]. Sriskandarajah and Sethi proposed a heuristic algorithm [11] to solve special flexible flow-shop problems in which only two machine centers exist and each machine center has the same number of homogenous machines. Each job must first be processed by the first machine center and then by the second machine center. Whenever a job arrives at a machine center, any free machine at that center can process it.

In the past, processing times for jobs have usually been assigned or estimated as fixed values. In many real-world applications, however, processing times may vary dynamically due to human factors or operating faults. The estimated processing times are then imprecise. Fuzzy set theory provides a good model for easily managing uncertain situations. Although fuzzy set concepts are mainly used in linguistic domains, they are also used in numerical domains, where each number is assigned a membership value.

In this paper, we use triangular membership functions for flexible flow shops with two-machine centers to examine processing-time uncertainties and to make scheduling more suitable for real applications. Since the processing time of each job is uncertain, the final completion time is also uncertain and can be represented by a triangular membership function. Triangular membership functions are used here to represent the fuzzy processing time of tasks. A triangular fuzzy membership function can be denoted by $A=(a, b, c)$, where $a \leq b \leq c$. The abscissa b represents the variable value with the maximal grade of membership value, i.e. $\mu_A(b)=1$; a and c are the lower and upper bounds of the available area. They are used to reflect the fuzziness of the data.

2 Review of Sriskandarajah and Sethi's Scheduling Algorithm

In [11], Sriskandarajah and Sethi proposed a heuristic algorithm for solving a special case of the flexible flow-shop problem. The special case is stated as follows. A factory has two machine centers. Each center contains the same number of homogeneous machines and each job consists of two tasks. The first task must be processed at the first machine center and the second task at the second machine center. Sriskandarajah and Sethi solved this problem obtaining nearly optimal final completion times.

Although the problem solved by Sriskandarajah and Sethi is only a special case of the flexible flow-shop problem, it is still quite complex. Sriskandarajah and Sethi decomposed the problem into three subproblems and solved each heuristically. The first subproblem is machine allocation; that is, each machine in center 1 is assigned to a machine in center 2 forming fixed pairs. The second subproblem is clustering; that is, jobs are clustered according to their processing times and the numbers of machines in each center. Each cluster of jobs is then allocated to a certain pair of machines. This is done using the LPT algorithm, which assigns jobs with the longest processing times to machines with the shortest completion times first. The third subproblem is sequencing and timing; that is, each cluster of jobs is allocated to a machine pair and the job sequences are determined. The start time for each job is also determined in this stage. This is done using the Johnson algorithm [9], which schedules jobs with shorter execution times on machine 1 than on machine 2 for execution first, and jobs with shorter execution times on machine 2 than on machine 1 for later execution. When a machine is free, the next unexecuted job is then assigned to it for execution.

3 Notation

Notation:
  m: *The number of machines in each machine center.*
  n: *The number of jobs to be scheduled.*
  mc_i: *The i-th machine center, $i = 1,2$.*
  m_{ij}: *The i-th machine of the j-th machine center.*
  F_i: *The i-th machine pair composed of one machine from each machine center (which can be thought of as a simple flow shop), $i = 1$ to m.*
  F_{ji}: *The j-th machine of the flow shop F_i, $j = 1, 2$.*
  f_i: *The fuzzy execution time of the i-th flow shop, F_i.*
  cf_{ij}: *The completion time $(a_{cfij}, b_{cfij}, c_{cfij})$ of the i-th machine in the j-th flow shop.*
  J_j: *The j-th job, $1 \le j \le n$.*
  J_{ij}: *The i-th task of job J_j, $i = 1, 2$.*

t_{ij}: *The fuzzy execution time* $\left(a_{tij}, b_{tij}, c_{tij}\right)$ *of the i-th task in* J_j, $i = 1, 2$.

tt_j: *The total fuzzy execution time* $\left(a_{ttj}, b_{ttj}, c_{ttj}\right)$ *of the j-th job* $(t_{1j} + t_{2j})$.

ff: *The final fuzzy completion time* $\left(a_{ff}, b_{ff}, c_{ff}\right)$ *of the entire schedule.*

4 Ranking Two Fuzzy Sets

Ranking numbers is quite important in a scheduling algorithm. Several fuzzy ranking methods have been proposed in the literature [1][2][3][4][6][10][12]. Below, the averaging ranking method is used as an example to show the proposed fuzzy scheduling algorithm. Note that other ranking methods can also be used in our fuzzy scheduling algorithm. Ranking using the averaging method is defined below.

The average height of a triangular membership function (a, b, c) is stated as follows.

$$ height = \frac{\int (\mu_A(x) \times x)dx}{\int \mu_A(x)dx} = \frac{\left| \int_a^b (\frac{x-a}{b-a}) \times xdx + \int_b^c (\frac{c-x}{c-b}) \times xdx \right|}{\left| \int_a^b (\frac{x-a}{b-a})dx + \int_b^c (\frac{x-a}{b-a})dx \right|} = \frac{1}{3}(a+b+c). $$

Let A and B be two triangular fuzzy sets. A and B can then be represented as follows:

$$ A = (a_A, b_A, c_A), $$
$$ B = (a_B, b_B, c_B). $$

Let

$$ h(A) = 1/3(a_A + b_A + c_A), $$
$$ h(B) = 1/3(a_B + b_B + c_B). $$

Using the average-height ranking method, we say $A > B$ if $h(A) > h(B)$.

5 The continuous fuzzy flexible flow-shop algorithm for two machine centers

The continuous fuzzy flexible flow-shop algorithm is based on Sriskandarajah and Sethi's algorithm [8] and uses triangular membership functions to manage uncertainty. The possible processing times for each task are represented by a triangular membership function. The proposed continuous fuzzy flexible flow-shop algorithm is shown below.

Continuous fuzzy flexible flow-shop algorithm for two machine centers
Input: A set of n jobs, each having two tasks with triangular fuzzy processing times, to be executed in turn by each of two machine centers with m homogenous machines.
Output: A schedule with a fuzzy completion time.

Part 1: Forming the machine groups.

Step 1: Form m machine pairs, each of which contains one machine from each machine center. Each machine pair can be thought of as a simple flow shop F_1, $F_2, \ldots, F_m$.

Step 2: Initialize the completion time of each flow shop $f_1, f_2, \ldots, f_m$ as a triangular membership function $(0, 0, 0)$. Its average height $ah(f) = \dfrac{1}{3}(a_{fi} + b_{fi} + c_{fi}) = 0$.

Step 3: For each job J_j, $1 \leq j \leq n$, find the total time $tt_j = t_{1j} + t_{2j}$ using the triangular fuzzy addition operation.

Part 2: Assigning jobs to machine groups.

Step 4: For each job J_j, find the average height $ah(tt_j)$; that is:

$$ah(tt_j) = \frac{1}{3}(a_{ttj} + b_{ttj} + c_{ttj}).$$

Step 5: Sort the jobs in descending order of average height $ah(tt_j)$; if any two jobs have the same $ah(tt_j)$ values, sort them in an arbitrary order.

Step 6: For each flow shop F_i ($i = 1$ to m), find the current average height $ah(f_i)$ of processing time f_i; that is:

$$ah(f_i) = \frac{1}{3}(a_{fi} + b_{fi} + c_{fi}).$$

Step 7: Find the flow shop F_i with the minimum average height $ah(f_i)$ among all the flow shops; if two flow shops have the same minimum $ah(f_i)$ value, choose an arbitrary one.

Step 8: Assign the first job J_j in the sorted list to the chosen flow shop, F_i, that has the minimum average height, $ah(f_i)$, among all m flow shops.

Step 9: Add the total time tt_j of job J_j to the completion time of the chosen flow shop, Fi, using the triangular fuzzy addition operation; that is

$$fi = fi + tt_j.$$

Step 10: Find the new average height $ah(f_i) = \dfrac{1}{3}(a_{fi} + b_{fi} + c_{fi})$, for the chosen flow shop, F_i, whose completion time has just been changed.

Step 11: Remove job J_j from the job list.

Step 12: Repeat Steps 7 to 11 until the job list is empty. After Step 12, jobs are clustered into m groups and are allocated to the m machine pairs (flow shops).

Part 3: Dealing with job sequencing in each flow shop.

Step 13: Find the average height for each task, t_{kj}, in each flow shop, F_i ($i = 1$ to m), using the following formula:

$$ah(t_{kj}) = \frac{1}{3}(a_{tkj} + b_{tkj} + c_{tkj}).$$

Step 14: Form the group of jobs U_{Fi} that takes less average processing time on the first machine center than on the second machine center for each flow shop F_i ($i = 1$ to m) with its allocated jobs.

Step 15: Form the group of jobs F_i that takes less or equal average time on the second machine center than on the first machine center for each flow shop F_i ($i = 1$ to m) with its allocated jobs.

Step 16: Sort the jobs in ascending order of the average height of the first tasks for each U_{Fi}.

Step 17: Sort the jobs in descending order of the average height of the second tasks for each V_{Fi}.

Step 18: Set the initial completion time of the first machine (cf_{1i}) and the second machine (cf_{2i}) to a triangular fuzzy membership function (0, 0, 0) in each flow shop F_i, with average height $ah(cf_{1i}) = ah(cf_{2i}) = 0$.

Step 19: Assign the first job J_j in the sorted U_{Fi} to the machines such that J_{1j} is assigned to F_{1i} and J_{2j} is assigned to F_{2i} for each flow shop F_i.

Step 20: Add the processing time t_{1j} to the completion time of the first machine cf_{1i} using the triangular fuzzy addition operation; that is:

$$cf_{1i} = cf_{1i} + t_{1j}.$$

Step 21: Set cf_{2i} = calculate-triangular-longer-time(cf_{1i}, cf_{2i})+ t_{2j}, where calculate-triangular-longer-time is a procedure given below.

Step 22: Remove job J_j from U_{Fi}.

Step 23: Repeat Steps 19 to 22 until U_{Fi} is empty.

Step 24: Schedule jobs in each V_{Fi} in a similar way (Steps 19 to 23).

Step 25: Set the final completion time of each flow shop f_i = the completion time of the second machine cf_{2i}.

Step 26: Use the calculate-triangular-longer-time procedure to find the final completion time ff with a triangular membership function among the completion times for all flow shops.

After Step 26, scheduling is finished and a triangular fuzzy completion time has been found. The calculate-triangular-longer-time procedure is designed in [7] and given below.

The calculate-triangular-longer-time procedure:

Input: m fuzzy triangular set $f_1 (a_1, b_1, c_1)$ to $f_m (a_m, b_m, c_m)$.

Output: the fuzzy maximum value $ff (a_{ff}, b_{ff}, c_{ff})$ for f_1 to f_m.

Procedure:

(I). *FOR* each f_i, i = 1 to m-1, *DO* the following:

(I-i). Set $a_{i+1} = max\{max(a_i, a_{i+1}), min(b_i, b_{i+1})\}$

(I-ii). Set $c_{i+1} = max(c_i, c_{i+1})$;

(I-iii). Set $b_{i+1} = \begin{cases} max(b_i, b_{i+1}); \text{if } max(b_i, b_{i+1}) \geq min(c_i, c_{i+1}) \\ \dfrac{c_i \times c_{i+1} - b_i \times b_{i+1}}{(c_i + c_{i+1}) - (b_i + b_{i+1})}; \text{if } max(b_i, b_{i+1}) < min(c_i, c_{i+1}) \end{cases}$

END FOR.

(II). Set $ff = (a_{ff}, b_{ff}, c_{ff}) = (max(a_m), max(b_m), max(c_m))$.

It is easily seen that the procedure is very simple and needs only little computation time. Details can be found in [7].

6. Conclusion

Scheduling jobs in flexible flow shops has long been known to be an NP-complete problem. Since task processing times in real applications are usually uncertain, in this paper, we have proposed a fuzzy scheduling algorithm for scheduling jobs in flexible flow shops with two machine centers. The scheduling results are a fuzzy set and can help system managers have broader views of scheduling and make good analysis. In the future, we will consider other task constraints, such as setup times, due dates, and priorities.

References

1. J. J. Buckley, "The multiple judge, multiple criteria ranking problem: A fuzzy set approach," Fuzzy Sets and Systems, Vol. 13, pp.25-37, 1984.
2. J. J. Buckley, "Ranking alternatives using fuzzy numbers," Fuzzy Sets and Systems, Vol. 15, pp.21-31, 1985.
3. J. J. Buckley and S. Chanas, "A fast method of ranking alternative using fuzzy numbers," Fuzzy Sets and Systems, Vol. 30, pp.337-338, 1989.
4. L. M. Campos Ibanez and A. Gonzalez Munoz, "A subjective approach for ranking fuzzy numbers," Fuzzy Sets and Systems, Vol. 29, pp. 145-153, 1989.
5. S. C. Chung and D. Y. Liao, "Scheduling flexible flow shops with no setup effects," The 1992 IEEE International. Conference on Robotics and Automation, pp.1179-1184, 1992.
6. M. Delgado, J.L. Verdegay and M.A. Vila, "A procedure for ranking fuzzy numbers using fuzzy relations," Fuzzy Sets and Systems, Vol. 26, pp. 49-62, 1988.
7. T. P. Hong and T. N. Chuang, "A new triangular fuzzy Johnson algorithm," accepted and to appear in *International Journal of Computers & Industrial Engineering*.
8. T. P. Hong and W. C. Chen, "Fuzzy flexible-flow shops at two machine centers," *Journal of Advanced Computational Intelligence,* Vol. 2, No. 4, 1998, pp. 142-149.
9. T. E. Morton and D. W. Pentico, Heuristic Scheduling Systems with Applications to Production Systems and Project Management, John Wiley & Sons Inc., New York, 1993.
10. G. Munoz, "A study of the ranking function approach through mean values," Fuzzy Sets and Systems, Vol. 35, pp. 29-41, 1990.
11. Sriskandarajah and S. P. Sethi, "Scheduling algorithms for flexible flow shops: worst and average case performance," European Journal of Operational Research, Vol. 43, pp. 143-160, 1989.
12. R. R. Yager, "A procedure for ordering fuzzy subsets of the unit interval," Inform Sci., vol. 24, pp. 143-161, 1981.

Testing the Performance of a Neural Networks-Based Adaptive Call Admission Controller with Multimedia Traffic

Yasser Dakroury, Ahmed Abd El Al, Osman Badr

Computer and Systems Engineering Dept., Faculty of Engineering, Ain Shams University, Cairo, Egypt.
{dakroury, amohamed, obadr}@shams.eun.eg

Abstract. We propose an Artificial Neural Network (ANN) based technique to be used for adaptive Call Admission Control (CAC) in Asynchronous Transfer Mode (ATM) networks. The mechanism does not depend on the traffic descriptors registered between the terminal and the network during connection setup time, thus inaccuracies in these descriptors will not affect the controller decisions. The controller monitors the current traffic state and adapts its decisions according to it. Also it takes into account the Quality of Service (QoS) of each traffic class separately as it bases the admission decision on the individual cell loss probability of each traffic class not on the average cell loss probability for mixed traffic. We modified the "Leaky Pattern Table" on-line training method to enable the NN to capture traffic variations more accurately. The proposed controller was tested with multimedia traffic as ATM is expected to be used for multimedia applications and we took into consideration the effect of the node buffer, otherwise the CAC mechanism will be too conservative. Reported results prove that this reactive controller is much more effective than other CAC mechanisms that depend on the traffic descriptors.

1 Introduction

Asynchronous Transfer Mode (ATM) is the technology recommended by ITU to provide voice, data, image and video services using a single integrated broadband network. The ATM solution is flexible enough to support a diverse mixture of multimedia traffic with different correlation's and burstiness properties. Not only does the different types of multimedia traffic differ in their statistical characteristics but also in their service requirements specified by the Quality of Service (QoS). The QoS defines a set of guaranteed performance measurement parameters (e.g., cell loss rate, delay, delay variability) that each type of the multimedia traffic requires from the network. Due to the absence of a channel structure in ATM-based networks, the user's traffic can be easily overwhelm the network resources leading to serious congestion problems. Thus congestion control mechanisms are required in ATM networks. Congestion control mechanisms proposed for ATM networks are classified into two categories: preventive control and reactive control [1], [2]. Preventive congestion control techniques attempt to prevent congestion by taking appropriate actions before they actually occur. However, it is generally agreed that preventive control techniques are not sufficient to eliminate congestion problems in ATM networks and that when congestion occurs it is necessary to react to the problem. Proposed reactive control techniques initiate the recovery from congested state. In a reactive scheme, the network is monitored for congestion. When congestion is detected, sources are requested to slow down or stop transmission for a while until congestion is cleared. Call Admission Control (CAC) is a preventive congestion control mechanism that decides to accept a new connection if the required QoS is guaranteed for the new connection while maintaining the QoS of the existing ones. Hence, at the call set-up phase, a traffic contract is "agreed-upon" between the network and the user. According to that contract, the CAC algorithm maintains a specific QoS for the user's connection as long as the user does not violate his contractual parameters (traffic descriptors) that characterize the traffic. The CAC mechanism should function in real-time and should attempt to maximize the utilization of network resources while meeting the user's QoS requirements.

CAC schemes may be classified as nonstatistical allocation and statistical allocation. Nonstatistical allocation allocates the peak rate to each traffic source independent of whether the source transmits continuously at the peak rate or not. Statistical allocation allocates bandwidth (called statistical bandwidth) that is greater than average rate and less than peak rate. As a result, the sum of all peak rates may be greater than the capacity of the output link. Different CAC algorithms have been proposed in literature for statistical allocation. One of these algorithms is "Mean Rate Allocation" in which the average transmission rate is allocated to each connection. Anther CAC algorithm is the "Equivalent Capacity" [3]. The equivalent capacity for a single source can be derived by assuming the source to be an on/off fluid source or, equivalently, an Interrupted Fluid Process (IFP). An IFP source can be completely characterized by the vector (R, r, b), where R is the peak rate, r the fraction of time the source is active, and b the mean duration of the active period. If the source feeds a finite capacity queue with size K and with constant service time, the service rate c that corresponds to a given cell loss ε was found to be as follows:

$$c = \frac{a - K + \sqrt{(a - K)^2 + 4Kar}}{2a} R \tag{1}$$

where a = $\ln(1/\varepsilon)$b(1 - r)R.

The problem with the previous CAC algorithms, that they depend on the traffic parameters declared by the user during call set-up and they use them to calculate the required bandwidth. This approach might not be effective since it is based upon simple parameters (e.g. peak, average bit rates) that are not accurate enough to characterize the correlations and burstiness properties of the traffic. Also due to real time constraints, which cause that not all the traffic characteristics might be known at call set-up with the required accuracy, there will be several inaccuracies in estimating these parameters, which in turn lead to similar inaccuracies in bandwidth allocation. The correct approach is to measure higher-order moments of the traffic such as the squared coefficient of variation. Of course, the problem in this solution is the difficulty in performing it in "real-time", since calculating these higher-order moments requires complex calculations at high rates. The problem will increase with the traffic of future services that are expected to have complex characteristics and require higher order moments in order to describe them, which will increase the possibility of traffic descriptor inaccuracies. Anther problem is that these mechanisms are static, so that once the connection is accepted they do not monitor the traffic and adapt their decisions according to the changes in network state.

One of the attractive tools for building an adaptive controller is the Neural Network (NN) because of its important features such as: adaptive learning, high computation rate, generalization from learning and high degree of robustness or fault tolerance. Artificial neural network is generally a multiple-input multiple-output nonlinear mapping circuit, which can learn an unknown non-linear input-output relation from a set of examples.

1.1 Proposed Model

In this paper, we propose a CAC mechanism that uses neural network (NN). The structure of the proposed controller is shown in Fig. 1. Although various ways have been proposed to define the buffer status, which will be used as the neural network input, the number of connections are used here since it is simple to handle by it. All connections are categorized into groups according to the traffic parameters and the numbers of connections in each group are counted. A simple way to define the groups is categorization by communication services used, such as audio, video, and data. Of course, in an actual network, these groups should be divided into much smaller groups according to the coding methods or cell generation characteristics. The output value of the neural network indicates a decision to accept the new connection request or not based on whether the required QoS for each class (which is cell loss rate in our case) is satisfied after the controller accepts all new requests or not. Reference [6] showed that there is a significant difference between cell loss probability calculated over all traffic classes and cell loss probability for each traffic class separately (individual cell loss probability).

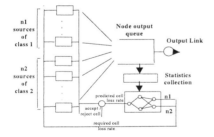

Fig. 1. Proposed NN based call admission controller

Section 2 describes the traffic model used. Section 3 describes the simulation environment and its parameters. Section 4 presents different simulation scenarios and obtained results whereas the conclusions are given in section 5.

2 Traffic Model

In our simulation we used two traffic classes, class 1 to represent video connections and class 2 to represent audio connections.

There are different video traffic models [7]. One of these models is the autoregressive stochastic model, which takes in account the Gaussian like pdf and the frame correlation properties in the video traffic. This model is useful to provide the bit rate variation with time but not useful for the analysis of statistical multiplexer, access control or admission control. Two other models can be used for analysis one of them is the Markov-modulated fluid flow based model and the other is the MMPP model.

A first order Autoregressive (AR) model was used in [5], [6] to approximate the video sources output rate. The autoregressive model is defined as:

$$\lambda(n) = \sum_{m=1}^{M} a_m \lambda(n - m) + b\omega(n) \tag{2}$$

where $\lambda(n)$ represents the source bit rate during the nth frame, M is the model order, $\omega(n)$ is a gaussian random process and a_m $(m=1,2,..M)$ and b are coefficients. The first order model is sufficient to match the mean and autocovariance function where $\omega(n)$ has a mean η and variance 1 and $|a_1|$ is less than 1. The values of coefficients a_1, b and mean η can be obtained by comparing these values with the corresponding values obtained from the actual video sequence as follows:

$$E(\lambda) = \frac{b\eta}{1 - a_1} \qquad |a_1| < 1 \tag{3}$$

$$\sigma^2_\lambda = C(0) = E(\lambda^2) - E^2(\lambda) = \frac{b^2}{1 - a_1^2} \tag{4}$$

$$C(n) = \sigma^2_\lambda |a_1|^n \qquad n \geq 0 \tag{5}$$

In order to obtain the values of coefficients a_1, b and mean η we compared equations (3), (4), (5) with the corresponding values obtained from a video sequence. The sample coded video sequence used was shown in [5] that represents a head of talking person. This video sequence is 300 frame long. Fig. 2 illustrates the coding bit rate of the captured sequence in (bits/pixel).

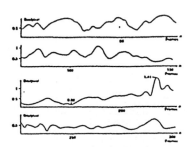

Fig. 2. Coding bit rate of the capture sequence (in bits/pixel)

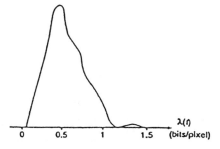

Fig. 3. Bit rate histogram

The minimum and maximum bit rates in the third strip of Fig. 2 are 0.08 bits/pixel and 1.41 bits/pixel respectively. From the corresponding bit rate appears in Fig. 3, we can notice its rough resemblance to a normal distribution. The average value λ is 0.52 bits/pixel and the standard deviation is $\sigma = 0.23$ bits/pixel.

Measurements made on the video sequence of Fig. 2 indicate that the covariance function for that sequence can be approximated fairly well by an exponential for values $n \leq 10$ frames.

$$C(n) = \sigma^2 e^{-\frac{kn}{30}} \tag{6}$$

From measurements, the covariance parameter k was found to be $k = 3.9$ sec., so that for the test sequence of Fig. 2,

$$C(n) = 0.0536\, e^{-0.13n} \tag{7}$$

By matching equations (3), (4) and (5) to the equivalent time (frame) averages of the actual video time sequence, we can find that:

$$a = e^{-0.13} = 0.878$$
$$b = 0.11$$
$$\eta = 0.58$$

Now the autoregressive process given in equation (2) can be used to evaluate the bit rate during the nth frame for the video sources. Parameters used in video sources modeling are shown in Table 1.

Table 1. Parameters used in video source simulation

Maximum number of sources	200
Average time between connections	3 sec.
Average call holding time	200 sec.

With respect to class 2 connections, that represent voice connections, the talk spurt duration and the silence duration were chosen as exponentially distributed. Average time between connections and average call holding time were also chosen as exponentially distributed [8]. Table 2 lists the traffic parameters for voice connections.

Table 2. Main characteristics of the voice traffic class used in simulation.

Maximum number of sources	200
Average time between connections	3 sec.
Average connection duration	200 sec.
Average talk spurt duration	0.35 sec.
Average silence duration	0.1167 sec.
Peak bit rate	2.12 Mbps
Mean bit rate	1.59 Mbps

3 Simulation Environment

We built a simulator to simulate an ATM network node, to which a number of traffic sources are attached. The number of active traffic sources varies with time. Cells from these sources are buffered and served by the link in FIFO discipline as sketched in Fig. 4.

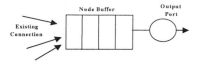

Fig. 4. An ATM node

The simulator was implemented using 'C'. We used some library functions that were used by SMPL simulator described in detail in [9]. Table 3 lists simulator parameters setting.

Table 3. Simulator parameters setting

Simulation duration	1000 – 1500 sec.
Sampling interval	1 sec.
Buffer size	50 cells
Link capacity	155.52 Mbps

Buffer size shown in Table 3 was determined as shown in [10], such that the maximum delay under FCFS discipline can be satisfied. That is, the output buffer size K and the maximum admissible delay T are assumed to satisfy the relation:

$$K = \frac{TC}{L} \qquad (8)$$

where C is the transmission link bandwidth and L is the cell size. Using K, C values as listed in Table 3 and $L = 53$ bytes. The maximum delay $T = 57.8$ msec. We studied the effect of dimensioning the buffer size, so as to obtain maximum delay = 25 msec., on our proposed CAC mechanism. Results showed that the corresponding reduction in the buffer size would not cause significant changes in results.

The NN was simulated using IBM Neural Network Utility/2 package [11]. This package provides a Common Application Interface (API) that was used by our simulator during on-line training.

3.1 Neural Network Off-Line Training

In order to prepare the data required for neural network off -line training, we built a traffic simulator that simulates different combinations of number of active sources in each traffic class and evaluates the cell loss rate for each traffic class separately (i.e. individual cell loss probability). Then it compares the cell loss rate for both classes with the target cell loss rate (chosen as 10^{-4} cells/sec.), if any of them exceeds the target cell loss rate the combination is recorded as high cell loss rate combination. This means that when this combination is presented to the neural network during operation, its correct decision is to reject new connections setup requests. Table 4 lists the parameters setting for the simulation.

Table 4. Parameters setting used for preparing the Off-Line training data

Maximum number of class 1 sources	300 sources
Maximum number of class 2 sources	100 sources
Simulation duration for each combination	50 sec.

The maximum number of sources in each traffic class were selected by assuming that the sources in each class are transmitting with their average bit rate and occupying the full link bandwidth. Fig. 5 shows the training data obtained from the simulation and used for neural network Off-Line training.

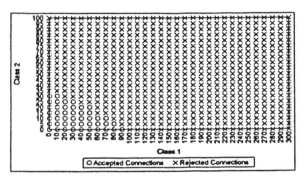

Fig. 5. Off-Line training data

In our work, back-propagation neural network was selected as the problem to be solved is a classification problem between low loss rate patterns and high loss rate patterns. The neural network has two inputs in the input layer, one for each traffic class and one neuron in the output layer to classify the input traffic pattern. The activation function of the NN was selected to be sigmoid. The number of hidden layer neurons, learning rate, and momentum, which control how the weight is updated by factoring in previous weight updates so as to reduce oscillation and help attain convergence, are selected by trail and error so as to achieve the lowest average RMS error. Other parameters were chosen as the default values of the model. The selected neural network parameters are listed in Table 5.

Table 5. Neural network parameters

Number of hidden layers	1
Number of hidden layer neurons	4
Learning rate	0.2
Momentum	0.9
Learning rate multiplier	1
Activation temperature	1
Tolerance	0.1

As Back propagation algorithm requires the data to be scaled to a small range from 0.0 to 1.0, the number of connections in each traffic class were scaled to this range by dividing the number of connections in each class by the maximum possible number of connections in this class before presenting them as an input to the neural network. Fig. 6 shows the average RMS error as a function of network epoch during off-line training.

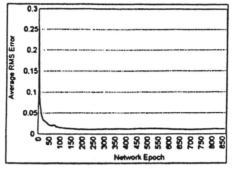

Fig. 6. Average RMS error versus network epoch

3.2 Neural Network On-line Training

After neural network off-line training, the trained neural network was embedded in the simulator and accessed using IBM NNU API. At the beginning the NNU API is initiated and the trained neural network is loaded. The on-line training method chosen is called "Leaky Pattern Selection Method" [4]. In this method, every training time, the current number of connections in each class as well as the cell loss probability for each traffic class since the last training time are computed. This training method requires two tables, one to hold low cell loss rate patterns and other for high cell loss rate patterns. These tables have fixed size = 100 entries each. The cell loss probabilities for both classes are compared to the required cell loss probability and accordingly the pattern is judged whether it is a high or low loss rate pattern. The pattern is inserted in the corresponding table and if the table is full it replaces the oldest pattern in the table. In order to decrease the frequency of training with old patterns, one pattern is discarded from pattern table with probability P_e. The pattern to be discarded is randomly selected from stored patterns. Since neural network requires training patterns distributed over the whole input pattern domain in random sequence, the training pattern is selected from high loss rate pattern table with probability P_1 and then the pattern is chosen from the selected table randomly. In our study, we have chosen $P_1 = 0.5$, $P_e = 0.03$ as in [4]. The pattern presented to the neural network for training consists of the number of connections in each traffic class scaled to range (0,1), by dividing it by the expected maximum number of connections in this class, as well as the required NN decision (i.e. Accept or Reject). Two breakpoints were used to prevent endless loop during training. The first breakpoint is that the average pattern error < 0.05, while the second is that the number of network epoch > 1000. The network stops training when either of them occurs. The original "leaky pattern table" method trains the network every fixed interval called "control cycle". This fixed training interval does not allow the NN to sense fast variations in the traffic characteristics due to the changes in the number of connections during this interval. In order to allow the NN network to capture network traffic variations more accurately, on-line training is performed every time an active connection ended or a new connection started or after 0.1 sec. if no changes occurred in the number of active connections during this interval. Our selection to this interval is based on compromising the load on the system due to training and at the same time to enable the NN to learn the traffic dynamics accurately, so we have chosen on-line training to be performed at intervals too small with respect to average time between connections for both traffic classes and in the same time adequate to collect statistics about cell loss probability.

4 Simulation Results

This section presents different simulation scenarios in order to validate our NN-based controller. Experiment 1 shows the effect of traffic descriptors inaccuracies on static CAC mechanisms. Experiment 2 tests the ability of our proposed NN CAC mechanism to achieve the required QoS. Experiment 3 tests our NN CAC mechanism in case of traffic descriptors inaccuracies.

4.1 Experiment 1:

This experiment tests the behavior of equivalent capacity allocation in case of traffic descriptors inaccuracies. We assume that traffic descriptors consist of peak rate, mean rate, and maximum burst duration. We represent the traffic descriptors inaccuracies by increasing class 1 peak rate 3 times the value specified in traffic descriptor for connections started after 250 sec. from starting the simulation.

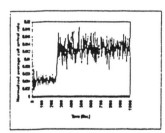 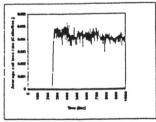 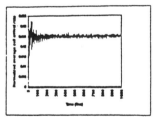

Fig. 7. Normalized average cell arrival rate for traffic class 1

Fig. 8. Average cell loss rate for traffic class 1

Fig. 9. Normalized average cell arrival rate for traffic class 2

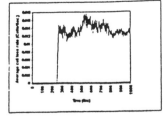 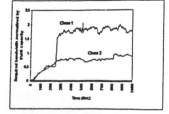

Fig. 10. Average cell loss rate for traffic class 2

Fig. 11. Bandwidth required by traffic classes normalized by trunk capacity

Previous figures show that after increasing the peak rate for class 1 connections the equivalent capacity allocation mechanism will use the peak rate value specified in the traffic descriptor to evaluate the required bandwidth for new connections. Thus it will allocate less bandwidth than actually required leading to an increase in average cell loss rate for traffic class 1. And because both classes share the same trunk, class 2 connections are also affected and their cell loss rate increased than the target value. We repeated the same experiment with other static CAC mechanisms (peak rate allocation, mean rate allocation) and we obtained similar results.

4.2 Experiment 2:

In experiment 2 we use the neural network after off-line training for call admission control.

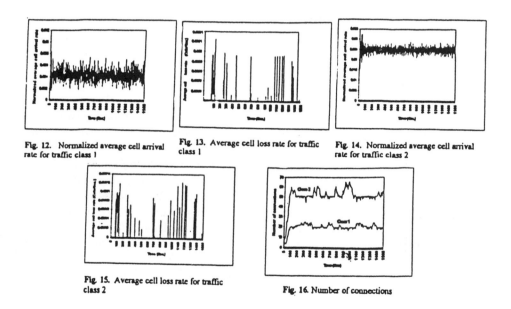

Fig. 12. Normalized average cell arrival rate for traffic class 1

Fig. 13. Average cell loss rate for traffic class 1

Fig. 14. Normalized average cell arrival rate for traffic class 2

Fig. 15. Average cell loss rate for traffic class 2

Fig. 16. Number of connections

Fig. 13 shows that the average cell loss rate for traffic class 1 is $\cong 1.05 * 10^{-5}$ cells/sec. Fig. 15 shows that traffic class 2 average cell loss rate is $\cong 2.2 * 10^{-5}$ cells/sec. This means that the neural network succeeded to keep the cell loss rate for both traffic classes below the target cell loss rate (10^{-4} cells/sec. in our case).

4.3 Experiment 3:

This experiment is similar to experiment 1, but here we used NN CAC mechanism instead of static CAC mechanisms.

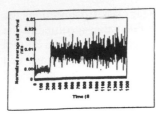

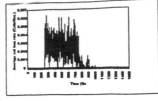

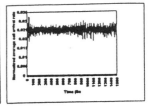

Fig. 17. Normalized average cell arrival rate for traffic class 1

Fig. 18. Average cell loss rate for traffic class 1

Fig. 19. Normalized average cell arrival rate for traffic class 2

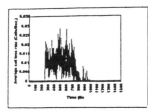

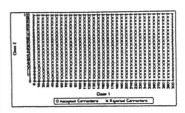

Fig. 20. Average cell loss rate for traffic class 2

Fig. 21. Number of connections versus time

Fig. 22. Neural network admission region after on-line training

It can be seen from Fig. 18 that the average cell loss rate during the first 250 sec. is $\cong 1.1 * 10^{-5}$ cells/sec., after that traffic class 1 peak cell rate increased by 3 times which cause increasing the cell loss rate. During about 500 sec., neural network was able to change its decision boundary by monitoring the generated traffic for each traffic class and on-line training. This new decision boundary returned the average cell loss rate to $\cong 5.4 * 10^{-6}$ cells/sec. Fig. 20 shows that during the first 250 sec. traffic class 2 average cell loss rate is $\cong 1.2 * 10^{-6}$ sec. After increasing traffic class 1 peak cell rate, the cell loss rate increases than the target value. Neural network new decision boundary returned the cell loss rate to average value $\cong 1.1 * 10^{-5}$ cells/sec. Fig. 21 shows how the number of connections for both classes changed in order to adapt to traffic state. Fig. 22 shows neural network admission region after the experiment.

The long response time taken by the NN to adapt its decision boundary can be justified that the simulated variations, increasing the traffic peak rate by 3 times the value in the traffic descriptors, are unrealistic and real world traffic variations are not so intense, so the neural network will be able to respond to them much faster.

Comparing Fig. 22 and Fig. 5 it can be seen that the neural network decreased its admission region in order to adapt with the increase in traffic class 1 peak cell rate so as to keep the cell loss rate for both traffic classes below the target value.

5 Conclusions

We propose an adaptive call admission controller using the capabilities of artificial neural network. The controller monitors the traffic state and the neural network adapts its decision boundary according to it, it does not depend on the traffic descriptors

91

registered between the terminal and the network during connection setup. In order to test our CAC mechanism we built a traffic simulator that generates traffic patterns representing video and voice connections. Simulation results show the effectiveness of the proposed mechanism while static CAC mechanisms fail to provide correct admission decisions when encountered with inaccurate traffic descriptors. Our study also shows the importance of taking in consideration the effect of the node buffer otherwise the CAC mechanism decisions will be too conservative. The size of this buffer can be dimensioned according to delay requirements of the most delay sensitive service expected to use the node. It is also very important to base the CAC decision on the individual cell loss probability for each traffic class not on the average cell loss probability for mixed traffic. Finally, it was shown that NN takes a long time to build its initial decision boundary, so this initial decision boundary can be built before embedding the NN in the application using a simple simulation model to the environment under consideration. Off-line training sets the decision boundary near optimal value. Then by on-line training NN decision boundary is adjusted to the optimal value and it is also changed according to the state of the environment. We modified the "leaky pattern table" learning method so as to allow the NN to capture traffic variations more accurately.

References

1. R. Onvural: Asynchronous Transfer Mode Network: Performance Issues, Artech House 1995
2. S. Yazid, H. T. Mouftah: Congestion Control Methods for B-ISDN, IEEE Commun. Mag., vol. 30, no. 7, pp. 42-47, Jul. 1992
3. R. Guerin, H. Ahmadi, M. Naghshinh: Equivalent Capacity and its Applications to Bandwidth Allocation in High-Speed Networks," IEEE J. Select. Areas Commun., vol. 9, no. 7, pp. 968-981, Sep. 1991
4. Hiramatsu: ATM Communications Network Control by Neural Networks, IEEE Trans. on Neural Networks, vol. 1, no. 1, pp. 122-130, Mar. 1990
5. Maglaris, D. Anastassiou, P. Sen, G. Karlson, J. D. Robbins: Performance Models of Statistical Multiplexing in Packet Video Communications", IEEE Trans. on Commun., Vol. 36, No. 7, Jul. 1988
6. T. Murase, et al.: A Call Admission Control Scheme for ATM Networks Using Simple Quality Estimate," IEEE J. Select Area Commun., vol. 9, no. 9, Apr. 1991
7. I. Habib, T. Saadawi: Multimedia Traffic Characteristics in Broadband Networks, IEEE Commun. Mag., vol. 30, no.7, pp. 48-54, Jul.1992
8. Y. Dakroury, A. Abd El Al, O. Badr: Neural Networks-Based Adaptive Call Admission Control in ATM Networks, in Proc. ICAIA 6, Cairo, Egypt, Feb. 1998
9. M. H. MacDougall: Simulating Computer Systems Techniques and Tools, MIT Press, 1987
10. H.Saito, K. Shimoto: Dynamic Call Admission Control in ATM Networks," IEEE J. Select. Areas Commun., vol. 9, no. 7, pp. 982-989, Sep. 1991
11. IBM, Neural Network Utility / 2: Programmers Reference, 1992

A Combined Neural Network and Mathematical Morphology Method for Computerized Detection of Microcalcifications

Daniel Manrique[1], Juan Ríos[1], Amparo Vilarrasa[2]

[1] Facultad de Informática, Universidad Politécnica de Madrid,
Campus de Montegancedo s/n, 28660 Boadilla del Monte, Madrid, Spain
jrios@fi.upm.es
[2] University Hospital 12 de Octubre, Servicio de Radiología
Ctra. Andalucía Km. 5,400, Madrid, Spain

Abstract. In this paper we present a new method for the automatic detection of microcalcifications combining morphological operations and artificial neural networks (ANN). The input chosen is a whole digitalized mammogram while the output of the algorithm is a new mammogram where microcalcifications (clustered or not) appear to be highlighted in white colour individually. Two new filters have been designed: one is based on mathematical morphology combined with other segmentation techniques and statistical methods, and the other one is made by training an ANN in order to be able to differentiate between those pixels belonging to a microcalcification and those being normal parenchymal patterns. Results for both algorithms are separately shown and how the combination of these two approaches improves the results.

1 Introduction

Breast cancer is one of the leading causes of death in women. Mammography has been proven to be the primary diagnostic procedure for the early detection of breast cancer. Between 30% and 50% of breast carcinomas demonstrate microcalcifications on mammograms, and between 60% and 80% of the carcinomas reveal microcalcifications upon histologic examination. This is the reason why using a good technique for computerized detection of microcalcifications can assist the radiologist in an early detection of breast cancer [1].

Machine recognition of forms has been the subject of extensive research in the last decade. Among the methods, techniques based on mathematical morphology have been the most used, a method based on the morphological inter-image substraction in order to extract user entered components from personal bankchecks is proposed in [2]. This technique requires a reference image without the components to be found which is not possible in our case of study because it is impossible to have the same mammogram twice, one of them with microcalcifications, and the other one without them. Morphological algorithms to analyse the features of the detected microcalcifications are also used in [3], or asynchronous dynamics for the iterative computation of morphological image filters to improve the convergence efficiency of such operators introduced in [4].

ANN is another of the most used techniques in pattern recognition and data classification in medical field [5]. They have been resulted to be a powerful tool for mammogram interpretation [6], to differenciate among patterns corresponding to various interstitial diseases in chest radiography [7], and to classify chest lesions [8].

We propose a robust algorithm for the automatic detection of microcalcifications, clustered or not, from a whole digitalized film mammogram. The output of the algorithm is the same image where the detected microcalcifications appear highlighted in white color. This method has been obtained combining a mathematical morphology based procedure [9] with an ANN based filter [10].

We retrospectively reviewed 154 mammographic studies showing micro-calcifications in 104 of them, for which surgical verification was available. All patients had standard mediolateral oblique and craneocaudal mammograms being obtained from the Department of Radiology of the University Hospital "12 de Octubre", Madrid, Spain and registered in a database implemented in our laboratory. The film screen mammograms were digitalized with a resolution of 1000 lpi, using 8 bits per pixel (256 grey levels). The minimum size in pixels of the digitalized images was 1800x1000. Eight bits per pixel were enough to obtain quite good results as can be seen in section 5 (Results) allowing also high speed processing which is required in real time systems.

Fig. 1. A detailed section of interest of a whole digitalized film mammogram.

In section 2 the procedure used to detect microcalcifications from morphological operations with other segmentation and statistical techniques is explained. Section 3 describes the building of the ANN based filter and how these two methods can be combined is exposed in section 4 improving its performance that is evaluated quantitatively in section 5 by means of the receiver operating characteristic (ROC) analysis [11].

It is shown throughout this work, as an illustrative example, the transformations made in the interesting region shown in figure 1, but bearing in mind that the input to the algorithm is always a complete mammogram.

2 Morphology Based Filter

Mathematical morphology based filter has two stages once the mammogram has been digitalized. Firstly, an initial filtering using the top-hat algorithm is applied, obtaining a new image where small areas appear to be highlighted over the rest of the pixels. Some of them are pixels belonging to microcalcifications. Secondly, a dynamic statistical method is employed to obtain the threshold value in order to decide which of the pixels of the image obtained in the previous step are really microcalcifications, and which are not.

2.1 Initial Filtering

When the background signal intensity of an image is constant, a simple thresholding may be used to detect features of interest. The top-hat algorithm provides a useful alternative when the background signal intensity is more variable, as is the case for stromal tissue on a mammogram (see figure 1) where microcalcifications are being searched. This method is based on morphological operations and consists of two steps.

Fig. 2. Results obtained in the region of interest shown in figure 1 after the top-hat algorithm has been applied. It can be seen small whitish areas, some of them corresponding to microcalcifications.

Firstly, a morphological opening is applied, comprising an erosion followed by a dilation. These operations require a kernel window (top-hat) systematically being placed on each pixel of the image. In our case we used a square kernel window of 21x21 pixels. During erosion, the pixel value at the center of the kernel is replaced by the minimum value of the neighborhood pixels. This operation reduces regions of high signal intensity in the digital image. After erosion, the complementary operation of dilation is applied. Now, the pixel value at the center of the window is replaced by the maximum value of the neighborhood pixels.

The second part of the top-hat algorithm involves the subtraction of the opened image from the original image.

2.2 Dynamic Thresholding

The purpose of this part of the algorithm is to distinguish between all the highlighted points that are microcalcifications, and the rest that are not which appear with a darker grey level. A dynamic grey-level thresholding segmentation algorithm is applied, and after obtaining the threshold value h, each point with a grey level value greater than it will be considered as part of a microcalcification, eliminating the rest which are lower than the threshold.

The threshold h is obtained dynamically depending on each image is being processed. For this purpose statistics methods have been used by calculation of the atypical values of a sample. First (q_1) and third (q_3) quartiles of the sample are calculated. Once these values have been obtained, it is statistically considered [12] that a point is atypical if it is greater than the upper limit U_1 or lower than the minimum limit L_1 using the formulas:

$$U_1 = q_3 + 1.5 (q_3 - q_1) \qquad\qquad L_1 = q_1 - 1.5(q_3 - q_1) \qquad\qquad (1)$$

Figure 3a. *Image obtained after the dynamic thresholding has been applied.*   **Figure 3b**. *Microcalcifications detected in the original image.*

Figure 3a shows the result after this stage of the algorithm has been applied in the region of interest of the figure 1, and how the microcalcifications appear to be highlighted in the original image (figure 3b) in order to the radiologist can give a final decision.

So, all the pixels with a grey level value greater than U_l ($h=U_l$) are considered to be microcalcifications. By doing so, a different threshold is obtained for each image depending on the grey-level values of its pixels. This feature makes also the algorithm independent from the parameters used in the digitalization of the mammogram.

As it can be seen from figure 2, the overwhelming majority of points have a grey level value equal or very near zero, which makes U_l has also a value very near zero. A great number of false microcalcifications would be obtained due to these points are greater than U_l.

To avoid this problem, a cut-off value C was taken in such a way that only those values greater than it are used to calculate U_l and L_l. Bearing that in mind, C cannot be greater than any pixel belonging to a microcalcification, because it would be eliminated from the set of possible points to study. The best results were obtained with a cut-off value of $C=25$.

3 Neural Network Based Filter

A filter for detection of microcalcifications in a digitalized mammogram using an ANN has been designed. The input to the network is a window of 9x9 pixels (81 input neurons) from the original digitalized image. The output of the network is 1 if the pixel value at the centre of the window belongs to a microcalcification, and 0 in other cases.

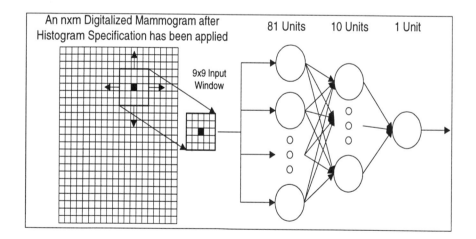

Figure 4. Neural network based filter architecture.

A total of 466 regions of interest (ROIs) of 9x9 pixels have been selected in order to train de ANN from the 154 cases of study. Histogram specification [13] has been employed to achieve some invariances to illumination shifts or gradients in all cases of study before extracting the training patterns. In 240 cases the pixel at the centre of the window belongs to a microcalcification (positive ROIs), and the other 226 are negative ROIs. 400 images have been used as training patterns and 66 to test the network.

We employed a full connected three layer, feed forward ANN. The number of the neurons for each layer was of 81-10-1. A back-propagation algorithm with generalized delta rule was employed in the training process. Sigmoid function was the activation function, defined as:

$$f(x) = \frac{1}{1 + e^{-x}} \tag{2}$$

The windows used as inputs to the network have 81 pixels with values in the range of [0,255] (256 gray-levels) which were standarized to [0,1]. The learning process was considered finished when the mean quadratic error reached 0.02.

Once the ANN has been trained, it is able to decide if the pixel at the centre of the 9x9 window introduced as its input belongs or not to a microcalcification.

The ANN based filter architecture is shown in figure 4 and works as follows: for each pixel in the original image, a 9x9 window is built and entered to the ANN. The output of the network will be a number $n \in [0,1]$. If n is greater than a certain threshold h, the pixel at the centre of the window belongs to a microcalcification. If the condition is not satisfied then the pixel is not a microcalcification. The best results for this filter were obtained giving h the value of 0.3.

4 The Combined Filter

Two different approaches to the computerized detection of microcalcifications from a digitalized mammography have been showed. Mathematical morphology based filter detects possible microcalcifications when sharp changes in the grey-level values of the pixels happen in areas of a size smaller than the size of the kernel window. This algorithm throws good results, but it gives too many false positives (non existent microcalcifications but detected by the algorithm) due to there are usually defects in the film mammogram as they can be artefacts, dust spots or even noisy points produced in the digitalization process.

In the other side, the ANN based filter can eliminate all those noisy points due to the training stage where patterns corresponding to this kind of noisy points are presented to the network. However, the only applying of an ANN to a whole digitalized image in order to know pixel by pixel if it is a microcalcification or not produces a great amount of errors. This is because there are many different kinds of patterns that can be in an image and presented to the network.

To avoid the inconvenients of both procedures and to get their advantages, this new algorithm is presented where both algorithms work together. Its working is explained as following and showed in figure 5.

The morphological scheme works as it was described in section 2, but instead of giving a new image as its output, when a pixel that could be a microcalcification is detected, a 9x9 windows is built with this pixel at the centre of it. This window will be the input to the ANN which will give the final output (this pixel belongs to a microcalcification or not). In the affirmative case, the pixel will appear in white color (grey-level value of 255) in the original image. In the negative case, the pixel will keep unchanged.

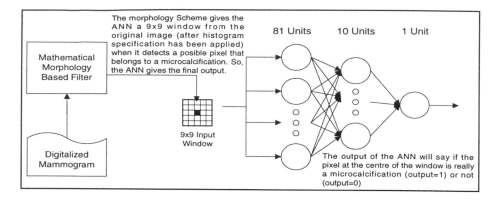

Figure 5. General scheme of the combined ANN and mathematical morphology method.

5 Results

The perfomance of each of three exposed algorithms has been evaluated quantitatively by means of the receiver operating characteristic (ROC) analysis in order to test that the combined algorithm has the best one. ROC curves have been constructed for each of the three methods. The area, Az, under an ROC curve measures the expected probability of the correct classification of a set of test cases.

To construct an ROC curve, the true positive fraction, tpf, is plotted against the false positive fraction, fpf, for various thresholds (in the case of the morphological filter, it was the cut-off parameter C, and the output given by the ANN in both ANN based and combined filter).

At a particular threshold, the true positive fraction may be defined as tpf=tp/(tp+fn), where tp is the number of true positives (a microcalcification exists, and it is detected by the algorithm), fn the number of false negatives (a microcalcification exists, but it is not detected by the algorithm). tp+fn is the number of malignant cases in the test set.

The false positive fraction is defined as fpf = fp/ (fp+tn) where fp is the number of false positives, tn the number of true negatives (benign tissue that has been correctly classified). fp+tn is the number of benign cases in the test set (there are not microcalcifications).

A value of Az near to 1 indicates that the algorithm can correctly detect microcalcifications with a small error rate. It can correctly classify those pixels belonging to a microcalcification, and those that do not. This fact can be obtained when small values of tpf (small number of false positives) give high values of tpf (small number of false negatives).

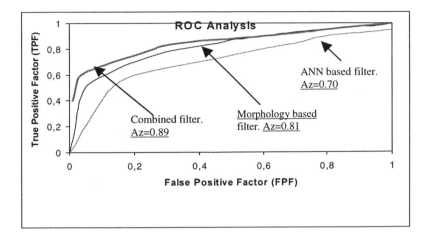

Fig. 6. ROC analysis of the detection of microcalcifications for each of three methods exposed.

Figure 6 shows three ROC curves for each of the exposed filters. It can be seen in this figure how the best results were obtained with the combined filter, with a computational load not very higher than the morphological filter. The morphological filter throws quite good results as well. The ANN based filter is the fastest, but its ability to classify correctly is very low.

6 Conclusions

Artificial Neural Networks appear to be useful in the detection of microcalcifications from digitalized mammograms. However, their only use does not give good results because a great amount of different patterns can be presented to their input. This fact produces some errors in the output given by the network. To solve this problem, it is needed to have a database that covers a large number of different patterns for the training of neural networks.

ANN better performs refining other procedures for microcalcifications detection as the one exposed here based on mathematical morphology. It was seen that that the area under the ROC curve improved from Az=0.81 to Az=0.89 by applying an ANN to the positive pixels given by the algorithm based on the mathematical morphology with a very low increase in the computational cost.

This improvement is made because ANN eliminates the false positives given by the morphological scheme. So, when film mammograms present defects or artefacts, the ANN can distinguish them from true microcalcifications, while morphological operations detect sharpen grey-level changes giving to its input a false positive.

It is also important to see how the only parameter which needs to be adjusted is the cut-off value C, while other detection algorithms have a great amount of factors and variables that make them difficult to tune.

Acknowledgments

Support for this study was provided by Technology and Science Commission (Comisión Interministerial De Ciencia y Tecnología CICYT). National Program of Information Technologies and Communications. Grant N° TIC-97/0956. Title "Automation and Real Time Detection of Breast Diseases".

References

1. Goerge, S.K., Evans, J., Cohen, G.P.B., McMillan, J.H.: Characteristics of Breast Carcinomas Missed by Screening Radiologists. Radiology Vol. 204, pp 131-135. 1997.
2. Okada, M., Shridhar, M.: Extraction of User Entered Components from a Personal Bankcheck Using Morphological Subtraction. International Journal of Pattern Recognition and Artificial Intelligence. Vol. 11 No 5, pp 699-715. 1997.
3. Betal, D.; Roberts, N.; Whitehouse, G. H.: Segmentation and Numerical Analysis of Microcalcifications on Mammograms Using Mathematical Morphology. The British Journal of Radiology, 70 pp 903-917. 1997.
4. Robin, F., Privat, G., Renaudin, M.: Asynchronous Relaxation of Morphological Operators: A Joint Algorithm-Architecture Perspective. International Journal of Pattern Recognition and Artificial Intelligence. Vol. 11 No 7, pp 1085-1094. 1997.
5. Shih-Chung, B.L., Heang-Ping, C., Jyh-Shyan L, et al.: Artificial Convolution Neural Network for Medical Image Pattern Recognition. Neural Networks, Vol. 8 No 7/8, pp 1201-1214. 1995.
6. Boone, J.M., Gross, G.W., Greco-Hunt, V.: Neural Networks in Radiologic Diagnosis: I. Introduction and Illustration. Ivest. Radiol. 25, pp 1012-1016. 1990.
7. Asada, N., Doi, K., MacMahon, H., Montner, S.M. et al.: Neural Network Approach for Differential Diagnosis of Interstitial Lung Diseases Proc. SPIE Med. Imag. 1233, pp 45-50. 1990.
8. Wu, Y., Giger, M.L., Nishikawa, M.: Computerized Detection of Clustered Microcalcifications in Digital Mammograms: Applications of Artificial Neural Networks. Medical Physics 19 (3), pp 555-560. 1992.
9. Crespo, J.: Morphological Connected Filters and Intra-Region Smoothing for Image Segmentation. PhD Thesis. Georgia Institute of Technology. 1993.
10. Maren, A., Harston, C., Pap, R.: Handbook of Neural Computing Applications. Academic Press. 1990.
11. Metz, C.E.: ROC Methodology in Radiologic Imaging. Investigative Radiology. Vol. 21, pp 720-733. 1886.
12. Peña Sánchez de Rivera, D. Estadística Modelos y Métodos. Alianza Universidad Textos. 2nd Ed. Madrid. 1991.
13. Gonzalo, R.C., Woods, R.E.: Digital Image Processing. Addison Wesley, 1993.

Study of Weight Importance in Neural Networks Working with Colineal Variables in Regression Problems*

A. Martínez[1], J. Castellanos[2], C. Hernández[2], and F. de Mingo[3]

[1] Dept. de Ciencias Básicas Aplicadas a la Ingeniería Forestal, EUIT Forestales,
Ciudad Universitaria, Madrid, Spain
amartin@forestales.upm.es
[2] Dept. de Inteligencia Artificial, Facultad de Informática, Campus de
Montegancedo, 28660 Boadilla del Monte, Madrid, Spain
jcastellanos@fi.upm.es
http://www.dia.fi.upm.es
[3] Dept. de Lenguajes, Proyectos y Sistemas Informáticos, EUIT de Informática, Cta.
de Valencia Km. 7, Madrid, Spain
lfmingo@eui.upm.es
http://www.lpsi.eui.upm.es

Abstract. This paper presents a new method that can be used for symbolic knowledge extraction from neural networks, once they have been trained with the desired performance. Weights are the basis for this method. This method allows knowledge extraction from neural networks with continuous inputs and outputs, more precisely in problems dealing with the general linear regression model where exists multicolineality among the input and output. An example of the application is showed by comparison of the results between the regression and the neural networks results, concernig the estimation that gasoline yields from crudes. This example is based on detecting the most important variables when there exists multicolineality.

1 Introduction

This paper shows the importance of the knowledge stored in the weights of a neural network. A trained neural network stores the acquired knowledge in numeric values that weights define [5]. The interpretation and extraction of such knowledge is a difficult task due to the special configuration of neural networks and to the wide domain of patterns [2, 3]. In a trained neural network is really difficult to understand the concepts that have been learnt, this is due to the fact that neural networks store the knowledge in the connection weights and therefore it is not easy to explain the concepts from which RNA outputs the desired response [7]. The kind of knowledge that this paper introduces is the relationship among inputs and outputs.

---

* Supported by INTAS 952

This paper proposes a method in order to detect the importance of the input variables. In multivariant analysis problems, when there exists conlineality among different variables of forecasting, the importance and the sequence when adding variables in linear regression model (analysis multivariant) can be detected, from the knowledge stored in the weight matrix [4]. The knowledge extraction method from the weights [6], or forecasting method, must be taken into account when the study of the correlation coefficients detect relationships among a set of variables.

Two examples are considered in order to compare the results using two forecasting methods: the general regression model and neural networks. The first one deals with weather variables with no relation at all among them (example of forecasting using regression with orthogonal variables, no colineality). In this example, neural networks and regression analysis output similar results. The second one deals with the Prater problem, that is, a set of data in order to forecast the petrol quality where N.H. Prater in 1956 developed a study of the general regression model provided that among the variables exists some correlation (colineality among forecast variables) [1]. In these problems is really difficult to detect the importance of input variables from the coefficients of the general regression model, and neural networks have some advantages over the regression model.

2 Method to Extract Knowledge

Tasks to follow in order to perform a study of the importance of input variables over output variables are the following ones:

1. Normalization of the input and output variables into the closed interval $[-1, 1]$.
2. Definition of the activation function (sigmoid function): $f(x) = \frac{1-e^{-x}}{1+e^{-x}}$
3. A neural network with n input neurones (forecasting variables) and one output neurone (variable to forecast) is used. The training algorithm considered is the backpropagation.
4. Division of the values associated to the variable to forecast into two intervals, the positive one with a positive output $[0, 1]$ and the negative interval with a negative output $[-1, 0)$. This way two independent neural networks are defined in order to be trained.
5. Established an error threshold for the forecasting process, each one of the two output classes of the variable to forecast (positive output values in the interval $[0, 1]$ and negative output values in the interval $[-1, 0)$) are divided into two new classes. For each one of the obtained classes (four classes) 4 neural networks are trained and the value of the weights is observed. If in these new obtained classes, the value of weights that are fixed after the training process is the same that the one obtained in the previous division, or is proportional, then go back to the previous division. If the value is not the same then this division is valid; therefore they will exist four neural

networks associated to the output intervals. This iterative division must go ahead until the weights of a new division will be the same of the previous division. When the weights are similar, then the successive divisions end. This process achieves a better error ratio, getting more powerful classification properties than classical nets, and this way a set of neural networks with their corresponding weights with the following information:

(a) The variable with the most influence over the variable to forecast will be the one with the highest absolute weight after the training process. These data must verify that the sign of the input variable multiplied by the sign of the weight must be equal to the sign of the variable to forecast.

(b) And if the relationship between the forccasting variable and the variable to forecast is a direct or inverse function, that is, if the sign of both variables are the same or not. If the output interval of the variable to forecast is a subinterval of interval $[0, 1]$ or a subinterval of interval $[-1, 0]$ and, if the domain of the forecasting variable (independent) multiplied by the corresponding weight is positive for a subinterval of the variable to forecast (dependent) of interval $[0, 1]$, we will say that the relationship is a direct one, other way it will be an inverse one; taking into account that the absolute value of the highest weight shows the importance of the forecasting variable over the variable to forecast. That is, the higher absolute value of the variable, the deeper influence in the output.

Different divisions of initial set of training data, obtained from the study of weights in the training subset, make that each one of the obtained training subset defines a different neural network to train in the whole subset. Each network, with its corresponding set of weights, denotes the importance of the forecasting variables over the variable to forecast.

(c) Besides extracting the importance of each variable in each output interval, for each one of the input variables it exits a network and a weight set that define the forecasting equation.

Therfore, the method is divided into two steps in order to better understand the two main process on it.

− The first step is used to classify using the bisection method the patterns of the initial set into several subsets, taking into account that this division is performed iteratively, studying the variation of the weights. When in a new division the weights do not change, then go back to the initial division.

− The second step is used once the initial pattern set is classify into several subsets and therefore into several neural networks. The importance of each input variable must be studied for each different network, taking into account the weight values, the variation domain of the input variable and the variation of the output; to study the influence over the variable to forecast. It must be considered:

1. The variables with the highest absolute weight.

2. Which of them verify that their variation domain for that input variable multiplied by its corresponding weight has the same sign of the variable to forecast according to the positive or negative interval $[-1, 0]$ or $[0, 1]$.

With these defined characteristics the most important forecasting variables are taken for each one of the intervals corresponding to the variable to forecast.

When the set of forecasting variables is a set of orthogonal variables (correlation among forecasting variables is null, that is, there is no correlation among the variables), then the proposed model and the general multivariant regression model detect in a similar way, the importance of the variables from the coefficients of the general regression model.

But the importance of the proposed model in this paper is when among the forecasting variables exist some kind of correlation, and the general regression model does not detect either the importance or the need of these variables. In these situations, the proposed method is more efficient that the general regression model.

3 Features of Forecasting Method

The classification of the patterns according to their outputs seems to be right. The learning of classes, that initially could be very difficult, is achieved if there exist enough patterns. That means that the patterns present some characteristics and relationships among the variables in the same class. When the correlation matrix is studied, it can be observed that these correlations exist and are varying into each interval, and the correlations are according to the weight values. The highest absolute weight is the variable with the highest correlation with the output variable.

The change of two variables that help a positive inference, is interrelationed with the change of another variable that goes against that inference, and they provide the variation intervals of such variables to belong to a given class.

The kind of problems that are considered presents multicolineality characteristics. Correlation between variables can be more or less strong, but it reflects the relationships that exist between the variables and with the variable to forecast, being these direct or inverse relationships. All these relationships are reflected by the weights, and that is the method proposed in this paper.

The first aim of the method is to obtain a good classification of the data set that is provided, studying the weights. With the iterative bisection process for the class division, an ideal classification is achieved for the data set. The classified data present a degree of homogeneity. This permits to extract characteristics for each one of the classes. Disjointless but different characteristics are obtained for each one of the several classes.

This method can be implemented with any set of variables who will be used to forecast a new variable with some relationship with the first ones, though these relationships are weak or are unknown, but could be interpreted heuristically.

This method has been achieved after the bisection process with a suitable learning rate and besides, this process has provided a set of weights that define the relationships among the variables, keeping away the interrelations or multicolinealities. This can not be computed with the general regression model. In such problems the coefficients of the lineal general regression model can not be used to obtain the importance of variables.

The proposed method is really efficient when dealing with variables that present colinelaity properties and there are a lot of data that are need in order not to lose information. Besides, the likely noise implemented by defected patterns, that could be present in the pattern set, when the amount of information is very high, is solved due to the generalization properties of neural networks.

First of all, the classes are obtained observing the change of weights along the different divisions in order to join the characteristics, these classes can not be obtained using statistical forecasting methods. Next, to obtain the characteristics and to obtain a prediction for the output. This method is compared with two different problems in sections 4 and 5.

4   Prediction without Colineality

A situation dealing with weather forecasting is analyzed in order to notice the characteristics of multicolineality, and another situation (section 5), developed by N.H. Pratter, is also studied. Given a regression model with two forecasting variables, if the coefficient corresponding to the simple correlation between two variables is zero, that is, the variables are orthogonal, then the effect of one of the variables over the response is measured in a total independent way of the other variable. If one or both forecasting variables are in the regression equation then the least squares estimation does not change its value.

Table 1. A sample of Weaher Forecasting data

Y	X_1	X_2	Regression	Neural Net
70	20	66	66.17	66.81
75	20	72	71.67	71.56
80	20	77	77.17	77.62
70	30	67	67.17	67.20
75	30	73	72.67	72.92
80	30	78	78.17	78.06
70	40	68	68.17	67.74
75	40	74	73.67	74.24
80	40	79	79.17	78.37

Data in table 1 were studied in order to show the effect of orthogonal variables, that refers to the feeling temperature Y, as a function of air temperature X_1 and the relative humidity X_2 [1]. The correlation coefficients corresponding to X_1 and X_2 are zero. Next, the following models are adjusted: $Y = \beta_0 + \beta_1 X_1 + \beta_2 X_2$, $Y = \beta_0 + \beta_2 X_2$, and $Y = \beta_0 + \beta_1 X_1$. The estimated coefficients for X_1 and X_2 are 1.1 and 0.1, regardless if the variable is present at the equation or not. The obtained results are the expected ones when dealing with orthogonal variables and there is not multicolineality.

This example clearly shows that the strength of variables is defined by the coefficients of the regression model and in a similar way by the weights of the trained neural networks, being the most important variable the one with the highest weight. This weight shows the influence degree of the variable over the output.

Table 2. Weights of the Neural Network after the Training Process

Bias	X_1	X_2
0.13	2.42	0.42

Table 3. Coefficients of Lineal Regression for the Weather Forecasting

β_0	β_1	β_2
−12.833	1.1	0.1

From tables 2 and 3, variable X_1 is the main one, so for the neural network as for the lineal regression model where variables are clearly orthogonal ones. The importance of variables and the output of both models are similar. The pattern does not contain a lot of data, this fact helps the regression model since there is not noise, and values are really homogeneous. However, neural network results have been better than regressions one concerning the weighted relationship between input variables and output one. The problem arises when this situation is not so favourable.

If the problem to solve is similar to Prater one (next section) with colineality among different variables, and when there is not possible to detect the influence of variables over the output using the general regression model then neural networks have shown to be very successful.

5 Prediction with Colineality

N.H. Prater (Estimate gasoline yields from crudes, Petroleum Refiner 35 (1956)). (Later Hydrocarbon Processing May (1956)) [1].

This section deals with the study of the problem known as the Prater problem, where the conlineality among different variables and their effect on the regression equation are shown; being these characteristics strongly jointed. The proposed neural networks method solves in an efficient way.

Two problems are solved in the developed example by Prater: the suitable way to find the individual effects of forecasting variables over the variable to forecast, based on the extras squares addition principle; and the way to find a set of forecasting variables to include in the regression equation, that is, identification of the better forecasting variable set or better regression equation.

Prater developed a regression equation to estimate the gasoline yields from crudes as a function of the distillation properties of some kind of crude. Four forecasting variables were identified: crude gravity $API(x_1)$; crude pressure $Psi(x_2)$; $10\% ASTM$ for crude $F(x_3)$ and $100\% ASTM$ for gasoline $F(x_4)$.

The aim of his study was in determining a regression equation for the gasoline production as a lineal function of the distillation properties of some kind of crude x_1, x_2, x_3, and of the final point desired for gasoline x_4 [1]. This problem is considered as belonging to the multiple lineal regression analysis.

The estimated regression equation is:

$$y = -6.82 + 0.23x_1 + 0.55x_2 - 0.15x_3 + 0.15x_4 \tag{1}$$

The multiple correlation coefficient is 0.9622, this implies that about 96% of total deviation of observation can be explained using the four forecasting variables included in the regression equation. But if the problem of Prater is studied [1], then there exists a little doubt about if the regression between the gasoline and the four variables is statistically important.

The addition of a given variable into a forecasting model does not implies that this variable will have an important effect over the response of the model, that is, if a researcher identifies a set of forecasting variables, he must check if they really affect the response. The appropiated method to find the isolated effects of forecasting variables is based on the extra squares addition principle (SCE).

From this point of view, it can be checked whether the contribution of a given regression coefficient is high enough to be included in the model or not. So, in some way, this method provides a measure over the individual effect of a given forecasting variable over a given response from the SCE. Then, applying this method, variables x_4, x_3, are the ones who have the most important effect over the variable to forecast.

It is used to obtain wrong conclusions with a casual point of view dealing with the application of a regression analysis, when there is not a complete view of problems that can arise. A frequent problem, in multiple lineal regression, is that some of the forecasting variables are correlated. If the correlation is small, then the consequences will be less important. However, if there is a high correlation between two or more forecasting variables, then regression results will be ambiguous, specially concerning the values of the estimated regression coefficients. A high correlation coefficient between two or more forecasting variables implies what is known as multicolineality. This problem is sometimes difficult to detect.

The forecasting equation, though not being exact in a physic way, must be a mean, empirical, to forecast the mean response against a condition of the forecasting variables. Multicolineality does avoid neither to have an adjustment nor the response to be forecasted belonging to the interval of observations. What happens is that when two or more forecasting variables have colineality, the estimated regression coefficients do not only measure the individual effects over the response, but also reflect a partial effect over the same one taking into account what happens with the other forecasting variables in the regression equation [1].

If the Prater data are considered, the regression equations including x_2 or x_3 will show that exists a multicolineality between x_2 and x_3. Moreover, the high correlation that exists between x_2 and x_3 has decreases in a drastic way the individual effect that x_3 with x_2 has over the response.

To show that there is a high correlation between x_2 and x_3, the matrix of correlation has to be computed for the four forecasting variables of the Prater data. The is a high lineal association between x_2 and x_3. This result is obvious if the data are examined. It can be checked that the higher presure the crude has x_2, the smaller x_3 $AST M 10\%$ and so on. These results imply the multicolineality. To solve the problem of multicolineality, the regression model without x_2 or x_3 has been examined. As a comparison the regression of (x_3) and (x_4).

5.1 Determination of the best set of forecasting variables

The determination of the best set of forecasting variables is similar to the previous problem, and is explained in order to compare with the results output by neural networks. That is, in the regression analysis, the determination of which variables must be included in the regression model to perform a forecasting output.

Let k the initial number of forecasting variables; the number of terms in the complete lineal model, including the fix term, is $m = k + 1$. A procedure that is highly recommended to determine the best set of forecasting variables to include in the regression equation is to compute and compare all the possible 2^k regression equations. This procedure provides the researcher the opportunity to evaluate and compare all the regression equations and, observing all the wrong ones, to extract the best equation. When k is high, it can not be very practical to determine and evaluate all the possible equations for all the variables. Again, Prater proposed a solution on the selection of forecasting variables that is used in this section and the neural network approach is compared to this one.

When dealing regression equations, the best equation to forecast the production of gasoline is an equation that contains x_3 and x_4.This result is very hard to obtain and can be simplified using neural networks. In the application of Neural Networks to the Prater problem, the data were normalized in interval $[-1, 1]$ with the sigmoid function as the activation function. When there is correlation between two variables of the general regression equation made up of four forecasting variables, the individual effect that x_3 with x_2 has over the response decreases drastically (in the equation the coefficients do not show the influence of variables over the output). This is not the case for a trained neural network.

Observing information at table 4, the most important variables are x_4, and x_3, that are obtained using Neural Networks or the method proposed by Prater.

Table 4. Weights of the Neural Network and Regression Coefficients

Bias	X_1	X_2	X_3	X_4	
−0.818	0.211	0.342	−1.058	2.155	Neural Network
−0.679	0.309	0.306	−1.018	1.974	Positive Net
−0.822	0.114	0.463	−0.933	2.104	Negative Net
−6.820	0.227	0.553	−0.149	0.154	Regression 4 Variables
18.468			−0.209	0.158	Regression 2 Variables
−0.973			−1.534	2.164	Net with 2 Variables

Previous table shows that the regression model does not compute the importance of variables when dealing with four of them. However, neural networks compute such importance by means of weights. If only two variables are taken to perform the study, then the regression coefficients are closer to the influence of variables over the output.

Besides, the regression analysis can be very complex if the importance of variables is tried to measure from the regression coefficients when there is a colineal property, as it can be read along the whole paper and along the Prater study. The information provided by the neural network has been easier to obtain and it has used the whole information. The most important thing is that neural networks obtain better results than the Prater model as in the previous case of weather forecasting.

6   Conclusions

Regression analysis also provides the equation to forecast a given variable. But it can not be forecasted which will be the optimum range for an output in a class. The proposed forecasting method that is related with usual forecasting methods based on multilineal regression models and correlation models, improves them in a successful way.

Neural networks are useful when dealing forecasting problems, especially in problem where there is multicolineality among the different input variables. First of all, because with new data the knowledge of the problem is increasing since neural networks adjust their weights to include such new patterns and the easy way to perform the training process. Moreover, when there is colineality among the variables, the knowledge of the relationship is obtained automatically by the training algorithm and the output of the net has a lower mean squared error than the regression analysis. It is important to note that weights of neural networks store the information about which is the most important variable, the importance of each variable, and more, characteristics that are very difficult to obtain via the regression analysis when there is multicolineality.

The problem that arises when there is multicolineality has been cited, and the identification of the best set of forecasting variables has also defined. The extraction of such knowledge via a regression analysis is very complex. This paper states that the knowledge of some set of data is store in the weights of a trained neural network, and that neural networks as a forecasting method can be used to explain why an output.

The proposed method also computes the forecasting value from the equation of weights. Two classes, positive output and negative output, are obtained, and two equation are used to forecast the output of the model. Due to the small number of patterns used in this example, the change of weights along different classes does involve similar trends. The proposed model takes into account the characteristics of forecasting variables could change from a different class to another, and that is the way it is necessary to use a division method, bisection method. This can be employed when dealing with a high number of patterns or to improve the error ratio.

The advantages of this method are the simplicity of itself. The weight matrix defines the most important forecasting variables, so as the equation to output a value. The only thing to do is to apply the bisection method to the data set and to train a neural network for each class identified by the algorithm. Results are very successful, since the method improves the regression analysis performance even when dealing with colineality properties.

References

1. Canavos, G.C.: Applied Probability and Statistical Methods. McGraw-Hill. Mexico. (1988).
2. Castellanos, J.; Castellanos, A.; Manrique, D.; Martínez, A.: A New Approach for Extracting Rules from a Trained Neural Network. Lectures Notes in Artificial Intelligence 1323. Applied Probability and Statistical Methods. Springer-Verlag, Pp. 297-302. (1997).
3. Castellanos, J.; Manrique, D.; Martínez, A.; Rios, J.: 4th World Congress on Expert Systems. Mexico. Pp. 598-603. (1998).
4. Cuadras, C.M.: Métodos de Análisis Multivariante. Eunibar. (1981).
5. Chang, B.L.; Hirsch, M.: Knowledge Acquisition and Knowledge Representation in a Rule-Based Expert System: Computers in Nursing. Volume 9, Number 5. Pp. 174-178. (1991).
6. Gail, A.C.; Tan, A.: Rule Extraction: From Neural Architecture to Symbolic Representation. Connection Science. Volume 7, Number 1. (1995).
7. Gallant, S.I.: Neural Networks, Learning and Expert Systems MIT Press, Massachusetts. (1993).

The Parallel Path Artificial Micronet

Gerard Murray, Tim Hendtlass and John Podlena

Center for Intelligent Systems
School of Biophysical Sciences and Electrical Engineering,
Swinburne University of technology,
PO Box 218 Hawthorn
AUSTRALIA 3122.
gmurray@swin.edu.au, thendtlass@swin.edu.au,
jpodlena@swin.edu.au

Abstract. Conventional artificial neural networks are based on greatly simplified models of biological neurons. In particular only one path exists to carry an input signal to the neuron body. The design of a parallel path artificial micro-net is described which can be used as the basic building block to construct networks. Preliminary training results are presented.

1 Introduction

Conventional artificial neural networks are based on greatly simplified models of biological neurons. Nonetheless, they have been successfully used for a large number of tasks. However, the standard model of an artificial neuron may be over simplified and, for some tasks, a more complex model may be required [4, 1]. In this paper a micro-net is described which provides added complexity by incorporating a commonly observed aspect in modern neurotransmission.

The axon of a biological neuron is joined to the dendrite of its neighbour at a synaptic junction. The amount of signal propagated is directly proportional to the amount of neurotransmitter released by the axon to bind to the dendritic receptors (figure 1.). This classical model of neurotransmission [3] can be described as a chemical vector, where the vector magnitude is the concentration of transmitter released. This is the state closely modeled by a single weight from an input to the body of the artificial neuron.

However, it has been suggested that some synaptic junctions may contain large, dense core, vesicles. In these vesicles a neurotransmitter co-exists with a neuropeptide (figure 2.). Here, both neurotransmitter and neuropeptide are involved in signal propagation, implying that more than one chemical pathway (or chemical vector) can exist in parallel.

A second example of parallelism in the biological synapse is the existence of multiple, post-synaptic receptor types modulated by a single classical neurotransmitter (figure 3.). It is possible to generate these parallel paths or chemical vector analogues in artificial networks by building more complex weight structures. This can yield new network architectures where the artificial neurons may

112

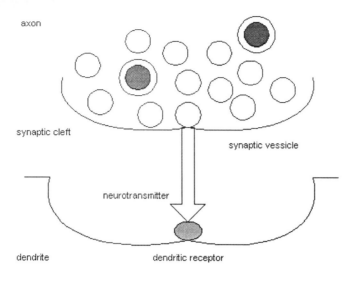

Fig. 1. Classical neurotransmission

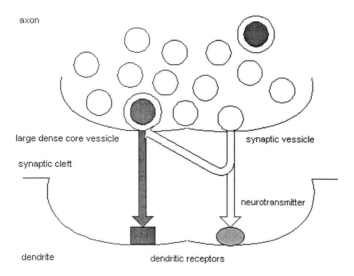

Fig. 2. Classical neurotransmitter in co-existence with a neuropeptide.

be 'super-maximally' connected. The use of different input paths under differing conditions can be likened to a collection of domain experts, each knowledgeable over a subset of input space (see for example [2]).

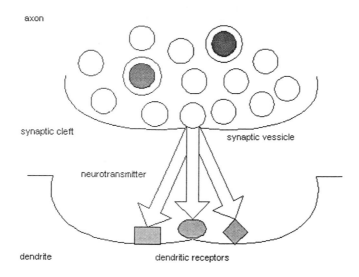

Fig. 3. Single neurotransmitter modulating multiple receptors

2 The Structure of a Micronet

A Micronet has N inputs and one output and consists of a tightly coupled pair of neurons, as shown in figure 4. Neuron one, the control neuron, receives signals from the N inputs via the conventional weights $W_{11}..W_{1N}$ and a bias signal via weight W_{1B}. Neuron two, the output neuron, receives input from the bias via the conventional weight (W_{2B}) and signals from the N inputs via N complex weights structures $S_1..S_N$. These complex weight structures also receive the output of the control neuron as an additional input. The output of each neuron is the hyperbolic tangent of the weighted sum of the signals arriving at that neuron.

Each complex weight structure contains a number of parallel paths (M) whose relative importance, in general, varies in response to the task being undertaken by the neuron.

Figure 5 shows a complex weight structure with two parallel paths. Each path consists of two components in series. One of these components is a conventional weight, the other is a control element and has a value determined by the output signal from the control node. The total effective weight of a single path is the product of the conventional weight and the control element for that path. The

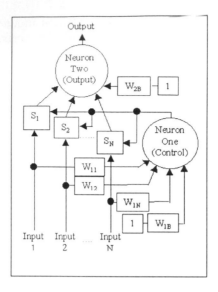

Fig. 4. A Micronet.

effective weight for M such paths in parallel is therefore:

$$W_{CN} = \sum_{i=1}^{M} W_{Ni} * C_{Ni}$$

The control elements are used to influence the relative importance of a particular path: the net importance of all these paths remains constant.

In the case of two parallel paths this is facilitated by the value of one control element being set to the output of the control node while the other control element is set to one minus the output of the control node.

3 Obtaining the Output from a Micronet

The output from the control neuron is calculated first. This output is the hyperbolic tangent of the weighted sum of the inputs to this neuron:

$$C_{NO} = Tanh(\sum_{i=1}^{N} (Input_i * W_{1i} + W_{1B}))$$

This output value is used to set the values of the control elements on the two parallel paths to

$$C_{N1} = W_F * C_{NO}$$

and

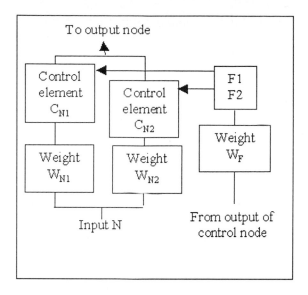

Fig. 5. Complex weight S_N.

$C_{N2} = W_F(1 - C_{NO})$.

The overall output is then calculated as

$$Tanh(\sum_{i=1}^{N}(\sum_{j=1}^{2}(W_{Nj} * C_{Nj} * InputN)) + W_{2B})$$

4 Training the Micronet

A variation of the back propagation algorithm is used to train the Micronet as follows. An estimate of the error in the internal activation of the output node is calculated from the product of the actual error at the output and the gradient of the output node transfer function as normal. Let this error estimate be E_{IAO}. This error is back propagated to estimate the error at the junction of C_{N1} and W_{N1} as $C_{N1} * E_{IAO}$. The signal at this junction is $InputN * W_{N1}$ and the estimate of the corrections that should be made to C_{N1} and W_{N1} are:

$\delta C_{N1} = LR_C * InputN * W_{N1} * E_{IAO}$

where LR_C is the learning rate for the control elements and

$\delta W_{N1} = LR_W * InputN * C_{N1} * E_{IAO}$

where LR_W is the learning rate for the conventional weights.

The conventional weights are updated directly from this result. However, as the control element values are directly determined by the control neuron, the δC_N values must be used to train the control neuron. The error at the output of the control neuron (E_{CNO}) is estimated (for the two parallel path case) as the sum of the four products $\delta C_N * W_F$. Hence the correction to W_F is estimated to be:

$$\delta W_F = LR_W * C_{NO} * E_{CNO}$$

The error in the internal activation of the control neuron and hence the estimate of the errors in the weights W_{11} to W_{1N} and W_{1B} are calculated as normal.

5 Results

The Micronet has initially been tested on the XOR and 3 bit parity problems. These problems are believed to suit a Micronet. For example, the only example in XOR that causes a learning problem is when both inputs are on / true but for which the output should be off / false. If the control node can identify this and switch to the second path for that example, XOR should be readily solvable.

In practice, the Micronet does solve this problem although the path switching used is of a more subtle and complex nature than that described above. The results for a series of tests on XOR are shown in table 1. The symmetric data set uses values of 0.3 and -0.3, while the asymmetric set uses values of 0.2 and 0.8. As might be expected, the symmetric set is easier for the Micronet to solve since the average output is zero and little change is required to the bias weights in the Micronet.

XOR data set	Control node learning rate	Output node learning rate	% of trials converged	Ave. number examples seen
symmetric	0.2	0.2	100%	580
asymmetric	0.2	0.2	81%	3844
symmetric	0.5	0.2	92%	480
symmetric	0.2	0.5	100%	360

Table 1: Results of training the Micronet on XOR data. In each case a minimum of 100 trials were done, using random initialization of the Micronet weights.

It can be seen that the error estimate for the control node weights is attenuated compared to the error of the output node weights owing to the extra back propagation step through the control node. As a result it might be expected that using a higher learning rate for the control node than for the output node would prove beneficial. As seen from table 1, this has proved to be the case, although other combinations of dissimilar learning rates can also produce good results (an example is shown in the bottom line of table 1).

Note that not all attempts are successful owing to the existence of a number of local minima in the error surface that the Micronet can get caught in.

A three input, two-path Micronet has also been tested on the three-bit parity problem. Again, not all attempts are successful, but for those that were the solution was quickly obtained.

6   Conclusion

Simple, minimalist, Micronets have been investigated and found to work on a limited range of problems. The smallest back propagation network that can solve the XOR problem contains six weights, a two input two path Micronet involves eight. This smallest back propagation network generally takes longer to learn this relationship. A three input, two path Micronet (10 weights) will learn the 3 bit parity relationship as will a 3-3-1 back propagation network (16 weights). Again the Micronet learns faster. The performance of Micronets with more than two paths and of networks of Micronets is currently being investigated.

References

1. Boers EJW and Kuiper H: Biological metaphors and the design of modular artificial neural networks. Masters' thesis, Leiden Univesity, Netherlands.
2. Jordan J.M. and Jacobs R.A: Hierarchical Mixtures of Experts and the EM Algorithm. Neural Computation, Vol 6. pp 181-214.
3. Lundberg, J.M. and Hokfelt, T.: Neurotransmitters in Action. Elsevier Biomedical Press, New York (1985), pp 113.
4. Rocha AF, Machado RJ and Gomide F.: Updating the Biology of the Artificial Neuron. Fuzzy Logic, Kluwer Academic Publishers, pp 237-249.

Acknowledgment: The authors gratefully acknowledge the many helpful discussions and suggestions made by Howard Copland.

A Generation Method to Produce GA with GP Capabilities for Signal Modeling

Ahmed Ezzat[1], Nobuhiro Inuzuka, and Hidenori Itoh

Department of Intelligence and Computer Science
Nagoya Institute of Technology
Gokiso-cho, Showa-ku, Nagoya 466-8555, JAPAN
{ezzat, inuzuka, itoh} @ics.nitech.ac.jp

Abstract. The present work is concerned with how to generate GA chromosomes having the capabilities of GP Chromosomes in signal modeling. Substructure formation genes are used to give the generated GA chromosomes the ability to build complex structures with complex substructures. To avoid any unnecessary structure formation and to save the time of the initial population generation phase, dynamic chromosome constructor is proposed. Hence a completed chromosome never scraped as in the conventional generation methods. Using this method, the generated chromosomes can represent very complicated structures with a relatively simple and direct dealing with the problem instead of dealing with complex programs.

1   Introduction

In system monitoring and control, system identification is an important task. As the systems became more complex, this task became more difficult. Therefore it is performed through signal modeling of measured signals or recorded data, [1-2]. For complex and multi input nonlinear systems, signal modeling is not easy. Hence GP is used, [3]. In GP, chromosomes are represented indirectly in the programming language that is implemented, [4]. Hence a capability of the programmer to deal with this problem is must. GP is using the mechanics of GA to solve the problem through the generation of various chromosomes in the form of various computer programs, [5]. So, it is logic to conclude that, if we succeeded to give the initial population in GA the same capabilities of that generated in GP, the same degree of success using GA with can be achieved. In addition, gaining the advantage of dealing with direct chromosome representation which lead to more simplicity in handling the problem can be achieved. In GA, chromosomes are generally represented in fixed length chromosomes with binary genes, [6]. So, it is difficult to use GA to deal with this kind of problem where complex and variable structures can not be represented, [3-4]. Therefore, searching for a new generation method to produce GA's chromosomes which can represent complex and variable structures seems must if we seeks the relative simplicity to deal directly with the problem rather than dealing with complex programs. Hence, substructure formation genes are used to construct complex and

---

[1] Research and teaching assistant, Mech. Eng. Dept., Assiut University, Assiut 71516, Egypt.

variable depth models that can be simply used to represent versatile and complex structures. Dynamic chromosome constructor is used to build, and monitor and correct any probable fault in the chromosome structure during its formation. Hence, the formation of legal chromosomes with necessary substructures along with saving the time of the initial generation phase is guaranteed by never scarping any chromosome. Otherwise, a completely constructed chromosome should be scraped and new one should be build from the beginning, [3-4].

2   Types of the used genes and its constraints

Since the chromosomes will represent the initial solutions of signal modeling problem, the genes must lead to representation of all possible parameters that can appear in a model. Hence, seven types of genes are used. Namely, constant valued genes, G1, input variable genes, G2, scale transformation genes, G3, trigonometric genes, G4, arithmetic operator genes, G5, substructure formation genes, G6, and end of chromosome gene, G7, Table.1. Based on gene's constraints, gene selection ability is determined, Table 1, columns number three and four.

In Table 1., column number two, G1 is to generate the model's constants in the next generations. Its number and values are user dependants. Guides may be extracted from the signal to be modeled. In harmonic wave, for example, it is possible to include the maximum and minimum values of the signal data to generate the amplitude and mean values in the next generations. This kind of genes can be selected if the constraints in column number three in Table 1 had been satisfied. The next gene can be one of the mentioned genes in column four of Table 1. G2 has the same constrains of G1 but the function is different, Table 1. G3 have the same constraints as G1 and G2 but its function is different. Guidelines for selecting values for genes in this type can be extracted from the nature of the included variables. As example, to deal with time or angle as inputs, 60 and π seem reasonable to be included.

G4 genes are necessary, at least one gene, if harmonics are there. The other functions can be generated through any phase shift generation in that function and the subsequent generations of these developed functions with the original one. G5 genes are necessary to perform the basic arithmetic operations. G6 genes were used to order the constructor to build in different depths to give the chromosome the required depth to represent variable and complicated structures. Constrains of these genes are mentioned in Table.1. G7 is the gene which terminate the chromosome growth at the previous selected gene. Selection ability of a gene depends on:

a- Chromosome minimum length and the current chromosome length,

b- Chromosome maximum length and the current chromosome length,

c- Maximum number of substructures and the current substructure number,

d- Maximum number of sub-substructures and the current sub-substructures number,

e- Depth at which building is carrying out,

f- Maximum substructure length and the current substructure length of the substructure under construction,

g- Maximum sub-substructure length and the current sub-substructure length of the sub-substructure under construction,

h- Number of the previous selected gene, and,

i- Type of the previous selected gene.

Table 1. Genes types and its functions and constrains

Gene Type		Gene function	Selection Constraints	Next Gene Type
G1. Constant valued genes		To generate the required models constants.	1- 1st selected gene, or, 2- Previous selected gene of type G5.	G5, G6 (go up), or, G7.
G2. Input variable genes		To represent the system inputs in the developed models.	Same as above.	Same as above.
G3. Scale transformation genes		To provide the required factors to transform from one unit to another.	Same as above.	Same as above.
G4. Trigonometric function genes		To produce harmonic waves.	1- 1st selected gene, or, 2- The previous selected gene is G5 or G6 (go down).	G6 (go down).
G5. Arithmetic operator genes		To perform the required arithmetic operations.	The previous selected gene is G1, G2, G3, or G6 (go up).	G1, G2, G3, G4 or G6 (go down).
G6. Sub-structure formation genes	Go down	To start building in substructure or sub-substructure on much deeper level.	1- 1st selected gene, or, 2- The previous selected gene is G5 or G4.	G1, G2, G3, G4, or G6 (go down).
	Go up	To end the building in the current sub-/ sub-substructure.	1- building takes place on depth different than zero, and the previously selected gene is G1, G2 or G3.	G7 (if building will be on zero depth), G5, G6 (go up, if building depth is none zero).
G7. End of chromosome gene		To end the tendency of the chromosome growth.	Building depth is zero and 1. Chromosome length reached the max., or 2. by nature if the min. chromosome length is exceeded.	None.

3   Dynamic chromosome constructor

Faulty chromosomes can be obtained if static chromosome constructor is used where this kind of constructors provides control only on how many gene are selected tell the chromosome end, [3-4]. Therefore, a fully controlled dynamic chromosome constructor is proposed.

To explain this constructor, the implemented definitions are given first:

Chromosome length: is the number of G5 genes on building depth equals zero. Using this definition, a wide variability even in chromosomes have the same lengths is provided.

Minimum chromosome length: is the chromosome length at which the selection switch of G7 gene will start to be on. This avoids chromosomes with lengths unable to represent dynamic systems.

Maximum chromosome length: It is the length at which all the selection switches of G5 genes will be off when the building depth equals zero.

Substructure: It is a structure started and ended on building depth one. It may contain sub-substructures on depths more than one.

Maximum substructure number: It is the maximum number of times the constructor can build separated substructures at depth one.

Substructure length: It is the number of G5 genes on building depth one.

Maximum substructure length: It is the maximum number of G5 genes on depth one inside the considered substructure.

Maximum sub-substructure number: It is the maximum number of sub-substructures can be contained in a substructure.

Maximum sub-substructure length: It is the maximum number of G5 genes in a sub-substructure lie on a depth equals the depth at which the sub-substructure is lie.

Selection switch of gene x: It is a switch related to gene x, which the constructor set it on or off based on the genes constraints.

The main features of the proposed constructor as shown in Fig. 1., are:
1. Input phase to define control parameters, available genes and its constraints. In this phase, minimum chromosome length, maximum number of substructures, maximum substructure length to avoid any endless substructure, maximum number of sub-substructures to avoid any endless substructure, and maximum sub-substructure length to avoid any endless sub-substructure are defined. Also the available genes and its constraints are defined.
2. Adjust the initial selection switches sittings based on the control parameters and the available genes and its constraints.
3. Select the first gene in the chromosome randomly. If the selection switch of this gene is off, reset and reselect randomly, else evaluate what is the selected gene.
4. Adjust the genes selection switches based on the first selected gene in the chromosome and the control parameters.
5. Select randomly a gene.
6. If selection switch of that gene is off, reset and reselect again. Else, evaluate the selected gene.
7. Check if the case is one of the cases that need to invoke the constructor checker. If yes and based on the case type the checker will work and will evaluate the substructure and its sub-substructures necessities and make the suitable remedy

action and go to the next step or close the current chromosome which proved to be legal and go to step 10.

8. Adjust the gene's selection switches based on the evaluation of the previous selected gene in the chromosome and the control parameters.

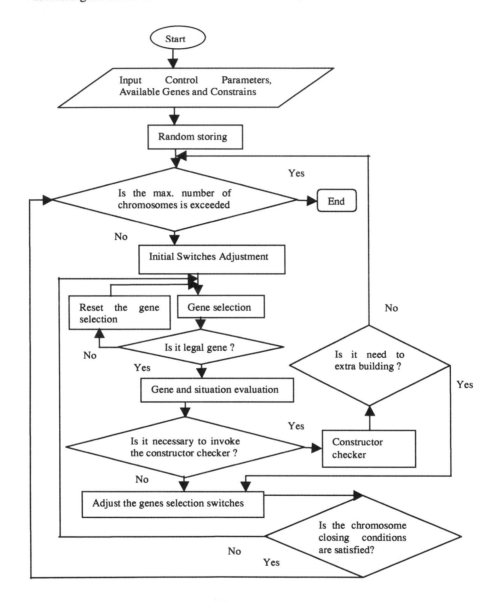

Fig. 1. Main features of the proposed dynamic chromosome constructor

9. Repeat steps 6 through 8 tell legal chromosome end is selected by nature.
10. Generate the next chromosome.

Mainly, producing legal chromosome without chromosome scraping depends on the constructor checker. Following, the explanation of this checker:
In step 7, if the case is G7 gene, the constructor checks if the previous gene is G6 (go up). If false, it goes to step 10. If true, the constructor will invoke its checker to check the necessity of that substructure. Here, the necessity of a substructure has the same meaning as it means to the computer. The necessity of substructure is defined as the necessity of that substructure to be at depth one as well as the maximum depth of its sub-substructures.

If a sub-substructure is proved to be unnecessary, the checker does the following:
1. Remove the G6 genes (go down and go up) of that sub-substructure.
2. Reorder the selected genes starting from the old position of the go down gene of that sub-substructure tell the chromosome end.
3. Unnecessary sub-substructure may lead to illegal sub-substructure in the concerned substructure if overlap was there between two sub-substructures because that one contains the unnecessary sub-substructure can possess, in this way, length greater than the maximum. Hence a truncation operation starting from the gene that follows the go down gene of that sub-substructure tell the end of the chromosome and rebuilding operation start from that gene.
4. Also, unnecessary sub-substructure inside a substructure can lead to illegal substructure which, by this way, can posses exceeded length than the maximum. In this case, a truncation start from the gene that follows the go down gene of this substructure tells its end and rebuilding operation takes place.

Unnecessary substructure may lead to illegal chromosome because the chromosome may posses extra length than the maximum. In this case, care will be paid to recount the chromosome length. If it exceeded the maximum, a truncation operation will start from the gene follows the G5 gene at which the chromosome exceeded the maximum length tell its end. The part starts from the chromosome beginning to that gene will be accepted as a legal chromosome.

Unnecessary substructure may lead to non-legal chromosome in different ways. The chromosome may posses extra number of substructures, then the maximum, resulted from the transfer of sub-substructures on depth equals two to be on depth equals one. This case is avoided by counting the sub-substructures became on depth equals one, keeping in mind the chromosome length. If the maximum substructure number is reached, truncation starts from the go down gene at which that sub-substructure is started, which became substructure, tell the end of the chromosome and rebuilding operation takes place, again, starting from that point.
If the considered substructure and its sub-substructure are proved to be necessary, the chromosome is accepted.

The second case to invoke the checker is that, the selected gene G5 gene and the previous gene is G6 gene (go up) of a substructure. In this case, the checker works in the previous mentioned manner but if all the substructure components are proved to be necessary, it returns to complete the chromosome building.

4 Experimental Work

To check the proposed method and the proposed dynamic constructor, the program is tested to generate ten chromosomes using the following data;

1. Five G1 genes to represent the constants C1, C2, C3, C4 and C5,
2. Three G2 genes to represent the input variables X1, X2 and X3,
3. Three G3 genes to represent 2, π and 60,
4. One G4 gene to represent the sin function,
5. Four G5 genes to represent *, /, + and – ,
6. Two G6 genes to represent the go up and go down building orders, and,
7. One G7 gene to represent the chromosome end.
8. Maximum substructures number in a chromosome is 5,
9. Maximum sub-substructures number in substructure is 3,
10. Maximum substructure length of a substructure is 6,
11. Maximum sub-substructure length of a sub-substructure is 4,
12. Maximum chromosome length is 11, and,
13. Minimum chromosome length is 2.

The generated chromosomes in the initial population were,

$$Y[1] = X3 - C5 * X3 \tag{1}$$

$$Y[2] = X3 / C1 - C1 - C1 * X1 \tag{2}$$

$$Y[3] = C1 * \sin(X3 * C4 + X3) + X1 - \pi / C4 - X1 / X2 * 2 / X2 * C2 / C4 + \pi \tag{3}$$

$$Y[4] = 60 + \pi / C1 / \pi + C2 \tag{4}$$

$$Y[5] = C3 - C3 + 2 - X3 + \pi \tag{5}$$

$$Y[6] = 2 - C4 * C5 + \sin(X2) + 60 - (C1 * C3 + X3 - X1 * C4 * \sin(60)) * C5 \tag{6}$$
$$* C3 / (60 - C5) - \sin(C1 - C5 * C4 - C4 * C5 + C3 * C1) * C4 + C1$$

$$Y[7] = C5 * X3 * 2 - 2 + C3 - C5 + C3 + C5 / C3 + C5 / X3 * X1 \tag{7}$$

$$Y[8] = \sin(C1 / C4 - X1 * X2 / \pi + X3) * \sin(\sin(C4 + C5 * X2 - C4 * C4) + \tag{8}$$
$$\pi + X3 - X1 * \pi + \sin(60 - X3 * 60)) + 60 / C4 / X3 / 2 * C4 * C1 / X2$$
$$* C4 * C2 + C4$$

$$Y[9] = C3 - C2 + X3 * C5 / C4 + X2 - X3 - \pi * \pi \tag{9}$$

$$Y[10] = X2 + 2 + C2 \tag{10}$$

It is evidence to conclude that, even in generating a small number of chromosomes, only ten, all the proposed genes were appeared in the initial population. In Fig. 2, the tree representation of chromosomes Y[1] and Y[6] is given.

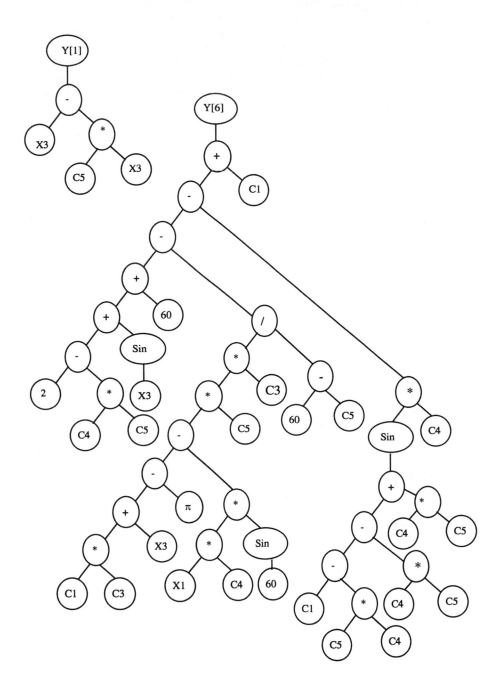

Fig. 2. Tree representation of chromosomes Y[1] and Y[6]

From Fig. 2, we can conclude that, very different chromosomes are generated, from simple models, such as Y[1], to complicated models, such as Y[6]. Comparing these initial chromosomes with those explained in [3], we could conclude that, the proposed method is efficient to give the GA's chromosomes the ability to represent any complicated structure that can generated using GP. Then, using the generation mechanics of the GA, which are same of GP, the same success of GP can be achieved with gaining the simplicity of dealing with simple and direct represented chromosomes, instead of complex programs. Also, because no chromosome scraping no wasting time can be involved in this phase.

5   Conclusion

A generation method to produce GA's chromosomes, which have the capabilities of GP's Chromosomes in signal modeling, is introduced. To perform this task, substructure formation genes are used to give the GA chromosomes the ability to build complex structures with complex substructures. Dynamic chromosome constructor is proposed to produce legal chromosomes without any chromosome scraping which lead to saving the time of the initial generation phase. The generated chromosomes can represent very complicated structures with the simplicity of dealing with the problem instead of dealing with complex programs.

6   Acknowledgement

This research is step in a research program supported by the Hori Information Science Promotion Foundation. Hence, the authors utilize the chance to express their appreciation to it.

References

1. Khajiavi, A.N., Komanduri, R.: Frequency and Time Domain Analyses of Sensor Signals in Drilling-I Correlation with Drill Wear. Int. J. Mach. Tools and Man., **35** (1995) 775-793
2. Lin, S.C., Yang, R.J.: Force-Based Model For Tool Wear Monitoring In Face Milling, Int. J. Mach. Tools and Man., **35** (1995) 1201-1211
3. Bastian, A.: Genetic Programming For Nonlinear Model Identification. Engineering Design and Automation, **3** (1995) 201-216
4. Koza, J.R.: Genetic Programming. MIT Press, Cambridge Massachusettes, (1992)
5. Langdon, W.B.: Genetic Programming and Data Structures. Kluwer Academic Publishers, Boston (1998)
6. Michalewicz, Z.: Genetic Algorithms + Data Structures = Evolution Programs. 3rd edn. Springer-Verlag, Berlin Heidelberg New York (1996)

Preventing Premature Convergence to Local Optima in Genetic Algorithms via Random Offspring Generation

Miguel Rocha and José Neves

Departamento de Informática
Universidade do Minho
Largo do Paço
4709 Braga Codex
PORTUGAL
Voice : 351 53 604466/70
Fax: 351 53 604471
mrocha@di.uminho.pt, jneves@di.uminho.pt
http://www.di.uminho.pt/~jneves
http://www.di.uminho.pt/~mrocha

Abstract. The *Genetic Algorithms (GAs)* paradigm is being used increasingly in search and optimization problems. The method has shown to be efficient and robust in a considerable number of scientific domains, where the complexity and cardinality of the problems considered elected themselves as key factors to be taken into account. However, there are still some insufficiencies; indeed, one of the major problems usually associated with the use of *GAs* is the premature convergence to solutions coding local optima of the objective function. The problem is tightly related with the loss of genetic diversity of the *GA*'s population, being the cause of a decrease on the quality of the solutions found. Out of question, this fact has lead to the development of different techniques aiming to solve, or at least to minimize the problem; traditional methods usually work to maintain a certain degree of genetic diversity on the target populations, without affecting the convergence process of the *GA*. In one's work, some of these techniques are compared and an innovative one, the *Random Offspring Generation*, is presented and evaluated in its merits. The *Traveling Salesman Problem* is used as a benchmark.

Keywords: Genetic Algorithms, Genetic Diversity, The Traveling Salesman Problem.

1 Introduction

Since its genesis, and in particular with the work by John Holland [Hol75], the *Genetic Algorithms (GAs)* paradigm has grown either in its areas of application or in its theoretical foundations. Indeed, the number of conferences and journals devoted to the subject is remarkable, as well as the amount of commercial software already available to tackle real-world problems.

In fact, the *GAs* are not too demanding, as could be natural to expect, in terms of their needs of computational power, being applied so far to a wide range of problems, going from the fields of *Combinatorial* or *Numeric Optimization* to *Image Processing* or even *Machine Learning*, showing to be efficient and robust.

However, and in spite of these achievements, the approach still suffers from a number of insufficiencies; or rather, one of the most harmful, and object of attention in this work, is the premature convergence of the *GA* to local optima of the objective function, defined for the target problem.

The significant influence of this anathema at the problem solving level, has promoted the development of techniques in order to overcome such an handicap. In this study, one pursues the purpose to hang in balance a set of such procedures, by evaluating their performance in tasks that may lead to renew the saga of the *Traveling Salesman Problem (TSP)*.

One's paper is organized as follows: it starts with a description of the problem at hand, the *TSP*. Then it endorses the *GA* paradigm in terms of its most important features. It continues addressing some techniques to avoid premature convergence to local optima and, finally, foregone conclusions and new directions for future work are reported.

2 The traveling salesman problem

The *TSP* is a classic, well known NP-hard problem in *Combinatorial Optimization*; given a set of n cities, and the costs associated with the travel between each pair, the objective is to find a roundtrip of minimal total cost (or length), visiting each city exactly once.

The problem is stated as a n-dimensional cost matrix of values d_{ij}, where the purpose of the exercise is to obtain a permutation of these values, such that the sum of the costs d_{ij}, for any i and j being i the precedent of j in the sequence, is minimal. There are problems in areas so distinct as *Computer Wiring*, *Wallpaper Cutting*, *Crystallography* or *Job Sequencing* that can be formulated and solved as instances of the *TSP*.

The *TSP* can be defined in terms of an *Integer Linear Programming* procedure and postulates (or restrictions) as follows [Lap91]:

$$Minimize: \quad \sum_{i=1}^{n} \sum_{j=1}^{n} d_{ij} x_{ij} \tag{1}$$

$$Subject\ to: \quad \sum_{j=1}^{n} x_{ij} = 1, \forall i \tag{2}$$

$$\sum_{i=1}^{n} x_{ij} = 1, \forall j \tag{3}$$

$$x_{ij} \in \{0, 1\}, \forall i, j \tag{4}$$

$$\sum_{i,j \in S} x_{ij} < |S|, \forall S \subset V, S \neq \emptyset \tag{5}$$

The first equation defines the cost function in terms of the values d_{ij} and of the decision variables, defined as binary variables in (4). When x_{ij} is one, the edge connecting i and j is in the solution, otherwise (x_{ij} is zero) the contemplated edge is not in the solution. The equations (2) and (3) define the constraints of only one edge entering and leaving a given node. In the last equation, V stands for the set of nodes in the instance and $|S|$ for the cardinality of S, defined to be a subset of V. The intended meaning of equation (5) is, therefore, to avoid feasible solutions containing cycles with length smaller than n.

3 The genetic algorithm

In the *GA* used, each individual (or chromosome), that makes the fixed population, codes a *TSP* valid tour. The *genotype* (the genetic constitution of an organism) of the individual is built on a sequence of n integers with no repeated values. The *phenotype* (the physical constitution of an organism as fixed by the interaction of its genetic structure with the environment) is fairly obvious, once the position of any allele on the chromosome determines the order by which the node that it codes is visited. An edge is assumed to connect the nodes that are given by the chromosome's last and first values.

This kind of genotypical representation is named *Order-Based Representation (OBR)*, and is in some ways quite different from the traditional *Binary-Based Representations (BBRs)*, where the order of the genes in the chromosome is not, or should not be, important to its phenotypical interpretation. The constraint on non-duplicates in *OBR*, and the dependence on the order of the genes, justifies the development of a whole new class of operators for the *crossover* and the *mutation* operations.

As far as a *TSP* instance is considered, the evaluation function will assign to each individual in the population a fitness value, a measure of the total cost of the solution coded by its *genotype*; it should also be noticed that the best individuals are the ones with the lowest fitness values.

The major structure of the *GA* used in one's approach is outlined in the pseudo-code of Figure 1, where *ps*, *nc* and *sr* stand, respectively, for the population size, the number of offspring generated per iteration, and the number of individuals replaced when moving from a generation to the next.

3.1 Selection

The *selection* operator is used to choose parents for reproductive trials, to pick the survivors from a generation to the next one, and to decide which of the offspring will be inserted into the population. The procedure used, in this study, is based on a stochastic process using a *Roulette-Wheel* scheme. The weight assigned to each individual is calculated to be inversely proportional to its ranking, considering the fitnesses of individuals on the whole population. The ranking is ascendent when the purpose is to select the best individuals, and descendent when one is looking to the worst of it.

3.2 Crossover

The *crossover* operator is defined based on the function

BEGIN
  Initialize time ($t = 0$).
  Generate, at random, *ps* individuals (initial population P_0) and out their evaluation.
  WHILE NOT (end test) DO
    Select from the present population (P_t) the individuals.
    Recombine these individuals to breed *nc* offspring and proceed on their evaluation.
    Select *sr* offspring to insert into the next population (P_{t+1}).
    Select *ps* − *sr* survivors from P_t to be inserted into P_{t+1}.
    Mutate P_{t+1} and re-evaluate mutated individuals.
    Increase current time ($t = t + 1$).
  END WHILE
END

Fig. 1. Structure of the *GA* used

$$Crossover : Individual \times Individual \times Parameters \mapsto Individual[\times Individual]$$

where the "$\times$" names Cartesian product, the "[" and "]" stand for an optional entity, and "$\mapsto$" (without danger of ambiguity) names "$\mapsto$".

In this study a number of different *crossover* operators were used, namely the so called *blind* operators, that recombine the genetic material of their ancestors without regard of the underlying solution; i.e., with no links to the target problem. There is also the ones known as *hybrids*, which are associated with operators that take advantage of the problem's specific knowledge; i.e., in the case of the *TSP*, the operators use the information in the cost matrix to guide the ancestors recombination process.

Below is a list of the *blind* operators implemented so far, as well as some references to their genesis [Sta91][RocNev98].

− *Uniform Order Preserving Crossover (UOPX)*
  The operator emphasizes the relative order of the genes in both ancestors, working with a randomly generated binary mask, with a size made equal to the genotype's length. It is the equivalent to the *Uniform Crossover* operator in *BBRs*, and some good results on its application to the *TSP* are being reported [Dav91].
− *EDGe Crossover (EDGX)*
  The edge family of crossover's operators is based on the principle of maintaining all possible pairs of adjacent genes (edges) on the chromosome. It was specially designed for the *TSP* problem [Sta91].
− *Maximum Preservative Crossover (MPX)*
  The *MPX* operator was designed by Mühlenbein [Muh91] with the purpose to tackle the *TSP* by preserving, in the offspring, subtours contained by both ancestors.

– *SCHleuter Crossover (SCHX)*
The *SCHX* [Sch89] is a variation of the *MPX*, with some features similar to the order preserving ones, and also contemplating the process of the inversion of partial tours.

When it comes to the *hybrid* operators, two were considered.

– *Greedy Crossover (GX)*
The *GX* was introduced by Grenfenstette [Gre85] and it is based on a simple rule, that states that among the existing edges, when recombining information from the ancestors to the offspring, one must choose the ones that carry the minimum cost to the solution, in each step, thus believing that local minimum sub-tours should lead to better global results than the use of the more expensive ones.
– *Half Greedy Crossover (HGX)*
The *HGX* operator was proposed in [Kur96]. The idea is to make the operator *less greedy*, delaying the convergence process with the purpose of preventing premature convergence to local optima.

3.3 Mutation

A mutation is an unary operator that can be defined as the function

$$Mutation : Individual \times Parameters \mapsto Individual$$

By analogy with what happens in nature, a mutation operation normally induces a small change to the genotype of the individual to which it applies, happening with a frequency, called *Mutation Rate (MR)*; *MR* defines the probability under which a mutation operator is applied, to a particular position of the genotype of an individual, in each iteration of the *GA*.

In this work, four different categories of mutation operators were considered, namely *adjacent swap*, *non-adjacent swap*, *sub-list scramble*, and *partial inversion*.

4 Preventing the premature convergence to local optima

In this section one aims to describe some of the techniques used to prevent the premature convergence to local optima. These methods work on to avoid the loss of genetic diversity of the whole population, and in principle, will not damage the convergence process.

4.1 Adaptive mutation rate

The mutation operator aims to introduce a random component into the search process, with the exploitation of new chunks of the solution space, thus promoting the increase of the genetic diversity of the population; i.e., it is not surprising to find out that one of the first steps to take in order to maintain the genetic diversity in a population is the *MR*'s increase.

However, a high value to this parameter introduces a certain degree of noise into the system, thus creating serious obstacles to the convergence process. Therefore, and in order to overcome this phenomenon, is was decided to change the value of the *MR*, with appeal to an adaptive strategy based on the population's genetic diversity, measured at regular spaces in time, being the standard deviation of the fitness values of the whole population used to estimate its diversity.

The process works as follows: one starts with an initial value for the *MR*, and at regular intervals in time, the value of the standard deviation is tested. If it is lower then a pre-defined limit, the *MR* is increased.

4.2   Social disasters technique

The *Social Disasters Technique (SDT)* was introduced by Kureichick and colleagues [Kur96] in order to avoid the premature convergence to local optima, when the *GAs* are applied to the *TSP*. The general idea is to diagnose the situations of loss of genetic diversity of the population, and in such a case to apply a *catastrophic operator* to it. These operators were defined with the purpose to return the population to an acceptable degree of genetic diversity, by replacing a number of selected individuals, by others, generated at random.

Two different operators were considered.

- *Packing.* Of all the individuals having the same fitness value, only one remains unchanged; all the others are fully randomized.
- *Judgment Day.* Only the individual with the best fitness value remains unchanged; all the others are fully randomized.

4.3   Random offspring generation

One of the features of a population converging to local optima is the large number of individuals sharing the same genetic material. But, when this situation occurs, there is a great probability that the crossover operator may receive as input two individuals with equal genotypes. In this case the recombination of their genetic material will be ineffective, since the offspring bred will simply be clone to their parents.

The idea behind the *Random Offspring Generation (ROG)* is to test the individual's genetic material, before the crossover operation, and if a situation as the one just referred is detected, the operation is not performed. Instead, one offspring, or even two, are randomly generated; i.e., their genotype will code a random solution on the problem's domain.

Two different strategies are possible, differing on the number of the random offspring created. With the former *(1-RO)*, the result is made of a random generated individual being the other one clonally obtained from their parents. With the latter *(2-RO)*, both descendents are randomly bred. When one uses a *hybrid* crossover operator, only the first strategy is applicable, due to the fact that it only generates one offspring per two parents.

5 Experimental results

The techniques described so far were applied to three *TSP* instances taken from the *TSPLIB95* [Rei95]. The problems are listed in Table 1, as well as the values of some of the relevant parameters for the *GA*. The instances referred can be classified as *Euclidean TSPs*, i.e., defined in a way that $\forall i, j, k \ d_{i,j} \leq d_{ik} + d_{kj}$. This is the case of all instances of the problem defined as a set of nodes characterized by their coordinates in a two-dimensional space.

Table 1. The TSP Instances

Problem	Nodes	Optimum	Population Size	Mutation Rates(%)
Eil51	51	430.0	100, 200	0.1 - 1
Eil76	76	545.4	150, 300	0.05 - 0.5
Eil101	101	642.3	200, 400	0.01 - 0.3

In the experiences conducted, several options regarding the policy used to prevent the loss of genetic diversity of the population were considered.

- *None.* No special technique was used to prevent the loss of diversity.
- *Adaptive Mutation Rate (AMR).* As described above considering the initial value of the *MR* to be 0, the value for each increment to be equal to 0.01%, 0.05% or 0.1%, the number of iterations between each test to be 25 or 50 iterations, and the minimum value for the standard deviation to be 1% or 2% of the smallest fitness value in the population.
- *Social Disasters Techniques (SDT)*, being considered two alternatives, the first one using the *Packing* operator *(SDT-P)*, and the latter using the *Judgment Day* operator *(SDT-J)*. A value of 0.5% of the smallest fitness value in the population was used as the minimum limit of the standard deviation, with tests every 100 generations.
- *Random Offspring Generation (ROG)*, being considered *1-RO* and *2-RO* strategies.

The results are given (Tables 2, 3 and 4), for each problem, in terms of the strategy applied to prevent loss of genetic diversity, and of the crossover operator used. The best result was obtained when combinations of other parameters, namely the *mutation rate*, the *population size* and the *mutation operator*, were considered. Each configuration was tested with 20 independent runs, being the result obtained as the average of the fitnesses of the best individuals in each run.

It is now possible to engage into some reflections:

Table 2. Experimental results for the problem *Eil51*

Crossover Operator	Premature Convergence Prevention Technique					
	None	*AMR*	*SDT-P*	*SDT-J*	*1-RO*	*2-RO*
UOPX	450.7	458.4	449.9	**441.8**	443.2	450.3
EDGX	443.9	452.6	451.5	445.6	447.0	**441.9**
MPX	438.3	448.9	442.9	442.9	**436.0**	455.9
SCHX	443.5	442.4	443.0	444.4	440.7	**437.8**
GX	434.3	436.5	436.2	435.5	**431.5**	-
HGX	434.1	**429.0**	429.5	429.5	**429.0**	-
Best	434.1	**429.0**	429.5	429.5	**429.0**	437.8

Table 3. Experimental results for the problem *Eil76*

Crossover Operator	Premature Convergence Prevention Technique					
	None	*AMR*	*SDT-P*	*SDT-J*	*1-RO*	*2-RO*
UOPX	592.0	**565.7**	578.6	573.5	582.3	580.6
EDGX	**569.3**	592.7	592.9	570.9	574.9	595.3
MPX	**590.1**	595.6	601.2	592.8	599.5	607.0
SCHX	577.8	565.8	569.5	591.7	**564.9**	575.8
GX	554.8	**550.6**	554.3	556.6	551.3	-
HGX	554.1	555.9	551.4	552.4	**551.1**	-
Best	554.1	**550.6**	551.4	552.4	551.1	575.8

- The *AMR* strategy leads to solutions with a similar degree of quality as the ones obtained via regular *GAs*. However, it must be stated that the use of an *AMR* strategy has an obvious advantage, in the sense that it releases the user from having to choose a specific value for the *MR's* parameter, an arduous task in many cases.
- The use of the *SDT* strategy, in either of its forms, does not seem to be an important factor to the improvement of the results so far obtained, although it may induce good solutions in some situations.
- The *ROG* strategy is, undoubtedly, the one that presents the best results, a fact that in itself is not completely surprising, since it is the approach that better takes care of the genetic diversity of the whole population.

6 Conclusions and future work

The data so far obtained, when the *GA* is applied to selected *TSP* instances, shows that the use of artifacts to prevent the loss of genetic diversity in the target population can improve significantly the quality of the results. In particular, the *ROG* strategy seems to be a simple, but powerful method to prevent premature convergence to local optima, and therefore improving the behavior of the *GA*.

Table 4. Experimental results for the problem *Eil101*

Crossover Operator	Premature Convergence Prevention Technique					
	None	*AMR*	*SDT-P*	*SDT-J*	*1-RO*	*2-RO*
UOPX	684.6	690.9	685.9	**669.0**	685.5	683.6
EDGX	721.5	715.5	710.4	693.5	**684.8**	708.5
MPX	708.8	716.3	719.9	718.2	711.2	**707.5**
SCHX	690.5	675.4	695.7	690.1	680.2	**674.1**
GX	652.7	664.7	**651.2**	669.8	653.2	-
HGX	646.3	653.4	653.4	654.0	**641.4**	-
Best	646.3	653.4	651.2	654.0	**641.4**	674.1

It must be mentioned that, unlike some other methods (eg. the *Crowding* scheme [DeJ75]), these techniques keep the selection procedures unchanged, and therefore their induced computational overheads may be disregarded.

Obviously, one does not intend to give the final solution to the problem of preventing the premature convergence and loss of genetic diversity that occur when using *GAs*. However, the results obtained by the *ROG's* scheme are encouraging, a reason to extend these tests to other domains, in order to evaluate the real usefulness of the method.

When one considers the way genetic diversity is created in nature, one comes to the conclusion that the spatial organization of the living species is crucial in their process of evolution. Therefore, it will not come as a surprise that the integration of these findings, as well as concepts such as diploid representations and dominance, along with the *GA's* machinery, must be object of consideration in any study that may be carried out on this arena.

References

[Dav91] Lawrence Davis ed. *Handbook of Genetic Algorithms*, Van Nostrand Reinhold, 1991.
[DeJ75] Kenneth A. DeJong, An Analysis of the Behavior of a Class of Genetic Adaptive Systems, Doctoral Dissertation, University of Michigan, 1975.
[Gre85] John J.Grenfenstette, Rajeev Gopal, Brian Rosmaita and Dirk Van Gucht, Genetic Algorithms for the Traveling Salesman Problem. In J.Grenfenstette ed. *Proceedings of the Second International Conference on Genetic Algorithms and their Applications*, MIT, Cambridge, July 1987, Lawrence Erlbaum Associates: Hillsdale, New Jersey.
[Hol75] J.H.Holland, Adaptation in Natural and Artificial Systems, University of Michigan Press, Ann Arbor, 1975.
[Kur96] V.Kureichick, A.N.Melikhov, V.V.Miaghick, O.V.Savelev and A.P.Topchy, Some New Features in the Genetic Solution of the Traveling Salesman Problem. In Ian Parmee and M.J.Denham eds. *Adaptive Computing in Engineering Design and Control 96(ACEDC'96), 2nd International Conference of the Integration of Genetic Algorithms and Neural Network Computing and Related Adaptive Computing with Current Engineering Practice*, Plymouth, UK, March 1996.

[Lap91] Gilbert Laporte, The Traveling Salesman Problem: An Overview of Exact and Approximate Algorithms, *European Journal of Operational Research*, 59:231-247, 1992.

[Muh91] H.Muhlenbein, Evolution in Time and Space - The Parallel Genetic Algorithm. In G.Rawlins ed. *Foundations of Genetic Algorithms*, pages 316-337, San Mateo, 1991. Morgan-Kaufmann.

[Rei95] G.Reinelt, TSPLIB95. Universitat Heidelberg, 1995.

[RocNev98] M.Rocha and J.Neves, An Analysis of Genetic Algorithms applied to the Traveling Salesman Problem. Technical Report, Universidade do Minho, 1998.

[Sch89] M.G.Schleuter, ASPARAGOS - An Asynchronous Parallel Genetic Optimization Strategy. In J.D.Schafer ed. *Proceedings of the Third International Conference on Genetic Algorithms*, George-Mason University, Morgan Kaufman, 1989.

[Sta91] T.Starkweather, S.McDaniel, K.Mathias, D.Whitley and C.Whitley, A Comparison of Genetic Sequencing Algorithms. In R.Belew and L.Booker ed. *Proceedings of the Fourth International Conference on Genetic Algorithms*, Morgan-Kaufmann, 1991.

Using Self Organizing Maps and Genetic Algorithms for Model Selection in Multilevel Optimization

Mohammed El-Beltagy and Andy Keane

{M.A.El-Beltagy,Andy.Keane}@soton.ac.uk

Evolutionary Optimization Group, Department of Mechanical Engineering
University of Southampton, Southampton SO17 1BJ, UK

Abstract. In Multilevel Optimization there is usually a choice to be made between different models when carrying out design evaluations. The choice is between accurate / computationally expensive evaluations and approximate/ computational cheap ones. Here, a strategy is sought for selecting between different models during the search. The focus of the paper is on preliminary work carried out using a self organizing map (SOM) for model selection.

1. Introduction

In engineering problems, it is often the case that a variety of ways may exist n which to model a particular problem. Some models may be quite elaborate, while others involve a simplification of the problem, with the former being more accurate but at the same time usually more computationally expensive than the latter.

The multiplicity of computational models for a given object of simulation may arise from at least three main causes. It could be due to different mathematical formulations being used to construct the model such as Euler and Navier Stokes approximations in computational fluid dynamics (CFD). It could also be due to different discretization limits within one formulation such as mesh densities in finite element analysis (FEA). Finally, it may come from the availability of approximate empirical models such as neural networks or response surfaces.

We use here the term "multilevel optimization" (MLO) to denote the process of optimizing such a multiplicity of models, where each level is essentially one of these models.

For the purposes of optimum design, it might not be the best strategy to use only one model throughout the optimization. A computationally expensive model would take an unreasonable amount of time to carry out an optimization. A cheap model on the other hand, might miss out on many details that only become evident in the expensive one and hence yield a poor optimum. Given the continuous improvement in design codes and the increasing computational burden they demand, this problem is not likely to disappear even with the advent of faster computers.

There is hence a strong need to develop an optimization strategy that integrates a large number of less accurate evaluations with few accurate ones to arrive at an optimum.

In contrast to the static optimisation problem $f(\vec{x}) \to opt, \ (\vec{x} \in M)$, the multilevel optimisation problem can be stated as $f_1(\vec{x}) \to opt, (\vec{x} \in M)$, where $f_1(\vec{x})$ is the most accurate function and there exist many $f_k(\vec{x})$ models where $k = 1..L$. The levels are such that $f_i(\vec{x})$ is more accurate and computationally expensive than $f_j(\vec{x})$ for $i < j$.

The self organizing map (SOM) [6] is an efficient method for vector quantization. Its power lies in capturing and representing, in a lower dimensional space, the important features from a high dimensional space. It can also be used to map the correlation between different models. Given a certain parameter vector $\vec{x}$ a well trained SOM may be used to select the appropriate model $f_j(\vec{x})$. An added benefit of the SOM is that it could be used subsequently for visualization and data mining purposes [5].

This paper is arranged as follows: In the next section a brief overview of multilevel optimizaion related work is presented. Section 3 outlines the essential elements of a typical SOM. Section 4 describes our proposed strategy for using a SOM with a GA. Section 5 details the test function we used and how we set it up to simulate a multilevel problem. Section 6 presents the experimental results obtained. The paper closes with a brief conclusion and discussion of future work

2. Related work

Dunham *et al.* were first to address the problem of multilevel optimization within an evolutionary optimization context [2]. They worked with a two level problem. In their study they used an approximate model most of the time, using the accurate/computationally expensive model only at the final stages of refinement.

Most recent efforts have focused on building variants of injection island genetic algorithms (iiGA) architectures for this sort of optimization [3][4][9]. The approach adopted was to have many islands using low accuracy/cheap evaluations that progressively pass on individuals to fewer islands using higher accuracy/expensive evaluations.

A case for the importance of multilevel optimization for aeronautical design was presented in [8]. [1] presents fundamental studies focusing on the use of a family of approximate fitness representations in GAs.

3. The Self Organizing Map

A self organizing map is a vector quantization algorithm that defines a mapping from a high dimensional input space $\Re^n$ to low dimensional array of nodes (typically a two dimensional array is used). It can been viewed as a "nonlinear projection" of the probability density function of a high-dimensional input data space.

Every node i in the SOM has a prametric reference vector $m_i \in \Re^n$ associated with it. During training an input vector x is compared with all map nodes to find the one that best matches it. Typically, the Euclidean distance measure $\|x - m_i\|$ is used. The winner node (the one most similar to x) is signified by the subscript c:

$$\|x - m_c\| = \min_i \{\|x - m_i\|\} \tag{1}$$

During training, nodes that are topologically close to the winner node learn from the same input. The learning rule is defined as:

$$m_i(t+1) = m_i(t) + \alpha(t)h_{ci}(t)[x(t) - m_i(t)] \tag{2}$$

In the above equation t is the discrete time variable, $h_{ci}(t)$ is the activation profile and $\alpha(t)$ is the learning rate. The activation profile defines to what degree the nodes closest to the winner node are activated; it is defined over the lattice points. Both the learning rate and activation profile are monotonically decreasing functions of time.

In our work we used a gaussian $h_{ci}(t)$

$$h_{ci}(t) = \exp\left(-\frac{\|r_c - r_i\|^2}{2\sigma^2(t)}\right) \tag{3}$$

Here r_i is the SOM lattice coordinate of node i, and σ determines the radius of the neighbourhood region.

4. Combining SOM and GA

As a first attempt to combine SOM with GA we considered a case of just two levels $f_e(\bar{x})$ and $f_a(\bar{x})$. Here $f_e(\bar{x})$ represents the exact and computationally expensive model and $f_a(\bar{x})$ the approximate and computationally cheap one.

The idea is to teach the SOM online to predict how the two models are related. Hence the vector used for training the SOM is

$$\vec{y} = \left[\vec{x}, \left|f_e(\vec{x}) - f_a(\vec{x})\right|\right]. \tag{4}$$

Here we use the SOM as estimator using autoassociative mapping [6, p.258].

After training, the SOM nodes are marked for evaluation using $f_e(\vec{x})$ according to their $\left|f_e(\vec{x}) - f_a(\vec{x})\right|$ value. A node having a high $\left|f_e(\vec{x}) - f_a(\vec{x})\right|$ would indicate that in its neighbourhood the two models don't match and hence it would be best to use $f_e(\vec{x})$. On other hand a node having a low $\left|f_e(\vec{x}) - f_a(\vec{x})\right|$ value would indicate that the approximate model is quite faithful and hence in its neighbourhood, evaluations using $f_a(\vec{x})$ would be sufficient. In our scheme we marked the top 20% of the nodes based on their $\left|f_e(\vec{x}) - f_a(\vec{x})\right|$ values.

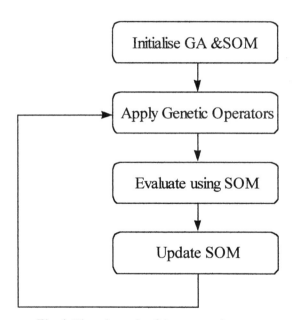

Fig. 1. The schematic of the proposed strategy

The strategy works as follows:

1. Initialize the population and evaluate it using both $f_e(\vec{x})$ and $f_a(\vec{x})$. Construct a training set using $\vec{x}$ and $\left|f_e(\vec{x}) - f_a(\vec{x})\right|$ and use this to train the SOM.

2. Do selection and apply the standard Genetic Operators to the population (mutation, crossover,etc.).
3. Evaluate using the SOM for model selection: If a member is best matched (using an Euclidean distance measure) to a node that is marked, $f_e(\vec{x})$ is used; otherwise $f_a(\vec{x})$ is used.
4. For those members evaluated using $f_e(\vec{x})$ evaluate $f_a(\vec{x})$ to get $|f_e(\vec{x}) - f_a(\vec{x})|$. Construct a training vector y and append it to the training set.
5. Delete the old SOM and construct a new one using the enlarged training set.
6. Repeat from step 2 until termination.

5. The test problem

The modified 'bump' test problem was used for this study [1]. It is defined as

$$\text{maximize} \frac{\text{abs}(\sum_{i=1}^{n} \cos^4(\alpha(x_i + \beta)) - 2\prod_{i=1}^{n} \cos^2(\alpha(x_i + \beta)))}{\sqrt{\sum_{i=1}^{n} i(x_i + \beta)^2}} \tag{5}$$

for

$$0 < x_i < 10 \quad i=1,\dots,n, \tag{6}$$

subject to

$$\prod_{i=1}^{n} x_i > 0.75 \quad \text{and} \quad \sum_{i=1}^{n} x_i < 15n/2 \tag{7}$$

Here the α and β parameters are used to describe the different levels. In this study $\alpha=1$ and $\beta=0$ were used for $f_e(\vec{x})$. For $f_a(\vec{x})$ we used $\alpha=1.1$ and $\beta=0.1$.

The α parameter spreads out the peaks ($\alpha < 1$) or makes them closer together ($\alpha > 1$), while β just shifts the peaks of bump in x_i.

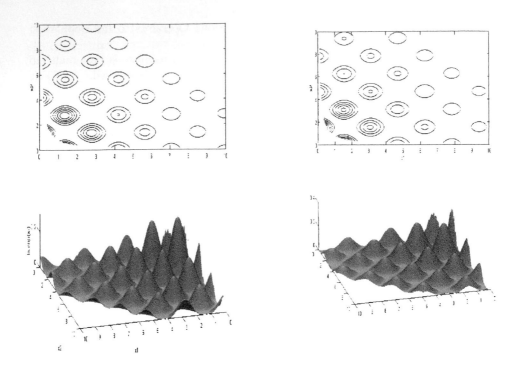

Fig. 2. On the left a contour map and a 3D plot a 2-D bump (n=2, α=1.1, β=0), on the right for (n=2, α=1, β=0)

6. Experimental Results

We tried the proposed strategy for a 20D, 10D and 5D bump. A real valued elitist GA incorporating linear scaling and tournament selection was used[1]. Its parameters are as shown below.

Population Size	300
Crossover Rate	0.9
Mutation Rate	0.05
Number of Generations	20

Table 1. GA parameter

---

[1] Matthew's GAlib C++ Library of Genetic Algorithm Components was used in developing the code used here. (http://lancet.mit.edu/ga/)

The SOM was trained using the SOM_PAK program package [7]. The map was intialized along a two-dimensional subspace spanned by the two principal eigenvectors of the input data vectors. It was trained for 50 epochs. Initial neighborhood radius was approximately half of the side length of the map. Initial value for the learning coefficient was 0.3 and this was decreased proportional to the inverse of time. A gaussian activation profile was used (see eqn. 3).

We tested the proposed strategy using a 10x10 SOM. For comparison the average number of $f_a(\vec{x})$ and $f_e(\vec{x})$ recorded was used in a sequential strategy GA [1]. In such a strategy the optimization is started using $f_a(\vec{x})$ and after a certain number of generations the GA is stopped and optimization is carried on using $f_e(\vec{x})$ but seeded with the final population of the low accuracy level. The scheme was set up such the average number of calls to $f_a(\vec{x})$ and $f_e(\vec{x})$ was the same in each strategy. The results were averaged over thirty runs and are as shown below.

	Average (SOMGA)	Average (Seq. GA)	Standard Deviation (SOMGA)	Standard Deviation (Seq. GA)
5D	0.483	0.480	0.047	0.054
10D	0.509	0.507	0.049	0.072
20D	0.424	0.403	0.036	0.051

Table 2. Results

The improvement obtained by the method is slight but not insignificant. Yet at the same time it is worthy noting that in our earlier work [1] we have shown that on a similar problem mixing between levels at random (which is equivalent to using an untrained map) always gave a *worse* result than the sequential method. So, it is clear that training the SOM has had a beneficial effect on the optimization efficiency.

7. Conclusion

We have made the case for multilevel optimisation and proposed a strategy for tackling it. In this embryonic work we have seen only slight improvement. It is hoped that further development of this approach will yield better results.

And added benefit of using the SOM-GA approach is that the constructed map cab be used latter for data mining and visualization.

Acknowledgements

This work was supported under EPSRC grant no GR/L04733

References

1. El-Beltagy MA, Keane AJ (1998). Optimization for Multilevel problems: A Comparison of Various Algorithms. *In: Parmee I (Ed.) Proc. of The Third International Conference on Adaptive Computing in Design and Manufacture (ACDM '98).* Springer, London

2, Dunham B, Fridshal D, Fridshal R, North JH (1963) Design by Natural Selection. Synthese 15:254-259

3. Eby D, Averill RC, Punch WF, Goodman ED (1998). Evaluation of Injection Island GA Performance on Flywheel Design Optimization. *In: Parmee I (Ed.) Proceedings of Third Conference on Adaptive Computing in Design and Manufacturing.* Springer Verlag, London

4. Goodman ED, Averill RC, Punch WF, Eby D (1997). Parallel Genetic Algorithms in the Optimization of Composite Structures. *In: Second World Conference on Soft Computing (WSC2).*,

5. Kaski S (1997) Data Exploration Using Self-Organizing Maps. Doctoral Thesis, Helsinki University of Technology.

6. Kohonen T (1997) Self-Organizing Maps, Second Edition edn. Springer-Verlag, Berlin

7. Kohonen T, Hynninen J, Kangas J, Laaksonen J (1996) SOM*PAK: The Self-Organizing Map Program Package. Technical Report A31 at Helesinki University of Technology, Laboratory of Computer and Information Science, 1996. Package available at (http://nucleus.hut.fi/nnrc/som*pak/).

8. Robinson GM, Keane AJ (1998). A Case for Multi-level Optimization in Aeronautical Design. *In: Proceedings of the Royal Aeronautical Society Conference on Multidisciplinary Design and Optimization.*,

9. Vekeria HD, Parmee I (1997). Co-operative Evolutionary Strategies for Single Component Design. *In: Bäck T (Ed.) Proceedings of The Seventh International Conference on Genetic Algorithms.* Morgan Kaufmann, San Fransico, CA

A Genetic Algorithm for Visualizing Networks of Association Rules

Fabrice Guillet[1], Pascale Kuntz[1], and Rémi Lehn[1,2]

[1] IRIN/IRESTE, Rue Christian Pauc 44300 Nantes, France,
[2] PerformanSE S.A., La Fleuriaye 44470 Carquefou, France
Remi.Lehn@irin.univ-nantes.fr

Abstract. In order to discover relevant information in a huge amount of data, a process commonly used in data mining consists in extracting logical association rules. As algorithms generally produce a large quantity of rules which hides the most interesting, it is essential to develop well-adapted rule mining tools which organize the rules and offer an intelligible representation of them.

In this paper, we focus on an approach based on the visualization of graphs modeling these association rule sets. The aesthetic criteria inherent to such representations are associated with combinatorial optimization problems unfortunately known to be NP hard. Moreover, in KDD applications it is necessary to introduce an additional criterion of stability when taking into account modifications in layout. We develop here a genetic algorithm for drawing association rule graphs which allows to satisfy readability constraints and to find very quickly new solutions close from the previous ones when slight modifications are inserted. Experimental results are presented, including the fitness function behavior with different GA parameters and graph sizes and a dynamic layout animation and an example on a corpus of real data is detailed.

Keywords : Knowledge Discovery, Association Rule Visualization, Genetic Algorithms.

1 Introduction

Since the beginning of the decade, Knowledge Discovery in Databases (KDD) has been an interdisciplinary field of growing interest aroused by firm needs in information processing. Its purpose is for decision makers to discover relevant information (e.g. regularities, exceptions, associations...) in a huge amount of data [4]. One common issue is to extract association rules, i.e. rules LHS $\Rightarrow$ RHS where LHS is an attribute conjunction and RHS is a single attribute [1]. But, in this case, a major problem comes from extraction algorithms : they generally produce a dramatically large quantity of rules, and those which are really interesting are consequently lost in the crowd. Although many heuristics have been developed for rule validation, the task of mining relevant rules is often left to the decision-maker. However, in practice, this evaluation is highly sensitive and requires a costly cognitive effort which may take quite a long time. Therefore, it is essential to develop well-adapted tools which first organize the rules, and second offer an intelligible representation of them. Until recently, research on association rules discovery have mainly focussed on automatic extraction and, the human-centered representation problem has been less studied. From the tradition of expert systems, rules are classically presented in textual form "if ... then ...". From the tradition of data analysis other studies have dealt with the representation of rule distribution (e.g. [6])

In this paper, we consider another approach based on the visualization of rule networks. Some papers are concerned with so-called "rule graphs" i.e. graphs such that nodes are single attributes and links are association rules (e.g. [12]). We extend such representations to rules with conjunction of attributes. As such rules define an oriented relation, they can naturally be represented by directed graphs. Adapted drawings of these graphs are very useful for reaching decisions as they present a well-structured and readable information. Indeed, graphic media can show a larger density of information than texts do while remaining easily understandable. We restrict ourselves here to two-dimensional representations as three-dimensional ones are just at their infancy and thus set, on the one hand, many algorithmic problems which are still completely open and, on the other hand, many questions on their interpretation.

1.1 Visualizing associative rules by graphs

A major difficulty is encountered in data visualization by graphs when formalizing æsthetic criteria which attempt to characterize readability. In order to highlight the underlying hierarchical relationships, directed acyclic graphs (digraph) are usually drawn in such a way that all the arcs flow in the same direction, e.g. from left to right. Moreover, when data can be partitioned in a *small* set of classes associated with a rank in a hierarchy – as it is the case with the partially ordered sets of rules treated here – it is convenient for the interpretation to constrain vertices to lie on a set of equally spaced vertical lines (e.g. [17]). Such graphs are called layered digraphs.

To favor the clarity of the drawing, one can add several constraints generally formulated by optimization criteria. A classical one is the minimization of crossing edges. But minimizing crossings for general layered digraphs is NP-hard even if there are only two layers [8]. Due to the importance of the problem for practical applications in various fields, many heuristics have been developed in the relevant literature (see [2] for a general overview).

However, for KDD applications, an important additional consideration must be taken into account : the graph layout must be dynamic, i.e. able to incorporate changes, according to the decision maker's point of view and to new data. And, these limited modifications should not entail too strong disturbances on the basic layout. On the contrary, it is of major importance for the interpretation of results that the dynamic layout preserves the user's mental map [3].

1.2 Genetic algorithms for dynamical graph drawing

In this context genetic algorithms (GA) appear as a promising field of research. Since the seminal work of Holland [11] on GA and Rechenberg [14] and Schwefel [16] on so-called evolutionary strategies, numerous papers have shown their relevance to tackle optimization problems for which standard methods such as gradient-based algorithms are not applicable [9]. Besides these properties for optimization, some recent studies on online control of process have underlined their interest for problem solving in dynamical environments [15]. In process control they have been shown to be not sufficient on their own because of the very strong industrial constraints. However, in our context, data evolve more smoothly and, by keeping track of several potential solutions, GAs allow to find new solutions very quickly, not too far from the previous ones, when *"small"* modifications appear. They can preserve the mental map's conservation without calculating explicitly distances between layouts as is often the case in incremental drawing [13]. In the optimization process, they can also take into account solutions proposed by the user by merely considering them as new individuals in the population of potential solutions.

In the first part of the paper, we detail our GA implementation; we introduce new operators and dynamical update of genotypes. Then, experimental results are presented, including the fitness function behavior with different GA parameters and graph sizes and a dynamic layout animation. These results confirm the interest of the approach. Some of our results are compared with results of other published algorithms. We conclude with a real-life application.

2 Description of the algorithm

We recall very briefly that the generic problem of genetic algorithms (GA) is to find the maximum value of a function f, called fitness function, in reference to the neo-Darwinian theory of evolution, from a set of potential solutions – individual – on $\mathcal{R}$. The optimization principle is based on the computation of successive generations composed of better and better adapted individuals i.e. with greater and greater fitness. Different stochastic rules – genetic operators – regulate this evolution (see [9], for details).

2.1 Fitness function

In the following, we consider that the partial order between association rules is associated with a directed graph $G = (V, A)$, with a vertex set V, of cardinality n, representing rules and an arc set A representing partial order relationships.

To simplify the graph layout problem, a transformation is previously made on A so that each arc is incident to the vertices placed on two adjacent layers only; an arc incident to vertices v_k and $v_{k'}$, respectively placed on layers k and k' so that $\mid k - k' \mid > 1$, is transformed into a path of $\mid k - k' \mid$ arcs using $\mid k - k' \mid -1$ virtual nodes.

Let $V_k \subset V$ be the subset of vertices placed on layer k. These subsets are here supposed to be previously computed with a ranking function. We denote by $c_{k-1,k}$ (resp. $l_{k-1,k}$) the number of crossings (resp. the sum of the length) of the arcs linking vertices of V_{k-1} and vertices of V_k. The number of potential places on each vertical layer is a discrete number bounded by a maximum height h_k.

In order to take into account the arc crossing and the arc length, the fitness function is here a function from $\mathcal{N}^n$ to the interval $[0..1]$ which maximize :

$$(1 - \frac{1}{\mathrm{Cross_{max}}} \sum_{k=2}^{K} c_{k-1,k})^p \times (1 - \frac{1}{\mathrm{Length_{max}}} \sum_{k=2}^{K} l_{k-1,k})^q$$

The constant $\mathrm{Cross_{max}}$ is a rough approximation of the maximum number of crossings for G^1 and $\mathrm{Length_{max}}$ is a boundary of the total length of the arcs[2]. These ratios are used to normalize the fitness and to make it not too sensitive to the addition or deletion of arcs from one generation to another. The exponents p and q tune the fitness function to focus rather on the minimization of crossings or rather on the arc length : p should be greater than q at the beginning of the layout process to get a suitable permutation of vertex places on each layer; then, q should be increased to optimize edge length.

2.2 Genotype space

One genotype codes the y-coordinates for each vertex of a possible layout. By convention, y-coordinates are stored, layer by layer, from left to right. Layers are then represented by contiguous bits in the genotype, but the order of the vertices in each of them is not important.

Each y-coordinate is binary-encoded with N bits (referred here by *gene*) on the genotype. N is $\mathrm{Max}_k n_k$, where n_k is the upper integer bound of $Log_2(h_k)$ and h_k, the height of the layer k (figure 1).

This encoding allows the representation of superposed vertices. A post-processing (during the genotype to phenotype decoding) is performed which shifts vertex positions with a minimal vertex distance when two or more vertices are superposed.

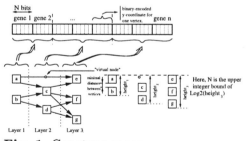

Fig. 1. Genotype space

2.3 Genetic operators

Selection : The selection scheme is the classical roulette wheel selection.

[1] $\sum_{k=2}^{K} \frac{|A_{k-1,k}| \cdot (|A_{k-1,k}| - 1)}{2}$

[2] product of the number of edges in G by the maximum h_k value.

Mutation : The mutation is the classical mutation operator too : a random bit is reversed according to a given probability. We add to the common mutation process a swapping function between genes. In our data structures, the number of crossings can often be decreased by simply swapping the vertices within a layer. As it is often more efficient to swap vertices which are close together, the swapped vertices are not chosen randomly. For each layer, vertex y-coordinates are sorted in a list. The first vertex to be swapped is chosen in this list randomly. Let i be its rank in the list. The second vertex has a rank $i \overset{+}{-} (int)Log_2(X)$ with X a random variable of uniform distribution on $[0..\frac{1}{2}]$.

Crossover : Crossover is here an operator from $\{0,1\}^{N \times n} \times \{0,1\}^{N \times n}$ into $\{0,1\}^{N \times n} \times \{0,1\}^{N \times n}$, i.e. two parents exchange genetic material to build up two offsprings. The crossover point can be chosen from two different methods according to a given probability : either on a random gene border or on a random internal bit. This allow to combine two objectives : finding a good vertex permutation on each layer and optimizing y-coordinates.

Population evolution : The next generation is a set of offsprings and parents. An offspring replaces a parent only if its fitness is better. To avoid always keeping the same individuals, we resort to the *ageing* technique : the fitness of each individual is multiplied by the reverse of the age of the individual, weighed by a parameter.

Meta-GA : In order to select all the parameters (probabilities of mutation, crossover, ageing weight, p and q of the fitness function, ...), we have resorted to another algorithm, called here *meta-GA*. Let Γ_i be a set of successive generations $\omega_1^i, ...\omega_t^i$ computed by our GA. Let $\Pi(\Gamma_i)$ be the set of parameters used in the computation of Γ_i. Let $\varphi(\Gamma_i) = \{f^\star(\omega_1^i), ..., f^\star(\omega_t^i)\}$, where $f^\star(\omega_k^i)$ is the best fitness for generation ω_k^i. The meta-GA computes for a set of Γ_i each associated difference $f^\star(\omega_t^i) - f^\star(\omega_1^i)$ (which characterize the fitness evolution) and retains the parameter set $\Pi(\Gamma_i)$ associated with the greatest difference. Although this meta-GA is very slow, it has allowed us to fix the initial values of the parameters cleverly -better than with a simple gradient- and, moreover it allows to adapt their values periodically to the evolution of the optimization process.

2.4 Dynamical layouts

It is very important in a KDD process to interact with decision makers. In order to take their point of view into account on the rule structuring, our software gives them the opportunity to add/remove vertices and edges and to move vertices.

When the user adds a vertex, a new place is allocated on every existing genotypes and a y-coordinate is randomly generated for the new vertex. Similarly, when a vertex is deleted, the associated gene is removed in every existing genotype. When a vertex is moved, the new *"user layout"* is encoded into a associated new genotype which replaces the genotype of the individual with the worst fitness value. For edge addition/deletion, the only change is concerned with the fitness whose value is calculated again.

3 Experimental results

3.1 Drawing quality

The quality of drawings is measured here by the fitness function which is a weighed product of two components : the number of edge crossings and the total arc length. For instance, on figure 6, the criterion of minimizing edge crossing is not completely satisfied (cf. the two long transitive arcs which are cut by another one); but to satisfy it, we should relax the other criterion ("arc length") and allow "very long" arcs. Such representations with long arcs are known to be hardly readable.

We compare here different implementations of GAs and follow the evolution of fitness for graphs of different sizes.

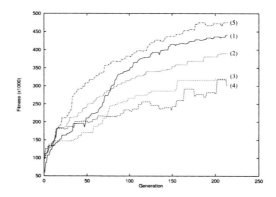

Fig. 2. Evolution of the fitness for 5 GA implementations

Comparison with different GAs' implementations We study here the influence of previously detailed genetic operators on the fitness. Four implementations are compared :

(1) Basic algorithm described in section 2 (an offspring replaces a parent only if its fitness is better, ageing is running).
(2) Basic algorithm, without ageing.
(3) Basic algorithm, and the offsprings always replace their parents.
(4) Basic algorithm, with ageing and the offsprings always replace their parents.

For all of these implementations, the size of the population has been set to 100 individuals. In order to understand the influence of this parameter on the optimization process, we have added a fifth comparison :

(5) Basic algorithm, with a population of 200 individuals.

Figure 2 confirms the relevance of our genetic operators. We see that the ageing process prevents the population from being stuck on local optima (see plateaus on curves (2) and (3)). Moreover, curves (3) and (4) show that our parent-offspring replacement strategy really speeds up the increasing of the fitness.

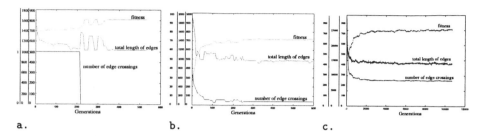

a.   b.   c.

Fig. 3. Evolution of the drawing for different digraphs **a.** (16 vertices, 21 edges), **b.** (26 vertices, 34 edges), **c.** (*jsort*, from AT&T Bell Labs' *graphviz*, 64 nodes, 85 edges)

Evolution of the fitness for different graph sizes We present here results of the influence of the graph size on the optimization process for three graphs : two "small" graphs with different complexities (figures **3a.** and **3b.**) and a complex benchmark, **jsort** (figure **3c.**).

These figures highlight how the algorithm proceeds to reach a balance between the two criteria considered here. For example, on figure **3a.**, an edge crossing is removed at generation #213, at the cost of longer arcs (see a *peak* of the *"total length"* curve when the edge crossing is removed). The vertex y-coordinates are adjusted afterwards (generations 215 – 350) to find shorter edged layouts again. This is not always feasible, for example, on figure **3b.**, the best layout (the higher fitness value) does not always correspond to the layout with the smallest number of edge crossings observed by the GA.

3.2 Dynamic layout

Unlike many graph drawing applications which are concerned with the representation of one complex graph given once and for all, we are interested here in a layout which evolves according to the user's requests. A usual pattern is described on figure 4 : the decision maker focuses on sub-populations having characteristics **a** and **e** and looks for the association rules; then he completes his knowledge step by step and the complexity of graphs representing the revealed rules increases with time. As the incremental drawing can be considered here as a real guide for the decision maker to keep mining relevant rules, it must retain some stability (i.e. not many changes between before and after updates on the graph).

152

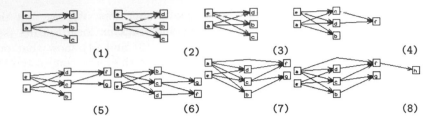

Fig. 4. Digraph updates and incremental layout

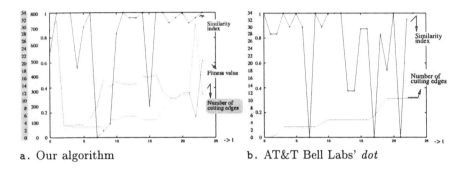

a. Our algorithm   b. AT&T Bell Labs' *dot*

Fig. 5. Evolution of the similarity index for an incremental drawing

In order to precise this notion, we introduce a similarity index $S(L_t, L_{t+1})$ to compare the layouts L_t and L_{t+1} of the graphs at time t and $t+1$. We take only into account the number of vertex permutations between two layouts as this criterion is the most sensitive one for the interpretation. Let C_k be the number of vertex couples on layer k whose order has changed between L_t and L_{t+1} and m_k be the number of vertex couple on layer k common to the two layouts. Then, $S(L_t, L_{t+1}) = 1 - \frac{1}{K} \sum_{k=1}^{K} \frac{C_k}{m_k}$.

Figure 5a. shows the evolution of S for an incremental drawing : some vertices and edges are randomly added at each time t to an initial vertex. Running time has been fixed to 0.5 second (user's time) which is a time comparable to "static" algorithms. Two phases appear. At the beginning, the graph is very small and its layout is very sensitive to changes : some permutations can improve the fitness very strongly. Then, changes in graphs have less disruptive effects on layouts. From time to time, important modifications are introduced when a new solution which is very different from the previous one has a significant better fitness value.

We have compared these results with a well known static algorithm (*dot*, from the Bell Laboratories, [7]) which has been applied to each graph generated at each time t. Figure 5b. shows that our algorithm is more robust to changes. We have also compared the quality of obtained drawings. But, as the optimization criterion in *dot* is different from our fitness value, we only retain here the number of cutting edges – which is a common criterion for the two algorithms – The values reached by our incremental algorithm are very close to those reached by *dot*.

Experiments confirm the interest of GA for such a problem of incremental optimization. However, undesirable switches between potential solutions L_t^i and L_t^j stored in the population may appear sometimes when their fitness are very close but are in reverse order in t and $t + 1$ ($f^\star(L_t^i) < f^\star(L_t^j)$ and $f^\star(L_t^i) > f^\star(L_t^j)$). This could be controlled by an additional local heuristic able to select some "suitable" solutions. Moreover, various tuneable parameters of the GA (ageing, mutation, ...) may be used to control the stability of the layout when updating the graph.

4   Applications

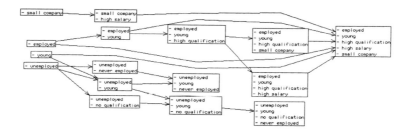

Fig. 6. Real-life application. Drawing elaborated in 121 generations (100 individuals), 3.5 sec. user time on an AMD-K6/2 300 MHz, Linux/jdk 1.1.5 with TYA JIT.

The drawing algorithm is included in a integrated KDD system which is composed of three major parts : a relational database which contains the data and the discovered rules, a user-driven mining algorithm for searching relevant rules associated with the user's requests in the database (see [10] for more details), and an interface to visualize knowledge as rules.

The interface offers on the one hand world wide web like navigation tools -based on HTML, Java and Apache- which communicate with the database via common SQL queries and, on the other hand the dynamic two-dimensional drawing described in this paper. This software has been tested on a real data base -from the French Agency for Employment- describing people who recovered a job. This database contains 6,200 rows of 150 attributes (company, salary, age, etc ...). Figure 6 gives an example of a drawing obtained after several user's requests. Vertices are descriptions (attribute conjunctions) of individuals in the database. And, the existence of a strong association rule between two descriptions is modeled here by an arc. The quality of rules is measured by the predictability and the intensity of implication [5]. This interface is able to visualize quickly and intelligibly states of knowledge selected by the user. It also allows to store the trail of his mining process on a structured way. Its limitations are due mainly to the size of handled graphs ; beyond 50 – 60 nodes the representation on a single screen is no more legible. In order to deal with such limitations we study at the moment a partitioning algorithm based on semantic constraints.

References

1. R. Agrawal, H. Manilla, R. Srikant, H. Toivonen and A. Inkeri Verkamo, *Fast Discovery of Association Rules*, in *Advances in Knowledge Discovery and Data Mining*, AAAI Press, (12):307–328, 1996.
2. G. Di Battista, P. Eades, R. Tamassia, and I. G. Tollis, *Algorithms for drawing graphs*, Mc Graw & Hill, 1998.
3. P. Eades, W. Lai, K. Misue and K. Sugiyama, *Preserving the mental map of a diagram*, Proc. Compugraphics 91, Portugal, 24–33, 1991.
4. V.M. Fayyad, G. Piatetsky-Shapiro, P. Smyth, R. Uthurusamy, *Advances in Knowledge Discovery and Data Mining*, the AAAI Press, Menlo Park, 1996.
5. L. Fleury, & Y. Masson, "The intensity of implication, a measurement for Machine Learning", The 8th Int. Conf. of Industrial and Engineering Applications of Artificial Intelligence and Expert Systems, IEA/AIE, Melbourne, 1995;
6. T. Fukuda, Y. Morimoto, S. Morishita, T. Tokuyama, *Data Mining using Two-dimensional Optimized Association Rules : Scheme, Algorithms and Visualization*, Proc. 1996 ACM SIGMOD, International Conference on Management of Data, Montreal, Quebec, Canada, 13–23, 1996.
7. E. R. Gansner, E. Koutsofios, S. C. North, K. Vo, *A Technique for Drawing Directed Graphs*, technical report, AT&T Bell Labs, available from http://www.research.att.com/sw/tools/graphviz 1993.
8. M.R. Garey, and D.S. Johnson, *Crossing number is NP-Complete*, SIAM J. algebraic and Discrete Methods, 4(3):312–316, 1983.
9. D.E. Goldberg, *Genetic Algorithms in search, optimization and machine learning*, Addison-Wesley Publishing Company, 1989.
10. S. Guillaume, F. Guillet, J. Philippé, Contribution of the integration of intensity of implication into the algorithm proposed by Agrawal. In *EMCSR'98, 14th European Meeting on Cybernetics and System Research*, Vienna, (2):805–810, 1998.
11. J. Holland, *Adaptation in natural and artificial systems*, University of Michigan Press, Ann Arbor, 1975.
12. M. Klemettinen, H. Manilla, P. Ronkainen, H. Toivonen and A. Inkeri Verkamo, *Finding interesting rules from large set of discovered association rules*, Proc. Third International Conference of Information and Knowledge Management, CIKM'94, 401–407, 1994.
13. S. C. North, *Incremental Layout in DynaDAG*, Tech. report, Software and System Research Center, AT&T Bell Laboratories, available from http://www.research.att.com/sw/tools/graphviz, 1998.
14. Rechenberg, *Evolutionsstrategies : Optimierung technicher System nach prinzipien der biologischen Evolution*, Fromman Holzboog, Verlag stuttgart, 1973.
15. J.M. Renders, *Genetic algorithms and neural networks*, (in French), Hermes, 1994.
16. H.P. Schwefel, *Numerical Optimization of Computer Models*, John Wiley & Sons, New-York, 2nd edition, 1981–1995.
17. K. Sugiyama, S. Tagawa, and M. Toda, *Methods for visual understanding of hierarchical systems*, IEEE Trans. on Systems, Man and Cybernetics, SMC-11(2):109–125, 1981.

Genetic Algorithms in Solving Graph Partitioning Problem

Sahar Shazely[1], Hoda Baraka[1], Ashraf Abdel-Wahab[2], and Hanan Kamal[3]

Computer Engineering Dept. Faculty of Engineering Cairo University, Egypt.
E-mail : {shazely_s@hotmail.com, hbaraka@idsc1.gov.eg}
Computers and Systems Dept. Electronics Research Institute Cairo, Egypt.
Email:ashraf@eri.sci.eg
Electrical Engineering Dept. Faculty of Engineering Cairo University, Egypt.
Email:hanan_kamal@hotmail.com

Abstract. The Graph Partitioning Problem (GPP) is one of the fundamental multimodal combinatorial problems that has many applications in computer science. Many algorithms have been devised to obtain a reasonable approximate solution for the GP problem. This paper applies different Genetic Algorithms in solving GP problem. In addition to using the Simple Genetic Algorithm (SGA), it introduces a new genetic algorithm named the Adaptive Population Genetic Algorithm (APGA) that overcomes the premature convergence of SGA. The paper also presents a new approach using niching methods for solving GPP as a multimodal optimization problem. The paper also presents a comparison between the four genetic algorithms; Simple Genetic Algorithm (SGA), Adaptive Population Genetic Algorithm (APGA) and the two niching methods; Sharing and Deterministic Crowding. when applied to the graph partitioning problem. Results proved the superiority of APGA over SGA and the ability of niching methods in obtaining a set of multiple good solutions.

1. Introduction

Graph Partitioning Problem (GPP) is a fundamental combinatorial optimization problem that has extensive applications in many areas, including scientific computing, VLSI design, parallel programming and Networks. This problem is NP-hard and many algorithms have been developed to obtain a reasonable approximate solution [7,8]. GPP can be considered as a multimodal optimization problem that has multiple good solutions. The choice of the suitable solution depends on the human experience. Genetic Algorithms (GAs) [2,4,6] are simple, robust and efficient search techniques that mimic natural genetics in natural evolution and selection. They have been applied to a wide range of practical optimization problems in a variety of domains. This paper introduces a new GA called "Adaptive Population Genetic Algorithm" (APGA) that overcomes some of the problems of SGA. In addition it presents a new approach in solving GPP as a multimodal optimization problem. Adaptive population GA (APGA) [12] is a new proposed genetic algorithm that overcomes the premature convergence of the conventional GA. APGA eliminates the premature convergence of SGA by replacing a set of either similar or undesirable weak individuals by a set of new generated individuals. Using this algorithm, solutions which is better than that of SGA are obtained. Niching methods [1,5,9,10,11] are techniques based upon Genetic Algorithms. They are used in order to obtain multiple good solutions. These methods have been used successfully in solving complex problems specially multimodal problems. Two niching methods; Sharing and Deterministic Crowding are used to obtain a set of multiple good solutions for the GP problem. Experiments are conducted to compare between four Genetic Algorithms when applied to problems of different sizes. These techniques are: Simple Genetic Algorithm, Adaptive Population GA, Sharing, Deterministic Crowding.

The paper is organized as follows: section 2 defines the Graph Partitioning Problem. Section 3 introduces the newly proposed GA "Adaptive Population GA". Section 4 presents a description of the niching methods. Problem representation and the used objective function is described in section 5. Section 6 presents the results of all the genetic algorithms. Section 7 presents the conclusion of this work.

2. Problem definition

A Graph can be represented as a set of nodes (V), and a set of edges (E). Each edge connects exactly two nodes. A Graph can be formulated as a relation between nodes and edges.

$$G = (V, E)$$
$$V = \{ v_1, v_2, \ldots\ldots, v_n \}$$
$$E = \{ e_1, e_2, \ldots\ldots, e_n \}$$
$$e_i = (v_j, v_k)$$

GPP is to partition a graph G into m subgraphs (g) such that the number of edges connecting nodes in different subgraphs is minimized, and the number of edges connecting nodes of the same subgraph is maximized.

$$G = \{ g_1, g_2, g_3, \ldots\ldots g_m \}$$

3. Adaptive Population GA (APGA)

APGA [12] is a new proposed genetic algorithm that overcomes the premature convergence of the conventional GA by replacing similar or extremely weak individuals by a set of new individuals generated by using high mutation rate for a set of unwanted individuals to produce new individuals that lead to better solutions. In genetic search, the process converges when most of the individuals of a population become identical, or nearly so. Unfortunately, this often occurs before a true optimum is reached, this behavior is called premature convergence. Eliminating the premature convergence depends on the efficiency of the used selection method.

Another disadvantage of SGA is that, in some cases, with a selection method that works close to the perfect sampling, a population may contain a large number of individuals that have low fitness. This situation may lead to undesirable solutions and hardly guide to further improvement of the solution within relatively large number of generations. In this case, the number of promising individuals produced by the crossover of two weak individuals is very low. The mutation operator reintroduces lost solutions, but high mutation rate increases the diversity of the search space at the cost of losing good or promising solutions. Low mutation rate is not able to introduce sufficient individuals that helps the population to replace its undesirable weak individuals. The pseudo code of the APGA is shown below. APGA module starts with multiple number of generations specified (G_{AP}) of a non improved solution.

Procedure APGA:
 - IF current generation = generation of best individual so far +
(APGA-Counter1 * G_{AP})
  - Sort the population by fitness.
  - Count the number of similar individuals of the highest fitness.
  - IF number of similar individual > accepted number of similarity
   - Reproduce new individuals by increasing the mutation rate
   of a number of similar individuals.
   - Calculate the fitness of new individuals.
  - Increment APGA-Counter1
   - IF current generation = generation of best individual so far
+
   (APGA-Counter2 * constant* G_{AP})
    - Reproduce new individuals by increasing the mutation
rate of a
    number of the weakest individuals.
   - Calculate the fitness of new individuals.
   - Increment APGA-Counter2
  - Shuffle the population
   - Apply GA operators (crossover and mutation)
 - IF any improvement to the solution occurred
  -Reset APGA-Counter1 and APGA-Counter2
End of APGA procedure

4. Niching methods

Niching methods [2,9,10,11] are efficient techniques that promote and maintain stable, diverse and final multiple solutions. The need of using niching methods in solving GPP is to obtain a set of multiple good solutions for the problem. SGA and APGA obtain only one suitable solution within the scope of one population. The strength of niching methods is to obtain a set of multiple good solutions within the cope of one population. In this paper two niching methods are used; Sharing and Deterministic Crowding.

4.1 Sharing

Sharing [1,5] derates the fitness of an individual by an amount related to the number of similar individuals that share the same niche. The new shared fitness of the individual, f' is equal to its old fitness f divided by its *niche count*. The niche count is calculated as the sum of the sharing function (sh). The sharing function depends on the distance d(i,j) between an individual and all other individuals in the population. It uses a constant threshold σ_{share} to determine whether the two individuals belong to the same niche or not. Both genotypic and phenotypic distance measures are used.

4.2 Deterministic Crowding (DC)

Deterministic crowding (DC) [9,10] is another niching method that operates as follows: First, all population elements are grouped into (population size/2) pairs. Then all pairs are crossed and optionally offsprings are mutated. Each offspring competes in a tournament against one of its parents by comparing their fitness (f). Two sets of parent-child tournaments are possible. DC holds the set of tournaments that forces the most similar elements to compete. Similarity can be measured using either genotypic (Hamming) or phenotypic (Euclidean) distance. Fig.4 illustrates the pseudo code of Deterministic Crowding.

5. Genetic representation for the GPP

This section presents the mapping of the GPP using GAs and the used objective function It also presents the used distance metric needed to measure the distance for niching methods.

5.1 Chromosome representation

The possible solutions are represented as fixed length chromosomes where :
Chromosome length = No of nodes.
Alleles are integers that represent the subgraph number which contain the node.
Interconnection between nodes are represented as real numbers.
An example of a chromosome representing a graph of 12 nodes is shown in fig. 1.

Subgraph number   Node number

| 2 | 3 | 0 | 1 | 1 | 0 | 2 | 2 | 3 | 1 | 3 | 0 |

Fig. 1. Chromosome representation of 12 nodes

5.2 Objective function

The objective function is to minimize the Performance Index (PI) of a graph. PI represents the lack of completeness and is defined as a simple description of the main two requirements of the objective function. The PI is defined as follows:

$$PI = \sum_K \left(\frac{N_k(N_k - 1)}{2} - A_{total} \right) + \sum_K External\_interconnections \tag{1}$$

K represents the maximum number of subgraphs and N_k represents the number of nodes of a subgraph k. The parameter A_{total} represents the total internal interconnections between nodes of a graph k. The summation over K indicates the summation over all subgraphs, the first term represents the lack of completeness of subgraph G_k. The second term represents the summation of external interconnections between all subgraphs. The fitness is calculated as :

$$Fitness\ function\ = \frac{1}{(PI + 1)} \tag{2}$$

5.3 Distance metric

Niching methods need a method to calculate distance between each two individuals. Both genotypic and phenotypic distance measure are used. Hamming distance is used as genotypic distance measure, While Euclidean distance is used as phenotypic distance measure. For two individuals of p dimensions x and y. $x = \{x_1, x_2, x_3, \ldots\ldots x_p\}; y = \{y_1, y_2, y_3, \ldots\ldots y_p\}$

$$Hamming\ distance\ H(x,y) = \sum_{i=1}^{p} d(x_i, y_i) \tag{3}$$

$$d(x_i, y_i) = \begin{cases} 1 & If\quad x_i \neq y_i \\ 0 & If\quad x_i = y_i \end{cases}$$

$$Euclidean\ distance\ E(x,y) = \sqrt{\sum_{i=1}^{p} (x_i - y_i)^2} \tag{4}$$

6. Experiments and Results

In order to select the proper set of GA operators adequate for the GPP, a number of experiments has been conducted with four different selection methods, and two crossover methods (single and double points). The selection methods are:Roulette Wheel, Tournament, Stochastic Remainder Sampling and Stochastic Universal Sampling (SUS). As a result of these experiments, tournament and SUS selection schemes are chosen to compare the performance and results of SGA with APGA. Tournament is chosen for its efficiency while SUS is chosen because it is the closest sampling selection method to the perfect sampling. Tournament selection is chosen to be used with Sharing.

6.1 Tuning parameters for APGA

A number of experiments are conducted in order to tune the parameters of the APGA. All the tuning experiments are applied on a graph of 23 nodes with a performance index equals to 20. The tuned parameters are: The percentage of accepted number of similar individuals (percentage of population size), the percentage of similar individuals to be replaced (percentage of similar individuals), the percentage of undesirable weak individuals to be replaced (percentage of population size) and the generation gap (percentage of the population size). Table 1 and figures (2 & 3) show that Tournament selection is sensitive to the change of the percentage of similar individuals to be replaced as Tournament selection leads to premature convergence by increasing the number of copies of the best fit individual. replaced.

Table 1. Codes of Parameters settings for APGA

Generation Gap (% of pop. size)	% of weak ind. to be replaced	% of similar ind. to be replaced	% of accepted similar ind. in a pop.	Exp. No.
0.0	0.1	0.5	0.1	1
0.0	0.1	0.8	0.2	2
0.0	0.2	0.5	0.1	3
0.02	0.1	0.5	0.1	4
0.02	0.2	0.5	0.1	5

6.2 Adaptive Population GA Experiments

Figures (4 - 7) show a comparison between SGA and APGA when applied to problems of different sizes 30 and 50 nodes with both Tournament and SUS selection methods. Results show the superiority of the proposed algorithm than SGA in obtaining better solution by preventing the premature convergence.

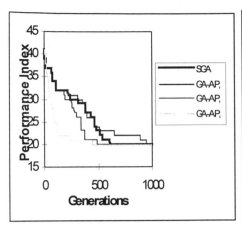

Fig. 2 - APGA - SUS

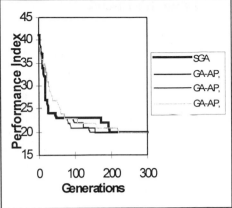

Fig. 3 - APGA - Tournament

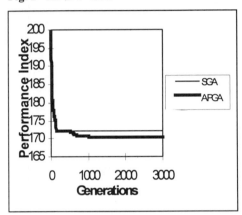

Fig. 4 - Graph of 30 nodes Tournament

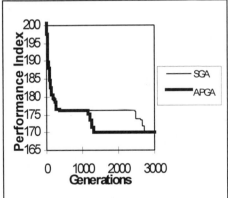

Fig. 5 - Graph of 30 nodes - SUS

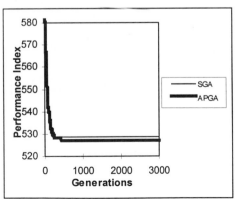

Fig. 6- Graph of 50 nodes Tour.

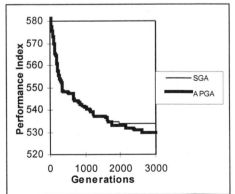

Fig. 7 - Graph of 50 nodes - SUS

162

6.3 Niching Experiments

Figures (8-11) show a comparison between SGA, and the two niching methods
Sharing and Deterministic Crowding when applied to problems of sizes 23, 30, 50 and
100 with Tournament selection method. A Sample of some solutions obtained by
niching methods are illustrated in table 2 and 3. Table 2 illustrates 10 solutions that
obtained by both sharing and DC of a graph of 23 nodes. These solutions are all
satisfying the objective function and have the same performance index but they are
completely different solutions. For example solution number 8 divided the problem of
23 nodes (n_i) into 7 different subgraphs (g_i) as follows: $g_1(n_1, n_5)$, $g_2(n_2, n_3, n_4)$, $g_3(n_6$
, n_7, n_8, $n_9)$, $g_4(n_{10}, n_{11}, n_{12})$, $g_5(n_{13}, n_{16})$, $g_6(n_{14}, n_{15}, n_{17}, n_{18})$ and $g_7(n_{19}, n_{20}$,
n_{21}, n_{22}, $n_{23})$. Solutions 1 and 2 divide them into 8 subgraphs but with different
arrangement. Solution 10 divide them into the minimum number of subgraphs.

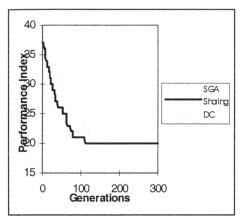

Fig. 8. Graph of 23 nodes

Fig. 9 . Graph of 30 nodes

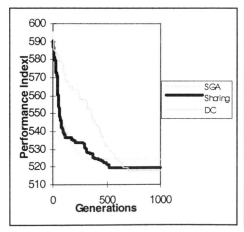

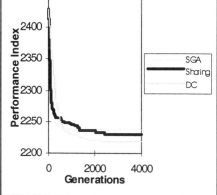

Fig. 10 - Graph of 50 nodes

Fig. 11 - Graph of 100 nodes

Table 2 - Example of 10 solutions of a graph of 23 nodes obtained by niching

Solution No.	Chromosome	Clusters	Best PI
1	0,1,1,1,1,2,2,2,2,3,3,3,4,4,5,5,6,6,7,7,7,7,7	8	20
2	0,1,1,1,0,2,2,2,2,3,3,3,4,4,5,5,6,6,7,7,7,7,7	8	20
3	0,0,0,0,0,1,1,1,1,2,2,2,3,3,4,4,5,5,6,6,6,6,6	7	20
4	0,1,1,1,1,0,2,2,2,3,3,3,4,4,5,5,6,6,7,7,7,7,7	8	20
5	0,1,1,1,1,2,2,2,2,3,3,3,4,4,5,5,5,5,6,6,6,6,6	7	20
6	0,1,1,1,1,2,2,2,2,3,3,3,4,4,5,5,5,5,6,6,6,6,6	7	20
7	0,0,0,0,0,1,1,1,1,2,2,2,3,3,4,4,4,4,5,5,5,5,5	6	20
8	0,1,1,1,0,2,2,2,2,3,3,3,4,5,5,4,5,5,6,6,6,6,6	7	20
9	0,1,1,1,1,2,2,2,2,3,3,3,4,4,4,4,4,4,5,5,5,5,5	6	20
10	0,0,0,0,0,1,1,1,1,2,2,2,3,3,3,3,3,3,4,4,4,4,4	5	20

Table 3 - PI for the best five solutions obtained by DC

Solution	Graph of 30 nodes	Graph of 50 nodes	Graph of 100 nodes
1	170.144	518.063	2218.109
2	170.83	519.550	2218.450
3	171.032	519.900	2219.843
4	171.66	520.321	2222.595
5	172.464	522.941	2222.735

7. Conclusions

The main objective of the research presented in this paper, is to assess the ability of GA-based techniques to solve GP problem and discover multiple solutions for this class of problems. Two approaches have been followed. The first approach is to find an optimal solution using GAs. The second approach is to consider GPP as a multimodal optimization problem that have multiple solutions. In order to find a good solution, a new proposed genetic algorithm, the Adaptive Population GA (APGA) has been formulated and tested. It has the advantage of eliminating premature convergence of SGA by replacing a number of either similar or undesirable weak individuals by new generated individuals that will be inserted in the population. Practical experiments were conducted to choose the best selection and crossover methods as well as to tune GA operators. Four selection schemes were compared together with both single point and double points crossover. Two selection methods were chosen to compare the performance of SGA and APGA with different problem sizes. Experiments with sample graphs of 30 and 50 nodes were used to show the ability of APGA to obtain better solutions than SGA. The second approach is to obtain a set of multiple solutions for GP problem. Two niching methods; Sharing and Deterministic Crowding were used to obtain multiple solutions for the GP problem.

Experiments were conducted on a variety of problems with different graph sizes ranging from 23 up to 100. Experiments showed that niching methods are able to obtain multiple solutions with better quality as compared to SGA. Comparing niching methods, DC is simple to implement, faster in convergence and it also maintain high level of stability. The solutions obtained by DC are of higher fitness than those obtained by Sharing. On the other hand, Sharing obtains large number of different solutions within the same population, although it may not converge to a global solution.

References

[1] Deb, K., & Goldberg, D.E., (1989). An investigation of niche and Species formation in genetic function optimization. Proceedings of the 3rd ICGA

[2] Deb, K. (1989). Genetic Algorithms in multimodal function optimization. Univ. of Alabama.

[3] Eid, S. A. & Fahmy A. H. (1978). "Optimal Partitioning of Large Scale Networks". International conference of measurement and control (MECO), Athens, Greece.

[4] Goldberg, D.E., (1989). Genetic algorithms in search, optimization, and machine learning. Addison-Wesley publicashing comp.

[5] Goldberg, D.E., and Richardson, J.(1987). Genetic algorithms with sharing for multimodal function optimization. Genetic algorithms and their applications: proceedings of the 2nd ICGA.

[6] Holland, J.H. (1975). Adaptation in natural and artificial systems. Ann Arbor: The University of Michigan press.

[7] Kernighan, B. W. & Lin S. (1970). "An efficient heuristic procedure for partitioning graphs" . Bell syst. Tech. J., Vol 49, pp 291 - 307.

[8] Laszewski G., A Collection of Graph Partitioning Algorithms. Northeast Parallel Architectures center, Syracuse University

[9] Mahfoud, S.W.(1995). A comparison of parallel and sequential Niching methods. Proceeding of 6th ICGA.

[10] Mahfoud, S.W. (1995). Niching methods for genetic algorithms (Doctoral dissertation / IlliGAL Rep. 95001). Urbana: u. of Illinois, Illinois Genetic Algorithms Lab.

[11] Mahfoud, S.W. (1993). Simple Analytical Models of Genetic Algorithms for Multimodal Function Optimization IlliGAL Rep. 93001). Urbana: u. of Illinois, Illinois Genetic Algorithms Lab.

[12] Shazely S., Eid, S., Baraka, H., Kamal H., 1998. "Solving Graph partitioning problem using Genetic Algorithms". M.Sc. thesis, Cairo University, Egypt.

Application of Genetic Algorithms in Power System Stabilizer Design

Ahmed I. Baaleh[1], Ahmed F. Sakr[1]

[1]Department of Electrical Power and Machines,
Faculty of Engineering, Cairo University, Giza, Egypt
abaaleh@hotmail.com,   sakr@starnet.com.eg

Abstract. The paper presents a new approach to design power system stabilizer via genetic algorithms. The optimum settings of the stabilizer parameters associated with selected set of grid points in the active/reactive power domain, are computed offline by minimizing an objective function. A single machine infinite bus system is considered to demonstrate the suggested technique.

1 Introduction

The problem of improving dynamic stability of power systems by means of supplementary stabilizers has received considerable attention during the last three decades. The supplementary feedback signal is introduced into the excitation and/or the governor loop of a synchronous generator to provide extra damping to the variations in frequency and terminal voltage [1-5].

It is important to recognize that machine parameters change with loading, making the dynamic behavior of the machine quite different at different operating points. Since these parameters change in a rather complex manner it is difficult to reach general conclusion from stabilizer parameter settings based only on one operating point. Therefore, to improve the damping characteristics of a power system over a wide range of operating points, various types of techniques have been proposed such as self tuning controllers [4], model reference adaptive control [8], variable structure power system stabilizer [3], and fuzzy logic controller [5]. A major disadvantage of the available techniques is the need of extensive on-line computations.

In this paper, a new approach is proposed to design a supplementary power system stabilizer (PSS) for synchronous machine infinite bus system. This approach is based on genetic algorithms (GA) to compute the optimization setting of fixed stabilizer parameters off-line. Genetic algorithms aim at finding the global optimum solution by using direct random search in any kind of system complexity. Mathematical properties such as differentiability, convexity, and linearity are of no concern for those algorithms. This is the biggest advantage of this search method over traditional optimization techniques.

2 Problem formulation

The mathematical model of a typical synchronous machine connected to infinite bus is described in Appendix A. this model can be linearized as follows:

$$\dot{X}_1(t) = A_1 X_1(t) + B_1 U_1(t) \tag{1}$$

Where $X_1(t) = [i_d(t) \quad i_F(t) \quad i_D(t) \quad i_q(t) \quad i_Q(t) \quad \delta(t) \quad \omega(t) \quad E_{FD}(t)]^T$

We suggest power system stabilizer as shown in Fig. (2), where the stabilizing signal is derived from the electrical torque signal ΔT_e and shaft speed variation $\Delta\omega$ [6][7]. The control signal U_{pss} can be expressed in frequency domain as:

$$U_{PSS} = \frac{S}{(S + \frac{1}{T})^2}(K_1\Delta\omega + K_2\Delta T_e) \tag{2}$$

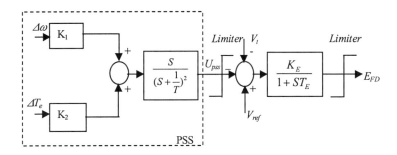

Fig. 2. Exciter-voltage regulator including PSS.

To linearize equation (2), the auxiliary state variables v_1 and v_2 are considered as:

$$\Delta v_1 = \frac{S}{S + \frac{1}{T}}(K_1\Delta\omega + K_2\Delta T_e) \tag{3}$$

$$\Delta v_2 = \frac{S}{S + \frac{1}{T}}\Delta v_1 \tag{4}$$

Equations (3) and (4) can be written in time domain as:

$$\Delta\dot{v}_1 = K_1\Delta\omega + K_2\Delta T_e - \frac{1}{T}\Delta v_1 \tag{5}$$

$$\Delta\dot{v}_2 = \Delta\dot{v}_1 - \frac{1}{T}\Delta v_2 \tag{6}$$

Dropping Δ and the addition of (5) and (6) to (1), the state equations for complete linear system including the PSS are given by:

$$\dot{X}(t) = AX(t) + BU(t) \tag{7}$$

Where

$$X(t) = [i_d(t) \ i_F(t) \ i_D(t) \ i_q(t) \ i_Q(t) \ \delta(t) \ \omega(t) \ E_{FD}(t) \ v_1(t) \ v_2(t)]^T \tag{8}$$

The problem is to determine the parameters of PSS i.e. K_1, K_2, T. In traditional methods these parameters are designed based on linear model that is calculated about one operating point. If the operating point is changed, new PSS parameters must be designed. While in the proposed approach, GA is used to design these parameters that simultaneously stabilize the family of N linear plants. Each plant is represented by linear model for every active (P) - reactive (Q) power combination. The linear models describing the N plants can be written as:

$$\dot{X}(t) = A_k X(t) + B_k U(t) \qquad k = 1,2,\ldots,N \tag{9}$$

Where N is the number of selected operating points corresponding to the set of grid points in the P-Q domain and $P \in [P_{min} \quad P_{max}]$, and $Q \in [Q_{min} \quad Q_{max}]$.

A necessary and sufficient condition for the set of plants described by equation (9) to be simultaneously table with supplementary signal is that the eigenvalues of the N closed loop systems lie in the left-hand side of the complex plane. This condition motivates the following approach for determining the parameters K_1, K_2, and T of the PSS.

Select K_1, K_2, and T to minimize the following objective function:

$$ObjV = \max\left(\frac{\max \operatorname{Re}(\lambda_{k,l})}{\operatorname{Im}(\lambda_{k,l})}\right) \qquad k = 1,\ldots,N; \ l = 1,\ldots,n \tag{10}$$

Where n=10 is the order of the complete linear model, and $\lambda_{k,l}$ is the lth closed-loop eigenvalues of the kth plant. The above objective function minimizes the damping ratio over the whole set of the N linear plants. The minimization is subject to the constraints that $|K_1| \leq a$ and $|K_2| \leq b$ and the time constant $|T| \leq c$ for appropriate pre-specified constants a, b, and c. Clearly if a solution is found such that $ObjV < 0$, then the resulting K_1, K_2, and T simultaneously stabilize the collection of plants. The existence of a solution is verified numerically by minimizing $ObjV$.

3 Design of PSS based on GA

The basic genetic algorithm methodology consist of the following broad steps [9-12]:

Step 1: Chromosomal representation: The first decision to be made regarding the GA is the representation of the parameter information. The chromosome structure is used to represent the parameters whose optimum values are being sought. Each solution in the population is represented by a real number string rather than as a binary string. In our case we are looking for three parameters that can be represented as follows $Z = [K_1\ K_2\ T]$. This vector represents one individual in the population. The chromosome is stored in a matrix of $N_{ind} \times N_{var}$ dimension where $N_{var}=3$ is the number of decision variables and N_{ind} is the number of individuals as follows:

$$Chrom = \begin{bmatrix} Z_{1,1} \\ Z_{2,1} \\ \vdots \\ Z_{Nind,1} \end{bmatrix} \quad \begin{array}{l} individual\ \ 1 \\ individual\ \ 2 \\ \quad \vdots \\ individual\ \ Nind \end{array} \tag{12}$$

This kind of chromosomal representation works much quicker than the binary one and it has two advantages. First, it guarantees that domain expertise embodied in the representation will be preserved. Second, the algorithm to be developed will feel natural to the designer.

The number of individuals N_{ind} plays an important role in GA. If the N_{ind} is small, GA has less genetic material to work with. This means less diversity in population and GA may not fully exploit the search space that can lead to sub-optimal solution. If N_{ind} is too large then the process takes very long time to evaluate and the progression towards a solution is slow. This is particularly acute when the fitness evaluation is complex where each individual must be tested at each generation. Therefore it is important to strike a balance between computation time and the population's information capacity.

For the problem at hand, $N_{ind} \approx 50$ is found to be a good compromise.

Step 2: Generating the initial population: In general the initial generations are chosen randomly; i.e. the genes that make up each chromosome are chosen at random until full population is created. This initial population is referred to as the zeroth generation.

Step 3: Evaluation of the performance of population: Each individual in the initial population has an associated objective function value. The objective function is used to evaluate the performance of chromosome in the problem domain. In our case the objective function is described by eq.(10). It is calculated for each individual as follow:

1. Select the individual $Z_{(1,i)} \in Chrom$ set.

2. For each active/reactive power (P and Q) combination, the eigenvalues that corresponds to maximum objective function is calculated.

3. For each one of the N linear systems we get a vector Λ containing N values of the objective function described by eq. (10) and associated to $Z_{(1,i)}$.

$$\Lambda = [\lambda_1 \ \lambda_2 \qquad \lambda_N] \tag{13}$$

4. The maximum value in the vector Λ is the objective function corresponds to the individual $Z_{(1,i)}$.

The above four steps are repeated for all individuals (N_{ind}). The result is stored in the *ObjV* vector as:

$$ObjV = \begin{bmatrix} f_1 \\ f_2 \\ \vdots \\ f_{Nind} \end{bmatrix} \begin{matrix} individual \ 1 \\ individual \ 2 \\ \vdots \\ individual \ Nind \end{matrix} \tag{14}$$

Step 4: Selection of candidates for reproduction: The selection of individuals to produce successive generations plays an extremely important role in genetic algorithm. A probabilistic selection is performed based upon individuals measured performance such that the better individuals have an increased chance of being selected. An individual in the population can be selected more than once with all individuals in the population having a chance of being selected to reproduce into the next generation.

Step 5: Generation of offspring: The process of generating offspring starts with the candidates who were selected for reproduction. The chromosomes of two parents are combined in order to produce new offspring using the crossover and mutation operators.

4 Simulation Studies

To test the efficacy of the suggested GA-based PSS, simulation tests on a real power system with the parameters given in Appendix B are carried out. The performance of the suggested GA-based PSS is compared to the performance of a conventional PSS based on pole placement. We consider symmetrical three-phase short circuit fault that occurs on one of the transmission lines.

The fault sequence is as follows:

1. **Stage one:** The system is in a pre-fault steady state, so the transmission lines can be represented by resistance R_e and reactance L_e.

2. **Stage two:** A fault occurs at the midpoint of one of the two transmission lines. The fault occurs at $t = 0$ second and is removed after four cycles by opening the breakers of the faulted line.

3. **Stage three**: The system is in a post fault-state (where R_e & L_e are doubled) i.e. the transmission lines can be represented by resistance $2R_e$ and reactance $2L_e$.

4. **Stage four**: The transmission line is restored at t = 3 second, and the system is in steady state, i.e. this stage is the same as stage one.

The four stages are shown in Fig. (3).

In the simulation the nonlinear mathematical model of the synchronous generator described in Appendix A is used.

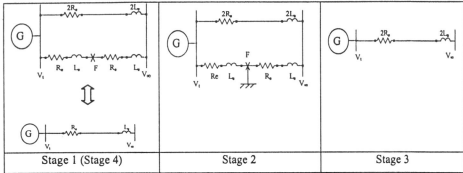

| Stage 1 (Stage 4) | Stage 2 | Stage 3 |

Fig. 3. The fault sequence stages

4.1 Design of PSS based on pole placement

For sake comparison, a conventional PSS is designed in the section based on the conventional pole placement technique. The two dominant eigenvalues of the linear system without PSS are (-0.65423±10.578I).

Fig. (4) shows the locations of dominant open loop eigenvalues versus the constant power lines and constant reactive power lines within the range of $P\in[0.7 \quad 1.2]$, $Q\in[0.5 \quad 0.9]$. It is clear that the system without PSS is always stable within the practical range of P and Q, while it may exhibit high oscillations. A PSS is designed to shift the dominant poles of the nominal system to the furthest location to the lift. The parameters of the PSS were found to be (K_1=131.06, K_2= -12.6, and T=0.001).

When the power system stabilizer added, the eigenvalues of the nominal system were (-1001.5 ± 56.656I, -13.572 ± 376.43I, -1.9989 ± 17.01I, -1.9989 ± 14.14I, -48.378, -46.04). Although the dominant eigenvalues, that were (-1.9989 ± 14.14I), has real part smaller than that of the dominant eigenvalues computed without PSS which makes the system more stable, but the simulation results illustrated by Fig. (9) shows that the system is unstable. This can be explained because some of the P/Q combinations make the system eigenvalues unstable as shown in Fig. (5).

4.2 PSS based on Genetic Algorithms (GA-PSS)

In this section a power system stabilizer PSS for the considered power system is designed based on genetic algorithms. The effect of adding nonlinear terms to the PSS is tested.

By applying the procedure given in section (3), the following PSS parameters were obtained: $K_1 = -178.54$, $K_2 = -15.7365$ and $T = 0.00105031$.

Fig. (7) illustrates the frequency and voltage response of the system when using the designed GA-PSS as compared to that without PSS. It is obvious from this figure that the power system response is greatly improved in the four fault stages.

The same procedure of GA is applied with different PSS structures that incorporate nonlinear terms in PSS parameters. Specifically; the PSS is selected to be in one of the following cases:

Case 1:
$$U_{PSS1} = \frac{S}{(S + \frac{1}{T})^2} (K_1 \Delta \omega + K_2 \Delta T_e + K_3 \Delta T_e^2) \qquad (15)$$

Case 2:
$$U_{PSS2} = \frac{S}{(S + \frac{1}{T})^2} (K_1 \Delta \omega + K_2 \sin(K_3 \Delta T_e)) \qquad (16)$$

For the structure depicted in equation (15), the optimum PSS parameters were found to be $K_1 = -200.36$, $K_2 = -14.8013$, $K_3 = 0.01$, $T = 0.00112351$. While for that depicted in equation (16) the optimum PSS parameters were found to be: $K_1 = -520.229$, $K_2 = -22.2177$, $K_3 = 0.723585$, and $T = 0.001$.

Fig. (8) shows the frequency and voltage response during the different fault stages for the system with different GA-PSS structures given by equations (2), (15) and (16). From this figure, it is clear that the GA-PSS with nonlinear terms provides better response, specifically the structure given by equation (16), which gives very good response.

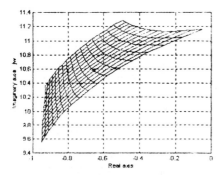

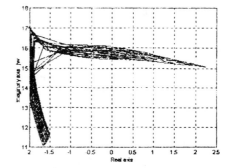

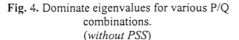

Fig. 4. Dominate eigenvalues for various P/Q combinations.
(*without PSS*)

Fig. 5. Dominate eigenvalues for various P/Q combinations.
(*with pole placement base PSS*)

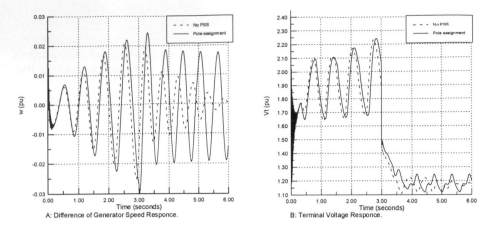

Fig. 6. Comparison of PSS based pole placement controller and without-PSS response.

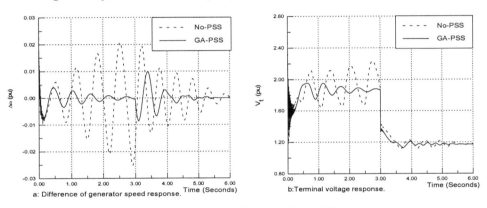

Fig. 7. Comparison of GA-PSS and without PSS response.

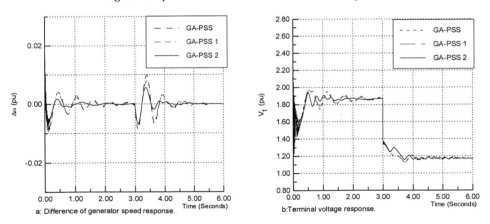

Fig. 8. Comparison of response for different GA-PSS structures.

5 Conclusion

In this paper, a GA based PSS approach is proposed. This approach provides a simultaneous stabilization of power systems over a wide range of operating conditions. The suggested design approach allows great freedom in selecting the desired objective function and the structure of the controller. Nonlinear control scheme can be designed easily via the proposed approach. Simulation results showed the effectiveness of the GA based PSS. It also showed that the damping of the frequency and voltage can be further improved by using a nonlinear PSS structure.

References

1. F. B. DeMello, and Concordia, " Concepts of synchronous machine stability as affected by excitation control", IEEE Trans. Power App. and Sys. PAS-88, 1969, pp. 316-329.
2. H. A. Moussa, and Y. N. Yu, "Optimal Power System Stabilization through Excitation and/or Governor Control", IEEE Trans. Power App. and Sys. PAS-91, 1972, pp. 1166-1174.
3. Abdel-Magid, Y.L. and Swift, G.W., "Variable structure power stabilizer to supplementary static-excitation systems", Proc. IEE, Vol. 123, No. 7, July 1976, pp. 697-701.
4. C. M. Lim, " A Self-Tuning Stabilizer for Excitation or Governor Control of Power System", IEEE Trans. Energy Conversion, Vol. 4, No. 2, 1989, pp. 152-159.
5. Y. Y. Hsu and C. H. Cheng, "A Fuzzy Controller for Generator Excitation Control", IEEE Trans. Sys. Man and Cyb. Vol. 23 No. 2, 1993, pp. 532-539.
6. P. M. Anderson, and A. A. Fouad, "Power System Control and Stability" Iowa State Univ. Press, Iowa, U.S., 1984.
7. P. S. Ray, P. B. Duttagupta, P. Bhakta, "Co-ordinated multimachine PSS design using both speed and electrical power", IEE Proc. C, Vol. 142, No. 5, 1995, pp. 503-510.
8. Y. Y. Hsu and C. H. Cheng, "Variable Structure and Adaptive Control of a Synchronous Generator", IEEE Trans. Vol. AES-24, No4. July 1988, pp. 337-344.
9. D. E. Goldberg, "Genetic Algorithm in Search, Optimization, and Machine learning", Addison Wesley, Reading, MA, 1989.
10. L. Davis (Editor), "Handbook of Genetic Algorithms", Van Nostrand Reinhold, New York, 1991.
11. K. Krishnakumar, D. E. Goldberg, "Control System Optimization Using Genetic Algorithms", Journal of Guidance, Control and Dynamics, Vol. 15 No. 3, May-June1992, pp.735-739.
12. C. J Downing, B. Byrne, K. Coveney, and W. P. Manane, "Controller Optimization and System Identification using Genetic Algorithms", Proceeding of Irish DSP and Control Colloquium, 1996, pp. 45-52.

Appendix A: A Mathematical Model of a Single Synchronous Generator

The system considered here is a synchronous machine connected to infinite bus through a double circuit transmission line as shown in Fig. (3, Stage 1). The dynamic behavior of the synchronous machine can be described by the following set of nonlinear differential equations [6]:

Mechanical Equations:

$$\dot{\delta} = (\omega - \omega_0)\omega_R \tag{A.1}$$

$$\dot{\omega} = \frac{1}{2H}(T_m - T_e - D\omega); \tag{A.2}$$

Where $T_e = \frac{1}{3}(L_d i_q i_d + kM_F i_q i_F + kM_D i_q i_D - L_q i_d i_q - kM_Q i_d i_Q)$

Electrical Equations:

$$v_d = -ri_d - \omega L_q i_q - \omega kM_Q i_Q - \frac{L_d}{\omega_R}\dot{i}_d - k\frac{M_F}{\omega_R}\dot{i}_F - k\frac{M_D}{\omega_R}\dot{i}_D \tag{A.3}$$

$$v_F = r_F i_F + k\frac{M_F}{\omega_R}\dot{i}_d + \frac{L_F}{\omega_R}\dot{i}_F + \frac{M_R}{\omega_R}\dot{i}_D \tag{A.4}$$

$$v_D = 0 = r_D i_D + k\frac{M_D}{\omega_R}\dot{i}_d + \frac{M_R}{\omega_R}\dot{i}_F + \frac{L_D}{\omega_R}\dot{i}_D \tag{A.5}$$

$$v_q = \omega L_d i_d - ri_q + \omega kM_F i_F + \omega kM_D i_D - \frac{L_q}{\omega_R}\dot{i}_q - k\frac{M_Q}{\omega_R}\dot{i}_Q \tag{A.6}$$

$$v_Q = 0 = r_Q i_Q + k\frac{M_Q}{\omega_R}\dot{i}_q + \frac{L_Q}{\omega_R}\dot{i}_Q \tag{A.7}$$

Network Equations. The equations for the transmission network with external resistance R_e and inductance L_e are

$$v_d = -\sqrt{3}\ V_\infty \sin(\delta - \alpha) + R_e i_d + \frac{L_e}{\omega_R}\dot{i}_d + \omega L_e i_q \tag{A.8}$$

$$v_q = \sqrt{3}V_\infty \cos(\delta - \alpha) + R_e i_q + \frac{L_e}{\omega_R}\dot{i}_q - \omega L_e i_d \tag{A.9}$$

The equation for the excitation system shown by block diagram in Fig. (1) is as follows:

$$T_e \dot{E}_{FD} = K_A(V_{ref} - V_t - U_{PSS}) - E_{FD} \tag{A.10}$$

Appendix B

The parameters of the system are as follow [6]:

1. Synchronous generator:
$\omega_R = 2\pi f = 377, \omega_0 = 1.0$ pu, $L_d = 1.70$ pu, $L_q = 1.64$ pu, $L_F = 1.65$ pu, $L_D = 1.605$ pu, $L_Q = 1.526$ pu, $kM_F = kM_D = M_R = 1.55$, $kM_Q = 1.49$ pu, $l_d = l_q = 0.15$ pu, $r = 0.001096$ pu, $r_F = 0.000742$ pu, $r_D = 0.0131$ pu, $r_Q = 0.054$ pu, $D = 0$, $R_e = 0.02$ pu, $L_e = 0.4$ pu, $H = 2.37s$.

2. Nominal load: P=1.0 pu, Q=0.62 pu.

3. Voltage Regulator and Exciter: $\tau_A = 0.05$ s, $K_A = 400$.

A Study of a Genetic Classifier System Based on the Pittsburgh Approach on a Medical Domain

Ester Bernadó i Mansilla[1], Abdelouahab Mekaouche[2], and Josep Maria Garrell i Guiu[1]

[1] Computer Science Department, Enginyeria i Arquitectura La Salle,
Ramon Llull University, P. Bonanova 8, 08022 Barcelona, Spain
{esterb,josepmg}@salleURL.edu
[2] IRIN, University of Nantes, 2 rue de la Houssinière,
BP 92208, 44322 Nantes Cedex 3, France
Abdel.Mekaouche@irin.univ-nantes.fr

Abstract. In this paper we present a classifier system based on Genetic Algorithms for a medical domain. The system evolves a set of rules, using the Pittsburgh approach. Therefore, each individual of the Genetic Algorithm codifies a complete set of rules. Our efforts have focused on the improvement of classification and prediction accuracy and the minimization of the number of rules required to describe the problem. In order to study the behaviour of our system in these areas, several experiments are presented.

1 Introduction

Genetic Algorithms (GAs) are based on a method inspired by the natural evolution of the species [6], [3]. They maintain a set (*population*) of potential solutions (*individuals*) to the problem to be solved. This population is evolved by imposing mechanisms of selective pressure and recombination of the best individuals. At the end of the process, the population tends to converge resulting in a "good solution", which is often the optimum or is very close to it.

We study the application of GAs in Machine Learning problems, which are called GBML (Genetic Based Machine Learning) Systems. This type of problem has been addressed mainly using two different approaches: the Michigan approach and the Pittsburgh approach, first exemplified by CS-1 [11] and LS-1 [5] respectively. In the Michigan approach, each individual of the population codifies one rule and the solution consists of all the members of the population. Some systems developed under this perspective are: SCS [3], Newboole [1], XCS [13], Alecsys [2]. In the Pittsburgh approach, each individual has a complete set of rules and the solution is the individual towards which the GA has converged. Some examples of this type of systems are GIL [7], GABIL [10], Samuel [4].

The Pittsburgh approach allows a more direct application of the Genetic Algorithm than the Michigan approach. In the former, the fitness of an individual is computed by testing his set of rules in the domain problem. On the contrary, in the Michigan approach, as one rule does not represent a complete solution, an additional algorithm is necessary to evaluate the contribution of each rule relative to the whole set of rules (e.g. Bucket Brigade Algorithm [5]) and the GA is limited to the search of new points of the space. Another difference between both systems is the solution that must be obtained. The Pittsburgh approach simply returns the individual with the highest performance, whereas the Michigan approach has to maintain a population of multiple solutions that globally represents the solution. This property is related to the systems with multimodal solutions, so the Genetic Algorithm has to provide some mechanism to preserve the diversity in the population.

Systems from both approaches have successfully been applied to a wide range of problems, such as learning sequential decision tasks for robots (Samuel, Alecsys, XCS) and classification in different domains: biological data sets (HDBPCS [9]), medical domains (Newboole, GABIL), etc.

In this paper, we study a GBML system, based on the Pittsburgh approach, in a medical domain: the diagnosis of breast cancer from mammary biopsy images.

In section 2 we describe in more details the Pittsburgh approach and the system we implemented. Section 3 describes our domain and presents our experiments. Finally we give our conclusions and describe the future work.

2 Description of the System

2.1 Representation

Each individual of the population has a set of rules of variable size. A rule consists of a condition part and a classification part: $condition \rightarrow classification$. The condition part is a conjunction of the tests on each attribute: $if\ T_1 \wedge T_2 \wedge ...T_n$, where T_i is the test of the ith attribute. A test on an attribute is performed for all the values of the attribute, allowing an internal disjunction. For example, if the attribute $Color$ can take the values $\{Blue, Green, Yellow\}$, a possible test can be: $if\ Color\ is\ Blue\ or\ Green$. The classification part of a rule gives the class to those examples that match the condition part.

The codification of the rules in the individuals can be done on a symbolic level (as in Samuel) or in a low level representation, in binary strings (e.g. GABIL). The first one allows a more natural representation but, on the other hand, genetic operators must be designed specifically. For that reason, we have chosen the binary codification, which allows the use of the classical crossover and mutation operators with minimum changes.

Each test on an attribute is codified by a binary string, whose length is the number of different nominal values of the attribute. For example, if the attribute $Color$ can have 3 nominal values, 3 bits are needed. The test $Color\ is$ $Blue\ or\ Green$ is codified as: 110, where each bit codifies one possible nominal value. Therefore, the condition part is a fixed-size binary string (whose length is the sum of the possible values of all the attributes). The classification part is also codified in a binary string (in our problem, as we performed a binary classification, only one bit was necessary).

In our problem, the attributes were described by real numbers, instead of nominal values. For this reason, prior to the codification, it was necessary to discretize all real values in intervals.

2.2 Genetic Operators

The crossover operator we used is *two-point crossover*, adapted to the variable size of individuals. The cut points are randomly generated and they can occur anywhere in the rule, not necessarily in a rule boundary. The process has been designed to preserve the consistency of the offspring in the following manner: first, we select the rules from each parent that are going to be swapped and then, we randomly choose the lower and upper bit of the cut point, which must be the same for both parents. For example, suppose that rules {2,4} are selected from parent 1 and {1,5} from parent 2, and the selected cut points are {3,6}. The swap will be performed between the substring {Rule2,Bit3 - Rule4,Bit6} from the first parent and the substring {Rule1,Bit3 - Rule5,Bit6} from the second one.

The strings that are swapped can have a different number of bits and therefore, length of individuals can be increased and decreased.

Mutation has been implemented as the classical bit-level mutation.

2.3 Evaluation Function

Individuals of the GA are evaluated by testing their sets of rules on the training set of examples. So, the fitness of each individual (I_i) can be computed as:

$$fitness(I_i) = \frac{\text{number of examples correctly classified}}{\text{total number of examples}} \qquad (1)$$

The evaluation function is a key point for guiding the GA towards a good set of rules. Function (1) tends to give good fitness to the individuals having consistent and complete rule sets.

Based on the Function 1, we have designed other evaluation functions which introduce a bias in order to obtain rule sets with the minimum number of rules. This type of sets are preferred because they keep a search space of reasonable size.

In Function (2) this bias is introduced by the parameter *Pen*: all rule sets with a number of rules greater than Pen, have their fitness decreased in a ratio $\frac{Pen}{NRules(I_i)}$, where $NRules(I_i)$ is the number of rules of the *ith* individual. Besides, a square was introduced in the percentage of correct classification (*P.Correct*), to give a differential fitness to those individuals with the highest percentage relative to the ones with low percentage.

$$fitness(I_i) = \begin{cases} (P.Correct)^2 & NRules(I_i) \leq Pen \\ (P.Correct)^2 \cdot \frac{Pen}{NRules(I_i)} & \text{otherwise} \end{cases} \qquad (2)$$

We have also experimented with the evaluation function (3), where the fitness of individuals is linearly decreased with an increasing number of rules:

$$fitness(I_i) = (P.Correct)^2 \cdot \left(1 - \frac{NRules(I_i)}{max}\right) \qquad (3)$$

where max is a user-defined parameter and establishes the maximum number of rules of an individual, where the fitness is 0.

3 Experiments

Our goal is to develop a system capable of learning a good set of rules, from a training set of examples, and then, the system must be able to predict (classify) new examples. For that reason, we have experimented with different strategies (the different evaluation functions, with different values for the parameters) in order to obtain a good set of rules in the training phase. This set of rules should be consistent, complete and with the minimum number of rules.

The measures used for the comparison between the different strategies are: the Classification Accuracy (CA), the Number of Rules (NR) of the obtained rule set and the Prediction Accuracy (PA). CA and PA are the percentage of correct classifications over the training and the test set of examples respectively.

3.1 Description of the Domain

Our database comes from mammary biopsy images with their correct classification made by the human expert. An example of an image is shown in Fig. 1. Each image is processed using mathematical morphology techniques [8], obtaining a set of 24 describing features. Therefore, our data set consists of 1028 examples, each one described by 24 real valued features and the solved classification, which is *cancerous* or *not*.

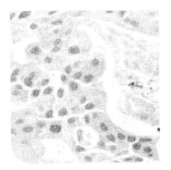

Fig. 1. Example of a mammary biopsy image

The set of examples are grouped in two sets: the training set and the test set. The GA is executed in two phases: in the first one, it learns the rule set from the training set of examples and two measures are obtained: the Classification Accuracy and the Number of Rules. The second phase consists of testing the Prediction Accuracy, which is measured by classifying the examples of the test set.

3.2 Parameter Set

Prior to the comparison between the different strategies, the GA parameters have been tuned and fixed for all the experiments. Their values are reported in Table 1.

Table 1. Genetic Algorithm parameters

Parameter	Value
Population Size	100
Crossover Probability	0.6
Mutation Probability	0.0006
Maximum iterations	15000
Elistism	Yes

3.3 Study of the Influence of the Different Fitness Functions on the System Performance

First, we compare the system performance with function (2) and (3), with different values of *Pen* and *max* respectively.

These experiments were performed on a reduced set of examples (10% of examples), due to CPU time. Results show the average over 5 different trials, each one for a different seed number for the GA.

Fitness Function 2. The first experiment demonstrates the necessity of imposing a restriction on the number of rules of individuals. The function used is (2), where *Pen* was set to a high value (*Pen* = 100), and the GA converged to an individual with 100 rules (see Table 2). Individuals always tend to grow until their number of rules is equal to *Pen*. If the number of rules of an individual is greater than this value, its fitness decreases and then, it is eliminated by the GA. Without this restriction on the number of rules, the individuals would increase until they exceed the memory capacity.

Table 2 shows the results obtained in learning performance for different values of *Pen*. A range for *Pen* from 5 to 30 gives similar performance, from which we can deduce that it is a good range for evolving good set of rules. For values under

Table 2. Comparison of different values of *Pen* in fitness function 2

Pen	Classification Accuracy	Number of Rules
2	85 %	2
5	90 %	5
9	92 %	9
15	90 %	15
25	90 %	25
30	94 %	30
100	84 %	100

5, the performance is worse and also for very high values of *Pen*. In the latter case, we can assume that the search space is too big for the GA, and population size should have been increased.

Fitness Function 3. Fitness function 3 was tested in order to see if better performance could be achieved. Results for different values of *max* are reported in Table 3.

Table 3. Comparison of different values of *max* in fitness function 3

max	Classification Accuracy	Number of Rules
50	90 %	8
70	92 %	15
90	92 %	16

From the Classification Accuracy point of view, there is not a clear difference between the two types of functions. The difference is in the way that the number of rules is controlled. In the first function, the parameter *Pen* establishes the maximum number of rules of the solution. For individuals having their number of rules greater than *Pen*, their evaluation is decreased in a ratio of $Pen/NRules(I_i)$. Therefore, these individuals with worse evaluation are not selected by the GA and they tend to disappear. For that reason, the selection of a good *Pen* parameter could be critical. On the contrary, in fitness function 3, evaluation is decreased linearly with the number of rules. Therefore, the choice of parameter *max* is not as critical. In Table 3, results show that with *max* 70 or 90, individuals are dynamically adjusted to 15 or 16 rules. This is not the minimum number of rules, but this method of adapting the number of rules could be more independent of the domain.

3.4 A Look Inside the Rule Sets

The experiments reported in Table 2 show a similar CA between the rule sets with 9 rules (9-RS) and rule sets with 25 or 30 rules (25/30-RS) which are quite greater. Our aim is to establish a more detailed comparison between these two type of rule sets (9-RS compared to 25/30-RS) in order to decide which one is better for our application.

First, we want to see if it is possible to eliminate some rules of the 25-RS or 30-RS, and still achieve the same CA. That is, if there were some rules that did not classify any example, we could eliminate them and obtain a reduced set. This process could be done at the end of the GA, once the GA had given the solution. This experiment was proved for the 30-RS and 25-RS and the rule sets did not present a high number of unused rules. As an example, Figure 2 shows the CA in a rule set of 25 rules, as rules were taken out from the rule set. The results show that 24 out of 25 rules were necessary, so the set could not be reduced significantly.

We also compared the Prediction Accuracy of the 30-RS and the 9-RS, and the results are slightly better for the 9-RS. Table 4 reports the PA of both Rule Sets. The Data Set is divided in 10 different groups: each group has a different percentage of examples in the training set and the test set. For example, the first group, 10%-90% contains 10% of examples for the learning phase, and 90% of examples are used for the test phase. The last group divides the examples in 90% for the learning stage and the rest (10%) for the test phase. In almost all cases, the PA in the 9-RS is better than the 30-RS (the PA mean is 7.4% higher). Therefore, the system can generalize better with fewer rules.

Table 4. Comparison in Prediction Accuracy between the 9-RS and 30-RS, for different Data Set groups

Training-Test set	PA(%) in 9-RS	PA(%) in 30-RS
10%-90%	53.9	55.3
20%-80%	62.1	50.7
30%-70%	70.5	62.4
40%-60%	72.5	57.7
50%-50%	71.7	63.1
60%-40%	73.1	67.4
70%-30%	75.3	71.0
80%-20%	73.5	66.7
90%-10%	85.8	77.1

In summary, rule sets with few rules are preferred because the learning time is better and the system can generalize more and obtain better PA results.

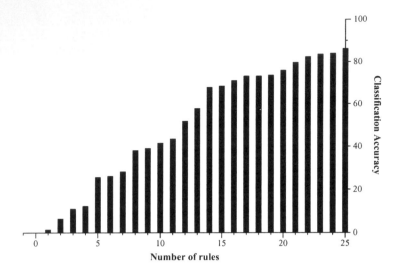

Fig. 2. Study of the influence on CA of each rule in a rule set of 25 rules

3.5 Performance in Test

Table 5 shows a summary of CA and PA results of the system (with fitness function 2 and $Pen = 9$). The dataset has been grouped into the same 10 groups described previously.

Table 5. Performance in training and test

Training-Test set	CA (%)	PA (%)
10%-90%	91.6	53.9
20%-80%	86.1	62.1
30%-70%	89.0	70.5
40%-60%	87.8	72.5
50%-50%	89.4	71.7
60%-40%	87.3	73.1
70%-30%	88.2	75.3
80%-20%	86.8	73.5
90%-10%	85.4	85.8

The CA ranges from 86.1% to 91.6%. The CA mean is about 88%. The percentage of correct classified examples in the test stage increases when more samples have been shown in the training phase, because there are more representative examples in the training phase for learning. The minimum value for Prediction Accuracy is reached in the 10%-90% group, with only 53.9% of correctly predicted examples, while the maximum value is obtained in the 90%-10% set, with a PA of 85.8%.

3.6 Comparison with a Non-GA Based Machine-Learning Method

In [12], a Neural Network is applied to the same data set. The maximum Prediction Accuracy result reported is achieved by a Backpropagation Neural Network (NN-24), with a mean PA value of 84.57%, in the 90%-10% training-test set. This result is improved using the same Neural Network with the data set reduced to 10 features (NN-10), which is obtained with a Non Linear Discriminant Analysis Network.

Table 6. PA obtained with the GA approach and a NN approach with 24 and 10 features (for the 90%-10% data set)

System	PA (%)
GA-approach	85.85
NN-24	84.57
NN-10	88.28

The GA-approach gives a similar result to the NN-approach with 24 features (see Table 6). The improvement in PA using an NN with a reduced set of 10 features leads us to consider performing a feature selection during the execution of the GA. This process would reduce the search space and could improve our results in two ways: a better PA and a reduction of learning time.

4 Conclusions and Future Work

In this paper, a system based on Genetic Algorithms in the Pittsburgh approach is studied for the automatic diagnosis of breast cancer. We have studied the behaviour of the system and our efforts have centred on the improvement of performance in two directions: obtaining better classification results and minimizing the learning time which has been related to the number of rules of our solutions.

The learning performance achieved is about 88% and the prediction accuracy depends on the set of examples shown in the learning phase. The maximun PA obtained is 85.85%. This result is slightly better than the Neural Network system (for the same number of features).

Our future work is focused on the improvement of PA by guiding the GA search. This goal could be achieved by performing feature selection, while the GA is running. We could also extend the idea of performing some type of rule selection, but during the evolution of the GA. Instead of limiting the number of rules by the evaluation function, there would be a specialized operator that could remove some rules from individuals: for example, those rules least used in the classification. These mechanisms would decrease the execution time and would guide the GA towards a more directed search.

We also intend to study the Michigan approach to see if better results could be achieved with less learning time.

Acknowledgements

Our thanks to Irin Laboratory, University of Nantes, and their members for their collaboration in the development of this work. We would also like to thank Enginyeria i Arquitectura La Salle, Ramon Llull University, for their support.

References

1. Bonelli, Pierre and Parodi, Alexandre: An Efficient Classifier System and its Experimental Comparison with two Representative learning methods on three methods on three medical domains. Fourth International Conference on Genetic Algorithms (ICGA'91). Morgan Kaufmann, (1991) 288-295
2. Dorigo, Marco: Alecsys and the AutonoMouse: Learning to Control a Real Robot by Distributed Classifier Systems. Machine Learning, 19, 3. (1995) 209-240
3. Goldberg, David E.: Genetic Algorithms in Search, Optimization and Machine Learning. Addison-Wesley Publishing Company, Inc. (1989)
4. Grefenstette, John J. and Connie Loggia Ramsey and Schultz, Alan C.: Learning Sequential Decision Rules Using Simulation Models and Competition. Machine Learning, 5(4), (1990) 355-381
5. Holland, John H.: Escaping Brittleness: The Possibilities of General Purpose Learning Algorithms Applied to Parallel Rule-Based Systems. Machine Learning: An Artificial Intelligence Approach, Vol. II , Morgan Kaufmann, (1986) 593-623
6. Holland, John H.: Adaptation in Natural and Artificial Systems: An Introductory Analysis with Applications to Biology, Control and Artificial Intelligence. MIT Press/ Bradford Books edition (1992)
7. Janikow, C.Z.: A Knowledge Intensive Genetic Algorithm for Supervised Learning. Machine Learning, 13, (1993) 198-228
8. Martinez Marroquin, E. and others, Morphological Analysis of Mammary Biopsy Images. Proceedings of Melecon'96, (1996) 1067-1070
9. Pei, Min, Goodman, Erik D., Punch, William F. and Ding, Ying: Genetic Algorithms for Classification and Feature Extraction. 1995 Annual Meeting, Classification Society of North America (1995)
10. Spears, William M. and Gordon, Diana F.: Adaptative Strategy Selection for Concept Learning. Proceedings of the First International Workshop on Multistrategy Learning (MSL-91). Harpers Ferry, MD, (1991) 231-246
11. Smith, S. F.: Flexible Learning of Problem Solving Heuristics through Adaptive Search. Proceedings of the 8th International Joint Conference on Artificial Intelligence, (1983) 422-425
12. Vos C.: Untersuchung von Biopsiegewebe zur Kalssifizierung von Brustkrebs mit neuronalen Netzen, Master's thesis, Universität Karlsruhe, Institut für Nachrichtentechnik, Institut für Automation und Robotik, (1996)
13. Wilson, Stewart W., Classifier Fitness Based on Accuracy. Evolutionary Computation, Volume 3, 2, (1995) 149-175

A New Gradient-Based Search Method: Grey-Gradient Search Method

Chin-Ming Hong[1], Chih-Ming Chen[2], and Heng-Kang Fan[3]

[1] Department of Industrial Education
National Taiwan Normal University

[2] Department of Electronic Engineering
National Taiwan University of Science and Technology

[3] Department of Electrical Engineering
Kuang-Ku Institute of Technology and Commerce

Abstract. Optimization theory and methods play very important role for engineering design and applications. In many domains of engineering applications, it is usually the most important process to find near optimal solution. The gradient-descent method is widely used to solve many engineering optimization problems. But the gradient-descent method has some disadvantages for searching optimal solution. Firstly, its convergent speed is very slowly and is easy to trap into local minimum in the applications of many actual problems. Secondly, the learning rate of gradient-descent method must been determined adequately for different engineering problem. If the learning rate set very small, the convergent speed will be very slowly. If the learning rate is set very large, the searching of solution is very easy to generate trashing or divergence. The main goal of this research is to propose a new method that is based on grey prediction theory to improve the gradient-descent method. We use the idea of grey prediction to speed up effectively the searching speed of gradient-descent method, and improve the drawback that gradient-descent method is very easy trap into local minimum. From the experimental results, we can show the workings of the proposed method that can speed up effectively the searching speed of gradient-descent method, and improve the drawback that gradient-descent method is easy to trapped into local minimum.
Keywords: Gradient-descent Method, Grey Prediction, Grey-Gradient Method, Optimization Method

1 Introduction

The optimization methods can be classified into three categories based on the type of information that must be supplied by the user : (1) Direct search methods, which use function values only. (2) Gradient methods, which require accurate values of the first derivative of function. (3) Second-order methods, which, in addition to the above, also use the second derivative of function [9]. None of the above methods can uniformly solve all problems with equal efficiency. The direct search method will take much calculating time for performance index during solution search, so we will not focus on direct search method. Optimization theory finds application in all branches of engineering in four primary areas [9]: (1) Design of components or retire systems. (2) Planning and analysis of existing operations. (3) Engineering analysis and data reduction. (4) Control of dynamic systems. The gradient-descent method is an optimization method that is used widely in many different engineering domains. For example on optimization control, the designer can use gradient-descent method to adjust and design the parameters of controller [8]. Furthermore, in machine learning, the Back Propagation learning algorithm for multiple-layer perceptrons also is based on gradient-descent method [7]. But the gradient-descent method has two main disadvantages: (1) the need to make an appropriate choice for learning rate, and the convergent speed of gradient-descent method is very slowly in the most real applications [9]. (2) It is very easy to trap into local minimum. So how to improve it is an interesting problem. In here, we joint the concept of short-term memory into our algorithm to record the prediction data during solution search, and use grey prediction to predict the direction and tendency of solution space. Because the grey prediction method has the following properties: (1) It is a precise prediction method and prediction does not need a large number of sample data [1][2][3][4]. (2) The calculation is simple [1][2][3][4]. Although the grey prediction us based on the high order analysis of mathematics, the process of calculation is simple and easy.

Prediction is to analyze the developing tendency in the future according to the past facts. Most of the prediction methods need a large number of history data, and use the statistic method to analyzes the characteristics of the system. Thus, this method is limited by the used data and its accuracy is low [2][4]. The grey theory do not requires lots of history data to make gery prediction. So the paper attempts to use grey prediction theory to predict the direction and tendency of optimization solution, and speeds up the performance of gradient-descent method.

2 Grey Prediction

The Grey System theory was first proposed in the paper, "Control Problems of Grey Systems" [1], proposed in the Journal of Systems and Control Letters by Dr. J. L. Deng in 1982. The Grey Prediction was an important section in the Grey System. The Grey Prediction can describe and analyze the future development according to the past and nowadays result. Because the Grey Prediction does not need a large number of sample data, the calculation is simple, it has been successfully used in the industrial, agriculture, commercial, military, economical, meteorological and civil engineering fields. The Grey Prediction has the following properties [1][2][3[4]:

(1) The Grey Prediction does not need a large number of sample data, it needs generally four sample datas.
(2) The calculation is simple. Although The Grey Prediction is based on the high order mathematics, the process of calculation is simple and easy.
(3) Generally, the Grey Prediction does not need a large number of relationship between each samples, so the sample data is easily got.

(4) The Grey Prediction can be used in near, middle and far prediction.
(5) The outcome of the Grey Prediction is more accurate than the other prediction methods.

According to the definition of Dr. Deng, the Grey Prediction is based on the GM(1,1) mathematics model [2].   The basic block diagram of the Grey Prediction is shown as Figure 1.

Grey Predictor

Figure.1. The basic block diagram of the Grey Prediction

The steps of the Grey Prediction are described as follow:
(1) Gather the original n sample data

$x^{(0)}(t)=\{x^{(0)}(1), x^{(0)}(2), x^{(0)}(3),...,x^{(0)}(n)\}$
where $x^{(i)}(t)$ represents the number t sample data of i times Accumulated Generating Operation.

(2) Apply Accumulated Generating Operation(AGO) by equation (1) below

$$x^{(1)}(k) = \sum_{i=1}^{k} x^{(0)}(i) \ , k=1,2,...,n \tag{1}$$

where $x^{(1)}(t)=\{\ x^{(1)}(1), x^{(1)}(2), x^{(1)}(3),..., x^{(1)}(n)\}$, $x^{(1)}(1)= x^{(0)}(1)$, $x^{(1)}(2)= x^{(0)}(1)+ x^{(0)}(2)$, $x^{(1)}(3)= x^{(0)}(1) + x^{(0)}(2)+x^{(0)}(3),...$, and $x^{(1)}(n)= x^{(0)}(1)+ x^{(0)}(2)+ x^{(0)}(3)+...+ x^{(0)}(n)$

(3) Create the Grey Differential Equation as show in equation (2)

Use the result of AGO to create the GM(1,1) grey differential equation.
$$\frac{dx^{(1)}(t)}{dt} + ax^{(1)}(t) = u \tag{2}$$
where the a and u are undecided coefficients of one order differentiable equation, we can get them that use mathematical method as shown below.
The equation (2) could be written as:
$$\lim_{\Delta t \to 0} \frac{x^{(1)}(k+1) - x^{(1)}(k)}{\Delta t} + a \cdot \frac{x^{(1)}(k+1) + x^{(1)}(k)}{2} = u \tag{3}$$
Let $\Delta t=1$ and $x^{(1)}(k+1)-x^{(1)}(k) = x^{(0)}(k+1)$, equation (3) will be

$$x^{(0)}(k+1)+a\cdot\frac{x^{(1)}(k+1)+x^{(1)}(k)}{2}=u \tag{4}$$

(4) Use the method of least squares estimation to get the parameters a and u

The equation (4) could be written as

$$[x^{(0)}(k+1)]=\left[-\frac{x^{(1)}(k+1)+x^{(1)}(k)}{2}\quad 1\right]\begin{bmatrix}a\\u\end{bmatrix} \tag{5}$$

Let $k=1,2,\dots,n-1$, and equation (5) will be:

$$\begin{bmatrix}x^{(0)}(2)\\x^{(0)}(3)\\\vdots\\x^{(0)}(n)\end{bmatrix}_{(n-1)\times1}=\begin{bmatrix}-\dfrac{x^{(1)}(1)+x^{(1)}(2)}{2}&1\\-\dfrac{x^{(1)}(2)+x^{(1)}(3)}{2}&1\\\vdots&\vdots\\-\dfrac{x^{(1)}(n-1)+x^{(1)}(n)}{2}&1\end{bmatrix}_{(n-1)\times2}\begin{bmatrix}a\\u\end{bmatrix}_{2\times1} \tag{6}$$

Equation (6) can be written as:

$$Y=\begin{bmatrix}x^{(0)}(2)\\x^{(0)}(3)\\\vdots\\x^{(0)}(n)\end{bmatrix}_{(n-1)\times1},\quad X=\begin{bmatrix}-\dfrac{x^{(1)}(1)+x^{(1)}(2)}{2}&1\\-\dfrac{x^{(1)}(2)+x^{(1)}(3)}{2}&1\\\vdots&\vdots\\-\dfrac{x^{(1)}(n-1)+x^{(1)}(n)}{2}&1\end{bmatrix}_{(n-1)\times2},\quad \Theta=\begin{bmatrix}a\\u\end{bmatrix}_{2\times1}$$

Use the method of least square estimation to get the estimated parameter, $\hat{\Theta}$:

$$\hat{\Theta}=\begin{bmatrix}a\\u\end{bmatrix}=(X^TX)^{-1}X^TY \tag{7}$$

(5) Solve the Grey Equation

Use equation (7) and equation (2), we can have:

$$\hat{x}^{(1)}(k+1)=\hat{x}_h+\hat{x}_p=\left(x^{(0)}(1)-\frac{u}{a}\right)e^{-ak}+\frac{u}{a} \tag{8}$$

where $\hat{x}_h=\left(x^{(0)}(1)-\dfrac{u}{a}\right)e^{-ak}$ is the homogeneous solution, $\hat{x}_p=\dfrac{u}{a}$ is the particular solution.

(6) Apply Inverse Accumulated Generating Operation(IAGO)

Because the Grey model is based on the AGO operation, we shall take the inverse accumulated generating operation to get the predict number.

$$\hat{x}^{(0)}(k) = \hat{x}^{(1)}(k) - \hat{x}^{(1)}(k-1) \tag{9}$$

3 The Grey-Gradient Search Method

The proposed grey gradient search method is a new search algorithm based on gradient-descent method. The algorithm records five history data that generated by original gradient-descent method, then applies one time Accumulated Generating Operation that detailed in section 2 to regularize the five history data. The main goal is to find the searching regular of gradient-descent method. Next, we use GM (1,1) model to create grey differential equation according to the result of one time Accumulated Generating Operation. The grey differential equation can build a prediction model to predict further searching direction of gradient-descent method. When we select different grey prediction step, we can get different prediction point. If we have already decided grey prediction step, we use Inverse Accumulated Generating Operation to get new prediction point, and using the new prediction point serves as the new searching point. The algorithm must detect stop criterion during the searching solution to decide if the algorithm continues searching or stops searching. In this algorithm, we define that $x^{(0)}$ is the initial searching point, the α is learning rate of gradient-descent method, k is the grey prediction step, the ε is stop criterion and GM(1,1) represents the grey prediction model of one time Accumulated Generating Operation and one order differential equation. About detailed procedure of the new algorithm is shown as follow:

Step 1. Execute the searching of gradient-descent method
   $x^{(n+1)} = x^{(n)} - \alpha \nabla f$ (gradient-descent method)
   Store all searching points of gradient-descent method
   Is accumulating five searching points?
   Yes: Go to Step 2 (make grey prediction)
   No: Go to Step 1

Step 2. Use GM(1,1) model to construct grey prediction model
   Use five searching points that step 2 stored to make Accumulated Generating Operation(AGO)
   Use GM(1,1) model to create grey prediction model
   Predict next searching point according to the prediction step (Solving the grey differential equation)
   Inverse Accumulated Generating Operation for new prediction point, and the result serve as the new searching point

Step 3. Detect stop criterion
   Is $\|\Delta x\| < \varepsilon$?

Yes: Go to Step 4
No: Go to Step 1

Step 4. Print results and stop

4 Experimental Results

To show the workings of proposed grey-gradient method, we use multiple variable functions as experimental target functions. In this experiment, we test and compare the properties between gradient-descent method and grey-gradient method. On the multiple variable experiment, we use Himmelblaus and Rosenbrock function. Figure 2 is the contour figure of Himmelblaus function. From this figure, we can see that the function has four local minimal values. The initial point is set as $x_1 = -3$ and $x_2 = 3$, and learning rate is set 0.001. Figure 3 and Figure 4 are the searching locus figure and function convergent figure for searching near local minimum by gradient-descent method. From the two figures, we can see that the near local minimum value is found after executing 37 epochs. The near local optimal solution is $x_1 = -2.8186$ and $x_2 = 3.126$. The function value of this point is 0.0072. Figure 5 and Figure 6 are the experimental results of grey-gradient method under the same initial condition and convergence criterion. The grey prediction step is set as 15. From the two figures, we can see that the grey-gradient just requires 15 epochs to find a near local minimum. The near local minimum value is $x_1 = -2.804$ and $x_2 = 3.1389$. The function value of this point is 0.0024. The grey-gradient method accelerates the searching speed of solution, and finds a better solution than gradient-descent method.

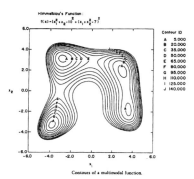

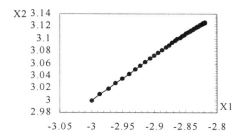

Figure.2. The contour figure of Himmelblaus function

Figure.3. The searching locus figure of Himmelblaus function using gradient-descent method

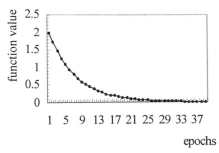

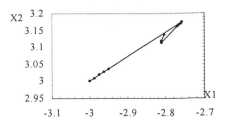

Figure.4. The convergent figure of Himmelblaus function using gradient-descent method

Figure.5. The searching locus figure of Himmelblaus function using grey-gradient method

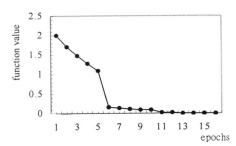

Figure.6. The convergent figure of Himmelblaus function using grey-gradient method

In the next, we will use a more complex Rosenbrock function of two variables to explain the properties of grey-gradient method. Figure 7 is contour figure of two variable Rosenbrock function. If the initial point is set as $x_1 = -1.2$ and $x_2 = 1.0$, learning rate is set 0.001. Figure 8 and Figure 9 are the searching locus figure and function convergent figure of gradient-descent method for searching local minimum. From the two figures, we can see that the gradient-descent method trap into a local minimum value, and the convergent error can not continue to reduce when epochs increase. The final convergent error is very large. The convergent point is $x_1 = -0.9743$ and $x_2 = 0.9593$. The function value of this point is 3.908. Under the same initial condition and stop criterion, Figure 10 and Figure 11 are the experimental results of grey-gradient method when the grey prediction step is set as 15. From the two figures, we can see that the grey-gradient method can escape the local minimum. It just needs 91 epochs to find a near global minimum. The near global minimum is $x_1 = 0.0479$ and $x_2 = 0.0297$. The function value of this point is 0.9816. The grey-gradient method speeds up the searching speed and finds a better solution than gradient-descent method.

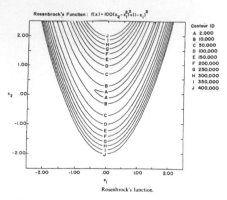

Figure.7. The contour figure of Rosenbrock function

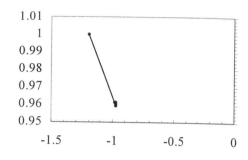

Figure.8. The searching locus figure of Rosenbrock function using gradient-descent method

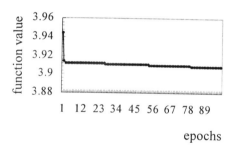

Figure.9. The function convergent figure of Rosenbrock function using gradient-descent method

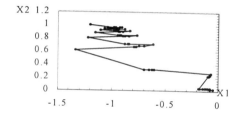

Figure.10. The searching locus figure of Rosenbrock function using grey-gradient method

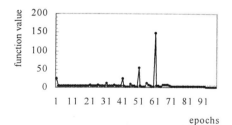

Figure.11. The searching locus figure of Rosenbrock function using grey-gradient method

5 Conclusions and Discussions

From the experimental results, we can conclude some properties about the grey-gradient method that shows as follow:

(1) The grey-gradient method can find the direction and developing tendency of solution according to the searching direction of gradient-descent method. It can predict the possible direction of solution with appropriate prediction step. The grey-gradient method has faster convergent speed than gradient-descent method, and it can speed up the searching speed of gradient-descent method.
(2) When we use appropriately grey prediction step, the grey-gradient method can escape effectively from the local minimal.
(3) For most engineering problems, the grey-gradient method will have more high probability to get global minimum solution than gradient-descent method does.
(4) Because the GM(1,1) model can be calculated at constant time, the grey-gradient method does not increase the time complexity too much.

Although the grey gradient search method can effectively improve the gradient-descent method, there are some issues that need to be discussed. They are:

(1) More mathematics analysis to prove the properties of grey gradient search method.
(2) More applications of grey gradient search method.

Acknowledgement

In acknowledgement of being funded by the National Science Council (title: Design of Genetic-Based Cerebellar Model Articulation Controller for DC Servo Motor Control project # NSC 87-2213-E-003-004)

References

[1] J. L. Deng ," Control problems of Grey System," Systems & Control Letters, Vol. 1, pp.288-294, 1982.
[2] J. L. Deng ," The Essential Methods of Grey System," Huazhong University of Science and Technology Press, pp. 1-137, 1987.
[3] J. L. Deng ," Grey System Control," Huazhong University of Science and Technology Press, pp. 25-134, 1985.
[4] J. L. Deng ," Introduction to Grey System Theory," The Journal of Grey System, Vol. 1, pp. 1-24, 1989.
[5] Chih-Ming Chen, Hong-Cen Lin and Chin-Ming Hong ," The Design and Application of a Genetic-Based Fuzzy Grey Prediction Controller," Journal of The Chinese Grey System Association, Vol.1, pp. 33-46, 1998.

[6]Chin-Ming Hong, Chin-Tang Chiang, Sinn-Cheng·Lin ," Control of Dynamic Systems by Fuzzy-based Grey Prediction Controller," The Journal of Grey System, Vol. 7, No. 1, pp. 23- 44, 1995.

[7]Jacobs, R. A. ," Increased Rates of Convergence Through Learning Rate Adaptation," Neural Networks, Vol. 1, pp. 295-307, 1988.

[8]Min-Shin Chen ," Control of Linear Time-Varying Systems by The Gradient Algorithm," IEEE Conference on Decision & Control, pp. 4549-4553, 1997.

[9]G. V. Reklaitis, A. Ravindran & K. M. Ragsdell ," Engineering Optimization Methods and Applications," A Wiley-Interscience Publication, 1983.

An Improvised A^* Algorithm for Mobile Robots to Find the Optimal Path in an Unknown Environment with Minimized Search Efforts

Douglas Antony Louis Piriyakumar and Paul Levi

Institute of Parallel and Distributed High-Performance Systems
University of Stuttgart
70565 Stuttgart
Germany
{piriyaku,levi}@informatik.uni-stuttgart.de

Abstract. An Algorithm for finding the optimal path given the initial and final positions of a cylinder-shaped mobile robot with the constraint on minimizing the search efforts is presented in this paper. This algorithm is based on the well known AI strategy A^* to get the optimal solution always. The strategy constructs a path from the available knowledge incrementally leading to the optimal path, and at the same time without any collision with the stationary obstacles. In this process, there are discontinuous "jumps" which demand the robot to move physically through the whole path traversing all along to reach the other end of the jump. Most of the times, the robot spends more time in traversing back and forth during the process of finding the optimal path. We have developed three new techniques, namely, *Petri expansion, Markovian cost function, and Retaining shortest path nodes* to minimize the search efforts in finding the optimal path. We have demonstrated the effectiveness of these techniques with ample examples. These three techniques always lead not only to the optimal path, but improves the search strategy by minimizing the "jumps" required. Computationally also, they reduce the number of nodes generated in the general A^* algorithm. The important issue is that the incorporation of these techniques in A^* is very easy with simple modifications and without any extra computational burden. ...

1 Introduction

To yield better products in higher efficiency in the modern Industry scenarios, the role of the mobile robots plays a key role. Most of the time, the mobile robot has to reach the point of service requirement at the earliest possible time. Sometimes, due to frequent modification suiting to the current need, this operation may have to be repeated many times coping with the new environment taking into care about slight modifications. This forces to find an optimal solution which perhaps may be repeatedly used to minimize each time spending lot of time in traversing through some path, rather than reaching every time through the optimal path minimizing the search effort.

The interesting problem of optimum path finding [6], [3] has been solved through several approaches such as graph methods [8], [4], some enumeration techniques [18], with maps [11] and possibly with refinement techniques by active vision also [20], [1]. An optimal and efficient path finding in partially known environments was formulated in [19]. A through analysis of Robot motion planning is given in [10], [9]. Recently, the researchers focussed their attention mainly on the sensor based navigation [5], [19], [8]. A dynamic graph search algorithm for motion planning in [2] describes a heuristically short motion in configuration space.

However, little attention was evinced to the efficiency of the search along the path. If a mobile robot has the required information about the environment apriori, it can precompute the optimal path performing searches in the computer memory itself using classical graph-search algorithms such as A^* [12], [14]. However, these algorithms are of less use as the mobile robot has to move physically to find the path in an unknown environment. During the course of finding the optimal path, given the initial and final positions, the robot has to make some "jumps" which are discontinuous in practical situations. Then, the robot has to spend more time in traversing the complicated path physically. The algorithms to minimize the search efforts are presented already in [14], [5], [18]. The algorithms [14], [5], make lot of search efforts leading to unbounded search in worst case [18]. In [18], the algorithm uses an A^* like behaviour to impose a bound on the depth of the search effort, and consequently to impose a bound on the search effort. It also used a DFS-like behaviour to minimize the search effort. Unlike the complicated hybrid strategy and cost propagation for updating the global costs of the nodes, we develop here there new techniques, namely, *Petri expansion, Markovian cost function, and Retaining shortest path nodes*. These techniques automatically leads to minimizing the search efforts within the framework of the A^* strategy itself with minor modifications.

Here, we combine the principles of petri nets, Markovian forgetfulness property, optimality in graph methods to get an improvised algorithm. This paper is divided into sections as follows. In section 2, the formulation of the problem with the required assumptions are specified. Section 3 explains the three new techniques and their applications. Section 4 illustrates the improvised algorithm. The analysis of the results and further extensions of the algorithm is sketched in section 5. Section 6 concludes the paper with a note on parallelization of the algorithm.

2   The Formulation of the Problem with Assumptions

A sensor-based cylinder-shaped mobile robot is considered here with the assumption that it has mechanism to distinguish the obstacle and locating the furthest point of visibility in the required directions. The obstacles are assumed to be stationary at some unknown locations which can be detected by the mobile robot using the sensors. The only moving object is the mobile robot with the known dimensions. Given is a car-like robot where the direction of motion is not impeded like car-like robot mentioned in [6]. The initial and final positions of the robot are known apriori. Given the initial and final position, the robot has to reach the final position by finding the optimal path without colliding with the obstacles and with the minimal search efforts. The focus is given to the minimal search efforts without sacrificing the optimality.

3 Three New Techniques to Reduce the Search Efforts in A^* Algorithm

The general A^* algorithm [12] is used here by the mobile robot to find the optimal path between the given points in an unknown environment. It may be recalled that the mobile robot standing at a point is required to go to the point which is selected for expansion by the A^* Algorithm. This discontinuous change of position is described as "jumps". The whole aim of the paper is to minimize the path traversed due to these jumps as the robot has to physically move. We have developed three new techniques which aim to reduce the search efforts during the process of finding the optimal path always. They are Petri Expansions, Markovian Cost Function and Retaining Shortest Path Nodes. We explain each one in the following subsections how the search effort could be minimized in each case.

3.1 Petri Expansions

The concept of Petri net is well known and widely discussed [13] and [16]. Succinctly explaining that a node is fired provided there is already one token in each of the incoming arc. This concept is transformed to suit to the need of the problem. The main idea is as follows, the search effort could be minimized if more information is available to the A^* algorithm to choose the next node to be expanded. There by unnecessary traversals can be easily eliminated. Now, the question boils down to how to give more information that too in an unknown environment. In fact, it is possible as most of the times two points P_1 and P_2 will be visible and P_1 will be expanded and P_2 waiting for its own turn for expansion. When the turn comes, the point P_2 being visible to the point P_1, will have one node with the last visited node being P_1. But, as we know, P_1 has been already expanded. Now, without wasting time, all points visible from P_1 can be included as nodes provided they are not visited earlier in the currently expanded node with the point P_2. These additional nodes can provide more information perhaps directly leading to the optimal path. Thereby, all intermediate nodes which were supposed to be visited will be not be visited in this case as a better choice is now available due to more information.

We have implemented this as a form of petri net. Whenever a point is expanded, it is noted as marked. During the expansion of some node, when the points which are marked are visible from the point of expansion, automatic petri expansion takes places, i.e., nodes are added with the nodes having their last point visited being visible to the marked points. This type of expansion happens as a firing as whenever the last node visited is already a marked node. This leads to lot of reduction in the search efforts as better node could be chosen for next viable expansion. The notable point is that there is very less additional effort required to do that. In fact, all these are incorporated in the A^* algorithm itself. Even though, there seems to be increase in the number of nodes generated, it is in due course lead to reduction in the total number of nodes generated due to the selection of better node for expansion apart from minimizing the search efforts.

3.2 Markovian Cost Function

The characteristics of a Markov process includes "forgetfulness" property [17], [7]. This forgetfulness property is used here while coining the cost function especially in g(x). In the general A^* algorithm, g(x) denotes the cost involved in coming to this node x from the initial node i.e., to this point from the initial position of the mobile robot. However, due to the physical movement of the mobile robot and because of jumps, the mobile robot may have to retraverse the path to reach a point. Normally in this case, all distances traversed starting from the initial position till this point will be included as the cost incurred to reach this point. Because of the complicated paths and jumps, this measure eventhough it is really the cost incurred to reach this point, does not give much hopes to go further. Mathematically, the cost function g(x) eventhough it represents the cost incurred to reach this point, does not help in getting a better node selected as the path in which it previously traversed becomes immaterial and only the distance counts. In this juncture, we introduce the concept of Markovian forgetfulness property to the cost function.

All that is needed is a good measure of g(x) which will help further to get the optimal path with minimal search efforts. So, instead of having g(x) as the cost involved in getting the mobile robot come to this point now, which is usually the case, we differ and introduce this property to forget the past and determine the future from this point. Mathematically, g(x) need not to be the total distance traversed to reach this point, but the minimal distance required to reach this point from the initial point with the available knowledge known so far. This gives better results and not only that we have ample examples to show that it drives the mobile robot exactly in the optimal path as required. Ofcourse, these may be rare examples. Yet these examples show concretely that such a possibilities are not only remote. By this proper choice of the next node to be expanded, the searching effort will be reduced to the large extent. It may be also recalled, no extra effort is needed as it is just the modification of the g(x) function in the A^* algorithm itself.

3.3 Retaining Shortest Path Nodes

This is one of the techniques in conjunction with other techniques yields highly appreciable results as far as the reduction of the search effort is concerned. The main idea is that what is the need of having a node with the last point P_l and another node with the same last point but a different path to reach P_l and it is shorter also. By expanding a node having the last point visited as P_l, and the distance from the initial point to P_l is not shorter compared to another node having the same last point visited as P_l and it is the shortest as per the knowledge at that moment of time available, is of no use as the path obtained as the solution can not be the optimal path because the path from the initial point to P_l is not optimal. This demands the node to be eliminated at the inception itself. This in turn will not only bound the explosion of the creation of new nodes, but also will reduce the search efforts.

The primary concern is to use this in conjunction with other techniques, eventhough individually it guarantees the optimal path also. The explosion of the creation of nodes are minimized as many nodes as which are not having the shortest path to the last node P_l are summarily eliminated. Thereby any possibility of jumping to these nodes are once far all eliminated from the search space of the A^* algorithm itself. As explained, this can be easily incorporated into the $A*$ algorithm, by just checking the f(x) value as the g(x) value is already modified to incorporate the Markovian cost function.

4   The Improvised A^* Algorithm

Before, we present the improvised A^* algorithm, few techniques introduced by us in [15] which reduces the space requirement and the computational time are briefed here as they are used in the algorithm.

4.1   Lower Bound

The lower bound is for the solution which is the minimum possible attainable solution. In the A^* algorithm, the algorithm has to continue even after finding a solution as it need not necessarily be optimal. Now the question lies how can it be proved that the given solution is the optimal solution so that the algorithm can be terminated atonce. The only possible way is that when the given solution is equal to the lower bound solution, obviously there could not be a better solution. Hence, the algorithm can terminate. LB = Euclidean distance between the initial position and final position of the mobile robot. One should be always careful that all feasible optimal solutions need not necessarily be lower bound solutions. The main advantage is that if the given problem has the lower bound solution, the algorithm terminates atonce it finds such solution, thereby reducing both the memory space required by the further expansions and the time to compute the same.

4.2   Upper Bound

The upper bound is a solution which the already available minimum solution. In the A^* algorithm, the algorithm has to evaluate the function f(x) at every node. Supposing that f(x) is greater than upper bound, that node need not to be expanded further. This will not affect the optimality as anyhow by expanding this node, the solution obtained will be more than that of the already available solution. However to start with, it is assigned a very high value for example say the product of the length and breath of the unknown field if it is known. However, once a solution is found first, the upper bound is set to be the solution. Further, whenever new solutions are found, it is updated provided it is better than the already available upper bound. So, using upper bound, the number of nodes generated are minimized thereby reducing the memory space and CPU time.

4.3 The Heuristics Function

The A^* strategy mainly depends on the effectiveness of the heuristic function. At node x, let there be P_1' points already visited. Then,
$g(x) = \Sigma \text{ Distance}(p_i, p_{i-1})$, $\forall p_i \in P_1'$, which are visited in the node x.

Now, to find the f(x) value, we need h(x), heuristic function. To produce always optimal solution, indeed $h^*(x)$ is required. The $h^*(x)$ is defined as, $h^*(x) = \text{Distance}(p_l, p_f)$ where p_l is the last point visited in the node x and p_f is the final position of the mobile robot, and Distance function calculates the Euclidean distance between the given points. In fact, it is easy to verify that $h^*(x) < h(x)$ to ascertain the optimality.

4.4 Our Algorithm for Finding the Optimal Path

1. Compute the lower bound solution, LB.
2. Set the upper bound UB as high value.
3. IF $(UB! = LB)$ THEN
4. c = 0 (* node count *).
5. Build the initial node N_0 with the initial point as first visited and insert it in the list with f(N_0) = LB.
6. REPEAT
7. Select the node N_k with smallest f value.
8. IF (N_k is not a solution) THEN
   (a) Generate the successors i.e., trying with all visible farthest points.
   (b) Do the following for each such points
       Include this point as the last point visited.
   (c) FOR each such visiting of points as N_i DO
       − Check for the duplication or shorter paths
       − IF (already available or not shorter) THEN
       Don't add the node
       ELSE
       Compute f(N_i) = g(N_i) + h(N_i) for this node N_i.
       IF (f(N_i) < UB)
       c = c + 1
       Insert it in the list
       IF (N_i is a solution) THEN

IF (f(N_i) = LB) THEN
Print the solution and quit.
IF (f(N_i) < UB) THEN
UB = f(N_i).
ENDIF
ENDIF
ENDIF
Start Petri expansion
ELSE
Prune the node N_i
ENDIF
ENDIF
ENDIF
ELSE
Print the solution and quit
9. UNTIL (N_k is solution OR list is empty).

5 Analysis of the Result and Future Work

To explain the effectiveness of the techniques, the simulations are carried out with various examples and few important cases are presented here to explain the salient features of the improvised algorithm Fig. 1-Fig. 5. For the sake of simplicity and explanation, linear obstacles are considered in these figures. Firstly, without these techniques the computations are made and then with techniques. They are tabulated in Table 1 and 2 respectively.

Table 1. The general A^* Algorithm

Ex	OP Dist	Nodes	AP Dist	Per Inc
Eg1	6.0	13		
Eg2	7.23	16	10.54	45.68
Eg3	7.47	23		
Eg4	9.54	21	19.54	104.86
Eg5	26.64	24		
Eg6	17.25	3267	1211.69	6923

It is very evident from the examples that our algorithm outperforms well in the complicated situations and performs equally well in simple situations and never worse than the general algorithm excepting for a marginal increase in the computation time due to the additions few extra nodes. Here, Ex denotes the example sets, OP Dist denotes the Optimal Path Distance, Nodes denotes the number of nodes generated as a measure of computational time, AP Dist denotes Actual Path Distance traversed, and Per Inc show the percentage of increase

Table 2. Our Improvised A^* Algorithm

Ex	OP Dist	Nodes	AP Dist	Per Inc
Eg1	6.0	22		
Eg2	7.23	19	10.54	45.68
Eg3	7.47	27		
Eg4	9.54	38	19.54	104.86
Eg5	26.64	27		
Eg6	17.25	280	73.21	324.35

between OP Dist and AP Dist. Whenever, AP Dist and Per Inc are not having values indicate that the path traversed is the optimal path and no extra distance is covered. It may be noted that it is same in table I and II as the same f(x) is used for the sake of comparison. The most interesting is the last case, where there is commendable achievement obtained by our improvised algorithm as it is easy to check that the search effort is minimized from 1211.69 to 73.21 and that too equally good reduction in Per Inc also. This evidently shows that the three techniques in the complicated situations reduces the search efforts enormously. In the cases of Eg1, Eg3, and Eg5, the exact path traversed is the optimal path and there is absolutely no extra distance traversed. This shows that the search efforts are even minimized to zero in some cases as evident from the examples shown here. This clearly demonstrates the effectiveness of the new techniques, especially Markovian cost function. As an easy extension, this algorithm can be modified either to stop at the first solution or any ϵ optimal solution taking into consideration of the lower bounds as the optimal solution.

6 Conclusion

In this paper, the algorithm for finding the optimal path given the initial and final positions of a cylinder-shaped mobile robot with the constraint on minimizing the search efforts is presented, which is based on the well known AI strategy A^* to assure the optimal solution always. Without any collision with the stationary obstacles, a path is constructed from the available knowledge incrementally leading to the optimal path. In this process, most of the time is spent by the robot in search of finding the optimal path. We have introduced three new techniques, namely, *Petri expansion, Markovian cost function, and Retaining shortest path nodes* to reduce the search efforts needed to find the optimal path in unknown environments. These three techniques always lead not only to the optimal path, but improves the search strategy by minimizing the "jumps" required. The reduction in the number of nodes generated comparing with the general A^* algorithm shows the efficiency of the techniques. The additional advantage of these techniques is that they can be easily incorporated in the A^* algorithm without much extra efforts. The Petri expansion and retaining the shortest path node exhibit enormous parallelism to be exploited. Thus, our algorithm paves way for efficient parallelization also.

References

1. Camillo, J.T. and David, J.K. "Vision-based motion planning and exploration algorithms for mobile robots", IEEE Trans. on Robotics and Automation, pp. 417-426, June, 1998.
2. Chen, P.C. and Hwang, Y.K. "SANDROS: A dynamic graph search algorithm for motion planning", IEEE Trans. on Robotics and Automation, pp. 390-403, vol. 14, no. 3, June, 1998.
3. Danny, Z.C. Robert, J.S. and John, J.U. "A framed-quadtree approach for determining Euclidean shortest paths in a 2D environment", IEEE Trans. on Robotics and Automation, pp. 668-681, October, 1997.
4. Fernandes, J.A. and Gonzalez, J. "Hierarchical graph search for mobile robot path planning", IEEE ICRA'98, pp. 656-661, April, 1998.
5. Foux, G. Heymann, M. and Bruckstein, A. "Two dimensional robot navigation among unknown stationary polygonal obstacles", IEEE Transactions on Robotics and Automation, vol. 9, no. 1, pp. 96-102, 1993.
6. Guy, D. and Francois, S. "An efficient algorithm to find a shortest path for a car-like robot", IEEE Trans. on Robotics and Automation, pp. 819-828, December, 1995.
7. Howard, R.A. "Dynamic programming and Markov processes", MIT Press, Cambridge Mass. 1960.
8. Hsu, J.Y.J. and Hwang, L.S. "A graph based exploration strategy of indoor environments by a autonomous mobile robot", IEEE ICRA'98, pp. 1262-1268, April, 1998.
9. Hwang, Y.K. and Ahuja, N. "Gross motion planning - A survey", ACM Computing Surveys, vol. 24, no.3, pp. 219-291, September, 1992.
10. Latombe, "Robot Motion Planning" Boston, M.A : Kluwer, 1991.
11. Lydia, E.K. Petr. S. Jean-claude, L. and Mark, H.O. "Probabilistic Road maps for path planning in High-dimensional configuration spaces", IEEE Trans. on Robotics and Automation, pp. 566-580, August, 1996.
12. Nilson, N.J. "Principles of Artificial Intelligence", Springer-Verlag, 1980.
13. Peterson, J.L. "Petri net theory and the modelling of systems", Prentice-Hall, Engelwood Cliffs, 1981.
14. Pearl, J. "Heuristics: Intelligent Search Strategies for Computer Problem Solving", Addison-Wesley, 1984.
15. Piriyakumar, D.A.L. Levi, P. "An efficient A^* based Algorithm for optimal Graph Matching applied to Computer Vision", The 5th German-Russian workshop on Pattern Recognition and Image Understanding, Munich, Germany, 1998.
16. Rozenburg. G. "Advances in Petri nets", Lecture Notes in Computer Science, Springer Verlag, Berlin, 1985.
17. Sharpe, M. "General theory of Markov processes", Academic Press, Boston, 1988.
18. Shmoulian L. and Rimon, E. "A_e^* DFS - an algorithm for minimizing search effort in sensor based mobile robot navigation", IEEE ICRA'98, pp. 356-362, April, 1998.
19. Stentz, A. "Optimal and efficient path finding for partially known environments", IEEE ICRA'94, pp. 3310-3317, May, 1994.
20. Wallner, F., Graf, R. and Dillman, R, "Realtime map refinement by fusing sonar and active stereo vision", IEEE ICRA'95, pp. 2968-2973, 1995.

204

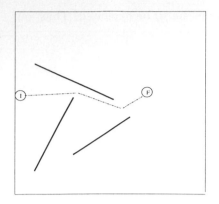

Fig. 1. Example1

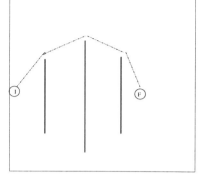

Fig. 3. Example4

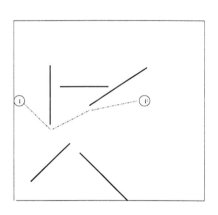

Fig. 2. Example3

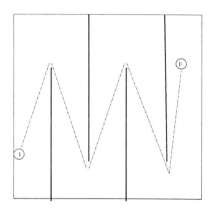

Fig. 4. Example5

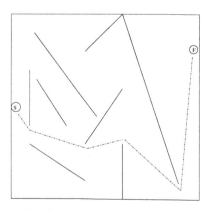

Fig. 5. Example6

Speeding the Vector Search Algorithm for Regional Color Channel Features Based Indexing and Retrieval Systems

Ahmed R. Appas[1]   Ahmed M. Darwish[2]*   Ayman I. El-Desouki[1]   Samir I. Shaheen[2]

1. Computer and Systems Department   2. Computer Engineering Department
Electronic Research Institute   Faculty of Engineering
Tahrir St., Dokki,   Cairo University
Giza 12411, EGYPT   Giza, 12613 EGYPT
arafat@eri.sci.eg   darwish@fru.eun.eg

Abstract. Color indexing is a technique by which images in the database could be retrieved on the bases of their color content. In this paper we propose a new algorithm for matching an input query with the color database images. The proposed algorithm is applied on the similarity retrieval system which was proposed in [Appas 99] to show the effect on the retrieval speed. The proposed algorithm adds one feature to an image feature vector. The added feature is obtained from the color channel features of the image regions. Comparing query hit percentage of image retrieval of the proposed algorithm in [Appas 99] with the new matching algorithm and with the normal exhausted matching algorithm shows that the performance of the retrieval system is nearly the same in both of cases but the retrieval speed increases specially with the large size databases when using the proposed matching algorithm.

1. Introduction

The great growth of digital image databases necessitates more than ever the automation of image storage and retrieval. New techniques are needed for effective searching and retrieval of images. Several systems have been created and developed for achieving this goal, but most of them are tailored to specific image ensembles for particular applications. Current research focuses on the choice of suitable features for image representation as well as the choice of parameters used to measure image similarity. The proposed algorithm in [Appas 99] used the exhaustive search method during matching the input query against the database images. The algorithm compares the query feature vector with all database image feature vectors which were stored in the database index. It chooses the nearest feature vectors to the query feature vector and considers their database images as the most similar images to this query.

In this paper we propose and evaluate a new algorithm for matching the query image with the database images. The new algorithm provides for efficient indexing, excellent retrieval performance and very fast retrieval response.

---

* To whom all correspondences should be addressed.

2. Literature Survey

Most of image retrieval approaches are based on extracting a feature vector from each image in the database and organize all feature vectors as a database index. At query time, a feature vector is extracted from the query image or the user-provided sketch and is matched against the feature vectors in the index. The main differences between the various systems lie in the features that are extracted and the algorithms that are used to compare feature vectors.

Moment based retrieval systems such as the systems in [Stricker 95], [Stricker 96] and [Appas 99] search for the best matches inside the whole database index. This means that for every query the query feature vector is compared with the whole database index. This will decrease the retrieval speed specially when the size of database becomes very large. In the following we will describe the retrieval system in [Appas 99] which uses the normal exhausted search. We will speedup the retrieval process by using the new matching algorithm.

2.1 Regional color channels features based image indexing and retrieval algorithm

The retrieval algorithm in [Appas 99] represents an image by fifteen features for five image regions, the center and the four corners. These features are stored in the image database index as a vector of size equal 15 floating point numbers (3 features x 5 image regions). The computed vector is attached to the image file or in a separate data relation with a field pointing to the image file (figure 1).

Image file name	$F_{0,1}, F_{0,2}, F_{0,3}$	.....	$F_{4,1}, F_{4,2}, F_{4,3}$

Fig. 1. A database index image vector

2.2 Image query and similarity function

To determine the similarity of two images at query time, the similarity between their indices is measured. Let Q and I be two color images with three color space channels. If the index entries of these images for their regions are $F_{l1,i}$, and $F_{l2,i}$ respectively. Then the similarity of the region $l1$ of Q and the $l2$ of I is defined as shown in equation (1).

$$d_{l1,l2}(Q,I) = \sum_{i=1}^{3} \left| F_{l1,i} - F_{l2,i} \right| \tag{1}$$

To compute the total similarity of the two images Q and I the central region is treated differently from the other regions. If the center region is R0 then the total similarity of the two images is defined as shown in equation (2).

$$d(Q,I) = S_0 d_{0,0}(Q,I) + \min_{f \in T90} \sum_{l1=1}^{4} S_{l1} d_{l1,f(l2)}(Q,I)$$

(2)

where f runs over the 4 rotations in $90°$ increments. The difference between the region $l1$ and every non-matched with before region $l2$ is computed. The minimum difference is chosen and the matched with region $l2$ is marked. Each corner in image Q is matched only with a corner in image I which has not been matched with any corner in image Q before and whose difference between the two regions is minimum. With the weights S_i ($i =0$, 1, 2, 3, 4) the user can specify how accurately each region of the query image should be matched. In contrast to the region weights, S_i, which could be used as a parameter when issuing query, the weights Wi can be fixed for all queries.

3. The new matching algorithm

The new algorithm extracts one region color feature from its three color channel features. Then an image color feature is extracted from its region color features. The image color feature is added to the feature vector which represents the image. The database index is ordered ascendingly according to the values of the image color feature of the stored images. The index is divided into groups. Each group contains 20 feature vectors. During query, a query image color feature is compared with image color feature of the starting feature vector of each group. The comparison is stopped when the query image color feature is less than one of the stored image color features which belongs to a feature vector at the start of one group. The feature vector which contains image color feature value larger than the value of the query image color feature is considered an edge. The 30 feature vectors before and the 10 feature vectors including and after that vector are considered the most similar 40 feature vectors to an input query feature vector. The normal matching algorithm is carried out on these 40 vectors and the most similar six images are displayed to the user. The new format of an image feature vector is shown in figure 2.

IF	Image File Name	$F_{0,1}, F_{0,2}, F_{0,3}$	$\cdots\cdots$	$F_{4,1}, F_{4,2}, F_{4,3}$

Fig. 2. A new database index image feature vector

3.1 The overall image color feature

The image color feature is obtained from the region color features of the image. The region color feature is extracted from the HSV color channel features of that region as shown in equation (3):

$$RF_i = 8Fh_i + 2Fv_i + Fs_i \tag{3}$$

Then the image color feature is computed as shown in equation (4):

$$IF = 12RF_0 + \sum_{i=1}^{4} RF_i \tag{4}$$

where RF_i = color feature of region i
   Fh_i, Fs_i, Fv_i = HSV color channel features for region I
   IF = image color feature

4. Results and analysis of experiments

Three color image databases that contain 109, 96 and 205 8-bit color bit-map images (320x200) are used to test the proposed algorithm in [Appas 99] with the new proposed matching algorithm and compare its performance with the normal exhausted matching algorithm. The first database is a database of aircraft images. The second database is a database of natural scenes. The third database is created from the first two databases to test the algorithm retrieval performance when the database contain images with different color distribution properties. To decouple the color channels at least partially, we perform all the tests in the HSV color space.

We have two sets of input query images which are used to test and compare the algorithms' performances. The first set contains query images extracted from some database images by rotation of the database images by 90 degrees. The second set of query images are new images which are not in the database nor extracted from the database images. We picked up twenty images from the third database and used them as the second set of query images so the database becomes 185 images. We list for every query image a group of database images which are similar to the query image in color distribution and shape. In the retrieval process, we consider the retrieval is succeeded if any one of the list of similar database images becomes retrieved in the first six images. Table 1 shows the results for 260 queries for 2 cases; the first with equal region weights (looking at the image in a global sense) and the second with the center region weight having a value greater than the 4 corner weight values (looking at a specific object located at the center region). The first case is more suitable for query in case of natural scenes while the second case is more suitable for query in case of aircraft images.

Table 1 shows the comparison between the results of the retrieval performance of the proposed algorithm in [Appas 99] with the new proposed matching algorithm and with the normal exhausted matching algorithm which was used in [Appas 99]. The results are shown in different classes of image query. Two sets of query images are used. We will discuss the results which were shown in these tables to extract and compare the retrieval performance with the two different matching algorithms in different query classes on different image databases.

Type of Database	[Appas 99] with the exhausted matching algorithm		[Appas 99] with the new matching algorithm	
	Equal region weight	Center region is zed	Equal region weight	Center region is emphasized
Aircraft DB 109 images	$\dfrac{14}{15}$	$\dfrac{15}{15}$	$\dfrac{14}{15}$	$\dfrac{15}{15}$
Natural scenes DB 96 images	$\dfrac{13}{15}$	$\dfrac{15}{15}$	$\dfrac{13}{15}$	$\dfrac{13}{15}$
Combined DB 205 images	$\dfrac{13}{15}$	$\dfrac{15}{15}$	$\dfrac{13}{15}$	$\dfrac{15}{15}$
Combined DB 185 images	$\dfrac{18}{20}$	$\dfrac{18}{20}$	$\dfrac{16}{20}$	$\dfrac{18}{20}$

Table 1. Comparison between the results of the retrieval performance of the proposed algorithm in [Appas 99] with the new proposed matching algorithm and with the normal exhausted matching algorithm which was used in [Appas 99].

If we consider that fetching, processing and sorting operations have an equal time then we can compute the system retrieval speed as shown in equation (5). How many times the retrieval speed is faster when the new matching algorithm is used than when the exhausted matching algorithm is used can be computed as shown in equations (6-7).

$$SRS = Ft + Pt + St \tag{5}$$

$$SPEED = \frac{SRSN}{SRSM} \tag{6}$$

$$SPEED = \frac{2(X * 20 * 15) + X * 20 * 1}{3(X * 1) + 2(40 * 15) + 40 * 1} = \frac{620X}{3X + 1240} \, times \tag{7}$$

Where   *SRS*   = System Retrieval Speed
         Ft   = Fetching Time
         Pt   = Processing Time
         St   = Sorting Time
         Speed = how many times the retrieval speed of the used system with the new
                 matching algorithm is faster than the normal system speed
         SRSN = System Retrieval Speed of the Normal system
         SRSM = System Retrieval Speed of the used system with the new matching
                 algorithm
         X   = number of groups of database images where each group contains 20
                 images at maximum

As an example :
       when X = 10 groups the speed = 5 times faster
       when X = 100 groups the speed = 40 times faster
       when X = 1000 groups the speed = 146 times faster

Figures 3 - 5 shows the retrieval performance of the proposed retrieval system in [Appas 99] with the new matching algorithm and with the exhausted matching algorithm. The shown examples are in both global (equal region weights) and local (center region is emphasized) query cases. The shown examples are retrieved from the combined DB 205 images.

From the results shown in table 1 and examples shown in figures 3-5 we can conclude that the retrieval performance with either of matching algorithms is nearly similar. The speed of retrieval with the new matching algorithm is nearly five times faster than the speed of the algorithm with the exhausted matching algorithm.

The reason of the nearly same retrieval performance is that the new image color feature represents the color content of the image accurately. The emphasizing of the role of H channel feature and the center region are the reasons of the accuracy of the new feature. This is because of that the hue is the most important feature of the color. Also the center region is less changes than the corner regions with respect to small luminance changes. So ordering the index according to the value of the new feature makes the nearly similar images are near in order in the index. This reduces the time spent in comparing the query image with all the index to obtain the nearly similar database images feature vectors. Hence the speed of retrieval is increased.

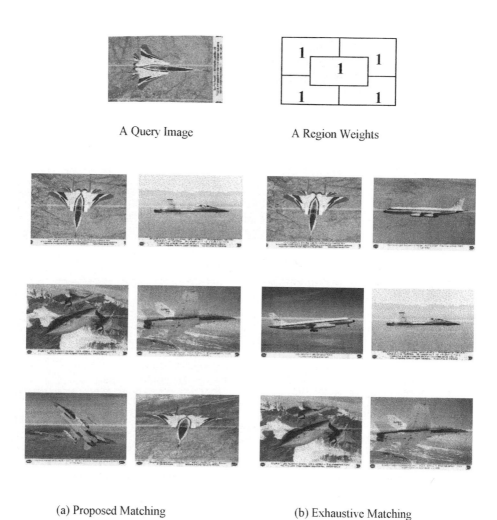

A Query Image   A Region Weights

(a) Proposed Matching   (b) Exhaustive Matching

Fig. 3. Retrieval performance of [Appas 99] with (a) Proposed matching algorithm and (b) Exhaustive matching in case of global query using an aircraft image.

For this example the correct image has been retrieved in the first order in both of (a) and (b).

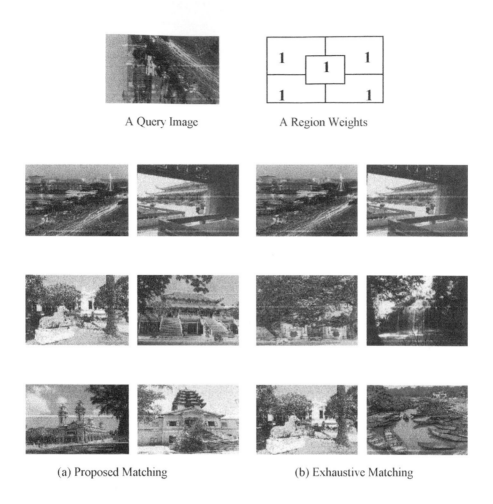

A Query Image   A Region Weights

(a) Proposed Matching   (b) Exhaustive Matching

Fig. 4. Retrieval performance of [Appas 99] with (a) Proposed matching algorithm and (b) Exhaustive matching algorithm in case of global query using a natural scene image.

For this example the correct image has been retrieved in the first order in both of (a) and (b).

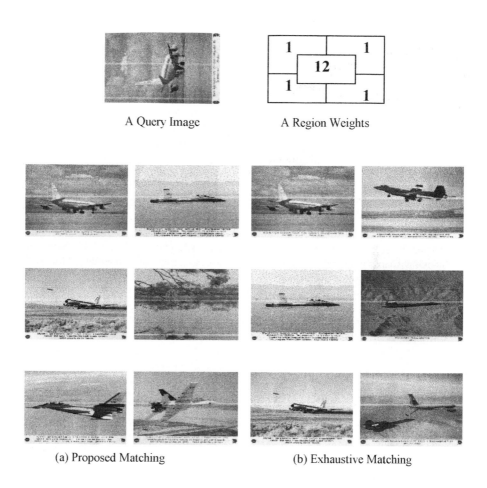

A Query Image   A Region Weights

(a) Proposed Matching   (b) Exhaustive Matching

Fig. 5. Retrieval performance of [Appas 99] with (a) Proposed matching algorithm and (b) Exhaustive matching algorithm in case of local query using an aircraft image.

For this example the correct image has been retrieved in the first order in both (a) and (b).

5. Conclusion

In this paper we present a new matching algorithm for the color based retrieval systems. The new matching algorithm adds one more feature to the image feature vector. The new feature is obtained from the regional color channel features of the image. The role of the center region color features is emphasized. The index is ordered ascendingly according to the value of the new overall image color feature. During matching the query image feature vector with the database index, the nearest forty images to the query image in the overall image color feature are chosen. The complete comparison between the query image features and the forty database images features is done. The nearest six images are retrieved to the user. The new matching algorithm speeds up the retrieval process. The difference in the retrieval speed between the normal exhausted matching algorithm and the proposed algorithm increases when the size of the database increases.

REFERENCES

1. Ahmed R. Appas, Ahmed M. Darwish, Ayman I. EL_Desouki, Samir I. Shaheen, " Image Indexing Using Regional Color Channels Features ", Electronic Imaging '99 IS&T/SPIE's 11th Annual Symposium, January 1999, Under Publishing
2. M. A. Stricker and M. Orengo, "Similarity of Color Images", Proc. SPIE, vol. 2420, 1995, pp. 381-392.
3. M. A. Stricker and A.Dimai, "Color Indexing with Weak Spatial Constraints", Proc. SPIE, vol. 2670, 1996, pp. 29-40.

Cost-Based Abduction Using Binary Decision Diagrams

Shohei Kato[†], Satoru Oono[‡], Hirohisa Seki[‡], and Hidenori Itoh[‡]

† Department of Electrical Engineering
Toyota National College of Technology, Eisei 2-1, Toyota, 471-8525, Japan
‡ Department of Intelligent and Computer Science
Nagoya Institute of Technology, Gkiso, Showa-ku, Nagoya, 466-8555, Japan
shohey@tctcc.cc.toyota-ct.ac.jp, {soono, seki, itoh}@ics.nitech.ac.jp

Abstract. This paper proposes an abductive reasoning system, which can find most preferable solution efficiently, using Binary Decision Diagrams. We propose a specialized BDD and its operation suitable for abductive reasoning: PBDD (Partial BDD) and GPC (Graft & Pruning Construction). We have implemented PBDD and GPC algorithm and built a cost-based abductive reasoning system which can find much more efficiently the most preferable explanation of a given observation. We have also made some experiments on the system with some diagnostic problems. Some good performance results are also shown.

1 Introduction

Abductive reasoning, a form of non-deductive inference which can reason suitably under uncertain knowledge, has attracted much attention in AI (e.g.,[Poo88]). Abductive reasoning has many interesting application areas such as diagnosis, scheduling and design. In general, abductive reasoning might find more than one solution. We, however, do not always require all the solutions. We often need the most preferable solution instead. Some work has been reported to solve such problem, by giving costs to hypotheses as the criterion judging which hypothesis to be selected preferably (e.g, [CS94,Poo93,KSI97]).

On the other hand, Binary Decision Diagrams (BDDs), which proposed by Akers [AS78] and developed by Bryant [Bry86], are well-known. BDD is the efficient representation and manipulation of Boolean functions and apply to many problems, such as computer-aided design for digital circuits, combinatorial problems, and so on.

In this paper, therefore, we aim to make abductive reasoning much more efficient by utilizing BDD. It ,however, causes some redundant computation to apply BDD to abductive reasoning; generating BDD for a given observation involves constructing BDDs for knowledge irrelevant to the observation, thus losing the goal-directedness. The problem of redundant computation of BDDs is essential for making abductive reasoning system much more efficient. We, therefore, propose a specialized BDD and its operation suitable for abductive reasoning: PBDD (Partial BDD) and GPC (Graft & Pruning Construction).

The organization of the paper is as follows. In section 2, we give a brief description of our framework of abductive reasoning. Section 3 explains a BDD and discusses abductive reasoning using BDD. In Section 4, we then proposes a Partial BDD. In section 5, we propose Graft & Pruning Construction algorithm so as to perform abductive reasoning suitably using PBDD. Section 6 shows some empirical results.

2   Cost-based abductive reasoning

The section describes our framework of abductive reasoning, cost-based abduction [KSI94]. We consider abductive reasoning to find the most preferable explanation of an observation, by giving each hypothesis a non-negative real number as its cost for the criterion of selection of preferable hypotheses.

Definition 2.1 Suppose that a set of propositional Horn clauses $\mathcal{F}$, called facts, and a set of atoms (ground unit clauses) $\mathcal{H}$, called the set of hypotheses, are given. Suppose further that an existentially quantified conjunction $\mathcal{O}$ of atoms, called an observation or simply a goal, is given. Then the most preferable (or an optimal) explanation h of $\mathcal{O}$ from $\mathcal{F} \cup \mathcal{H}$ is a subset of $\mathcal{H}$ such that

$\mathcal{F} \cup h \vdash \mathcal{O}$ ($\mathcal{O}$ can be proved from $\mathcal{F} \cup h$)   (AR1)
$\mathcal{F} \cup h \not\vdash false$ ($\mathcal{F} \cup h$ is consistent)   (AR2)
$cost(h) \leq cost(D)$ for all D: $\mathcal{F} \cup D \vdash \mathcal{O}$, $\mathcal{F} \cup D \not\vdash false$
    ($cost(h)$ is minimum among all sets of
    hypotheses which satisfy AR1 and AR2)   (AR3)

where $cost(D)$ is the sum of costs of hypotheses in D.   ∎

Abductive reasoning is defined to be a task of finding an optimal explanation h of $\mathcal{O}$ from $\mathcal{F} \cup \mathcal{H}$. In this framework, $\mathcal{F}$ is assumed to be consistent and treated as always true. However, there is a possibility of hypotheses D being inconsistent with $\mathcal{F}$.

Definition 2.2 A headless clause in $\mathcal{F}$ is called a *consistency condition*. A consistency condition is denoted by " $false \leftarrow A_1, \cdots, A_n$", where $A_i(1 \leq i \leq n$, $n \geq 1)$ is a hypothesis and $false$ designates falsity.   ∎

A consistency condition, " $false \leftarrow A_1, \cdots, A_n$" means that it causes inconsistency if all of $A_1, \cdots, A_n$ are assumed to be true.

Example **2.1** Figure 1 shows an simple example P_{ex}. Suppose that an observation "C" is given. Knowledge base P_{ex} and goal "$\leftarrow$ C" correspond to $\mathcal{F} \cup \mathcal{H}$ and $\mathcal{O}$ respectively. We know that $\{b\}$ becomes the most preferable explanation of $\mathcal{O}$ since the cost of $\{b\}$ is minimal among all sets of hypotheses (e.g., $\{b\}$, $\{a, c, e, g\}$, $\{a, d, e, g\}$) obtained from cost-based abductive reasoning, while maintaining the consistency.   ∎

Facts($\mathcal{F}$)		Hypotheses($\mathcal{H}$)	Costs
$A \leftarrow (c \wedge d) \vee (c \wedge e) \vee (d \wedge e)$	(2.1)	a	3
$B \leftarrow f \wedge A$	(2.2)	b	2
$C \leftarrow (a \wedge g \wedge A) \vee b$	(2.3)	c	5
$false \leftarrow a \wedge b$	(2.4)	d	2
$false \leftarrow c \wedge d$	(2.5)	e	1
$false \leftarrow e \wedge f$	(2.6)	f	1
		g	1

Fig.1: An example P_{ex}

3 BDD approach to Abductive Reasoning

3.1 BDD

BDD, a directed acyclic graph representation of a Boolean function, is derived by reducing a binary tree graph representing the recursive execution of Shannon's expansion. BDD has two terminal nodes labeled 0 and 1 representing the constant function 0 and 1 (called 0-node and 1-node respectively). Each non-terminal node is labeled with a variable name v and has two edges labeled 1(or then) and 0(or else). Each non-terminal node represents the Boolean function corresponding to its 1 edge if $v = 1$, or the Boolean function corresponding to its 0 edge if $v = 0$ as to Shannon's decomposition principle (see [Bry86,BRB90] for details).

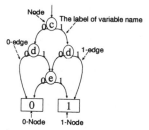

Fig.2: BDD for $(c \wedge d) \vee (c \wedge e) \vee (d \wedge e)$

Figure 2 shows BDD for $(c \wedge d) \vee (c \wedge e) \vee (d \wedge e)$. The variable order have $c \succ d \succ e$. BDD for $f(x)$ means BDD representing the logical formula $f(x)$. In this paper, we refer to non-terminal nodes as nodes. We also refer to the left edges as the edge labeled "0" (called 0-edge), and to the right edge as the edge with "1" (called 1-edge). We consider utilizing BDD, having the above properties, for abductive reasoning.

218

3.2   Cost-based abductive reasoning using BDD

We assign a cost, corresponding to a cost of a hypothesis, to each 1-edge of a node labeled with the hypothesis.[1] And we suppose that a cost of a path in BDD is the sum of the cost of edges in the path. A cost of an explanation, then, is the sum of the costs of 1-edges included in a path from root node to 1-Node.

By utilizing the BDD, abductive reasoning is performed with the following steps, when a knowledge-base $\mathcal{F} \cup \mathcal{H}$ and an observation O are given.

i) **Generating BDDs.**   Generate BDDs for each of Horn clauses in $\mathcal{F} \cup \mathcal{H}$ and O. In this step, we adopt *completion* of programs [Llo84], and transform a propositional Horn clause "$f \leftarrow a_1, a_2, \ldots, a_n$" to a logical formula "$f = a_1, a_2, \ldots, a_n$".[2] More in detail of generating BDDs, refer to [Bry86] and [Min93].

ii) **Consistency Checking.**   For each path in the BDDs generated above step, extract all nodes whose 1-edge are included in the path, and collect all labels of the nodes. Then, check if some of the labels represent all hypotheses in any *consistency condition* in $\mathcal{F} \cup \mathcal{H}$. The path satisfying the condition, making contradiction to facts, is excepted from the BDDs.

iii) **Finding Explanation(s).**   Collect paths, called *successful paths*, from root to 1-node from the BDD for O, and select a path, called *an optimal successful path*, whose cost is minimal from the successful paths. Then, extract nodes such that: 1-edge of the node is included in the path, and the node is labeled with hypotheses.

Abductive reasoning, finally, finds the hypotheses as an explanation of the observation in the BDD for O (i.e., target BDD).

Example **3.1** We consider abductive reasoning using BDD for example P_{ex} with an observation "C" (shown in Figure 1). In the example, we suppose that variable order is $a \succ b \succ c \succ d \succ e \succ f \succ g$. Figure 3 shows the result of step ii). In the figure, a path marked with "✕" indicates that it makes contradiction to facts in P_{ex}. We know, through the step iii), that {b} becomes the most preferable explanation of Osince the cost of {b} is minimal among all sets of hypotheses (e.g., {b}, {a, c, e, g} and {a, d, e, g}) obtained from all successful paths in the BDD for $\mathcal{O}$, while maintaining the consistency. ∎

The above mentioned procedure, however, has two crucial problems. Figure 3 shows that the BDDs have lots of unnecessary paths so as to find an optimal successful path. A BDD for an observation generated by the above steps includes all successful paths; all explanations of the observation can be obtained from the paths. However, it should be noticed that we need an optimal explanation only. We, therefore, introduce a heuristic search control technique into generation of BDD.

[1] We suppose that 1-edge of a node labeled with a fact in knowledge is given zero as its cost.
[2] = designates *equivalent* and not *unifiable*.

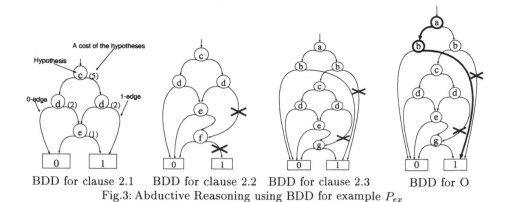

BDD for clause 2.1   BDD for clause 2.2   BDD for clause 2.3   BDD for O

Fig.3: Abductive Reasoning using BDD for example P_{ex}

In another problem, paying attention to the BDD for clause 2.2 and the BDD for O in the figure, we know, as a result of generation of the BDD for O, that the BDD for clause 2.2 did not have to be generated, since it is independent on the BDD for O; thereby, the BDD for clause 2.2 is irrelevant to the BDD for O. The problem is that it is impossible to check the relevancy between a BDD and target BDD. A static analysis strategy of given knowledge-base may be solve this problem. Overhead of the analysis, however, can not be negligible, especially for propositional logical formulas as knowledge-base. On our strategy, generation of a BDD pauses on an intermediate phase, and it is completed after the BDD turns out to be involved with a target BDD.

In the next section, PBDD (Partial BDD) is defined so as to represent a BDD under generation. We, then, propose GPC (Graft & Pruning Construction) algorithm incorporated with heuristic search strategies, which can efficiently generate a target BDD utilizing PBDD, in Section 5.

4   Partial BDD

We now discuss constructing BDD for a propositional logical formula "$f = v_1 \cdot v_2 \cdot \cdots \cdot v_n$" ($\cdot$ designates a logical operator), taking into consideration with the complexity of the formula. It is quite efficient to construct the BDD if f is composed of only atoms (i.e., v_i is an atom). the BDD for f, then, can be very compact. The efficiency, however, may not hold in case that f nests other formulas (i.e., $v_i = w_{i1} \cdot w_{i2} \cdot \cdots \cdot w_{im}$). this is ineludible for constructing a whole BDD. From the viewpoint of applying BDD to a kind of constraint satisfaction problem such as abduction and deduction, the above anxiety can be avoidable, since abduction and deduction do not require the whole truth value table for f, they need only truth values of v_i such that f becomes true, instead. This means that it is unnecessary to constructing BDD for $v_i = w_{i1} \cdot w_{i2} \cdot \cdots \cdot w_{im}$ if truth-value of v_i is resultingly irrelevant to that of f. In general, it is impossible to judge the relevancy of v_i to f without any pre-analysis. In this paper, we

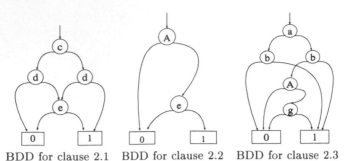

BDD for clause 2.1   BDD for clause 2.2   BDD for clause 2.3

Fig.4: PBDDs for Horn clauses in Example P_{ex}

propose a method; construction of BDD for f is partially done by treating v_i as like an atom, and it is, then, completed after the BDD for v_i turns out to be necessary for the BDD for f. We, thus, modify the original BDD as follows.

Definition 4.1 Let $\bullet$ be a logical operator, and $f = v_1 \bullet v_2 \bullet \cdots \bullet v_n$ be a propositional logical formula, and F be a set of fs. Then, PBDD (Partial BDD) for f in F is defined as a BDD for f constructed such that;

- Each of propositional variables v_i $(1 \le i \le n)$ in f is decomposed by Shannon's decomposition even if it is not an atom, that is, a formula $v_i = w_1 \bullet w_2 \bullet \cdots \bullet w_m a$ exists in F.
- The order of variable f is the highest in all orders of variables v_is, that is, the node labeled with f is positioned higher than all nodes labeled with v_is in the BDD. ∎

Example **4.1** Suppose we construct PBDDs for Horn clauses in knowledge base in Figure 1, where the variable order is given as $C \succ B \succ a \succ b \succ A \succ c \succ d \succ e \succ f \succ g$ by definition 4.1.[3] Figure 4 shows PBDDs for the Horn clauses in Figure 3. By comparison with BDDs, representing same clauses, in Figure 3, the figure shows that PBDD is easier and smaller than BDD. PBDD structures the same diagram with BDD if a given Horn clause is composed of only atoms (see PBDD for clause 2.1 in the figure). ∎

PBDD is characterized by as follows. In case that constructing PBDD for a logical formula requires another PBDD to apply some logical operation, it suspends the applying, while constructing BDD dose apply the operation by bottom-up computation without lookover for the relevancy. In our method, Graft & Pruning Construction algorithm (shown in the next section) enables this applying to be done by top-down computation after its necessity turns out.

---

[3] In general, definition 4.1 may allow more than one order. In this example, so as to compare with BDDs in Figure 3, we take same order with that in *Example* 3.1 with respect to atoms "a, b, c, d, e, f, g".

5 Cost-based Abductive Reasoning using PBDD

The Section describes our cost-based abductive reasoning, which can find an optimal explanation efficiently, by using PBDD with heuristic search control technique. By utilizing the PBDD, abductive reasoning is performed with the following steps, when a knowledge-base $\mathcal{F} \cup \mathcal{H}$ and an observation O are given.

i) Generation of PBDDs:
Based on Definition 4.1, decide the order of propositional variables in $\mathcal{F} \cup \mathcal{H}$ and O. and then, generate PBDDs for each of Horn clauses in $\mathcal{F} \cup \mathcal{H}$ and O.

ii) Graft & Pruning Construction (GPC):
Construct BDD for the observation in the manner of GPC algorithm (see section 5.1) so as to find an optimal successful path in BDD for O. Consistency checking of the path is included in the GPC.

iii) Finding an Optimal Explanation:
Extract nodes from the path such that: 1-edge of the node is included in the path, and the node is labeled with hypotheses.

Abductive reasoning, finally, finds the hypotheses as an optimal explanation of the observation in the BDD for O.

5.1 Graft & Pruning Construction

We propose Graft & Pruning Construction (GPC) algorithm, which can efficiently construct an optimal successful path from a PBDD. GPC enables constructing a target BDD to avoid involving BDD for knowledge irrelevant to the target BDD, and to avoid expanding nodes irrelevant to a given observation. GPC thus makes abductive reasoning much more efficient.

Figure 7 shows the algorithm of GPC. In the figure, F_m indicates a conjunction of logical formula, and N_m indicates a node in PBDD. *Open* and *Closed* indicate sets of nodes.

N_m is composed of four elements: the label, F_m, $Hypo_m$ and $\hat{h}(N_m)$, where F_m means the logical formula should be represented by sub-PBDD under N_m, $Hypo_m$ means the set of hypotheses already assumed to be true at the path form root to N_m, and $\hat{h}(N_m)$ means evaluation value of N_m.

The algorithm executes the iterations of series of three procedures, "node generation", "consistency checking", and "path selection". The iterations are done in order to find an optimal successful path. At the m-th iteration, the GPC algorithm behaves intuitively as follows, where N_m is in the form of $(null, f(h_j, h_{j+1}, \cdots, h_n), Hypo_m, \hat{h}(N_m))$, and $,h_j$ has the highest position in the variable order for variables in formula $f(h_j, h_{j+1}, \cdots, h_n)$.

node generation procedure assigns h_j to node N_m as its label, and then, generates descendant nodes by either of following three ways according to the label h_j:
in case h_j =1-Node, GPC finds an optimal successful path. GPC, thus, terminates, in success, with out put $Hypo_m$. (see lines 16-17 in Figure 7)

in case $h_j =$ is a hypothesis, by Shannon's decomposition principle, formula $f(h_j, h_{j+1}, \cdots, h_n)$ can be decomposed into disjunction of $h_j \wedge f(1, h_{j+1}, \cdots, h_n)$ or $\bar{h}_j \wedge f(0, h_{j+1}, \ldots, h_n)$. The descendant nodes N_{m_1} and N_{m_0} are constructed in the forms of $(null, f(1, h_{j+1}, \ldots, h_n), Hypo_m \cap h_j, \hat{h}(N_m) + cost(h_j))$ and $(null, f(0, h_{j+1}, \cdots, h_n), Hypo_m, \hat{h}(N_m))$, under 1-edge and 0-edge of N_m, respectively. (see lines 18-28 in Figure 7)

in case h_j is consequence of Horn clause $h_j \leftarrow H_j$, propositional variable h_J is replaced with propositional formula H_j. And then, by Shannon's decomposition principle, $f(H_J, h_{j+1}, \ldots, h_n)$ can be decomposed into disjunction of $H_j \wedge f(1, h_{j+1}, \cdots, h_n)$ or $\bar{H}_j \wedge f(0, h_{j+1}, \cdots, h_n)$. The descendant nodes N_{m_1} and N_{m_0} are constructed in the forms of $(null, H_j \wedge f(1, h_{j+1}, \cdots, h_n), Hypo_m, \hat{h}(N_m))$ and $(null, \bar{H}_j \wedge f(0, h_{j+1}, \ldots, h_n), Hypo_m, \hat{h}(N_m))$, under 1-edge and 0-edge of N_m, respectively. the above replacement means that constructing PBDD resumes applying logical operation involving another PBDD. The operation is done by top-down computation; PBDD for H_j is grafted onto descendant node of N_m. It follows from Horn clause $h_j \leftarrow H_j$ that PBDD for H_j is relevant to a target BDD including the node labeled with h_j. (see lines 29-39 in Figure 7)

It should be noticed that Shannon's decomposition can be done easily using PBDDs already constructed.

consistency checking procedure checks the consistency of nodes generated by the above procedure. For a node $(label, f(h_j, \cdots, h_n), Hypo_m, \hat{h}(N_m))$, if $Hypo_m$ includes all hypotheses in any *consistency condition*, the node is excepted from the PBDD. (see lines 41-42 in Figure 7)

node selection procedure, in the best-first manner, updates *Open* and *Closed* with generated nodes and expanded node respectively. It, then, selects node N_{m+1} whose evaluation value $\hat{h}(N_{m+1})$ is minimal from *Open* for the next iteration. If $Open = \phi$, GPC terminates in failure. (see lines 43-51 in Figure 7)

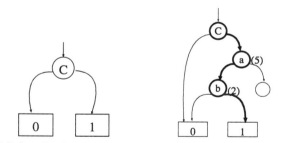

PBDD for the Observation $\mathcal{O}$ A Result of GPC Algorithm

Fig.5: An Input and An Output of GPC Algorithm

Example **5.1** Abductive reasoning for the example knowledge-base shown in Figure 1 is performed by the following steps. Firstly, PBDDs are constructed in the

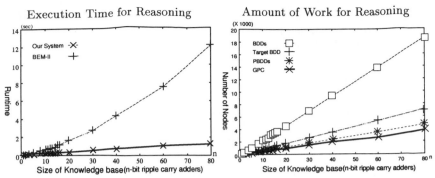

Fig.6: Experimental Results

way of Example 4.1, and PBDD for O, shown in Figure 5, is inputted to GPC algorithm. GPC is executed and then finds an optimal successful path shown in Figure 5. Abductive reasoning, thus, can find $\{b\}$ as an optimal explanation, since 1-edge of the node labeled with b is included in the optimal successful path. In Figure 5, a node having no label shows that the node is not expanded. It should be noticed that GPC uses only PBDD for O and PBDD for clause 2.3. This means that GPC can avoid involving PBDDs irrelevant to given observation and expanding nodes irrelevant to an optimal successful path (PBDD for clause 2.2 and PBDD for clause 2.1 are corresponding to this case). ∎

6 Some Empirical Results

We have made some experiments. We have considered a diagnostic problem to diagnose an n-bit ripple carry adder circuit. We have represented each hypothesis as a state in which a gate x is (i.e. okay$_x$, on$_x$ or off$_x$), and have given each hypothesis $|log_e \mathcal{P}(state_x)|$ as its cost, where $\mathcal{P}(state_x)$ is a probability of x being in $state$[4]. Figure 6 shows the experimental results on the problem, by changing the size n of the circuit. Our experimental system is written in C and all runtimes are average CPU times of a hundred experiments on SPARCStation5/85Hz with 64MB memory. We have compared our system with BEM-II [Min93] in the experiments. In this particular example, the results indicate that our system is about $1.2 \sim 9.9$ times faster than BEM-II as the problem becomes larger.

We have also measured amount of work for reasoning on the problem. In the figure, BDDs indicates the total amount of BDDs generated by BEM-II, and PBDDs indicates the total amount of PBDDs generated by our system. Target BDD indicates the size of a BDD for given observation (target BDD), and GPC indicates the size of a BDD constructed by GPC algorithm. Results on the problem with its size $n = 80$ (diagnosis of 80-bit ripple carry adder circuit)

---

[4] It follows from $0 \leq \mathcal{P}_i \leq 1$ and $\prod_i \mathcal{P}_i = exp(\sum_i |log_e \mathcal{P}_i|)$ that $\prod_i \mathcal{P}_i$ is maximum if $\sum_i |log_e \mathcal{P}_i|$ is minimum.

shows that our system can involve only about 25% of total amount of work by BEM-II. The result also shows that our system can find an optimal solution with a half sized target BDD, in comparison with BEM-II.

7 Conclusion

We proposed an abductive reasoning system which can find most preferable solution efficiently, by utilizing BDD with heuristic search control technique. In this paper, a specialized BDD, Partial BDD, was defined and its operation suitable for abductive reasoning, Graft & Pruning Construction, was proposed. PBDD is effective to maintain the goal-directedness of reasoning, and GPC algorithm can avoid generating some parts of BDDs which are irrelevant to the most preferable solution.

We implemented PBDD and GPC algorithm and built a cost-based abductive reasoning system which can find much more efficiently the most preferable explanation of a given observation. We also made some experiments on the system with some diagnostic problems. The results show that our system gives good performance comparable to a system using an existent efficient implementation of BDD. Our proposed BDD operation incorporated with heuristic search control would be interesting also from the viewpoint of BDD applications for constraint satisfaction problems, since it gives a way of controlling top-down construction.

References

[AS78]   Akers and S.B. Binary Decision Diagrams. *IEEE Trans. on Computers*, C-27(6):509–516, June 1978.

[BRB90]   K.S. Brace, R.L. Rudell, and R.E. Brayton. Efficient Implementation of a BDD Package. In *ACM/IEEE Proc. 27th DAC*, pages 40–45, 1990.

[Bry86]   Randal E. Bryant. Graph-Based Algorithms for Boolean Function Manipulation. *IEEE Trans. on Computers*, C-35(8):677–691, August 1986.

[CS94]   E. Charniak and S. E. Shimony. Cost-based abduction and MAP explanation. *Artificial Intelligence*, 66:345–374, 1994.

[KSI94]   S. Kato, H. Seki, and H. Itoh. Cost-based Horn Abduction to Focus on the Most Probable Diagnosis. In *Proc. of the 5th Intl. Workshop on Principles of Diagnosis*, pages 148–152, New Paltz, NY, October 1994.

[KSI97]   S. Kato, H. Seki, and H. Itoh. A Parallel Implementation of Cost-Based Abductive Reasoning. In *Proc. of the second Intl. Symp. on Parallel Symbolic Computation (PASCO'97)*, pages 111–118, Maui, Hawaii, July 1997. ACM press.

[Llo84]   J. W. Lloyd. *Foundations of Logic Programming*. Springer, 1984. Second, extended edition, 1987.

[Min93]   S. Minato. BEM-II: An arithmetic Boolean expression manipulator using BDDs. *IEICE Trans. of Fundamentals*, E76-A(10):1721–1729, 1993.

[Poo88]   D. Poole. A Logical Framework for Default Reasoning. *Artificial Intelligence*, 36:27–47, 1988.

[Poo93]   D. Poole. Probabilistic Horn abduction and Bayesian networks. *Artificial Intelligence*, 64:81–129, 1993.

Graft & Pruning Construction

```
1  begin
2     F_0 := O; Hypo_0 := φ; ĥ(N_0) := 0; N_0 = (F_0, F_0, Hypo_0, ĥ(N_0)); Open := {N_0};
3     Closed := { 0-Node }; finish := false; PBDD := {PBDDs for Horn clauses in F ∪ H};
4     while (Open ≠ φ and not finish)
5     begin
6        Children := φ;
7        N_m := (h_j, X_{m0} ∧ ⋯ ∧ X_{mn}, Hypo_m, ĥ(N_m)) ∈ Open | for all N_k ∈ Open: ĥ(N_m) ≤ ĥ(N_k);
8        if h_j = null then
9        begin
10          h_j := a lowest variable in the variable order;
11          for each X_{mi} of X_{m0}, …, X_{mn}
12             for all pbdd ∈ PBDD
13                if (∃(h_{ji}, X_{mi, -, -}) ∈ pbdd and variable order of h_{ji} is higher than h_j) then h_j := h_{ji};
14       end
15       case of h_j
16          1-Node:                          % an optimal successful path is found.
17             Hypo := Hypo_m;   finish := true;
18          Hypothesis:                       % one step expansion of h_j.
19             F_{m_0} := 1; F_{m_1} := 1;
20             for each X_{mi} of X_{m0}, …, X_{mn}
21             begin
22                decompose X_{mi} into (h̄_j ∧ X^0_{mi}) ∨ (h_j ∧ X^1_{mi});
23                F_{m_0} := F_{m_0} ∧ X^0_{mi};   F_{m_1} := F_{m_1} ∧ X^1_{mi};
24             end
25             Hypo_{m_1} := Hypo_m ∪ h_j;
26             generate node N_{m_0} in the form of (null, F_{m_0}, Hypo_m, ĥ(N_m));
27             generate node N_{m_1} in the form of (null, F_{m_1}, Hypo_{m_1}, ĥ(N_m) + cost(h_j));
28             Children := {N_{m_1}, N_{m_0}};
29          Consequence of Horn clause :     % graft another PBDD onto h_j.
30             H_j := a_0, …, a_l | ∃ h_j ← a_0, …, a_l ∈ F ∪ H;
31             F_{m_{h̄_j}} := H̄_j; F_{m_{h_j}} := H_j;
32             for each X_{mi} of X_{m0}, …, X_{mn}
33             begin
34                decompose X_{mi} into (h̄_j ∧ X^0_{mi}) ∨ (h_j ∧ X^1_{mi});
35                F_{m_{h̄_j}} := F_{m_{h̄_j}} ∧ X^0_{mi};   F_{m_{h_j}} := F_{m_{h_j}} ∧ X^1_{mi};
36             end
37             generate node N_{h̄_j} in the form of (null, F_{m_{h̄_j}}, Hypo_m, ĥ(N_m));
38             generate node N_{h_j} in the form of (null, F_{m_{h_j}}, Hypo_m, ĥ(N_m));
39             Children := {N_{h̄_j}, N_{h_j}};
40       end case
41       if (∃ Hypo_{m_1}) then                    % cheking the consistency
42          if (Hypo_{m_1} ∧ F → false) then N_{m_1} := 0-Node;
43       for ∀N ∈ Children | N = (_, F, _, ĥ(N))     % in the best-first manner.
44          if N ≠ 0-Node then
45          begin
46             if (N' ∉ Open ∪ Closed | N' = (_, F, _, ĥ(N'))) then    Open := Open ∪ N;
47             if (N' ∈ Open | N' = (_, F, _, ĥ(N')) and ĥ(N) ≤ ĥ(N')) then
48                Open := Open ∪ {N} \ {N'};
49             if (N' ∈ Closed | N' = (_, F, _, ĥ(N')) and ĥ(N) ≤ ĥ(N')) then
50                Closed := Closed ∪ {N} \ {N'};
51          end
52       Closed := Closed ∪ N_m;
53    end while
54    if (Open = φ) then return false;
55    return Hypo;
56 end.
```

†Shannon's decomposition in line 20-24,32-36 is executed easily by inputted **PBDD**.

Fig.7: Algorithm of Graft & Pruning Construction

New Directions in Debugging Hardware Designs *

Franz Wotawa

Technische Universität Wien, Institut für Informationssysteme, Paniglgasse 16, A-1040 Wien, Austria, E-Mail wotawa@dbai.tuwien.ac.at

Abstract. This paper introduces a new approach in the debugging of hardware designs. The design is given as a VHDL program and converted in a component connection model. The conversion is similar to the synthesis of register transfer into gate level programs. The resulting model is directly used for locating faults within the design. To do this, we propose the application of model-based diagnosis. The advantage of this approach is its degree of automation and that it can be applied even on today's mid-size to large size programs.

1 Introduction

Since hardware designs continue to become larger and more complex the verification task becomes a crucial factor. In order to reduce time to market and the overall design costs, formal verification techniques and extensive simulations are currently used during the design cycle. However, currently there is almost no help for the designer in fixing the faults thus found. Only traditional debuggers allowing to go step by step through the code are available, making fault localization and correction a time consuming and difficult task.

Because a large amount of time is spent in fault localization and correction, automation of this task helps to improve the design process. Work has recently been done in this direction include [CWH94,FSW96,SW98]. In this paper we describe a new approach for diagnosing hardware designs. Similar to previous work we convert VHDL programs (see [Nav93,VHD88] for an introduction into VHDL) into a declarative model describing the programs behavior. Such a model consists out of component models, associated with statements or expressions, and connections, associated with signals and variables used in the program. By applying standard diagnosis techniques (see [Rei87]) we can locate faulty components in the model and map them back to the associated statements or expressions within the program. This debugging approach is not restricted to VHDL.

To be general applicable the model has to be automatically derived from the program. The main part of the conversion process is to map parts of the programs into components. Which parts are considered will influence the granularity of the diagnosis results. If statements are directly mapped to components, bugs can only be localized on the statement level. Faults inside a statement can not be diagnosed with such a model. This must not be a drawback since diagnosis time depends on the number of diagnosis

---

* The work presented in this paper has been partially funded by Siemens Austria under research grant DDV GR 21/96106/4 and the Austrian Science Fund Project N Z29-INF.

components. On the other hand, if a converted program consists only out of one component, diagnosis does not provide any discrimination. Therefore the granularity should be carefully chosen.

In [FSW96] VHDL concurrent statements were mapped to diagnosis components. Since the number of concurrent statements varies from 1 to about 1,000 with an average value of about 20, this diagnosis granularity is good enough to provide substantial informations in most cases. Additionally, an extended model (see [SW98]) allows to apply hierarchical diagnosis on the contents of a concurrent process statement. The hierarchical diagnosis process is done as follows. First, the bug is located at the level of concurrent statements. If one VHDL process statement can be identified as source of an error, a model using the sequential statements in a process as components, allows to locate the bug within the process.

The approach used in this paper differs from the previous in the following point. Instead of using a hierarchical diagnosis approach, we convert the program into a flat structure. Statements and expressions are converted to (diagnosis) components. This is similar to the synthesis process (see [VHD98]) where (VHDL) programs are directly converted into hardware, implementing the same functionality. However, integers and other data types are not mapped into boolean values in the manner usual for synthesis tools. One advantage of this approach is that the resulting system does not need to take account of the VHDL operational semantics speeding up computation. Of course it is guaranteed that the diagnosis system implements the same functionality as the program. In summary the proposed approach has the following benefits: (1) A model used for diagnosis is automatically derived from the given program. (2) Diagnosis is done using standard model-based diagnosis techniques. (3) Because of the syntactical restrictions the approach can be applied even to large size designs. (4) The approach is applicable to VHDL register transfer programs. Most of todays designs are written using the VHDL register transfer subset.

This paper is organized as follows. In section 2 the approach is introduced using a small example. It is followed by a description of the conversion procedure. Section 4 shows how the model can be used for debugging. A section about related research and future directions concludes the paper.

2   A basic example

We illustrate our approach using a small VHDL program implementing a 2 bit counter. The counter is reseted by a positive value on the RESET input and counts up every time a positive edge of the clock input CLK is detected. Figure 1 shows the VHDL program consisting of an entity COUNTER and describes the inputs (CLK,RESET) and the outputs (O1,O2) of the device. The behavior of the counter is given by the architecture BEHAV. Two processes (MEM and COMB_IN) implement the desired functionality. The behavior of the counter is shown in figure 2 (a).

The program behavior (given by the VHDL semantics) is described such as follows: Processes are only executed after detecting an event on a input signal at the current simulation time. In our case an event occurs if at least one signal of the process sensitivity list changes the value. The sensitivity list of the MEM process has two signals: CLK

```
entity COUNTER is
   port ( CLK, RESET : in STD_LOGIC;
      O1, O2 : out STD_LOGIC );
end COUNTER;
architecture BEHAV of COUNTER is
   signal D1,D2 : STD_LOGIC ;
begin
   mem: process ( CLK, RESET )
   begin
      if RESET = '1' then
         O1 <= '0'; O2 <= '0';
      else if CLK = '1' and CLK'EVENT then
         O1 <= D1; O2 <= D2;
      end if; end if;
   end process mem;
   comb_in : process ( O1, O2 )
      variable V: STD_LOGIC;
   begin
      V := not(O1); D1 <= V;
      D2 <= not((O1 and O2) or (V and not(O2)));
   end process comb_in;
end BEHAV;
```

Fig. 1. The VHDL COUNTER(BEHAV) program

and RESET. So if RESET is set to '1' the process is executed. The signals O1 and O2 are set to '0'. Since both signals occur in the sensitivity list of COMB_IN, it is executed and D1 is set to '1' and D2 to '0'. If RESET is set to '0' the counter starts counting up every time a positive edge on CLK is detected. In the first step O1 is set to '1' and O2 to '0'. Then the process COMB_IN is executed leading to D1 = '0' and D2 = '1'.

The same functionality as described above can be implemented using hardware. [VHD98] describes a VHDL subset together with a conversion algorithm for mapping programs into a gate level representation. The gate level view of the COUNTER(BEHAV) program corresponds directly to hardware. In our approach we use the basic idea of RTL synthesis. However, we do not restrict ourself to such a small class of RTL programs as synthesis does. See [VHD98] for a description of how VHDL process must be organized to be mapped to flip flops, latches and other digital gates. Instead the only restriction we currently apply is that wait statements are not allowed within processes[1] and that delay times are ignored. Expressions, conditionals, assignments, loops, etc. are converted to simple components while signals and variables are converted to connections.

3 Modeling programs

In this section we show the conversion of programs into diagnosis systems. For presentation purposes we restrict VHDL to a subset. However, the used techniques can be easily applied to other VHDL constructs namely component instantiation statements, loops, and others. The overall conversion process can be summarized as follows:

1. Convert the program into a component connection model. This representation may contain cycles.

---

[1] Usually the wait statement can be replaced by a corresponding process sensitivity list.

2. Make the model acyclic. In the hardware context this step creates a combinational circuit from the sequential circuit.

3. Do diagnosis using the model and observations. In debugging observations come from an oracle (a formal specification or the programmer).

Before discussing the program conversion process we define a component connection model. A component is an object with ports and a specified behavior. Connections are associated to ports. We introduce a function p for every port of the given component returning the associated connection, i.e., $p : COMPS \mapsto CONNS$. For example, given a *NOT* component C with two ports *in* and *out* we assume the existence of the functions $in(C)$ and $out(C)$. We further introduce a function $ports : COMPS \mapsto 2^{PORTS}$ returning a set of ports for a given component. For component C we have $ports(C) = \{in, out\}$. The behavior of a component can be obtained using the *behavior* function. For simplicity we introduce a function $connPorts$ returning all ports connected with the given signal. The model for a program is given by a set of components and the connections can be computed from the underlying information. This model forms the system description used in model-based diagnosis to determine faulty components. In our approach we convert programs into models, compute diagnoses, and map the diagnoses back to the associated parts of the programs.

VHDL programs are given by an entity and an architecture[2]. While entities specify the interface, i.e., those signals accessible from the outside, architectures implement the (maybe faulty) behavior. Therefore only architectures must be converted into diagnosable systems. For simplicity we assume that an architecture consists out of processes. That is no semantic restriction because every concurrent statement relevant for simulation that is not a single process can be rewritten in this manner.

We further assume a function $conn : (SIGNALS \times STATEMENTS) \cup VARIABLES \mapsto CONNS$ mapping driving signals and variables to connections, and a function $signalConn : SIGNALS \mapsto CONNS$ associating signals to connection. Note, that a driver of a signal is generated whenever a process assigns a value to a signal. The driver itself is a signal that determines the signal value according to the VHDL semantics. The mappings of signals are never changed during conversion. They are initially given by associating a new connection to every signal driver occurring in the VHDL program (for $conn$) and by associating a connection to every signal(for $signalConn$). For every variable we also initialize $conn$ with a new connection. However, the associated connection may change during conversion. This distinction is necessary because of the semantically different handling of signals and variables in a process. Details about the handling of resolution functions used to determine a value in case several driving signals are used can be found in [Wot99]

The conversion algorithm converts VHDL architectures by successively converting their concurrent statements, i.e., processes. Processes are converted by converting the sequential part. The conversion of the sequential part is done statement by statement in the same order they appear in the source code. Because of limited space most of the formal definition of this conversion together with the conversion of expressions is

---

[2] Configurations are ignored for our purposes because it is always possible to automatically generate a program with an equal behavior without using configurations.

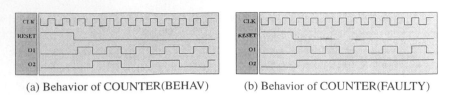

| (a) Behavior of COUNTER(BEHAV) | (b) Behavior of COUNTER(FAULTY) |

Fig. 2. The behavior of the two COUNTER programs

omitted. The whole algorithm is given in [Wot99]. To show the basic principles we define the conversion of VHDL assignments.

Signal, Variable Assignments are converted to a component $C_{assignment}$ with one input in and one output out. The behavior is simple. If the assignment is assumed to be correct the input value is converted to the output and vice versa, i.e., $assignment(C) \wedge \neg ab(C) \Rightarrow in(C) = out(C)$. The following algorithm converts a signal assignment to a component oriented view.

convert$(T <= E, comp, conn, signalConn, P)$
  $c = $ **new** $C_{assignment}$ $comp = comp \cup \{c\}$
  $out(c) = conn((T, P))$ $in(c) = $ **convert**$(E, comp, conn, signalConn)$

As stated previously the execution of the variable assignment immediately changes the variable value. We model this by changing the associated connection.

convert$(T := E, comp, conn, signalConn, P)$
  $c = $ **new** $C_{assignment}$ $comp = comp \cup \{c\}$
  $in(c) = $ **convertExpr**$(E, comp, conn, signalConn)$
  $conn(T) = $ **new connection** $out(c) = conn(T)$

The **convert** algorithm defines the conversion of programs into a component oriented view. Since this representation may contain cycles we must eliminate them. Therefore we first convert the program into a functional dependency graph and remove the signal vertices eliminating all cycles. In [SW95] the conversion of VHDL programs into a functional dependency graph is shown.

Let $\{S_1, \ldots, S_n\}$ be a minimal set of signals that, when removed, eliminates all cycles for a given program. We adapt the component connection model described above by introducing new connections $C_1, \ldots, C_n$ (associated with $S_1, \ldots, S_n$) and replacing every connection $conn(S_i)$ connected with a port p with the new connection C_i. This final model represents the system description used for diagnosis. Figure 3 shows the final system description for the example program COUNTER(BEHAV).

It remains to give a condition specifying those signals and variables that must be observed in order to compute a diagnosis. This set is given by all signals and variables where their associated connection is only used in exactly one port. All such signals and variables are either inputs or outputs.

In our case we have to give the values for the inputs CLK, $CLK'EVENT$, $RESET$, $Q1(IN)$, $Q2(IN)$ and the outputs $Q1$ and $Q2$. The connection $Q1(IN)$ is associated with signal $Q1$ and the connection $Q2(IN)$ is associated with signal $Q2$.

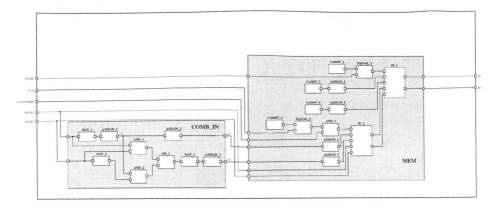

Fig. 3. The final diagnosis system for COUNTER(BEHAV)

The name is only used to distinguish between the input and the output value for $Q1$ ($Q2$). From the semantics point of view the value of $Q1(IN)$ at time $t \neq 0$ is given by the value of $Q1$ at the immediately preceding time point $t - 1$. At time $t = 0$ the value of $Q1(IN)$ is equal to the initial value.

4   Using the model for program debugging

Until now we have shown the conversion of VHDL programs into system descriptions for model-based diagnosis and specified the set of signals and variables whose values must be known to find the location of a bug. We use the example from section 2 to demonstrate how bugs can be located. Therefore we introduce a (single) bug into the COUNTER(BEHAV) program. Instead of " D2 <= **not**((O1 **and** O2) **or** (V **and not**(O2))); " in the correct program we assume that the faulty architecture (named FAULTY) has the assignment " D2 <= **not**((O1 **and** O2) **and** (V **and not**(O2))); ".

The faulty program COUNTER(FAULTY) now produces a behavior as given in figure 2 (b). The Comparison of this waveform trace with the original one (figure 2 (a)) leads to the detection of discrepancies. In the next step we use the specified behavior together with the computed system description. Because of the kind of bug the structure of the system description of both programs is equal. Therefore we can use the schematics depicted in figure 3 for diagnosis. We only have to assume that component OR_1 behaves like an *AND* component. The specified behavior (given in figure 2 (a)) can be viewed as observations for the diagnosis system. Table 1 shows the list of observations. The columns of the table are the test vectors for diagnosis. Vectors leading to contradictions are denoted with a star (*) on the right side of the table.

At this point we are interested in locating the bug. Since we use standard model-based diagnosis techniques this means that we are searching for a subset of the component set, that when assumed to behave incorrect, does not anymore lead to a contradiction. See [Rei87] for a formal definition. In the context of model-based diagnosis, debugging is viewed as searching for components not leading to a contradiction. For example if we assume that IF_1 is faulty, i.e., $ab(\text{IF}\_1)$, and that all other components

Table 1. List of observations for diagnosis

test	RESET	CLK	CLK'EVENT	Q1(IN)	Q2(IN)	Q1	Q2	
1	'1'	'0'	true	'0'	'0'	'0'	'0'	
2	'1'	'1'	true	'0'	'0'	'0'	'0'	
...								
7	'0'	'1'	true	'0'	'0'	'1'	'0'	*1
8	'0'	'0'	true	'1'	'0'	'1'	'0'	
9	'0'	'1'	true	'1'	'0'	'0'	'1'	
...								
13	'0'	'1'	true	'1'	'1'	'0'	'0'	*
...								

$C \in COMPS$ behave correct, i.e., $\neg ab(C)$ then it can be proven that no contradiction occurs. So we have found a diagnosis $\{ab(\text{IF\_1})\}$ which can be mapped back to the first conditional statement of the process MEM. In summary we get 12 diagnoses: $\{ab(\text{AND\_1})\}$, $\{ab(\text{OR\_1})\}$, $\{ab(\text{NOT\_3})\}$, $\{ab(\text{ASSING\_3})\}$, $\{ab(\text{ASSIGN\_7})\}$, $\{ab(\text{IF\_2})\}$, $\{ab(\text{CONST\_2})\}$, $\{ab(\text{EQUAL\_2})\}$, $\{ab(\text{AND\_3})\}$, $\{ab(\text{CONST\_1})\}$, $\{ab(\text{EQUAL\_1})\}$, $\{ab(\text{IF\_1})\}$.

This result is good in the sense that the number of possible fault locations is reduced. Only 12 of original 22 components can cause the faulty behavior. However, it does not improve the results obtained using much simpler approaches such as [FSW96], where the two processes MEM and COMB_IN would be diagnosis candidates. Although the result of the new model is more accurate it does not distinguish between the two processes. In order to improve diagnosis using the component connection model we have to take a closer look at the involved components, their behavior modes, and the possible corrections.

Conditional statements have three behavior modes: $\neg ab(C)$, $ab(C)$, and $wrong(C)$ to indicate correct behavior, incorrect behavior, and a fault within the condition causing the execution of the wrong path. The $ab(C)$ mode is the most general one. Nothing about the real behavior is specified leading to the conclusion that everything is possible. Therefore this mode should only be taken into account if nothing else (e.g., $wrong(C)$) explains to a wrong behavior. Correcting a faulty conditional statement with an $ab(C)$ mode is difficult. Several corrections are possible including removing the conditional, adding lines of code before or after the conditional statement, among others. The $wrong(C)$ mode gives more information about the bug location than the $ab(C)$ mode because it focuses the attention only to the condition. The correction of a faulty conditional statement with a $wrong(C)$ behavior has to be done by correcting the condition.

Assignments have two modes, one for the correct behavior ($\neg ab(C)$) and one indicating a fault ($ab(C)$). A $ab(C)$ gives no exact hint about the kind of fault. Only two possible faults can be excluded, a faulty expression and a wrong target signal. The wrong expression is handled by the connected components associated with the expression. The wrong target signal can only be handled by assuming a structural fault in the converted system. A possible correction for a faulty assignment is to add new statements before or after it. By assuming that only few changes are necessary to correct the fault we can ignore assignments as diagnosis unless nothing else is a diagnosis.

Constants, Functions can be correct $(ab(C))$ or faulty $(\neg ab(C))$. Faulty constants or functions can be corrected by replacing them with another constant, function or variable (see [SW98]).

Filtering diagnosis candidates is done using 3 rules:
(1) Remove all diagnoses where components mapped to a condition are included.
(2) Remove candidates where conditional statements C are assumed to behave incorrectly $ab(C)$ but where we know that the condition is not faulty $(\neg wrong(C))$.
(3) Eliminate diagnoses that includes faulty assignments.

Note, that the rules (2), (3) are only applicable while at least one diagnosis remains. Rule (1) is not allowed to be used if diagnosis focus on finding the bug in a conditional. We illustrate the filtering criteria by using our running example where only 3 diagnoses remain: $\{ab(\text{AND}\_1)\}$, $\{ab(\text{OR}\_1)\}$, $\{ab(\text{NOT}\_3)\}$.

This result can be further improved by using a technique from [SW98] where multiple test cases are used. Assume for example that diagnosis $\{ab(\text{NOT}\_3)\}$ is correct. In this case we can compute the following input/output vectors. For test 7 we get $(in(\text{NOT}\_3), out(\text{NOT}\_3)) = ('0','0')$, for test 9 and 11 $('0','1')$, and for test 13 $('0','0')$. We see that with the same input value, i.e., '0', different output values should be returned. Assuming the working hypothesis, that the correct program is a small variant of the wrong one, leads to the conclusion that NOT_3 can not be a single diagnosis. Note, that if this working hypothesis is not used, then replacing a *NOT* by an *AND* (with one additional input) might be applicable and we cannot reject this diagnosis. With this technique diagnosis $\{ab(\text{AND}\_1)\}$ can also be rejected. Hence, finally only the diagnosis $\{ab(\text{OR}\_1)\}$ remains. Because the described diagnosis and filtering steps are performed automatically using the new model, the debugging tool can find the bug automatically.

Although, the techniques described in the paper are not general applicable, they can be used in a wide variety of examples with similar results. In some examples where the simplifying assumptions can not be used, we can find the bug locations by using measurement selection [dKW87] and oracle calls, i.e., user calls or a formal specification. However, the results computed remain correct and provide a reduction of the initial search space. However, there is a problem with the model. In the current version it can not handle bugs related to the use of a wrong variable. Finding such a fault is related with a structural change in the model. Similar problems occur in hardware diagnosis, e.g., bridge faults [Böt95].

In the remainder of this section we argue that the model can be used for debugging of medium-sized and large VHDL programs. Because we have not yet implemented the proposed approach we give an estimation for the diagnosis of single bugs. This is done by comparing state of the art diagnosis algorithms with the expected number of generated components per K-Bytes of design size. Table 2 (a) gives the performance results of the DRUM-2 algorithm (see [FN97]) for combinational circuits.

Assuming independence of involved random variables and equal probability, the expected diagnosis time per gate is 0.007751 seconds. In hardware design, typical sim-

Table 2. Empirical informations of the performance of diagnosis algorithms and real-world designs

Circuit	Gates	Diag. Time [sec]	Time per Gate [sec]
c499	202	0.13	0.000644
c880	383	0.02	0.000052
c1355	547	1.78	0.003254
c2670	1193	6.43	0.005390
c3540	1669	1.42	0.000851
c5315	2307	0.31	0.000134
c6288	2406	118.59	0.049289
c7552	3512	8.4	0.002392

(a) Performance of DRUM-2 (from [FN97])

Design	Size [k-Byte]	Gates	Gates per k-Byte
D-01	2,553	17,188	6.73
D-02	952	1,030	1.08
D-03	48	1,335	27.81
D-04	18	809	44.94
D-05	13	220	16.92
D-06	12	284	23.67
D-07	8	156	19.50

(b) Approximated number of gates for real-world designs

ulation runs last from several minutes to several days. If we want to restrict diagnosis time by an upper limit of 10 minutes (that is convenience to hardware designers) we can diagnose systems with up to 77,400 components. Within 10 seconds, smaller programs with up to 1,300 components can be diagnosed in the average case. Note, that this estimation is based on the assumption that the time for computing single diagnoses is linear in the number of components.

Finally, we combine the performance results with the expected number of components per k-Byte of source code. This number is taken from real-world examples (see figure 2 (b)). Assuming statistical independence we get that about 20 components are created for 1 k-Byte of source code. Note, that we do not restrict components (gates) to be digital. Therefore, the given number of components in figure 2 (b) is smaller than the number of digital components obtained after applying synthesis, i.e., converting VHDL programs into hardware.

Hence, we can expect to diagnose programs with up to 3.8 MB of source code within 10 minutes and programs with up to 65 k-Byte of source code within 10 seconds. Although this estimation is very rough it provides a strong evidence for the usefulness of the proposed method.

5 Conclusion and related research

Over the past years several approaches for debugging hardware designs have been proposed. [CWH94] introduces algorithms for locating and correcting faults in gate level designs where only boolean values and functions were used. Our approach on the other hand is not restricted to boolean values. Instead arbitrary data types and their functions can be used. Applying model-based diagnosis to software debugging especially for the hardware description language VHDL is not new. [FSW96] has proposed an abstract model in order to handle very large programs with up to 10 MB source code. This approach is similar to the program slicing [Wei84] but extents it in some directions. In [SW98] a program description for VHDL process statements handling the full (sequential) VHDL semantics was introduced. This paper extents previous work in the VHDL domain by using a concise model for the concurrent statements. Since not all parts of

VHDL are utilized it is not general usable such as [FSW96]. However, we can handle those parts of VHDL that are mostly used in todays hardware designs.

In summary we use the following restrictions for VHDL programs: (1) Only the predefined data types including IEEE standard logic are allowed. (2) Neither loops nor recursive function calls are allowed. Code with loops that can be translated into a loop free version can be supported with the proposed technique. (3) No delay times must be specified. (4) Only one resolution function (for standard logic) is supported. For RTL synthesis the above restrictions must be also applied. But some synthesis restrictions are not used because a possible hardware implementation must not be taken into account. Future work and research in this area should include: (1) A proof that the above model really does not change the VHDL semantics for the used subset. (2) An implementation together with a detailed analysis. Performance and an evaluation of the diagnosis results should be done. (3) Extending the VHDL subset to the more general case including user defined data types and recursive function calls.

References

[Böt95]   Claudia Böttcher. No faults in structure? How to diagnose hidden interaction. In *Proc. IJCAI*, Montreal, August 1995.

[CWH94]   Pi-Yu Chung, Yi-Min Wang, and Ibrahim N. Hajj. Logic design error diagnosis and correction. *IEEE Transactions on Very Large Scale Integration (VLSI) Systems*, 2:320–332, 1994.

[dKW87]   Johan de Kleer and Brian C. Williams. Diagnosing multiple faults. *Artificial Intelligence*, 32(1):97–130, 1987.

[FN97]   Peter Fröhlich and Wolfgang Nejdl. A Static Model-Based Engine for Model-Based Reasoning. In *Proceedings 15th International Joint Conf. on Artificial Intelligence*, Nagoya, Japan, August 1997.

[FSW96]   Gerhard Friedrich, Markus Stumptner, and Franz Wotawa. Model-based diagnosis of hardware designs. In *Proc. ECAI*, Budapest, August 1996.

[Nav93]   Zainalabedin Navabi. *VHDL Analysis and Modeling of Digital Systems*. McGraw-Hill, 1993.

[Rei87]   Raymond Reiter. A theory of diagnosis from first principles. *Artificial Intelligence*, 32(1):57–95, 1987.

[SW95]   Markus Stumptner and Franz Wotawa. Modeling VHDL Programs for Diagnosis with Linear Computational Complexity. Technical Report DBAI-MBD-TR-95-03, Technische Universität Wien, June 1995.

[SW98]   Markus Stumptner and Franz Wotawa. Model-based debugging of functional programs. In *Proc. DX'98 Workshop*, Cape Cod, May 1998.

[VHD88]   IEEE Standard VHDL Language Reference Manual LRM Std 1076-1987, 1988.

[VHD98]   IEEE P1076.6/D1.12 Draft Standard For VHDL Register Transfer Level Synthesis, 1998.

[Wei84]   Mark Weiser. Program slicing. *IEEE Transactions on Software Engineering*, 10(4):352–357, July 1984.

[Wot99]   Franz Wotawa. New Directions in Debugging Hardware Designs. Technical Report DBAI-TR-99-24, Technische Universität Wien, 1999.

Reasoning with Diagrams:
The Semantics of Arrows

Gérard Ligozat

LIMSI/CNRS & Université Paris-Sud
P.O. Box 133, 91403 Orsay, France
ligozat@limsi.fr

Abstract. This paper describes the use of diagrams as pictorial, non analogical representations in mathematics. Three interpretations of diagrams, of increasing complexity, are discussed: commutative diagrams, exact sequences, universal diagrams. Typically, reasoning with diagrams (in the case under scrutiny, geometrical structures of arrows) involves three steps: representation, construction, inspection and interpretation. We show how reasoning with diagrams makes a metaphoric use of the properties of the representations and suggest how extensions of the existing paradigms can enrich the emergent domain of hybrid reasoning.
We think that much insight on how to use picture-based reasoning can be gained by analyzing the way diagrammatic reasoning *is* used by humans in existing fields of research. This study complements the more familiar topic of common-sense reasoning with pictures and diagrams: here, expert knowledge, rather than common-sense knowledge, is involved.

1   Introduction

The use of pictorial representations in mathematics is widespread in geometry and topology. In classical Euclidean geometry, analogical representations of the geometric objects themselves are used as auxiliary tools for conceptualization and reasoning. Typically, this will involve three steps:

Representation A figure is drawn to represent the problem. For instance, to take the geometric proof of Pythagoras' theorem (see [11]), an arbitrary right triangle is given.

Construction This part is concerned with introducing suitable objects in the universe of discourse. For example, in the proof of Pythagoras' theorem, two squares are constructed out of four copies of the original triangle.

Inspection and interpretation Knowledge about the interpretations of the objects is used. For example, the formulas which give the area of a square are used for both squares, yielding the desired result after some algebraic calculation.

Another, simpler example is the proof that all three mediators in a triangle intersect. The construction part would consist in drawing two mediators and taking their intersection. The inspection and interpretation part would use the fact that this point is equidistant from the three vertices, hence also lies on the third mediator.

In [11] a formal language for expressing diagrammatic proofs is investigated. We will not pursue this topic here, but rather consider informal uses of reasoning with diagrams, as used by mathematicians.

It is fair to say that figures in geometry are also commonly held in suspicion as imperfect representations of the pure, mathematical structures. Geometry itself has been described as "the art of reasoning correctly about incorrect figures". Figures are analogical, imperfect representations of the intended *bona fide* mathematical objects, such as circles, lines, or triangles. Apart from their ability to suggest facts that might prove to be false because of the approximate nature of the representation, figures also hide implicit assumptions: for instance, in the proof about mediators, we implicitly assumed that two mediators in a triangle do in fact intersect. Figures in geometry, then, tend to be assigned a subsidiary role: although they are aids for intuition, they cannot by themselves give any validity to a geometrical proof.

An analogous situation holds in topology: As good supports for intuition, figures are also commonly used. Since many objects considered in topology are not embeddable in two or three dimensional Euclidean space, analogical representations are used which sometimes leave out some of the properties of the real objects: for example, the "Klein bottle", a non orientable surface which is not embeddable in 3-D space, is drawn as a self intersecting surface. Very useful representations of topological objects are in terms of quotient objects: for instance, all compact, connected smooth surfaces can be represented as constructed from a polygon by identifying suitable pairs of sides (see e.g. [7]). Figure 1 represents the construction of a torus and the construction of the Klein bottle from a square by identifying opposite sides of the square in the way indicated by the arrows.

This last class of pictorial representations is interesting in a new way. Those representations are still partly analogical and still useful as a way of visualizing the object, but they are representations of constructions of the object rather than of the object itself. Hence, they leave out *in a principled way* some (topological) properties of the represented object; for instance, the representation of a surface has boundaries which the intended object does not have. Those "cut and glue" representations have a *procedural* aspect to them. We will encounter analogous cases in discussing diagrams.

A last example of using analogous representations in mathematics is in the area of mathematical logic. In their paper about heterogeneous logic, Barwise and Etchemendy [1] cite Tennant [10] to the effect that "[the diagram] has no proper place in the proof as such", and contend that "this dogma is misguided". The example they use for logics uses a universe which is a spatial interpretation of a logical language. Hence, in a way, their example still belongs to the class of analogical representations. The same phenomenon we observed for topology also appears here: some properties of the representation have to be interpreted on another level. For instance, the fact that a block object lies outside the grid

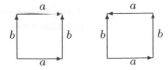

Fig. 1. Constructing the torus and the Klein bottle

in the image has to be interpreted as knowing about the existence of the object, but having no information about its location.

One of the aims of this paper is to defend the view that the usefulness of pictorial representations in mathematics is not limited to *analogical* representations, by considering the uses of diagrams in algebraically oriented areas of mathematics. The representations we will discuss cannot be viewed as analogical ones, except from the fact that they represent the linear structure of composition of arrows in graph-like structures.

We show how some diagrammatic devices used in mathematics allow to transport some of the reasoning procedures described above to abstract domains.

We will consider three instances of this kind of diagrammatic reasoning, with increasing complexity in their semantic interpretations:

- Commutative diagrams.
- Commutative diagrams and exact sequences
- Universal properties.

2   Representing composition: commutative diagrams

Reasoning with maps often involves expressing that composing two maps $f : A \to B$ and $g : B \to C$ to get $(g \circ f) : A \to C$ gives the same result as $h : A \to C$. A diagrammatic way of expressing this is to say that the triangle in the left part of Fig. 2 is *commutative*, i.e. arrows in it can be followed in any way to give the same result. In the same way, the commutative square in Fig.2 expresses the fact that $g \circ f = k \circ h$. An interesting property of this representation is the fact that this amounts to saying that both triangles in the square of Fig. 2 are separately commutative. Hence the representation has a "block like" property of allowing to reason from the figure itself by assembling block-level information into bigger units.

Reasoning with commutative diagrams typically involves the three steps described in the preceding section: representation, construction, inspection and interpretation. For instance, to prove that a given diagram is commutative, we may construct a bigger diagram which contains it, inspect and use information about subparts of the new diagram, and conclude by assembling the information obtained in that way.

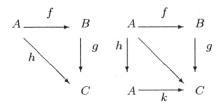

Fig. 2. Commutative diagrams

In reasoning about commutative diagrams, the semantics we use is basically path semantics: paths on a graph are interpreted as composition of maps. Commutativity means that two different paths with the same source and target represent the same map. We now consider adding more semantics to diagrams.

3 Reasoning about input/output properties: exact sequences

A common situation in mathematics is decomposing objects into simpler objects. For instance, consider the additive group A of integers modulo 6, whose elements are 0, 1, ..., 5. The subset $B = \{0, 2, 4\}$ is a three element cyclic subgroup in A. This means that the natural map $i : B \to A$ is in fact a group homomorphism. Now consider the map j defined by $j(x) = 3x$, which is also a group homomorphism. It sends A unto the subgroup $C = \{0, 3\}$ of A. Hence $j : A \to C$ is a surjective group homomorphism. We get the following sequence:

$$B \xrightarrow{i} A \xrightarrow{j} C$$

Notice that applying first i, then j, (this is denoted by $j \circ i$) always gives the unit element 0 of the group. In other terms, j "kills" all elements input by i. The subset of elements killed by a group morphism f is called the *kernel* of f, and written $Ker(f)$. Using the usual notation $Im(f)$ for the image of a map f, we may express what happens by saying that in the sequence $B \to A \to C$ the kernel of j contains the image of i.

In fact, the kernel of j and the image of i coincide. This is expressed by saying that the sequence is an *exact* sequence (at A). In this particular example, i is injective, and j surjective. This amounts to saying that the sequences $0 \to B \to A$ and $A \to C \to 0$ are both exact (where the group homomorphisms from and onto 0 are the only possible ones).

Putting things together, we get the *short exact sequence*:

$$0 \longrightarrow B \xrightarrow{i} A \xrightarrow{j} C \longrightarrow 0$$

In many cases, this is the best we can hope to get to express that the object A can in some sense be decomposed into B and C. In our case, the exact sequence even *splits*, which means that we also can find a map from C back to A, which is a group homomorphism, and such that "pulling back" elements from C and applying j gives the identity map: we only have to choose the canonical injection of $C = \{0, 3\}$ into A. This splitting property expresses the fact that A is basically the direct sum of A, a 3-element group, and C, a 2-element group. However, in many cases, the existence of an exact sequence is the best one can hope for.

The use of exact sequences was initiated in algebraic topology by Eilenberg and Steenrod [3]. It illustrates a basic notational device which, as we will see, has grown into a whole (informal) mechanism of proof in this particular field of mathematics, as well as in category theory, algebraic geometry and formal logics.

More generally, many situations give rise to *long exact sequences*, that is, sequences such that $Im(\alpha_{i-1}) = Ker(\alpha_i)$ for all i:

$$\cdots \longrightarrow A_{i-1} \xrightarrow{\alpha_{i-1}} A_i \xrightarrow{\alpha_i} A_{i+1} \longrightarrow \cdots$$

4 Using compiled knowledge about sequences: some lemmas

In many situations in algebraic topology, algebraic geometry, and other algebraically oriented domains of mathematics, complex diagrams are used to represent and reason about the properties of algebraic objects attached to topological spaces, varieties or other entities. Because of their frequent uses, some results associated to particular patterns in the diagrams have emerged as lemmas. We give some typical examples of such lemmas in what follows. Algebraists call "diagram chasing" the mathematical activity involved in following paths along the diagrams. Numerous examples can be found in standard textbooks such as [2, 8, 9]).

4.1 The five lemma

The *five lemma* is used in the situation represented in Fig. 3. It asserts that, if:

- the diagram is commutative;
- the upper and the lower sequence are exact;
- α, β, δ, ϵ are isomorphisms,

then γ itself is an isomorphism.

In the common case where A, A', E and E' are 0, the result boils down to the fact that C and C', which are "composed" of isomorphic objects are themselves isomorphic.

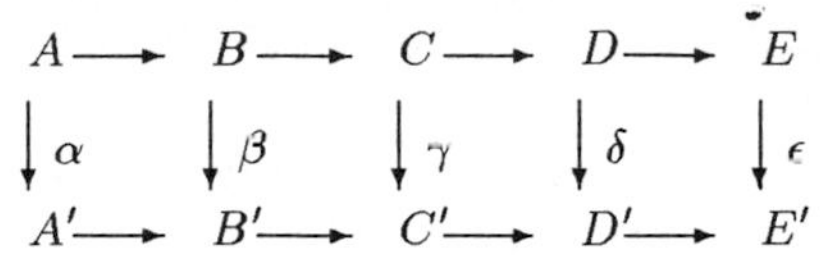

$$A \longrightarrow B \longrightarrow C \longrightarrow D \longrightarrow E$$

$$\downarrow \alpha \quad\quad \downarrow \beta \quad\quad \downarrow \gamma \quad\quad \downarrow \delta \quad\quad \downarrow \epsilon$$

$$A' \longrightarrow B' \longrightarrow C' \longrightarrow D' \longrightarrow E'$$

Fig. 3. The five lemma

4.2 The snake lemma

The snake lemma deals with two short exact sequences as in Fig. 4. Suppose that there exist α, β, γ such that the diagram is commutative. Then the lemma asserts the existence of a canonical exact sequence:

$$0 \to Ker(\alpha) \to Ker(\beta) \to Ker(\gamma) \to A'/Im(\alpha) \to B'/Im(\beta) \to C'/Im(\gamma) \to 0$$

The name of the lemma is best understood by looking at Fig. 5.

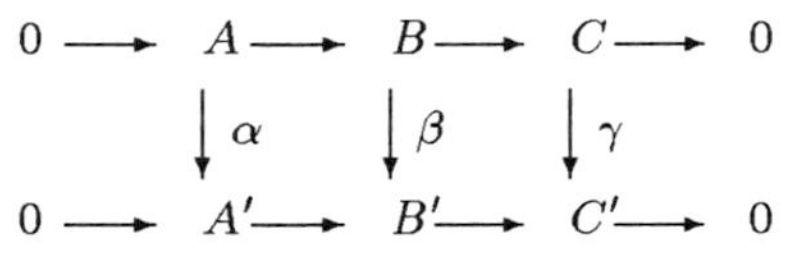

$$0 \longrightarrow A \longrightarrow B \longrightarrow C \longrightarrow 0$$

$$\downarrow \alpha \quad\quad \downarrow \beta \quad\quad \downarrow \gamma$$

$$0 \longrightarrow A' \longrightarrow B' \longrightarrow C' \longrightarrow 0$$

Fig. 4. The snake lemma: what is given

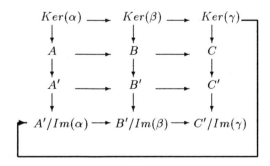

Fig. 5. The snake lemma: what is deduced

4.3   The nine lemma

The nine lemma deals with nine objects as in Fig. 6. It asserts that, if the diagram
is commutative, if all columns are exact, and if the middle row is exact, then, if
either of the upper or the lower sequence is exact, then both are.

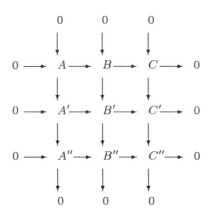

Fig. 6. The nine lemma

4.4   Discussion

As the preceding examples show, reasoning with exact sequences in diagrams
uses the spatial properties of diagrams in two dimensions. Other more complex
examples can use other geometric figures (the hexagonal lemma for example) or
virtual three dimensional diagrams by considering e.g. commutative cubes. The
general pattern of proof is still in terms of three steps: representation, construction, inspection and interpretation.

   The new fact is that the language of representation itself supports a degree of
independently compiled knowledge. Typically, in a proof, both the assumptions
and conclusions of a diagrammatic lemma are in terms of the diagram structure
itself. The fact that the class of results we are considering are called lemmas
rather than theorems shows that they are not considered as fully mathematical
in content, but rather as intermediate knowledge resulting from the diagram
chasing activity. Diagram chasing is not considered as really meaningful in itself,
but rather as operating on the representation itself. The critical point here is
that this representation has spatial properties which are used in the proving
activity. To get a feeling of what the diagrammatic structure brings about, we
have only to try to express the same lemmas without using any figures.

5 Procedural aspects of diagrammatic reasoning: universal constructs

The last sections described how the semantics of the basic language of paths could be enriched by reading off extra properties in terms of exact sequences. In this section, we briefly describe a class of interpretations of diagrams which further complexifies the interpretation by assigning implicit procedural properties to the diagrams.

This new class is somewhat analogous to the polygonal representations of surfaces in topology we discussed above.

Consider the diagram in Fig. 7. The object C is a sum of A and B if this diagram has the following *universal* property: α and β are injections (or monomorphisms in a general category), and for all pairs of arrows $u : A \to D$, $v : B \to D$, there exists a unique arrow $w : C \to D$ such that the diagram commutes. Reversing all arrows gives the definition of a product. Many useful constructs, such as projective and injective limits (products and sums are special cases) can be expressed in terms of universal properties of diagrams.

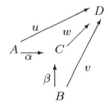

Fig. 7. A sum object in a category

Reasoning with universal properties makes use of the potential generative power of universal definitions. As a typical example, take the proof of the fact that the sum of two objects is unique up to isomorphism. Suppose that C' is another sum of A and B, with injections α' and β'. Since C is a sum, there must exist $w : C \to C'$ such that α' and β' factor through it: $\alpha' = w \circ \alpha'$, and $\beta' = w \circ \beta'$. Since C' is also a sum, there exists $w' : C' \to C$ such that α and β factor through it. This easily implies that w and w' are inverse isomorphisms.

Universal properties are quite extensively used in category theory [6] where they allow to describe many important constructs along a basic uniform line: initial and final objects, injective and projective limits, and many more.

6 Diagrams as objects: from diagrammatic reasoning to reasoning about diagrams

Category theory gives a way of describing commonalities in many fields of mathematics in terms of arrows and objects. In fact, topos theory, which is part of category theory, can be used to replace set theory as an alternative foundation for mathematics.

The existence of the whole field of category theory is possible because of the fact that many fields of mathematics (if not all of them) can be expressed in terms of arrows and objects. A standard example is the definition of group objects: the axioms of a group can be expressed using diagrams. For example, the property expressing that group multiplication in a group G is associative can be expressed by the commutative diagram in Fig. 8, where id is the identity and $\mu : G \times G \to G$ is group multiplication. Similar diagrams can be used to express other group axioms.

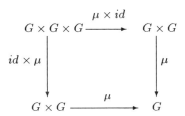

Fig. 8. Group multiplication is associative

An often tedious but necessary part of reasoning in category theory is concerned with routine diagram chasing. Part of it is considered as expressing rather "shallow" knowledge, and is often referred to somewhat disparagingly as "general nonsense".

The use of diagrams in category theory also illustrates how diagrams can be further enriched by endowing them with a richer semantics. In fact, there, diagrams live in categories, and the existence of particular kinds of diagrams expresses deep structural properties of the category itself.

7 Conclusions and perspectives

We have considered the informal non analogical use of diagrams in mathematics, in three domains of increasing semantic complexity. Although this use does not rely on a similarity of structure between the objects of study and the intended mathematical objects, contrary to what is the case in geometry or topology, analogous methods are used for reasoning in both cases, which exploit the immediate access to structure allowed by diagrammatic representations. Starting with the simple geometry of composition in commutative diagrams, we described how diagrams are used to express the input/output ratios in exact sequences, and also to represent encapsulated procedural knowledge when describing universal properties of objects.

The use of diagrams in mathematics sheds an interesting light on the potential fruitfulness of developing new paradigms for diagrammatic representations. In particular, many areas of reasoning where the intended interpretations are not of a spatial nature might be accessible to spatial varieties of representation. For instance, in the domain of temporal and spatial reasoning, the topology of temporal and spatial relations, of which the notion of conceptual neighborhood is an important aspect, can be shown to be representable by spatial structures in Euclidean space [5]. Another case in point is qualitative physics, where spatial regions can be used to represent configuration spaces [4].

We think that progress in the use of sophisticated spatial and diagrammatic representations in Artificial Intelligence will benefit from the study of widely used, if not widely known, similar enterprises in other fields of science. Both the study of informal methods of proof, as considered here, and of formalizations of reasoning techniques, as in [11] are worthy of further consideration and development.

References

[1] J. Barwise and J. Etchemendy. Heterogeneous logic. In J. Glasgow, N. Hari Narayanan, and B. Chandrasekaran, editors, *Diagrammatic Reasoning: Cognitive and Computational Perspectives*, pages 211–234. AAAI Press/ The MIT Press, Cambridge MA, 1995.

[2] G. Bredon. *Topology and Geometry.* Springer-Verlag, 1993.

[3] S. Eilenberg and N. Steenrod. *Foundations of algebraic topology.* Princeton University Press, Princeton, NJ, 1952.

[4] B. Faltings. Qualitative Spatial Reasoning Using Algebraic Topology. In *Proc. of COSIT'95*, LNCS 988, pages 17–30. Springer Verlag, 1995.

[5] G. Ligozat. Towards a general characterization of conceptual neighborhoods in temporal and spatial reasoning. In *AAAI-94 Workshop on Spatial and Temporal Reasoning*, Seattle, WA, 1994.

[6] S. MacLane. *Categories for the Working Mathematician.* Springer Verlag, 1971.

[7] W. S. Massey. *Algebraic topology: an introduction.* Springer Verlag, 1967.

[8] J.R. Munkres. *Elements of Algebraic Topology.* Addison-Wesley, 1984.

[9] E.E. Spanier. *Algebraic Topology.* Mc Graw-Hill, 1966.

[10] N. Tennant. The Withering Away of Formal Semantics. *Mind and Language*, 1, 1986.

[11] D. Wang, J. Lee, and H. Zeevat. Reasoning with diagrammatic representations. In J. Glasgow, N. Hari Narayanan, and B. Chandrasekaran, editors, *Diagrammatic Reasoning: Cognitive and Computational Perspectives*, pages 339–393. AAAI Press/ The MIT Press, Cambridge MA, 1995.

MAD: A Real World Application of Qualitative Model-Based Decision Tree Generation for Diagnosis

Heiko Milde, Lothar Hotz, Jörg Kahl, Bernd Neumann, and Stephanie Wessel

Laboratory for Artificial Intelligence, University of Hamburg
Vogt-Koelln-Str. 30, 22527 Hamburg, Germany
milde@kogs.informatik.uni-hamburg.de

Abstract. Computer diagnosis systems grounded on hand-crafted decision trees are wide-spread in industrial practice. Since the complexity of technical system increases and innovation cycles are shortened, the need for systematic decision tree generation and maintenance arises. In this paper, the MAD system is introduced which generates decision trees based on qualitative device models. Existing resources such as design data and expert design know-how as well as decision trees and diagnosis knowledge can easily be reused and integrated into decision tree generation. Since decision tree generation is based on device models, applying MAD reduces average fault identification cost and facilitates quality management of diagnosis equipment. Furthermore, cost of diagnosis system generation, modification and maintenance is reduced. We have successfully evaluated the MAD system in cooperation with the german forklift manufacturer STILL GmbH Hamburg.

1 Introduction

More than 100.000 forklifts made by the german company STILL GmbH Hamburg are in daily use all over Europe. In order to reduce forklift downtimes, approximately 1100 STILL service workshop trucks utilize decision-tree-based computer diagnosis systems for off-line diagnosis. Due to the complexity of the electrical circuits employed in forklifts, decision trees may consist of more than 5000 objects. When forklift model ranges are modified or new model ranges are released, decision trees are manually generated or adapted by service engineers who apply detailed expert knowledge concerning faults and their effects. Obviously, this practice is costly and quality management is difficult. Furthermore, average cost of decision-tree-based fault identification may be unnecessarily high because decision trees are not optimized. Hence, there is a need for computer methods to support systematic generation, modifications and optimization of diagnosis systems. The introduction of new diagnosis techniques, however, raises challenges.

- First, it is essential to integrate innovative with established concepts. A total redesign of existing diagnosis systems is usually unacceptable for economical reasons. In particular, for STILL, abandoning decision trees was not acceptable.
- Second, it is essential to utilize available resources such as expert knowledge and computer-based product data for diagnosis system generation. This way, the cost of diagnosis systems can be reduced and the trustworthiness of diagnosis data can be improved.

Model-based decision tree generation is a promising answer to the challenges noted above. In particular, model-based techniques facilitate the integration of available resources into the diagnosis equipment. Furthermore, grounding diagnosis systems on a model provides a systematic way for modification, reuse and optimization.

In our application, model-based approaches have to deal with electrical circuits of the automotive domain. These circuits usually consist of components which show a variety of different behavior types, such as analog, digital, static, dynamic, linear, nonlinear and software-controlled behavior. In principle, model-based techniques provide a systematic way for predicting the behavior of electrical circuits, including faulty behavior. However, adequate modeling of heterogeneous circuits is still a challenge.

In the STILL application scenario, diagnosis follows the branches of a decision tree. Nodes of a decision tree represent fault sets, edges are labeled by the tests (involving measurements, observations, display values and error codes) which must be carried out to verify the corresponding child node. Although the basic concepts of model-based generation of such decision trees are already described in [2] and [5], for the reader's convenience, we briefly outline the main ideas of the approach in the following. Due to the STILL application scenario, we focus on the electrical domain, although, in principle, dealing with devices of different technical domains such as hydraulics or mechanics is feasible.

The first step of model-based decision tree generation is to model a device. This step is supported by a component library and a device model archive (see Figure 1). Design data and knowledge from the design process (knowledge concerning intended device behavior, expected faults, and available measurements) are exploited in this step. In a second step, ok and faulty device behavior is predicted automatically by evaluating the device model. The third step is to build decision trees from behavior predictions. This step is supported by a decision tree archive and a cost model for the tests which can be performed. Decision tree generation can be performed automatically or guided by service know-how, i.e. knowledge concerning preferable decision-tree topologies and fault probabilities.

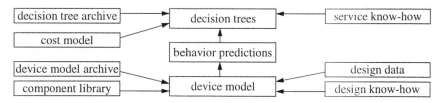

Figure 1. Basic concepts of model-based decision tree generation

In order to realize these concepts, we implemented the MAD system (Modeling, Analyzing and Diagnosing) whose main parts are described in this paper. Section 2 describes MAD's user interface which facilitates adequate device modeling. Furthermore, the internal representation of electrical circuitry is explained. In Section 3, MAD's model-based behavior prediction is described. Section 4 outlines the decision tree generation. The evaluation of the MAD system described in Section 5 was performed in cooperation with the STILL GmbH Hamburg.

2 Device Modeling

In this section, COMEDI, the user interface of the MAD system is presented and MAD's internal representation of electrical circuits is described.

2.1 COMEDI

COMEDI (COmponent Modeling EDItor), the user interface of MAD is similar to a CAD tool which hopefully assures a high degree of acceptance in industry (see Figure 2). For device modeling, COMEDI provides two different libraries, a device model archive and a component library (see Figure 1). The device model archive allows systematic reuse and modification of device models which were created during former modeling sessions. The component library contains different qualitative models of electrical components. A simplified COMEDI model of a forklift frontlight and backlight circuit is shown in Figure 2.

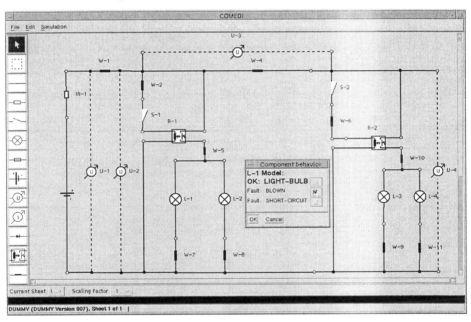

Figure 2. COMEDI model of forklift frontlight and backlight circuit

COMEDI models of simple components represent a single ok behavior mode and optionally several fault behavior modes. Modes of behavior are explicitly marked as correct or faulty. For instance, a COMEDI light bulb model consists of three different behavior modes, i.e. *ok: light-bulb, fault: light-bulb-blown*, and *fault: light-bulb-short-circuit* (see Figure 2).

There are electrical components with complex behavior, i.e. components with several operating modes. For instance, hand-operated switches and relays can be open or closed. In COMEDI, each operating mode of a component is described by a distinct model. Each model consists of a single ok behavior mode and (some) fault modes. Additionally, attached to each model, there is a model condition defining requirements under which the corresponding model holds. Alternative models of switches and relays and the corresponding model conditions are shown in Table 1. In this simple example, each component model shows only one faulty behavior mode.

Table 1. Alternative models of switches and relays and corresponding model conditions

component model	ok behavior mode	fault behavior mode	condition
switch model 1	ok: open	fault: stuck-closed	opened-manually
switch model 2	ok: closed	fault: stuck-open	closed-manually
relay model 1	ok: switch-open	fault: switch-stuck-closed	coil-passive
relay model 2	ok: switch-closed	fault: switch-stuck-open	coil-active

Model conditions are of two different types, i.e. internal conditions and input conditions. Input conditions relate to inputs of the investigated technical device. Hence, in order to model certain device input the user of COMEDI can choose the corresponding component model. For instance, in order to model a hand-operated closed switch, a CO-MEDI user selects switch model 2 (see Table 1).

Internal model conditions relate to internal parameters of the device. For instance, the internal relay model condition *coil-active* (see Table 1) relates to the current through the relay coil. Section 3.2 outlines MAD's automatic behavior prediction and its treatment of alternative behavior models with associated internal conditions.

Since in COMEDI, component behavior is described in a language similar to the way engineer's think about component behavior, design experts can handle the modeling task. This reflects the insight that the design of modern technical systems and of appropriate innovative diagnosis systems is inseparable. In particular, given certain components or subcircuits, knowledge concerning ignorable physical effects as well as know-how about intended behavior is essential to model devices at an adequate degree of abstraction.

Qualitative modeling is adequate because, usually, faults and symptoms are described qualitatively in this domain. Furthermore, qualitative techniques allow to handle parametric variants of a device without changing the device model and, thus, the complexity of diagnosis equipment is reduced. As another point, qualitative modeling reduces the number of fault models because, often, a class of different faults is represented by only one qualitative model. In MAD, dealing with a small number of different fault models is essential because the number of faults determines the size of the decision tree and the computational efforts to generate it. Additionally, using qualitative techniques, electrical circuits can be modeled at an adequate degree of abstraction what is necessary to deal with complex circuits that consist of a large number of components. Thus, for model-based decision tree generation, quantitative network analysis such as SPICE [1] seems to be problematic.

For modeling devices, in COMEDI, component models can be easily combined because of their local internal behavior descriptions (no-function-in-structure principle, [4]) presented in the following subsection.

2.2 Standard Components

Internally, COMEDI models are mapped to formalized standard components showing well defined and idealized behavior. MAD provides four different standard components, i.e. idealized voltage sources, consumers, conductors and barriers. The behavior of idealized voltage sources is well-known from electrical engineering. Consumers are passive and their current/voltage characteristic is monotonous. Idealized conductors do not allow any voltage drop while idealized barriers do not allow any current. Standard components can be connected in combinations of series, parallel, star and delta groupings. This simple internal representation of electrical circuits is sufficient for the following reasons.

- In STILL service workshops, only steady state diagnosis of electrical circuits is performed. Therefore, only steady state behavior of physical components has to be represented in component models. In particular, an explicit representation of temporal dependencies is not necessary.
- A small number of qualitative standard components suffices, because, often, different physical components show similar electrical behavior, i.e. their current/voltage characteristics differ only slightly. Qualitative versions of these current/voltage characteristics are frequently identical.
- MAD's standard components are deliberately selected so that important behavior classes of the application domain can be represented adequately.

Due to analogies between electrics, mechanics and hydraulics, the internal MAD representation is, in principle, also adequate for other technical domains.

2.3 Qualitative Parameter Representation

In electrical circuits, faults may modify component behavior or may even change circuit structures. Hence, heterogeneous symptoms, such as slight deviations of parameter values or total loss of functionality may occur. In general, any circuit behavior that is different from the expected behavior can be a fault symptom. Thus, representing actual parameter values as well as deviations from reference values is helpful to characterize faults and symptoms adequately.

However, MAD's qualitative parameter representation consists of *three* attributes, i.e. actual value, deviation value and reference value. At first sight, this representation may seem to be redundant because *actual value = reference value + deviation* holds. However, for qualitative value spaces this is not necessarily true since a certain qualitative deviation may lead to more than one possible actual value. For example, consider the situation that the reference value of a certain parameter is known to be *positive* and the deviation is *negative*. In this case, the corresponding actual value may be *positive, zero* or *negative*. Hence, qualitative computations can be sharpened if all three attributes are carried along. In [7], we elaborate on this topic.

Table 2 and 3 show attributes and corresponding qualitative value sets of currents and voltages. The semantics of the qualitative values should be obvious. Note that in MAD's internal models of electrical devices, infinite current values may occur because MAD provides idealized voltage sources and idealized conductors as standard components. MAD's set of standard components does not include idealized current sources. Thus, in MAD's internal device models, voltages show certain limits and voltage values beyond these limits can be considered as *impossible* values.

Table 2. Qualitative representation of currents

attributes	qualitative values
actual value / reference value	negative-infinite, negative, zero, positive, positive-infinite
deviation value	negative, zero, positive

Table 3. Qualitative representation of voltages

attributes	qualitative values
actual value / reference value	negative-infinite, negative-impossible, negative-maximum, negative-between, zero, positive-between, positive-maximum, positive-impossible, positive-infinite
deviation value	negative, zero, positive

Due to the MAD's explicit representation of voltage limits, in principle, dealing with logical circuits is possible. For instance, logical values (*low, high*) can be mapped to MAD's voltage values *zero* and *positive-maximum*. Furthermore, MAD's qualitative voltage representation allows to handle electrical devices showing more than one source. In particular, the representation of impossible voltage values paves the way to define a qualitative version of the superposition principle well-known from electrical engineering. Dealing with logical values as well as handling multiple sources is the basis for dealing with hybrid systems consisting of both analog and digital subsystems.

3 Automated Behavior Prediction

In this section, MAD's computation of qualitative values is briefly described. A detailed description can be found in [7]. Furthermore, in this section, MAD's generation of fault-symptom tables is summarized.

3.1 Computation of Qualitative Values

In order to compute qualitative current and voltage values, local propagation methods have been investigated [9]. Since detailed studies proved that local propagation in electrical networks is inappropriate, we follow a different approach first presented in [6]. Networks are transformed into trees representing the network structure. In particular, series, parallel, star and delta groupings are represented explicitly. Exploiting these structure trees, qualitative behavior can in fact be computed by local propagation. Unlike other approaches such as QCAT [8], the SPS method [6], and the Connectivity method [10], MAD offers certain features to improve the accuracy of qualitative prediction. This is explained in the following.

- First of all, rather than relying on qualitative versions of basic arithmetics, MAD computes qualitative values for currents and voltages by a set of qualitative operators which are qualitative versions of complex quantitative equations. In effect, these equations describe behavior of series, parallel, star and delta groupings. Utilization of complex operators avoids multiple applications of simple operators and, thus, avoids spurious predictions. For instance, a voltage divider operator is invoked to compute qualitative voltage values instead of determining current values first and computing voltage values from current values in a second step. In principle, for network analysis, a limited number of operators suffices because MAD's internal representation of electrical circuits offers a limited number of standard components and elementary network structures.

- Second, qualitative operators are defined by applying the corresponding quantitative equation to the interval boundaries which represent actual values and reference values of input parameters. Resulting boundaries represent the corresponding qualitative values of output parameters. Qualitative deviation values are computed from actual and reference values. Additionally, output deviation values are inferred from input deviation values, assuming that parameter dependencies are monotonous. Operators are represented by a set of tables comprising more than 30.000 entries which had to be generated by computer in order to secure reliability. Due to the properties of this qualitative calculus, spurious solutions do not occur at all if the network can be structured into series and parallel groupings of standard components.

- Third, in addition to local propagation of qualitative values, MAD globally analyses network structures and structure trees in order to eliminate (some) spurious predictions. For instance, a global analysis of the network structure allows to determine current directions. Knowledge about current directions can be used to eliminate certain qualitative current values. Therefore, global network analysis may prevent spurious current predictions.

3.2 Dealing with Complex Component Behavior

As stated in Section 2, there are electrical components whose behavior depends on internal parameter values, e.g. the behavior of a relay switch (open / closed) depends on the current through the relay coil. In COMEDI, these components are described by sets of alternative behavior models with associated internal conditions. MAD's dealing with these components is similar to QCAT. In a first step, for each of these components, one of its alternative models is instantiated. Second, qualitative voltage and current values are computed. Third, internal conditions are verified. If an internal condition is violated, one component model is changed and computation of qualitative values is restarted. If all internal conditions are fulfilled, the steady state behavior prediction with the chosen set of models is successful. Steady state behavior prediction fails if all possible combinations of alternative models lead to violated model conditions as may happen in the case of instable behavior. This case is explicitly reported to MAD users.

3.3 Generation of Fault-symptom Tables

In order to generate decision trees, behavior predictions are performed for all operating modes, faults, and fault combination for which diagnosis support is required. For each operating mode and fault assumption, all symptoms (measurements, observations, error codes, display values) are computed which are in principle available for diagnosis. The output of the prediction step is model-based diagnosis knowledge in form of an extensive table of fault-symptom associations. This table is the basis for decision tree generation. For the forklift frontlight and backlight circuit, MAD generates the fault-symptom table shown in Figure 3.

Behavior modes	Measurements: U-1_S1-G-S2-G	U-4_S1-G-S2-G	U-3_S1-G-S2-G	U-2_S1-G-S2-G
IR-1-is_BATTERY-LOSSY	L_PB_US6 [0 4]	L_PB_US6 [0 4]	N_0_US5 0	L_PB_US6 [0 4]
W-1-is_RESISTOR-TOO-HIGH	H_PB_US6 [3 5]	L_PB_US6 [0 4]	N_0_US5 0	L_PB_US6 [0 4]
W-2-is_RESISTOR-TOO-HIGH	N_PB_US6 [3 4]	N_PB_US6 [3 4]	N_0_US5 0	N_PB_US6 [3 4]
W-5-is-BROKEN-WIRE	N_PB_US6 [3 4]	N_PB_US6 [3 4]	N_0_US5 0	N_PB_US6 [3 4]
W-5-is_RESISTOR-TOO-HIGH	N_PB_US6 [3 4]	N_PB_US6 [3 4]	N_0_US5 0	N_PB_US6 [3 4]
L-1-is_LAMP-BLOWN	N_PB_US6 [3 4]	N_PB_US6 [3 4]	N_0_US5 0	N_PB_US6 [3 4]
L-2-is_LAMP-BLOWN	N_PB_US6 [3 4]	N_PB_US6 [3 4]	N_0_US5 0	N_PB_US6 [3 4]
W-7-is-BROKEN-WIRE	N_PB_US6 [3 4]	N_PB_US6 [3 4]	N_0_US5 0	N_PB_US6 [3 4]
W-7-is_RESISTOR-TOO-HIGH	N_PB_US6 [3 4]	N_PB_US6 [3 4]	N_0_US5 0	N_PB_US6 [3 4]
W-8-is-BROKEN-WIRE	N_PB_US6 [3 4]	N_PB_US6 [3 4]	N_0_US5 0	N_PB_US6 [3 4]
W-8-is_RESISTOR-TOO-HIGH	N_PB_US6 [3 4]	N_PB_US6 [3 4]	N_0_US5 0	N_PB_US6 [3 4]
W-4-is_RESISTOR-TOO-HIGH	H_PB_US6 [3 5]	L_PB_US6 [0 4]	H_PB_US5 [3 5]	H_PB_US6 [3 5]
W-6-is_RESISTOR-TOO-HIGH	H_PB_US6 [3 5]	H_PB_US6 [3 5]	N_0_US5 0	H_PB_US6 [3 5]
W-10-is_RESISTOR-TOO-HIGH	H_PB_US6 [3 5]	H_PB_US6 [3 5]	N_0_US5 0	H_PB_US6 [3 5]
L-3-is_LAMP-BLOWN	H_PB_US6 [3 5]	H_PB_US6 [3 5]	N_0_US5 0	H_PB_US6 [3 5]

Figure 3. Fault-symptom associations for forklift frontlight and backlight circuit

4 Decision Tree Generation

MAD offers three different possibilities to generate decision trees. First, based on fault-symptom tables, decision trees can be created fully automatically. Second, decision trees from archives can be reused. Third, in order to permit manual adaption and modification of decision trees, MAD offers basic editing operations, such as moving a certain fault from one fault set to another and recomputing the corresponding tests. In the following, automated decision tree generation is presented in more detail. One can choose from the following criteria to guide decision tree generation.

- *Minimization of average diagnosis cost.* Automated decision tree generation uses the well-known A*-algorithm [3] to select the tests minimizing the average diagnosis cost according to a cost model which specifies the cost for each test.
- *Grouping by observations, error codes, display values.* Decision trees are generated such that subsets of faults correspond to a prespecified symptom. For instance, all faults are grouped together which cause the frontlights not to shine correctly.
- *Grouping by aggregate structure.* If the aggregate structure of the device is known, decision trees can be generated such that subsets of faults correspond to the same physical aggregate. For instance, faults occurring on a certain board may be grouped together.

Figure 4 shows a decision tree for the forklift frontlight and backlight circuit. This decision tree was generated automatically, guided by the criterion *minimization of average diagnosis cost.* Model-based prediction and automated decision tree generation guarantee, that decision trees are correct and complete with respect to the underlying device model. All faults considered in the device model occur in the generated decision tree, and tests are selected correctly to discriminate fault sets. This holds even if decision trees are modified manually because the editor enforces complete coverage of all faults and correct test assignments. Furthermore, average diagnosis cost is minimal within the constraints imposed by a prespecified decision tree structure.

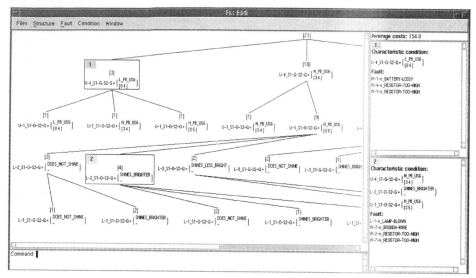

Figure 4. Decision tree for forklift frontlight and backlight circuit

5 Evaluation and Conclusions

The MAD system generates diagnostic decision trees based on a new method for qualitative electrical network analysis which allows accurate behavior predictions for the following reasons. First, MAD's internal standard components represent important behavior types of the electrical domain. Furthermore, since qualitative values describe actual values as well as deviations from reference values, faults and symptoms can be adequately characterized. As another point, exploitation of network structures and certain features to avoid spurious solutions (see Section 2 and 3) assure precise behavior predictions. Using MAD, existing resources such as design data and expert design know-how as well as decision trees and diagnosis knowledge can easily be reused and integrated into decision tree generation.

In cooperation with the STILL GmbH Hamburg, we have evaluated the MAD system in the application scenario and found that using the modeling techniques of MAD with some extensions regarding electronic control units, more than 90% of the faults of the current hand-crafted diagnosis system can be handled successfully. The prototypical implementation allows model-based behavior prediction and automatic generation as well as manual modification of decision trees. Furthermore, we successfully integrated these decision trees into existing STILL diagnosis systems.

Computer-based decision tree generation is a challenging task, because decision trees are wide-spread in industry. With model-based decision tree generation a systematic way for diagnosis system generation has been developed providing the benefits of reduced cost for diagnosis system generation, modification, and maintenance, improved quality management and cost optimal fault identification.

Acknowledgments

We would like to thank our partners in INDIA for their valuable contributions. This work has been supported by the Bundesministerium für Bildung, Wissenschaft, Forschung und Technologie (BMBF) under the grant 01 IN 509 D 0, INDIA.

References

1. Banzhaf, W.: Computer-aided circuit analysis using SPICE, Prentice Hall, Englewood Cliffs, New Jersey (1989)
2. Friedrich, G., Nejdl, W.: Generating Efficient Diagnostic Decision Tree Procedures for Model-Based Reasoning Systems, in: Proc. DX-89, International Workshop on Model-based Systems, Paris, France (1989)
3. Hart, P. E., Nilsson, N. J., Raphael, B.: A formal basis for the heuristic determination of minimum cost paths, in: IEEE Transactions on Systems Science and Cybernetics, SSC-4(2): 100 - 107 (1968)
4. de Kleer, J., Brown, J. S.: A Qualitative Physics Based on Confluences, in: AI Journal (1984)
5. Mauss, J.: Analyse kompositionaler Modelle durch Serien-Parallel-Stern Aggregation, DISKI 183, Dissertationen zur Künstlichen Intelligenz (1998)
6. Mauss, J., Neumann, B.: Qualitative Reasoning about Electrical Circuits using Series-Parallel-Star Trees, in: Proc. QR'96, 10th International Workshop on Qualitative Reasoning about Physical Systems (1996)
7. Milde, H., Hotz, L., Kahl, J., Wessel, S.: Qualitative Analysis of Electrical Circuits for Computer-based Diagnostic Decision Tree Generation, submitted to DX-99, 10th International Workshop on Principles of Diagnosis (1999)
8. Pugh, d., Snooke, N.: Dynamic Analysis of Qualitative Circuits, in: Proc. of Annual Reliability and Maintainability Symposium, 37 - 42, IEEE Press (1996)
9. Struss, P.: Problems of Interval-based Qualitative Reasoning, in: Readings in Qualitative Reasoning about Physical Systems, Weld, D., de Kleer, J. (Eds.), Morgan Kaufmann (1990)
10. Struss, P. Malik, A., Sachenbacher, M.: Qualitative Modeling is the Key, in: Proc. DX-95, 6th International Workshop on Principles of Diagnosis (1995)

WebMaster: Knowledge-Based Verification of Web-Pages

Frank van Harmelen[1] and Jos van der Meer[2]

[1] AIdministrator & Vrije Universiteit Amsterdam, frankh@cs.vu.nl
[2] AIdministrator, Jos.van.der.Meer@aidministrator.nl

Abstract. Maintaining contents of Web sites is an open and urgent problem on the current World Wide Web. Although many current tools deal with problems such as broken links and missing images, very few solutions exist for maintaining the *contents* of Web sites. We present a *knowledge-based approach to the verification of Web-page contents*. The user exploits semantic markup in Web-pages to formulate rules and constraints that must hold on the information in a site. An inference engine subsequently uses these rules to categorise Web-pages in an ontology of pages, while the constraints are used to define categories of pages which contain errors. We have constructed WebMaster, a software tool for knowledge-based verification of Web-pages. WebMaster allows the user to define rules and constraints in a graphical format, and is then able to use these rules to detect outdated, inconsistent and incomplete information in Web-pages. In this paper, we describe the various options for semantic markup on the Web, we define a precise logical and graphical format for rules and constraints, and we report on our practical experiences with WebMaster.

1 Introduction

Maintaining contents of Web-pages is an obvious open problem. Anybody who has used the WWW has experienced the amounts of outdated, missing, and inconsistent information on many Web sites, even on those sites that are of crucial importance to individuals, companies or organisations. Websites are large, frequently updated, and constructed by multiple authors. All this makes it impossible to do manual maintenance on contents of Websites. In this paper we describe a software tool (provisionally named WebMaster) which supports an important aspect of Web maintenance, namely *verification*: the location of errors in Web-page contents. The functionality of WebMaster is in sharp contrast with most existing Web site maintenance tools. These existing tools deal with problems such as broken links, missing images, incorrect HTML etc, but unlike WebMaster, they do not analyse the semantic contents of the site.

---

[0] The work reported in this paper has only been possible with the contributions from all current and past members of the WebMaster team at AIdministrator: Jan Bakker, Chris Fluit, Herko ter Horst, Walter van Iterson, Arjohn Kampman and Gert-Jan van de Streek.

A complication is that the data typically found on Web sites is only weakly structured. This makes computer support for Web-site maintenance hard to provide. For databases, methods have been developed to maintain the quality of high volumes of fast changing information from multiple providers, but the key to these methods is the very strict structure that a database imposes on its contents. These methods are not applicable to weakly structured data. In this paper we present an approach which *does* apply to weakly structured data.

WebMaster is intended to solve three types of problems that occur in Web sites: outdated information, missing information, and inconsistent (i.e. contradicting) information. A category of problems that is missing from this list concerns *incorrect information*. We have deliberately omitted it from the list of error-categories that WebMaster can locate. The reason for this is as follows: Each of the three above error-categories can be identified on the basis of knowledge about the Web site alone: inconsistent information can be detected by comparing different locations within the same site; missing information can be detected on the basis of rules stating which types of information must occur within a site; and outdated information can be identified by comparing temporal statements in the Web site with the external time. Because of this, each of these categories can indeed be identified effectively, as we will show in the remainder of this paper. For "incorrect information" however, it would be necessary to compare the contents of a Web site with the actual state of affairs in the external world described by the site. This would require a full "world-model" of the world described by the site, while the other three categories only require reasoning about a model of the Web site contents, and not of the external world

Summarising, the aim of WebMaster is to support the maintenance of the contents of Web sites by verifying the contents of sites for outdated, missing and inconsistent information.

This approach is indeed "knowledge-based": the information providers use their domain specific knowledge to express which constraints should be imposed on the Web site contents. As usual in knowledge-based approaches, a strong point is that these integrity constraints can be highly domain specific since they are provided by domain experts, such as the information provider themselves.

2 Semantic Markup

In order to express integrity constraints on the contents of Web pages, an information provider must be able to refer to the semantic contents of such Web pages in a machine-accessible way. As has been pointed out by many authors [11, 4, 10], HTML pages as currently encountered on the Web are unsuitable for this purpose.

One of the results of a general push towards more semantic structure on the Web has been the development of the XML markup language XML allows Web-page creators to use their own set of markup-tags. These tags can be chosen to reflect the domain specific semantics of the information, rather than merely its lay-out. Fig 2 shows the same piece of information in HTML-markup (fig. 2a) and in semantically well chosen XML (fig. 2b).

From the example it is clear that in XML we can now recognise pieces of information

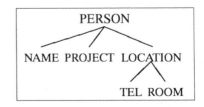

Fig. 1. XML markup as a labelled tree

such as a person's name or telephone number. In essence, XML allows us to structure Web-pages as labelled trees, where the labels can be chosen by the information provider to reflect as much of the documents semantics as is required. The labelled tree for fig. 2b is shown in figure 1.

```
<H3>F. van Harmelen</H3>
works for project
<B>WebMaster</B>
and can be reached at


tel. <I>47731</I>
or in room <I>T3.57</I>
```

```
<PERSON>
 <NAME>F. van Harmelen</NAME>
 works for project
 <PROJECT>WebMaster</PROJECT>
 and can be reached at
 <LOCATION>
  tel. <TEL>47731</TEL>
  or in room <ROOM>T3.57</ROOM>
 </LOCATION>
</PERSON>
```

Fig. 2. HTML and XML markup

Summarising, we can say that markup to indicate the semantics of Web-page contents is a necessary requirement for expressing rules and constraints on this contents. Such semantic markup can be expressed in XML (which has been designed with this specific purpose in mind).

3 Ontologies: types and constraints

Now that we know how to express semantic markup in Web-pages, the next step in the knowledge-based verification of Web-page contents is to express rules and constraints on this contents. These rules and constraints will capture the users knowledge on the required contents of these pages, and will be used by an inference engine to determine potential errors (constraint violations) in the site. In this section, we will describe the formalism we have developed for expressing rules (to categorise pages into types) and constraints (to determine potential errors in pages).

3.1 Types: describing an ontology of Web-pages

As the first step towards identifying errors in a Web site, it is useful to divide Web-pages into categories, where pages within a given category share certain properties. By organising such categories in a hierarchy of subcategories, we get a type-hierarchy (or: ontology). An example is shown in figure 3. Such

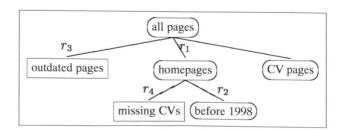

Fig. 3. An example page-ontology

ontologies are well understood and often used as modelling devices in fields such as Knowledge Engineering and Software Engineering.

The top of WebMaster's ontologies is always the type of "all pages" in the site (i.e. the universal type). Ontologies must be a tree (i.e. no multiple supertypes are allowed). To allow maximal flexibility in ontological modelling, we do *not* require subtypes to be either exhaustive or exclusive.

3.2 Constraint-types: describing categories of errors

Constraint types are special types meant to indicate error-categories. Whereas normal types group together all pages that share a given property, constraint types group together all pages that *fail* to satisfy a given property, where this property is again specified in the rule defining the constraint type. For example in fig. 3, the type "missing CVs" consists of all homepages (ie. pages satisfying rule r_1) which *fail* to satisfy rule r_4. In fig. 3, constraint types are indicated by rectangles and normal types by rounded boxes. Typically, normal types are used to group pages into meaningful categories, while constraint types are used to collect pages that contain a particular type of error.

Both normal types and constraint types can be further divided into subtypes. This is shown in fig. 3 for normal types, but it is often also useful for constraint types. This allows to subdivide error-types into gradually more refined and smaller types of errors.

4 Rules

As mentioned above, both types and constraint-types are defined intensionally by rules that express which properties must hold (or fail to hold) on a page. In this section we will describe the formalism that is used in WebMaster for expressing such rules. As already mentioned earlier, these rules will refer to the semantic markup of the pages that must be categorised. More precisely, the rules will be phrased in terms of the labelled tree structure of these pages (fig. 1).

A trade-off must be struck between the expressiveness of these rules (to allow the information providers to express powerful constraints) and the efficiency with which the rules can be tested on specific Web-pages by the inference engine. Following a suggestion in [19], we have chosen the following general logical form for our rules and constraints:

$$\forall x [\exists y \bigwedge_i P_i(x_k, y_l)] \rightarrow [\exists z \bigwedge_j Q_j(x_k, z_m)] \tag{1}$$

where the x, y and z are sets of variables, and each of the P_i and Q_j are binary predicates. The variables may be quantified over a given type T, for which we will use the notation $\forall x_k \in T$ (and similar for $\exists$). The binary predicates P_i and Q_j can express one of the following types of relations:

- *Arbitrary nesting of tags:* The predicate $descendant(\langle \text{TAG} \rangle x \langle /\text{TAG} \rangle, y)$ is true *iff* the tagged structure
  $\langle \text{TAG} \rangle x \langle /\text{TAG} \rangle$ occurs somewhere within y. For example, if we take for y the XML text of fig. 2b, then $descendant(\langle \text{TEL} \rangle 47731 \langle /\text{TEL} \rangle, y)$ is true.
- *Direct nesting of tags:* The predicate $child(\langle \text{TAG} \rangle x \langle /\text{TAG} \rangle, y)$ is true *iff* the tagged structure $\langle \text{TAG} \rangle x \langle /\text{TAG} \rangle$ is one of the direct children of y. If we again take for y the text of fig. 2b, then $child(\langle \text{NAME} \rangle \text{F. van Harmelen} \langle /\text{NAME} \rangle, y)$ is true, but not $child(\langle \text{TEL} \rangle 47731 \langle /\text{TEL} \rangle, y)$

– *Simple binary operations:* We will also need simple binary tests on tag-contents or on entire page. We will use the following in the remainder of this paper:
  - string-tests on tag-content such as string equality, substring, initial-substring, etc.
  - comparisons on ordered types such as integers, clock- and calendar-times, etc.
  - tests on links between pages such as direct and indirect links between pages.

This class of formulae is less expressive than full first order logic over the predicates P_i and Q_j (because of the limited nesting of the quantifiers), but is more expressive than Horn Logic (because of the existential quantifier in the right-hand side of the implication). As examples of such rules, we will now give some of the rule-definitions required for the ontology of fig. 3.

Rule r_1: Homepages. As the simplest example possible, let us assume that homepages can be identified simply because they contain the tag $\langle \text{HOMEPAGE} \rangle$ somewhere inside:

$$\forall x \in \textit{all-pages} : \top \rightarrow \exists z : descendant(\langle \text{HOMEPAGE} \rangle z \langle /\text{HOMEPAGE} \rangle, x)$$

This rule simply demands the presence of the $\langle \text{HOMEPAGE} \rangle$-tag anywhere in the page. According to fig. 3, any page fulfilling this demand will be a member of the type *homepages*. In terms of the general schema above, the sets x and z consist of just a single variable, the sets y and P_i are empty (so the left-hand side of the implication is trivially true, indicated by $\top$), and only one Q_i predicate is used, namely $descendant(\cdot)$.

Clearly, this example rule is so simple that it could still have been performed by a plain text-search engine (simply searching for the string "$\langle \text{HOMEPAGE} \rangle$"). The second example already goes beyond the capabilities of a text-based search engine:

Rule r_2: Homepages before 1998. Any homepage which is last modified before 1 Jan 1998 can be identified easily, assuming that homepages contain a $\langle \text{MODIFIED} \rangle$-tag mentioning the last modification date of the page:

$$\forall x \in \textit{home-pages} : \top \rightarrow \exists d : descendant(\langle \text{MODIFIED} \rangle d \langle /\text{MODIFIED} \rangle, x) \wedge \\ d < \texttt{01-01-1998}$$

Since the test on the date $d < \texttt{01-01-1998}$ should only be applied to dates that appear in the context of a $\langle \text{MODIFIED} \rangle$, this simple rule already goes beyond the capabilities of a text-based search engine.

Rule r_3: Outdated pages. Obviously, dates mentioned in announcements must be in the future, or more precisely: *if* any page contains anywhere an $\langle \text{ANNOUNCE} \rangle$-tag, *and if* that $\langle \text{ANNOUNCE} \rangle$-tag directly contains a $\langle \text{DATE} \rangle$-tag, *then* that date must be in the future:

$$\forall x \in \textit{all-pages} \, \forall d : [\exists a : descendant(\langle \text{ANNOUNCE} \rangle a \langle /\text{ANNOUNCE} \rangle, x) \wedge \\ child(\langle \text{DATE} \rangle d \langle /\text{DATE} \rangle, a)] \\ \rightarrow d > today().$$

Since this definition concerns a constraint-type in figure 3, any page that fails to satisfy the above demand on announcement-dates belongs to the constraint-type of *outdated-pages*

Rule r_4: Missing CV-page. As discussed earlier, we might require that every homepage must link to the CV page of the corresponding employee. This property of home-pages can be enforced by the following constraint rule:

$$\forall h \in \text{home-pages} \, \forall n : \quad [\exists p : \quad descendant(\langle \text{PERSON} \rangle p \langle / \text{PERSON} \rangle, h) \wedge$$
$$child(\langle \text{NAME} \rangle n \langle / \text{NAME} \rangle, p)]$$
$$\rightarrow$$
$$\exists c \in CV\text{-pages}$$
$$\exists p' \, \exists n' : \quad [descendant(\langle \text{PERSON} \rangle p' \langle / \text{PERSON} \rangle, c) \wedge$$
$$child(\langle \text{NAME} \rangle n' \langle / \text{NAME} \rangle, c) \wedge$$
$$n = n' \wedge links\text{-}to(h, c)]$$

This rule states that *if* a homepage contains anywhere within it a $\langle \text{NAME} \rangle$-tag occurring directly inside a $\langle \text{PERSON} \rangle$-tag (i.e. the left-hand side of the rule), *then* (i) there should exist a CV-page[1] which (ii) should contain a $\langle \text{NAME} \rangle$ tag which directly inside a $\langle \text{PERSON} \rangle$-tag, (iii), the names appearing in both pages should be equal, and (iv) there should be a link from the home-page to the CV-page. Again, this rule defines a constraint type in fig. 3, so any homepage violating this constraint will belong to the constraint-type *missing-CVs*.

Rule r_5: link back to root. A final example shows that our rules and constraints can also be used to check the connectivity in a site. Consider the following example:

$$\forall p \in all \; pages : links\text{-}to(p, \texttt{index.html}).$$

Used as a normal rule, this defines the type of all pages which have a direct link back to the root-page of the site (i.e. a "back to root" button). Used as a constraint, it states that all pages *must* have such a "back to root" button, and collects all pages that fail to satisfy this requirement.

Clearly, we cannot expect the average Web-page builder to be able to express such logical formulae. Instead, we have developed a graphical notation for this purpose. This allows expressions like the above to be stated in a diagrammatic form which is intuitive for the information provider. These diagrams abstract from logical details, and yet are 1-to-1 translatable with the underlying logical formulae. In this paper we do not describe this graphical notation, but an example of this notation is shown in figure 4. This figure shows the graphical representation of rule r_4.

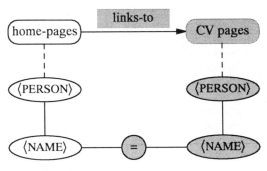

Fig. 4. Graphical notation of rule r_4.

5 Inference Engine

Of course the whole point of having an intensionally defined ontology as in fig. 3, defined by rules and constraints as defined above is to have these rules and constraints

---

[1] This rule assumes an existing definition for the type of CV-page's.

executed by an inference engine. This will deduce which Web-pages belong to which types. Type membership of constraint types is particularly relevant, since such constraint types indicate pages with errors.

We have constructed such an inference engine for the rule format defined in the previous section. Our current experience indicates that for this rule format, no sophisticated deduction-technology is required, and a simple generate-and-test engine (with some optimisation for the interleaving of the generate- and test-steps) is sufficient for practical ontologies on sites of up to a thousand pages.

6 Practical experience

6.1 Implementation

We have implemented a version of WebMaster which realises the full verification process. It enables users to graphically construct an ontology as in figure 3, allowing for both normal types and constraint types. For each of these types, the user can graphically define the rule or constraint that must be used to determine the type-membership. Once these rules have been defined, the inference engine can be called to determine type-membership either for indicated types or for the entire ontology. The system displays a graphical map of the site that is being verified, and uses this map to indicate which pages belong to a given type. Figure 5 shows a screen-snapshot of WebMaster. It displays a graphical map of the site showing pages and the links between them (top-left window), a type-hierarchy of pages (broad center window), and a rule used to define one of the types in the hierarchy (to the right of the site-map).

6.2 Re-engineering Ontobroker

As described above, our knowledge-based approach to verification of Web-sites works best on Web-pages with semantic markup. (Whether this markup is expressed in XML or in HTML is irrelevant). However, such semantically marked-up pages are hard to find on the current Web. An important exception is the $(KA)^2$-Ontobroker project [1, 11]. In this project, Knowledge Engineering researchers have developed an ontology to describe their own community, and have annotated a set of existing Web-pages (typically home-pages of individual researchers, research groups and research projects), using an HTML-compliant syntax which allows expressions from F-logic [13] (objects, classes and attribute-value pairs). These annotated pages are subsequently used to answer queries posed to an F-logic inference engine that has access to both the ontology and the pages.

We have used WebMaster's rule-language to recreate much of the $(KA)^2$ ontology. Membership of ontological classes in Ontobroker is mostly specified extensionally, so defining the ontological classes was simply a matter of recognising the relevant Ontobroker markup. Some of the $(KA)^2$ classes are defined intensionally via rules. Most of these could again be modelled using WebMaster's rule-formalism. Finally, we reconstructed many of the Ontobroker queries from the Ontobroker home-page[2]. We formulated each Ontobroker query as a WebMaster rule, after which Ontobroker's answer-set to the query corresponded to the WebMaster type defined by the rule.

---

[2] http://www.aifb.uni-karlsruhe.de/WBS/broker/help.html

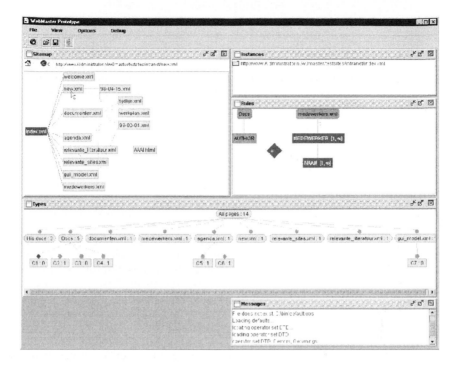

Fig. 5. Screen snapshot of WebMaster

6.3   Application to plain HTML

Given the lack of Web-pages with semantic markup, we have also experimented with the application of WebMaster to plain HTML-pages, which only contain structural or layout markup but no semantic markup. What did come as a surprise to us was the amount of useful rules and constraints that we were able to formulate on plain HTML-pages, not using any semantic markup whatsoever. Constraints concerning reachability of pages (e.g. distance from root), layout of pages (e.g. conformance to house style), HTML usage (e.g. frames or scripts) and page-updates (e.g. find all recently updated pages) could all be expressed in simple WebMaster rules.

7   Related Work

In this section we discuss a number of related projects, which all exploit machine-accessible semantics in Web-pages for various purposes.

7.1   SHOE

SHOE ("Simple HTML Ontology Extension", [15]), is an extension of HTML in which one can define ontologies and use these ontologies to annotate HTML pages. There are a number of important differences with WebMaster's notion of ontology. Firstly, WebMaster derives type-membership on the basis of required properties as stated in rules while in SHOE type-membership is mostly stated extensionally. Secondly, type objects in SHOE can be parts of an HTML page, while in WebMaster, only entire pages are members of types. Thirdly, SHOE's annotation language allows for attribute-value pairs to be associated with objects, while no such notion exists in WebMaster. Fourthly, although the SHOE markup language does allow rules to be stated for intensionally defined type-membership (in the form of negation-free Horn clauses), this notion is currently hardly used in SHOE, while in WebMaster, these rules are central to the entire approach.

7.2 Strudel

In section 1 we distinguished three different approaches to the problem of Web-site maintenance. One of these was to construct sites from information residing in data-bases, thereby reducing the problem of Web-site maintenance to that of data-base main-tenance. Strudel [12] is a system which takes exactly this approach. In Strudel, infor-mation is taken from a variety of (possibly weakly-structured) data-sources. Such infor-mation is put into a structured format using specialised wrappers for each information source. Queries in a specialised query-language are then formulated on this structured information. These queries compute the required combinations of information-elements that must be presented in the Web-site. Finally, the results of these queries are turned into HTML-presentations using predefined HTML-templates.

The approach as taken Strudel does indeed guarantee the correct *structure* of Web-sites, since this structure is machine-generated on the basis of user-specified queries. However, the *contents* of the information in the site is not covered in the Strudel ap-proach: the information is taken at face value from the various data-sources. In general, these data-sources are weakly structured and very heterogeneous, so verification of a combination of such information sources is a serious problem.

8 Summary and conclusions

Outdated, inconsistent and incomplete information is all too frequently encountered on the current WWW. In this paper, we have argued that knowledge-based verification techniques can be used to locate such errors. The knowledge-based approach exploits has a number of advantages over other potential approaches: in contrast with guarded updates, it leaves does not constraint the freedom of the page-author. Compared with generating Web-sites from a database, the knowledge-based approach can deal with much less strictly structured information, which is an important advantage in the context of the WWW.

Our knowledge-based verification of Web-sites relies on rules and constraints that are formulated by information-providers. WebMaster uses the rules to group Web-pages into ontological categories. It uses the constraints to define categories of pages that con-tain errors. We have defined a logical formalism for such rules and constraints. Because information providers cannot be expected to express themselves in a formal language, we have designed a graphical notation for rules and constraints which closely reflects the structure of annotated Web-pages (for the benefits of the human users) and which can be 1-1 translated into our logical formalism (for the benefits of WebMaster's infer-ence engine).

Our experiences with WebMaster indicate that its rule-language is sufficiently ex-pressive to capture almost any constraint occurring in practical Web-site verification. WebMaster performed well on semantically annotated pages that were constructed by others in an independent project. Somewhat to our surprise, rich semantic markup is not necessary precondition for WebMaster's knowledge-based approach. Already on plain HTML pages it is possible to formulate a rich variety of very useful rules and constraints. The possibility to incrementally add semantic markup to HTML pages makes it possible to gradually migrate from a traditional layout-oriented Web-site to a semantically rich site where AI technologies such as the one described in this paper can be used to their full potential.

References

1. R. Benjamins and D. Fensel. The ontological engineering initiative (KA)^2. In N. Guarino, editor, *Proceedings of the International Conference on Formal Ontologies in Information Systems (FOIS'98)*, Frontiers in Artificial Intelligence and Applications. IOS-Press, June 1998.
2. M. Blazquez, M. Fernandez, J. Garcia-Pinar, and A. Gomez-Perez. Building ontologies at the knowledge level using the Ontology Design Environment. In *Proceedings of the Eleventh Workshop on Knowledge Acquisition for Knowledge-Based Systems (KAW'98)*, Banff, Alberta, 1998.
3. R. Brachman, D. McGuinness, P. Patel-Schneider, L. Alperin Resnick, and A. Borgida. Living with CLASSIC: When and how to use a KL-ONE-like language. In J. Sowa, editor, *Principles of Semantic Networks: Explorations in the representation of knowledge*, pages 401–456. Morgan Kaufman, 1991.
4. T. Bray, J. Paoli, and C. Sperberg-McQueen. Extensible Markup Language (XML) 1.0. W3C Recommendation, February 1998. http://www.w3.org/TR/1998/REC-xml-19980210.
5. C. Breiteneder, M. Hitz, and A. Mueck. Metadata mining in legacy data sets. In *First IEEE Metadata Conference*, 1996.
6. J. Clark and S. Deach. Extensible Stylesheet Language (XSL) version 1.0. World Wide Web Consortium Working Draft, August 1998. http://www.w3.org/TR/WD-xsl.
7. G. Crowder and Ch. Nicholas. Using statistical properties of text to create metadata. In *First IEEE Metadata Conference*, 1996.
8. A. Deutsch, M. Fernandez, D. Florescu, A. Levy, and D. Suciu. XML-QL: A query language for XML. Submission to the W3C, August 1998. http://www.w3.org/TR/NOTE-xml-ql/.
9. L. Wood et al. Document object model (DOM) level 1 specification version 1.0. W3C Recommendation, October 1998. http://www.w3.org/TR/REC-DOM-Level-1/.
10. O. Etzioni. Adaptive Web sites: an ai challenge. In *IJCAI'97*, 1997.
11. D. Fensel, S. Decker, M. Erdmann, and R. Studer. Ontobroker: The very high idea. In *In Proceedings of the 11th International Flairs Conference (FLAIRS'98)*, 1998.
12. M. Fernandez, D. Florescu, A. Levy, and D. Suciu. Web-site management: The Strudel approach. *Data Engineering Bulletin*, 21(2):14–20, 1998.
13. M. Kifer, G. Lausen, and J. Wu. Logical foundations of object-oriented and frame-based languages. *Journal of the ACM*, 42, 1995.
14. N. Kushmerick, D. Weld, and R. Doorenbos. Wrapper induction for information extraction. In *IJCAI'97*, 1997.
15. S. Luke, L. Spector, D. Rager, and J. Hendler. Ontology-based Web agents. In *Proceedings of First International Conference on Autonomous Agents (AA'97)*, 1997.
16. R. MacGregor. A description classifier for the predicate calculus. In *AAAI'94*, pages 213–220, 1994.
17. E. Maler and S.DeRose. XML linking language (XLink). World Wide Web Consortium Working Draft, March 1998. http://www.w3.org/TR/1998/WD-xlink-19980303.
18. E. Maler and S.DeRose. XML pointer language (XPointer). World Wide Web Consortium Working Draft, March 1998. http://www.w3.org/TR/1998/WD-xptr-19980303.
19. M.-C. Rousset. Verifying the World Wide Web: a position statement. In F. van Harmelen and J. van Thienen, editors, *Proceedings of the Fourth European Symposium on the Validation and Verification of Knowledge Based Systems (EUROVAV'97)*, 1997.

Towards Task-Oriented User Support for Failure Mode and Effects Analysis

Gerhard Peter[1] and Dietmar Rösner[2]

[1] Research Institute for Applied Knowledge Processing
at the University of Ulm (FAW), P.O. Box 20 60, 89010 Ulm, Germany
gerhard@faw.uni-ulm.de
[2] Faculty of Computer Science, Otto-von-Guericke-University,
P.O. Box 41 20, 39016 Magdeburg, Germany
roesner@iws.cs.uni-magdeburg.de

Abstract. Failure mode and effects analysis (FMEA) is an important method for preventive quality management. However, the issue of user support has largely been ignored. We have developed an approach for a task-oriented user support. Its goals are: (1) to aid a user in getting her work organized; (2) to improve FMEA results by providing suitable evaluation functions; and (3) to support the search for relevant information. In this paper, we focus on the first two goals. The support given is situation-specific, i.e., the state of the current task and the role of the user requesting advice, are taken into account. The major resource is an explicit model of the task at hand. FMEA is representative of information-intensive and support-intensive tasks which can be described by a hierarchical task model.

1 Introduction

Failure mode and effects analysis (FMEA) is an important method for preventive quality management. It involves investigating and assessing all causes and effects of all potential failure modes on a system in the earliest development phases [1]. An FMEA is typically conducted by a team of specialists from varying departments (e.g. design or production).

If an FMEA is done properly, the resulting documents contain a lot of knowledge about systems and production processes in a company. Therefore, it is a valuable source of know-how of a company. Furthermore, because it supports the early detection of weaknesses of a design, a reduction of development costs and less changes during series production are expected.

But even 30 years after its introduction in the aerospace industry and despite more than 10 years of experience in using this method in development, FMEA is still a challenge for many companies. Engineers consider FMEA as laborious and time consuming (and thus expensive) to carry out. Despite the major effort, however, the results are rather poor (cf. [2]): Descriptions of systems and functions are often incomplete and inconsistent, the knowledge in the FMEAs is hardly reusable, and FMEA is treated as a stand alone technique, i.e., it is integrated neither with the design process nor with other methods of quality management (e.g. quality function deployment). Hence, acceptance and motivation are low, which presents a major problem considering that the cost of errors (e.g. omitted failure modes) can be immense as recall actions of various car manufacturers have demonstrated.

There have been various research efforts to resolve these problems. For example, the FLAME system [12] fully automates an FMEA for electrical circuits and thus reduces the effort considerably. In [7] an approach for maintaining FMEA information is discussed and [11, 10] describe how to integrate FMEA with other methods (diagnosis and quality function deployment respectively). However, the issue of user support has largely been ignored. This comes as a surprise for the following reasons:

- An FMEA is a complex task, i.e., there are many threads of activity; various knowledge sources have to be accessed; an increasing amount of information has to be managed; and there is no standard way of performing the task, i.e., the subtasks, the interdependencies and the sequence of execution cannot be determined in advance.
- In general, an FMEA is conducted by specialists from various departments in one or more meetings. However, among the participants exist considerable differences regarding what they know about the FMEA task and with respect to the vocabulary used.

Our system actively supports a user in goal establishment (i.e. what task to perform next) and the evaluation of (preliminary) results according to various criteria (e.g. consistency or completeness). It also suggests how to resolve potential problems. The main tool is an *agenda*, which provides access and different views on a task. So far, support is mainly aimed at individual users. To facilitate cooperation among the members of a team we propose to use an *electronic blackboard* that functions like a newsgroup. On this blackboard users can post questions in natural language regarding their tasks at hand.

The approach draws on three resources: an explicit model of the task and its subtasks; a model that describes the users, especially the different *roles* they can assume when carrying out an FMEA; and a library of problem-solving methods that are responsible for providing appropriate answers to user requests. User model and task model provide the context for the advice-giving process.

Although an FMEA is essentially a team effort we focus on supporting an individual team member. This is motivated by the fact that in order to make team sessions more effective tasks are increasingly assigned to individuals; only the results are then discussed in the team.

The remainder of the paper is organized as follows: The following section describes related work. Then the features of an FMEA task and the requirements for user support are derived. Next, the underlying knowledge bases are described and the approach is presented in detail. Finally, the approach is discussed and possible directions of future research are pointed out.

2 Related Work

Up to date there hardly exists user support for FMEA. A notable exception is the intelligent, multimodal interface employed in XFMEA.

The user support in XFMEA [13] is based on a multilayer blackboard architecture in which knowledge sources in the lower layer provide interaction knowledge while those in the top layer contain learning knowledge. It learns how to interact with a user and aids a user when she does not know about performing a particular task. The blackboard utilizes a number of knowledge sources, e.g., knowledge of the user, the computer, the working environment, interaction modes, the user tasks etc. Compared to our system much more knowledge has to be modeled before the system can be utilized. Its learning facilities definitely reduce the modelling effort required. However, if a system changes continuously the user may experience problems developing a *mental model* of the application.

3 Features of FMEA Tasks

We consider an FMEA primarily to be a design task. Its most prominent features and consequences for user support in our domain are (cf. [5]):

Distribution of information: Both, goal state and start state are not fully specified and the transformation function from the start to goal states is completely unspecified. Thus, providing generic advice on how to get to goal states does not suffice; the advice has to be adapted to the current situation (i.e. the state of the task).

Size and complexity of problems: A print-out version of an FMEA can easily consist of several hundred pages. The size and complexity render it difficult to reuse FMEA results. Hence the design rationale should be captured. Additionally, as long as a product is maintained by a company, an FMEA cannot be considered as completed. Nowadays, engineers only spend a fraction of their time with FMEA-related activities. This may lead to lengthy intervals between team meetings. The user should be able to inform herself quickly about the state of an FMEA.

Right and wrong answers: There are only better or worse answers. A user should have access to heuristics that allow her to evaluate the FMEA results she has achieved so far and thus to come to better answers. What is considered to be better or worse may change over time (e.g. through new safety regulations). Hence, there is a need for users to be able to adapt these heuristics.

Personalized evaluation functions: Because there are no right or wrong answers and direct feedback is lacking, the evaluation functions that designers use will be derived from personal experience and immersion in the profession. Again, it is likely that these evaluation functions will change over time and support for adaptation is needed. Other designers also benefit from the description of such evaluation functions.

Team effort: An FMEA is essentially a team effort. Although most work is performed in team meetings, there is a growing tendency to assign tasks to individuals. Thus, a system should support an individual user in organizing her work, e.g., provide advice on what to do next.

In a nutshell, the support should be adaptive, i.e., it should take the current situation into account and the system should also be adaptable [4], i.e., a user should be able to adapt the system (e.g. heuristics). However, supporting adaptability and representing design rationale are subject to further research.

4   Knowledge Representation

User support draws on knowledge concerning tasks, users (and their roles), and domain knowledge. We also employ a library of problem-solving methods that are applied to provide appropriate answers to user requests.

A *task* describes the problem type to be solved. A *method* is a way of accomplishing a task. In general, a task can be accomplished by alternative methods. Both, application tasks (e.g. FMEA) and advice giving tasks, are modelled as *task-method-hierarchies*. A task-method hierarchy is a tree of tasks, methods, and subtasks applied recursively until *primitive* methods are reached. A primitive method does not contain further subtasks. All subtasks of a method are executed. The basic notion of task and (problem-solving) method, and their embedding into a task-method-hierarchy are concepts which are nowadays shared among most of the knowledge engineering methodologies (e.g. KADS [16]).

We have developed a task-method-hierarchy for FMEA. A part of it is shown in Fig. 1 (rectangles denote methods and ovals denote tasks). It is based on an analysis of various FMEA standards (e.g. [1]) and on experiences gained in two projects [17, 10] while observing FMEA sessions and through interviewing their participants.

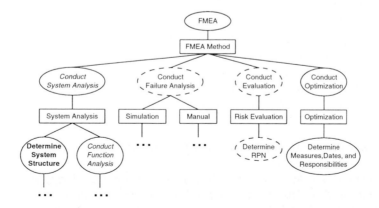

Fig. 1. Task-method hierarchy for FMEA

The applicability of a method is determined by one or more preconditions. In the case that several methods can be applied to an advice giving task (i.e. fulfil the preconditions), one is picked at random. Methods for application tasks are chosen by a user. For example, the task `Conduct Failure Analysis` (see Fig. 1) can either be accomplished by carrying out a simulation method (preferred for electrical devices; cf. [12]) or manually, i.e., an FMEA team itself determines the causes and effects of a potential failure.

Besides administrative information, a user model contains information on the user's experience (previous roles and tasks carried out), on individual preferences (e.g. viewing style), and on the tasks she is currently involved in. Each user can inspect her model and adapt it. A role is represented by a *stereotype* [14]. A role defines the access rights and the preferences a role inhabitant is assumed to have. Fig. 2 shows an example of a user model. Mrs. Jones is a `Designer` and has experience in system analysis. In contrast to the preferences given in the stereotype, Mrs. Jones prefers brief explanations. Currently, she is involved in system analysis and failure analysis. The moderator has also granted her for these tasks the additional access right to view any QFD-data related to the current FMEA.

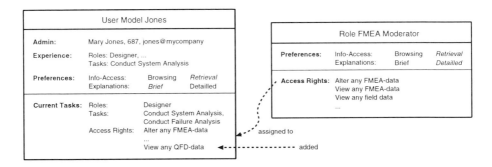

Fig. 2. User model and stereotype

The stereotypes form an specialization hierarchy. For example, the role `Designer` is a subtype of `TeamMember`. Other team members (e.g. for QFD tasks) can thus possess different access rights and preferences. The structure of a user model for FMEA participants and the stereotype hierarchy for roles are also results of the projects mentioned above.

5 Task-oriented Approach to User Support

In this section, we describe how the system supports goal establishment and the evaluation of results. Both functions are mainly aimed at individual users. An electronic blackboard is a first means to facilitate cooperation among team members.

We also discuss the question whether the system should provide only passive help or also active help, i.e., whether it can interrupt the user and if yes when.

Goal establishment A user can turn to the system to ask for some advice on what task to perform next. Let the current state of an FMEA be depicted in Fig. 1. Overdue tasks appear in italics (e.g. Conduct Function Analysis) and completed tasks are written in bold characters (Determine System Structure). Tasks that have been started but are not finished yet are surrounded by a dotted oval. Such a visualization in an agenda will give a user a first impression of the state of task execution.

The task-method structure of our system for selecting the next task is depicted in Fig. 3. The underlying algorithm can be summarized as follows: The user request What next? is mapped to the predefinded task Select Task. There are alternative methods to solve this task: Shallow Selection will be applied, for example, if a user does not want any evaluation to be performed by the system; i.e. the task Evaluation in Fig. 3 is not carried out. The system retrieves all unfinished tasks depending on the user's role and the user chooses one of them.

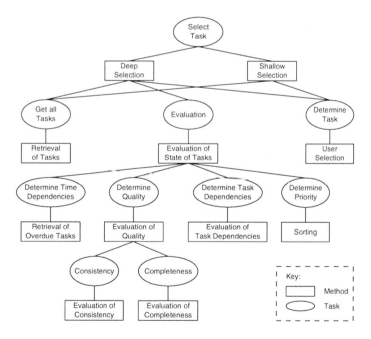

Fig. 3. Task-method hierarchy of the task What next?

Deep Selection, on the other hand, also evaluates the available tasks. Several criteria are applied (cf. Fig. 3):

– All overdue tasks are retrieved (Determine Time Dependencies).

- The *quality* of the tasks is evaluated (Determine Quality), i.e., all tasks are collected that are either incomplete, inconsistent, or both (see below).
- Finally, all tasks are retrieved that will be overdue soon and on whose results other users of the group depend (Determine Task Dependencies).

The result of Determine Priority is a sorted list presented to the user. The sorting criteria determine the sequence of the tasks in the list. As a default, overdue tasks are presented first.

A task-method hierarchy is processed in a depth-first manner. Each task is stored in a *conflict set* which holds all tasks that have not been accomplished. A problem-solving component determines the sequence of their execution. Currently, the most recent posted task is treated first. The problem-solving component is also responsible for retracting tasks if a subtask failed. In case a task cannot be accomplished, the problem-solving component tries to execute an alternative method that matches the given preconditions. To determine if a task fails or not is context-dependent. For example, if the method Get all Tasks returns an empty set then no FMEA task can be selected. On the other hand, if Retrieval of Overdue Tasks returns an empty set the task Determine Time Dependencies is still accomplished.

In the task What next? several retrieval methods are employed. The retrieval methods form an abstraction hierarchy, e.g.:

```
retrieve (object)
    retrieve (task)
        retrieve (application task)
    retrieve (role)
    retrieve (date)
    . . .
```

The hierarchy depends on the generality of the objects as specified in the data model. For example, object is more general than task which in turn is more general than application task. The latter consists only of the FMEA tasks and none from the problem-solving library.

When modeling a task suitable methods are selected from the abstraction hierarchy and included into the task model. If, for example, no specific retrieval method is available then at least the most general retrieval method (retrieve (object)) can be included. Executing retrieve (object) starts the retrieval interface and the user can formulate the search request by herself. In contrast, retrieve (task) automatically returns all subtasks of a given task; no user interaction is required. Note that retrieve (object) does not retrieve all objects of a knowledge base as this would cause an information overload.

Evaluation It has been mentioned above that there are no right or wrong answers; there are only better or worse answers. We employ heuristics, i.e., rules associated with tasks that evaluate FMEA results from different perspectives. Currently, the system checks for completeness and consistency of FMEA results.

A user decides by selecting a task from a task model what heuristics are applied. For example, selecting `Conduct Function Analysis` causes all technical functions to be checked for completeness and consistency. Incomplete or inconsistent functions are pointed out to the user (e.g. each function should be assigned to a component of a device). Moving up a task hierarchy will cause more heuristics to be applied to the results of an FMEA, moving down in less. Thus, a user can determine quite clearly on what aspect of the overall task she wants advice. She can only select those tasks she is responsible for (as indicated in her user model).

The system can also provide help on how to resolve potential problems, i.e., in the case of an incomplete technical function it suggests potential components the function can be assigned to. The domain model of our system can be considered to be one big hypertext. Thus, in one strategy, it first follows the `is-Subfunction-of`-link to the superfunction an then the `is-realized-by`-link to the component realizing the superfunction. Potential candidates are its sub-components. On request, the system offers a limited justification of its suggestion by displaying a textual explanation of the appropriate heuristic ("canned text"). An explanation can either be brief or more detailled according to the preference given in the user model.

We take the notion seriously that a user should not be distracted from carrying out an FMEA too often. We also have to take into account that there are no fixed criteria when an FMEA task has been completed. Thus, we employ the following rule of thumb: each time FMEA-related information (e.g. a functional model) is saved, its completeness and consistency are also automatically evaluated. Each heuristic carries the additional information whether the advice has to be given directly to the user (and thus to interrupt her) or not. In the latter case the advice is added to the respective task in the agenda.

However, a user can decide to switch off evaluating information when it is saved. Then, evaluation is only started on her request (see above) or when a task in the agenda is marked as completed. In that case, all advice is given directly to the user. Although, such after task support is considered to be less useful than immediate support it is pointed out in [15] that it becomes more useful as task difficulty increases.

Note that a user is always free to reject an advice and to proceed as she wishes. If a user wants to process an advice at a later point in time the advice is added to the respective task in an agenda. However, unprocessed advice by the user regarding inconsistent or incomplete functions, failures, or components also marks the respective task to be incomplete or "inconsistent", respectively.

Electronic Blackboard It is a shared workspace accessible to all people logged into the system. Thus, it can be accessed by team members and non-team members alike. There are different views on questions and their respective answers: the task in whose context the question originated, the role of the user who posed the question, and the user herself.

The system also automatically attaches a note to the respective task in the agenda that there is a pending request. As long as the request has not been answered, i.e., the note has not been removed from the task, the task is again considered to be incomplete.

6   Discussion

In this paper, we proposed an approach for a task-oriented user support. We consider its main advantage that a user is relieved from a lot of routine work: FMEA tasks are managed automatically in an agenda; a user can get advice on what to do next; and by looking at an agenda a user can also get an immediate understanding of the state of a task. Hence, she can concentrate on carrying out an FMEA. Additionally, FMEA results are improved by heuristics which not only highlight potential problems but also recommend how to resolve them. Finally, an electronic blackboard facilitates cooperation among team members.

An explicit task model is the main resource for user support. We assume that such a model remains fairly stable over time. A library of problem-solving methods is applied to provide appropriate answers to user requests. Such a library has been successfully applied in supporting human-computer interaction in the context of real-time decision for management problems (e.g. traffic control) [6] and in a framework that guides a user in breaking down a task of knowledge discovery in databases [3].

The task model and information on the user, especially her role in the task, are employed to adapt the advice to the current situation. In the approach no strict guidance is intended.

FMEA is representative of information-intensive and advice-intensive tasks which can be described by a hierarchical task model. One of our next steps will be to transfer our results to a new application. An initial estimate is that most of the problem-solving methods can be reused. For example, the result of `What next?` depends on the context and that is the user's role and her task. It has to be investigated to what extent existing role descriptions and parts of the FMEA task model can be reused as well. Alternatively, for each new application a task model has to be developed and additional roles have to be described.

It has also to be investigated, how such a system can be employed in a team session. A smooth integration is a major determinant regarding user acceptance.

Currently, we are developing a first prototype as an extension to IPQM [10], an existing prototypical system for preventive quality management developed at FAW Ulm. The prototype is intended to provide a proper basis for field tests. We have begun to implement the problem-solving methods using the NéOpus system, a first-order inference engine embedded in Smalltalk [8]. One of its prominent features is the declarative specification of control with *metarules* [9].

Acknowledgements

The first author's work is supported by a doctoral grant from the Research Institute for Applied Knowledge Processing at the University of Ulm (FAW),

Germany. Thanks are due to Brigitte Grote and Holger Knublauch for valuable comments on a draft version of the paper.

References

1. British Standards Institution. Reliability of systems, equipment and components – Part 5. Guide to failure modes, effects and criticality analysis (FMEA and FMECA). British Standard BS 5760 (1991)
2. Dale, B.G., Shaw, P.: Failure mode and effects analysis in the U.K. motor industry – a state-of-the-art study. Qual. Reliab. Engng. Int. **6** (1990) 179–188
3. Engels, R.: Planning tasks for knowledge discovery in databases; performing task-oriented user-guidance. In: Proc. of the 2nd International Conference on Knowledge Discovery and Data Mining (KDD '96). AAAI Press, Menlo Park (1996) 170–175
4. Fischer, G., Reeves, B.N.: Beyond intelligent interfaces: Exploring, analyzing and creating success models of cooperative problem solving. Applied Intelligence **1** (1992) 311–332
5. Goel, V., Pirolli, P.: The structure of design problem spaces. Cognitive Science **16** (1992) 395–429
6. Hernández, J., Molina, M.: Advanced human-computer interaction for decision support systems using knowledge modeling techniques. In: Proc. 15th IFIP World Computer Congress, Information Technology and Knowledge Systems (IT & KNOWS). Wien Budapest (1998)
7. Krämer, A., Peter, G.: Using change notification for maintaining an FMEA database. In: Proc. of the Third International Conference on Concurrent Engineering (CE '96). Toronto (1996)
8. Pachet, F.: On the embeddability of production rules in object-oriented languages. Journal of Object-Oriented Programming **8** (1995) 4 19–24
9. Pachet, F., Perrot, J.-F.: Rule firing with metarules. In: Proc. Software Engineering and Knowledge Engineering (SEKE '94). Jurmala Lettonia (1994) 322–329
10. Peter, G., Berthold, B., Krämer, A., Rupprecht, C.: Knowledge-based techniques and applications for preventive quality management. In: Leondes, C.T. (ed): Expert Systems Techniques and Applications. Gordon and Breach. Newark (1999) Forthcoming
11. Price, C., Taylor, N.: Multiple fault diagnosis from FMEA. In: Proc. 9th Conference on Innovative Applications of Artificial Intelligence (IAAI '97). AAAI Press. Menlo Park (1997)
12. Price, C.J., Pugh, D.R., Wilson, M.S., Snooke, N.: The Flame system: Automating electrical failure mode & effects analysis (FMEA). In: Proc. Annual Reliability and Maintainability Symposium. IEEE Press. (1995) 90–95
13. Puerta, A.R., Bonnell, R.D.: A blackboard model for a self-improving intelligent user interface. In Proc. Annual Conference International Association of Knowledge Engineers. College Park (1989) 619–635
14. Rich, E.: User modeling via stereotypes. Cognitive Science **3** (1979) 329–354
15. Silverman, B.G., Wenig, R.G.: Engineering expert critics for cooperative systems. The Knowledge Engineering Review. **8** (1993) 4 309–328
16. Wielinga, B., Schreiber, A.Th., Breuker, J.A.: KADS: a modelling approach to knowledge engineering. Knowledge Acquisition. **4** (1992) 1 5–53
17. Wirth, R., Berthold, B., Krämer, A., Peter, G.: Knowledge-based support of system analysis for failure modes and effects. Engineering Applications of Artificial Intelligence. **9** (1996) 3 219–229

Incremental and Integrated Evaluation of Rule-Based Systems

P. G. Chander[*1],

R. Shinghal[2], and T. Radhakrishnan[2]

[1] Bell Laboratories, Business Communications Systems R&D, Lucent Technologies,
Denver, CO 80234,
gokul3@bell-labs.com,
[2] Concordia University, Dept. of Computer Science, Montreal, QC H3G 1M8,
shinghal@cs.concordia.ca, krishnan@cs.concordia.ca

Abstract. Rule-based systems that are easily testable are required for high reliability applications. However, as a rule base evolves, developers prefer incremental evaluation owing to the high cost of regression testing. For quality and reliability improvement, researchers advocate that the evaluation phase be integrated with development: thus, incremental evaluation becomes more important in this context. In this paper, we propose a three-tiered life-cycle model for integrating evaluation in a rule-based system life cycle. We then outline its use to facilitate knowledge acquisition using "goal specification" and its realization in a rule base using "paths." Path-based validation has been well-accepted in the literature as a reliable method for structural testing. However, extracting paths for every rule base modification for its evaluation incurs an enormous effort. In this paper, we identify situations that can help in incremental path extraction and present some issues that are important for integrating evaluation into the life-cycle of rule-based systems. We also outline how our approach can facilitate handling these issues.

Key words: Expert Systems, Verification & Validation.

Introduction and Motivation

One of the primary reasons that makes the evaluation of rule-based systems harder is the lack of a well defined link between system conception and its realization. In fact, in many cases, there does not even seem to be a design level to weigh the various compromises with which domain knowledge and constraints can be represented in the rule base. For rule base designs that integrate verification and validation (V&V) processes with ease, we need a well defined link that connects the following three stages in its development: (1) the functional requirements of the system, (2) the rule base design constraints, and (3) the

---

[*] The work reported in this paper was done while the author was at Concordia University, Montreal.

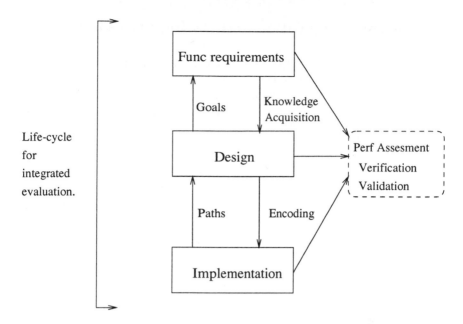

Fig. 1. A simple software development perspective for rule-based systems for viewing evaluation as part of all the phases of a system's life-cycle.

implementation of the rule base (which, in turn, determines the extent of evaluation that can be performed on the system). This is portrayed in Figure 1. It shows a development perspective for rule-based systems.

Our motivation for the work reported in this paper arises partly from our experience in the design and evaluation of rule-based systems, and from the increasing emphasis on integrating evaluation and on reducing evaluation costs [22, 20, 10]. Contemporary researchers believe that quality and reliability improvement for rule-based systems can be obtained through formal approaches to system construction and integrated evaluation [23, 1, 20]. Integrating evaluation in a system's life-cycle, however, is non-trivial as costs can be prohibitive if all tests are automatically repeated for every modification to the system. Based upon these observations, we believe that integrating evaluation in a system's life-cycle should emphasize incremental evaluation procedures and should take into account the role of the structure of a system for its evaluation [7].

The paper is organized as follows. In section 2, we describe our approach to knowledge acquisition called **goal specification**, and explain how goals are realized in a rule base using **paths**. In section 3, we describe a method for incremental path extraction. In section 4, we identify several issues that are pertinent for integrated evaluation, and outline how goal specification and paths can play a useful role in tackling such issues. Section 5 provides concluding remarks.

1 Knowledge Acquisition and Its Representation

In our approach towards knowledge acquisition, the domain expert specifies a set of significant states that need to be reached to solve problems in the domain. For example, in a medical diagnosis domain, inferring a `liver-disease` is a significant state toward inferring `liver-cirrhosis` as a final diagnosis. Typically, the domain expert specifies concepts associated with the domain that serve as significant states for problem solving, and the knowledge engineer translates these concepts into a representation language, say a conjunction of first-order logic atoms that capture the intent of the domain expert. Such states are called *goals*. In addition, the domain expert also specifies *inviolables*, which are constraints associated with the domain; an inviolable is a conjunction of atoms that should never be true, for example, $MALE(x) \land PREGNANT(x)$; obviously, no goal should contain an inviolable.

Definition 1 (Goal Specification) *The set of goals and inviolables of a domain constitutes the goal specification of that domain.*

Every goal in a goal specification, when translated into a first order logic formula, consists of a conjunction of hypotheses. The hypotheses that are used as goal compositions are called *goal atoms*, in order to contrast them with the other hypotheses in the system that may be needed (for rule base coding) called *non-goal atoms*. Solutions to problems in the domain are also clearly demarcated at the time of goal specification. Thus, it is possible to partition the goal specification into two goal classes: *intermediate* goals and *final* goals. Typically, the intermediate goals are those that are achieved in order to infer a final goal.

A rule base will consist of a set of rules and hypotheses. The hypotheses are atoms of first order logic that capture a concept, or inference associated with the domain. For example, the designation of "professor in a university" is typically captured as $PROFESSOR(x, y)$, where x refers to the professors name, and y to the university. However, this alone does not portray the importance of this concept (hypothesis) in relation to solving problems in its domain. Thus, capturing knowledge in terms of atoms for rule encoding is necessary, but not sufficient. The insufficiency arises because in diagnostic domains where rule-based systems are typically employed, the required knowledge should capture the *progress* of problem solving in the domain through a set of concepts/hypotheses so that the rule base can be designed to reflect this progress through well defined rule sequences.

A rule base constructed based on a given goal specification of a domain implements the problem solving by rule sequences (called *paths*) that progress from goal(s) to goal [3]. The complete knowledge of the domain is represented in a system via the goals and the paths inferring these goals causing a progression in problem solving. The extent to which a given rule base realizes the acquired knowledge of goal inference is reflected by the paths in the rule base; they are collectively said to portray the **structure** of the rule base [14]. Figure 2 shows a sample rule base, goals and paths.

Definition 2 (Rule Base Structure) *The structure of a rule base (or, simply structure) is defined as $< \mathcal{G}, \Sigma >$ where $\mathcal{G}$ is the goal specification of the domain, and Σ is a set of paths in the rule base such that,*

$$(\forall \Phi \in \Sigma)(\exists g)(g \in \mathcal{G}) \; G \wedge \Phi \vdash g$$

where G is the set of goals that are required for path Φ.

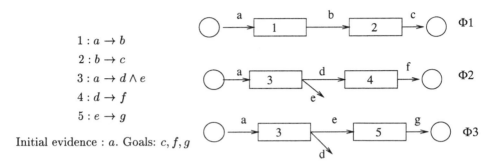

$1 : a \rightarrow b$

$2 : b \rightarrow c$

$3 : a \rightarrow d \wedge e$

$4 : d \rightarrow f$

$5 : e \rightarrow g$

Initial evidence : a. Goals: c, f, g

Fig. 2. An example of a rule base, goals, and paths. In this model, the boxes represents rules and a directed edge from r_1 to r_2 represents unification between an atom in r_1's consequent and r_2's antecedent. Rule r_2 is also said to be *reachable* from r_1. For more details, refer to [8].

Paths in a rule base have been the basis for a variety of evaluation processes [11, 19, 14]. However, the relation of the paths to the acquired domain knowledge is implicit. By making this link explicit via the goals (cf. the above definition of rule base structure), we capture the rule interactions that pertain to problem solving in a meaningful manner because these rule interactions can be mapped to inferring goals. In addition, rule interactions in a path are completely *localized*: by its definition, all rules in a path are ready to fire once the goals required by the path are inferred.

A tool called Path Hunter has been developed that can extract the paths from a rule base given its goal specification [14]. Path Hunter has been shown to be effective in extracting and enumerating paths for large rule bases [15]. However, using Path Hunter for path re-extraction, when modifications are made to the rule base as it evolves, however, is practically infeasible since it extracts all the paths as opposed to those paths that are affected by the change. This incurs huge computational cost and effort. In the next section, we remedy this defect by outlining our method for incremental path extraction to facilitate integrated evaluation [5].

2   Incremental Evaluation to Reduce Costs

Rule bases evolve by test results, by feedback of users on test operation, and by the changing requirements of a domain. Every evolution should, of course, be tested to confirm its correct operation with respect to the change(s) in the system. For example, adding one more rule could result in changed paths, thus altering the goal relevancy status of the system. Hence, all the paths should be extracted to determine the effect of the change. However, extracting all the paths once again even for a small change (that is, a regression testing approach) can entail a large cost [22] as path extraction has an exponential complexity in the worst case [14]. In this section, we state the problem of incremental evaluation, and outline a method for incremental path extraction.

More generally, incremental evaluation can be viewed as an optimization problem, where we wish to minimize the cost of evaluation while evaluating the system due to some modifications without compromising on the quality and reliability of test results.

Problem Statement 1 (Incremental Evaluation) *Let S and R be the system specification and rule base respectively. Let the operator Δ denote a small change of an entity. Let C be the cost of testing the system so that ΔS and ΔR represent changes in system specification and rule base modifications, respectively. Determine an evaluation strategy that is based upon ΔR and ΔS so that the cost for evaluation is comparable to that change (that is, it should be ΔC, as opposed to C). For example, if the change to a rule base is 10% (with respect to some reference), then the cost of evaluation should be of the order to 10% of C so that the <u>total</u> cost for testing taking the modifications into account should be $1.1C$ ($C + 0.1C$) as opposed to $2C$ (repeating the tests).*

In our framework, all our system evaluation perspectives (verification, validation, and performance and quality assessment) are based upon paths and goals [10]. Hence, the above problem transforms to incremental path extraction: that is, identify and extract only those paths that are affected by a change in the rule base. In the description below, we assume that the goal specification has not changed, and modifications involve only non-goal atoms. The procedure taking into account changed goal specification as well as rule base modifications is complicated, and is currently under our investigation.

Given the changes to a rule base ΔR, to identify paths that require re-extraction, we must first identify paths that are potentially affected by that change $\Delta \Phi$, where Φ is the set of paths in a rule base. From $\Delta \Phi$, we then obtain $\Delta \Phi'$, the set of paths that require re-extraction. In the worst case, (for regression testing) $\Delta \Phi' \equiv \Delta \Phi \equiv \Phi$. Let $R = \{r_1, r_2, \ldots r_n\}$ and $\Delta R = \{r'_1, r'_2, \ldots r'_m\}$, $0 \leq |\Delta R| \leq |R|$, where we use the notation $|S|$ to denote the size of a set S.

Since the goal specification is unchanged, the changes to a rule cannot affect existing goals; thus, $\Delta \Phi$ can be easily computed ($O(|\Phi| * |\Delta R|)$) by noting every path that contain a rule from ΔR. The path requires re-extraction, iff the rule dependency in the path has changed. Thus, we need to re-compute only the rule dependency between rules in ΔR and the other rules in the rule base. If this remains the same, no path re-extraction is required. If it is different, then we need to focus only on the subset of paths that are affected by the change in the dependency. In Figure 3, we consider four cases: addition (deletion) of atoms in a rule consequent (antecedent). It outlines the heuristics for quick path extraction to facilitate testing whenever changes involve non-goal atoms (the underlying assumption being that non-goal atoms would be added/deleted more frequently than atoms used to infer goals).

---

1. **Addition of an atom in a rule antecedent** Every path that contains the rule requires re-extraction as the reachability of this rule has changed. However, if the atom added in the antecedent is inferred in that path, do not re-extract. If the atom is an initial evidence or if it is a goal atom contained in the goals required for this path, then no re-computation is required. For example, addition of atom e in rule 4 does not cause any path re-extraction.

2. **Deletion of an atom in a rule antecedent** For saving computation, paths that contain this rule r need not be re computed as deletion does not affect the reachability of this rule from the other rules in the path (but the deleted atom and dependency due to this atom in the path should be noted). Of course, if the deletion causes a rule antecedent to become empty, or a path to get disconnected, report this as potential redundancy. For example, deletion of b in the antecedent of rule 2 makes path $\Phi 1$ disconnected. This should be reported, but the path need not be re-extracted (since its goal would be inferred any way), but rule 1 is now redundant as it makes a useless inference.

3. **Addition of an atom A in a rule consequent** If the rules reachable from this rule r are unchanged, then no computation is required for path extraction. Otherwise, for every new rule r' in path Φ' now reachable from r, one new path inferring the same goal g' as Φ' is added. In the new path r replaces a rule (or a set of rules) from Φ' that infers atom A. Note, adding new paths entail little computation: copy paths such as Φ' and simply perform a rule replacement. For example, adding d to the consequent of rule 1 adds a path inferring goal f containing rule 1 replacing rule 3 (in path $\Phi 2$).

4. **Deleting an atom in the consequent of a rule** If the rule dependency of a path is unaffected, do not re-compute. Otherwise, re-extract every path that contains this rule (because it can cause path disconnection). For example, deleting e from rule 3 causes a re-extraction of only path $\Phi 3$ and a subsequent report that rule 5 is redundant due to this change.

Fig. 3. Incremental Path Extraction for changes involving addition or deletion of non-goal atoms. Examples correspond to the rule base shown in Figure 2.

All test conclusions based upon the above changes would be correct since they preserve the rule dependency in a path (warning the developer as appropriate), and ensure that all rules in a path are ready to fire once the goals required by a path are inferred. A general method for path extraction involving unconstrained rule modification that also takes into account modifications to the goal specification is an open problem.

Incremental evaluation methods play a crucial role for Integrating evaluation in a rule-based system life-cycle. However, integrated evaluation for rule-based systems is a research problem that is much broader in its scope. In the next section, we identify several key issues that influence the design of rule-based systems for integrated evaluation [1, 5].

3 Design Issues for Integrated Evaluation

A design that allows integrated verification and validation should emphasize rule sequences rather than individual rules. The intent is that the design captures the rule interactions in the form of rule sequences rather than individual (or pairwise) rules so that the role of every rule in problem solving can be explicitly mapped to the acquired knowledge. An ideal model would capture every possible rule sequence without losing computational tractability, but owing to the exponential complexity associated with rule sequence enumeration, one can only hope to be closer towards the ideal [24, 17].

Goal specification can be used to extract paths and can also be used to control the computation for path extraction. This allows one to speculate on its role for integrated evaluation. Below, we outline issues that confront integrating evaluation in a system's life-cycle and explore how goal specification based on rule base designs can *facilitate* integrating evaluation into a system's life-cycle. We do not claim that goal specification and path-based evaluation methods [10] are panacea to the problem of efficiently integrating evaluation in a system life-cycle, but merely point out many of its features that support this objective.

Structure extraction for V&V As defined earlier, the structure of the system in our case is a static part of the system which is a set of rule sequences that move from goal-to-goal resulting in a progression of problem solving. The structure of a rule base forms the basis for structural validation [19, 11, 27, 14]. The extraction of paths, however, is a non trivial issue due to the combinatorial explosion that arises while trying to enumerate the paths in a system, and goal specification can be used as "meta knowledge" of the domain to cut down the computation required to extract the paths involved in goal-to-goal progressions in the system. A software tool, called Path Hunter, has also been developed to extract the paths in a rule base given the goal specification [14].

Behavior understandability The use of appropriate dynamic coverage measures is an important issue in integrated evaluation [1]. The understandability of behavior thus implies the ability to map a run trace unambiguously to a set of paths thereby knowing the actual goal-to-goal progression that occurred while solving a given problem. Goal specification helps one to understand the role of a fired rule by mapping this rule to a path.

Performance evaluation One of the problems that is currently faced by researchers in performance evaluation is that of defining a "good" criteria to assess performance [16]. In our case, the goal specification approach incorporates the notion of a goal, captured during knowledge acquisition, as a unit of work done by the system in solving a problem. For more details, see [6].

Incremental Evaluation One of the merits of integrated evaluation is its inherent ability to facilitate incremental testing to reduce evaluation costs [22]. Evaluation based upon goals and paths provide support to this objective in the sense that incremental path extraction is possible subject to some limitations (based on the type of modifications effected on a goal specification and its associated rule base). A method for incremental path extraction was presented in section 3.

Test case generation The ease with which test cases can be generated to test a given rule base is important to control evaluation costs. In our case, sequences of paths enumerated from permissible initial evidence to final goals can help identify initial evidence that can be input selectively to the system [9]. Automated test-case generating tools, similar in spirit to those described in [2], can easily integrate with our approach.

Ease of analysis of test case coverage The paths extracted from a goal specification based design can help measure given test case coverage of the rule base by means of a set of criteria that allows how many paths are exercised and the extent to which a given rule sequence is exercised for a test run; a software tool, called Path Tracer, has also been developed by us to analyze the run trace of a system to measure rule sequence coverage [25].

Comprehensive verification of rule base anomalies Inference chains are widely used as a basis for comprehensive verification schemes [12, 13, 26, 21]. Paths are generalized inference chains compared to the linear chains used for computing labels in the above schemes. A set of criteria has also been developed for comprehensive rule base verification by spotting certain path combinations called "rule aberrations" [8] to detect rule base anomalies.

Quality Assurance Goal-based design scheme provide several metrics to assess the "goodness" of knowledge representation, and implementation [4]. It is also relatively easy to track these metrics as a rule base evolves. In addition, paths can provide several implementation-specific metrics to assess the various qualities of a rule base such as its complexity, verifiability, etc [24].

As a final note, all the above aspects of verification, validation, and performance evaluation and tools aimed towards these objectives can be used in any part of the incremental life cycle in a goal specification based design: all that is required is to (incrementally) extract the current set of paths, and apply verification and validation procedures. Indeed, by recording the goal and rule sequences over several versions, this incremental verification capability is similar to the approach in [22], and can also help in maintenance over several versions by using paths for rule grouping similar to the approach in [18].

4 Summary & Conclusion

The design and development of rule-based systems often cause anomalies in the rule base: not only the development tends to be error prone, but there is confusion in applying evaluation processes for these systems [1, 17]. In this regard, our work emphasizes the following: (1) the system structure should play a major role in system evaluation; and (2) integrating evaluation should exploit incremental evaluation to cut cost and effort for testing without compromising on the quality of testing. We also outlined several issues that should be considered as part of integrating evaluation in a rule-based system life-cycle, and how our approach can help a developer in tackling these issues.

References

1. Ed P. Andert Jr. Integrated Design and V&V of Knowledge-Based Systems. In *Notes of the Workshop on Validation and Verification of Knowledge-Based Systems (Eleventh National Conference on Artificial Intelligence)*, pages 127–128, Washington D.C., July 1993.
2. Marc Ayel and Laurence Vignollet. SYCOJET and SACCO: Two Tools for Verifying Expert Systems. *International Journal of Expert Systems*, 6(3):273–298, 1993.
3. P. G. Chander, T. Radhakrishnan, and R. Shinghal. Quality Issues in Designing and Evaluating Rule-based Systems. In *Notes of the Workshop on Verification & Validation of Knowledge-Based Systems (Thirteenth National Conference on Artificial Intelligence)*, pages 33–42, Portland, Oregon, August 1996.
4. P. G. Chander, T. Radhakrishnan, and R. Shinghal. Design Schemes for Rule-based Systems. *International Journal of Expert Systems: Research and Applications*, 10(1):1–36, November 1997.
5. P. G. Chander, T. Radhakrishnan, and R. Shinghal. Issues in Designing Rule-based Systems for Integrated Evaluation. In *Notes of the Workshop on Verification & Validation of Knowledge-Based Systems (Fifteenth National Conference on Artificial Intelligence AAAI-98)*, pages 18–24, Wisconsin, Madison, July 1998.
6. P. G. Chander, R. Shinghal, B.C. Desai, and T. Radhakrishnan. An Expert System for Cataloging and Searching Digital Libraries. *Expert Systems with Applications*, 12(4):405–416, 1997.
7. P. G. Chander, R. Shinghal, and T. Radhakrishnan. Performance Assesment and Incremental Evaluation of Rule-based Systems. In *Notes of the Workshop on Verification & Validation of Knowledge-Based Systems (Fourteenth National Conference on Artificial Intelligence AAAI-97)*, pages 40–46, Providence, Rhode Island, July 1997.
8. P. G. Chander, R. Shinghal, and T. Radhakrishnan. Using Goals to Design and Verify Rule Bases. *Decision Support Systems*, 21(4):281–305, 1997.
9. P. Gokul Chander, R. Shinghal, and T. Radhakrishnan. Static Determination of Dynamic Functional Attributes in Rule-based Systems. In *Proceedings of the 1994 International Conference on Systems Research, Informatics and Cybernetics, AI Symposium (ICSRIC 94)*, pages 79–84, Baden Baden, Germany, August 1994.
10. Prabhakar Gokul Chander. *On the Design and Evaluation of Rule-based Systems*. PhD thesis, Department of Computer Science, Concordia University, Montreal, May 1996.
11. C. L. Chang, J. B. Combs, and R. A. Stachowitz. A Report on the Expert Systems Validation Associate (EVA). *Expert Systems with Applications*, 1(3):217–230, 1990.
12. Allen Ginsberg. Knowledge-Base Reduction: A New Approach to Checking Knowledge Bases for Inconsistency & Redundancy. In *Proceedings of the 7th National Conference on Artificial Intelligence (AAAI 88)*, volume 2, pages 585–589, St. Paul, Minnesota, August 1988.

13. Allen Ginsberg and Keith Williamson.   Checking for Quasi First-Order-Logic Knowledge Bases. *Expert Systems with Applications*, 6(3):321–340, 1993.

14. C. Grossner, P. Gokulchander, A. Preece, and T. Radhakrishnan. Revealing the Structure of Rule-Based Systems. *International Journal of Expert Systems: Research and Applications*, 9(2):255–278, 1996.

15. C. Grossner, A. Preece, P. Gokulchander, T. Radhakrishnan, and C.Y. Suen. Exploring the Structure of Rule Based Systems. In *Proceedings of the 11th National Conference on Artificial Intelligence (AAAI 93)*, pages 704–709, Washington D.C., 1993.

16. Giovanni Guida and Giancarlo Mauri. Evaluating Performance and Quality of Knowledge-Based Systems: Foundation and Methodology. *IEEE transactions in Knowledge and Data engineering*, 5(2):204–224, Apr 1993.

17. David Hamilton, Keith Kelley, and Chris Culbert.   State-of-the-Practice in Knowledge-based System Verification and Validation. *Expert Systems with Applications*, 3(3):403–410, 1991.

18. Robert J. K. Jacob and Judith N. Froscher. A Software Engineering Methodology for Rule-Based Systems. *IEEE Transactions on Knowledge and Data Engineering*, 2(2):173–189, 1990.

19. James D. Kiper. Structural Testing of Rule-Based Expert Systems. *ACM Transactions on Software Engineering and Methodology*, 1(2):168–187, 1992.

20. Sunro Lee and Robert M. O'Keefe. Developing a Strategy for Expert System Verification and Validation. *IEEE Transactions on Systems, Man, and Cybernetics*, 24(4):643–655, Apr 1994.

21. Stephane Loiseau and Marie-Christine Rousset. Formal Verification of Knowledge Bases Focused on Consistency: Two Experiments Based on ATMS Techniques. *International Journal of Expert Systems*, 6(3):273–298, 1993.

22. Pedro Meseguer. Incremental Verification of Rule-based Expert Systems. In B. Neumann, editor, *10th European Conference on Artificial Intelligence*, pages 829–834, Vienna, Austria, 1992.

23. Robert T. Plant. Expert System Development and Testing: A Knowledge Engineer's Perspective. *Journal of Systems Software*, 19(2):141–146, Oct 1992.

24. A. Preece, P. Gokulchander, C. Grossner, and T. Radhakrishnan. Modeling Rule Base Structure for Expert System Quality Assurance. In *Notes of the Workshop on Validation of Knowledge-Based Systems (Thirteenth International Joint Conference on Artificial Intelligence)*, pages 37–50, Savoie, France, August 1993.

25. A. Preece, C. Grossner, P. Gokulchander, and T. Radhakrishnan. Structural Validation of Expert Systems: Experience Using a Formal Model. In *Notes of the Workshop on Validation and Verification of Knowledge-Based Systems (Eleventh National Conference on Artificial Intelligence)*, pages 19–26, Washington D.C., July 1993.

26. Marie-Christine Rousset. On the Consistency of Knowledge Bases: The COVADIS System. *Computational Intelligence*, 4(2):166–170, May 1988. Also in *ECAI 88, Proc. European Conference on AI* (Munich, August 1–5, 1988), pages 79–84.

27. John Rushby and Judith Crow. Evaluation of an Expert System for Fault Detection, Isolation, and Recovery in the Manned Maneuvering Unit. NASA Contractor Report CR-187466, SRI International, Menlo Park CA, February 1990.

A Model of Reflection for Developing AI Applications

Mamdouh H. Ibrahim and Fred A. Cummins

EDS/Leading Technologies and Methods, 5555 New King, Troy, MI 48007
Mamdouh.ibrahim@eds.com   and   Fred.Cummins@eds.com

Abstract. Computational systems that allow their programs to reason about themselves and reflect on their computations are fundamental to delivering true Artificial Intelligence applications. This paper introduces our definition of full reflection of programming languages and environments, and presents a unified model of reflection in a class-based object-oriented environment that is complete and consistent with the stated definition. This model subsumes a number of other models of reflection. An implementation of the model and how it is used to develop an integrated, multi AI programming environment is described to illustrate the validity of the model.

1   Introduction

In considering reflection in computer languages, the tendency is to focus on applications of reflection and, in particular, the ability to analyze, monitor, or alter the actions of an active application process. This leads to a view of reflection which is subject to the particular interest of the researcher. A more consistent and universally accepted definition of reflection should be sought by returning to the reflection hypothesis defined by Brian Smith [17] as follows:

"In as much as a computational process can be constructed to reason about an external world in virtue of comprising an ingredient process (interpreter) formally manipulating representations of that world, so too a computational process could be made to reason about itself in virtue of comprising an ingredient process (interpreter) formally manipulating representations of its own operations and structures."

Brian Smith

Here we see reflection as the ability to step back from a principle activity and consider the state of that activity and how it is being conducted. This is the essence of reflective thought, where we think about what we are thinking and analyze our assumptions and techniques.

In this paper we present an object-oriented model of reflection that achieves both structural and computational reflection, as defined in the literature [5]. The model achieves structural reflection through the representation of all entities, including procedural behaviors, instance variables, and executable expressions of the language, as objects.

Computational reflection has been demonstrated using the metaobject model [15]. We will show that a simple trap mechanism, implemented as part of the model, can provide similar computational reflection. However, we believe that computational reflection does not stop at monitoring, multiple-inheritance, and statistic gathering. Computational reflection should be more consistent with the reflection hypothesis definition by including the state of computation, which is not fully realized by the meta object model.

To address this issue, our model includes objects representing the context in which execution is interpreted and the global and meta state of the environment. These extensions allow more computational reflection than can be achieved by the metaobject model or the structural model alone.

The model presented here, for the most part, has been implemented in KSL [3], [9], a reflective object-oriented language. Like Smalltalk [7] and Java [8], KSL provides class-based inheritance, where objects are described by classes. In KSL, everything is an object, including classes, behaviors, and program instructions; thus KSL can operate on itself. This quality enables classes, behaviors, and instructions to be created and modified programmatically. It also enables objects to be both components of the problem model and active elements in the program execution process, thus allowing dynamic modification of a program.

In the following section we propose conceptual and operational definitions of full reflection that satisfy Smith's reflection hypothesis. We then focus our attention to how reflection can be implemented in a computer system, particularly in an object-oriented environment. We begin with a unified, basic object-oriented model of reflection. We then present extensions for the basic model necessary to incorporate access to the state of the environment, perform ad hoc reflective operations, and capture the meta state of computation. We conclude by discussing the limitations of reflection and directions for future work.

2. Reflection: Definitions

Conceptual Definition. The reflection hypothesis provides fundamental concepts in reflection that must be included in any consistent and universally accepted definition. Based on this hypothesis, we can conceptually think of reflection as the process by which a system operates on itself. This definition was further refined by Smith, during the discussions of the Workshop on Reflection and Metalevel Architectures held in conjunction with OOPSLA '90 [12], as follows:

"Reflection: an entity's integral ability to represent, operate on, and otherwise deal with its self in the same way that it represents, operates on, and deals with its primary subject matter."

The definition suggests that to cover all aspects of a system operating on itself, it is not enough to examine the system's own activities, but it must also be capable of altering its execution course and state. This has been referred to as the "causal link" which is considered a condition for reflection [15].

Operational Definition. Once we have a conceptual definition of reflection, we need a consistent operational definition that will allow us to achieve reflection in computer systems, particularly in an object-oriented environment. Computer systems consist of programs written in a particular programming language with an interpreter (which may be the hardware) that executes these programs within an environment. Our operational definition consists of the following two necessary and sufficient conditions.

1) The language must incorporate internal structures that represent the language itself and the programs written in it in the same way they represent the external domain.
2) The interpreter of such a language must allow programs written in the language to access and manipulate the structures that implement the language and its programs, as well as the state of computation.

In object-oriented programming, objects represent the entities of the problem domain. For an object-oriented language to be reflective, the entities of the object-oriented paradigm must be also represented with objects. With such a model, a system can reflect on object-oriented activity through the object-oriented paradigm. The sections that follow will present such a model in stages, beginning with a basic model and then extending it to incorporate additional aspects of reflection appropriate to an object-oriented environment.

3   The Basic Model

Fundamental to the basic model is the representation of all entities in the environment as objects that are instances of classes. The reflective model involves four principle classes of objects: class, metaclass, behavior, and expression. Figure 1 illustrates these principle classes which represent object structure and behavior.

Target Object Class. A target object is defined by its class. The class defines the structure and the behavior of its instances. Each class identifies its super-class from which the initial definitions are inherited. It also incorporates local method and instance-variable definitions which may override or add to the behaviors inherited by the class. Most class-based object-oriented environments use this conceptual form of target object class.

289

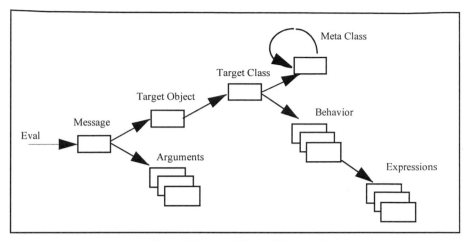

Figure 1. Structural View of the Basic Model

Meta Class. Classes are instances of a class which is an instance of itself. The meta class defines behaviors on classes. The meta class could be specialized to define new types of classes.

Behaviors. Behaviors describe the mechanism of response to specific message types. There are two fundamentally different behaviors, (1) an "instance-variable access" behavior that implicitly defines an instance variable and may perform integrity checks or propagation of effects when a variable value is assigned, and (2) a method which may perform various computations and send messages to other objects. In our model, as in our KSL language, both methods and instance variables are specified with behavior objects since both are accessed through message sends. Local behavior objects are components of the target class definition. Figure 2 presents the class hierarchy of KSL behavior classes (the TrapObject class will be discussed later).

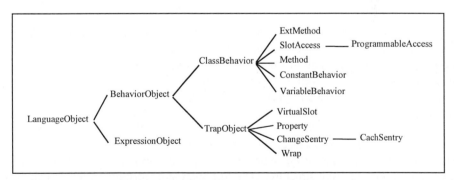

Figure 2. The Class Inheritance of the KSL Behavior Objects

Expressions. Operations performed by behaviors are specified by expression objects. The fundamental expression object is a message, but there are, necessarily, other classes of expressions which perform such actions as local variable references and assignments and execution flow control. An expression object is executed by sending it an evaluation message. Figure 3 illustrates the expression classes in KSL.

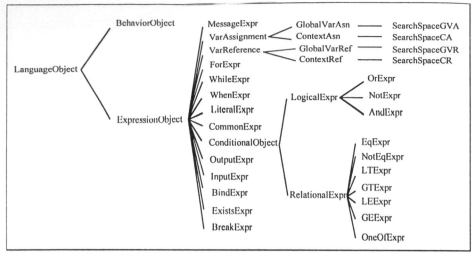

Figure 3. The Class Hierarchy of KSL Expression Objects

Object Interactions. The roles and relationships of the objects in this model can be better understood by examination of their interactions in the resolution of a message to the target object, as depicted in Figure 1. The target object contains the identity of its class. The behaviors defined on the class are searched for a selector that matches the message selector. When the matching behavior is found, an evaluation message is sent to it. The behavior, assuming a method, evaluates its executable expressions in sequence to perform the desired operation and then returns a result.

An instance variable access behavior has the additional role of retrieving the value of an instance variable. This is performed through a primitive behavior on the target class that is capable of accessing the appropriate element of the target object structure.

Note that in this model, the first condition of our definition of reflection is realized as a result of representing all the language constructs as objects. This gives us access to the programs (methods) and instructions of the language (expressions) including the messages themselves, as well as object structure definitions (classes). This concept was utilized to extend KSL to provide an integrated AI programming paradigm that includes both rule-based and logic-based programming. A *RuleSet* class is created as a specialization of the *Method* class. *RuleSet* inherits all the *Method* class behavior; however, its list of expressions contains rule instances which are instantiations of the *Rule* class. The *Rule* class is implemented as a specialization of the *ExpressionObject* class that defines the rule structure. The left hand side attribute of a rule contains a KSL condition expression and the right hand side attribute contains a list of action expressions. *RuleSet* objects contain an inferencing control object which provides the inference engine behavior. Behaviors defined on these control objects determine the type of inferencing to be performed, i.e., backward chaining or forward chaining.

Thus the running program can change the behavior of a *RuleSet* by associating the *RuleSet* with a different inference control object.

Similarly, KSL/Logic was developed as an extension of KSL that integrates logic programming with objects and rule-based programming using the same basic model of reflection described above. Through extensions to the class hierarchy, logic expressions are implemented as specializations of the *ExpressionObject* class, while a *Predicate* class was introduced as a specialization of the *Method* class to represent first order logic predicates. New variables that allow dynamic binding and backtracking were implemented as specialization of *VarAssinment and VarReference classes*, respectively. Further discussion can be found in [11]

4. Access to State of the Computation

The basic model described above provides access to the definitions of objects but not to the state of the environment. This requires representation of message-sending and variable binding with a context class.

Context Objects. A context object contains an association of local variable names with values. It is essentially the working memory of a method and corresponds to a stack entry in conventional function-call processing. In KSL, the current context is passed as an argument of the evaluation message to any expression. If the expression is a variable reference, the value is retrieved from the context. If the expression is a variable assignment, the value is associated with the variable name in the context. Each context also identifies the previous context that was in effect when it was created.

Message Resolution. The creation of a context occurs as the result of evaluation of a message expression. The message evaluation behavior first evaluates the component expressions of the message to obtain the actual, as opposed to symbolic, identities of the target object and message arguments. The message selector, target object, and message arguments then become arguments in a message-resolution message to the target object class. The class message-resolution behavior determines the appropriate target behavior to be executed. The message-resolution message is then forwarded to the target behavior. If the target behavior is a simple instance variable access behavior, the appropriate access is performed and a value returned. If the target behavior is a method, a context is created to associate the target object and arguments with the method parameters. The local variables and default values are also specified in the context. The currently active context pointer is assigned to an instance variable of the new context; this implements the stack structure of the processing environment. The new context is then used in the evaluation of the method's executable expressions.

Note that in this extended model, the message resolution mechanism and the structure access mechanism (and thereby the underlying implementation of object structure) are behaviors of the class and as such are both defined by the meta class, the class that defines classes.

Global Variables. In addition to parameters and local variables, an environment requires global variables as initial points of reference to its representations. In this model, and in KSL, global variables are represented as values associated with symbol objects. Thus a global variable reference is a reference to an instance variable of a symbol object. This defines the global "context".

This extension provides the model with the necessary mechanism to allow its interpreter to access the structures that represent the state of execution.

5   Ad hoc Reflective Operations

So far, the extended model provides the ability to reflect on the current state of the system and the code being executed, but it does not provide a good mechanism for introducing ad hoc side effects to invoke reflective operations at key points in the processing. We will refer to these reflective operations here as monitoring.

Traps. Monitoring is accomplished by extending our model with another class of objects, traps, that replace the target object reference to its class. When a message is sent to a trapped target object, the message evaluator then sends the message-resolution message to the trap instead of to the class. The trap message-resolution behavior then performs special processing for the particular instance. This is illustrated in Figure 4.

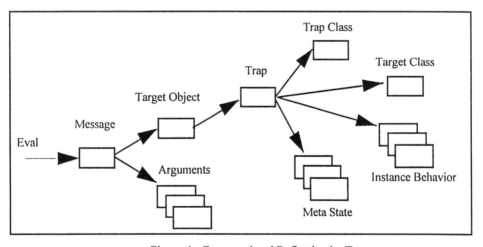

Figure 4. Computational Reflection by Traps

The trap provides a mechanism for instance specialization; in other words, it supports addition of local behaviors and side effects on the particular instance. Instance behaviors are attached to the trap. When a message is received, the trap message-resolution behavior determines if there is an instance behavior with a matching selector, i.e., if the message is to be intercepted by an instance behavior. If there is a matching behavior the message is forwarded to it. If there is no matching instance behavior, the message is forwarded to the class of the target object to be processed in the normal manner. Instance behaviors may, in turn, forward the message to the class for normal processing before and/or after producing special side effects.

Compiled vs. Interpreted Behaviors. Another form of ad hoc reflection is replacement of a compiled behavior with its object representation so that method resolution shifts to reflective (interpreted) mode to allow stepping, tracing and break-point setting.

KSL uses these extensions to implement several reflective facilities. For example, traps are used to incorporate access-oriented programming. Monitoring of object behaviors is achieved by trapping appropriate messages to a target object and invoking corresponding behaviors attached to the trap. Debugging, tracing, and stepping, all reflective facilities, are implemented in KSL through posting traps on the language expression objects to intercept their evaluation messages during method execution.

6   The Meta State

The reflection model is still not complete unless there is a mechanism for recalling how the current state of an object was reached. Reflection must be able to analyze events leading up to the current situation.   In order to explain the current state, "meta-state" information must be captured each time the state of an object changes. Each event should be captured in a meta-state object which identifies the source of the event, the prior state, and the next-previous meta-state object.

It is not practical nor necessary to recall everything, only those things that might need to be explained or retracted. Consequently, the capture of meta-state information should not be programmed into the class of a target object. Instead, capture of meta-state information should be performed by message resolution on a trap object (see Figure 4). An instance variable on the trap object can provide a link to the latest meta-state object. The trap message resolution can recognize special messages for access to the meta-state information. Additional capture logic should determine if and when old meta-state objects should be discarded as no longer relevant.

KSL uses this concept to achieve inference and domain explanation for its rule-based facilities [10]. The explanation facilities determine what rule events occurred to bring about the current state. As inferencing proceeds, the inference engine generates event objects which constitute a trail of the sequence of events. The explanation information is captured in the domain model by attaching events with traps. For example, if a slot (instance variable) defined with an "explainable" access behavior is active, then any change to the slot value will cause an event object to be attached to explain how the new slot value was obtained. In addition, a record of the inferencing activities is stored on the current session object instantiated at the beginning of the inferencing process. An event browser is developed as a network display facility to examine the event trail.

7   Relation to other Reflection Models

An excellent background on Computational Reflection and its implementation is presented by Malenfant et. al. in Reflection'96 [16]. However, most reflection models in the literature address certain aspects of reflection, but not all. The best summary of such models was given by Ferber [5]. He identified three models, the metaclass, metaobject, and meta-communications models, to deal with structural reflection [4] and computational reflection [15].

In our basic model, structural reflection is achieved through the representation of all entities as objects. Other object-oriented languages and systems achieve partial reflection by adopting a limited form of this approach. For example, Smalltalk [7] represents classes and methods as objects, and OBJVLISP [2] represents entities as instances of classes that are instances of other classes called metaclasses. Metaclasses in OBJVLISP are subclasses of the class Class that is an instance of itself. Our basic model extends both Smalltalk and OBJVLISP models to represent the procedural behaviors, instance variables, and executable expressions of the language as objects. This allows access to all aspects of the language structure thus providing greater structural reflection.

Computational reflection as defined by Maes [15] was implemented using the metaobject model. Several systems were implemented to provide this type of reflection [15], [18]. The approach these systems adopted for achieving computational reflection focused on introducing a metaobject (an instance of the metaobject class) for each target object to intercept its messages and perform desired reflective operations. The simple trap mechanism, implemented as part of our model, can provide such computational reflection. However, we extended our basic model to include objects that represent the global state of the environment and the context in which execution is interpreted. These extensions are unique to our unified model and allow more computational reflection than can be achieved by the metaobject model alone.

The model also lends itself to representing features of other object-oriented languages. For example, behaviors may be selected based on argument classes to implement the multi-method capabilities of CLOS [14]. Furthermore, a specialized form of trap can be used to implement delegation for a prototyping form of representation [13], and a specialization of the trap class can also provide a dispatcher to implement concurrent object-oriented programming a la ACTORS [1].

8   Conclusions and Future Work

In this paper we proposed conceptual and operational definitions of full reflection that satisfy Smith's reflection hypothesis. We presented our view of a unified object-oriented model of reflection that is necessary to achieve full reflection, and showed that it subsumes most of the existing models of reflection. The model, for the most part, has been implemented in KSL, but the model is also suitable for representation of other object-oriented languages.

The principle challenge for the future is to provide reflection with performance. Research in this area should focus on two main issues: (1) implementation of the non-reflective underlying code, and (2) the ability to transition code that is compiled (non-reflective) to reflective mode allowing selective but ad hoc reflection. The model described here should provide an appropriate framework for this work.

References

1. Agha, G., ACTORS - A Model of Concurrent Computation in Distributed Systems, The MIT Press, 1986.
2. Briot, J-P and Cointe, P., "A Uniform Model for Object-Oriented Languages Using the Class Abstraction," Proceedings of IJCAI-87, August 1987.
3. Cummins, F., Bejcek, W., Ibrahim, M., O'Leary, D., and Woyak, S., "EDS/OWL Reference Manual," Internal Technical Report, EDS R&D, Troy, MI, 1987.
4. Cointe, P., "MetaClasses Are First Class Objects: The OBJVLISP Model," Proceedings of OOPSLA-87, 1987.
5. Ferber, J., "Computational Reflection in Class based Object Oriented Languages," Proceedings of OOPSLA-89, 1989.
6. Friedman, D. P., and Wand, M., "Reification: Reflection without Metaphysics," Proceedings of the 1984 ACM Symposium on Lisp and Functional Programming, 1984.
7. Goldberg, A., and Robson, D., Smalltalk-80 -- The Language and its Implementation, Addison-Wesley, Reading, MA, 1983.
8. Gosling, J., Joy, B., and Steele G., The Java Language Specification, Addison-Wesley, 1996.
9. Ibrahim, M. and Cummins, F., "KSL: A Reflective Object-Oriented Programming Language," Proceedings of the IEEE Computer Society International Conference on Computer Languages, 1988.
10. Ibrahim, M. and Woyak, S., "An Object-Oriented Environment for Multiple AI Paradigms," Proceedings of the IEEE Computer Society Conference on Tools for Artificial Intelligence, 1990.
11. Ibrahim, M. and Cummins, F, "Objects with Logic," Proceedings of ACM CSC '90, 1990.
12. Ibrahim, M., "Report on the OOPSLA/ECOOP '90 Workshop on Reflection and Metalevel Architectures in Object-Oriented Programming," OOPSLA/ECOOP '90 Addendum to Proceedings, SIGPLAN Notices Special Issue, 1991.
13. Ibrahim, M., Bejcek, W., and Cummins, F., "Instance Specialization without Delegation," Journal of Object-Oriented Programming, Vol. 4, No. 3, June 1991, pp. 53-56.
14. Keene, S. E., Object-Oriented Programming in Common Lisp: A Programmer's Introduction to CLOS, Addison-Wesley, 1989.
15. Maes, P., "Concepts and Experiments in Computational Reflection," Proceedings of OOPSLA-87, 1987.
16. Malenfant, J., Jacques, M., and Demers, F-N., "A Tutorial on Behavioral Reflection and its Implementation," Proceedings of Reflection '96, 1996.
17. Smith, B. C., "Prologue to Reflection 5 and Semantics in a Procedural Language," in Readings in Knowledge Representation, Ronald J. Brachman and Hector J. Levesque (Eds.), Morgan Kaufmann Publishers, Los Altos, CA, 1985.
18. Watanabe, T., and Yonezawa, A., "Reflection in an Object-Oriented Concurrent Language," Proceedings of OOPSLA-88, 1988.

A Compositional Process Control Model and Its Application to Biochemical Processes

Catholijn M. Jonker, Jan Treur

Vrije Universiteit Amsterdam, Department of Artificial Intelligence
De Boelelaan 1081a, 1081 HV Amsterdam, The Netherlands.
URL:http://www.cs.vu.nl/~{jonker,treur}.   Email: {jonker,treur}@cs.vu.nl

Abstract. A compositional generic process control model is presented which has been applied to control enzymatic biochemical processes. The model has been designed at a conceptual and formal level using the compositional development method DESIRE, and includes processes for analysis, planning and simulation. It integrates qualitative and quantitative techniques. Its application to enzymatic chemical processes is described.

1 Introduction

Process control is a task that has many application domains, like production processes in industry (e.g., chemical industry, car industry), any automated process that uses a conveyor belt or assembly-line, but also in hospitals (e.g., brain-scanners, intensive-care) and remote robot control (e.g., space shuttles, the docking of space-crafts, nuclear reactors, deep-sea exploratory vessels) for environments that are hostile for human beings or for situations in which humans are not capable of receiving and interpreting sensory information quickly enough to make the right decisions.

Applications for process control often are developed in an ad hoc manner, with no explicit specification at a conceptual level or built in facilities for reuse or verification. In this paper, a reusable model for process control is described which has been designed using the compositional development method DESIRE (cf. [1], [2]). The model covers analysis of the current state (and history) and the possibility to simulate a possible plan before actually selecting and executing it.

The model is generic in two senses: it is generic with respect to the processes or tasks, and it is generic with respect to the information structures and knowledge. Genericity with respect to processes or tasks refers to the level of process abstraction: a generic model abstracts from processes at lower levels. A more specific model with respect to processes is a model within which a number of more specific processes, at a lower level of process abstraction are distinguished. This type of refinement is called *specialisation*. Genericity with respect to knowledge refers to levels of knowledge abstraction: a generic model abstracts from more specific information structures and knowledge. Refinement of a model with respect to the knowledge in specific domains of application, is refinement in which knowledge at a lower level of knowledge abstraction is explicitly included. This type of refinement is called *instantiation*. Reuse of such a generic model can take place by

- adding domain-specific information structures and knowledge (instantiation)
- adding more specific sub-processes within the processes defined by the generic model (specialisation)
- adding or deleting components (reconfiguration)

In addition to the possibility to reconfigure, also verification is supported by the compositional structure of the design.

The process control model presented has been reused in the domain of enzymatic reactions, in particular for penicillin production processes. The prototype implementation developed integrates qualitative methods acquired in the form of expert knowledge, and quantitative techniques (a numerical simulation model).

In Section 2 the problem of process control is discussed, and the example domain of application is presented. In Section 3 the processes in the generic model are presented, together with their composition relation. In Section 4 the generic information types are presented. In Section 5 it is shown how it has been instantiated by information structures and knowledge on the specific application domain.

2 Problem Description

For effective control of a process, a good understanding of the current situation and history of that process is vital. Often it is also important to make predictions of future situations of the process. By undertaking proper actions it might be possible to prevent undesired (predicted) situations. Process control can be used to keep the process within acceptable bounds, but also to optimize a process.

The two basic generic information elements for process control are observations and actions. Observation information (for example, acquired by sensors) is needed to assess the current situation. Based on the assessment of the situation (but also previous situations), actions must be performed to control the process. Process control can be performed with or without *simulation* of the plans that are determined to control the process.

2.1 Domain of Application: Enzymatic Biochemical Reactions

In chemical industry more and more production processes for medicins are based on enzymatic reactions. For example, benzylpenicillin is an antibiotic that is directly used as a medicin. It can be produced from 6-amino penicillin acid (6-APA) and phenyl acetate. The reaction is described by the following:

benzylpenicillin $\leftrightarrow$ 6-amino penicillin acid + phenyl acetate

The reaction is a balance reaction, where the balance is determined by the pH of the mixture. The reaction takes place in water until an equilibrium is reached at a certain pH, depending on the starting concentrations. Since it is an enzymatic reaction, the mixture needs to contain penicillin amidase for the reaction to take place.

To produce benzylpenicillin the mixture needs to contain the same amount of 6-APA as it contains phenyl acetate. Furthermore, the mixture must contain so much phenyl acetate that it has a pH-degree lower than 5. An example: if pH=4.4, then 88% of the 6-APA is transformed into benzylpenicillin. Because the mixture will contain less and less acetate by this reaction, the pH will rise (i.e., the mixture will become less acid). Therefore, if the production is to continue, the pH must be kept low, so phenyl acetate is to be added. Furthermore, if phenyl acetate is added, the same amount of 6-APA must be added as well.

The enzyme is very sensitive to acids, it deteriorates rapidly if the pH drops below 4.3. Furthermore, the enzyme only functions good if the temperature is close to 25°C. To monitor and control the production process correctly, there are two thermometers (one for the temperature in the kettle, and one for the temperature of the surroundings), one pH-electrode (to measure the pH in the kettle), a dial to set the heating of the kettle, four smaller kettles containing 6-APA, enzyme, acetate and sodium hydroxide

(NaOH) respectively. The sodium hydroxide is a base, and can therefore be used to raise the pH if necessary. Each of these kettles can be made to release a standard amount of material.

2.2 The Requirements

The process control system has to be able to analyse the state of a process in terms of the assessments specified for the domain of application. On the basis of these assessments the system is to determine a plan of actions with which the process is to be controlled. These plans must be tested first before being applied to the process in the external world. Therefore, the process control system has to contain a simulation of the world process.

The task of analysing (or monitoring) the process must be exercised on the process running in the external world as well as on the simulated process. Generated plans are tested on the simulation, and the simulation results have to be analysed in order to adapt the plan before executing it on the process in the external world. Given that for complex processes observation might be costly and/or time consuming, the system has to determine when and which observations are to be performed both on the process in the external world as well as on the simulated process.

The simulation of the process has to be quick enough so that the plan is still useful for the process in the real world. On the other hand it has to give a reasonable prognosis of the effect of the plan on the process running in the external world. A problem is that very accurate simulations are (often) time consuming. These two constraints have to be balanced within the process control system.

3 A Compositional Generic Model for Process Control

Within the generic model different levels of process abstraction are considered. The process composition relation (cf. [2]) defines how the behaviour of a component emerges from the behaviours of its sub-components at the next lower level of process abstraction. The definition of a process composition relation consists of a static part (the information links) and a dynamic part (task control).

At the top level, two components are modelled, see Fig. 1: process control task and external world. They interact with each other in a bidirectional manner. The information on observation results is transferred from the external world to the process control task by the link world observation information. The information on the observations and actions to be performed is transferred from the process control task to the world by the link selected actions and observations.

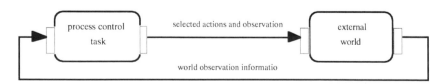

Fig. 1 Top level process composition: information links

The component process control task is composed of the components process analysis, simulated world processes, and plan determination, see Fig. 2.

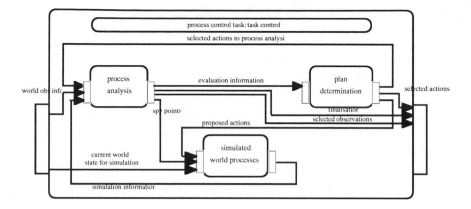

Fig. 2 Process composition of the process control task: information links

The component process analysis within the process control task is composed of two components: process evaluation and determine observations, see Fig. 3.

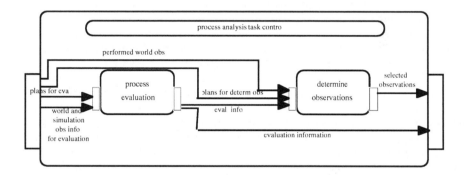

Fig. 3 Process composition of process analysis: information links

The information that can be obtained from observations using the sensors is modelled by the information type observation result info (in the example: observation information on temperature and pressure). The information type observation result info is used both in the output interface of the component external world and in the input interface of the component process control task.

The information types observation info and action info model information on the observations and actions to be performed in the external world. These information types are used in the output interface of process control task and the input interface of external world. Within the component process control task information types are defined that are used by its three internal components. The information type assessments is used in the output interface of the component process analysis and in the input interface of plan determination. Assessments represent information on the state of the process. The information type observation result info is used as an input of simulated external world to callibrate the simulation models used. Note that two types of observations can be performed: *incidental observations* that return an observation result for only the current point in time, and *continuous observations* that continuously return all updated observation results as soon as changes in the world occur.

Task control within the process control system is only discussed globally. All processes and information links at the top level are awake, i.e., both the external world and the process control task process information as soon it arrives. This enables the process control task to interpret new observation result information as soon as possible. Furthermore, plans can be determined as quickly as possible, and long term effects can be predicted by use of the simulation task.

Within process control task a cycle of process activations takes place continuously. In each cycle, first process analysis checks the effects on the process running in the external world of the plan that is currently being executed and determines which incidental observations have to be performed. If this check is satisfactory, the simulation is activated, to give a prognosis of the rest of the current plan with respect to the last observation result information from the external world, after which a new cycle begins. If, however, the current plan is no longer satisfactory, component plan determination is activated with task control focus new plan. Component plan determination comes up with a new plan, sends it to the simulation and the analysis components. If plan determination finished determination of a new plan, process analysis is activated. Component process analysis determines which observations have to be performed on the simulation that are relevant for the new plan. Then the simulation is activated with the new plan and the latest observation results. Component simulated world processes provides simulation results after which component process analysis is activated again. Component process analysis check the new plan, if it is satisfactory it makes the new plan the current plan and allows it to be executed in the external world and a new cycle begins. If the check is unsatisfactory, component plan determination is activated, starting a new sub-cycle for the production of a satisfactory new plan. Information links are activated as relevant for the above cycle. However, the information link world obs info to component process analysis is made awake to enable the component to react directly to the latest information (to be able to react to emergency situations at any time). The processes within process analysis are activated in a row, first proces evaluation, then observation determination. All information links are awake.

4   Knowledge Composition: Generic Information Types

In this section the generic information types used in the process control model are briefly discussed. Based on the *observation result information* the process controller is to decide which actions are to be performed. The process controller receives statements about what has been observed of the state of the process. The generic information type observation results is used to express observation result information.

The aim is to express statements like "the observation result is that it is true that the pressure is low" and "the observation result is that it is false that the pressure is high". The information type observation results includes the sorts INFO ELEMENT and SIGN. Terms of the sort INFO ELEMENT refer to statements in the language defined by the domain specific information type domain info. In the generic model the information type domain info remains empty. If the model is applied this information type is instantiated with domain-specific information structures.

The information type action info makes use of the generic information type actions to be performed and is meant to enable the reasoning about actions. In the information type actions to be performed the sort ACTION is introduced and the relation to be performed is added to be able to reason about actions. This information type is generic; no reference is made to domain specific information types. The information type action info is composed of this generic information type and the domain specific information type domain actions. In the generic model this information type remains empty. If the model is applied, this information type is instantiated with the domain-specific action names. For example, the atom to_be_performed(add(enzyme)) refers to one of the actions used in the application domain addressed in Section 5. Similarly the other information types have been specified. For example, the information type assessments defines the relation assessment that is used to express output of the analysis process.

5 Application Domain Specific Knowledge

In this section the relevant knowledge in the application domain of enzymatic reactions is described, and related to the generic model.

5.1 Domain Specific Knowledge used in Analysis

In enzymatic reactions the following observations can be made:
* pH
* T, the (internal) kettle temperature
* Tlab, the temperature of the laboratory
* heater

Given this observation information from the component external world the task of the component process_analysis is to analyse the process. For the domain of benzylpenicillin a knowledge base has been acquired from the domain expert which provides assessments of the process in terms of:
* kettle temperature is too high
* kettle temperature is high, but not too high
* kettle temperature is optimal
* kettle temperature is too low
* kettle temperature is low, but not too low
* pH is too high
* pH is high, but not too high
* pH is optimal
* pH is too low
* pH is low, but not too low
* no enzyme in the mixture
* enzyme in the mixture
* no reaction is taking place
* reaction is taking place
* not enough 6-APA in the mixture
* enough 6-APA in the mixture

5.2 Domain Specific Knowledge used in Planning

The process can be influenced in the following manners:
* addition of: acid, base, 6-APA, or enzyme
* changing the heater.

Given the information from the component process_analysis the component plan_determination is to produce a plan to correct the process. For the domain of benzylpenicillin the component plan_determination can have to correct the following situations:

- kettle temperature above optimum but not too high
- kettle temperature too high
- kettle temperature below optimum but not too low
- kettle temperature too low
- pH above optimum but not too high
- pH too high
- pH below optimum but not too low
- pH too low
- no enzyme in the mixture
- no reaction is taking place
- not enough 6-APA in the mixture

The following table (acquired from the domain expert) represents the knowledge used for action selection:

react ion	enzy me	kettle temp	pH	action type			
				temp	pH	enzyme	apa
+	+	above optimum	above optimum	decrease	acid	naught	naught
			ok	decrease	naught	naught	naught
			below optimum	decrease	base	naught	naught
		ok	above optimum	nil	acid	naught	naught
			ok	nil	naught	naught	naught
			below optimum	nil	base	naught	naught
		below optimum	above optimum	increase	acid	naught	naught
			ok	increase	naught	naught	naught
			below optimum	increase	base	naught	naught
-	+	above optimum	above optimum	decrease	acid	naught	naught
			ok	decrease	naught	naught	naught
			below optimum	decrease	base	naught	naught
		ok	above optimum	nil	acid	naught	naught
			ok	nil	naught	naught	add
			below optimum	nil	base	naught	naught
		below optimum	above optimum	increase	acid	naught	naught
			ok	increase	naught	naught	naught
			below optimum	increase	base	naught	naught
-	-	above optimum	above optimum	decrease	acid	naught	naught
			ok	decrease	naught	naught	naught
			below optimum	decrease	base	naught	naught
		ok	above optimum	nil	acid	naught	naught
			ok	nil	naught	add	naught
			below optimum	nil	base	naught	naught
		below optimum	above optimum	increase	acid	naught	naught
			ok	increase	naught	naught	naught
			below optimum	increase	base	naught	naught

The combination reaction + and enzyme - is impossible.

6   The Simulation Model

The enzymatic reaction for the production of benzylpenicillin is as follows:

$$E + A \underset{k_1}{\overset{k_{-1}}{\rightleftarrows}} EA \underset{k_2}{\overset{k_{-2}}{\rightleftarrows}} E + P + Q$$

Notation	Entity	Measure
E	enzym: penicillin amidase	mol / liter (concentration)
A	benzylpenicillin	mol / liter
EA	intermediate result	mol / liter
P	phenyl acetic acid	mol / liter
Q	6-APA	mol / liter
k	reaction rate constant	/ sec

From the reaction specification it is clear that four reaction rate constants play a role. However, an additional deactivation rate constant k_3 plays a role, due to the deterioration of the enzyme according to temperature and pH.

For the computation of the concentration of the different substances in the solution, the following difference equations hold:

$$c_A(t + dt) \; = \; c_A(t) + (- k_1 c_E c_A + k_{-1} c_{EA}) dt$$

$$c_P(t + dt) \; = \; c_P(t) + (- k_{-2} c_E c_P c_Q + k_2 c_{EA}) dt$$

$$c_Q(t + dt) \; = \; c_Q(t) + (- k_{-2} c_E c_P c_Q + k_2 c_{EA}) dt$$

$$c_{EA}(t + dt) \; = \; c_{EA}(t) + (k_1 c_E c_A - k_{-1} c_{EA} + k_{-2} c_E c_P c_Q - k_2 c_{EA}) dt$$

$$c_E(t + dt) \; = \; c_E(t) + (-k_1 c_E c_A + k_{-1} c_{EA} - k_{-2} c_E c_P c_Q + k_2 c_{EA} - k_3 c_E) dt$$

where dt is the duration of 1 step measured in minutes.

Experimental results show that, if the temperature is between 5°C and 30°C, the reaction rate constants behave approximately linear with respect to temperature. The reaction rates are 0 at 0°C, as the mixture in the kettle then freezes. Therefore, we chose to use the following linear equations for the reaction rate constants:

$$k_1 \; = \; 0.04 * c * (T - 273)$$
$$k_{-1} \; = \; 5000 * c * (T - 273)$$
$$k_{-2} \; = \; 6 * c * (T - 273)$$
$$k_2 \; = \; 1 * c * (T - 273)$$

where the rates correspond to reaction rates per minute. Although it seems as if k_{-1} is much higher than the others, it's effect is less than that of k_{-2}. Both have more effect than k_1 and k_2, which is as one would hope if penicilline is to be produced.

The case i = 3 is special: the enzym deteriorates rapidly if the pH leaves the vicinity of 4.4 and if the temperature within the kettle rises above 25°C. This was modelled by:

$$k_3 \; = \; c \left(1 - \frac{1}{1 + z(pH - 4.4)^2}\right) + c \frac{1}{2} \left(1 + \frac{T - 303}{\sqrt{1 + (T - 303)^2}}\right)$$

The parameter z determines how fast the deterioration goes. For this application the value 25 for the parameter z was estimated. The current pH is given by:

$$pH \quad = \quad - \log(\sqrt{(cP1.7 \times 10^{-5})})$$

Computing backwards from the optimum pH of 4.4, we see that then c_p should be 0.000093229 mol/liter. Since these are unpleasant figures to work with, we normalize the computation of pH in such a way that if we choose $c_p = 100$, then pH = 4.4:

$$pH \quad = \quad - \log(\sqrt{(0.00000093229 \times cP \times 1.7 \times 10^{-5})})$$

The change in the kettle temperature T per dt depends on the temperature of the laboratory and the state of the heater:

$$\Delta T / dt \quad = \quad c * \beta * heater - c * \alpha * (T - Tlab)$$

In the current model we estimated β to be 300000, and α to be 50000. The heater can have the following values: 1, 2, and 3. The temperature of the laboratory fluctuates, but is assumed to be somewhere between 10°C and 30°C.

7 Conclusions

The generic model for process control presented in this paper was designed on the basis of earlier experiences in the control of ship building processes. On the basis of the generic model, the application to the control of enzymatic reactions, in particular in antibiotics production was designed in a relatively short time; most of the effort was spent in building the simulation model. The application integrates qualitative methods (acquired from our domain expert) and quantitative techniques (the simulation model based on differential equations). The prototype implementation that was automatically created on the basis of the design, using the DESIRE software environment, has been tested in a simulated environment, but not yet in the real environment.

This project has shown that the generic model for process control indeed provides a strong form of reusability, and improves the efficiency of the development process of applications to a large extent. The generic and compositional nature of the process control model supports reusability of the model as a whole, but also of separate components within the model.

To prove that an application with this model works properly, the compositional verification method introduced in [4] can be used, in a similar manner as how this has been done for a generic model of diagnosis; see [3]. This compositional verification method relates dynamic properties of a system as a whole to properties of system components, and properties of components to sub-components, and so on. Finally the dynamics of the system as a whole can related to properties of the knowledge used to specify the primitive components, and environmental and domain assumptions. The formulation of these properties and the proofs of their relations can be performed in a generic manner. The generic and compositional structure of the model presented here provides an appropriate basis for this, in addition to existing techniques, for example in specification and verification of reactive systems (cf. [5]). Behavioural properties of the compositional process control model such as reactiveness and pro-activeness can also be studied in the context of intelligent agents; cf. [4], [6], [7].

Acknowledgements

The authors have learned a lot on process control of enzymatic reactions from the domain expert Carla van Wees (Technical University Delft and Lonsa, Switserland). Frances Brazier, David de Klerk and Pieter van Langen contributed to the development of the model in the context of co-ordination of ship building projects. Moreover, Wieke de Vries, Lourens van der Mey, and the students of the course 'Design of Multi-Agent Systems' in the year 1997/1998 provided support and feedback on the model and the application.

References

1.   Brazier, F.M.T., Dunin-Keplicz, B., Jennings, N.R. and Treur, J., Formal specification of Multi-Agent Systems: a real-world case. In: V. Lesser (Ed.), *Proceedings of the First International Conference on Multi-Agent Systems, ICMAS-95*, MIT Press, Cambridge, MA, 1995, pp. 25-32. Extended version in: International Journal of Cooperative Information Systems, M. Huhns, M. Singh, (Eds.), Special issue on Formal Methods in Cooperative Information Systems: Multi-Agent Systems, vol. 6, 1997, pp. 67-94.

2.   Brazier, F.M.T., Jonker, C.M., and Treur, J., Principles of Compositional Multi-agent System Development. In: J. Cuena (ed.), Proceedings of the 15th IFIP World Computer Congress, WCC'98, Conference on Information Technology and Knowledge Systems, IT&KNOWS'98, 1998, pp. 347-360.

3.   Cornelissen, F., Jonker, C.M., Treur, J., Compositional verification of knowledge-based systems: a case study in diagnostic reasoning. In: E. Plaza, R. Benjamins (eds.), *Knowledge Acquisition, Modelling and Management, Proceedings of the 10th EKAW'97*, Lecture Notes in AI, vol. 1319, Springer Verlag, 1997, pp. 65-80.

4.   Jonker, C.M., Treur, J., Compositional Verification of Multi-Agent Systems: a Formal Analysis of Pro-activeness and Reactiveness. In: W.P. de Roever, H. Langmaack, A. Pnueli (eds.), *Proceedings of the International Workshop on Compositionality, COMPOS'97*. Lecture Notes in Computer Science, vol. 1536, Springer Verlag, 1998, pp. 350-380.

5.   Manna, Z., Pnueli, A., Temporal Verification of Reactive Systems: Safety. Springer Verlag, Berlin, 1995.

6.   Wooldridge, M., Jennings, N.R. (eds.). *Intelligent Agents,* Lecture Notes in Artificial Intelligence, Vol. 890, Springer Verlag, Berlin, 1995.

7.   Wooldridge, M., Jennings, N.R., Agent theories, architectures, and languages: a survey. In: [6], 1995, pp. 1-39.

Visual and Textual Knowledge Representation in DESIRE

Catholijn M. Jonker[1], Rob Kremer[2], Pim van Leeuwen[1], Dong Pan[2], Jan Treur[1]

[1] Vrije Universiteit Amsterdam, Department of Artificial Intelligence
De Boelelaan 1081a, 1081 HV, Amsterdam, The Netherlands
Email: {jonker, treur}@cs.vu.nl, URL: http://www.cs.vu.nl/{~jonker,~treur}

[2] University of Calgary, Software Engineering Research Network
2500 University Drive NW, Calgary, Alberta T2N 1N4, Canada
Email: {kremer, pand}@cpsc.ucalgary.ca

Abstract. In this paper, graphical, conceptual graph-based representations for knowledge structures in the compositional development method DESIRE for knowledge-based and multi-agent systems are presented, together with a graphical editor based on the Constraint Graph environment. Moreover, a translator is described which translates these graphical representations to textual representations in DESIRE. The strength of the combined environment is a powerful -- yet easy-to-use -- framework to support the development of knowledge based and multi-agent systems. Finally, a mapping is presented from DESIRE, that is based on order sorted predicate logic, to Conceptual Graphs.

1 Introduction

Most languages for knowledge acquisition, elicitation, and reasoning result in specifications in pure text format. Textual representation is easier for a computer program to process. However, textual representation is not an easily understandable form, especially for those domain experts who are not familiar with computer programming. Visual representation of knowledge relies on graphics rather than text. Visual representations are more understandable and transparent than textual representations [7].

DESIRE (DEsign and Specification of Interacting REasoning components) [2] is a compositional [3] development method used for the design of knowledge-based and multi-agent systems. DESIRE supports designers during the entire design process: from knowledge acquisition to automated prototype generation. DESIRE uses composition of processes and of knowledge composition to enhance transparency of the system and the knowledge used therein.

Originally, a textual knowledge representation language was used in DESIRE that is based on order sorted predicate logic. Recently, as a continuation of the work presented in [6] a graphical representation method for knowledge structures has been developed, based on conceptual graphs [8].

Constraint Graphs [5] is a concept mapping "meta-language" that allows the visual definition of any number of target concept mapping languages. Once a target language is defined (for example, the DESIRE's graphical representation language) the constraint graphs program can emulate a graphical editor for the language as though it were custom build for the target language. This "custom" graphical editor can prevent the user from making syntactically illegal constructs and dynamically constraints the choices of the user to those allowed by the syntax. Constraint Graph's graphical environment is used to present knowledge in a way that corresponds closely to the graphical representation language for knowledge that is used in DESIRE. A translator is described that bridges the gap between the graphical representation and the textual representation language in DESIRE.

Another well-known knowledge representation language, that also makes use of graphical notations, is Conceptual Graphs [8]. Knowledge presented in Conceptual Graphs can also be represented in predicate logic. Since DESIRE is based on order sorted predicate logic, such knowledge can also be represented in DESIRE. In this paper, a mapping is given from DESIRE to Conceptual Graphs, thus bringing DESIRE closer to that representation language.

2 Graphical Knowledge Representation in DESIRE

In this section both graphical and textual representations and their relations are presented for the specification of knowledge structures in DESIRE [2]. Knowledge structures in DESIRE consist of information types and knowledge bases. In Sections 2.1 and 2.2 graphical and textual representations of information types are discussed. In Section 2.3 representations of knowledge bases are discussed.

2.1 Basic Concepts in Information Types

Information types provide the ontology for the languages used in components of the system, knowledge bases and information links between components. In information type specifications the following concepts are used: sorts, sub-sorts, objects, relations, functions, references, and meta-descriptions. For the graphical specification of information types, the icons in Fig. 1 are used.

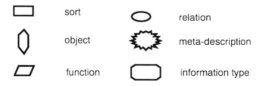

Fig. 1. Information types: legenda

A sort can be viewed as a representation of a part of the domain. The set of sorts categorizes the objects and terms of the domain into groups. All objects used in a specification have to be typed, i.e., assigned to a sort. Terms are either objects, variables, or function applications. Each term belongs to a certain sort. The specification of a function consists of a name and information regarding the sorts that form the domain and the sort that forms the co-domain of the function. The function name in combination with instantiated function arguments forms a term. The term is of the sort that forms the co-domain of the function. Relations are the concepts needed to make statements. Relations are defined on a list of arguments that belong to certain sorts. If the list is empty, the relation is a nullary relation, also called a propositional atom. The information type birds is an example information type specifying sorts, objects, functions and atoms with which some knowledge concerning birds can be specified. The information type is specified in Fig. 2. With information type birds it is, for example, possible to express statements like "Tweety is of the type that it prefers vegetarian food": is_of_type(tweety, food_preference(vegetarian)).

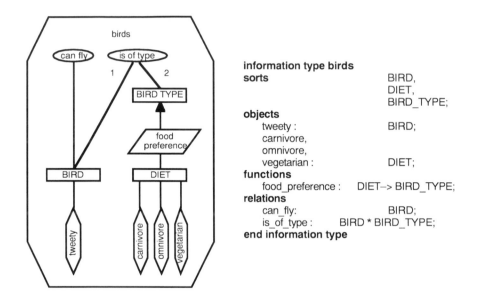

Fig. 2. Information type: **birds**

Note that being able to express a statement does not mean that the statement is true, it could be false.

2.2 Compositionality of Information Types

Compositionality of knowledge structures is important for the transparency and reusability of specifications. In DESIRE two features enable compositionality with respect to information types: information type references, and meta-descriptions. By means of information type references it is possible to import one (or more) information type(s) into another. For example, information type birds above can be used in an information type that specifies an extended language for specifying knowledge that compares birds.

Example 1

```
information type compare_birds
    information types  birds;
    relations          same_type:                          BIRD * BIRD;
end information type
```

The second feature supporting compositional design of information types is the meta-description representation facility. The value of distinguishing meta-level knowledge from object level knowledge is well recognized. For meta-level reasoning a meta-language needs to be specified. It is possible to specify information types that describe the meta-language of already existing languages. As an example, a meta-information type, called about_birds, is constructed using a meta-description of the information type birds (see Fig. 3). The meta-description of information type birds connected to sort BIRD_ATOM ensures that every atom of information type birds is available as a term of sort BIRD_ATOM. Using information type about_birds it is possible to express that it has to be discovered whether bird Tweety can fly (to_be_discovered(can_fly(tweety))).

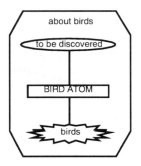

Fig. 3. Meta-descriptions: **about birds**

2.3 Knowledge Bases

Knowledge bases express relationships between, for example, domain specific concepts. Reasoning processes use these relationships to derive explicit additional information.

Example 2

```
knowledge base birds_kbs
    information types compare_birds;
    contents
        if             has_type(X: BIRD, Y: BIRD_TYPE)
        and            has_type(Z: BIRD, Y: BIRD_TYPE)
        then           same_type(X: BIRD, Z: BIRD);

        if             has_type(X: BIRD, type(Y: DIET, flying, Z: HABITAT))
        then           flies(X: BIRD);

        has_type(tweety, type(vegetarian, flying, hot));
end knowledge base
```

The knowledge base birds kbs specified in Example 2 expresses which birds are of the same type, and which birds fly. Although, knowledge bases can be represented graphically as well, examples have been omitted from this paper.

Finally, a knowledge base can reference several other knowledge bases. The knowledge base elements of knowledge bases to which the specification refers are also used to deduce information (an example has been omitted).

3 Constraint Graphs

Constraint graphs is a concept mapping "meta-language" that allows one to visually define any number of target concept mapping languages. Once a target language is defined (for example, the DESIRE knowledge representation language) the constraint graphs program can emulate a graphical editor for the language as though it were custom build for the target language. This "custom" graphical editor can prevent the user making synactically illegal constructs. Furthermore, the editor dynamically constraints user choices to those allowed by the syntax.

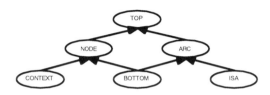

Fig. 4. The base type lattice for Constraint Graphs

In order to accommodate a large number of visual languages, constraint graphs must make as few assumptions about concept mapping languages as possible. To this end, constraint graphs defines only four base components: node, arc, context, and isa (see Fig. 4). Nodes and arcs are mutually exclusive, where nodes are the vertices from graph theory, and arcs interconnect other components, and are analogous to edges in graph theory. Both nodes and arcs may (or may not) be labeled, typed, and visual distinguished by color, shape, style, etc. Contexts are a sub-type of node and may contain a partition of the graph. Isa arcs are a sub-type of arc and are used by the system to define the sub-type relation: one defines one component to the be a sub-type of another component merely by drawing an isa arc from the sub-type to the supertype.

Futhermore, the generality requirement of constraint graphs dictates that arcs are not always binary, but may also be unary or of any arbitrary arity greater than 1 (i.e., trinary and n-ary arcs are allowed). For example, the between relation puts a trinary arc to good use. Constraint graphs arcs may interconnect not only nodes but other arcs as well. This is not only useful, but necessary because all sub-type and instance-of relations are defined using an isa arc, arcs between arcs are required to define the type of any arc. Finally, within constraint graphs no hard distinctions are made between types and instances, but rather, the object-delegation model [1] is followed where any object can function as a class or type.

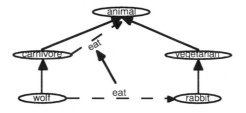

Fig. 5. An example Constraint Graphs definition

To illustrate some of the above points, Fig. 5 shows a simple definition. Here, the fat, directed arcs are the constraint graphs isa arcs and define carnivore and vegetarian to be sub-types of animal, wolf as a sub-type (or instance-of) of carnivore, and rabbit as a sub-type (or instance-of) of vegetarian. Furthermore the eat binary relation (dashed arc) is defined and starts on carnivore and terminates on animal. These terminals are important: the components at the terminals constrain all sub-types of eat to also terminate at some sub-type of carnivore and animal respectively. The second eat arc is defined (by the fat isa arc between it's label and the first eat arc's label) to be a sub-type of the first eat arc. It is therefore legally drawn between wolf and rabbit, but the editor would refuse to let it be drawn in the reverse direction: the eat definition says that rabbits can't eat wolves.

4 The Translator

In Constraint Graphs, three basic types of objects exist: nodes, arcs and contexts. The elements of the language to be expressed in the Constraint Graphs' environment therefore need to be mapped onto these basic types. Table 1 below shows the mapping between DESIRE's knowledge elements and nodes, arcs and contexts.

Object	Sort	Subsort	Meta description	Function	Relation	Information type	Knowledge Base
NODE	NODE	ARC	ARC	ARC	ARC	CONTEXT	CONTEXT

Table 1. Mapping between DESIRE and Contraint Graphs

Constraint Graphs allows the user to further constrain the language definition in by, for example, restricting the shapes and connector types of the nodes and arcs the language elements are mapped onto. In our case, we restrict the shape of node Sort to a rectangle, and the shape of Object to a diamond. Furthermore, sub-sorts, meta-descriptions and relations will be represented as directed labeled arcs, where the label takes the shape of an ellipse. Moreover, functions will be depicted as directed labeled arcs as well, but the label will be a parallelogram. Finally, information types and knowledge bases are mapped onto contexts, and the shape of these contexts will be the default: a rectangle.

Fig. 6 below gives an impression of a specification of the DESIRE information type birds (compare to Fig. 2) in Constraint Graphs.

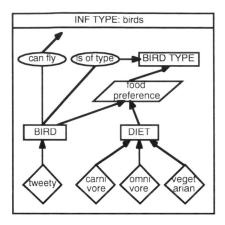

Fig. 6. Example of a DESIRE information type represented in Constraint Graphs

5 Relation to Conceptual Graphs

In this paper, graphical notations for knowledge in DESIRE are presented, as well as a translator which translates specifications of these notations in a graphical environment called Constraint Graphs to the textual DESIRE representation. Having this graphical interface brings the knowledge modelling in DESIRE closer to other well-known knowledge representation languages, such as Conceptual Graphs [8] because a dedicated interchange procedure could be added to the software. The relation between Conceptual Graphs and predicate logic is well-known. The fact that DESIRE is vased on order sorted predicate logic, and the possibility to represent different meta-levels of information within DESIRE, ensures that all knowledge represented in Conceptual Graphs can also be represented in DESIRE. In this section a translation from representations in DESIRE to representations in Conceptual Graphs is defined.

A conceptual graph is a finite, connected, bipartite graph, which consists of two kinds of nodes: concepts and conceptual relations. Concepts are denoted by a rectangle, with the name of the concept within this rectangle, and a conceptual relation is represented as an ellipse, with one or more arcs, each of which must be linked to some concept. Fig. 7 below shows an example conceptual graph, representing the episodic knowledge that a girl, Sue, is eating pie fast.

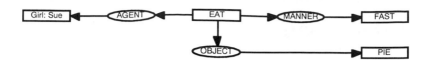

Fig. 7. An example Conceptual Graph

When comparing conceptual graphs with the graphical notations for DESIRE, many similarities become apparent. For instance, DESIRE's relations are denoted by ellipses, like conceptual relations, and sorts appear as rectangles, like concepts. Other elements however, are harder to translate to a Conceptual Graph notation. Table 2 provides an overview of the translation of DESIRE elements to Conceptual Graphs. Part of this is discussed in some detail.

Objects
Objects in DESIRE are instances of a sort. In Conceptual Graphs (CG) these instances are represented by individual concepts, i.e., concepts with an individual marker following the concept name. For example, the object tweety of sort BIRD in DESIRE is represented by [BIRD: tweety] in CG. Also anonymous individuals can be translated, e.g., the DESIRE variable X:BIRD is translated into [BIRD: *x] of CG which means that it is known that an individual of type BIRD exists, but it is unknown which individual.

Functions
In DESIRE, functions group sorts together by mapping them onto another sort. Functions can be regarded to be sub-types of a general CG concept FUNCTION, which takes one or more arguments and produces a result. In DESIRE functions act as a named placeholder for an object of its result, in which the argument(s) and the name of the function ensure the placeholder's uniqueness. Function food_preference, for example of Fig. 2, can be represented by the following Conceptual Graph:
    [DIET]<-(ARG)<-[food_preference]-(RSLT)-[BIRD_TYPE].

Relations
Relations in DESIRE can be classified according to their arity. This arity determines the mapping to Conceptual Graphs. 0-ary relations in DESIRE will have to be translated to concepts; concepts in Conceptual Graphs form a graph in itself, like nullary relations form a DESIRE atom in DESIRE. Relations with an arity greater than zero can be translated into either a conceptual relation with the same arity or a combination of a concept and (an)other conceptual relation(s). For example, the relation between: space * brick * brick in DESIRE could be translated into the following Conceptual Graph:

    [SPACE] - (BETW) - [BRICK]
                      -[BRICK]

This graph is a triadic relation, which could be read as "a space is between a brick and a brick". Relation is_traveling_from_to: person * origin * destination however could be translated into the graph

```
[TRAVEL]   -
           (AGNT) - [PERSON]
           (ORG) - [ORIGIN]
           (DEST) - [DESTINATION]
```

Sub-sorts

In DESIRE, hierarchical relations between sorts are allowed. Sub-sorts in DESIRE correspond to the type hierarchy of concepts in Conceptual Graphs. In Conceptual Graphs, hierarchies of both concepts and conceptual relations are possible, but these hierarchical is-a relations are kept in a separate semantic net from other relations that exist in the domain.

Desire Element	Graphical Equivalent in Constraint Graphs	Equivalent in Conceptual Graphs
Object	diamond	individual concept
Sort	rectangle	generic concept
Sub-sort	rectangle connected to super-sort by instance-of arrow	type hierarchy of concepts
Meta-Description	dashed arrow from information type to sort	conceptual relation -(METALEVEL)-
Function	parallelogram	concept FUNCTION
Relation	ellipse	conceptual relation or concept and conceptual relation(s)
Information type	context-box labeled SIG	context
Knowledge Base	context box labeled KB	context
Antecedent	context-box labeled ANT	context
Consequent	context-box labeled CONS	context
NOT-context	context-box labeled NOT	negative context
Information type Reference to information type	arrow between information types	context enclosed in another context
Knowledge base Reference to KB	arrow between knowledge base contexts	context enclosed in another context
KB reference to information type	arrow from kb to information type	comparable to first three and last component in a canon
Rule	arrow labeled "implies" between antecedent and consequent	conceptual relation -(IMP)-

Table 2. DESIRE, Constraint Graphs, and Conceptual Graphs

6 Conclusion

In this paper, graphical representations for knowledge structures in DESIRE [2] have been presented, together with a graphical editor based on the Constraint Graph environment [5]. Moreover, a translator has been described which translates these graphical representations to textual representations in DESIRE. This software environment can be regarded as a graphical design tool for knowledge in DESIRE, an interface which offers many advantages to a textual interface. First, Constraint Graphs can be used to specify knowledge structures, allowing the user to work with a mouse, pull-down menu's and windows instead of typing the specification conform the textual DESIRE syntax. Second, the graphical representation of knowledge structures (supported by the software environment for Constraint Graphs) offers a clear visual representation, facilitating communication between domain expert and knowledge engineer in the development process. Third, the graphical representations bring DESIRE closer to other knowledge representation languages, such as Conceptual Graphs [8], by defining a mapping from DESIRE to Conceptual Graphs (the other direction was already covered). In conclusion, the strengths of the Constraint Graphs environment as an easy to use representation tool in combination with the DESIRE environment allows for a powerful framework to support the development of knowledge based or multi-agent systems.

References

1. Abadi, M., and Cardelli, L. *A Theory of Object*, Springer, New York, 1996.
2. Brazier, F.M.T., Dunin-Keplicz, B., Jennings, N.R., and Treur, J., Formal specification of Multi-Agent Systems: a real-world case. In: V. Lesser (Ed.), *Proceedings of the First International Conference on Multi-Agent Systems*, ICMAS-95, MIT Press, Cambridge, MA,, 1995, pp. 25-32. Extended version in: *International Journal of Cooperative Information Systems*, M. Huhns, M. Singh, (Eds.), special issue on Formal Methods in Cooperative Information Systems: Multi-Agent Systems, vol. 6, 1997, pp. 67-94.
3. Brazier, F.M.T., Jonker, C.M., and Treur, J., Principles of Compositional Multi-Agent System Development, In: J. Cuena (ed.), *Proceedings of the IFIP World Computer Congress*, WCC'98, Conference on Information Technologies and Knowledge Systems, IT&KNOWS'98, 1998.
4. Gamma, E., Helm, R., Johnson, R., and Vlissides, J., *Design Patterns: Elements of Reusable Object-Oriented Software*, Addison-Wesley, Reading, Mass., 1994.
5. Kremer, R., Constraint Graphs: *A Constraint Graphs Meta-Language*, PhD Dissertation, Department of Computer Science, University of Calgary, 1997.
6. Moeller J.U., and Willems M. CG-DESIRE: Formal Specification Using Conceptual Graphs; In: Gaines, B.R. and Musen, M.A. (eds), *Proceedings of the 9th Banff Knowledge Acquisition for Knowledge-Based Systems Workshop* KAW-95, Calgary, 1995, pp. 25/1 - 25/20.
7. Nosek, J. T., and Roth, I., A Comparison of Formal Knowledge Representation Schemes as Communication Tools: Predicate Logic vs. Semantic Network, In: *International Journal of Man-Machine Studies*, vol. 33, 1990, pp. 227-239.
8. Sowa, J.F., *Conceptual Structures: Information Processing in Mind and Machine*, Addison-Wesley, Reading, Mass., 1984.

Verification of Knowledge Based-Systems for Power System Control Centres

Jorge Santos[1], Luiz Faria[1], Carlos Ramos[1], Zita A. Vale[2], Albino Marques[3]

[1]Polytechnic Institute of Porto, Institute of Engineering
Department of Computer Engineering
Rua de S. Tomé, 4200 Porto, Portugal
{jsantos | lff | csr}@ dei.isep.ipp.pt

[2]Polytechnic Institute of Porto, Institute of Engineering
Department of Electrical Engineering
Rua de S. Tomé, 4200 Porto, Portugal
zav@dee.isep.ipp.pt

[3]REN – Portuguese Transmission Network, EDP Group
Apartado 3, 4471 Maia Codex, Portugal

Abstract. During the last years, electrical utilities began to install intelligent applications in order to assist Control Centres operators. The Verification and Validation (V&V) process intends to assure the reliability of these applications, even under incident conditions.
This paper addresses the Validation and Verification of Knowledge-Based Systems (KBS) in general, focussing particularly on the V&V of SPARSE, a KBS used in the Portuguese Transmission Network for operator assistance in incident analysis and power restoration.
VERITAS is a verification tool developed to verify SPARSE Knowledge Base. This tool performs knowledge base structural analysis allowing knowledge anomalies detection.

1 Introduction

Nowadays, Control Centres (CC) are of high importance for the operation of electrical networks. These Centres receive real-time information about the state of the network and Control Centre operators must take decisions according to this information.

Under incident conditions, a huge volume of information may arrive to these Centres, making its correct and efficient interpretation by a human operator almost impossible. In order to solve this problem, some years ago, electrical utilities began to install intelligent applications in their Control Centres. These applications are usually Knowledge-Based Systems (KBS) and are mainly intended to provide operators with assistance, especially in critical situations.

The correct and efficient performance of such applications must be guarantied through Verification and Validation (V&V). V&V of KBS are not so usual as desirable but are usually undertaken in a non-systematic way. The systematic use of formal V&V techniques is a key for making end-users more confident about KBS, especially when critical applications are considered.

This paper addresses the Validation and Verification of Knowledge-Based Systems in general, focussing particularly on the V&V of SPARSE, a KBS to assist operators of Portuguese Transmission Control Centres in incident analysis and power restoration.

It is known that knowledge maintenance is an essential issue for the success of a KBS but it must be guaranteed that the modified KB remains consistent and will not make the KBS incorrect or inefficient. There is no general agreement on the meaning of these terms. For the remaining of this paper, the following definitions will be used:

- Validation - Allows to assure that the KBS provides solutions that present a confidence level as high as the ones provided by the expert(s). Validation is then based on tests, desirably in the real environment and under real circumstances. During these tests, the KBS is considered as a "black box" and only the input and the output are really considered important.

- Verification - Allows to assure that the KBS has been correctly conceived and implemented and does not contain technical errors. Verification is intended to examine the interior of the KBS and find any possible errors.

Most KBS are only validated and verified. Although validation process can guarantee that when the system is deployed, its performance is correct, the existing problems may arise when there is a need to change the Rule Base.

Verification should rely on formal methods requiring the development of tools to implement these methods. Although there are already some available verification tools in the market, specific needs of Power System applications usually require the development of specific tools. As formal methods of verification rely on mathematical foundations, they are able to detect a large number of possible problems. In this way, it is possible to guarantee that a KBS that has passed through a verification phase is correct and efficient. Moreover, it is possible to assure that it will provide correct performance with examples that have not been considered in the validation phase.

The present section focus the mains aspects related with KBS knowledge maintenance, stressing its relation with V&V stages.

Section 2 describes the SPARSE's characteristics, namely, architecture, reasoning model, rule selection mechanism and its implications for Verification and Validation work.

Section 3 describes the Validation stage of SPARSE development, especially the field tests and the need of applying formal methods in SPARSE's V&V.

Section 4 presents VERITAS, a verification tool based on formal methods. This tool has been successfully applied to several KBS: SPARSE; ARCA, an expert system applied to Cardiology diseases diagnosis; and another expert system created to assist in Otology diseases diagnosis and therapy. Finally, section 5 presents some conclusions and future work.

2 SPARSE

SPARSE is a KBS developed for the Control Centres of the Portuguese Transmission network, owned and operated by REN[1]. This KBS assists Control Centres operators in incident analysis and power restoration [7] [8] [9].

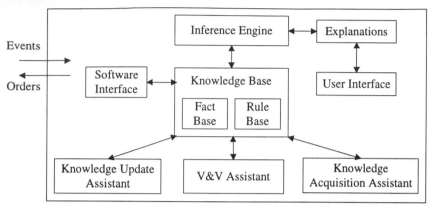

Fig. 1 - SPARSE Architecture

SPARSE (see: Fig. 1) has been developed using PROLOG and C language and runs on-line in a DECstation 5000/240 under ULTRIX operating system. This machine is connected, through a Local Area Network Ethernet of duplicate configuration, with two MicroVAX II machines that support SCADA (Supervisory Control And Data Acquisition) functions in the Control Centre.

SPARSE presents some features that make the verification work more difficult than for most KBS. These features include nonmonotonic behavior, temporal reasoning and the meta-rules used in rule triggering. Considering the following rule:

```
rule xx : 'EXAMPLE' :
[
 [C1 and C2 and C3]
 or
 [C4 and C5]
]
==>
[A1,A2].
```

The conditions considered in the LHS (Left Hand Side) (C1 to C5 in this example) may be of one of the following types:

– A fact which truth must be proved (normally these facts are time-tagged);

– A temporal condition;

The actions/conclusions to be taken (A1 to A2 in this example) may be of one of the following types:

– Assertion of facts (conclusions to be inserted to the knowledge base);

---

[1] REN is the Portuguese Transmission Network

– Retraction of facts (conclusions to be deleted from the knowledge base);

– Interaction with the user interface.

The rule selection mechanism uses facts with the following structure:

```
trigger(NF,NR,T1,T2)
```

This fact means that rule number NR should be triggered, until it is successful, between instant T1 and T2, because of the arrived fact NF.

SPARSE has passed through a validation phase and is presently installed in one of the two Control Centres of REN - Vermoim Control Centre, providing real-time assistance to operators.

SPARSE's Validation and Verification have been especially important for the success of this project, namely in what concerns knowledge updating.

3 Validation

The process of Verification and Validation should start as early as possible during the development of the application. The SPARSE V&V have been considered since the very beginning and special arrangements have been made in order to provide conditions for this process performing.

The project team aimed to perform the validation of SPARSE using examples as close as possible to the ones that the application should face in the real environment. According to this, it was considered that validation should be based mainly on real information about the network.

Another important aspect that has been considered since an early stage of development is the software required to interface SPARSE with SCADA applications used in the Control Centre. In fact, it was realised that some limitations imposed by SCADA should be considered since the very beginning in order to allow to take them into account during the development of the prototype, namely during the knowledge acquisition phase.

When integration issues are not addressed in an early phase of the project, the changes that are required when the system is integrated in the real environment may be very significant and impose almost a complete rebuilding of the system. The experts should namely, consider these issues during the knowledge acquisition phase.

REN's staff developed an application named TTLOGW [5] to acquire real-time information from SCADA and to send it to SPARSE. It acquires information related to the state of electrical network equipment, which is used to generate material for SPARSE's validation.

This application acquires the information related to the state of the equipment of the electrical network.

In this way, files concerning real incidents have been obtained and have been used in order to validate SPARSE conclusions. Experts involved in the project commented these conclusions and corrections in the Knowledge Base were made whenever necessary.

New validation techniques need to be applied after SPARSE was first installed in the control centre, since it now received real-time information from TTLOGW. The validation of SPARSE considering real-time information was very important due to several reasons:

- Temporal reasoning should be tested under real situations in order to assure its correction;

- Consideration of multiple faults is an important aspect of SPARSE performance that is very dependent from the way information flows;

- Processing times should be tested in order to guarantee real-time performance, even under incident conditions.

As nowadays electrical networks are very reliable it was not possible to completely validate SPARSE with real incidents. A large number of different types of incidents had to be simulated to allow validation. As this simulation should be as accurate as possible, two different techniques have been used:

(1) Simulation of incidents by operators located in chosen substations

(2) Simulation of incidents using a programmable impulse generator and a Remote Terminal Unit (RTU).

These two techniques complement each other, allowing a complete validation.

The simulation of incidents by operators allowed to obtain real-time information that was forced to be generated but presenting exactly the same characteristics as the information obtained during a real incident. During these tests, operators, making the whole system act as if a real incident was taking place simulated the behaviour of the protection equipment. In this way, the information, used by SPARSE was generated, as it would be under a real incident.

Due to the difficulties of co-ordinating operators in several substations, the simulation is not always correct and the whole process may have to be repeated several times in order to obtain a good test case.

In spite of all the difficulties and costs involved, this kind of tests has been considered absolutely essential for the validation of SPARSE, allowing to increase the confidence in its real-time behaviour.

In order to undertake a complete set of tests without the extremely high costs required by this technique, a different technique of test has also been used. This technique involves the use of a Remote Terminal Unit (RTU) and of a programmable impulse generator (PIG). The PIG generates impulses in order to force the alarm messages creation by the SCADA system. This technique was used to simulate a wide set of incidents allowing a more complete SPARSE Knowledge Base validation with reduced costs.

These methods of validation have been considered sufficient to put SPARSE in service, without the need to undertake formal verification of SPARSE Knowledge Base. However, when a Knowledge-Based System, as SPARSE, is in continuous use, the necessity to make changes in the Rule Base arises sooner or later. In the case of the Portuguese Transmission network, the introduction of new substations, with different types of operation or layout, has already imposed some modifications. Under these circumstances, it is not possible to accept the need to undertake complete validation tests, as the ones described before. Even if the costs are acceptable, the required time would oblige the Knowledge-Based System to be either out of service or to be in service without a validated Rule Base for longer than desirable. This problem must be addressed with a verification tool using formal methods. The use of this kind of tools to detect possible problems in the modified Rule Base allows to reduce the time required in Verification and validation process.

4 VERITAS, a verification tool

In what concerns SPARSE, there were two major reasons to start the verification work. First, the SPARSE team carried out a set of tests (see section Validation) in order to assure the quality of the answers of SPARSE to a set of real and simulated cases. Considering the expected high reliability and confidence of the tools to be applied in power systems area, it was decided to develop a verification tool to perform anomaly detection in SPARSE KB, assuring the consistency of the represented knowledge. On the other hand, tests applied in the Validation phase, namely the field tests, are very expensive because during it was necessary to assign a lot of technical personnel and physical resources for their execution (e.g. transmission lines). It seems obvious that it is impossible to carry out those tests after each knowledge updating so the developed verification tool offers an easy and inexpensive way to assure the knowledge quality maintenance.

A specific tool, named VERITAS [6] (see: Fig. 2) has been developed to be used in the verification of the SPARSE, performing structural analysis allowing to detect knowledge anomalies.

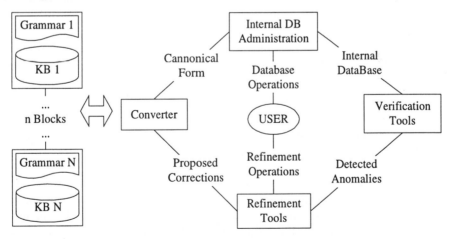

Fig. 2 - VERITAS Architecture

VERITAS is knowledge-domain and rule-grammar independent. It has been developed with an open and modular architecture (Fig. 2) allowing user-interaction along all the verification process. Since the tool is independent of KB grammar, theoretically any rule-based system can be analysed by VERITAS.

The Converter module allows the representation of external rules in an internal canonical form that is recognised by the other modules. Notice that this module works in two directions. It can also convert the canonical form into an external KB, generating new rules during knowledge updating, after anomaly detection, using an external grammar.

The Internal DB Administration module is responsible for the extraction and classification of all the information needed during the anomaly detection phase. In the first step all literals extracted from rules are classified according to the following schema:

- Fact – if it just appears in rule antecedents;
- Conclusion – if it just appears in rule consequents;
- Hypotheses – if it appears in both sides of the rules.

Notice that this classification is domain independent and just makes sense for verification procedures. This classification offers the advantages of a more compact knowledge representation and the reduction of the complexity of the rule expansion generating process. As it will be described later, this process corresponds to the analytical calculation of all possible inference chains.

In the second step, the Internal DB Administration module generates useful information about existing relations between literals (previously obtained). That information will be used not just to make the expansions generation process faster but also in the automatic detection of Single Value Constraints. VERITAS considers some type of constraints already described in literature [10]. Considered constraints can be classified in the following classes:

- Semantic Constraints – this type of impermissible set is formed by literals that cannot be present at the same time in the KB. Semantic constraints have to be introduced by the user.
- Logical Constraints – there are just two types of logical constraints: A and not(A) (where A stands for a literal); A and notPhysical(A); this designation is obtained by analogy with logical negation and allows to represent the constraint defined by a literal and by its retraction from the KB.
- Single Value Constraints – this type of impermissible set is formed by only one literal but considering different values of its parameters. Notice that those potential constraints are automatically detected. After this, the constraint can be either confirmed or changed by users.

The anomaly detection module (included in the Verification Tools) works in an autonomous way with no user interaction (i.e. it can run in batch mode). Presently this module can be used integrated with a developed tool (Knowledge Update Assistant) that, among other functions, allows rule edition. This functionality shows the existing relations between the rules that are to be modified and the remaining existing knowledge in the KB. This information is supplied in a graphical interface using a graph type representation. Moreover, it is possible to verify the rule in question immediately and to assure the KB consistency after the insertion of that rule.

When the verified Knowledge Base has large dimensions according to the number of rules and inference chains, the information generated during anomaly detection can be huge.

The detected anomalies have to be reported using a form suitable for easing its analysis. Special care has been put in this task, in order to reduce the time needed for the information analysis, so, it is possible to aggregate or select information by type of anomaly, number of rule and literal identification.

The anomaly detection relies on the rule expansions and constraint analysis. This method is also used by some well known V&V tools, as KB-REDUCER [1] and COVER [3]. As it has been described before, SPARSE has some specific features, due to these features the used technique is a variation of common ATMS (Assumption-based Truth Maintenance System) [2]. Namely, the knowledge represented in the meta-rules had to be considered in rule expansion generation.

VERITAS allows the rule expansion generation to be done in two different modes (see: Table 1): normal or exhaustive.

As an example, consider the following KB:

```
r1:  t(X) and r(a) → s(a)
r2:  f(a) → t(a)
r3:  f(b) → t(b)
r4:  h(a) → r(a)
r5:  j(a) → r(a)
```

Table 1 – Rule Expansions Calculation

Normal Mode	Exhaustive Mode
t(X) and h(a)→ s(a)	f(a) and h(a)→ s(a)
t(X) and j(a)→ s(a)	f(a) and j(a)→ s(a)
	f(b) and h(a)→ s(a)
	f(b) and j(a)→ s(a)

It is possible to notice that "normal mode" generates fewer expansions but, on the other hand, the information obtained after anomaly detection is more useful. The "exhaustive mode" wastes a lot of time generating the rule expansions implying also more wasted time to analyse them, but, in principle, it will be possible to detect more potential errors.

The detected anomalies could be grouped in three major classes: redundancy, circularity and inconsistency (see: Fig. 3). There is another type of anomaly that is not, yet, detected by VERITAS, named deficiency. To detect this anomaly it is not enough to know the KB and its syntax, since deficiency detection requires that all inputs and outputs to/from the system are known. For the SPARSE system this work can be done using all types of SCADA messages.

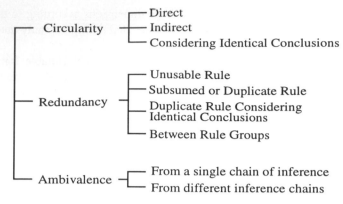

Fig. 3 - Anomaly Classification

This classification is based on Preece classification [4] with some modifications. First, the matching values are considered in rule analysis, meaning that a new set of anomalies will arise. Considering the following circular rules:

```
r1: t(a) and r(X) → s(a)
r2: s(a) → r(a)
```

For X=a some inference engines could start an infinite loop.

Another situation concerns to redundancy between groups of rules. In the following example:

```
r1: a and b and c → z
r2: not a and c → z
r3: not b and c → z
```

rules r1, r2 and r3 could be replaced by rx rule:

```
rx: a and b and c or not a and c or not b and c → z
```

Applying logical simplifications to rule rx, it is possible to obtain the following rule:

```
rx': c → z
```

Redundancy between groups of rules is a generalisation of the unused literal situation already studied by Alun de Preece [4]. Notice that this type of redundancy could be desirable. VERITAS can detect these situations using an improved Quine-McCluskey method for logical expression simplification.

5 Conclusions

This paper dealt with some important aspects for the practical use of KBS in Control Centres, namely knowledge maintenance and its relation to the Verification and Validation process.

The systematic use of Verification and Validation methods is very important for the acceptance of Knowledge-Based Systems by their end-users, especially when critical applications are considered. The use of Verification tools, based on formal methods, increases the confidence of the user and eases the process of changing KB, reducing the testing costs and the time needed to implement them.

This paper described SPARSE's V&V process, focusing on field-tests and techniques used during the validation phase. For the verification of SPARSE it was decided to implement a tool using a formal verification method.

VERITAS is a verification tool that performs the structural analysis in order to detect knowledge anomalies. We argue that the usefulness of VERITAS increases proportionally with KB size and the number of knowledge modifications, which must be undertaken.

Presently, VERITAS is being improved in order to allow the detection of anomalies related to temporal and nonmonotonic reasoning. We are also envisaging the use of VERITAS in verification of knowledge generated by Data Mining applications.

References

[1] Ginsberg, A. 1987. A new approach to checking knowledge bases for inconsistency and redundancy. In *Proceedings of the 3rd Annual Expert Systems in Government Conference.* 102-111. Washington, D.C., IEEE Computer Society.
[2] Kleer, J. 1986. An assumption-based TMS. *Artificial Intelligence* (Holland). 28(2):127-162
[3] Preece, A. 1990. Towards a methodology for evaluating expert systems. *Expert Systems* (UK). 7(4):215-223.
[4] Preece, A; and Shinghal, R.1994 Foundation and Application of Knowledge Base Verification. *Intelligence Systems.* 9:683-701.
[5] Rosado, C. 1993. Process TTLOGW. EDP Technical Report, RESP/SCDS 20/93, Electricidade de Portugal
[6] Santos, J. 1997. Verificação e Validação de Sistemas Baseados em Conhecimento – VERITAS, uma Ferramenta de Verificação. MSc Thesis diss., Dept. de Engenharia Electrotecnica e Computadores, Faculdade de Engenharia do Porto.
[7] Vale, Z. and Moura, A. 1993. An Expert System with Temporal Reasoning for Alarm Processing in Power System Control Centres. *IEEE Transactions on Power Systems* 8(3):1307-1314.
[8] Vale, Z.; Faria, L.; Ramos, C.; Fernandes, M.; and Marques, A. 1996. Towards More Intelligent and Adaptive User Interfaces for Control Centre Applications. In *Proceedings of the International Conference on Intelligent Systems Applications to Power Systems (ISAP'96).* 2-6. Orlando, Florida.
[9] Vale, Z.; Moura, A.; Fernandes, M.; and Marques, A. 1994. SPARSE - An Expert System for Alarm Processing and Operator Assistance in Substations Control Centres. *Applied Computing Review.* 2(2):18-26. ACM Press.
[10] Zlatareva, N.; and Preece, A. 1994. An Effective Logical Framework for Knowledge-Based Systems Verification. *International Journal of Expert Systems.* 7(3):239-260.

Self-Crossover and Its Application to the Traveling Salesman Problem

Malay K. Kundu and Nikhil R. Pal

Machine Intelligence Unit
Indian Statistical Institute
203 B. T. Road,
Calcutta 700035, INDIA
e-mail: { malay,nikhil}@isical.ac.in

Abstract. Crossover is an important genetic operation that helps in random recombination of structured information to locate new points in the search space, in order to achieve a good solution to an optimization problem. The conventional crossover operation when applied on a pair of binary strings will usually not retain the total number of 1's in the offsprings to be the same as that of their parents. But there are many optimization problems which require such a constraint. In this article, we propose a new crossover technique called, "self-crossover", which satisfies this constraint as well as retains the stochastic and evolutionary characteristics of genetic algorithms. We have also shown that this new operator serves the combined role of crossover and mutation. We have proved that self-crossover can generate any permutation of a given string. As an illustration, the effectiveness of this new operator has been demonstrated in solving the traveling salesman problem (TSP) using GA. This new technique is best suited for path representation of tours and performs better for TSP with large number of cities. Performance of the proposed scheme is compared with that of ordered crossover (OC) scheme.
Keywords : Genetic Algorithms, Heuristic Searching, Self-crossover, Traveling Salesman Problem(TSP), Ordered Crossover.

1 Introduction

The TSP, a well-known NP-hard problem, can be easily stated as follows : A traveling salesman must visit every city in his territory exactly once and then return to the starting point. An itinerary has to be found such that the total distance traversed in the tour is minimum. Mathematically, given a sequence of cities $c_1, c_2, \ldots, c_n$ and intercity distances $d(c_i, c_j)$, TSP finds a permutation π of the cities that minimizes the sum of distances

$$\text{TC(tour)} = \sum_{i=1}^{n-1} d(c_{\pi(i)}, c_{\pi(i+1)}) + d(c_{\pi(n)}, c_{\pi(1)}). \tag{1}$$

We considered, $d(c_i, c_j) = d(c_j, c_i)$ for $1 \leq i, j \leq n$.

During the last decades, several algorithms emerged like nearest neighbor, greedy algorithm, minimum spanning tree [4] to approximate the optimal solutions for problems. Another group of algorithms (like 2-opt algorithm, Lin-Kernighan algorithm) aims at a local optimization - an improvement of a tour by local perturbations. But none of the heuristics is well suited for large TSPs (*i.e.* TSP with large number of cities).

Genetic algorithms (GAs) are probabilistic heuristic search processes based on natural genetic system. They are capable of solving a wide range of complex optimization problems using three simple genetic operations (selection/ reproduction, crossover and mutation) on coded solutions (strings/ chromosomes) for the parameter set, not the parameters themselves, in an iterative fashion. There are several interesting features which made GA very popular. GAs consider several points in the search space simultaneously, which reduces the chance of convergence to a local optima. GAs use only the payoff or penalty function (objective function) called, the fitness function and do not need any other auxiliary information.

Eventually, TSP also became a target for GA applications. TSP tours can have various representations. Some of them are 1) adjacency 2) ordinal 3) path and 4) binary matrix representations. Based on different representations and choices of genetic operators several GA-based algorithms have already been reported [2, 3].

In this note we propose a new genetic operator, called *self-crossover* (SC). This operator is capable of randomly permuting the entries of a GA chromosome. If the chromosome is a binary string, in absence of mutation, this operator is able to retain the number of 1's in the string same, before and after the self-crossover operation. So, this operator can be successfully applied to a group of problems like selection of a fixed number of features, selection of a fixed number of prototypes for designing a nearest neighbor (NN) classifier where the constraint on total number of 1's in the chromosome string is very important. We have shown that self-crossover can produce any arbitrary string from any arbitrary starting chromosome. Hence self-crossover alone (*i.e.*without mutation) is sufficient for GA to be applicable to TSP and some other problems[7].

2 Review of GAs for TSP

In our investigation, we use the path representation of tours. In path representation, a tour $5 - 1 - 7 - 8 - 9 - 4 - 6 - 2 - 3$ is represented simply as (5 1 7 8 9 4 6 2 3).

There are several crossover techniques applicable to path representation. Order crossover (OC) of Davis [5], builds offspring by choosing a subsequence of a tour from one parent and preserving the relative order of cities from the other parent.

Recently a non-vector representation scheme for TSP has been evolved [1]. In this scheme, tours are represented by a *binary matrix* M. Matrix element m_{ij} contains a 1 if and only if the tour goes from city i directly to city j. This means that there is only one non-zero entry for each row and each column in the matrix. This representation avoids the problem of specifying the starting city. However, not every matrix with these constraints would represent a single valid tour.

Homaifar and Guan [1] proposed a new crossover technique based on matrix representation, called matrix crossover (MC). MC exchange all entries of the two parent matrices after a crossover point (which represents a column number of matrices). An additional "repair algorithm" is run to ensure that each row and each column has precisely single 1; and to cut and connect sub-tours to produce a single legal tour. The first step of the"repair algorithm" moves some 1's in the matrix to satisfy the row and column constraints. The cut and connect phase takes into account the existing edges in the original parents, preserving as many of the existing edges from the parents as possible. Homaifar and Guan [1] incorporated *inversion* operator to bring the effect of mutation in the GA. Inversion operation needs the path representation of the tours and selects an arbitrary subsequence of cities from the parent tour. The order of this subsequence of cities is reversed producing a new tour.

Though it is true that the order of cities (not the positions of the cities) are important for tours, matrix crossover using matrix representation is not an elegant way to evolve new tours. For mending the illegal tours some repair algorithms are needed. Effectively, efficiency of those repair algorithms will determine the efficiency of GA-based heuristic algorithm for solving TSP.

MC alone is not able to evolve each and every corner of the total tour space. So some additional operators are needed to blend the flavor of mutation with it. In [1] authors considered inversion to play this role. For that they need to switch over to path representation from matrix representation and vice versa.

3 Self-Crossover(SC) : A new genetic operator

Unlike Conventional crossover mechanism, self-crossover mechanism alters the genetic information within a *single* potential string selected **randomly** from the mating pool to produce an offspring. This is done in such a manner that the stochastic and evolutionary characteristics of GAs are preserved.

Let $S = 00010010011001011011$ be a string of length 20 selected from the mating pool. For self-crossover, first we select a random position p $(0 < p < L)$ and generate two substrings s_1 and s_2 : $s_1 =$ bits 1 through p of S and $s_2 =$ bits $p + 1$ through L of S. Now we select two random positions p_1, $0 \leq p_1 \leq p$ and p_2, $0 \leq p_2 \leq (L - p)$. Then four substrings are generated as follows :

$$s_{11} = \text{bits 1 through } p - p_1 \text{ of } s_1$$
$$s_{12} = \text{bits } (p - p_1 + 1) \text{ through } p \text{ of } s_1$$
$$s_{21} = \text{bits 1 through } L - p - p_2 \text{ of } s_2$$
$$s_{22} = \text{bits } (L - p_2 + 1) \text{ through } L \text{ of } s_2$$

Using operations similar to crossover we generate $S^1 = s_{11} \mid s_{22}$ and $S^2 = s_{21} \mid s_{12}$. Finally, the self-crossovered offspring of S is generated as $S_1 = S^1 \mid S^2$. It is easy to see that number of 1's in S and S_1 is the same. We now explain it with the example string S of length 20.

$$S = 00010010011001011011$$

A random position, $p = 9$, is selected for splitting the string into two substrings (s_1, s_2) as follows : $s_1 = 000100100$ and $s_2 = 11001011011$

Now two random positions, $p_1 = 4$ and $p_2 = 7$, are selected for s_1 and s_2 respectively. After splitting s_1 and s_2 at $4th$ and $7th$ position, respectively we get,

$s_{11} = 00010$, $s_{12} = 0100$, $s_{21} = 1100$, and $s_{22} = 1011011$.
The two new substrings S^1 and S^2 are then obtained as :
$S^1 = 000101011011$ and $S^2 = 11000100$.
Finally, the offspring (S_1) is generated by concatenating S^1 and S^2 as :
$S_1 = 00010101101111000100$

Thus, self-crossover exchanges substrings s_{12} and s_{22}. If the parent string consists of all 0's or all 1's, the offspring generated through self-crossover will resemble its parent because of the underlying constraint on the total number of 1's in the string. It is also clear that if we do not start GA with a all '1' or all '0' string, GA with self-crossover technique, will never generate such strings as offsprings. So, self-crossover will generate new offsprings as iterations go on.

We can see very well that mutation is not effective in producing such constrained offsprings. But self-crossover can regenerate any lost genetic information. So we may not need mutation when we use the new technique in constrained GA applications.

Next we show through a Lemma that self-crossover (without mutation) can generate any target string.

Lemma : Given a string of symbols, self-crossover operations can generate any arbitrary permutation of the symbols.[7]

Proof : We can represent any arbitrary string P by $P = S_1 \mid S_2 \mid S_3 \mid S_4$ where S_i represents a subsequence of symbols. S_i can be empty sequence as well. Now if we are able to prove that a parent string $P = S_1 \mid S_2 \mid S_3 \mid S_4$ can produce an offspring $O = S_1 \mid S_3 \mid S_2 \mid S_4$ using a finite number of self-crossover operations, then we can iterate the process to cook up a sequence of self-crossover operations to reach any target offspring.

The position for splitting P is chosen such that two subsequence s_1 and s_2 are formed as $s_1 = S_1 \mid S_2$ and $s_2 = S_3 \mid S_4$. Now two random positions for s_1 and s_2 are selected such that after splitting of s_1 and s_2 at these two positions we get

$s_{11} = S_1$, $s_{12} = S_2$, $s_{21} = S_3$, and $s_{22} = S_4$. So the resultant child is obtained as
$P_1 = S_1 \mid S_4 \mid S_3 \mid S_2$.

We apply once more the self-crossover operation on intermediate child P_1. Now the random position for splitting P_1 is chosen such that
$s_1 = S_1 \mid S_4$ and, $s_2 = S_3 \mid S_2$.

Again we select two random positions for s_1 and s_2 such that after splitting of s_1 and s_2 at these two positions we get

$s_{11} = S_1$, $s_{12} = S_4$, $s_{21} = S_3 \mid S_2$, and s_{22} – empty sequence.
The offspring now becomes

$O = S_1 \mid S_3 \mid S_2 \mid S_4$ which is nothing but what we wanted to produce.

Since S_1 could be a null string, so any symbol from the parent string can be brought at the beginning of the offspring through substring S_3 by two successive self-crossover operations. The lemma also ensures that any substring consisting of symbols starting from the beginning of a parent string can be preserved in the child through substring S_1. Hence, any target permutation can be grown from the left side. Proceeding this way in the terminal phase of the process S_2 and S_4 will be empty; S_1 will contain the entire target substring except the last symbol which will be in S_3.

Note that, the lemma does *not* say that there is no more need for mutation in GA with self-crossover technique. It simply says that for combinatorial problems like TSP, use of self-crossover without mutation can generate all possible valid solution strings. For problems like feature selection[7], data editing for NN classifier where we want to select the best subset of features or data points of a prefixed cardinality, self-crossover without mutation is sufficient. In fact, conventional mutation for such problems may produce invalid solutions, *i.e.*, it may generate a substring of arbitrary cardinality, not equal to the prefixed cardinality.

At the first sight, it might appear that self-crossover is nothing but a parallel random search, but this is not the case because of two reasons. Self-crossover is done only on a randomly selected subset of strings and self-crossover does not alter the substring s_{11}. It exchanges, only s_{22} and s_{12}. Consequently, the evolutionary characteristics of GA are preserved. The similarity between the parents and offsprings will be more if we take $p_1 = p_2 = p'$ (say) = a random number selected between 1 and $Min\,(p, L-p)$; i.e., $0 < p_1 = p_2 = p' < Min\,(p, L-p)$. Here, the bits in positions 0 through p' and in positions $p+1$ through $L - p'$ will remain unaltered. Consequently, the evolution pressure will be high.

4 TSP with self - crossover(SC)

It is seen that the SC operator plays the combined role of conventional crossover and mutation. So, we are getting two-in-one effect with the help of this operator. Because of the simplicity of this operator, our algorithm becomes fast. Here we use path representation of tours. Since the SC operator can generate any permutation of a given string, as already discussed, given a valid tour, it will generate only valid tours. Now we describe the algorithm for solving TSP. The objective here is to minimize the total cost of a tour, TC(tour). To convert it to a maximization problem, we take the fitness function

$$f(tour) = - \ TC(tour).$$

1. Start with the population of initial valid tours(a set of integer strings/ chromosomes).
2. Evaluate fitness of every tour in the current population.
3. Generate new mating pool .

4. Each tour of the current population is self-crossovered and copied to the new population.

5. Repetition of steps 2 through 4 until the system ceases to improve or some stopping criterion is reached.

5 Results and Discussion

The results obtained for TSPs of different sizes are depicted in table I. For each problem we synthetically generated a cost matrix with known cost for the optimum tour. The initial population is generated randomly, consisting of only valid tours. We adopted partially disruptive selection strategy *i.e.*, at each iteration we kept few best candidates and chose the rest of the population randomly.

Table : I

Result with SC				
Cities	Population size	Generations	Best tour-cost	Optimum cost
10	10	2876	49.45	49.45
20	20	99943	36.74	36.74
30	30	151987	35.349	34.349
50	50	201000	7.609	5.98

From table I we found that GA equipped with SC can determine exact optimum solution for TSPs of sizes 10 and 20 within few generations. For TSP with 30 cities, our scheme can find a solution within 3% of the optimum solution.

Note that, in table I we used different population sizes for different problems of different sizes. For TSP the population size should be chosen carefully, it should be dependent on the size of TSP; larger the size of TSP, larger should be the population size.

SC(Self-crossover) operation does not involve any comparison operation which is an explosive cpu operation. It needs to choose 3 random numbers and a few string copy and string concatenation operations. But OC(Ordered-crossover) operation needs 2 string comparisons per iteration. String comparison operation cost is once again proportional to the string length. Unlike SC, number of OC operations per iteration is proportional to the square of the population size since OC involves a pair of strings (Number of all possible pairs $=n*(n-1)/2$ where n represents the population size). As a result the time complexity for GA with OC is much more than GA with SC. The time complexity for GA with OC increases even more prominently for higher number of cities, compared to GA with SC. As an illustration, OC needs 17 minutes for 5000 iterations for 20 cities, whereas SC requires only 50 seconds for 5000 iterations for same problem.

6 Conclusions

There is a class of optimization problem which require the number of '1's in the strings to be constant. Conventional crossover / mutation does not guarantee this. We have introduced a new crossover operation named *self-crossover* which preserve this constraint. GA with SC has been successfully used for selection of a fixed number of good features for pattern recognition[7]. Here GA with SC is used for study the TSP. SC has some distinct advantages over other GA based approaches for TSP. For example SC produces only legal tours and prevents generation of duplicate tours. Moreover, this new operator plays the combined role of mutation and as well as conventional crossover operator. GA with SC is found to be quite successful for TSP of moderate size. However experiments with problems of much bigger sizes are needed to make finer conclusion.

7 References

1. Homaifar, A. Guan, S. and Liepins, G. E. : A New Approach on the Traveling Salesman Problem by Genetic Algorithms, 460-466. To appear in *Complex Systems.*
2. Grefenstette, J. J., *et al.*, Genetic Algorithm for the TSP, in Grefenstette (ed), *Proceedings of an International Conference of Genetic Algorithm and their Applications,* Texas Instrument and U.S. Navy Center for Applied Research and Artificial Intelligence, 154-159, (1985).
3. Suh, *et al.*, The Effects of population size, Heuristic Crossover and Local Improvement on a GA for the TSP, in the J.D.Schaffer (ed), *Proceedings of the Third International conference for Genetic Algorithms*, Morgan Kaufmann Publishers, Inc., 110-115, (1989).
4. Johnson, D. S., Local Optimization and Traveling Salesman Problem, in M.S. Paterson (Editor), *Proceedings of the 17th Colloquium on Automata, Languages, and Programming, Springer-Verlag, Lecture Notes in Computer Science,* **443**, 446-461, (1990).
5. Davis, L., Applying Adaptive Algorithms to Epistatic Domains, *Proceedings of the International Joint Conference on Artificial Intelligence,* 162-164, (1985).
6. Lawler, E. *et al.*, The Traveling Salesman Problem, *John Wiley and Sons,* New York.
7. , Pal, N. R. and Nandi, S. and Kundu, M. K. : Self Crossover: a new genetic operator and its application to feature selection, *International Journal of System Sciences,* **29**, no. 2, 207-212, (1998).

Enabling Client-Server Explanation Facilities in a Real-Time Expert System

Nuno Malheiro [1], Zita A. Vale [2], Carlos Ramos [1], Jorge Santos [1] and Albino Marques [3]

[1] Polytechnic Institute of Porto (IPP) , Institute of Engineering,
Department of Computer Engineering
Rua de S. Tomé - 4200 Porto – Portugal,
Fax: 351-2-8321159, Phone: 351-2-8340500
{ ntm l csr l jsantos}@dei.isep.ipp.pt
[2] Polytechnic Institute of Porto (IPP), Institute of Engineering,
Department of Electrical Engineering
Rua de S. Tomé - 4200 Porto – Portugal,
Fax: 351-2-8321159, Phone: 351-2-8340500
zav@dee.isep.ipp.pt
[3] REN - Portuguese Transmission Network (EDP Group)
Apartado 3 – 4471 Maia Codex – Portugal
Fax: 351-2-9486758 Phone: 351-2-9448132

Abstract. Although expert systems have been increasingly applied for solving power system problems in the past decade, their capabilities have not been fully used. Their ability to provide explanations concerning their knowledge and inference process is one good example of misuse of expert system possibilities. In the case of expert systems designed for real-time operation, the introduction of effective explanation features poses important constraints that must be taken into account. This paper presents an explanation mechanism that has been designed and implemented to be used by SPARSE, an expert system for intelligent alarm processing and power restoration aid which is running on-line in a Portuguese transmission control center.

Keywords: Control Center, Expert Systems, Explanation, Intelligent Tutor

1 Introduction

Expert systems have been increasingly used in the last few years to address a wide range of problems in power systems. [1] [2]. An important part of these systems are intended to work in real time as on-line applications in power system control centers. These applications address issues such as intelligent alarm processing, power restoration aid and power dispatch.

Although expert systems have been considered an adequate approach to several problems in the power systems area, the total number of successful projects is rather small. In fact, the use of expert systems for real-time applications in power system control centers is highly demanding due to the characteristics of these applications. Real time performance, huge amounts of information and temporal and nonmonotonic reasoning requirements can be identified as the most difficult issues that have to be addressed [3].

On one hand, the process of knowledge acquisition is complex and very time consuming, imposing long development periods. On the other hand, difficulties in knowledge maintenance make the life of some deployed expert systems very short [4].

Some alternative techniques have also been used from which artificial neural networks (ANN) and genetic algorithms (GA) are good example [2]. Although these techniques are very useful in some situations, expert systems have a unique characteristic that can justify the preference towards their use. In fact, knowledge-based systems and expert systems are the only techniques, among intelligent applications, that can provide a real understanding of the knowledge encoded [5] [6] [7] [8] [9]. This point can be very important for the future use of expert systems in real applications but has, somehow, caught little attention till the present.

In fact, explanation capabilities have not been fully used in power system applications although they can significantly contribute to their acceptance and their effective use. However, the production of explanations for on-time applications, with very restrictive real-time constraints, imposes important difficulties.

The authors of this paper have been involved in the team responsible for the development of SPARSE, an expert system for intelligent alarm processing and power restoration assistance [10] [11]. SPARSE is presently working as an on-line application in Vermoim Control Center of the Portuguese transmission network.

The SPARSE team is presently involved in some research projects whose main goals are the following:

- To adapt SPARSE in order to be integrated in the software and hardware acquired by the Portuguese transmission utility (REN) to be fully operational in middle 1999
- To provide SPARSE with additional features to enhance its performance and functionality
- To use the knowledge acquired during the development of SPARSE to develop some intelligent applications to be integrated in Portuguese distribution control centers.

These goals led us to revise the philosophy of SPARSE's architecture and to undertake a modification process, which is intended to continue till the end of 2000. Presently, an important part of this process is already accomplished and some new applications are already available.

The full use of explanation capabilities has been considered an important point in this process, as a means to make the use of SPARSE more effective. As explanation capabilities could not put in danger the efficient real-time performance of the expert system, a special mechanism has been designed and implemented.

This paper addresses this mechanism that enables real-time performance of the expert systems and the use of explanation capabilities for several purposes. The authors discuss the effectiveness of such mechanism and how its use can really improve expert system performance.

In our opinion, the use of efficient explanation mechanisms makes expert system applications in power systems much more promising and effective.

2 SPARSE'S EXPLANATIONS

2.1 Overview

Explanations have been used in SPARSE's development since the very beginning. In fact, explanations have been very useful in the knowledge acquisition phase [10].

In an early phase of the project, explanations were internally generated by SPARSE and could be seen in a window included in its user interface. This architecture was due to the coexistence in the same process, in an early stage of the project, of the expert system and its graphical interface.

With this kind of architecture, real-time performance of the expert system could be endangered by the use of the explanation module. In fact, users should not overload the expert system by asking for explanations under incident situations.

Changes in SPARSE software architecture led to the separation of the expert system and the user interface into individual processes communicating with each other [12]. Presently, this philosophy is used for most of the components of the system and has proved to be quite efficient even under incident conditions.

2.2 The new approach: A Client – Server philosophy

In a multiagent environment, which is presently SPARSE's case, where more than one process needs the availability of a functionality, a Client – Server philosophy can be used.

In this case, SPARSE, as the Server, will be able to supply explanations to any process (which becomes SPARSE's Client) that questions it. From this point forward, the SPARSE designation can be used to identify the whole system or just the expert system, depending on the context. The acronym ESES, which means Expert System Explanation Server, will also be used.

A certain process becomes an ESES's client by sending it an initial message that contains the following data:

- Client's identification
- Client's means of contact (message queue, socket, file, ...)
- Client's means of contact identification (message queue key, socket port, file name, ...).

This initial message registers the Client in the ESES's knowledge base, allowing it to use the explanation production mechanisms.

The same process questions ESES, sending it a message that contains the following data:

- Client's identification
- Explanations chaining method
- Explanation depth
- Fact to explain.

These clients can be running either in the same machine or in several machines, connected through a Local Area Network (LAN) or a Wide Area Network (WAN).

This methodology presents the following characteristics:

- It allows more than one process to question ESES at the same time, handling and scheduling these requests
- The explanation production mechanism is independent from the client
- ESES operates with high granularity in explanation prodution, controlling the time spent in explanation production and delivery
- The Explanation Server schedules its activity to the periods when the Expert System is free
- It makes the explanations available when they are needed by each Client
- The requests can be interactive (one level of explanation depth at a time) or full (explanation until the SCADA message level is reached)
- The operation is very flexible being the form of explanation request and presentation defined by the Client.

The Explanation Server has been integrated in SPARSE with minimum modification in the already existing software. The Explanation Server is presently embedded in the expert system. Other modules, such as the user interface and the intelligent tutor, which exist as independent processes, can use this new functionality.

2.3 Available Types of Explanations

2.3.1 "How" Explanations
The "how" explanations exist to supply to the user of a knowledge based system a description of the way that has been used to infer a certain result. This kind of explanation has several uses:
- To the system builder's, in a non automatic verification and validation perspective
- To help experienced users in knowledge maintenance
- As a tutor to inexperienced users.

In the case of SPARSE, the "how" explanation starts in any conclusion drawn by SPARSE, basing its fundament on the rule which succeeded to obtain that conclusion and on the facts used as its premises. These premises can, eventually, be explained by rules that fired to create them and their own facts used as premises. This chaining will always end in SCADA messages, which are the lowest level of abstraction that the system can reach.

2.3.2 "Inference" explanations
The "inference" explanations show, in natural language, the way that the forward chaining expert system (SPARSE) works. They start in a SCADA message or in an intermediate conclusion drawn by the system, showing all the information that was inferred with it.

This particular type of explanation is intended to explain the reasoning of the system in a language that the user can understand. Its obvious use is the validation of the system, supplying the information needed to verify if the system not only draws the right conclusions, but also that it draws all the possible conclusions with the available information.

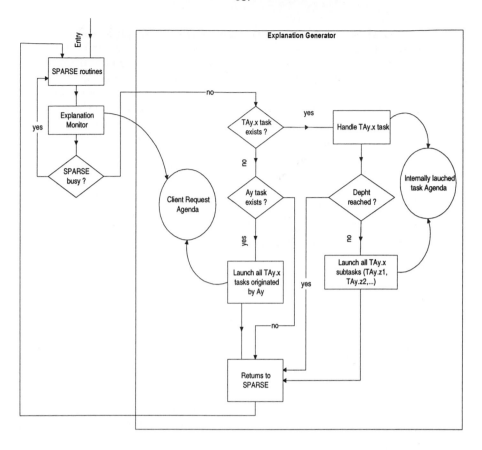

Fig. 1. Explanation Server flowchart.

2.4 Knowledge modeling

It's quite intuitive that a larger quantity of information is necessary to produce explanations about a certain conclusion, than the amount of information needed to infer that conclusion.

In order to provide explanations it is necessary to have:

- The information used to infer
- Additional information about the information used to the inference (e.g. natural language translation)
- Information about the inference process itself (how the information owned by the system is used to infer new information).

It is very important to limit the quantity of the information required to produce any type of explanations in order to make knowledge maintenance effort relatively low. For this purpose, it is desirable to organize the required information in an effective way.

In the case of SPARSE, there are internal facts, obtained through SCADA messages, which correspond to the status of devices. These facts can comprehend several devices, each of which with several possible states. In a first approach, the translation of these facts to natural language, we could consider them in an independent form. This would be a very redundant form of modeling since the grouping of their common properties can significantly reduce the dimension of the required knowledge base. In this case, it is only necessary to store the essential information: the types of devices that are considered by the system, the possible status for each type and how can a generic fact of this kind be explained.

In what concerns the information concerning the inference process, it is required to encode knowledge about the kind of production rules used, which includes the explanation model for a generic rule. It is also needed to encode knowledge about the characteristics of the inference engine considering its type of inference mechanism (forward chaining mechanism in the case of SPARSE) and specific features such as temporal and non monotonic reasoning.

2.5 Server and explanation generation mechanism structure analysis

The explanation Server is divided in two modules:
- Monitor
- Explanation Generator.

The Monitor's function is to receive the messages that arrive to the Server, handling them in the shortest possible time. It is the Monitor's job to update the Client registry and the Explanation Generator Agenda, considering Client requests.

The Explanation Generator Agenda is a Server's crucial issue. It contains the tasks that the Generator must supply to the Clients, ordered in a First Come, First Served (FCFS) basis.

The Explanation Generator works in a very different way: it is a data-driven mechanism that only operates when new information arrives at its Agenda. It is also restricted by the fact that it can only work if SPARSE expert system is fully stopped. As SPARSE is itself a data-driven application, we are uncertain about the time of information arrival, its quantity or even how much internal activity it will provoke (rule firing). Because of this, the Explanation Generator must work in a very granular form. Complex tasks have to be decomposed into small subtasks, whose processing time is small enough to consider the SPARSE system as being active at all times. Each of these subtasks must be handled one at a time, verifying the SPARSE's activity at its completion and acting accordingly.

The explanation generation module works as follows:
- The Explanation Generator verifies if it can be activated (if the SPARSE expert system is stopped)
- It verifies the existence of a TAy.x entry (task previously scheduled, as sub-task x of the task y, requested by a Client) in the internally launched task Agenda
- If a TAy.x task does exist, it handles it, eventually launching new internal tasks (if the explanation depth demands it) and returns control over to SPARSE
- It determines the oldest Ay entry in the Server's Agenda
- If Ay does not exist, it returns control over to SPARSE

- It transforms task Ay into internal TAy.x sub-tasks and returns control over to SPARSE.

The Explanation Server control flow is presented in Fig. 1.

Each TAy.x task can be linked to a fact known by the system, which is relevant to the requested explanation. The explanation of that fact can be done in two possible ways: either the fact is a SCADA message and, as such, explained by being system acquired data or it is an inferred fact, explainable by the information needed to perform its inference.

The handling of a new Az task is only undertaken when all TAy.x subtasks of the previously initiated Ay have been concluded, and its results sent to the requesting Client.

The handling of each TAy.x is done in a way that, at most, only a production rule or a SCADA message is explained, thus limiting the time spent in a task and succeeding to achieve the desired granularity.

2.6 Server-Client interaction

Any process can become an explanation client as long as it has:
- Ability to communicate trough message queues
- Knowledge about the communication protocol
- Knowledge about SPARSE's inferred conclusions.

In order to have the ability to request explanations, a process needs to have knowledge about the system's conclusions so that it registers and actually requests the desired information.

The expected Client must be prepared to graphically present the explanations using a graph form because the information supplied to clients is intended to be presented under this form. In this graph, each node represents a SPARSE rule firing, including all the facts that were used as premises, all the concluded facts and/or applicable constraints. The user interface can allow direct and inverse navigation capabilities, based in the "how" (backward chaining) and "inference" (forward chaining) explanation types. When navigation is demanded to a unknown node, the information about that node can be simply requested from the Server and then presented.

The Server can read messages sent to the message queue with a specific key. The Client must use this key to communicate with the Server.

So that the communication between the Client and ESES can be achieved, the following procedure must used:
- Process registration as an ESES's Explanation Client:

A message with the following format is used:

r_id(MyID, MyCommType , MyCommInfo)

where:

MyID is the process identification

MyCommType is the Client's preferred means of communication (message queue, socket or file)

MyCommInfo is the information needed by the ESES to communicate to the Client in its preferred form (message queue key, socket's machine and port or file name).
- Sending to ESES an explanation request message with the following format:

msg(MyID,ExplType,ExplDepth,Fact)

where:

MyID is the identification with which the Client was registered

ExplType is the type of the desired explanation (how or forward)

ExplDepth is the wanted explanation depth level

Fact is the fact to explain; It can be supplied as a conclusion (conclusion's text as produced by SPARSE), or through SPARSE's internal fact numbering.

• Sending the messages that compose the explanation to the client.

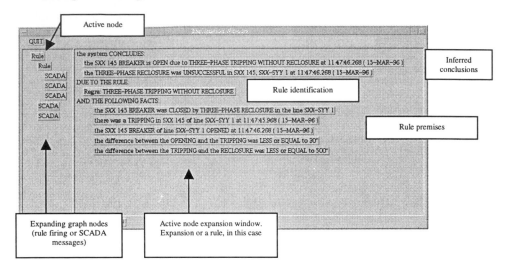

Fig. 2. Application

3 USE OF THE EXPLANATION MECHANISM

Explanations are especially intended for the following purposes:

• Knowledge validation

• Intelligent tutor

• Real-time operator assistance.

From these applications, only the last one presents real-time requirements. Fig. 2 shows the interface designed for this kind of application.

In the existing application, the window on the left shows the graph nodes. Its interconnections are visualized by indentation. Navigation is possible in the window on the left, clicking the mouse on the nodes.

In the right window we can see the expanded active node. This expansion is, itself, interactive. If a browser premise is clicked, it will try to find (if it exists) the node that concluded it (and possibly other facts). Given the case of a conclusion being clicked, the browser will activate the node where that particular conclusion was used as a premise.

Let us consider the following explanation request: "explain how you arrived at the conclusion 'the SXX 145 BREAKER is OPEN due to THREE-PHASE TRIPPING

WITHOUT RECLOSURE at 11:47:46.268 (15-MAR-96)' ", with one depth level only. The answer that is provided by the explanation generator is presented in Fig. 3. The protocol used to produce each of the lines shown in Fig. 3 is of use to the client application so that it can inter-relate the separated messages and present the information in the way it desires. This data has been obtained using data from a real incident.

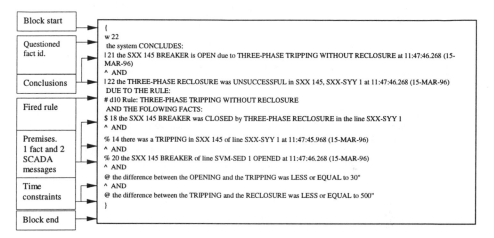

Fig. 3. Example of a "how" explanation with one depth level.

4. CONCLUSIONS

Explanation capabilities are a very important characteristic of expert systems, which can make their use much more effective. In the case of expert systems developed for power system on-line applications, severe real-time constraints make the introduction of effective explanation capabilities a difficult matter.

This paper presents an explanation production mechanism that has been designed and implemented to be used by SPARSE. SPARSE is an expert system for intelligent alarm processing and power restoration aid that is presently running on-line in a Portuguese transmission control center. This mechanism is based on a client-server philosophy and is especially designed to guarantee real-time performance of the expert system.

An intelligent tutor that allows control center operators to undertake training sessions during their normal activities in the control center presently uses the explanation server. This server is also being used to provide real-time explanations required by control center operators.

The integration of this explanation mechanism with SPARSE has enabled a much richer use of its capabilities.

REFERENCES

1. R. Bayada, "A Knowledge-Based Multimedia Tool Providing Explanations to Power Plant Operators", First International Conference on Sucesses and Failures of KBS in Real-World Applications, pp. 112-116,Bangkok, Thailand, 28-30 October, 1996
2. B. Chandrasekaran, W. Swartout, "Explanations in Knowledge Systems – The Role of Explicit Representation of Design Knowledge", IEEE Expert, pp. 47-49, June 1991
3. CIGRE - Working Group 38.06.03, "Expert Systems: Development Experience and User Requirements", Electra, n. 146, pp. 29-67, February 1993
4. J. Durkin, "Expert Systems: A View of the Field", IEEE Expert, February 1996
5. C.C. Liu, D. Pierce, H. Song, "Intelligent System Applications to Power Systems", IEEE Computer Applications in Power, vol. 10, no. 4, pp. 21-2, October 1997
6. W. Swartout et al., "Design for Explainable Expert Systems", IEEE Expert, pp. 58-64, June 1991
7. Z. A. Vale and A. M. Moura, "An Expert System with Temporal Reasoning for Alarm Processing in Power System Control Centers", IEEE Transactions on Power Systems, vol. 8, No. 3, pp. 1307-1314, August 1993
8. Zita A. Vale, M. Fernanda Fernandes, Couto Rosado, Albino Marques, Carlos Ramos, Luiz Faria, "Better KBS for Real-time Applications in Power System Control Centres: What can be learned by experience?", First International Conference on Successes and Failures of Knowledge-Based Systems in Real-World Applications, Bangkok, Thailand, 28-30 October, 1996
9. Zita A. Vale, A. Machado e Moura, M. Fernanda Fernandes, Albino Marques, Couto Rosado, Carlos Ramos, "SPARSE: An Intelligent Alarm Processor and Operator Assistant", IEEE Expert, vol.12, no. 3, Special Track on AI Applications in the Electric Power Industry, pp. 86-93, May/June, 1997
10. Zita A. Vale, Carlos Ramos, Luiz Faria, "User Interfaces for Control Center Applications", The 1997 International Conference on Intelligent Systems Applications To Power Systems (ISAP'97), Seoul, Korea, pp. 14-18, 6-10 July, 1997
11. Zita A. Vale, Carlos Ramos, Luiz Faria, Jorge Santos, M. Fernanda Fernandes,, Couto Rosado, Albino Marques, "Knowledge-Based Systems for Power System Control Centers: Is Knowledge the Problem?", The 1997 International Conference on Intelligent Systems Applications To Power Systems (ISAP'97), pp. 231-235, Seoul, Korea, 6-10 de July, 1997
12. I. Zukerman, R. Mconachy, "Generating concise Discourse that Addresses a User's Inference", 13th International Joint Conference on Artificial Intelligence (IJCAI'93), pp. 1202-1207, Chambéry, France, 28 August-3 September, 1993

Using Extended Logic Programming for Alarm-Correlation in Cellular Phone Networks

Peter Fröhlich[1] *, Wolfgang Nejdl[1], Michael Schroeder[1,2]
Carlos Damásio[3], and Luis Moniz Pereira[3]

[1] Universität Hannover, Germany (froehlic, nejdl)@kbs.uni-hannover.de
[2] City University, London, msch@cs.city.ac.uk
[3] Centria, Universidade Nova de Lisboa, Portugal (cd,lmp)@di.fct.unl.pt

Abstract. In this paper, we describe how to realize alarm-correlation in cellular phone networks using extended logic programming which provides integrity constraints, implicit, and explicit negation.

1 Introduction

Though cellular phone networks already contain intelligent network elements diagnosing local faults, alarm bursts caused by network element failures cannot be handled properly by this technology [5]. To deal with alarm bursts alarm correlation systems are required to filter and condense the incoming alarms to meaningful high-level alarms and diagnoses. We review the application described in [5] and show how the problem is modelled and solved with extended logic programming which provides integrity constraints, implicit and explicit negation. Such a model-based approach allows to diagnose a system by constraining observed and predicted behaviour of the system. Consider the diagram below.

On the right we have observations of the actual system and on the left the model of the system with its predicted behaviour. If the constraints that observed and predicted behaviour should not differ is violated we can adapt the model's assumptions in order to satisfy the constraints and thus compute a diagnosis.

Mobile networks can be divided into three parts: the mobile station (MS), the access network with the base station transceiver (BTS) consisting of antennas, radio transceivers, cross connect systems (CC) and microwave (ML) or cable links (CL) and the base station controller (BSC), and the switched network, which is connected to the access network by the BSC's. The BSC provides the radio resource management, which serves the control and selection of appropriate radio channels to interconnect the MS and the switched network. The switched network interconnects the MS to the communication partner, which might be another MS or an ISDN subscriber [5].

---

* Peter Fröhlich works now at ABB, Heidelberg

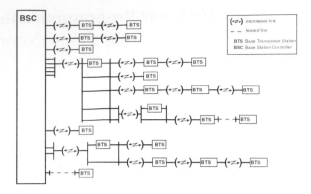

Fig. 1. Star-configuration of a base station subsystem [5].

The rest of the paper is organised as follows. First we introduce extended logic programs and diagnosis, then we show how to model cellular phone networks and alarm-correlation as extended logic programs, and finally we show how the alarm-correlation is realised in REVISE [3], a system for contradiction removal of extended logic programs.

2 Extended Logic Programming and Diagnosis

Since Prolog became a standard in logic programming much research has been devoted to the semantics of logic programs. In particular, Prolog's unsatisfactory treatment of negation as finite failure led to many innovations. Well-founded semantics [6] turned out to be a promising approach to cope with negation by default. Subsequent work extended well-founded semantics with a form of explicit negation and constraints and showed that the richer language, called WFSX, is appropriate for a spate of knowledge representation and reasoning forms [2].

Definition 1. An *extended logic program* is a (possibly infinite) set of rules of the form $L_0 \leftarrow L_1, \ldots, L_m, notL_{m+1}, \ldots, notL_n$ $(0 \leq m \leq n)$, where each L_i is an objective literal $(0 \leq i \leq n)$. An objective literal is either an atom A or its explicit negation $\neg A$.[1] Literals of the form $notL$ are called default literals. Literals are either objective or default ones.

The behaviour of the system to be diagnosed is coded as an extended logic program. To express the assumption that the system works correctly by default we use negation by default.

---

[1] Note that the coherence principle relates explicit and default, or implicit, negation: $\neg L$ implies $notL$ for every objective literal L.

Example 2. Consider the cellular phone network in Fig. 1. Assumption a microwave link is ok, it properly propagates signals:

$$
\begin{aligned}
&signal(NE, up, Sender, Signal) \leftarrow \\
&\quad not\ ab(NE), \\
&\quad type(NE, ml), type(Sender, bts), \\
&\quad class(Signal, farend\_or\_status\_signal), \\
&\quad signal(NE, down, Sender, Signal).
\end{aligned}
$$

Rules as the one above allow one to predict the behaviour of the system to be diagnosed in case it is working fine. To express that normality assumptions may lead to contradictions between predictions and actual observations we introduce integrity constraints.

Definition 3. An *integrity constraint* has the form $\perp \leftarrow L_1, \ldots, L_m, not L_{m+1}, \ldots, not L_n$ ($0 \leq m \leq n$) where each L_i is an objective literal ($0 \leq i \leq n$), and $\perp$ stands for false.

 Syntactically, the only difference between the program rules and the integrity constraints is the head. A rule's head is an objective literal, whereas the constraint's head is $\perp$, the symbol for false. Semantically the difference is that program rules open the solution space, whereas constraints limit it, as indicated on the left.

Example 4. Integrity Constraint

Now we can express that a contradiction arises if predictions and observations differ. In the setting of alarm-correlation we use for example the constraint

$$\perp \leftarrow \neg signal(bsc, down, Sender, alive), signal(bsc, down, Sender, alive).$$

to express that it is contradictory for the BSC to allegedly have received an alive signal of a BTS and to know at the same time that it has not.

A contradiction is always based on the assumption that the components work fine, i.e. the default literals are false. In general, we can remove a contradiction by partially dropping some closed world assumptions. Technically, we achieve this by adding a minimal set of revisable facts to the initially contradictory program:

Definition 5. The *revisables R* of a program *P* are a subset of the default negated literals which do not occur as rule heads in *P*. The set $R' \subseteq R$ is called a *revision* if it is a minimal set such that $P \cup R'$ is free of contradiction , i.e. $P \cup R' \not\models_{WFSX} \perp$

For details on the definition of the inference operator $\models_{WFSX}$ see e.g. [2].

Example 6. In the example above *not ab(NE)* is a revisable.

The limitation of revisability to default literals which do not occur as rule heads is adopted for efficiency reasons, but without loss of generality. We want to guarantee that the truth value of revisables is independent of any rules. Thus we can change the truth value of a revisable whenever necessary without considering an expensive derivation of the default literal's truth value.

In the next section we will show how to model alarm-correlation as extended logic programs.

3 Modelling Cellular Phone Networks

The cellular phone networks are configured in a star topology (see Fig. 1) with exactly one path from a BTS to the BSC. Since such networks are highly dynamic an explicit model of the network is a necessary prerequisite for alarm correlation. Consider the path depicted in Fig. 2. Its topology is modelled by facts on the components' types and connections (see Fig. 3). The fact $conn(ml16, up, bsc, down)$ states, for example, that the up-stream port of microwave link $ml16$ is connected to the down-stream port of the BSC.

Fig. 2. Network's topology where microwave link $ml18$ is faulty [5].

$type(ml16, ml)$. $conn(ml16, up, bsc, down)$.
$type(ml17, ml)$. $conn(ml18, up, ml16, down)$.
$type(ml18, ml)$. $conn(bts18, up, ml17, down)$.
$\ldots$ $\ldots$

Fig. 3. Facts for the network's topology.

The network elements are intelligent and perform local diagnosis resulting in alarm messages which are sent to the BSC. We distinguish three classes for the alarm messages:

(1) Farend alarms are generated by a BTS if the the components farther away from the BSC are not reachable anymore. (2) BTS-failure alarms are generated directly in the BSC, when it detects that a BTS is not reachable anymore. (3) The status signal *alive* is sent from each BTS on a periodical polling of the BSC. It is used to generate appropriate alarms in the BSC in case it gets lost on the way from the BTS to the BSC. The message is not physically present.

The alarms are classified by the facts *class* and *bts_failure_alarm* as shown in Fig. 4. The generation and suppression of alarms is captured by the rules shown in Fig. 5. The rules on the left-hand side state that each BTS sends an alive message and an alarm message in case of a farend alarm, respectively. The rules on the right-hand side express that there cannot be an alive message at the BSC if there is a BTS-failure or a farend alarm. Here explicit negation ($\neg signal$) proves to be very useful to get a compact model.

In Fig. 6 we formalize how signals are propagated over connections and through components (BTS or ML). While the BTS cannot fail, the microwave link may be faulty, though we assume by default that it is not abnormal ($not\ ab(NE)$). This default literal is a revisable whose truth value may be changed to satisfy the constraints.

Finally, we have to specify the integrity constraints that it is contradictory to have and not to have an alive message at the BSC and that there has to be either an alive message, or a BTS failure alarm, or the message was lost by the BTS (see Fig. 7). The

class(bts_omu_link_fail, bts_failure_signal). class(lapd_link_failure, bts_failure_signal).
class(bcch_missing, bts_failure_signal). class(farend_alarm_1, farend_signal).
class(available_traffic, bts_failure_signal). class(alive, status_signal).

bts_failure_alarm(Sender) ← bts_failure_alarm(Sender) ←
    alarm(Sender, bts_omu_link_fail),   alarm(Sender, available_traffic),
    type(Sender, bts).   type(Sender, bts).
bts_failure_alarm(Sender) ← bts_failure_alarm(Sender) ←
    alarm(Sender, bcch_missing),   alarm(Sender, lapd_link_failure),
    type(Sender, bts).   type(Sender, bts).

Fig. 4. Alarm classes.

signal(Sender, up, Sender, alive) ← ¬signal(bsc, down, Sender, alive) ←
    type(Sender, bts).   bts_failure_alarm(Sender),
                                        type(Sender, bts).
signal(Sender, up, Sender, Signal) ← ¬signal(bsc, down, Sender, alive) ←
    alarm(Sender, Signal),   alarm(Sender, Signal),
    class(Signal, farend_signal),   class(Signal, farend_signal),
    type(Sender, bts).   type(Sender, bts).

Fig. 5. Signal generation and suppression.

literal *message_lost* is a revisable which is assumed false, but may be changed to satisfy the constraints. Lost messages are more likely to occur that abnormal microwave links which is specified in Fig. 7.

The model of the network is consistent as long as there are no alarms. If alarms are generated and received by the BSC the constraints are violated and satisfying them yields the correct diagnoses of the problem.

Example 7. Consider the alarms in Fig. 8. The burst of messages is difficult to survey for a human operator, though the most probable explanation by revision is fairly easy: microwave link *ml*16 is abnormal.

To correlate the alarms, or to speak more generally, to compute revisions of contradictory extended logic programs, we use the REVISE system [3] which implements an adaption of the hitting-set algorithm [8,7] suitable for extended logic programming.

signal(NE_2, down, Sender, Signal) ←
    type(Sender, bts), class(Signal, farend_or_status_signal),
    conn(NE_1, up, NE_2, down), signal(NE_1, up, Sender, Signal).

signal(NE, up, Sender, Signal) ← signal(NE, up, Sender, Signal) ←
    class(Signal, farend_or_status_signal),   not ab(NE),
    type(NE, bts), type(Sender, bts),   type(NE, ml), type(Sender, bts),
    NE ≠ Sender,   class(Signal, farend_or_status_signal),
    signal(NE, down, Sender, Signal).   signal(NE, down, Sender, Signal).

Fig. 6. Signal propagation.

$\perp \leftarrow type(Sender, bts),$
  $not\ signal(bsc, down, Sender, alive),$
  $not\ bts\_failure\_alarm(Sender),$
  $not\ message\_lost(Sender).$

$\perp \leftarrow \neg signal(bsc, down, Sender, alive),$
  $signal(bsc, down, Sender, alive).$
$probability(ab(\_), 0.001).$
$probability(message\_lost(\_), 0.1).$

Fig. 7. Constraints for signals and a-priori probabilities of revisables.

$alarm(bts17, bcch\_missing).$
$alarm(bts17, bcf\_bie\_alarm\_in).$
$alarm(bts17, pcm\_fail).$
$alarm(bts17, bts\_omu\_link\_fail).$
$alarm(bts18, bcch\_missing).$
$alarm(bts18, pcm\_failure).$
$alarm(bts18, bts\_omu\_link\_fail).$
$alarm(bts19, bcf\_bie\_alarm\_in).$

$alarm(bts19, pcm\_failure).$
$alarm(bts19, bts\_omu\_link\_fail).$
$alarm(bts20, bcch\_missing).$
$alarm(bts20, pcm\_fail).$
$alarm(bts20, bts\_omu\_link\_fail).$
$alarm(bts21, bcch\_missing).$
$alarm(bts21, pcm\_fail).$
$alarm(bts21, bts\_omu\_link\_fail).$

Fig. 8. Alarms.

4 Computing the Revisions

Before we show how the revisions are computed, we need some defintions. Conflicts are sets of default assumptions that lead to a contradiction.

Definition 8. Let P be an extended logic program with default literals D. Then $C \subset D$ is a *conflict* iff $P \cup \{\neg c \mid not\ c \in C\} \models \perp$

To compute revisions, we have to change default assumptions so that all conflicts are covered. Such a cover is called hitting set, since all conflicts involved are hit.

Definition 9. A *hitting set* for a collection of sets C is a set $H \subseteq \bigcup_{S \in C} S$ such that $H \cap S \neq \{\}$ for each $S \in C$. A hitting set is minimal iff no proper subset of it is a hitting set for C.

Theorem 10. *[8] Let P be a program. Then R is a revision of P iff R is a minimal hitting set for the collection of conflicts for P.*

Theorem 10 states that revisions can be computed from conflicts and hitting sets which can be obtained from hitting set trees [8]:

Definition 11. Let C be a collection of sets. An HS-tree for C, call it T, is a smallest edge-labeled and node-labeled tree with the following properties:

1. The root is labeled $\sqrt{}$ if C is empty. Otherwise the root is labeled by an arbitrary set of C.
2. For each node n of T, let $H(n)$ be the set of edge labels on the path in T from the root node to n. The label for n is any set $\Sigma \in C$ such that $\Sigma \cap H(n) = \{\}$, if such a set Σ exists. Otherwise, the label for n is $\sqrt{}$. If n is labeled by the set Σ, then for each $\sigma \in \Sigma$, n has a successor, n_σ, joined to n by an edge labeled by σ.

To compute hitting set trees, Reiter proposed an algorithm [8] which was corrected in [7]. For the sake of brevity and clarity we explain the algorithm with its adaption to extended logic programs informally and by means of an extensive example.

The derivation tree on the right of Figure 9 show only the most relevant literals, the actual proof tree is based on the SLX proof-procedure for WFSX described in [1]. Since programs may be contradictory the paraconsistent version of WFSX is used. The top-down characterization of WFSX relies on the construction of two types of AND-trees (T and TU-trees), whose nodes are either assigned the status successful or failed. T-trees compute whether a literal is true; TU-trees whether it is true or undefined. A successful (resp. failed) tree is one whose root is successful (resp. failed). If a literal L has a successful T-tree rooted in it then it belongs to the paraconsistent well-founded model of the program (WFM_p); otherwise, i.e. if all T-trees for L are failed, L does not belong to the WFM_p. Accordingly, failure does not mean falsity, but simply failure to prove verity.

A T-tree is constructed as an ordinary SLDNF-tree. However, when a $notL$ goal is found the subsidiary tree for L is constructed as a TU-tree: $notL$ is true if the attempt to prove L true or undefined fails. To enforce the coherence principle we add in TU-trees the literal $not\neg L$ to the resolvant, whenever the objective literal $\neg L$ is selected for expansion. When a $notL$ goal is found in a TU-tree the subsidiary tree for L is constructed as a T-tree: $notL$ is true or undefined if the attempt to prove L true fails. Besides these differences, the construction of TU-trees proceeds as for SLDNF-trees.

Apart from floundering, the main issues in defining top-down procedures for Well-Founded Semantics are infinite positive recursion, and infinite recursion through negation by default. The former gives rise to the truth value false (the query L fails and the query $not\ L$ succeeds when L is involved in the recursion), and the latter to the truth value undefined (both L and $not\ L$ fail). Cyclic infinite positive recursion is detected locally in T-trees and TU-trees by checking if a literal L depends on itself. A list of local ancestors is maintained to implement this pruning rule. For cyclic infinite negative recursion detection a set of global ancestors is kept. If one is expanding, in a T-tree, a literal L which already appears in an ancestor T-tree then the former T-tree is failed. If one is expanding, in a TU-tree, a literal L which already appears in an ancestor TU-tree then the literal L is successful in the former TU-tree. The SLX proof-procedure has been extended in the REVISE implementation, and named SLXA, to be able to return the conflicts supporting $\bot$.

The SLXA procedure makes sure in T-trees that a revisable is false; in TU-trees it assumes that every revisable found is true. The revisables discovered are collected and returned to the invoker. Mark the similarities with Eshghi and Kowalski's abductive procedure where there is a consistency phase and an abductive phase [4].

The calls to SLXA are driven by the REVISE engine. Its main data structure is the hitting-set tree. The construction of the hitting-set tree is started on candidate {}, meaning that the revisables initially have their default value. We say that the node {} has been expanded when the SLXA procedure is called to determine one conflict. If there is none, then the program is non-contradictory and the revision process is finished. Otherwise, the REVISE engine computes all the minimal ways of satisfying the conflicted integrity constraint returned by SLXA, i.e. the sets of revisables which have to be added to program in order to remove that particular conflict. For each of these sets of revisables, a child node of {} is created. If there is no way of satisfying the conflicted integrity the program is contradictory. Else the REVISE engine selects a node to expand according to some preference criterium and cycles: it determines a new conflict, it expands that node with the revisables which remove the conflict, etc.... When a solution is found, i.e. there is no conflict, it is kept in a table for pruning the revision tree, by removing any nodes which contain some solution, and have been selected according to the preference criterium.

The order in which the nodes of the revision tree are expanded is important to obtain minimal solutions first. To abstract the preference ordering we assign to every node a key and a code. The code is a representation of the node in the lattice induced by the preference ordering $\preceq$. The key is a natural number with the property that given two elements e_1 and e_2 in the preference ordering $\preceq$ such that the corresponding key_1 and key_2 obey $key_1 \leq key_2$ then $e_2 \not\prec e_1$. This guarantees that if we expand first the nodes with smallest key we find the minimal solutions first. Furthermore, two nodes with the same key are incomparable. For the case of minimality by set-inclusion, the key corresponds to the cardinality of the candidate set for revision.

Minimality criteria, as for example minimality by cardinality, minimality by set-inclusion or minimality by probability, are implemented as an abstract data type with three functions: empty_code, add_new_rev, and smaller_code. The first returns the code associated with the candidate set {}. By definition, the key of this candidate set is 0. Given a new literal which has to be added to a node, the second function computes the new code and new key of the new candidate set. The latter function implements the subsumption test of two codes with respect to the preference ordering.

Example 12. Consider the description of the phone network of the previous section. Given the two alarm $\{alarm(bts20, bcch\_missing), alarm(bts20, lapd\_link\_failure)\}$ the most probable solution is that microwave 19 is faulty and a message from $bts21$ was lost. To compute these two revisions, REVISE proceeds as shown in Figure 9. The left column shows the expanding hitting set tree and the right the inference of contradictions with the responsible conflict. For clarity's sake, the inference tree contains only the most relevant literals such as *signal*, etc. Assuming nothing (2), the base station controller explicitly derives from the given alarms that there cannot be an alive signal of $bts20$. However, since all components are assumed to be working correctly, the BSC also derives that there should be an alive signal of BTS20. The components assumed fault-free and involved in the propagation are microwave links $ml16, ml18, ml19$. They form the first conflict.

In (3) these three microwave links form the first conflict, the root of the hitting set tree. The tree is now step by step expanded. First, it is assumed that $ml16$ is abnormal (4). A new conflict is derived since the *bsc* does not have an alive signal from $bts17$, neither a failure alarm, nor was a message lost. The latter is revisable, i.e. we may assume that messages get lost, and thus the conflict $msg\_lost(bts17)$ is returned. The hitting set tree is updated accordingly (5) and in particular, the nodes are re-ordered according to their probability. While the branches for microwave links 18 and 19 have a probability of 0.001, the branch of $ab(ml16), msg\_lost(bts17)$ has only a probability of 0.0001.

Similar to (4), $ml18$ is now assumed abnormal (6), leading to a new conflict involving $msg\_lost(bts19)$. This process of changing assumptions, deriving a contradiction with associated conflict and updating the hitting-set tree accordingly until finally, assuming $ml19$ abnormal and the message from $bts21$ lost, there is no further conflict. Due to the constant re-ordering of the hitting-set tree to expand the most probable candidate next, we know that we found the most probable solution and can terminate the search.

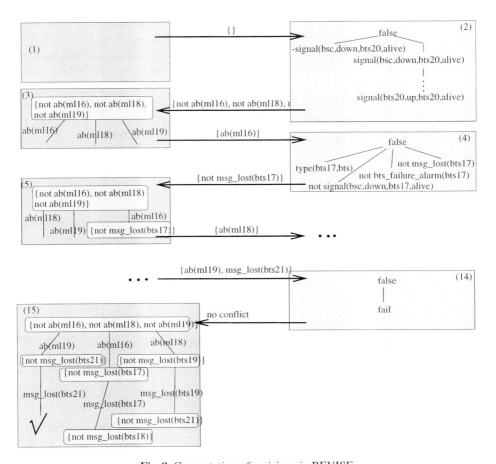

Fig. 9. Computation of revisions in REVISE.

Conclusions

In cellular phone networks faults of microwave links are likely to cause alarm showers, which put the system operator to a hard test. Alarm correlation tools are needed, which support the operators by identifying the cause of a large number of alarm messages.

The main contribution of this paper is twofold: We have shown how to model alarm-correlation in cellular-phone networks declaratively as extended logic programs and how to diagnose these networks by satisfying violated integrity constraints using the REVISE[2] system. The integrity constraints and implicit negation are used to constrain predictions and observations and express default assumptions about the components states, respectively.

Many practical problems like alarm-correlation require a logical explanation of observations and therefore abductive capabilities as used in our approach. When compared with other approaches such as neural networks [9], there are three major benefits:

1. A very small and maintainable system description, that separates structural from behavioural components and thus makes changes of the network topology easy.
2. A propagation model which allows to correctly diagnose unforeseen errors as well as multiple faults.
3. Failure probability estimates (see Fig. 7), which lead to correct diagnoses even on noisy data, where alarm messages have been lost.

Acknowledgements We'd like to thank ICCTI-BMFT.

References

1. J. J. Alferes, C. V. Damásio, and L. M. Pereira. A logic programming system for non-monotonic reasoning. *Journal of Automated Reasoning*, 14(1):93–147, 1995.
2. J. J. Alferes and L. M. Pereira. *Reasoning with Logic Programming*. (LNAI 1111), Springer-Verlag, 1996.
3. C. V/ Damásio, L. M. Pereira, and M. Schroeder. REVISE: Logic programming and diagnosis. In *Proc. of LPNMR97*. LNAI 1265, Springer–Verlag, 1997.
4. K. Eshghi and R. Kowalski. Abduction compared with negation by failure. In *6th Int. Conf. on LP*. MIT Press, 1989.
5. P. Fröhlich, W. Nejdl, K. Jobmann, and H. Wietgrefe. Model-based alarm correlation in cellular phone networks. In *Proc. of MASCOTS97*, 1997.
6. A. Van Gelder, K. Ross, and J. S. Schlipf. Unfounded sets and well-founded semantics for general logic programs. In *Proc. of Symp. on Principles of Database Systems*, 1988.
7. R. Greiner, B. A. Smith, and R. W. Wilkerson. A correction of the algorithm in reiter's theory of diagnosis. *Artificial Intelligence*, 41(1):79–88, 1989.
8. R. Reiter. A theory of diagnosis from first principles. *Artificial Intelligence*, 32(1):57–96, 1987.
9. H. Wietgrefe et al. Using neural networks for alarm correlation in cellular phone networks. In *Proc. of Ws on Applications of Neural Networks in Telecommunications*, 1997.

---

[2] REVISE is online at www.soi.city.ac.uk/˜msch.

Knowledge Acquisition Based on Semantic Balance of Internal and External Knowledge

Vagan Y. Terziyan, Seppo Puuronen

Department of Computer Science and Information Systems,
University of Jyvaskyla, P.O.Box 35, FIN-40351, Jyvaskyla, Finland
e-mail: {vagan, sepi}@jytko.jyu.fi

Abstract. This paper presents a strategy to handle incomplete knowledge during acquisition process. The goal of this research is to develop formal tools that benefit the law of semantic balance. The assumption is used that a situation inside the object's boundary in some world should be in balance with a situation outside it. It means that continuous cognition of an object aspires to a complete knowledge about it and knowledge about internal structure of the object will be in balance with knowledge about relationships of the object with other objects in its environment. It is supposed that one way to discover incompleteness of knowledge about some object is to measure and compare knowledge about its internal and external structures in an environment. If there exist differences between the internal and the external semantics of an object, then these differences can be used to derive more knowledge about the object to make knowledge complete. The knowledge refinement process is done step-by-step as a continuous evolution of a knowledge base. Each step consists first automatic analysis of semantic balance which is then followed by attempts to derive knowledge that will balance differences between internal and external semantics of the object. This paper describes an algebra that is used to describe the internal and external semantics of an object and to derive unknown part of it. The results presented are mostly theoretical ones.

1   Introduction

This paper deals with a cognition strategy based on semantic model of world. It describes one refinement technique to handle incompleteness of knowledge in acquisition process. Knowledge base refinement is now one of the central problems of expert systems [12]. It needs a fundamental research using basic concepts of philosophy and cognitive science. The main focus of this paper is to describe and apply one of the fundamental philosophic principles - "Balance in Nature" in terms of semantic networks to define the strategy of improving knowledge during a cognition process. The goal of this research is to develop formal metasemantic algebra that benefits the law of semantic balance, i.e. there should be balance between the internal and external semantics of an object in the possible world (*WORLD* in short onwards in this paper). If this semantic law holds in the *WORLD* and there exists any difference between the internal and the external semantics of an object, then this difference can be used to acquire more knowledge about the object. The refinement proceeds step-by-step as a continuous evolution of knowledge base, where each step includes two substeps: first substep makes automatic analysis of semantic balance and if the situation is not in balance then the second substep attempts to derive knowledge that will reestablish balance.

Knowledge base refinement as a method to improve an incorrect, inconsistent, and incomplete domain theory has also been suggested in [12]. His ODYSSEUS system refines knowledge bases of advanced rule-based systems. It learns by watching apprentice. His refinement program tries to construct an explanation of an observed action of an expert. Context of explanation allows to generate candidate of knowledge base repairs. ODYSSEUS system is designed for use with heuristic classification using hypothesis-directed reasoning. A processing stage prior to apprenticeship learning removes an inconsistent knowledge from the domain theory, which is responsible for deterioration of the performance of the system due to sociopathic interactions between elements of the domain theory. Sociopathicity implies that some kind of global refinement for the acquired knowledge is essential for machine learning.

Current books in formal semantics widely use approaches based on fundamental conceptual research in philosophy and cognitive psychology. For example Larson and Segal [6] give equal weight to philosophical, empirical, and formal discussions. They study a particular human cognitive competence governing the meanings of words and phrases. They argue that speakers have unconscious knowledge of the semantic rules of their language. Knowledge of meanings is both the semantics of domain attributes (properties and relations) and learning technology how to derive semantics of inconsistent and incomplete meanings.

During last several years one can see the growth of interest to semantic models of *World* [7]. The reason seems to be in extremely fast development of global information networks. Study of large domains with numerous objects and groups of objects with relations requires possibilities to have closer considerations inside objects (their properties), outside objects (their external semantics), and both inside and outside considerations also for groups of objects. This kind of situations arises for example with WWW, the organization of which requires net-based semantic models and good technology of self-organization to handle problems of their complexity [4]. One can interpret acquired knowledge only if "internal" part of it is in a conformity with "external" one. In other words these parts have to be in "balance". Phenomena of balance is very important in understanding problems related to knowledge [10]. It was used in [5] to minimise incompleteness of internal and external knowledge represented in neural networks. Balance has to be taken into account in cooperative modeling and machine learning, according to [8, 2], in systems control according to [11].

The main focus of this paper is to describe in formal way and apply the fundamental philosophic principle of balance between internal and external semantics of a domain object. We use and further develop the formalism of metasemantic algebra [9,1,3] to describe internal and external semantics of any single or compound object in a network and the formal use of the law of semantic balance during the cognition process. Chapter 2 of this paper gives a short introduction to the metasemantic algebra. In chapter 3 the ways to formulate the internal and external semantics of an object is presented and chapter 4 introduces the law of semantic balance between the internal and external semantics of an object. Chapter 5 describes the stepwise process of knowledge refinement utilizing unbalanced situations and chapter 6 concludes with further research suggestions.

2 A Metasemantic Algebra

The metasemantic algebra was proposed in [9] and further developed in [1]. The basic elements of the algebra are objects and their relations with special operations upon semantic meanings of relations. In this chapter we will introduce the basic elements of semantic network and semantic operations.

2.1 A Semantic Network

Let A_i be an atomic *object*. It can also have its internal structure but in relations it is considered as an atomic object. Let L_k be a *semantic meaning* of *relation* between two objects or one object with itself. The second one corresponds to the *property* of the object. The *semantic predicate P* is:

$$P(A_i, L_k, A_j) = \begin{cases} 1, \textit{ if there is relation between} \\ \quad A_i \textit{ and } A_j \textit{ with meaning } L_k; \\ 0, \textit{ otherwise.} \end{cases} \tag{1}$$

Semantic network S is: $S = \bigwedge_{i,j,k} P(A_i, L_k, A_j)$, where A_i is the source of the relation, A_j is the object of the relation, and L_k is the semantic meaning of the relation between those objects.

2.2 Semantic Constants and Operations

There are two semantic constants:

- semantic *ZERO* (notation - *IGN*): (it means a total ignorance about relationship between the source object and the target object):

$$\forall (A_i, A_j) \neg \exists L_k (P(A_i, L_k, A_j)) \Rightarrow P(A_i, IGN, A_j) ; \tag{2}$$

- semantic *UNIVERSE* (notation *SAME*): (it means a total knowledge about relationship between the source object and the target object).

There are two special relations *HAS_PART* and *PART_OF* which have their ordinary meanings. If it is true that $P(A_i, HAS\_PART, A_j)$ or

$P(A_j, PART\_OF, A_i)$ then object A_j is included in the object A_i. In the special case when an object is not part of any other object we call it as a (possible) *World*, i.e. $\forall A_i \neg \exists A_j (P(A_i, PART\_OF, A_j)) \Rightarrow A_i = World$ and in the special case when an object has no other object that is part of it, we call it as *Atom*, i.e.: $\forall A_i \neg \exists A_j (P(A_i, HAS\_PART, A_j)) \Rightarrow A_i = Atom$.

The ordinary semantic operations are:

- semantic inversion: $P(A_i, L_k, A_j) \equiv P(A_j, \tilde{L}_k, A_i)$;
- semantic addition: $P(A_i, L_k, A_j) \wedge P(A_i, L_n, A_j) \equiv P(A_i, L_k + L_n, A_j)$;
- semantic multiplication: $P(A_i, L_k, A_m) \wedge P(A_m, L_n, A_j) \equiv P(A_i, L_k * L_n, A_j)$, where $i \neq j \neq m$.

In the graphical representation, objects are described by circles and relations with directed arcs leading from the source object to the target object. Object and relations that form an internal semantic structure of an object are presented inside the circle of that object. When the internal structure of an object is not under consideration it is not necessarily shown. The *WORLD* under consideration is presented by the outermost circle as in Figure 1 where the *WORLD* is *W*. It includes an object *A* which internal structure is also presented in the figure. The object *A* includes an object A_2 which internal structure is not presented. There is a relation L_1 between the objects A_3 and A_1 and the object A_2 has relationship with itself.

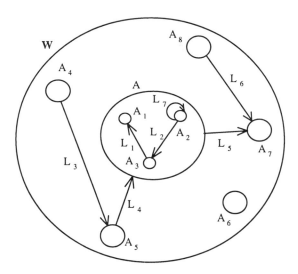

Fig. 1. The world *W* with its objects and relations.

3 The Internal and External Semantics

In this chapter we will derive the formulas of the semantic algebra which present the internal and external semantics of an object.

3.1 The Internal Semantics of an Object

We suppose that the internal semantics of an object can be defined using the components of the object and the relationships between those components. We further suppose that the quality of internal semantics behaves in monotonous way. It means that if any additional knowledge about the internal structure (components and their relations) of the object is achieved then the quality of the formulated internal semantics is never going worse. Thus by acquiring more and more knowledge about the object (its internal structure) we are able to achieve more and more complete internal semantics of this object.

We define that the internal semantics of an object A_i is the semantic sum over all the possible paths between any pairs of objects (A_j, A_k) included in the object A_i plus the paths from each included object to itself. A path between any pair of objects (A_j, A_k) includes successive relations (or their inverse) from the object A_j to the object A_k so that no relation (or its inverse) is taken twice. The only path where the same object is visited twice is the path from the object to itself. This later one guaranty that the properties of the included objects will be taken into account. Thus for a given object A_i its internal semantics $E_{in}(A_i)$ is:

$$E_{in}(A_i) = \sum_{\substack{\forall j,k,j \le k, \\ P(A_i, HAS\_PART, A_j) \\ P(A_i, HAS\_PART, A_k)}} L_{A_j - A_k} . \tag{3}$$

where $L_{A_j - A_k}$ is a path from A_j to A_k.

In the case of Fig. 1 the internal semantics of the object A is:

$$E_{in}(A) = \tilde{L}_1 * \tilde{L}_2 + L_2 + \tilde{L}_1 + L_7 . \tag{4}$$

The internal semantics corresponds the knowledge seen in Fig. 2a.

3.2 The External Semantics of an Object

We suppose that the external semantics of an object A can be defined using the components of the possible world outside the object and the relationships between those components. We further suppose that the quality of external semantics behaves in monotonous way. It means that if any additional knowledge about the external structure (components and their relations) of the object is achieved then the quality of the formulated external semantics is never going worse. Thus by acquiring more and more knowledge about the world outside the object (its interaction with its environment) we are able to achieve more and more complete external semantics of this object.

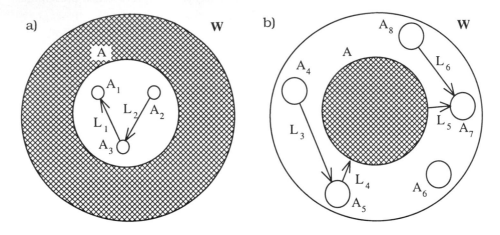

Fig. 2.   a) The internal semantics of the object A in the possible world W,
b) The external semantics of the object A in the possible world W.

We define that the external semantics of an object A_i is the semantic sum over all the possible paths between any pairs of objects (A_j, A_k) included in the world outside the object A_i plus the paths from each included object to itself. The object A_i belongs to the objects that participate in pairs but the path from the object A_i to A_i is not included. For a given object A_i its external semantics is:

$$E_{ex}(A_i) = \sum_{\substack{\forall j,k,j\leq k,j\neq k\neq i, \\ (A_j \in W/A_i)\vee(A_j = A_i), \\ (A_k \in W/A_i)\vee(A_k = A_i)}} L_{A_j - A_k} . \tag{5}$$

On the other hand $E_{ex}(A_i)$ is the internal semantics of the *World* when A_i is taken as *Atom* (without noticing its internal structure). This gives a formula:

$$E_{ex}(A_i) = E_{in}(World \,/\, E_{in}(A_i)) . \tag{6}$$

In the case of Fig. 1 the external semantics of the object A is:

$$E_{ex}(A) = L_3 * L_4 + L_3 * L_4 * L_5 * \tilde{L}_6 + \tag{7}$$
$$+ L_3 * L_4 * L_5 + L_4 + L_4 * L_5 * \tilde{L}_6 +$$
$$+ L_4 * L_5 + L_5 * \tilde{L}_6 + L_5 + L_3 + \tilde{L}_6 .$$

This external semantics corresponds the knowledge seen in Fig. 2b.

4 The Law of Semantic Balance

Let us suppose that there exists a possible world where the ideal situation for an object A_i is that its internal semantics (i.e. its internal structure = objects and their relations) and its external semantics (i.e. its properties when it interacts its environment) are in balance. In this ideal situation the law of semantic balance holds: $E_{in}(A_i) = E_{ex}(A_i)$. Usually, especially with knowledge bases, the ideal situation has not been achieved. Human knowledge about objects is almost always incomplete, and the knowledge base usually includes incompleteness, inconsistencies, and incorrectness. Sometimes we know more about the structure of an object than its external properties and sometimes on the contrary. Let $ign_{in}^t(A_i)$ be our ignorance about the internal semantics of the object A_i at the time t and let $ign_{ex}^t(A_i)$ be our ignorance about the external semantics of the object A_i at the time t then according the law of semantic balance we can write:

$$E_{in}^{(t)}(A_i) + ign_{in}^{(t)} = E_{ex}^{(t)}(A_i) + ign_{ex}^{(t)} . \tag{8}$$

5 Strategy of Knowledge Refinement

In this chapter we consider a strategy of improving incomplete or incorrect knowledge about some object during the acquisition process using the law of semantic balance. The strategy is shown in Fig. 3.

Let us suppose that we have acquired some knowledge $E_{in}^1(A_i)$ about the internal semantics of the object A_i and that we have acquired some knowledge $E_{ex}^1(A_i)$ about the external semantics of the object A_i. Let us assume that these two semantics are not in balance. Then we can try to make them in balance trying to remove some part of ignorance from either or both sides of the formula:

$$E_{in}^{(1)}(A_i) + ign_{in}^{(1)} = E_{ex}^{(1)}(A_i) + ign_{ex}^{(1)} . \tag{9}$$

If this equation succeeds, at least partially, then we receive another amount of knowledge $E_{in}^2(A_i)$ about the internal semantics of the object A_i and another amount of knowledge $E_{ex}^2(A_i)$ about the external semantics of the object A_i. If these two semantics are not in balance or if some outer knowledge source gives extra knowledge that makes them unbalance again, then we try to make them in balance trying to remove some part of ignorance using the same formula as above and so on (Fig. 3).

360

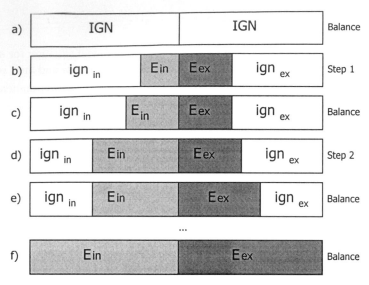

Fig. 3. The strategy of step-by-step knowledge refinement using the law of semantic balance.

6   Conclusion

We have discussed some aspects of using semantic balance in knowledge acquisition. It uses as a basic measure of unbalance the difference between the internal and external semantics of an object. These semantics are expressed using formal metasemantic algebra. The solution of equations of this algebra needs further development. Nevertheless the principle of semantic balance seems to be a good way to understand the dynamics of knowledge.

References

1. Bondarenko, M., Grebenyuk, V., Terziyan, V.: Reasoning Based on the Algebra of Semantic Relation. Pattern Recognition and Image Analysis 3(4) (1993) 488-499
2. DeJong, G., Bennett, S.: Permissive Planning: A Machine Learning Approach to Linking Internal and External Worlds. AAAI Press/MIT Press, Menlo Park (1993) 508-513
3. Grebenyuk, V., Kaikova, H., Terziyan, V., Puuronen, S.: The Law of Semantic Balance and its Use in Modeling Possible Worlds. In: STeP-96 – Genes, Nets and Symbols, Publ. of the Finnish AI Society (1996) 97-103
4. Heylighen, F., Bollen, J.: The World Wide Web as a Super-Brain: From Metaphor to Model. In: Proceedings of the 13-th European Meeting on Cybernetics and Systems Research, Austrian Society for Cybernetic Studies (1996) 917-922
5. Kamimura, R.: Generation of Inhibitory Connections to Minimize Internal and External Entropy. In: Alexander, I., Taylor, J. (eds.): Artificial Neural Networks 2, North Holland, Amsterdam (1992) 969-972

6. Larson, R., Segal, G.: Knowledge of Meaning. An Introduction to Semantic Theory. Mass.: A Bradford Book (1995)
7. Li, L.: Possible World Semantics and Autoepistemic Reasoning. Artificial Intelligence 71(2) (1994)
8. Morik, K.: Balanced Cooperative Modeling. Machine Learning 11(2/3) (1993)
9. Puuronen, S., Terziyan, V.: A Metasemantic Network. In: STeP-92 – New Directions in Artificial Intelligence, Publ. of the Finnish AI Society (1992) 136-143
10. Schultz, T. R., Mareschal, D., Schmidt, W. C.: Modeling Cognitive Development on Balance Scale Phenomena. Machine Learning 16(1/2) (1994)
11. Sen, M. de la, Jugo, J.: Robust Control for a Class of Systems with Internal and External Point Delays, Cybernetics and Systems '94, World Scientific Publishing (1994) 191-198
12. Willkins, D. C.: Knowledge Base Refinement as Improving an Incorrect and Incomplete Domain Theory. In: Buchanan B. G., Wilkins D. C. (eds.): Readings in Knowledge Acquisition and Learning, Morgan Kaufmann (1993) 631-641

Alex: A Computer Aid for Treating Alexithymia

Raoul N. Smith[1], William J. Frawley[2]

[1]College of Computer Science, Northeastern University, Boston, MA 02115 USA
rnsmith@ccs.neu.edu
[2]Department of Linguistics and Program in Cognitive Science, University of Delaware,
Newark, DE 19711 USA
billf@copland.udel.edu

Abstract. One of the principal skills that a patient in psychotherapy must acquire is to be aware of and be able to express his or her feelings and emotions. Many patients, however, especially at the beginning of their therapy, have a great deal of trouble doing this. They are able to articulate "I feel bad" but nothing more precise. One reason for this is that they may be suffering from a cognitive-affective disorder called alexithymia.[†] This paper describes a system called Alex that we have built to help treat alexithymia. This system, currently a prototype, consists of a query processor, a pattern matcher, and a text generator written in C++.[1]

1 Introduction

Alexithymia is a cognitive-affective disturbance that affects the way people experience emotions and express them [20, 21]. The term was originally coined by Sifneos [16] from the Greek _a_ meaning 'lack of', _lexis_ meaning 'word', and _thymos_ meaning 'emotion' or 'mood.' It can be most generally characterized by the inability to verbalize affect. Krystal [8] claimed this disturbance as the most important single factor impeding the success of psychoanalysis and of psychodynamic psychotherapy.[2]

[†] The authors would like to thank Carroll Izard, John T. Murray, Eric Schuster, and Mary H. Smith for their thoughtful comments on an earlier draft of this paper. It has been much improved by our having incorporated many of their suggestions. We also wish to thank Chantana Chantrapornchai for programming a large part of the initial system.

[1] A complete version of Alex will require a very large affect base. We are currently seeking funding to build such a large database of emotions. It will be based on a database of 7500 emotion terms that we have compiled

[2] Alexithymia may also be one of the single largest indirect causes in the increased cost in health care. Because alexithymics have difficulty in verbalizing feelings and distinguishing emotion states, they often express their psychic problems by means of physical complaints. Their physicians look at these complaints as suggesting some organic pathology. The physician orders extensive texts, none of which disclose an organic problem. Since at least

2 Characteristics of Alexithymia

Characteristics of alexithymia were first observed among patients who expressed classical psychosomatic disorders. Ruesch [13] referred to these patients as 'infantile personalities.' Freedman and Sweet [5] referred to them as 'emotional illiterates.' Later, Krystal [6] observed similar symptoms in patients suffering from post-traumatic stress. Rybakowski, Ziolkowski, Zasadska, and Brzezinski [14] observed them in alcoholics and drug addicts.

Alexithymia is manifested mainly in communicative style. Its most characteristic features are:

- a striking inability in recognizing and verbally describing one's own emotions
- markedly reduced or absent symbolic thinking[3] so that inner attitudes, feelings, wishes, and drives are not expressed
- inability to use feelings as signs of emotional problems
- thinking that is literal, utilitarian, and concerned with the minutiae of external events [20]
- few dreams and, when they do occur, they lack color, bizarreness, and symbolism (ibid.)
- difficulty in discriminating between emotional states and bodily sensations [11]
- stiff posture [10]
- lack of expressive facial expressions [7]
- impaired capacity for empathy (ibid.).

Clearly alexithymics have a deficiency in the symbolic processing required in identifying and communicating emotions.[4]

Consequently, when an alexithymic is asked about his[5] feelings about some highly charged event, he will respond either by describing physical symptoms or by not understanding the question [12]. Our goal in building this system is to break through the alexithymia barrier. We see this as one tool in a set of tools that can help a patient in his recovery from addiction and subsequent growth. The set of interrelated

---

30% of the population is alexithymic [17], this results in a huge expenditure to the health care delivery system.

[3]This does put into question the present approach and that of [4] but sub-symbolic, neural network approaches that might be suggested are, at the moment, not at all clear.

[4]Carroll Izard (personal communication) has an interesting interpretation of alexithymia. If one recognizes the relative independence of the emotion and cognitive systems, then alexithymia can be described as a function of dissociation between emotion states and cognitive processes.

[5]We use the masculine pronoun because men are roughly four times as likely to suffer from alexithymia as women [1].

computer tools that we envisage is a component that accepts characteristic behaviors and returns a description of the emotion that characterizes it (the subject of the present paper), a set of affirmations specific to that emotion, a narrative/story retrieval system for retrieving stories with that emotion in it (a version of this is described in [19]), on-line discrimination exercises as in gestalt therapy to help the patient, and a network used for on-line support

Many if not all, forms of therapy have teaching as their most consistent method. For patients with alexithymia, they have to be taught to identify and differentiate among their feelings. For example, the alexithymic patient can be made aware that, when one is shamed, it is normal to blush, to feel flushed, to lower one's eyes. These sensations and behaviors can be called "embarrassment," "shamefulness." Patients must be taught to pay attention to their bodily sensations and to connect these to the notion of experiencing a feeling. This is the basis for our approach. It is supported by the work of Lesser [9]. Lesser states that traditional psychotherapy with alexithymics is difficult and a more instructional, reality-based approach should be taken rather than the more traditional conversational approach. In addition, a very supportive approach is needed because even though alexithymic patients may not be articulate in describing their feelings, it does not mean that their feelings are not real. And these feelings may be painful.[6]

3 The System

The system we designed is meant to be accessible by the patient on his own or, preferably, as part of a therapy or counseling session with a professional. This professional could be a nurse practitioner at an HMO who would be working with a group of patients working at terminals. It could also be someone overseeing the recovery of an employee as part of an Employee Assistance Program. It would be used after an initial diagnosis of alexithymia and before extensive therapy with a professional therapist or in conjunction with on-going therapy. We see this being of long-term use. The three principal components of the systems are a query processor, a pattern matcher, and a text generator.

3.1 The Query Processor

The purpose of the query processor is to present to the patient a series of questions. The responses to the questions are accumulated. These are then used to match against a database of frames representing emotions. The frame that is the closest match (the

---

[6]This is consonant with Izard's view. The feelings are there but there is a repression or dissociation of cognitive content.

greatest number of matched hits) is assumed to be the emotion that the patient is exemplifying.

There are problems with implementing this, however. First, the problem with open-ended questions, such as "Why do you think that happened?," is that the subsequent responses are difficult to analyze. Also, "why" questions are often unanswerable. In addition, the answer can be almost anything including a long monologue. We know that task and domain severely restrict the syntax and lexicon in natural language processing [18]. But even in this application task, a cause for experiencing an emotion can be a very long open list. In some cases that may be alright. For example, the first author and some of his students developed a system called FOCUS that can be used in cognitive psychotherapy to teach patients new patterns of growth. In FOCUS the answer does not matter, as long as the person responds. The question can be quite generic and only locally dependent on responses[7].

As an example, we present here our description of the emotion of shame. Shame is perhaps the most "complex" or "basic" emotion. It seems to "underlie" or motivate many other emotions. Because of this, it may appear adventuresome on our part to model it. But we have focused on it *because* of its importance. Bradshaw [2] refers to shame as "the master emotion." It plays an important role in the progression of the disease of alcoholism, an application area that we have addressed elsewhere. Also it is known that alcoholics form a large portion of the population which suffers from alexithymia[8].

The object structure we have designed is a frame-like object. A partially filled *shame* object has the following structure:

---

[7]FOCUS was developed in conjunction with Dr. Michael Pearlman. It existed in two versions, one written in VP-Expert and the other in Hypercard.

[8]Alcoholics have that shame reduced when they accept the fact that alcoholism is a disease over which they have no control. They are not flawed just because they have a disease. Still another correlation between shame and alcoholism exists with gays and lesbians. Kominars 1989 claims that 10% of the US population is alcoholic or addict (and this estimate may be very low) and in the gay and lesbian community that number increases to 30%. It should also be noted that the number of gays and lesbians who suffer from alexithymia may therefore also be higher than the straight community. Furthermore, shame plays an important role in the field of literary criticism based on "queer performativity." This branch of literary criticism focuses on the identification of shame in literary texts and uses it as the prime analytical tool. (See, for example, [3] and [15]). We would like to see someone apply our shame description in the analysis of such literary texts.

Slot Name	Filler
name	shame
instance-of	emotion
may-lead-to	aggression ∨ addiction ∨ obsession ∨ narcissism ∨ depression ∨ perfectionism ∨withdrawal ∨diffidence ∨ combativeness
provenience	child rearing practices
time-acquired	early
source	rebukes ∨ warnings ∨ teasing ∨ ridicule ∨ ostracism ∨ neglect ∨ abuse∨
caused-by	perception of self ∨unconscious feeling of unworthiness
typical-examples	telling a joke that causes offense ∨ overstaying your welcome
mutually-disjoint-with	pride
Syn	being caught and wanting to hide
Time$_{onset}$	sudden
Time$_{duration}$	range (short → long)
yields	fear of humiliation
feelings	ugly∨stupid∨impotent∨unmanly∨unwomanly∨phony∨ grasping∨boring∨cheap∨insignificant∨immature∨ unable to love ∨ unlovable ∨weak ∨needy ∨ rejected ∨abandoned ∨ dismissed

behavioral characteristics	expect disappointment ∨refuse to see reality ∨ look away when embarrassed ∨ avoid eye contact ∨control interactions with people ∨pay great attention to looks ∨ pay great attention to behavior ∨control own behavior∨ try to be right ∨avoid negative comments of others ∨ perfectionist ∨blame others ∨do not articulate needs ∨ do not articulate wants ∨cover up mistakes ∨shame others when they are wrong ∨ do not trust
cognitive characteristics	think that you are flawed ∨think that you are a mistake∨think that you are worthless∨think you should not think the way you do∨think that you should not want what you want∨think you should not imagine what you do

Since people with alexithymia have a great deal of trouble identifying their emotions and feelings, the emotion objects we store in the database cannot refer to these. We therefore elicit the emotions via a set of generated questions based on the contents of only the slots labeled behavioral characteristics and cognitive characteristics in our frame structure. A portion of the questions includes:

Do you think that you are a mistake?
Do you think that you are flawed?
Do you blush when you are embarrassed?
Do you look away under those circumstances?
Do you avoid eye contact then?

As questions get answered, a skeleton structure is built up. That structure is that of a typical emotion frame. Both *yes* and *no* responses are important so they are both saved.[9] The skeleton structure is as large as the union of the slot fillers. This may mean two hundred or more filled slots. Once the skeleton is filled (or almost filled), it is compared to all the emotion frames in the database. The emotion frame that most closely matches the skeleton structure is retrieved from the database and is passed to the text generator for presentation. The process of pattern matching is described next.

---

[9]An intermediate structure is built with the number of the question and the response. The responses are then matched against the phrases in a table. The appropriately spelled phrase is then stored in the skeleton structure.

3.2 The Pattern Matcher

After the patient has responded to all the questions, the query program returns the answer pattern in the form of a packed array. Each element of the array is one bit representing *yes* or *no*. This is the input to the pattern matcher. The role of the pattern matcher is to match the patient's answer pattern to the most probable emotion description in the affect base. The algorithm we use, nearest neighbor is also that used in Dr. Bob [19].[10] For example, *shame* and *anger* might have the following answer patterns:

> *Shame*: 10110011101
>
> *Anger*: 01001110110

If the first question is "Do you think you are a mistake?", the answer would probably be yes (1) for someone suffering from *shame* and no (0) for someone experiencing anger. The ordered set of answers is then matched against the answers to the prototypical responses for all the emotions in the affect base. The closest match is then returned to the text generator.

3.3 The Text Generator

The input to the text generator is the set of slot fillers of a particular frame retrieved by the query processor from the affect base. The text generator is a

---

[10]In the full system we intend to implement a more sophisticated pattern matcher. In that system there is an external file containing the answer pattern for each emotion. The file has the following format:

> question-weights   range-values   emotion-name

The first field contains the weight of each question. We want to introduce weighting because the same answer might be appropriate for a variety of questions. This field then contains several values, one for each of the questions. Each is a real number whose position is determined by the order of the question presented by the query processor. Its value is the weight of the corresponding question as it relates to the emotion listed in the emotion-name. For instance, some answer may be weighted differently depending on the emotion of which it is symptomatic. Suppose that the question illustrated above is:

> Do you think you are flawed in some way?

Clearly, since that cognitive characteristic is highly symptomatic of the emotion of *shame*, a *yes* answer to this question should be given more weight than to *anger* where a *yes* might also be acceptable but less likely.

grammar that consists of a set of sentence frames and a set of rules for merging the slot fillers into the slots of the sentence frames. For each sentence there is more than one version. The reason for this is to have a variety of outputs. If this were not available, the output would be very repetitive and therefore potentially boring. The grammar includes the following:

$$\text{Text} \rightarrow P_1 + S_{\text{lead-to}} + P_3 + S_{\text{feelings}} + S_{\text{behavior}}$$

$$P_1 \rightarrow S_{\text{emot}} + S_{\text{time-acquired}} + S_{\text{provenience}} + S_{\text{source}} + S_{\text{caused-by}} + S_{\text{syn}} + S_{\text{yield}} + S_{\text{mutually-disjoint-with}}$$

$$S_{\text{emot}} \rightarrow \text{The emotion you are describing is probably } \underline{\quad}.$$

...

A much abridged sample text of sentence frames with emotion slot fillers merged with them follows. The slot fillers are underlined in the paragraph. The emotion is shame.

> The emotion you are describing is probably <u>shame.</u> This emotion is acquired <u>early</u> and is due to <u>child rearing practices.</u> Its source can be traced to <u>rebukes, warnings, teasing, ridicule, ostracism, neglect,</u> or <u>abuse.</u> These sources lead to <u>a perception of the self</u> as <u>unworthy.</u> It is sort of like <u>being caught and wanting to hide.</u> What it yields is <u>a fear of humiliation.</u> Its opposite is <u>pride.</u>
>
> <u>Shame</u> can lead to a variety of behaviors. These include <u>aggression, addiction, obsession narcissism, depression, perfectionism, withdrawal, diffidence,</u> or <u>combativeness.</u>

4 Future Plans

Being asked a long series of questions, many of which might be repeated at subsequent sessions, can become tedious to a patient. We are currently working on a tree-pruning algorithm to help speed up the questioning process by eliminating some questions which are not appropriate to the set of emotions characteristic of the emotions represented by the behavioral and cognitive features identified up to that point in the session. This presupposes a tree structuring of the emotions as well, a very interesting theoretical question with many implications for the underlying organization of emotions.

5 Summary and Conclusions

We have built a system for helping in the treatment of alexithymia. Alex presents a set of questions to the patient. The responses are collected and matched against the characteristics of emotions recorded in the database. The emotion with the closest match is returned to a text generator. This component generates a description of the emotion, based on paragraph frames that match the slot names in the emotion data structure. This text is presented to the patient. This presentation of a detailed description of his emotion helps the patient become aware of his feelings and articulate them. This will greatly assist the patient in his work, especially in the early stages of recovery.

References

1. Blanchard, E.B., Arena, J.G., and Pallmeyer, T.P.: Psychosomatic Properties of a Scale to Measure Alexithymia. Psychotherapy Psychosoma 35 (1981) 64-71

2. Bradshaw, J.: Healing the Shame that Binds You. Deerfield Beach, FL: Health Communications, Inc. (1988)

3. Butler, J.: Critically Queer. Gay and Lesbian Quarterly 1 (1993) 17-32

4. Epstein, P. S.: A Semiotic Approach to Alexithymia. In Litowitz, B. E., and Epstein, P.S., Semiotic Perspectives on Clinical Theory and Practice: Medicine, Neuropsychiatry, and Psychoanalysis. New York: Mouton de Gruyter (1991)

5. Freedman, M.B., and Sweet, B.S.: Some Specific Features of Group Psychotherapy and Their Implications for Selection of Patients. International Journal of Group Psychotherapy 4 (1954) 355-368

6. Krystal, H.: Massive Psychic Trauma. New York: International Universities Press (1968)

7. _____ Alexithymia and Psychotherapy. American Journal of Psychotherapy 33 (1979) 17-31

8. _____ Alexithymia and the Effectiveness of Psychoanalytic Treatment. International Journal of Psychoanalytic Psychotherapy 9 (1982) 353-378

9. Lesser, I. M.: Current Concepts in Psychiatry. The New England Journal of Medicine 312.11 (1985) 690-692

10. Nemiah, J.C.: Alexithymia and Psychosomatic Illness. Journal of Continuing Education in Psychiatry 39 (1978) 25-37

11. _____, Freyberger, H., and Sifneos, P. E.: Alexithymia: A View of the Psychosomatic Process. In Hill, O. (ed.): Modern Trends in Psychosomatic Medicine, Vol. 3. London: Butterworths (1976)

12. _____ and Sifneos, P.E.: Affect and Fantasy in Patients. In Hill, O. (ed.): Psychosomatic Medicine. Butterworths, London (1970) 26-34

13. Ruesch, J.: The Infantile Personality. Psychosomatic Medicine 10 (1948) 134-144

14. Rybakowski, J., Ziolkowski, M., Zasadska, T., and Brzezinski, R.: High Prevalence of Alexithymia in Male Patients with Alcohol Dependence. Drug and Alcohol Dependence 21 (1988) 133-136

15. Sedgwick. E.K.: Queer Performativity. Gay and Lesbian Quarterly 1(1993) 1-16

16. Sifneos, P.: Short-Term Psychotherapy and Emotional Crisis. Harvard University Press, Cambridge, MA (1972)

17. Smith, G. R.: Alexithymia in Medical Patients Referred to a Consultation/Liaison Service. American Journal of Psychiatry 140 (1983) 99-101

18. Smith, R. N., Bienstock, D., and Housman, E.: A Collocational Model of Information Transfer. Information Interaction: Proceedings of the Annual Meeting of the American Society of Information Science 19 (1982) 281-284

19. Smith, R. N., Chen, C.C., Feng, F.F., and Gomez-Gauchia, H.: A Massively Parallel Memory-based Story System for Psychotherapy. Computers and Biomedical Research 26 (1993) 415-423

20. Taylor, G. J.: Alexithymia: Concept, Measurement, and Implications for Treatment. The American Journal of Psychiatry, 141 (1984) 725-732

21. Taylor, G. J., Bagby, R. M., and Parker, J.D.A.: Disorders of Affect Regulation. Cambridge University Press (1997)

Mechanizing Proofs of Integrity Constraints in the Situation Calculus

KOUNALIS Emmanuel, URSO Pascal

Université de Nice - Sophia Antipolis, Laboratoire I3S, Dpt. Informatique, Parc Valrose,
06108 Nice CEDEX 2, France
kounalis@essi.fr, urso@essi.fr

Abstract. We address the problem of proving automatically integrity constraints in databases represented within the situation calculus. Integrity constraints specify what counts as a legal database state; it is a property that every database state must satisfy. Reiter ([REI, 91] and [REI, 93]) has presented an approach to database integrity constraints in which the concept of a database satisfying its constraints is defined in terms of inductive entailment from the database, induction axioms on database states, and other axioms of the situation calculus. In this paper we develop and implement a framework for proving integrity constraints in databases. We show how our induction theorem prover that includes strategies for finding suitable induction schemes and inference rules for automatically generalizing conjectures can be used to prove many integrity constraints completely automatically from an update database specification alone. We provide computer applications to a toy database (education) to illustrate our framework.

1 Introduction

1.1 Motivation

The *situation calculus* [McC, 69] is a well-known formalism for representing temporal domains. In the situation calculus, the basic notions are *actions* and *situations*. A situation may be viewed as a time period. The set of fluents that holds an instance of the time is the state of the world at that instance. Actions are the cause of situation transitions: a result function is used to map pairs (state, situation) to the new situation resulting from the execution of that action in that situation. Reiter [REI, 91], [REI, 93] and [REI, 93a] has described how one may represent databases and their update transactions within the situation calculus. The basic idea consists in realizing that a database evolves in time, so that updateable relations should have an explicitly state argument representing the current database state. This argument records the sequences of update transactions that the database has undergone thus far. Database transactions are treated as functions, and the effect of a transaction is to map the current database state into a successor state. These two features - an explicit argument for updateable relations, and functions for transactions - lead him to specify databases and their update transactions within the language of situation calculus. In this setting, the basic approach to specify database transactions consists of an axiomatic system that contains the following information: 1) state independent knowledge about objects of the database; 2) knowledge about the state of the database at the initial situation; 3) preconditions for performing the different actions; 4) effects of actions, when they are possible, in terms of the fluents whose truth values are known to be changed by the execution of the action. A consequence of the previous idea is that the only way the database evolves is through transactions. Accordingly properties of the form "no matter how the database evolves from the initial state, such-and-such must be true in all database features" are of particular importance. It is then natural to represent them as universally quantified sentences over states (*integrity constraints*). Now, since constraints are sentences quantified over states, and states change only by virtue of transaction "occurrences" it is clear that mathematical induction is required for reasoning about them.

The need to be able to reason about integrity constraints in a rigorous formal way is self-evident. Formal methods are more and more frequently adopted by industry for hardware and software verification. They require efficient automatic tools to relieve designers and programmers of related proof obligations. Traditional database management systems do not provide facilities for proving that the integrity constraint of the database cannot be violated as a result of performing transactions. More often, these systems provide support for testing the integrity of the database whenever a transaction takes place. *The motivation for this paper is to add to the understanding of formal reasoning about databases using equational logic, by providing a formal framework for proving automatically integrity constraints from the updateable database specification alone.*

1.2 The Problem

The language for writing down updateable database specification is first-order, many sorted, with sorts *state* for situations, *action* for actions, and *object* for everything else. L has the following domain independent predicates and functions: a constant S_0 of sort *state* denoting the initial situation; a binary function $do(a,S)$ denoting the situation resulting from performing the action a in the situation S; a binary predicate $Poss(a,S)$ meaning that the action a is possible (executable) in situation S; and an unary predicate *Reach* to identify the subset of states reachable by executing actions possible in other reachable states. L also has a finite number of *state independent* predicates with arity $object^n$, $n \cdot 0$, a finite number of *state independent* functions $object^n \rightarrow object$ and a finite number of fluents which are predicate symbols of arity $object^n \times state$, $n \cdot 0$. Using L, one may write down axioms that specify databases and their update transactions. These axioms may express the *transaction preconditions* of actions, i.e., they specify the sufficient conditions which the current state of a database must specify before the transaction can be performed in this state. They may also express the *effect* of all transactions on all updateable database relations. Finally they may express of what is true of the initial state S_0 of the database. All these axioms are arbitrary first-order formulas. However, from a theorem proving point of view, these formulas are not amenable for automation since they are quantified. Fortunately, under reasonable conditions, they can be translated in the language of equational logic. The essential idea for this translation is that one may use rewrite rule based systems to carry out automatic proofs.

We assume familiarity with the basic notions of equational logic and rewrite systems (see for instance, [DJ, 90]). Let $L(F,X)$ denote the many-sorted language of (first-order) *terms* built out of function symbols taken from the finite vocabulary F and a denumerable set X of variables. Let A be a set of *conditional equations*, i.e., expressions of the form $e_1 \wedge ... \wedge e_n \Rightarrow e_{n+1}$, where e_i are equations which are usually written as $t = s$. $e_1, ..., e_n$ are the *premises* and e_{n+1} is the *conclusion* of $e_1 \wedge ... \wedge e_n \Rightarrow e_{n+1}$. A clause is an expression of the form $\neg e_1 \vee ... \vee \neg e_n \vee e'_1 \vee ... \vee e'_m$. We shall sometimes write this expression in the following equivalent way: $e_1 \wedge ... \wedge e_n \Rightarrow e'_1 \vee ... \vee e'_m$. We identify a conditional equation and its corresponding representation as a Horn clause. A clause $\phi := \neg e_1 \vee ... \vee \neg e_n \vee e'_1 \vee ... \vee e'_m$ is an *inductive consequence (inductively valid)* of a set A of axioms (written as $A \models_{ind} \neg e_1 \vee ... \vee \neg e_n \vee e'_1 \vee ... \vee e'_m$) if and only if for any ground substitution σ, (for all i: $A \models e_i \sigma$) implies (there exists j such that $A \models e_j \sigma$). Since ground terms can easily be well ordered, induction can be used as a natural technique to prove inductive clauses. In general, to establish inductive consequences, one requires "eureka steps" such as finding a suitable variable to induce upon, additional lemmas, generalizations, or case analysis for the proof to go through. In some cases, a combination of these concepts is needed.

In this paper we deal with the following problem:

- **Input**: A set of axioms A over the language $L(F,X)$ that represents an updateable database specification and a formula ϕ that represents an integrity constraint in the database.
- **Output**: Is $A \models_{ind} \phi$? In other words, is ϕ true in all database states?

1.3 Aims of the Paper and Related Works

This paper deals with the problem of proving automatically integrity constraints in databases represented within the situation calculus. Our solution consists of two steps: at the first step a Horn-clause like axiomatization is derived from the situation calculus database definition. This translation can be carried out automatically. At the second step an inductive proof of the integrity constraint is performed in this axiomatization. The basic framework of this second step is a set of transition rules suitable for database axiomatization. It contains: 1) a set of classical inference rules suitable for induction (see for details [KR, 90], and [BKR, 95]) 2) a general method for finding out good induction schemes, and 3) an inference rule for automatically generalizing conjectures that is particularly useful in the context of proofs of integrity constraints. The proposed framework has been implemented in C. The initial experimentation is very pleasing. For instance, the method can be employed to automatically establish proofs of integrity constraints without any user interaction as in [BPSKS, 96].

2 Outline of our Approach: An Example

In this section we give the essential ideas underlying the proofs of integrity constraints. We illustrate them using a simplified version of the *education* updateable database definition ([REI, 93]). This education database involves two relations:
1. *enrolled(student ,course, s)*: Student *student* is enrolled in course *course* when the database is in state *s*.
2. *prerequ(pre,course)*: *pre* is a prerequisite course for course *course*.

The update transactions involved in the database are the following:
1. *register(student, course)*: Register student *student* in course *course*.
2. *drop(student, course)*: Student *student* drops course *course*.

As we pointed out, transactions have preconditions that must be satisfied by the current database before the transaction can be "executed". In general, these preconditions are quantified formulas. For instance, the register precondition *"It is possible to register a student in a course if and only if he is registered in all prerequisites for the course"* can be formulated as:

$$\text{Poss (register(st,co),s)} \equiv \forall \text{co' \{prerequ(co',co)} \Rightarrow \text{enrolled(st,co',s)\}} \qquad \textbf{(1)}$$

Further, the effects of all transactions on all updateable database relations are specified by the successor state axioms. For instance, *"A student is enrolled in a database state if and only if he has been registered by the last transaction or was enrolled in the previous state and has not dropped the course in the last transaction"* is a successor axiom for the relation *enrolled* and can be written as:

$$\text{enrolled(st,co,do(a,s))} \equiv \text{a=register(st,co)} \lor \text{enrolled(st,co,s)} \land \text{a} \bullet \text{drop(st,s)} \qquad \textbf{(2)}$$

Since this database evolves from state to state, a constraint that must be satisfied is the following: *"If a student is enrolled in a course, then he is enrolled in each prerequisite of this course"*. This can be formulated as:

$$\text{reach(s)} \land \text{enrolled(st,co,s)} \land \text{prerequ(co1,co)} => \text{enrolled(st,co1,s)} \qquad \textbf{(IC)}$$

To perform an automatic proof of this constraint in the education database we first translate the Reiter-like axiomatization in the language of conditional equations. The main problem of such a translation is the transformation of the quantified first-order axioms into equivalent quantifier-free formulas. In general, this translation is not possible. However, there are two points of interest here: The first point is the finite nature of the domain of objects (*domain*) under consideration in a database application; the second point is that the only way the database evolves is through transactions. Consequently, we may use new predicates and explicitly building domain to eliminate all quantifiers (see the details of this construction [KU, 98]). For instance, the assertion (1) can be translated as:

```
(15) poss(register(xst,xco),xs) = fall(xst,xco,xs,dom(xs))
(16) fall(xst,xco,xs,nothing) = True
(17) fall(xst,xco,xs,c(st(x),xdom)) = fall(xst,xco,xs,xdom)
(18) prerequ(xco1,xco) = False =>
fall(xst,xco,xs,c(ct(xco1),xdom)) = fall(xst,xco,xs,xdom)
(19) prerequ(xco1,xco) = True, enrolled(xst,xco1,xs) = False =>
fall(xst,xco,xs,c(ct(xco1),xdom)) = False
(20) prerequ(xco1,xco) = True, enrolled(xst,xco1,xs) = True =>
fall(xst,xco,xs,c(ct(xco1),xdom)) = fall(xst,xco,xs,xdom)
```

This translation is obtained by introducing the new function symbol *fall* over *dom(s)* in order to replace the $\forall$-quantifier. *dom* is the function (see for its definition below) which describes the *domain* of objects occurring in a state of the database ([BPSKS, 96]):

```
(26) dom(s0) = nothing
(27) dom(do(register(xst,xco),xs)) =
c(st(xst),c(ct(xco),dom(xs)))
(28) dom(do(drop(xst,xco),xs)) = c(st(xst),c(ct(xco),dom(xs)))
```

The range of *dom(s)* contains all objects that exist in *s*. Thus, the *domain* will contain the objects in the initial state plus those that appear as arguments in actions that have been executed. This is presented by a list that is computed by using the free constructors *nothing* and *c*. *nothing* (resp. *c*) corresponds to the LISP-function *nil* (resp. *cons*).

Having transformed all quantified formulas into quantify-free, we use simple rules of first-order logic to get the final equational axiomatization. For example, unquantified formulas of the form $P \Rightarrow (A \Leftrightarrow B \wedge C)$ can be translated to the language of equational logic as $P = true \wedge C = true \Rightarrow A = B$ and $P = true \wedge C = false \Rightarrow A = false$. Hence, axioms like

$$\text{Reach(s0), Reach (do(action, state))} \equiv \text{Poss (action, state)} \wedge \text{Reach (state)} \qquad (3)$$

are formulated as follows:

```
(11) reach(s0) = True
(12) poss(xa,xs) = True => reach(do(xa,xs)) = reach(xs)
(13) poss(xa,xs) = False => reach(do(xa,xs)) = False
```

Similarly, the conditional equations associated to assertion (2) are as follows:

```
(21) eqAct(xa,register(xst,xco)) = True =>
enrolled(xst,xco,do(xa,xs)) = True
(22) eqAct(xa,register(xst,xco)) = False,
eqAct(xa,drop(xst,xco)) = True => enrolled(xst,xco,do(xa,xs)) =
False
(23) eqAct(xa,register(xst,xco)) = False,
eqAct(xa,drop(xst,xco)) = False => enrolled(xst,xco,do(xa,xs)) =
enrolled(xst,xco,xs)
(24) enrolled(xst,xco,s0) = False
```

The following conditional equations complete the education example that has been given to our prover:

```
(1) True = False =>
```

```
(2)  eqAct(x,x) = True
(3)  x1 = x2, y1 = y2 => eqAct(drop(x1,y1),drop(x2,y2)) = True
(4)  x1 = x2, y1 != y2 => eqAct(drop(x1,y1),drop(x2,y2)) = False
(5)  x1 != x2 => eqAct(drop(x1,y1),drop(x2,y2)) = False
(6)  x1 = x2, y1 = y2 => eqAct(register(x1,y1),register(x2,y2)) =
True
(7)  x1 = x2, y1 != y2 => eqAct(register(x1,y1),register(x2,y2))
= False
(8)  x1 != x2 => eqAct(register(x1,y1),register(x2,y2)) = False
(9)  eqAct(drop(x1,y1),register(x2,x3)) = False
(10) eqAct(register(x1,x2),drop(x3,x4)) = False
(14) poss(drop(xst,xco),xs) = False
(25) prerequ(xco,xco) = False
```

Notice that equations 1 to 10 express the definition of the equality relation on transactions (see for details [Lloyd, 87 p. 79]). Equations 14 and 25 are part of our database specification.

Having completed the translation of the updateable specification we are ready to verify whether the database satisfies given constraints. In general, our prover performs a proof of a formula ϕ (integrity constraint) from a set A of conditional equations using a *reduction ordering* > (i.e, a stable and monotonic well-founded relation > on terms), and a *test set* TS(A) for A (i.e, a finite set of terms which describes the states, the actions and the objects of a database). For instance, the set *{s0, do(xa, s), drop(student, course), register(student, course), True, False}* is a test set for the education database.

The proof strategy first instantiates *induction* variables in the conjecture with the members of the test set, and then applies *rewriting* to simplify the resulting expression. This rewriting uses any of the axioms, lemmas and induction hypotheses provided they are smaller (w.r.t. the reduction order >) than the current conjecture. There are two fundamental ideas behind our proof strategy. The first idea consists of applying the generate rule (see section 3.1) on an induction variable as fixed in definition 3. This rule produces the induction schemes needed for the proof of a goal. The second idea consists of using the GT-rule (see definition 4) just before another induction is attempted. This rule produces a generalization of a goal by abstracting well-defined subterms in the goal in order to prevent failure.

For instance, the constraint (IC) is given to our prover for a proof by induction:
```
enrolled(xst,xco,s) = True, prerequ(xco1,xco) = True, reach(s) =
True => enrolled(xst,xco1,s) = True
```

3 Automatic Proofs of Integrity Constraints: The Machinery

The induction procedure is formalized by a set of transitions rules [BKR, 95], [BR, 95] applied to pairs of the form (E, H), where E is the set of clauses which are conjectures to be proved and H the set of inductive hypotheses. The major benefit of this procedure is that it allows proofs by mutual induction, that is, proofs of multiple conjectures that use each other in their proofs in a mutually recursive fashion.

3.1 Transition Rules

The following transition rules is a part of the inference system used to prove integrity constraints (see [BKR, 95], and [BR, 95] for the theoretical justification):

Generate: $(E \cup \{C\}, H) \vdash_1 (E \cup (\cup_\sigma E_\sigma, H \cup \{C\})$ for all test set substitution σ

Case Simplify: $(E \cup \{C\}, H) \vdash_1 (E \cup E', H)$ if $E' = $ Case analysis(C)

Simplify: $(E \cup \{C\}, H) \vdash_1 (E \cup \{C'\}, H)$ if $C \rightarrow_{R<H\cup E>} C'$.

Subsume: $(E \cup \{C\}, H) \vdash_1 (E, H)$ if C is subsumed by a clause of $R \cup H \cup E$

Delete: $(E \cup \{C\}, H) \vdash_1 (E, H)$ if C is a tautology.

Fail: $(E \cup \{C\}, H) \vdash_1$ // if no condition of the previous rules holds for C.

A success-full derivation is a sequence of transitions $(E, H) \vdash (E1, H1) \vdash .... \vdash (En, Hn)$ such that $En = \emptyset$. The fail rule is a kind of conjecture disprover. It is very important, when a convergent set of axioms exists for ground terms, to guard against false lemmas.

To simplify goals we use axioms in A, induction hypotheses and other conjectures which are not yet proved during simplification. In particular, *Case Analysis* simplifies a conjecture with conditional rules provided that the disjunction of the preconditions is inductively valid in R. For instance, consider the following clause C that has been generated by our system when proving the initial integrity constraint:
```
enrolled(xst,xco,do(register(x1,y2),x3)) = False,
prerequ(xco1,xco) = False, reach(do(register(x1,y2),x3)) = False
```
The Case Simplify rule can be used to simplify C as follows:
```
## Case Simplify:
[by 21] eqAct(register(x1,y2), register(xst,xco)) = False, True
= False, prerequ(xco1,xco) = False,
reach(do(register(x1,y2),x3)) = False
[by 22] eqAct(register(x1,y2),drop(xst,xco)) = False,
eqAct(register(x1,y2), register(xst,xco)) = True, False = False,
prerequ(xco1,xco) = False, reach(do(register(x1,y2),x3)) = False
[by 23] eqAct(register(x1,y2),drop(xst,xco)) = True,
eqAct(register(x1,y2), register(xst,xco)) = True,
enrolled(xst,xco,x3) = False, prerequ(xco1,xco) = False,
reach(do(register(x1,y2),x3)) = False
```

3.2 Finding Good Induction Schemes for Integrity Constraints

The generate rule captures the essence of induction. It allows obtaining the induction schemes that entail the proof of the integrity constraints under consideration. An induction scheme must be formed in the way that has a good chance of success. In some sense, the success of a proof strongly depends to what variable the generate rule applies. Then a question naturally arises: *What is the best variable to replace in order to be able to apply a definition and eventually the induction hypothesis?*

Definition 1. (inductive position set). If R is a set of axioms and f a function symbol in F, then $IP(R,f) \equiv \{p \mid p \bullet \varepsilon$ and $\exists (P \Rightarrow l \rightarrow r)$ in R such that $l(\varepsilon) \equiv f$ and $l(p) \in F\}$ is the inductive position set of function f w.r.t. R. IP(R) denotes the set {IP(R,f)| f∈ F} and is the inductive position set of R.

Example. The inductive position set for the education example as computed by our prover is the following: {*reach : [1], poss : [1], enrolled : [3], prerequ : [], eqAct : [1 2], dom : [1], fall : [4]* }

Definition 1 allows selecting among the set of variables of an equation, a subset which is suitable for the application of generate rule. However, we may cut down this set by discarding variables judged useless since they lead to goals to which definitions fail to apply. Let $C \equiv \neg a_1 = b_1 \vee ... \vee \neg a_n = b_n \vee a_{n+1} = b_{n+1} \vee ... \vee a_m = b_m$. We denote by *comp(C)* the set of atoms of C: {t | t is either a_i or b_i for i = 1, 2, ..., m}. Since we are working with database axiomatizations that are left linear (no variable occurs more than once in the left-hand side of the conclusion of a conditional equation), we have the following:

Definition 2. Suppose R is a left-linear set of conditional rules, and C is a clause. $U_x(t)$ is defined to be the set of positions u in t labeled by x such that whenever there is a rule $P \Rightarrow l \rightarrow r$ in R such that l unifies with t/v for some prefix v of u then, |u/v| < D(R) and there is a position p in IP(R,l(ε)) such that p is a prefix of u}. The multiset $U_x(C)$ of variable x in C is then defined as {{ u | u ∈ $U_x(t)$ and t ∈ comp(C)}}.

The main reason for considering the set $U_x(C)$ is just to select the best variables in a clause to induce upon. The following defines formally the notion of induction variables in our setting:

Definition 3 (induction variables): Variable x in C is an induction variable if for all variables y in C either x is a generalized variable and y is not a generalized variable or $|U_y(C)| < |U_x(C)|$.

A variable x is said to be *generalized* if it has been introduced by the GT-rule below, i.e. if x is the variable obtained by replacing common non-trivial subterms in a clause by x. Notice that this requirement is consistent with definition 2 since the GT-rule generalizes subterms that are at positions suffixes of positions in the induction position set of a function. Notice also that in the case when $|U_y(C)| = |U_x(C)|$ either x or y may be used to induce upon.

Example:
– Consider the clause C: reach (s) = True, prerequ (xco1, xco) = True, enrolled (xst, xco, s) = True => enrolled (xst, xco1, s) = True. Since $|U_y(C)| < |U_s(C)|$ for all variables y in C we get:
```
## Generate on {s}: reach(s) = True, prerequ(xco1,xco) = True,
enrolled(xst,xco,s) = True => enrolled(xst,xco1,s) = True
```

3.2 Finding Generalizations of Integrity Constraints

We now introduce an inference rule that turns out to be an indispensable part for automating the proofs of integrity constraints. The essential idea behind this rule is to propose a generalized form of the conclusion just before another application of

```
## Deleted:
    enrolled(xst,x3,x4) = True, prerequ(x3,xco) = False,
prerequ(x3,y2) = True, prerequ(xco1,xco) = False, reach(x4) =
False, ballreg(x1,y2,x4,y5) = False, False = False

## Conjectures :

 ######## FINISH ########
    reach(xs) = True, prerequ(xco1,xco) = True,
enrolled(xst,xco,xs) = True => enrolled(xst,xco1,xs) = True
Are all induction properties.
```

References

[BPSKS, 96] L. Bertossi, J. Pinto, P. Saez, D. Kapur, and M. Subramaniam. Automated proofs of integrity constraints in situation calculus. ISMIS'96, p. 212-222 (1996).

[BR, 95] Bouhoula, A., Rusinowitch, M. "Implicit Induction in Conditional Theories", Journal of Automated Reasoning, 14(2):189-235, (1995).

[BKR, 95] Bouhoula, A., Kounalis E., Rusinowitch M. "Automated Mathematical Induction", Journal of Logic and Computation, 5(5):631-668 (1995).

[DJ, 90] Dershowitz N., Jouannaud J.P. (1991). "Rewriting systems", Handbook of Theoretical Computer Science.

[KR, 90] Kounalis, E. and Rusinowitch, M. "Mechanizing Inductive Reasoning", Proc. *8th* (AAAI-90) . p. 240-245, (1990).

[KU, 98] Kounalis, E. And Urso, P. "Mechanizing Proofs of Integrity Constraints in the Situation Calculus", Technical Report. Laboratoire I3S UNSA (1998).

[KU, 99] Kounalis, E. And Urso, P. "Generalization Discovery for Proofs by Induction in Conditional Theories", FLAIRS'99. AAAI Press, to appear (1999).

[LLO, 87] Lloyd, J.W. *Foundations of Logic Programming*. Springer Verlag, (1987).

[McC, 69] J. McCarthy and P. Hayes. "Some philosophical problems from the standpoint of the artificial intelligence. Machine Intelligence, Vol. 4, p. 463-502, (1969).

[REI, 91] R. Reiter. "The Frame Problem in the Situation Calculus: a simple solution (sometimes) and a completeness result for goal regression". A.I and M.T.C: Papers in Honor of J. McCarthy, p. 359-380 (1991).

[REI, 93] R. Reiter. "Formalizing database evolution in the situation calculus". In Proc. of 5th Generation Computer Systems, pages 600-609, Tokyo, Japan, (1992).

[REI, 93a] R. Reiter. "Proving properties of states in the situation calculus". In Artificial Intelligence 64, p. 337-351, Dec. (1993).

generate rule is attempted. The GT-rule below provides a way to transform conjectures into more general ones by abstracting a common non-trivial subterm that is a suffix of an inductive position. A term is *trivial* if it is a linear variable in a clause. The GT-rule works in two steps: at the first step, it transforms a positive literal of a clause B into a literal that both sides share a common non-trivial subterm. This common subterm should be, in both sides, at positions that are suffixes of inductive positions. At the second step it replaces these common subterms with a fresh variable. To create common subterms, GT-rule uses the clause A whose a test set instance of it has been reduced to B. The following definition captures this discussion.

Definition 4 (Generalized Transform): Let $A \equiv P[x] \vee (l[x] = r[x])$, where x is an induction variable in A. Let B be a clause that derived from $P[x/t] \vee (l[x/t] = r[x/t])$ using the generate rule with term t and the simplification inference rules (case simplify or simplify). Let B be of the form $Q[s] \vee Q'[s] \vee (a[s]_p = b)$. If

1. s is a non-trivial term in B at position p such that p is a suffix of a position in IP(R),
2. there is a substitution $\eta\theta$ such that $l\eta\theta \equiv d[s]_q$, for some context $d[]$, $P\eta\theta$ subsumes $Q[s] \vee Q'[s]$ and $b \equiv r\eta$.
3. q is a suffix of a position in IP(R),
4. $(l\eta\theta = r\eta\theta) <_c (l[x/t] = r[x/t])$.
5. $d[s]_q \bullet a[s]_p$
6. $Q[g] \vee Q'[g] \vee (a[g]_p = d[g]_q)$ is inductively valid for any ground term in TS(R), then, the clause $Q[W] \vee Q'[W] \vee (a[W]p = d[W]q)$ is said to be a generalized transform of B, and variable W is said to be the generalized variable in B.

GT-Rule : $(E \cup \{B\}, H) \vdash_\neg (E \cup \{D\}, H)$ if D is a generalized transform of B.

Let us illustrate the use of GT-rule on an example. Notice that without this rule all integrity constraints proofs goes with failure. In [BPSKS, 96] approach all these kind of lemmas are provided by the user.

Example: Let A be the formula below in the "Hypotheses" and B the formula in the "Conjectures".

```
## Hypotheses : enrolled(xst,xco1,x3) = True, reach(x3) = False,
prerequ(xco1,xco) = False, fall(xst,xco,x3,dom(x3)) = False,
## Conjectures : fall(xst,xco,do(register(x1,y2),x4),dom(x4)) =
False, prerequ(xco1,xco) = False, reach(x4) = False,
enrolled(xst,xco1,x4) = True, fall(x1,y2,x4,dom(x4)) = False,
prerequ(y2,xco) = True
```

Then by using the GT-rule we get:

```
## GT-Rule      : prerequ(y2,xco) = False, fall(x1,y2,x4,W0) =
True, enrolled(xst,xco1,x4) = False, reach(x4) = True,
prerequ(xco1,xco) = True =>
fall(xst,xco,do(register(x1,y2),x4),W0) = fall(xst,xco,x4,W0)
```

After simplification of the formulas obtained from application of the generate rule to the previous clause we get:

Using Cases for Process Modelling: An Example from the Water Supply Industry

Guy Saward

University of Hertfordshire
College Lane, Hatfield,
Herts AL10 9AB, UK
e-mail g.r.saward@herts.ac.uk

Abstract. This paper describes the use of cases as a process representation technique and shows how case base reasoning can be used to navigate through, or execute a complex control process. The work is based on a prototype decision support application for a UK water utility company in which two processes were modelled - one diagnostic and one operational. A brief description is given of how processes are represented as cases and how those cases are are used to animate the business process.

1 Introduction

Case base reasoning is traditionally seen as a problem solving technique to be applied where an effective model cannot be created [1,2]. A prototype decision support application for a UK water utility company, provisionally titled Cascade (Cbr ASsisted Customer ADvicE), demonstrates how an existing process model can be captured as cases. Each path through the process model can be viewed as a script comprising a series of questions and actions, as described in section 4, which is akin to a procedural schemata [3]. The system iteratively matches a given process scenario against the set of all possible scripts, resulting both in the selection of the appropriate script and the incremental execution of the script as described in section 5.

Cascade differs from traditional CBR where the search process is driven by the data and is either programmed as part of a matching algorithm, or is generated from the data. Nearest neighbourhood algorithms are an example of the former where a (partial or complete) set of attribute values is matched, using a static algorithm, against existing attribute value sets. Inductive techniques such as ID3 are an example of the latter where the resulting decision tree matching process is focussed on partitioning the cases for fast retrieval.

In Cascade, a process model is used to generate the data for the cases. The resulting cases can be viewed as generic or model cases, as opposed to the concrete, or instance cases created from experimental, observed or theoretical data. Although the engineering of model cases is not a new concept [4], the concept of engineering a matching process is new and is akin to the knowledge engineering employed in traditional rule or model based systems [5].

2 Significance

Separating the decision making process or business logic from a complex system implementation is critical for a number of reasons. For any utility market in the UK (water, gas or electricity), consistency of execution and application of regulatory principles provide strong evidence of compliance with industry regulations. The separation of application and business process also allows business processes to be altered without changing the system implementation.

The engineering of generic process cases retains the utility of cases as episodic knowledge which is one of the key benefits of CBR [6,1]. It has also been argued that cases are a good starting point for heuristic model building [7]. The Cascade application supports this and allows the argument to be taken one stage further by suggesting that cases are a better representation than rules. This is based on the view that cases do not force an unnatural decompilation of knowledge into rule sets. Cases provide a "vertical", end-to-end view of the problem solving process incontrast to rule sets that typically implement horizontal slices through an inference model.

The need for knowledge engineering raises the spectre of the Knowledge acquisition bottle-neck. However, the prevalence of process modelling techniques in industry means that the knowledge engineering can be limited to facilitating and validating the construction of process models. The simple translation of process models into implemented cases should also reduce the requirement for knowledge engineering per se.

3 Process Overview

Two processes were mapped as part of the Cascade prototype: a diagnostic process concerned with water quality and an operational process related to billing. Existing flow charts were used to facilitate the construction of more detailed process models. The process models include additional steps that were used to drive the end user GUI. The knowledge for the processes was elicited in four 2 hour workshops and validated in a further two workshops.

Table 1. Process Summary Statistics

Process Attributes	Process 1	Process 2
Sub-processes	4	8
Decision points / actions	37	110
Routes through sub-process	27	114
Total routes	720	43 million
Longest path (steps)	32	60
Cases	26	55
Longest case (# of steps)	8	12

Each process was split into a number of sub-processes. Each path through the sub-process was implemented as a case. The cases were created using Inference's CBR3 Professional Author tool. Rules were used to link the sub-processes given the relatively small size of the problem domain and the project time scales. Meta-cases could have been used to link the sub-processes [8, 9] although this would have involved recursive use of the CBR3 Search Engine. The number of elements in each process is shown in table 1.

4 Representing Processes as Cases

Figure 1 shows part of a generic contact screening process derived from an application built for a UK water utility company. A flow chart notation is used as it is easy for business analysts and users to understand. Diamonds represent decision points with the arrows representing potential answers. The rounded boxes represent links to other parts of the process.

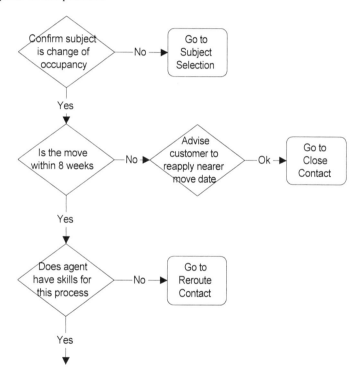

Fig1. Process fragment showing part of the
operational procedure for change of occupancy

To represent the process fragment in figure 1 as cases, each path through the process model is represented as a single case in which questions become case attributes, and answers become attribute/value pairs. This results in the four cases shown below in table 2.

Table 2. Example cases derived from Figure 1.

	Attributes	Case Attribute Values			
#	Decision Point Description	Case 1	Case 2	Case 3	Case 4
1	Subject confirmed	No	Yes	Yes	Yes
2	Move in 8 weeks		No	Yes	Yes
3	Advise reapply		Ok		
4	Appropriate skills			No	Yes

Representing the process as cases could be viewed as slicing the process vertically into inference chains, in contrast to a fine grained rule-based approach which would slice the process up horizontally and focus on individual inferences. In defining each case, only those attributes that have a value are associated with the case as shown by Case 2 in figure 2. This case has been authored using Inference's CBR3 tool in which attributes are stored as questions and values as answers to the questions.

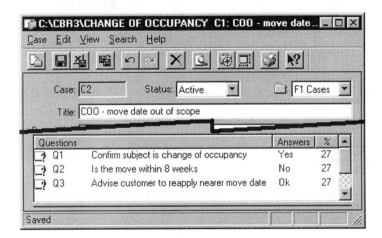

Fig2. Edited case window showing Case 2 from table 1.

5 Case Base Process Animation

The Case-Based Process is animated by an iterative case-base search algorithm. In this way, Case-Based Processes combine the knowledge engineering approaches of rule-based systems with the simple, generic retrieval algorithms of case-base reasoning. The search algorithm used is based on iterative case retrieval using a nearest neighbour algorithm. The value for attribute 1 is determined, either through user intervention or some external system interaction, and all cases are scored and ranked. The highest scoring case is used to determine the next attribute value. The attribute order, which can be varied on a case by case basis, is used to select the next unfilled attribute. A value is then determined for this attribute, the cases are re-scored and re-ranked and the process repeated.

For the cases in table 1, if the subject is confirmed as change of occupancy (attribute 1), cases 2,3 and 4 are equally ranked and the system will determine whether the move is within 8 weeks (attribute 2). If it is (attribute 2 = "Yes"), then the search moves on to determine whether the agent has the skills appropriate to work through this process.

The incremental case retrieval for Case-Based Processes (CBP) is analogous to the use of decision trees built on top of case bases [10,11] but differs in that:

- the sequence of decision points in decision trees are derived from structure of the data underlying the cases, while cases in CBP are engineered to implement a particular sequence of inferences;
- traversing a decision tree leads to specific cases which are the solution, while the process of selecting cases in CBP is actually animating the decision making process - there is no solution to be enacted once a case has been selected;
- decision trees are just that – they are limited to tree structures, while CBP allows for network (non-tree) structures.

Although knowledge engineering has been applied to CBR [7] the idea of engineering the search or retrieval process rather than the cases themselves represents a change in emphasis.

Where a process requires the execution of an action or procedure by an external agent, it is included in the CBP as an attribute or decision point with an appropriately worded title, e.g. "Advise customer to reapply nearer move date". The possible responses to this direction depend on the possible outcomes to the CBP invoked action.

6   Conclusions

The figures from table 1 illustrate how a relative small number of cases can be used to model complex processes. . The technique described is applicable to a wide range of industries. This case study, along with research based on another 2 studies is the basis of on-going research investigating the use of cases as a representation technique. This compares the use of cases with other techniques such as rule-based systems and looks at the issues of process decomposition. Other results focus on the use of cases as a single technique for integrating model-based reasoning and experiential data.

References

1. Kolodner, J., Case-Based Reasoning, Morgan Kaufmann, 1993
2. Watson, I., Applying Case-Based Reasoning: Techniques for Enterprise Systems, Morgan Kaufmann, 1997

3. Turner, R.M., Adaptive Reasoning for Real World Problems: A Schema-Based Approach, Lawrence Erlbaum, 1994

4. Bartsch-Sporl, B., Towards the Integration of Case-Based, Schema-Based and Model-Based Reasoning for Supporting Complex Design Tasks, Proceedings ICCBR-95, Springer, 1995

5. Wielinga, B.J., Schreiber, A. T., Breuker, J.A., KADS: A Modelling Approach to Knowledge Engineering, Knowledge Acquisition, 4, 1992

6. Simoudis, F., Using Case-Based Reasoning for Customer Technical Support, IEEE Expert, 7(5) , 1992

7. Strube, G., Enzinger A., Janetzko, D., Knauff, M., Knowledge Engineering from a Cognitive Science Perspective, Proceedings ICCBR-95, Springer, 1995

8. Bergmann, R., Wilke, W., On the Role of Abstraction in Case-Based Reasoning, in Proceedings EWCBR-96, Springer, 1996

9. Saward G., Shankararaman, V., Process Modelling Using Cases, University of Hertfordshire Technical Report 357, in preparation

10. Quinlan, R. 1986. Induction of Decision Trees. *Machine Learning 1*, 81-106.

11. Quinlan, R., 1993. *C4.5: Programs for Machine Learning*. San Mateo, CA: Morgan Kaufmann.

Financial Analysis by Case Based Reasoning

Esma Aïmeur and Kamel Boudina

University of Montreal
Departement d'Informatique et de Recherche Operationnelle
C.P. 6128, succursale Centre-ville
Montreal, Canada H3C 3J7

{aimeur,boudina}@iro.umontreal.ca
Tel: (514) 343-6794
Fax: (514) 343-5834

Abstract. Financial analysis is based on complex concepts and rules; its goal is to propose problem-adapted solutions. The evaluation of a particular financial situation has to consider human factors like savers risk tolerance and consumers behavior. It also has to consider political factors like interest rates variations and currency policy. The financial planner has to analyze the client's financial situation to elaborate a financial portfolio adapted to his or her needs. On the grounds of the nature and the diversity of the parameters describing a client's financial profile, we need tools that will memorize and reuse this information in different situations. In order to provide training and evaluation tools for financial analysts, we propose a system called FIPS (Financial Planification System) using Case-Based Reasoning. In FIPS, case-based reasoning is used in the case retrieval process, and also in a reflexive way during the adaptation stage. FIPS proposes to the learner the client's data like financial goals, acceptable risk, income, etc., and expects a balanced financial portfolio suggested by the student. It uses old cases, already treated and memorized, to propose an adapted solution that is compared to the learner's solution. The learner's evaluation is based on the distance between the solution provided by the system and the solution suggested by the learner.

Keywords: Self-training, Evaluation, Case based reasoning system, Reflexive approach, Financial situation.

1. Introduction

There are different approaches in the training domain. In this paper we are interested in the Case Based Reasoning (CBR) [6] approach in the training and the evaluation of learners. CBR systems attempt to adopt a pragmatic approach, based on the experience elaborated on the solved problems, exactly like a human expert develops experience and becomes subtler in his reasoning.

In the financial analysis domain, the nature of the problems to solve is not adapted to simple application of a set of general rules. In fact, the client's financial situations are very different and specific for each one. So the blind application of rules to treat those situations is not an appropriate approach. The financial expert has to build a case study for each new client and try to adapt the financial portfolio in order to answer to the client's needs. Because of the differences between the cases, we adopt a CBR approach to develop a software tool called FIPS for the training and the evaluation of financial experts.

The next section presents some tutorial systems that use the CBR approach. After that, we describe our prototype implementation. We also explain the structures defined for the case memory organization, the similarity measure, the indexation and the adaptation. In the third section, we expose our approach for the training and the evaluation process. The last section presents the conclusions of this work, the future perspectives and the possible enhancements of our system.

2. Case-based tutoring systems

There are several CBR systems that provide training in different knowledge domains [8]. The following case-based tutoring systems are used to help training and evaluating learners [9]. The system DECIDER [4] helps students understand or resolve a pedagogical problem by selecting and presenting appropriate cases from a database that respond to the student's goal. The system HYPO [1] is a case-based tutoring system for law students. The system is used to generate fresh cases for analysis in response to a particular issue of interest as identified by a tutor. We also have other examples of tutorial systems, like GuSS [5] that provide a training of complex social tasks like how to sell products or services. In the next section, we present our prototype structure and the approach adopted to implement it.

3. Prototype realization

The goal of this work is to develop a system, that proposes to a learner the information describing a client's profile, and to evaluate if the financial portfolio suggested by the student is adapted to that client. The system uses a case base with an efficient classification of its data. As it adapts past solutions, it builds an adaptation case base. The system's architecture is presented in [2].

3.1   Representation

The FIPS system is built on five major modules: indexation, retrieval, adaptation, evaluation and input interface. The different modules are described in the following sections. In the current section, we present the case and solution structures. We also describe, the functional structure of the FIPS system and the memory organization used to represent the case base.

Case structure. In the client's description, we have different information such as annual income, financial goals, fortune, etc. We also have a significant client characteristic, which is his capacity to manage the risk of his investment. After the evaluation of that client's capacity, the financial expert tries to affect a numerical value between 1 and 10 for his risk tolerance.

The case structure and its representation in the system are shown in Figure 1.

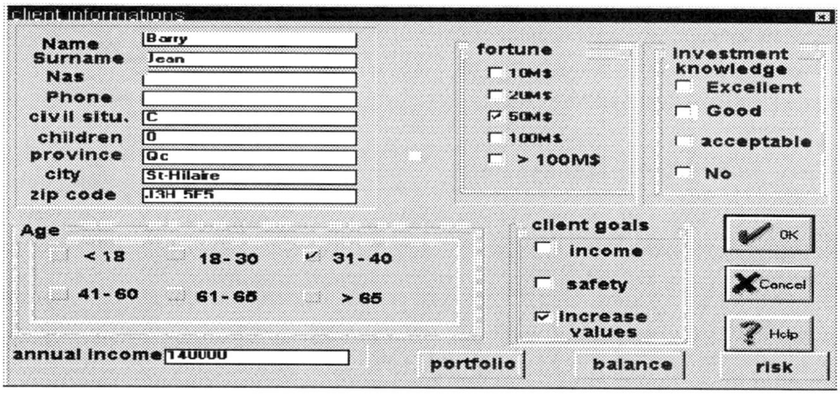

Fig. 1. Case Description

Description of the solution components. The system proposes different financial portfolios like solutions for the problems to solve. The combination of the different assets has to take into account the client's goals and the current economical situation. Before the introduction of the rules that will serve to adapt a solution for a particular client, we will introduce some notions linked to the financial analysis. A financial portfolio is composed of three great categories; shares, fixed yield values (bonds, debentures) and specie or quasi-specie (treasury bills) [3]. For each category, we'll see the related factors:

Shares: To evaluate the future share values, we use four great indicator categories. The fundamental indicators which are represented by the companies profits, the technical indicators which are related to the curve of the stock indicators (DowJones, TSE300...), the economical indicators which are the GNP, retail business, unemployment rate, etc.

Fixed yield values: The fixed yield values are divided in three major categories: the long, short and medium term values. To evaluate the future productivity of the fixed yield values, we must analyze the future tendency of the interest rates.

Specie and quasi-specie: The specie productivity is evaluated depending on the anticipated interest rates at the time of the fixed yield values analysis.

3.2 Indexation in FIPS

The case base is represented by a tree. The tree leaves are the pages making up the case base. Each node points to another node or directly to a page. A virtual address assigns a page number to each node. The system evaluates the virtual address for each new case and the hash-code function returns its page number. In the FIPS system, each field describing a case has a weight, which represents its importance. The consequence of the case virtual addressing is the creation of a hierarchy of indexes. The index with the highest level corresponds to the field with the highest weight.

3.3 Case retrieval

In the FIPS system, each field describing a case presents a neighborhood expressed in percentage. In other words, for an attribute value vi, all the values vj with a distance form vi lower than a certain level δ_i are in the neighborhood of vi. The work of Wess and Ritcher [11] and the work of Wess and Globig [10] inspire our approach. The distance between the values vi and vj is the absolute value of the arithmetic difference. If the values of the different attributes describing two cases are in the same neighborhood then the cases are members of the same class.

However, it is possible to have two cases with just a subset of their attributes in the same neighborhood. In this situation, we have to take into account the attribute weights. It follows that the distance between a case c and another one q is calculated like shown below:

$$dis(c,q) = \Sigma a=1,n \ wa * disa \ (qa , ca)$$

wa :Weight of the attribute a. $disa$: Local distance for the attribute a.

Retrieval case algorithm. The retrieval case algorithm is described as follows:

- Read the case q and choose from each field (attribute) a number of bits proportional to the attribute's weight in order to obtain a virtual address
- Determine in the tree, the node N which contains similar cases to the case q
- Read the physical page number p stored in the node N

- Load *p* from the disk
- Read the first case *c* in the page *p*
- Initialize *retrieved_case* to *c* and *max_similarity* to 0
- **Do while** *Not (end of page)*
  Compute the similarity degree between *c* and *q* from the distance between *c* and *q* :
  Similarity(c,q) = 1/dis(c,q)
  If *similarity(c,q) > max_similarity*
  Then *max_similarity = similarity(c,q)*
  Retrieved_case = c
  Endif
  EndDo
- Read the solution *S* corresponding to the case *retrieved_case*
- Adapt the solution *S. (*see adaptation algorithm *)*

3.4   Case adaptation

In this section, we present the rule and case based adaptation process. The originality in the approach of the system FIPS is the use of case-based reasoning in a reflexive way. In fact, the CBR approach is used to retrieve similar cases and the old adapted solutions. In other words, the system keeps a trace of the solution transformations in order to reuse them [7].

Adaptation case base. The case base in the system FIPS is a set of pairs $<C_i, S_i>$, C_i are the cases and S_i are the associated solutions. To treat a new case, the system will find in the base, the most similar case C_i and will adapt the solution S_i using the rules presented in 3.4.3. The transformation of the solution S_i *will* give us the solution $S_i`$ which is the solution proposed by the system for the new case. Finally, the system will store the pair $<S_i, S_i'>$ in the adaptation case base.

Indexation and distance of the adaptation cases. In the adaptation base, the pairs $<S_i, S_i'>$, are indexed on the different fields of the solution S_i . There is a field for each type of financial value. However, there is an additional field corresponding to the distance between the case C_i and the new case to solve. Like in the case base, the fields of the elements in the adaptation base have different weighing and we also have a hierarchy of indexes. The same approach is adopted for the field's neighborhood in the adaptation base. Like in the case base, the distance between two cases in the adaptation base is the absolute value of the arithmetical difference.

Adaptation rules in the system FIPS. To elaborate a financial portfolio, we have to take into account two major aspects. The first one is the financial situation and the financial goals for the client, and the second one is the overall economic climate. The adaptation of the retrieved solution will be done in a first step using the rules related to the overall economic climate and in a second step, using the rules related to the distance between the retrieved case and the current one.

Example of economic climate rules

> **R1: _If_** Increasing interest rates **_then_** reduce the term of the fixed yield values **_endif_**

Example of rules related to the client financial situation

> **R2: _If_** the client goals are the safety and the income **_then_** reduce the percentage of shares and increase the percentage of fixed yield values **_endif_**

The rules presented before are used by the system FIPS for the first adaptations. The results of those adaptations are stored in the adaptation base in order to be reused.

Adaptation Algorithm. The adaptation algorithm is the following:

- Read the case *newC* and retrieve the nearest case *Ck*
- Read the solution *Sk* related to the case *Ck* and search for an adaptation rule in the rule set *RS*
- **If** there is a rule corresponding to the difference between *newC* and *Ck*
  Then Apply rule to *Sk* and obtain *Sk`*
    Insert the pair (*Sk* , *Sk`*) in the adaptation base (*AB*) and return the adapted solution *Sk`*
  Else Retrieve the solution *Sk`* which is the closest to the solution *Sk* in the *AB*
    Take and return the solution *Sk``* from the pair (*Sk`* , *Sk``*)
  Endif

Training algorithm. The goal of our approach is to develop the learner's capabilities of memorization and solution adaptation. The training algorithm is composed of several steps:

- •   (a) Propose a case q from the case base to the learner
- •   (b) Retrieve in the case base the solution S' associated to the case q
- •   (c) Read the solution S suggested by the learner
- •   (d) Evaluate the distance dis between S and S'
- •   (e) Evaluate and display the similarity degree sim from the distance dis.
      ($sim = 1/ (dis+1)$)
- •   **if** $sim < Acceptable\_Rate$ **then** Goto (a) **Endif**

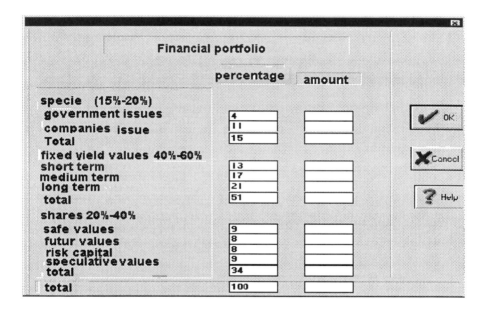

Fig. 2. Example of solution suggested by a learner

Evaluation algorithm.   After several tests are successfully passed, the system proposes an opposite approach when it provides a solution and expects from the learner a profile description. If the last test is successful then the learner evaluation is positive and the training process was well assimilated by the learner.

> - (1) **Do while** #*Step* < *StepNumber*
>   Training case (see the training algorithm)
>   **EndDo**
> (2) Propose a solution *S* to the learner
> - (3) **Read the case *q* suggested by the learner**
> - (4) **Retrieve in the case base the case *q'***
> - (5) **Evaluate the distance *dis* between *q* and *q'***
> - (6) **Evaluate he similarity degree** *similarity*(**from the distance *dis*. (*sim* = 1/ (*dis*+1))**
> - if *sim* < *Acceptable_Rate* then **Goto (a)** Endif

3.5 Example

In this section, we see a training and an evaluating example to illustrate what we presented previously. Let *C* be a case described by the attributes:

$$C = \{ Age,\ civil\ situation,\ Salary,\ risk\ tolerance,\ financial\ goals \}$$

Let *S* be a solution:

$$S = \{ Specie,\ short\ term,\ medium\ term,\ long\ term,\ safe\ values,\ future\ values,\ risk\ capital,\ speculative\ values \}$$

Let *Case1* be a case to propose to a learner (**step(a)** in the training algorithm):

$$Case1 = \{ 38,\ single,\ 120000,\ 60\%,\ safety \}$$

Let **S1** be the solution retrieved by the system for the case *Case1* : (**step(b)**)

$$S1 = \{ 14\%,\ 11\%,\ 19\%,\ 21\%,\ 8\%,\ 9\%,\ 9\%,\ 9\% \}$$

Let *S2* be the solution suggested by the learner : (**step(c)**)

$$S2 = \{ 12\%,\ 13\%,\ 17\%,\ 23\%,\ 7\%,\ 8\%,\ 11\%,\ 9\% \}$$

The distance between *S1* and *S2* will be evaluated by the system : (**step(d)**)

$$dis\ (S1, S2) = \Sigma_{a=1,8}\ wa * disa\ (S1a, S2a) = 12.5\ (2+2+2+2+1+1+2+0) = 150$$

The evaluation of the student's solution is the value of the similarity degree between *S1* and *S2*:

$$sim = 1/ (1 + dis\ (S1, S2)) = 1/(1+150) = 66\%\ (\textbf{step(e)}).$$

The system proposes *n* different cases (this number depends on the learner's profile) to the learner and evaluates his solutions as shown for the solution *S2*. If the value of the learner's solutions is greater than an acceptable rate (for example 65%), the system skips to the learner's evaluation level (**step(2)** in the evaluation algorithm), and proposes a solution (for example *S1*) and checks if the case proposed by the learner (**step(3)**) is similar to the case *Case1* (**steps(4,5,6)**). Depending on the answer provided by the learner, the system evaluates the success of the training process (**step(7)**).

4. Conclusion

A rule-based system is powerless in front of any non-planned problem. The set of rules guiding an expert system are fixed and offer no evolution whatsoever. In other words, this type of system has no capacity to go beyond its predefined rules and enhance its knowledge of the application domain in which it operates. Contrary to a rule based system, a case based system is capable of learning by rendering the solution available for use in any future problem, thus adding a learning mechanism to the process of problem solving. This technique offers a certain advantage over the rule-based approach.

The degree of local similarity between two values of an attribute figuring in two cases reflects the local distance separating these two values. The similarity between the two cases is translated into a composed (rather than a single) similarity, which takes into account the local attribute's similarity as well as the weight associated to them.

The originality of this work is the recursive use of the CBR approach. In fact, the CBR mechanism is used to retrieve similar cases and the old adapted solutions; words, for each solution adaptation, the system keeps in memory a trace of that operation in order to reuse it in the future. Therefore the adaptation has a mixed approach with the use of a Case-Based Reasoning and a rule based process.

The approach of FIPS for the learner training is based on the weights of the attributes in the description of the solutions. Therefore, we have a good memory organization with a hierarchical structure and a mechanism built on attributes with a strong power of discrimination. The case memory is dynamically modified while the system evolves. The case based adaptation process gives the system the possibility to learn more on how to adapt its solutions after each new adaptation.

On the other hand, an interesting enhancement for FIPS could be the addition of a base of student's profiles. For a new learner, the system can use a CBR approach to retrieve an old similar student profile in order to know what kind of examples can be used to provide a training adapted to the new student.

FIPS has been tested successfully with a base of one hundred cases and gives good results in financial analysis. We conclude by highlighting the fact that all the modules in FIPS are easily reusable in other application domains, like medical diagnostic and mechanical-failure detection.

5. References

1. Aleven, V., Ashley, K. D., "Automated Generation of examples for a tutorial in Case-Based Argumentation", Proceedings *of The Second International Conference on Intelligent Tutoring Systems (ITS 92), 575-584*. Edit. By C. Frasson; G. Gauthier; G.L. McCallan. Berlin: Springer-Verlag. Montréal, 1992.
2. Boudina, K. "Le raisonnement à base de cas dans la planification financière", Master Thesis in Computer Science, the University of Montreal, 1998
3. CCVM "Cours sur le Commerce des Valeurs Mobilières au Canada", *Institut Canadien des Valeurs Mobilières, 1995.*
4. Farrel, R., "Intelligent Case Selection and Presentation", *Proceeding of the Tenth International Joint Conference on Artificial Intelligence.* IJCAI-87,1,74-76. Milan, Italy, 1987.

5. Kass, A., Burke, R. D., Blevis, E., Williamson, M., "The GuSS project: Integrating instruction and practice through guided social simulation", Northwestern University, Institute for the Learning Sciences Technical Report no.34, 1992.
6. Kolodner, J. L., "Case-Based Reasoning". Morgan Kaufmann Publishers, 1993.
7. Leake, D. B., Kinley, A.,Wilson , D., "Learning to Improve Case Adaptation by Introspective Reasoning and CBR". Case-Based Reasoning. Experiences, Lessons and Future Directions. 185-197. Ed. Leake, D. B., AAAI Press/ MIT Press, 1996.
8. Shiri, M., Aïmeur, E., Frasson, C. "Student Modelling by Case Based Reasoning" *ITS-98, Fourth International Conference on Intelligent Tutoring Systems*, Lecture Notes in Computer Sciences, no. 1452, B.P. Goettl, H.M. Halff, C.L. Redfield, V.J. Shute Editors, Springer Verlag, San Antonio, Texas, pp. 394-403, August 1998.
9. Watson, I., "Applying Case-Based Reasoning: Techniques for Entreprise Systems", Morgan Kaufmann Publishers, San Fransisco, California. University of Salford. United kingdom, 1997.
10. Wess, S., Globig, C., "Case-Based and Symbolic Classification Algorithms", University of Kaiserslautern, Germany,.1994
11. Wess, S., Ritcher, M., "Similarity, Uncertainty and Case-Based Reasoning in PADTEX". University of Kaiserslautern, 1993.

Training of the Learner in Criminal Law by Case-Based Reasoning

Simon Bélanger, Marc-André Thibodeau, Esma Aïmeur

University of Montreal
Department of Computer Science and Operational Research
C.P. 6128, succ. Centre-ville
Montreal, Canada H3C 3J7
Tel: (514) 343-6794; Fax: (514) 343-5834
[belanger, thibodea, aimeur]@iro.montreal.ca

Abstract. FORSETI is a system that uses Case-Based Reasoning applied to Criminal law. It has two functional modes: the **expert** mode and the **tutorial** mode. The first mode makes use of its knowledge in order to resolve new cases, i.e. to determine a sentence (imprisonment, delay before request for parole, etc.). Canadian jurisprudence of similar cases to the one presented constitutes the basis for FORSETI's judgement. An expert may also consult the case base in order to corroborate the result obtained and supply the base with new cases. The second mode is an educational tool for professionals in Criminal law. In virtue of this mode, two types of exercises are possible. The first one focuses on developing the user's judgement in determining sentences and the second one on improving his jurisprudence analysis. FORSETI uses an experimental method for adaptation that we call *planar interpolation*. The goal of this method is to improve the level of consistency in the knowledge base and produce significant results in adaptation of new cases.

Keywords: Training, Case-Based Reasoning, Tutorial systems, Criminal Law, Sentence.

1. Introduction

In the course of his everyday life, a human being is often confronted with problems that he can solve by using his memory by looking into his past to find situations that are similar to the one to which he is confronted. These situations all contain, at least in part, knowledge required to the resolution of a new problem. This illustrates the basic idea of a relatively new field of interest in artificial intelligence research called *Case-Based Reasoning* (or CBR) [4]. Its goal is to resolve problems in a particular field of activity. The two basic problems in CBR are the indexing and the definition of an appropriate heuristic for estimating the similarity between two or more cases. To this end, it requires a base of cases from the past linked to their solutions. When it is confronted with a new problem, it searches in this base for the cases that contain the most similarities, then attempts to adapt their solutions in order to create one which will satisfy the aspects of the new situation.

It is possible to apply this Case-Based Reasoning to an array of different fields. Law (JUDGE) [2], medicine (CASEY) [5], architecture (ARCHIE) [7], design (NIRMANI) [8] and military planning (Battle Planner) [3] are only a few examples. Their development has been growing rapidly, as is shown by the systems described by [11]. FORSETI[1], the system which we have conceived, is specialized in the field of Criminal law. Its first objective is to determine sentences for crimes on the basis of cases similar in jurisprudence. Its second, is to help lawyers further their analysis of criminal cases in a continuing development of Criminal law decision-making. FORSETI uses Case-Based Reasoning to these purposes.

Our work has consisted in conceiving and developing the FORSETI system and to establish a case base linked to jurisprudence. First, this document presents the basic concepts of the theory on Case-Based Reasoning, such as representation of cases, indexation and adaptation of new cases presented to the system. This last method which is adapted presents a new resolution approach called planar interpolation. Secondly, it is an incursion into the field of Criminal law. Finally, we will discuss with more emphasis the tutorial function of FORSETI.

2. The elaboration of FORSETI

This section presents the basic principles which have guided the elaboration of FORSETI and explains why Case-Based Reasoning has been chosen for its realization. This segment ends with a brief presentation of Criminal law in Canada.

2.1 Use of CBR for training

Jean Piaget, an epistemologist, was the one to unite the two great philosophies of innatism and empiricism in laying the foundation of the theory of Constructivism [10]. Today, the theory of Constructivism is the most credible, well-rounded and accepted theory in the field of education. It integrates the new concept of construction of knowledge. For Piaget, it is by his interaction with his environment that a human being elaborates the mental structures which permit him to assimilate the reality that surrounds him and enable him to reflect on this reality. Therefore, forming an individual implies putting him in contact with a problematic situation, then bringing him to integrate this situation in his memory (assimilation) and to modify his internal cognitive structures (accommodation).

[1]In Scandinavian mythology, Forseti is the God of Justice

This conception of learning is exactly the one adopted by Case-Based Reasoning. It aims at permitting the system learning by the construction of knowledge base. This construction is realized by presenting new situations to the system to which it must react conveniently. If let doesn't do so, then the system must adapt its internal structures. It finally integrates this situation and the newly acquired knowledge in its database in order to be able to use them in the future, when other situations occur.

For example, the intelligent tutorial systems HYPO [1] and SARA [9] use Case-Based Reasoning for training. The first is used for the training of students while the other is applied to the teaching of mathematics.

2.2   Principles behind CBR

Representation of cases: A case is, conceptually, a piece of knowledge representing a certain experience. Typically, a case should contain:
- The state of the problem in its original context
- The solution to the problem or the new situation of the case when the problem has been resolved.

Cases can be represented in a variety of forms including frames, objects, predicates, semantic networks and rules.

Indexation: Each case posseses different attributes. Some are used to index the case in order to limit and accelerate research of similar cases, whereas others are used to characterize the cases. *Indexation based on difference* and *indexation based on explanation* are two examples of methods that are commonly used.

Search of similar cases: To extract similar cases from the database, two general methods exist: *algorithm of the nearest neighbor* and *method by induction*.

Adaptation: It is a central step of Case-Based Reasoning. It consists of finding a solution for a new case by modifying the solution of a similar case. The manner in which the adaptation is realized can become a difficult problem. It is in fact one of the major stakes of CBR for which there has been enormous research. David Leake is one of the researchers brought attention to the study of adaptation. Substitution and transformation of solutions are methods used for adaptation [6].

2.3   Criminal law

Criminal law aims at the repression of certain conducts, law enforcement and the protection of public wealth. It is characterized by a consequence; a sentence which can be a fine, an interdiction or an ordinance.

FORSETI is applied to Criminal law, more precisely to crimes against the person under the Canadian Criminal code. All cases used for the FORSETI system have been extracted from the Quebec Penal law jurisprudence index which provides abstracts of decisions in Quebec Penal law. The justice system must conserve a coherent link between present and past sentencing. Therefore it is of most importance to analyze the judicial system in order to identify the most influent criteria that lead to the judgement of a case. These are the seven principles that must guide the courts in the imposition of a sentence in Canada:
- Effective gravity of an infraction with regard to the law,
- Effective gravity of the infraction,

♦ Subjective gravity of the infraction, criminal record, age and reputation of the accused.
♦ Frequency of the same category of crime,
♦ Circumstances extenuating and increase,
♦ Rehabilitation of the accused,
♦ Salutary or exemplary effect of the imposed appeal.

3.   The FORSETI system

Java, an Object-oriented programming language, was used in the realization of the FORSETI system. Object-oriented programming permits a better structure and helps maintain systems which use a high complexity level. Furthermore, Java brings the advantage of portability to FORSETI, permitting it to be executed on any given platform.
The description of a case is constituted by thirteen descriptive attributes, two indexation attributes (the type of crime and the plea) and four solution-oriented attributes (the sentence, the delay, the extra fine and the interdiction). The attributes of age, sentence and delay are the only numeric attributes, the others being hierarchic (for example, the precision «spouse» concerning the attribute of victim can only exist if the precision «known» is already attributed to the victim). Furthermore, certain attributes can be of different levels of importance.

3.1   Calculating similarity

The algorithm used to calculate the degree of similarity between two cases is a variation of the *closest neighbor algorithm*. It consists of comparing the value of the descriptive attributes of two cases while taking note of the different levels of importance given to each attribute. These levels of importance have been determined with the collaboration of a law expert and reflect the importance of the attribute in the decision of the court (see section 2.3).

3.2   Adaptation

The adaptation of new cases consists of determining the value of the four solution-oriented attributes of a new case by examining similar cases in the database.

Planar interpolation The idea is to interpolate a numeric value by the equation of a plane in a tridimensional space. Finding such an equation implies the knowledge of the coordinates (x, y, z) of three non-aligned locations in a given space. These three locations will each correspond to a case chosen for the adaptation of the solution. The method for determining the coordinates of a location based on a case is relatively simple. The user that has just entered a new case to adapt chooses two attributes that he judges the most important, and that should most influence the decision of the judge in this particular case. It should be noted that it is possible to make another choice if the adaptation obtained is not satisfactory. These two attributes become x and y. It is then necessary to calculate the value of these variables x and y, for each case, according to the value of these attributes. To this end, we have determined, with the assistance of a lawyer, a scale of «gravity» which indicates the relative gravity of different possible values for the four most important attributes: circumstances, motive, consequences on the victim and character of the victim. For example, an «accidental» circumstance will be much less important on the scale of gravity than the «planning of a crime».

Then, we must determine the z axis, the one relative to the attribute for which we have decided to interpolate the value. For example, we can interpolate the length of imprisonment. In this case, the number of months of imprisonment in similar cases will be the z coordinate of the corresponding points.

Example: If we take three locations (3,10,80), (8,8,90) and (10,5,80). The obtained equation is $50x + 70y - 11z = -30$ and the corresponding interpolation plan is presented by Figure 1.

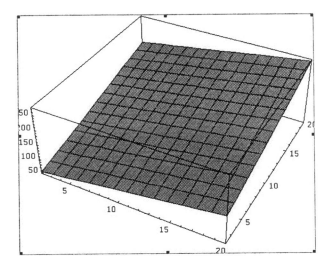

Fig. 1. Interpolation plan

The result obtained for the value of the attribute of the new case also considers the value of this attribute in the three most similar cases, instead of only one case, as in JUDGE. This solution cannot contradict those of the three cases used for interpolation, thus enhancing the coherence of the decisions taken.

It is also interesting to observe the potential of this method. Two important improvements could be made. First, the number of cases used for interpolation could be increased arbitrarily. Many interpolation methods that are more complex are known to use more locations. It is for instance the case of the *Collocation surfaces method* which consists of finding a polynomial whose degree varies according to the number of locations. This polynomial determines a surface which passes through all the locations and gives an interpolation whose realism increases with the number of locations used.

A surface obtained with this method for nine locations can look for example like Figure 2.

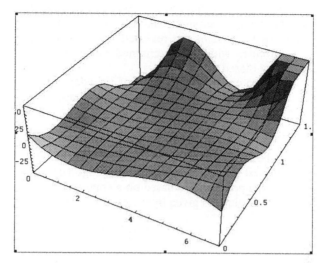

Fig. 2. Collocation surface calculated with nine points

It would be possible to use all similar cases in interpolation, maximizing at the same time the coherence of the derived solution with the one of similar cases. But the problem is as follows: implementing the Collocation surfaces method in a non-functional programming language such as Java is a complex task which has not yet been realized. It would certainly be an interesting extension to our research.

The second improvement that could be made would be to use more attributes. This would translate into increasing the dimension of the interpolation space. For example, we could, with four axes, interpolate a value in a space with four dimensions by using three attributes instead of two. More locations would obviously be necessary, as well as more similar cases for each equation. This can again be generalized at n dimensions. The implementation of the «spatial» interpolation which would result would be of a certain interest, since the use of all the attributes for the adaptation would certainly give even superior results. Nonetheless, this goes beyond the objective of this research.

Adjusting parameters When the attribute of the solution to be adapted possesses a numeric value, but that there are less than three similar cases found, then an adjustment method of parameters is launched. This method considers four factors we have already submitted. The degree of gravity is calculated for each of those four attributes by using the same scales as for planar interpolation. The gravity of each attribute is then compared to the corresponding attribute in the most similar case. The parameter is increased if the gravity is superior, decreased if the gravity is inferior.

3.3 Validation of the system

The quality of the results obtained with FORSETI has been measured qualitatively by tests made on a database of a hundred cases. The first test aimed at verifying the calculating of similarity. It consisted in calculating the similarity of each case with all the other cases of the database. This test revealed that a relatively small importance should be given to attributes with non-determined values so that the value of the calculated similarity is not biased.

The second test was used to evaluate the quality of the adaptation. We have adapted the solution of each case of the database as if it was a new case by using other cases of the base as reference. We then compared the solution obtained with the real solution, for each case. The principal information furnished by the test is an indication on the adjustment of attributes and mainly, on the importance in the choice of locations for the planar interpolation method. It is important to choose, as locations, attributes that for the adapted case have played a determining role in the judge's decision-making.

An evaluation of the system performance based on a comparison between the solutions given by FORSETI and the solutions given by the expert is underway.

4. FORSETI as tutorial

FORSETI also aims at bringing the user that is not an expert to exercise his judgement and to help him in understanding jurisprudence. The lawyer, in the exercise of his work, must frequently search for similar cases to the one he is defending. It is of great importance for him to develop an ability to rapidly extract permanent information from jurisprudence. Furthermore, he must train to evaluate cases in order to determine which sentence would be acceptable for each of them. FORSETI possesses all the information necessary to help the training of a law professional. As such, in tutorial mode, this system offers two types of exercises to the user.

4.1 First type of exercise

FORSETI presents jurisprudence to the learner as it could be found in indexes. The learner is then invited to analyze texts taken from jurisprudence and to determine which attributes are most important, by filling in a form (Figure 3). Once filled, the form is assimilated by the system as a new case. This case is compared to the case corresponding to the text which has been presented. This comparison is made possible by calculating the similarity between the cases. If the similarity obtained is high, then the description created by the user is close to the one made by the expert and which is contained in the system. The user is finally brought to reflect on each difference encountered between the two cases. As such, the system questions the learner on the value that he has given to litigious attributes. Has he given this value simply because he has seen a certain term used in a case? Has he considered the context in which the term has been used in the text? Could he have regrouped a few terms by a more general term, for example «troubles-personality(aggressive) + troubles-personality(dangerous)» by «troubles-personality(criminal)? Or on the other hand, should he have given more previsions, for example by replacing the value «physical-sequels» by «physical-sequels(major-injuries)» for the attribute of «consequence»? In these terms, FORSETI's goal is to bring the learner to reflect on his mistakes.

Furthermore, the system displays the percentage of efficiency of the learner so that he can be aware of his level of learning and of his progression. This percentage of efficiency is calculated simply: S being the maximal similarity level that can be obtained and s the level of similarity between the case and the description given by the user, then the percentage of efficiency is:

$$T = s/S * 100$$

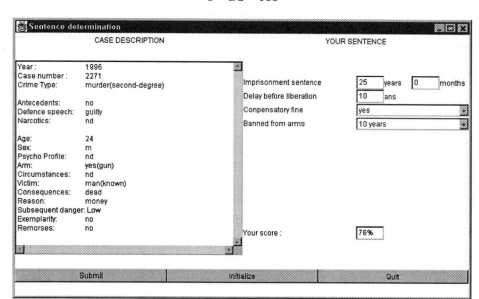

Fig. 3. First type of exercise

4.2 Second type of exercise

FORSETI presents cases chosen randomly in a base of unresolved exercises. Theses cases possess solutions, contained in a «solutions» folder, but are not part of the database from which FORSETI works. The learner must then, after analyzing the attributes of the case, determine a solution, i.e. enter a punishment that he finds appropriate (Figure 4). The system then tries itself to complete the exercise. It proceeds to the adaptation of the case, as if it was a new case, and also determines a solution. The three solutions are then presented to the learner: his own, FORSETI's, and the expert's (the real judgement that was given for the case, as described in the «solutions» folder). The success obtained by the system is an indication of the difficulty of the exercise and the efficiency of the user.

In fact, if FORSETI determines a solution which is far from the real solution, it is either that there are only a few similar cases in its database, or that the case is exceptional, having obtained a judgement that contradicts those of similar cases (in this case, the level of difficulty of the exercise is very high). The learner should be encouraged if his solution is closer to the real solution than the one FORSETI has put forward. This brings us to determining a method for measuring the difference δ between two solutions. We calculate this difference by a function:

$$\delta = w_p \max(p_1, p_2)/\min(p_1, p_2) + w_d \max(d_1,d_2)/\min(d_1,d_2) + sw_s + iw_i$$

where w_k determines the relative importance of each attribute in a solution, according to the expert. p_k represent the values determined for the sentence. d_k are values determined for the delay before request for parole. s is the difference between the value for the estimated compensatory extra fine and the real value (s equals 0 if this value has been determined correctly, 1 if not) and i is the difference between the values of interdiction to possess firearms (i = number of levels of difference for the interdiction. For example, «life» and «5 years» have two levels of difference, so i equals 2).

As such, a value of δ is calculated for the difference between FORSETI's solution and the real solution, and another for the difference between the real solution and the learner's.

The learner's efficiency rate is in this case calculated as such:

$$T = \delta_F/(\delta_F + \delta_a) * 100$$

where δ_F is the error in FORSETI'S solution and δ_a the error in the learner's solution. A high value for δ_F (and so of the difficulty of the exercise) will bring the T rate closer to the maximum (100%) while a high value of δ_a will have the contrary effect. With this formula, an exact answer from the learner gives a rate of 100%. The T rate considers simultaneously the learner's answer and the level of difficulty of the exercise.

As in the first type of exercise, the learner is brought to reflect on his mistakes. This time, the tutorial gives him the opportunity to consult similar cases extracted from the database during adaptation. As such, the user can see if the judgement he has made corroborates the reality of jurisprudence. He is also questioned on the reasons for his false reasoning. Each learning session can be saved at any time by the user so that he can pursue it at another moment.

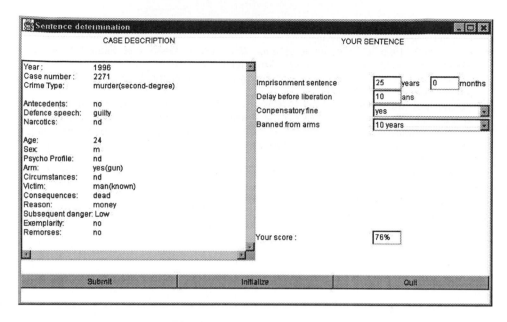

Fig. 4. Second type of exercise

5 Conclusion

FORSETI is a system that uses Case-Based Reasoning capable of determining sentences in cases of Criminal law. The cases presented in the system are those from 1984 to 1997. Its second function is the development, by exercises, of the ability of law professionals to analyze jurisprudence.

The system possesses characteristics which could be the object of future improvements, in particular with respect to the automation of the determination of the importance of the attributes. Nonetheless, FORSETI innovates by using an experimental method of adaptation for the parameters of new cases. This method helps the system conserve a high level of consistency and coherence in decision-making with prior cases. It could be generalized to be used for other fields of law.

6 References

1   Aleven, V., and Ashley, K.D. 1992. Automate Generation of Examples for a Tutorial in Case-Based Argumentation. In Proceedings of the Second International Conference on Intelligent Tutoring Systems. Berlin. Springer-Verlag.
2   Bain, W. 1989. Inside Case-Based Reasoning. Northvale, ed. C.K. Riesbeck and R.C. Schank.
3   Goodman, M. 1989. CBR in Battle Planning. In Proceedings : Workshop on Case-Based Reasoning (DARPA). pp.576-584, San Mateo, California.: Morgan Kaufmann.
4   Kolodner, J. 1993. Case-Based Reasoning. SanMateo. : Morgan Kaufmann Publishers.

5   Koton, P. 1989. Using Experience in Learning and Problem Solving. Ph.D. diss., Dept. of Computer Science, MIT, Boston.

6   Leake, D.B. 1996. Case-Based Reasoning, Experiences, Lessons & Futures Directions. Manlo Park. : American Association for Artificial Intelligence.

7   Pearce, M., Ashok, K.G., Kolodner, J., Zimring, C., and Billington, R. 1992. A Case Study in Architectural Design., IEEE Expert.

8   Perera, R.S., and Watson, I. 1995. A Case-Based Design Approach for the Integration of Design and Estimating, In Progress in Case-Based Reasoning. Berlin. : Springer-Verlag.

9   Shiri, M., Aïmeur, E., and Frasson, C. 1998. Student Modelling by Case-Based Reasoning, ITS 98 Conference, 4th International Conference on Intelligent Tutoring Systems. pp.394-403, San Antonio, Texas, 1998.

10 Wadsworth B.J. 1989. Piaget's Theory of Cognitive and Affective Developmen. New York. : Longman.

11 Watson, I. 1997. Applying Case-Based Reasoning : Techniques for Enterprise Systems. San Francisco, California. : Morgan Kaufmann Publishers, Inc.

Constraint-Based Agent Specification for a Multi-agent Stock Brokering System

Boon Hua Ooi and Aditya K. Ghose
Decision Systems Lab
Dept. of Business Systems
University of Wollongong
NSW 2522 Australia
email: {boono, aditya}@uow.edu.au

Abstract. This paper outlines a constraint-based agent specification framework and demonstrates its utility in designing and implementing a Multi-Agent Stock Brokering System (*MABS*). We present first an extension of the BDI agent programming language AgentSpeak(L) which incorporates constraints as first-class objects. We then describe the specification (and implementation) of *MABS* using this new language, AgentSpeak(L)+C. Our results suggest that the integration of constraints in a high-level agent specification language yields significant advantages in terms of both expressivity and efficiency.

1. Introduction and Motivation

The design of complex multi-agent systems requires the use of expressive high-level specification languages which eventually translate into efficient implementations. This paper suggests that constraint-based agent specification meets these requirements. To support this thesis, we present an augmentation of the BDI agent programming language AgentSpeak(L) [6] with constraints and describe an implementation of a Mult-Agent Stock Brokering System (*MABS*) in which agents are specified in this new agent programming language. The proposed language, called AgentSpeak(L)+C, improves over AgentSpeak(L) in a manner parallel to the gains achieved by integrating constraints in a logic programming framework to obtain *constraint logic programming* [3], in terms of both expressivity and efficiency. We present some preliminary observations on BDI agent architectures and constrain-based reasoning in Section 2. In Section 3, we present the syntax of AgentSpeak(L)+C. Sections 4 and 5 describe the design and implementation, respectively, of the *MABS* system. Section 6 presents an example of *MABS* agent specification using AgentSpeak(L)+C, while Section 7 concludes the discussion.

2. Preliminaries

2.1 BDI Agent Architectures

In the context of a BDI architecture [1][5][7], the information, motivation and decisions/actions of an agent are modeled using the mental attitudes of *belief*, *desire* and *intention*. Typically, a BDI architecture consists of four data structures, one each for beliefs, desires and intentions and an additional data structure called the *plan library*. Each plan specifies a course of action that may be followed to achieve certain intentions, as well as the pre-conditions that must hold for the plan to be undertaken and the post-conditions that would hold after plan execution (specified in terms of the intentions that would be achieved).

2.2 Constraint-Based Reasoning

Constraint-based reasoning involves an expressive knowledge representation language in which problems are formulated as a set of variables (with associated domains) for which value assignments are sought and which respect a specified set of *constraints* on the allowed combinations of values of these variables. The power of the constraint-based reasoning derives from both the simple formulation of problems and from the wide array of methods for efficiently solving such problems. Constraint-based reasoning is a popular approach to solving a wide variety of scheduling, configuration and optimization problems and has been successfully integrated with logic programming and database technology to yield *constraint logic programming* [3] and *constraint database* toolkits.

3. AgentSpeak(L)+C Syntax

AgentSpeak(L), developed by Rao [6] and further elaborated by D'Inverno [2], is a popular BDI agent programming language. We outline below the syntax of AgentSpeak(L)+C, an agent programming language which augments AgentSpeak(L) with constraints from a constraint domain C. Note that the actual instance of AgentSpeak(L)+C is parameterized by the actual choice of constraint domain C (reals, finite domains etc.).

Definition 1 : An *agent program* in AgentSpeak(L)+C is a triple *<Belief, Goal, Plan>*.

Definition 2 : A *belief* is an assertion of the form $b(t)$ [] $c_1(t)$,..., $c_n(t)$ or $\{b(t)\}$ or $\{c(t)\}$ where b is a belief symbol, t is a vector of terms from constraint domain C and each of $c_1,..., c_n$ are valid constraint symbols from C. *Ground belief* (*base belief*) will be in the form of $b(t)$ or $c(t)$ where t contains no variables.

Definition 3 : A goal is an assertion which can take one of the two possible forms : $!g(t)$ or $?g(t)$ where g is a goal symbol and t is a vector of terms.

Definition 4 : If $b(t)$ [] $c_1(t)$,..., $c_n(t)$ or $b(t)$ are *beliefs* and $!g(t)$ or $?g(t)$ are goals, then $\pm b(t)[]c_1(t)$,..., $c_n(t)$, $\pm b(t)$, $\pm!g(t)$, $\pm?g(t)$ are *triggering events*.

Definition 5 : If a is an action symbol and $t_1,...,t_n$ are terms, then $a(t_1,...,t_n)$ is an *action*.

Definition 6 : If e is a triggering event, $\{b(t)\}$ or $\{c(t)\}$ are beliefs, and $h_1,...,h_m$ are goals or actions, then a plan will be in one of the following form :

Plan 1   - $e : \{b_1(t)\},...,\{b_n(s)\} \Leftarrow h_1;...;h_m$

   or

Plan 2   - $e : \{c_1(t)\},...,\{c_n(s)\} \Leftarrow h_1;...;h_m$

Plan 1 can be reduced or reorganized to the representation in plan 2 base on ground beliefs expressed in the form of :

$$\{b_1(t)[]c_1(t),...,c_n(t)\},...,\{b_2(s)[]c_1(s),...,c_n(s)\}.$$

The following are the typical specifications of beliefs and plan in AgentSpeak(L).

Beliefs :	client (john, valid).
	fund (john, sufficient).
Plan :	+!buy (X, Qty, Stk, Prc) : *client (X, valid) &*
	fund (X, sufficient)
	$\Leftarrow$ submit (order).

In the above plan, the head is a combination of an addition of a new goal (invocation) and 2 conditional context beliefs. The primitive action of Submit (Order) will be carried out if the relevant context is derivable from the base beliefs, which in this case if X is a valid client and there is sufficient fund available to make the purchase.

The original processing of the above declarative specifications in AgentSpeak(L) will be based on the unification process which is specified in detail in [6]. A constraint directed improvisation (as in AgentSpeak(L)+C) could be incorporated into the computation strategy employed during the interpretation process. The improvised approach introduces a constraint system into the declarative semantics of AgentSpeak(L)+C whereby the basic elements of BDI architecture are constructed based on constraint clauses. Richer data structure introduces by the constraint system will allow beliefs and goals of a BDI agent to be directly expressed and manipulated. Constraint solving will subsequently be applied to select the satisfied plan to be fired. The extension preserves the simple, direct approach of AgentSpeak(L) in agent specification and at the same time maintains the established constructs of BDI architecture.

For instance, the above beliefs and plan can be re-specified using AgentSpeak(L)+C in a more expressive and precise manner as follows:

Beliefs :	Validity = true.
	Fund = 100000.00
Plan :	+ !buy (X, Qty, Stk, Prc) : *Fund = 100000.00 &*
	Validity = true &
	Client = X & (Qty * Prc < Fund)
	$\Leftarrow$ submit (order).

By embracing constraint directed technology within the BDI architecture, it will be able to have a better ***understanding*** and ***control*** over the behavior of an agent as the beliefs and goals can now be specified in a more precise, expressive and unambiguous terms. Corollary issues on performance, real-time reactive behavior and processing efficiency of an agent will be made less discouraging as these problems can now be tackled by manipulating the constraint that controls the mental states and beliefs of the agent. In short, it will be much more comfortable to handle and manipulate constraint to achieve the desired result.

4. The Multi-Agent Stock Brokering System (MABS) : Design

4.1 Agent Design and Specification

The *MABS* is an object-oriented software implemented on a Java interpreter designed for a declarative agent programming and specification language - AgentSpeak(L)+C . The Java interpreter translates the declarative semantics and operational semantics of AgentSpeak(L)+C into an object-oriented model by exploring on the well established and matured technology of object-oriented development strategy.

Under this approach, the entire system is mapped onto a hierarchical structure which identifies the functional roles and the relationships between the respective agents within the system. This mapping will isolate the functions performed by each of the autonomous agent and at the same time define the basic, fundamental functions that form the core engine of the entire software. An abstract, high level modeling of the *MABS* is made up of the followings :

- 2 Interfaces that provide the templates for the behavior of BDI Engine and constraint engine to interpret AgentSpeak(L)+C.
- A number of high level classes that implement the 2 interfaces and other top level common features of the agents.
- Individual agent classes that implement different agents in MABS.

4.2 Agent Model

There are two distinct sets of agents deployed in this model – the *static* agent set and the *dynamic* agent set. Static agents are instantiated at system initialization and remain 'alive' throughout the run-time. Dynamic agents are agents that may be instantiated during run-time and will vanish from the system when the tasks assigned to it have been completed.

Communication and collaboration between the agents are performed via message passing from one agent object to another to invoke the relevant methods or to instantiate the required run-time agent objects to perform specific function within the systems. Each of the agent object will be able to respond immediately to all triggering events by relying on its current beliefs set and plan set, irrespect of whether the triggering event is from neighbouring agent or external source. This will enable the agents to display reactive behaviour in a real-time environment. *Fig. 1* depicts the overall layout of the *system architecture* at run-time.

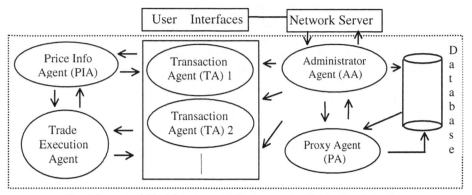

Fig. 1 Layout of *MABS* system architecture and collaboration behaviors between agents

4.3 Member Agents

The run-time of MABS is handled by a number of *daemon threads* and *process threads*. Daemon threads are background thread for static agent that constantly listens for triggering events, monitors on activities performed on the contents of relevant registry within an agent etc. They are initiated at the 'startup' of the system and remain active throughout the run-time.

Process threads are a multiplicity of threads that are spawned each time when a dynamic agent is instantiated (e.g. when a transaction agent is created to handle the processing for a particular transaction within a specified time constraint). Each process thread remains active for the specified period of time and will be removed when it expires.

The member agents of MABS are : (*Fig.2* depicts a Java based interpreter for a BDI architecture)

- *Transaction Agent (TA) – Transaction agent* is an agent object that is instantiated with an initial belief set whenever a new distinct, transaction is entered into the system. It will vanish at the end of transaction cycle. During its life span, its belief set will change with addition or deletion of beliefs based on various states of its life cycle. Every *transaction agent* will be assigned a unique identification tag and its computation state is maintained via the assigned tag. The state of each of the transaction agents will be constantly monitored by the *administrator agent* which will act accordingly based on its current beliefs on each of the *transaction agents*. *Transaction agent* is a dynamic agent and there can be a set of active *transaction agents* at run-time. It is able to authorize and to submit trade for execution by the *trade execution agent*. It also has the ability to spawn new *transaction agent* (e.g. to perform additional buy or sell, to arbitrage) based on its existing beliefs and plans. Through the collaboration effort with the *price info agent*, it will enable the *transaction agent* to behave reactively and to facilitate the trading strategy of the trader in a limited fashion.

- *Administrator Agent (AA)* – It adopts an interface role that receives instructions from trader, pass over the request to *proxy agent* for authorization and subsequently instantiate a new *transaction agent* if the request is approved. *Administrator agent* is a static agent that maintains a registry of all the active *transaction agents* whether they are instantiated by itself or are spawned by another active *transaction agent*. It has its own belief set that reflects the state information for all *transaction agents* that are still active. It will 'talk' to them in order to change its belief about the current state of each *transaction agent*. At regular time interval, belief information at designated state will be downloaded to a database for permanent storage.

- *Proxy Agent (PA)* – *Proxy agent* is a static agent that receives query from *administrator agent* to verify and to validate all trade requests from the traders. These verification and validating process is carried out based on the beliefs it has for each of the trader. The universal set of the belief terms for the entire client (trader) base is stored in a database that is accessible to the *proxy agent*. The information verified and validated by the *proxy agent* are trader's status and trader's particulars. The *proxy agent* will respond to each of the query by returning a 'not approved' reply or an 'approved' reply together with relevant transaction information (e.g. fund available) for the *administrator agent* to act on.

- *Price Info Agent (PIA)* – A static agent that maintains a real-time registry with the pricing information for all the stock counters. Its belief set consists of tuples of **<counter, requestor, price requested>** information (e.g. price limit for arbitrage, requested price) about trades waiting for arbitrage opportunities and trades in the on-line registry of *trade execution agent* waiting to be executed. It will notify the relevant *transaction agent* when the right price for arbitrage is offered in the market and it will also announce to the *trade execution agent* all the prices for the trades that are still outstanding and due to be executed.

- *Trade Execution Agent (TEA)* – It plays the role of a scheduler that submits trade for execution and maintains an on-line registry which keeps track of all the outstanding trades waiting to be executed. A set of **<counter, requestor, price requested>** information is relayed to the *price info agent* to facilitate the price monitoring process. *Trade execution agent* is a static agent that will relieve the trader from tedious effort of constantly monitoring the price changes. It will react to price changes in accordance to the instructions given by the *transaction agent*. Online price information provided by *price info agent* enables it to immediately submit trade for execution or hold back and wait for the right timing before it acts. Its behavior is determined by the beliefs set it holds for each of the outstanding trades on its registry. There are a set of plans (e.g. submit trade immediately, hold back until specified date, hold back until price is above or below specified limit etc.) that will be fired accordingly based on the different beliefs for different trade. Executed trade will be updated to the relevant *transaction agent* and hence change its belief set with the addition of a new belief.

5. The Multi-Agent Brokering System (MABS) : Implementation

The MABS is designed using the BDI (Belief-Desire-Intention) agent framework in [1][4][5][7], whereby agents constantly monitor the environment around them and behave accordingly based on the three mental attitudes of *belief*, *desire* and *intention*. In the BDI perspective, this framework provides an efficient maneuver of agent's behavior necessary to respond reactively based on changes in beliefs and work autonomously to achieve goals through the use of plans.

The three major components of MABS are a set of run-time libraries, a BDI package and a set of user application package.

- The run-time class libraries provide and implement the basic run-time behavior of MABS. There are two distinct packages within the run-time libraries :
  i.   BDI engine, an implementation of a BDI architecture.
  ii. Constraint engine, an implementation of constraints driven computation within the BDI architecture.
- The BDI package is an abstraction layer between the MABS application code and the run-time packages. It provides a standard framework to organize the belief set and plan set that specify the intended behavior of an agent.
- The user (MABS) application package deals with tasks of creating, initializing and controlling the behavior of individual agents. Program package in this library will determine the functionaries and capabilities of the MABS system.

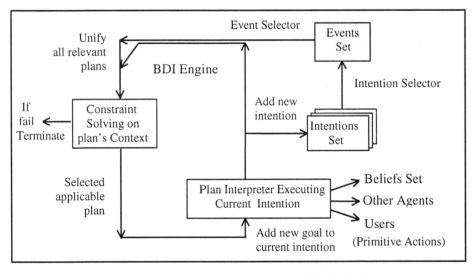

Fig. 2 Java based interpreter for a BDI architecture augmented by
constraint solving

The summarized sequential transition state for an execution cycle that takes place within the above architectural design are as follows :

i.   Creation (Instantiation) of Event Selector and Intention Selector (class objects)
ii. Selection of an triggering event (from events set)
iii. Compilation of all relevant plans
iv. Perform constraint solving on relevant plans set to identify an appropriate plan
v.   Apply goals and actions of plan body

There are a number of system parameters that can be preset and used to control the performance of the run-time system based on the available computing resources. For instance, it should be able to control or limit the validity period (in a flexible manner) for a particular transaction specified by an investor. This measure will help to reduce the number of unproductive, redundant transactions 'hanging around' in the system (taking up valuable resources) which prevent other genuine investors who intend to trade.

6.   A MABS Specification Example

The BDI framework employed in MABS is implemented with an improvised computation strategies - a synergy of *unification* [6] and *constraint solving* [3][8][9]. The improvisation applies constraint directed solving on the context section of a BDI agent's plan specification in order to determine an application plan to fire. This modification takes advantage of the constraint solving mechanism to improve on its performance capabilities based on 2 distinctive properties exist in the working context of MABS :

* Looking at the fact that real number computation forms the major bulk of computation involved in MABS, constraint solving naturally becomes an efficient solution for this purpose, i.e. the constraint domain in this instance of AgentSpeak(L)+C is real arithmetic (R).
* There is a possibility of having one or a range solution for each computation process. Constraint solving approach takes care of this elegantly by precisely define problem at hand using a set of constraint clauses.

The constraint system introduced into the BDI framework maintains a constraint store that collects a set of constraints that augment the beliefs of an agent. Before any constraint solving is being performed, the constraint store is first initialized the relevant constraint beliefs. Existing satisfied constraints set is then enhanced with new constraints (additional beliefs) when selected applicable plan was executed. These incremental constraints collecting process will eventually lead to a final beliefs set whereby relevant solution can be derived if required and the consistent constraints set itself becomes the most updated base beliefs of the agent concern. During the process of generating an applicable plan in respond to a trigger, constraint solving is applied to determine whether the constraint(s) in the context of plan is(are) consistent with the constraints already collected in the store. New constraints are added to the store and goals or actions are activated to initiate the next cycle of processing if the constraint solver returns a true value after the consistency checking process.

The following example demonstrates a Transaction Agent that receives a goal, generates an order, confirms trade done and arbitrages the trade. The Transaction

Agent is instantiated by a trigger event - a goal to *BUY* a certain *Quantity(Q_R)* of *Stock(STK_X)* @ *Price(P_R)* for *Client X*. Other variables involved in the constraint expressions are as indicated in the base beliefs set below (1 to 8), in which S represent sale, B is bank balance, P is outstanding purchases, M is margin amount available, R is the rate of return expected and ED is the arbitrage expiry date.

Relevant triggering events with unifiers $\sigma1$, $\sigma2$ and $\sigma3$ would invoke the following plans (AA, BB, CC) respectively and their context can be reformulated (partly with the assistance of current beliefs) into a more expressive and explicit form shown in 13, 14 and 15.

(AA) $+\,!\,buy\,(\,X, STK_X\,, Q_R\,, P_R\,)$	:	*valid_client (X)*
	$\leftarrow$	create (order).
(BB) $+\,done\,(\,X, STK_D, Q_D, P_D\,)$	:	*buy (X, STK_X , Q_R , P_R) &*
		equal (STK_D , STK_X) &
		equal (Q_D , Q_R) &
		equal_less(P_D , P_R)
	$\leftarrow$	notify (X).
(CC) $+\,return\,(\,CurPrice, R_{\%})$:	*arbitrage (buy (X, STK_X , Q_R , P_R)) &*	
	equal_less (R , $R_{\%}$)	
$\leftarrow$	$+\,!\,sell\,(STK_R\,, Q_R\,, CurPrice)$	

Trigger with rel. unifiers $\sigma1$ = {X/john, STK_X/abc co., Q_R/2000, P_R/3.00}

Trigger with rel. unifiers $\sigma2$ = {X/john, STK_D/abc co., Q_D/2000, P_D/2.95}

Trigger with rel. unifiers $\sigma3$ = {CurPrice/3.25, $R_{\%}$/10%}

---

1. X = john.
2. Validity = true.
3. Status = active.
4. S = 20000.
5. B = 80000.
6. P = 30000.
7. M = 80000.
8. R = 10.
9. ED = 18 Jun 1999
10. valid_client (john) [] X = john, Validity = true, Status = active,
$$(S + B - P > Q_R * P_R).$$
11. arbitrage (buy(X, STK_X , Q_R , P_R)) [] buy (X, STK_X , Q_R , P_R).
12. buy (john, abc co., 2000, 3.00) [] X = john , STK_X = abc co., Q_R = 2000,
$$P_R = 3.00.$$

13. $+!\,buy\,(X, STK_X\,, Q_R\,, P_R)$	:	*Validity = true & Status = active &*
		*(S + B - P > Q_R * P_R)*
	$\leftarrow$	create (order).
14. $+\,done\,(X, STK_D, Q_D, P_D)$	:	*X = john & STK_X = abc co. &*
		Q_R = 2000 & P_R = 3.00 &
		STK_D = STK_X & Q_D = Q_R & (P_D ≤ P_R)
	$\leftarrow$	notify (john).

15. + return (CurPrice, R $_{\%}$) : $X = john$ & $STK_X = abc\ co.$ &
            $Q_R = 2000$ & $P_R = 3.00$ & $(R \leq R_{\%})$ &
            $(current\_date \leq ED)$.
        $\leftarrow$ +! sell (ABC Co., 2000, 3.25).

16. +! buy (john, jkl co., Q_R, 9.00) : $Validity = true$ & $Status = active$ &
            $(S_X + B_X - P_X > Q_R * P_R)$
        $\leftarrow$ create (order).

17. +! buy (john, jkl co., 8000, 10.00) : $Validity = true$ & $Status = active$ &
            $(S + B - P > Q_R * P_R)$
        $\leftarrow$ create (order).

18. +! buy (X, jkl co., 8000, 10.00) : $Validity = true$ & $Status = active$ &
            $(S + B + M - P > Q_R * P_R)$
        $\leftarrow$ create (order) &
        holdamount ((S + B + M - P) - ($Q_R * P_R$)).

In the above example, initial *ground primitive beliefs* are made up of 1 to 9. During the processing of each plan, new constraints will be added incrementally to augment the existing constraint store. Constraint solving applies to the context of each plan will determine whether the relevant plan selected is the applicable plan to be fired. Beside its use to ensure constraints set is always consistent and to pick the right plan of action, constraint solving performs on the context of an applicable plan is also used to generate values (solutions) for subsequent action. For instance in plan 15, constraint solving based on the existing belief set plus the new constraints in the context of plan 15 will generate an answer of *true* to enable subsequent arbitrage action (SELL transaction) to take place. In another instance (plan 16), constraint computation will assist in generating an allowable value (Q_R) for the primitive action - create a buy order with Q_R = 7000 (rounded down to nearest thousand). In plan 17 and 18, Constraint computation is used to determine selective action to be performed. In this case, Plan 17 stops a buy order based on insufficient cash fund at hand. However plan 18 has provided an alternative to plan 17 by taking into consideration fund available in the margin account (M), hence it enables the 2 primitive actions of *create* and *holdamount* to take place.

7. Conclusion

This paper has presented a general overview and informal discussion of the concept of incorporating constraint-based processing into the specification of an agent. The idea represents a confluence of :

- AgentSpeak(L) for programming and specification of a BDI agent.
- Constraint directed computation with improved expressive power and preciseness.

The AgentSpeak(L)+C is an evolution from the above with the capabilities of providing richer data structures for programming a BDI agent. The work on MABS has shown the value added brought by AgentSpeak(L)+C in term of understanding and control of multi-agent behaviour.

References

[1] M. d'Inverno, D. Kinny, M. Luck, and M. Wooldridge. A formal spec. of dMARS, *Tech. Rep. 72, Australian Art. Intell. Inst.*, Mel, Aust., Nov 1997.

[2] M. d'Inverno and M. Luck. A formal specification of AgentSpeak(L). *Journal of Logic and Computation*, 8(3) :233-260, 1998.

[3] J. Jaffar and M.J. Maher. Constraint logic programming: A survey. *Journal of Logic Programming*, 19,20:503-581, 1994.

[4] D. Kinny and M. Georgeff. Modeling and design of multi-agent systems. In *Intelligent Agents III*, Lecture Notes in Artificial Intelligence. Springer, Berlin, 1997.

[5] D. Morley, Semantics of BDI agents and their environment. *Tech. Rep. 74, Australian Art. Intell. Institute*, Mel., Aust., May 1996.

[6] A.S.Rao. AgentSpeak(L): BDI agents speak out in a logical computable language. In *Agents Breaking Away: Proceedings of the 7th European WS on Modelling Autonomous Agents in a Multi-Agent World, (LNAI Vol 1038)*, pg42-55, Springer-Verlag: Heidelberg, Germany, 1996.

[7] A.S. Rao. and M. Georgeff. BDI Agents: from theory to practice. In *Proceedings of First International Conference on Multi-Agent Systems (ICMAS-95)*, pages 312-319, Sna Francisco, CA, June 1995.

[8] P. van Hentenryck. Constraint Satisfaction Using Constraint Logic Programming, *Artificial Intelligence*, 58:113-159, 1992.

[9] P. van Hentenryck and Y. Deville. Operational semantics of constraint logic programming over finite domains. In *Proceedings Symposium on Programming Language Implementation and Logic Programming, (LNCS 528)*, pages 395-406, 1991.

A Communication Language and the Design of a Diagnosis Agent – Towards a Framework for Mobile Diagnosis Agents[*]

Christian Piccardi and Franz Wotawa

Technische Universität Wien, Institut für Informationssysteme, Paniglgasse 16, A-1040 Wien, Austria, E-Mail {piccardi,wotawa}@dbai.tuwien.ac.at

Abstract. In this article we present a communication language and a design concept for a diagnosis agent. The communication language is divided in two parts. One part contains primitives necessary to control and query the agent. The other provides constructs for formulating the structure and behavior of systems to be diagnosed, and for observations determining the current state of the system. Additionally we give an overview of a WWW-based architecture for accessing an (diagnosis) agent's functionality. This approach is intended to make diagnosis tools accessible from todays WWW browsers.

1 Introduction

Model-based diagnosis (MBD) [Rei87,dKW87] has been successfully applied to several domains including the automotive industry for on- and off-board diagnosis of cars [MS96,PS97], for software debugging [FSW96,SW98], and tutoring systems [dKB98] among others. Only little effort has been spent so far on the design and implementation of diagnosis agents accessible using the WWW infrastructure. Currently most diagnosis systems use their own way to store, load, and process system models and observations, thus the prerequisite for the implementation of a WWW-based agent is a commonly agreed language for describing the behavior and structure of diagnosis systems, i.e., the system model, and the observations. A mobile diagnosis agent can be used in various domains ranging from academic institutions for educational purposes or in industry allowing easy maintenance and extendibility.

In this paper we describe the first part of the *MObile DIagnosiS (MODIS)* project. The objectives of MODIS are the design of a general framework for diagnosis, the exchange of diagnosis informations between agents, and the improvement of diagnosis algorithms to allow effective and efficient use in mobile computers such as personal digital assistants (PDAs) in case online communication is not possible. Our interest is to improve the algorithms to diagnose mid-size systems with up to 1,000 components using a PDA with limited computational power and memory.The objectives of the first part of MODIS are the design and implementation of a diagnosis agent for the WWW

---

[*] The work presented in this paper has been funded by grant H-00031/97 from the Hochschulju-biläumsstiftung der Stadt Wien. Christan Piccardi was supported by the Austrian Science Fund (FWF) under project grant P12344-INF.

using a communication language allowing to exchange diagnosis knowledge such as system descriptions and observations. The diagnosis agent itself has been implemented in Smalltalk. The client, i.e., the interface agent, has to be implemented using HTML and a Java applet. TCP/IP sockets are used as the underlying communication protocol.

During the last years several researchers have proposed different architectures of diagnosis agents. See for example [SDP96] or more recently [HLW98]. In [SDP96] an architecture of a diagnosis agent based on a non-monotonic inference systems is described. [HLW98] uses the communication language KQML ([FLM97]) and CORBA for communication. In contrast to previous approaches we do not base our system on one inference system. Instead, by using a communication module as interface between the diagnosis engine and requests from outside, the actual implementation of the diagnosis engine becomes less important, although the diagnosis engine has to implement all functions accessible from other agents. Additionally, we propose the use of standard WWW technology as the underlying communication architecture because of its availability on most of todays computer systems.

This paper is organized as follows. First, we give an introduction into model-based diagnosis followed by a section describing the agent's architecture and its design. Section 4 gives an overview of the communication language and the diagnosis functionality. A concise introduction into the used system description language for describing the behavior and the structure of a diagnosis system and the observations is given in section 5. A conclusion and future research directions are presented at the end of this paper.

2   Model-based diagnosis

In model-based diagnosis [Rei87,dKW87] a model of the *correct* behavior of a system or device is used to find a subset of system components that cause a detected misbehavior. This subset is termed a diagnosis. In contrast to other approaches, e.g., rule-based expert systems, no explicit effect-to-cause relationships have to be given to describe specific failure occurrences (although the model can be extended to describe typical incorrect behaviors). Very importantly, models can be composed of model fragments, i.e., behavioral descriptions of individual components such as an *AND* gate, merely by specifying the types of components and their interconnections. The model is thus easily maintainable. However, model-based diagnosis is only applicable if a model is available, which fortunately is the case in almost all technical domains.

A diagnosis is defined for a system (a device), consisting of a behavioral description SD and a set of components $COMP$, and a set of observations OBS. Formally (see [Rei87]), a set $\Delta \subseteq COMP$ is a diagnosis iff

$$SD \cup OBS \cup \{\neg ab(C) | C \in COMP \setminus \Delta\} \cup \{ab(C) | C \in \Delta\}$$

is consistent. The predicate $ab(C)$ is used to indicate that a component C is behaving abnormally. A correctly behaving component C is therefore described by $\neg ab(C)$.

We illustrate this diagnosis definition using the full adder example from figure 1. The behavior of the full adder is given by the behavior of its subcomponents (in this case, two inverters, two AND gates, one OR gate) and their interconnections. The correct behavior of an *AND* gate C can be formalized as follows:

$and(C) \Rightarrow (\neg ab(C) \rightarrow out(C) = andBehav(in_1(C), in_2(C)))$
where $andBehav(true, true)$ = $true$, $andBehav(false, X)$ = $false$,
$andBehav(X, false) = false$.

Similar rules can be given for the XOR and the OR gate. The connection part of the system description is specified by:

$in_1(X2) = c_n$. $in_2(X2) = out(X1)$. $out(X2) = q_{n+1}$
$in_1(X1) = a_n$. $in_2(X1) = b_n$.
$in_1(A1) = b_n$. $in_2(A1) = a_n$. $out(A1) = in_2(O1)$.
$in_1(A2) = a_n$. $in_2(A2) = out(X1)$. $out(A1) = in_1(O1)$.
$out(O1) = c_{n+1}$.

Now assume the input values a_n = $false, b_n$ = $true, c_n$ = $true$ have been used and the output values q_{n+1} = $true, c_{n+1}$ = $true$ have been observed. Assuming the correct behavior of all components, i.e., $\neg ab(A1), \neg ab(A2), \neg ab(O1), \neg ab(X1), \neg ab(X2)$, the system description contradicts the observations. A model-based diagnosis algorithm such as the Hitting Set algorithm described in [Rei87], computes only one single diagnosis $ab(X2)$ (i.e., the failure of $X2$ alone would explain the observations). All other diagnoses consist of more than one faulty component.

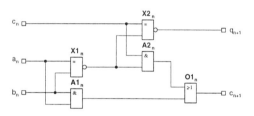

Fig. 1. A schematic of a full adder

In summary, the model-based diagnosis approach provides: (1) a precise diagnosis definition, (2) a clear separation between the knowledge about a system and the diagnosis algorithms, (3) a component oriented view that allows reuse of knowledge and easy maintenance, and (4) the use of multiple models. The only prerequisite for using model-based diagnosis is the existence of a component oriented system model. The diagnosis time depends on the number of system components. To give an upper bound for the diagnosis time we assume a model expressible in propositional logic where the size is a multiple of the number of components. The consistency of such a model can be checked in $o(|COMP|)$ time [Min88]. Searching only for single diagnoses results in at most $|COMP|$ consistency checks. Therefore the maximum time for computing all single diagnoses is of order $o(|COMP|^2)$. However, the expected runtime for diagnosis does not reach this bound in practice. The best diagnosis algorithms available today can diagnose systems with up to 10,000 components in less than one minute [FN97,SW97].

3 The diagnosis agent architecture

In order to provide easy access to an agent's capabilities the use of the WWW is recommended because of the provided functionality and the widespread use. We therefore introduce a global architecture for mobile diagnosis agents based on the WWW. The agent, that implements the diagnosis functionality, is executed on a dedicated server. The interface agent implemented as a Java applet embedded in a HTML page is executed on the client side. TCP/IP sockets are used for communication between them. Although the functionality of an applet is restricted for security purposes, socket communication between the applet and the server distributing the applet is allowed. Figure 2 gives an overview of the global architecture.

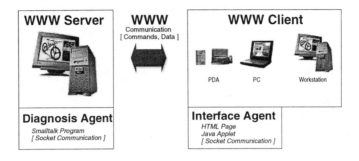

Fig. 2. The MODIS WWW architecture

As a consequence of the definition of model-based diagnosis, the diagnosis agent must include at least a diagnosis engine and a theorem prover. For interaction with other agents we additionally need a communication module and an user interface. Figure 3 (a) shows the agent's internal architecture.

The communication module controls the information exchange between agents, i.e., interface or other diagnosis agents. Its task is to parse the incoming messages and call the associated (diagnosis specific) functions together with their parameters. The result is coded using the communication language and sent back to the sender of the original message.

The diagnosis engine implements the whole diagnosis functionality such as diagnosis, probing (measurement selection), among others. Information about the behavior of the currently used diagnosis system and the corresponding observations are located within the theorem prover which is used for checking consistency while computing all diagnoses. Diagnosis parameters such as the maximum number of faulty components or the maximum number of diagnoses to be computed together with different diagnosis algorithms are part of the engine.

From the object-oriented point of view we have to deal with objects rather than modules. We illustrate the agent design by specifying some objects and their relations necessary for implementing the diagnosis engine. The most important entity for diagnosis is a system storing components, connections, observations, results and parame-

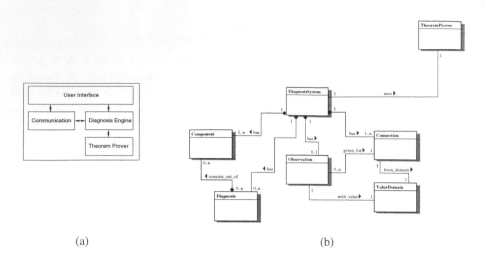

(a)   (b)

Fig. 3. Diagnosis agent architecture (a) and diagnosis system design (b)

ters. Components of a specific type have an unique behavior. Values associated with connections must be from a given domain. Additionally, for diagnosis we need theorem provers which are specific for a given system. From this verbal description we can directly derive the objects DiagnosisSystem, Component, Connection, Observation, ValueDomain, Diagnosis, TheoremProver with the relationships given in figure 3 (b).

Because DiagnosisSystem has access to all diagnosis related informations we assume that it implements the whole diagnosis functionality, e.g., diagnosis or measurement selection. So a diagnosis engine can be implemented using DiagnosisSystems.

4 The communication and diagnosis requirements

One of the subtasks of MODIS has been the development of a language for the interchange of diagnosis informations between agents. In our view agents are programs implementing functionality accessible from the outside. The state of the agent is given by its internal data and is not necessarily accessible from outside. This approach is similar to [ESP98] and has the advantage of separating communication properties from their implementation. However, our approach is simpler but adequate for our purposes.

The communication language is divided in two parts. The first part contains those constructs necessary for communication functionality, i.e., controlling and querying the agent. The second part called *system description language (SDL)* is used to describe diagnosis systems and observations. Therefore the application domain of SDL is not restricted to agent communication. Instead it can also be used for storing diagnosis informations, e.g., building a model library. SDL is concisely described in section 5.

We assume that an agent $A_i = (i, F_i)$ has an unique identifier i and a set of functions F_i accessible by other agents. Communication is done by sending a function expression to an agent using the underlying communication protocol, i.e., sockets. It is assumed, that every time a function expression $f(a_1, \ldots, a_n), f \in F_k$ of arity n is sent to an agent A_k, a reply from A_k is sent back. This reply contains the message $f(a_1, \ldots, a_n) = Result$ where $Result$ represents the resulting value of the function call. For example $loadSD(\ldots)$ will return $loadSD(\ldots) = done$ when the given system description has been successfully loaded. This hand-shake mechanism ensures proper synchronization between sender and receiver.

Using the above formalization a basic diagnosis agent D_i is defined by the tuple $(i, \{startDiagnosis, startProbing, loadSD, loadOBS, getSD, getOBS,$ $getDiagnoses, getProbes, resetOBS, reply\})$. The functions have the following meaning:

Diagnosis At least diagnosis and measurement selection should be provided.

   startDiagnosis() starts the diagnosis process using previously received informations about the system description and observations.

   startProbing() runs the measurement selection routine. The system description, observations, and the derivable diagnoses must be available.

Access The system description, observations, diagnosis and measurement selection result must be accessible from the outside.

   loadSD(SystemDescription) loads the system description into the diagnosis agent.

   loadOBS(Observations) loads the observations into the diagnosis agent.

   getSD() asks for the currently used system description.

   getOBS() asks for the currently used observations.

   getDiagnoses() asks for the last computed diagnoses.

   getProbes() asks for the result of the measurement selection function.

   resetOBS() removes the observations together with the diagnoses and measurement selection result.

A more general diagnosis agent can provide functions for setting and removing a component focus, or for computing informations based on system descriptions, e.g., a failure-mode and effects analysis (FMEA). In the next section the second part of the communication language is introduced that is used for exchanging diagnosis informations.

Since our diagnosis agent is intended to be used for educational purposes, i.e., teaching students model-based diagnosis, we add the functions for accessing predefined system descriptions and observations to the diagnosis agent:

allSDid() returns a list of names of system descriptions stored by the diagnosis agent.

getSD(SDIdentifier) returns a system description with the name *SDIdentifier*.

allOBS(SDIdentifier) returns a list of names of observations for the given system description named by *SDIdentifier*.

getOBS(SDIdentifier, OBSIdentifier) returns the observations named *OBSIdentifier* for a system *SDIdentifier*.

5 The system description language

The system description language (SDL) (see [Pic98] for the full syntax and semantics of SDL) is used for describing the properties of components and their interconnection.

It distinguishes between atomic and hierarchic components. In both cases it is necessary to specify the associated input, output or bidirectional ports which act as interfaces to other components. For atomic components the associated behavior modes are defined, whereas the behavior of hierarchical components is derived by their subcomponents and the connections between them.

<AtomicComponent> →
  component <ComponentName> <PortDefinitions> [<VariableDefinitions>]
  [<AxiomDefinitions>] <BehaviorDefinitions> **end component**
<HierarchicComponent> →
  component <ComponentName> <PortDefinitions> <SubcomponentDefinitions>
  <ConnectionDefinitions> **end component**

Each port of a component is associated with a type, determining the possible values that can be assigned to it. Predefined types are `integer, float, string, character` and `boolean`. Own types can easily be added by enumerating the corresponding elements or using the set-theoretic union of already existing types or arbitrary sets of elements.

Variables are used to allow for compact behavior definitions and are defined by associating a variable name with a set of elements. SDL only allows finite sets to be associated with variables.

Axioms are formulated in terms of predicate expressions, defining relations between objects for a given component. During the evaluation of a behavior mode, decisions based on components' axioms can be made. Axioms have a dynamic character and can be added or removed during mode evaluation thus having direct impact on a components behavior. Like the introduction of variables, the use of axioms allows for more efficient and compact behavior definitions.

The heart of each component is its behavioral model, stating how it behaves under certain circumstances. Each component in a system knows one or more different behavioral modes based on different assumptions. A behavior mode definition for an atomic component is given by the following syntax:

<BehaviorDefinition> →
  [**default**] **behavior** <BehaviorName> [**prob**:<Probability>] [**cost**:<RepairCosts>]
  [**action**:<RepairAction>] <RuleList> **end behavior**

For each behavior mode we know a probability stating how probable the occurence of this behavior is, an associated cost value telling how expensive the necessary repairs are and an action description stating which steps have to be taken in order to repair the component. The <RuleList>-production is used to formulate the causal dependencies and legal inference steps within the component.

The behavior of hierarchic components is derived from the behavior of their subcomponents and is based on the connections between them. For this reason a mechanism is needed to instantiate components within the current component definition, which is provided by the <SubcomponentsDefinition>-production. Names are associated with already defined components and can then be used to specify the connections between them.

<ConnectionDefinition> →
  <ConnectionName> [**observable**:{**true** | **false**}]
  [**cost**: <ObservationCost>] <ComponentPort> -> <ComponentPortList>.

Connections may or may not be observable and can have an associated cost value determining how expensive an observation would be. Each connection has its origin at the port of one component (<ComponentPort>) and can branch to many others (<ComponentPortList>).

As mentioned above the causal dependencies and legal inference steps within a component are formulated using rules. Each rule consists of a prerequisite and a consequence. Intuitively, if the prerequisite is true then the consequence can be infered. SDL uses two inference operators to allow for compact rule definitions.

<Rule> → [<Sentence>] =: <Sentence>. | <Sentence> :=: <Sentence>.

In the first case the right-hand side of the rule can be infered iff the left-hand side is true. In the second case inference is possible in both directions. If in the first case the left-hand side is empty the consequence can always be infered. Rules are defined in terms of sentences.

SDL distinguishes four types of sentences: *predicate sentences, conjunctive sentences, implication sentences* and *quantified sentences*.

<PredicateSentence> → <PredicateName>(<ArgumentList>)
<ConjunctiveSentence> → <Sentence>, <Sentence>
<ImplicationSentence> → <Sentence> -: <Sentence>
<QuantifiedSentence> → {**forall** | **exists**} <VariableList> : <Sentence>

<ArgumentList> is a possibly empty list of constants, ports, function expressions and variables.

When used in the prerequisite of a rule, we have to assign truth values to sentences, determining wheter the consequence can be infered. This is done as follows. viewing a predicate as a set of tuples, we define the truth value for a predicate sentence as *true* if and only if the tuple of objects determined by the arguments are an element of this set, otherwise as *false*. Given a conjunction the truth value is defined as true if and only if the truth value of every conjunct is true. For an implication s_1 -:s_2 the truth value is defined as follows: If s_1 is true then the truthvalue of the implication is true if and only if s_2 is true. If s_1 is false then the implication is true, regardless of the truthvalue of s_2 (ex falso quod libet). The truth-value for an universally quantified sentence is defined as true if and only if for all possible variable assignments the truth value of the sentence following the colon is true. Given an existentially quantified sentence the truth value is defined as true if and only if for at least one possible variable assignment the truth value of the sentence following the colon is true.

In order to provide a diagnosis tool with reference values for certain ports, we need a way to specify observations. We do this by specifying the observations and mapping them to a specific diagnosis component.

<ObservationDefinition> → **observation** <ObservationName> <PortAssignments>.
<PortAssignment> → <PortName> = <Value>.
<DiagnoseDefinition> → **diagnose**(<ComponentName>,<ObservationName>)

```
component XorGate
    input in1, in2 : bool.
    output out: bool.
    default behavior
        Neq(in1, in2) :=: Eq(out, true).
        Eq(in1, in2) :=: Eq(out, false).
    end behavior
end component
```

```
component AndGate
    input in1, in2 : bool.
    output out: bool.
    var Port: {in1, in2}.
    default behavior
        forall Port: Eq(Port, true) :=: Eq(out, true).
        exists Port: Eq(Port, false) =: Eq(out, false).
    end behavior
end component
```

```
component OrGate
    input in1, in2 : bool.
    output out: bool.
    var Port: {in1, in2}.
    default behavior
        exists Port: Eq(Port, true) =: Eq(out, true).
        forall Port: Eq(Port, false) :=: Eq(out, false).
    end behavior
end component
```

```
component FullAdder
    input a,b,c : bool.
    output q,z: bool.
    subcomponent
        and1, and2: AndGate.
        or1: OrGate.
        xor1, xor2: XorGate.
    connection
        s1: a -> xor1(in1), and1(in2).
        s2: b -> xor1(in2), and1(in1).
        s3: c -> and2(in1), xor2(in1).
        s4: xor1(out) -> and2(in2), xor2(in2).
        s5: and1(out) ->or1(in2).
        s6: and2(out) ->or1(in1).
        s7: xor2(out) -> q.
        s8: or1(out) -> z.
    end component
```

Fig. 4. The SDL program implementing a full adder

Finally, we illustrate the capabilities of SDL using the full adder example depicted in figure 1. The full adder uses three types of boolean gates: a XorGate, an AndGate, and an OrGate. The SDL program describing the behavior of the three gates and the full adder consisting out of the gates is given in figure 4.

6 Conclusion

In this paper we have presented the architecture and design of a diagnosis agent accessible using a standard WWW browser. For communication purposes we have introduced a communication language consisting out of two parts. One for sending queries and requests to the agent and one for describing diagnosis systems and observations. Currently, the system description language and the diagnosis agent and parts of the WWW interface have been implemented.

In summary the paper provides: (1) a general architecture for diagnosis agents, (2) a description of a WWW interface for a diagnosis agent, and (3) a communication language for controlling and querying an agent, and exchanging and storing diagnosis informations. Future research in this domain include the improvement of diagnosis algorithms, their implementation within the introduced framework, and the design of a framework for a graphical user interface especially for diagnosis.

References

[dKB98] Kees de Koning and Bert Bredeweg. Using GDE in Educational Systems. In *Proceedings of the European Conference on Artificial Intelligence (ECAI)*, pages 279–283, Brighton, UK, August 1998.

[dKW87] Johan de Kleer and Brian C. Williams. Diagnosing multiple faults. *Artificial Intelligence*, 32(1):97–130, 1987.

[ESP98] Thomas Eiter, V.S. Subrahmanian, and George Pick. Heterogeneous active agents. Technical report, Institut für Informatik, Justus-Liebig-Universität Gießen, Germany, IFIG Research Report 9802, March 1998.

[FLM97] Tim Finin, Yannis Labrou, and James Mayfield. KQML as an Agent Communication Language. In Jeffrey M. Bradshaw, editor, *Software Agents*, pages 291–317. AAAI Press / The MIT Press, 1997.

[FN97] Peter Fröhlich and Wolfgang Nejdl. A Static Model-Based Engine for Model-Based Reasoning. In *Proceedings 15th International Joint Conf. on Artificial Intelligence*, Nagoya, Japan, August 1997.

[FSW96] Gerhard Friedrich, Markus Stumptner, and Franz Wotawa. Model-based diagnosis of hardware designs. In *Proceedings of the European Conference on Artificial Intelligence (ECAI)*, Budapest, August 1996.

[HLW98] Florentin Heck, Thomas Laengle, and Heinz Woern. A Multi-Agent Based Monitoring and Diagnosis System for Industrial Components. In *Proceedings of the Ninth International Workshop on Principles of Diagnosis*, pages 63–69, Cape Cod, Massachusetts, USA, May 1998.

[Min88] Michel Minoux. LTUR: A Simplified Linear-time Unit Resolution Algorithm for Horn Formulae and Computer Implementation. *Information Processing Letters*, 29:1–12, 1988.

[MS96] A. Malik and P. Struss. Diagnosis of dynamic systems does not necessarily require simulation. In *Proceedings of the Seventh International Workshop on Principles of Diagnosis*, 1996.

[Pic98] Christian Piccardi. AD^{2}L An Abstract Modelling Language for Diagnosis Systems. Master's thesis, TU Vienna, 1998.

[PS97] Chris Price and Neal Snooke. Challenges for qualitative electrical reasoning in circuit simulation. In *Proceedings of the 11th International Workshop on Qualitative Reasoning*, 1997.

[Rei87] Raymond Reiter. A theory of diagnosis from first principles. *Artificial Intelligence*, 32(1):57–95, 1987.

[SDP96] M. Schroeder, C.V. Damasio, and L.M. Pereira. REVISE Report: An Architecture for a Diagnosis Agent. In *ECAI-96 Workshop on Integrating Non-Monotonicity into Automated Reasoning Systems*, Budapest, Hungaria, August 1996.

[SW97] Markus Stumptner and Franz Wotawa. Diagnosing tree-structured systems. In *Proceedings 15th International Joint Conf. on Artificial Intelligence*, Nagoya, Japan, 1997.

[SW98] Markus Stumptner and Franz Wotawa. VHDLDIAG+:Value-level Diagnosis of VHDL Programs. In *Proceedings of the Ninth International Workshop on Principles of Diagnosis*, Cape Cod, May 1998.

Information Broker Agents in Intelligent Websites

Catholijn M. Jonker, Jan Treur

Vrije Universiteit Amsterdam, Department of Artificial Intelligence
De Boelelaan 1081a, 1081 HV Amsterdam, The Netherlands
URL: http://www.cs.vu.nl/~{jonker,treur}. Email: {jonker,treur}@cs.vu.nl

Abstract. In this paper a generic information broker agent for intelligent Websites is introduced. The agent architecture has been designed using the compositional development method for multi-agent systems DESIRE. The use of the architecture is illustrated in an Electronic Commerce application for a department store.

1 Introduction

Most current business Websites are mainly based on navigation through the available information across hyperlinks. A closer analysis of such conventional Websites reveals some of their shortcomings. For example, customer relations experts may be disappointed about the *unpersonal treatment* of customers at the Website; customers are wandering around anonymously in an unpersonal virtual environment and do not feel supported by anyone. It is as if customers are visiting the physical environment of a shop (that has been virtualised), without any serving personnel.

Marketing experts may also not be satisfied by the Website; they may disappointed in the lack of facilities to support *one-to-one marketing*. In a conventional Website only a limited number of possibilities are provided to announce new products and special offers in such a manner that all relevant customers learn about them. Moreover, often Websites do not acquire information on the amounts of articles sold (sales statistics). It is possible to build in monitoring facilities with respect to the amount of products sold over time, but also the number of times a request is put forward on a product (demand statistics). If for some articles a decreasing trend is observed, then the Website could even warn employees so that these trends can be taken into account in the marketing strategy. If on these aspects a more active role would be taken by the Website, the marketing qualities could be improved.

The analysis from the two perspectives (marketing and customer relations) suggests that Websites should become more active and personalised, just as in the traditional case where contacts were based on humans. Intelligent agents provide the possibility to reflect at least a number of aspects of the traditional situation in a simulated form, and, in addition, enables to use new opportunities for, e.g., one-to-one marketing, integrated in the Website.

In this paper it is shown how a generic broker agent architecture can be exploited to design an intelligent Website for a department store. In Section 2 the application domain is discussed; two types of information agents participating in the application are distinguished. In Section 3 their characteristics and required properties are discussed. In Section 4 the generic broker agent architecture is described and applied to obtain the internal structure of the agents involved in the application.

2 Application: an Intelligent Website for a Department Store

The application addresses the design of an active, intelligent Website for a chain of department stores. The system should support customers that order articles via the Internet. Each of the department stores sells articles according to departments such as car accessoires, audio and video, computer hardware and software, food, clothing, books and magazines, music, household goods, and so on. Each of these departments has autonomy to a large extent; the departments consider themselves small shops (as part of a larger market). This suggests a multi-agent perspective based on the separate departments and the customers. Four types of agents are distinguished:

- *customers* (human agents)
- *Personal Assistant agents* (an own software agent for each user)
- *Department Agents* (software agents within the department store's Website)
- *employees* (human agents)

A Personal Assistant agent serves as an interface agent for the customer. As soon as a customer visits the Website, this agent is offered and instantiated to the customer. The Personal Assistant is involved in communication to both its own user and all Website agents. From the user it can receive information about his or her profile, and it can provide him or her with information assumed interesting. Moreover, it can receive information from any of the Website agents, and it can ask them for specific information. The Website agents communicate not only with all Personal Assistants, but also with each other and with employees. The customer only communicates with his or her own Personal Assistant.

3 Requirements for the Department Agents

The departments should relate to customers like small shops with personal relationships to customers. The idea is that customers know at least somebody (a Department Agent) related to a department, as a representative of the department and, moreover, this agente knows specific information on the customer. Viewed from outside the basic agent behaviours *autonomy, responsiveness, pro-activeness* and *social behaviour* such as discussed, for example in [10] provide a means to characterise the agents (see Table 3). In addition the interaction characteristics as shown in Tables 1 and 2 have been specified.

Interaction with the world	Department Agent
observation passive	- its own part of the Website - product information - presence of customers/PAs visiting the Website
observation active	- economic information - products and prices of competitors - focussing on what a specific customer or PA does - search for new products on the market
performing actions	- making modifications in the Website (e.g., change prices) - showing Web-pages to a customer and PA - creating (personal or general) special offers - modification of assortment

Table 1. World interaction characteristics for a Department Agent

Communication	Department Agent
incoming	*from PA*: - request for information - request to buy an article - paying information - customer profile information - customer privacy constraints *from Employee*: - requests for information on figures of sold articles - new product information - proposals for special offers and price changes - confirmation of proposed marketing actions - confirmation of proposed assortment modifications - proposals for marketing actions - proposals for assortment modifications *from other DA*: - info on assortment scopes - customer info
outgoing	*to PA*: - asking whether DA can help - providing information on products - providing information on special offers - special (personal or general) offers *to Employee*: - figures of articles sold (sales statistics) - analyses of sales statistics - numbers of requests for articles (demand statistics) - proposals for special offers - proposals for assortment modifications *to other DA*: - info on assortment scopes - customer info

Table 2. Communication characteristics for a Department Agent

The following requirements have been imposed on the Department Agents:
- *personal approach; informed behaviour with respect to customer*
  In the Website each department shall be represented by an agent with a name and face, and who knows the customer and his or her characteristics, and remembers what this customer bought previous times.
- *being helpful*
  Customers entering some area of the Website shall be contacted by the agent of the department related to this area, and asked whether he or she wants some help. If the customer explicitly indicates that he or she only wants to look around without getting help, the customer shall be left alone. Otherwise, the agent takes responsibility to serve this customer until the customer has no wishes anymore that relate to the agent's department. The conventional Website can be used by the Department Agents to point at some of the articles that are relevant (according to their dialogue) to the customer.
- *refer customers to appropriate colleague Department Agents*
  A customer which is served at a department and was finished at that department can only be left alone if he or she has explicitly indicated to have no further wishes within the context of the entire department store. Otherwise the agent shall find

out in which other department the customer may have an interest and the customer shall be referred to the agent representing this other department.
- *be able to provide product and special offer information*
  For example, if a client communicates a need, then a product is offered fulfilling this need (strictly or approximately), and, if available a special offer.
- *dedicated announcement*
  New products and special offers shall be announced as soon as available to all relevant (on the basis of their profiles) customers, not only if they initiate a contact with the store, but they also shall be contacted by the store in case they do not contact the store

Basic types of behaviour	Department Agent
Autonomy	- functions autonomously, especially when no employees are available (e.g., at night)
Responsiveness	- responds to requests from Personal Assistants - responds to input from Employees - triggers on decreasing trends in selling and demands
Pro-activeness	- takes initiative to contact Personal Assistants - takes initiative to propose special offers to customers - creates and initiates proposals for marketing actions and assortment modifications
Social behaviour	- cooperation with Employees, Personal Assistants, and other Department Agents

Table 3. Basic types of behaviour of a Department Agent

- *analyses for marketing*
  The Department Agents shall monitor the amounts of articles sold (sales stastitics), communicate them to Employees (e.g., every week) and warn if substantially decreasing trends are observed. For example, if the figures of an article sold decrease during a period of 3 weeks, thenmarketing actions or assortment modifications shall be proposed.
- *actions for marketing*
  Each Department Agent shall maintain the (history of) the transactions of each of ·the customers within its department, and shall be willing to perform one to one marketing to relevant customers, if requested. The Employees shall be able to communicate to the relevant Department Agents that they have to perform a marketing campaign. The agent shall propose marketing actions to Employees.
- *privacy*
  No profile is maintained without explicit agreement with the customer. The customer has access to the maintained profile.

3.2 Characteristics and Requirements for the Personal Assistants

For the Personal Assistants the interaction characteristics are given in Table 4, and their basic types of behaviour in Table 5. The following requirements can be imposed on the Personal Assistants:
- *support communication on behalf of the customer*
  Each customer shall be supported by his or her own Personal Assistant agent, who serves as an interface for the communication with the Department Agents.

- *only provide information within scope of interest of customer*
  A customer shall not be bothered by information that is not within his or her scope of interest. A special offer that has been communicated by a Department Agent leads to a proposal to the Customer, if it fits in the profile, and at the moment when the Customer wants such information
- *sensitive profiling*
  Customers are relevant for a special offer if they have bought a related article in the past, or if the offer fits in their profile as known to the Personal Assistant.
- *providing customer information for Department Agents*
  every week the relevant parts of the profile of the Customer is communicated to the Department Agent, if the Customer agrees.
- *privacy*
  The Personal Assistant shall protect and respect the desired privacy of the customer. Only parts of the profile information agreed upon are communicated.

Interaction characteristics	Personal Assistant
A. Interaction with the world	
observation passive observation active	- notice changes and special offers at the Website - look at Website for articles within the customer needs
performing actions	
B. Communication with other agents	
incoming	*from Department Agent*: - product info - special (personal and general) offers *from Customer*: - customer needs and preferences - agreement to buy - privacy constraints
outgoing	*to Department Agent*: - customer needs - payment information - profile information *to Customer*: - product information - special offers

Table 4. Interaction characteristics for the Personal Assistant

Basic types of behaviour	Personal Assistant
Autonomy	autonomous in dealing with DAs on behalf of customer
Responsiveness	responsive on needs communicated by customer
Pro-activeness	initiative to find and present special offers to customer
Social behaviour	with customer and DAs

Table 5. Basic types of behaviour for the Personal Assistant

4 The Internal Design of the Information Broker Agents

The agents in the application presented in the previous sections have been designed on the basis of a generic model for a broker agent. The process of brokering as often occurs as a mediating process in electronic commerce involves a number of activities. For example, responding to customer requests for products with certain properties, maintaining information on customers, building customer profiles on the basis of such customer information, maintaining information on products, maintaining provider profiles, matching customer requests and product information (in a strict or soft manner), searching for information on the WWW, and responding to new offers of products by informing customers for whom these offers fit their profile. In this section a generic broker agent architecture is presented that supports such activities. This generic model has been used as a basis for both the Department Agents and the Personal Assistant agents.

4.1 A Generic Broker Agent Architecture

For the design of the generic broker agent the following main aspects are considered: process composition, knowledge composition, and relations between knowledge and process composition, as discussed in Section 2. A compositional generic agent model (introduced in [2]), supporting the weak agency notion (cf. [10]) is used; see Fig. 1. At the highest abstraction level within an agent, a number of processes can be distinguished that support interaction with the other agents. First, a process that manages communication with other agents, modelled by the component agent interaction management in Fig. 1. This component analyses incoming information and determines which other processes within the agent need the communicated information. Moreover, outgoing communication is prepared. Next, the agent needs to maintain information on the other agents with which it co-operates: maintenance of agent information. The component maintenance of world information is included to store the world information (e.g., information on attributes of products). The process own process control defines different characteristics of the agent and determines foci for behaviour. The component world interaction management is included to model interaction with the world (with the World Wide Web world, in the example application): initiating observations and receiving observation results.

The agent processes discussed above are generic agent processes. Many agents perform these processes. In addition, often agent-specific processes are needed: to perform tasks specific to one agent, for example directly related to a specific domain of application. The broker agent may have to determine proposals for other agents. In this process, information on available products (communicated by information providing agents and kept in the component maintenance of world information), and about the scopes of interests of agents (kept in the component maintenance of agent information), is combined to determine which agents might be interested in which products. Fig. 1 depicts how the broker agent is composed of its components.

Part of the exchange of information within the generic broker agent model can be described as follows. The broker agent needs input about scopes of interests put forward by agents and information about attributes of available products that are communicated by information providing agents. It produces output for other agents about proposed products and the attributes of these products. Moreover, it produces output for information providers about interests. In the information types that express communication information, the subject information of the communication and the

436

agent from or to whom the communication is directed are expressed. This means that communication information consists of statements about the subject statements that are communicated.

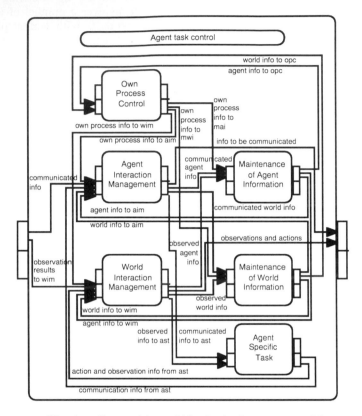

Fig. 1. Composition within the broker agent model

Within the broker agent, the component own process control uses as input belief info, i.e., information on the world and other agents, and generates focus information: to focus on a scope of interest to be given a preferential treatment, i.e., pro-active behaviour will be shown with respect to this focus. The component agent interaction management has the same input information as the agent (incoming communication), extended with belief info and focus info. The output generated includes part of the output for the agent as a whole (outgoing communication), extended with maintenance info (information on the world and other agents that is to be stored within the agent), which is used to prepare the storage of communicated world and agent information.

Information on attributes of products is stored in the component maintenance of world information. In the same manner, the beliefs of the agent with respect to other agents' profiles (provider attribute info and interests) are stored in maintenance of agent information. The agent specific task uses information on product attributes and agent interests as input to generate proposals as output. For reasons of space limitation the generic and domain-specific information types within the agent model are not presented; for more details; see [5].

4.2 The Department Agent: Internal Design

The broker agent architecture provides an appropriate means to establish the internal design of the two types of agents involved.

For the Department Agent, the internal storage and updating of information on the world and on other agents (the beliefs of the agent) is performed by the two components maintenance of world information and maintenance of agent information. In Table 6 it is specified which types of information are used in these components. Profile information on customers is obtained from Personal Assistants, and maintained with the customer's permission. Also identified behaviour instances of the Personal Assistants can give input to the profile. Profile information can be abstracted from specific demands, using existing datamining techniques.

Maintenance of Information	Department Agent
world information	- info on products within the DA's assortment - info on special offers
agent information	- info on customer profiles - info on customer privacy constraints - info on MA's needs for figures - info on customer preferences in communication - info on which products belong to which other DA's assortments - info on providers of products

Table 6. Maintenance information for the Department Agent

The component agent interaction management identifies the information in incoming communication and generates outgoing communication on the basis of internal information. For example, if a Personal Assistant agent communicates its interests, then this information is identified as new agent interest information that is believed and has to be stored, so that it can be recalled later.

In the component agent specific task specific knowledge is used such as, for example:

- if the selling numbers for an article decrease for 3 weeks, then make a special offer with lower price, taking into account the right season
- if a customer asks for a particular cheap product, and there is a special offer, then this is proposed
- if an article is not sold enough over a longer period, then take it out of the assortment

Within this component non-strict (or soft) matching techniques can be employed to relate demands and offers.

4.3 The Personal Assistant: Internal Design

In this section some of the components of the Personal Assistant are briefly discussed. For the Personal Assistant, as for the Department Agent, the internal storage and updating of information on the world and on other agents is performed by the two components maintenance of world information and maintenance of agent information. In Table 7 it is specified which types of information are used in these components.

Maintenance of Information	Personal Assistant
world information	- product information - special offers
agent information	- customer needs and profile - customer privacy constraints - offers personal to the customer - DAs assortment scopes

Table 7. Maintenance information for a Personal Assistant

As in the previous section, the component agent interaction management identifies the information in incoming communication and generates outgoing communication on the basis of internal information. For example, if a Department Agent communicates a special offer, then this information is identified as new agent information that is believed and has to be stored, so that it can be recalled later. Moreover, in the same communication process, information about the product to which the special offer refers can be included; this is identified and stored as world information.

6 Discussion

In this paper a multi-agent architecture for intelligent Websites is proposed, based on information broker agents. A Website, supported by this architecture has a more personal look and feel than the usual Websites. Within the architecture, also negotation facilities (e.g., as in [8]) can be incorporated. The broker agent architecture has also been applied within a project on virtual market environments in Electronic Commerce, in co-operation with the Internet application company Crisp.

Applications of broker agents (addressed in, e.g., [3], [4], [6], [7], [8], [9]), often are not implemented in a principled manner: without an explicit design at a conceptual level, and without support maintenance. The broker agent architecture introduced here was designed and implemented in a principled manner, using the compositional development method for multi-agent systems DESIRE [1]. Due to its compositional structure it supports reuse and maintenance; a flexible, easily adaptable architecture results. Moreover, within the agent model facilities can be integrated that provide automated support of the agent's own maintenance [5]. Therefore, the agent is not only easily adaptable, but, if in such a manner adaptation is automated, it shows adaptive behaviour to meet new requirements (either in reaction to communication with a maintenance agent, or fully autonomous).

Acknowledgements

Mehrzad Kharami and Pascal van Eck (Vrije Universiteit Amsterdam) supported experiments with some variants of the broker agent model. Working on a design for the department store application with various employees of the software company CMG provided constructive feedback on the architecture introduced.

References

1. Brazier, F.M.T., Dunin-Keplicz, B., Jennings, N.R., and Treur, J., Formal specification of Multi-Agent Systems: a real-world case. In: V. Lesser (ed.), *Proceedings of the First International Conference on Multi-Agent Systems, ICMAS'95*, MIT Press, Cambridge, MA, 1995, pp. 25-32. Extended version in: *International Journal of Cooperative Information Systems*, M. Huhns, M. Singh, (eds.), special issue on Formal Methods in Cooperative Information Systems: Multi-Agent Systems, vol. 6, 1997, pp. 67-94.

2. Brazier, F.M.T., Jonker, C.M., and Treur, J., Formalisation of a cooperation model based on joint intentions. In: J.P. Müller, M.J. Wooldridge, N.R. Jennings (eds.), *Intelligent Agents III (Proceedings of the Third International Workshop on Agent Theories, Architectures and Languages, ATAL'96)*, Lecture Notes in AI, volume 1193, Springer Verlag, 1997, pp. 141-155.

3. Chavez, A., and Maes, P., Kasbah: An Agent Marketplace for Buying and Selling goods. In: *Proceedings of the First International Conference on the Practical Application of Intelligent Agents and Multi-Agent Technology, PAAM'96*, The Practical Application Company Ltd, Blackpool, 1996, pp. 75-90.

4. Chavez, A., Dreilinger, D., Gutman, R., and Maes, P., A Real-Life Experiment in Creating an Agent Market Place. In: *Proceedings of the Second International Conference on the Practical Application of Intelligent Agents and Multi-Agent Technology, PAAM'97*, The Practical Application Company Ltd, Blackpool, 1997, pp. 159-178.

5. Jonker, C.M., and Treur, J., *Compositional Design and Maintenance of Broker Agents*. Technical Report, Vrije Universiteit Amsterdam, Department of Mathematics and Computer Science, 1998.

6. Kuokka, D., and Harada, L., On Using KQML for Matchmaking. In: V. Lesser (ed.), *Proceedings of the First International Conference on Multi-Agent Systems, ICMAS'95*, MIT Press, Cambridge, MA, 1995, pp. 239-245.

7. Martin, D., Moran, D., Oohama, H., and Cheyer, A., Information Brokering in an Agent Architecture. In: *Proceedings of the Second International Conference on the Practical Application of Intelligent Agents and Multi-Agent Technology, PAAM'97*, The Practical Application Company Ltd, Blackpool, 1997, pp. 467-486.

8. Sandholm, T., and Lesser, V., Issues in Automated Negotiation and Electronic Commerce: Extending the Contract Network. In: V. Lesser (ed.), *Proceedings of the First International Conference on Multi-Agent Systems, ICMAS'95*, MIT Press, Cambridge, MA, 1995, pp. 328-335.

9. Tsvetovatyy, M., and Gini, M., Toward a Virtual Marketplace: Architectures and Strategies. In: *Proceedings of the First International Conference on the Practical Application of Intelligent Agents and Multi-Agent Technology, PAAM'96*, The Practical Application Company Ltd, Blackpool, 1996, pp. 597-613.

10. Wooldridge, M., and Jennings, N.R., Agent theories, architectures, and languages: a survey. In: [11], pp. 1-39.

11. Wooldridge, M., and Jennings, N.R. (eds.), *Intelligent Agents, Proceedings of the First International Workshop on Agent Theories, Architectures and Languages, ATAL'94*, Lecture Notes in AI, vol. 890, Springer Verlag, 1995.

Learner-Model Approach to Multi-agent Intelligent Distance Learning System for Program Testing

Tatiana Gavrilova, Alexander Voinov, and Irina Lescheva

Institute for High Performance Computing and Databases,
194291, P.O. Box 71, St. Petersburg, Russia,
gavr@limtu.spb.su,
WWW home page: http://www.csa.ru/Inst/gorb_dep/artific

Abstract. A project of an intelligent multi-agent system for distance learning, which is currently being developed in the Institute for High Performance Computing and Databases, is described. The system's adaptability, i.e. its ability to adjust itself to the personal characteristics and preferences of its user, is based on User Model, which is initiated according to the results of prior user testing and is adjusted dynamically during the user's session. As an example of an application domain the program testing is chosen - the domain, which is one of the weakest points in the software engineering practice in Russia. The described system is intended for using via Internet by common WWW-browsers.

1 Introduction

An idea to apply the technology of multi-agent systems to the computer aided learning (CAL) is currently being developed by many authors in many research groups (see, e.g., [17]). The theoretical foundations of the design and architecture of distributed learning systems are, however, on their early stage of development.

It is generally agreed (see e.g. [7] [12]), that at least three important issues should constitute the basis for such systems: they are expected to be distributed, intelligent and adaptive.

The adaptability of DL systems, in all meanings of this concept, is announced in most of currently developed projects, but usually weakly implemented in practice. As emphasized in [17], adaptation and learning in multi-agent systems establishes a relatively new but significant topic in Artificial Intelligence (AI). Multi-agent systems typically are very complex and hard to specify in their behavior.

It is therefore broadly agreed in both the Distributed AI and the related communities that there is the need to endow these systems with the ability to adapt and learn, that is, to self-improve their future performance. Despite this agreement, however, adaptation and learning in multi-agent systems has been widely neglected in AI until a few years ago. On the one hand, work in Distributed AI mainly concentrated on developing multi-agent systems whose activity repertoires and coordination mechanisms are more or less fixed and thus less robust and effective particularly in changing environments. On the other hand, work in related fields, such as Machine Learning, mainly concentrated on learning techniques and methods in single-agent or isolated-system settings. Today this situation has changed considerably, and there is an increasing number of researchers focusing on the intersection of Distributed AI and related topics (such as machine learning).

There may be at least two different kinds of adaptability: purely engineering one, related to the situation, as mentioned above, when a program is able to change its behavior in changing context regardless of the nature of the latter; and one arising from Human-Computer Interaction (HCI) issues [9] [6], which concern a program adjusting its behavior to meet the user's expectations and peculiarities according to some reasonable strategy.

The second, HCI-related kind of adaptability is crucial for any tutoring system. It is evident, that a qualified tutor works differently with different students, regardless of possibly equal "marks" of the latter. This issue is even more important for distance learning systems, in which the online contact with human tutor is lost.

Therefore, one of the main directions of the described project is to investigate possible strategies of the automatic adaptation of a DL system, based on the concept of Student Model [3] [5] [6] and the corresponding Learning Process Model. In contrast to cited papers, together with [10], these models are supposed not to function separately, but to form an "agent model" in the meaning of [15].

These investigations are accompanied by the development of a prototype of an adaptive DL system IDLE (Intelligent Distance Learning Environment). The technology of program testing, which is one of the weakest components of the Russian software engineering, is taken as a sample domain to be tutored by IDLE.

2 DL Systems in Internet

WWW unifies several existing Internet protocols (such as ftp, http, wais, etc.) and one new (http) around the concept of hypertext. The role of WWW w.r.t. Internet may be compared with that of a windowing system w.r.t. its underlying OS. Using one universal visual language, any user, both programmer and non-programmer, gains control over all the technological features of Internet.

- Plain distribution of volumes of learning material. These include both online tutorials in standard formats (.ps, .hlp, .rtf, etc.) and some special interactive courses, intended to be run as local applications.
- Collaborative Learning in the network (both locally in Intranet, and essentially distant).
- Interactive online courses with immediate access via HTML browsers (possibly with custom extensions, see below).
- Implementation of the second and third methods of DL requires special Internet programming tools and programmer's qualification.

Having analyzed some existing DL systems, which function in WWW, one may draw its typical structure. Usually the following active components which may be represented either by "real" human persons or special programs (here we enter the multi-agency) are found in such systems:

1. Tutor, which forms and presents learning material. It may be either a human being or a computer program (intellectual agent).
2. Supervisor, which watches and controls the learning process. Again, it may be either human person, or an special (agent) program.
3. Assistant, which tries to help student in various aspects of learning process. The fields of assistance may include domain knowledge, adaptation of interface, Internet usage, etc.

The other usual components of DL systems include

1. Learning Material. It may be both hypertext and special training programs.
2. External Data Sources. Everything not supported explicitly by the system, but required or recommended during education (hardcopy tutorials, video cassettes, etc.).
3. Auxiliary Tools. This includes various computer techniques, which out of the scope of the system, but are required for it to function properly (such as communication programs).

Such a typical structure may be implemented differently, as illustrated by the existing DL systems [13].

The simple overview show that there are two opposite approaches to the organization of distance learning in WWW. The first of them uses on-line mode, the second one - off-line. In the first case only a standard WWW-browser is required, while in the second case auxiliary software is necessary on the client host. Both of the systems, however, function in the framework of the client-server technology and use CGI interface, which is common for most of such systems.

It is worth mention that difference in these approaches is well correlated with the complexity of the corresponding learning material. For now, it is very difficult and at least inefficient to simulate complex processes via standard HTML (even with Java applets), therefore the use of special client software is justified.

Currently, the following methods of distance learning in Internet are well studied and widely used in practice:

- WWW as a data source without any efforts to maintain a DL system.;
- Server-hosted software development;
- Auxiliary client-hosted software development.

The most important directions of further development of Internet-based technologies which would help in maintaining DL systems, are (except Java or other script language applets):

- HTML extensions for CCI (Client Communication Interface);
- Synchronous conversation applications for WWW (analogs of Unix 'talk' or Windows 'chat').
- Multimedia newsgroups.
- Virtual reality.

3   The Architecture of IDLE

According to the existing taxonomy the software implementation of the IDLE prototype may be regarded as: multi-agent, portable, access-restricted and using multimedia effects.

System contains both the modules of traditional (isolated) architecture, which examples are tutor's and system administrator's workbenches, and the modules of multi-agent architecture, which implement e.g. system supervising, immediate control over learning.

Among the agents comprising the system, are those, which may be regarded as "classic" [17]:

Expert Tutor.

Expert Tutor controls the learning process, applies different education strategies according to the cognitive and personal features of the student, together with his/her educational "progress". Interacts with Interface Engineer (see below) on adaptation and user modeling issues.

Interface Engineer.

Interface Engineer maintains User Model (see below) for current student (both by the preliminary testing and dynamically during the student's working with the system) and provides it to the human users and to the other agents. In particular this information is used in choosing the scenario of the presentation of courseware and evaluating the student progress and his/her assessment.

Domain Expert (Domain Assistant).

Domain Expert accumulates and provides knowledge on subject domain, maintains testing exercises, provides these data to its users and other agents, analyses the students "feedback" in domain terms.

Such system architecture as presented here may be regarded as a development of a traditional for AI idea of "active knowledge bases": each of the knowledge base components is access via corresponding agent, which adapts knowledge to the context of usage. Furthermore, a close analog of the agent-mediated knowledge bases is seen in the works of [8] where a knowledge base with reflection was developed. It has both domain knowledge and some meta-knowledge about its integrity, and could apply this in reasoning. Certainly some close analogs of these techniques are seen also in the field of databases, especially object-oriented ones [2].

Behavior of expert tutor is influenced by the hypertext model of presentation of learning material. In contrast to the traditional mode of hypertext navigation, when the default routes are embedded into the page, in the described system the navigation is controlled by the expert tutor, which restricts the navigation freedom, and puts "barriers". It does not allow student to move to those places of course, where his/her visit is unreasonable (according to the strategy and the target of learning).

For this, all the domain material is stratified into "clusters", containing one or more hypertext nodes and representing "learning units". The set of clusters, which nodes were visited by the student during all his/her learning sessions, is called "scope of visibility" and occurs a union of both already and currently

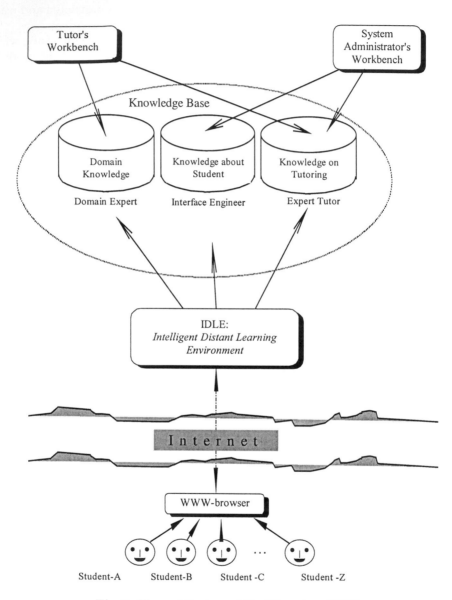

Fig. 1. The architecture of the DL system IDLE.

learned material. The navigation within the scope of visibility is unrestricted. To widen the scope of visibility, the student should overcome a "barrier", which provides some examination test.

4 Learner Modeling: the Background for Interface Engineer

Different factors, which affect the learner's productivity in a concrete computer application environment, may be grouped into several classes. Within each class only those factors, which allow for a more or less formal estimation, are emphasized.

Demographic factors

They comprise such essential user's parameters as age, gender, first language, place of birth, social and cultural peculiarities.

Professional factors

They tie such user's features as position, professional experience, computer skills into integral "portrait" of his/her expertise. In particular, they include the expertise level [1] as one of the most important characteristics that an adaptive system takes into account. The special expertise tests are developed both for professional skills in knowledge engineering and computer operating. The educational parameter also is important.

Physiological factors

From the point of view of physiology the factors which could influence the user's productivity are mostly grouped around reaction, workability, attention, etc. One of the possible taxonomies may be descried as follows:

- Perception: global, peripheral and spatial perception;
- Memory: operative, iconic memory, long-term memory;
- Dynamic Motorics: response time, response orientation, senso-motor performance, etc;
- Neuro-physiology: functional cerebral asymmetry;
- Cognitive functions: estimating speed and distance, long-term reactive stress tolerance, numerical intelligence, tracking;
- Attention: attention in a monotonous environment, focal attention, vigilance.

Psychological factors

They can be defined through the two dominant strata [4]: communicative stratum and cognitive stratum.

The communicative stratum deals with the area of communications between the user and the computer and it comprises the problems of perception, understanding, clarity, usability, handiness and some other shallow visible features of man-machine interactions. The problem is how to gain the user's sympathy and confidence, taking into account the temper and intelligent peculiarities.

The cognitive stratum comprises more deep problems related to human mind and memory. In the recent years the interest to cognitive modeling has grown significantly. But still every novice feels the cognitive dissonance between his/her expectations and the interface layout and behavior.

An extremely important factor is personal cognitive style, which considerably influences the problem solving way. Very important is also the style of logical mentality or deductive/inductive strategies. Users using deduction always perform their cognitive activity with the top-down strategy from the higher level of abstraction to more and more detailed schema. On the contrary, in the variant of induction the users ascend from the unconnected elementary concepts to meta-concepts.

An adequate formalization of the factors mentioned above, which is necessary both for theoretical investigations and the practical system design, may be achieved via the concept of Learner/user model (LM) [14] [18].

A frame structure for LM may look like:

LM:
     demographic:
       age: young | medium | old,
       education: higher | medium | none,

       ....

     professional:
       experience: value

       ....

     physiological:
       cerebral_asymm: left | right | balanced,

       ....

     psychological:
       assertiveness: high | medium | low

       ....

Here, a conventional pseudo-syntax is read as follows: the pairs A:B denote the slots together with the types and/or sets of possible values, indentation is used to group the sub-frames. The symbol '—' is used to separate the (constant) alternatives and may be read as "or".

This information is maintained by the interface engineer agent and kept in its internal database which is used to adapt personal interfaces. Such features of interface design as - layout, color, font size, speed, navigation, functionality and help instructions - are the matter of adjustment and adaptation according to learner model.

The experimental framework for acquiring this information is supported by specially designed software system TOPOS [5], which, on the current stage of the project, is represented by a stand-alone (isolated) program, constituting a component of the system administrator workbench (see above).

TOPOS is aimed to provide user-friendly interface for all the procedures related to psycho-physiological experiments measuring user characteristics together with some other related data.

Its functionality includes:

- Management of the database, containing the respondents' test and personal data.
- Performing questionnaires.
- Performing game-like non-verbal and indirect (so called "projective") tests.
- Performing special test on syllogistic reasoning.

These data are then processed and interpreted in TOPOS with the use of a Prolog inference engine, which set of built-in predicates is enriched by some special routines, providing such services as mathematical statistics and database access (dBase and Paradox). Apart from intelligent analysis of data processing, this component of the system provides a possibility for the quickly building of questionnaires for arbitrary domains without programming: the first versions of TOPOS tests were prototyped using this possibility.

As mentioned before, other agents may request the LM information via inter-agent communication protocol and then use it in their specific tasks.

5 Subject Domain: the Program Testing

As a subject domain for the first prototype of the DL system IDLE the technology (and art) of program testing is chosen [11]. Testing is understood here as a program execution aimed at revealing of errors. This does not mean that "tested" program (i.e. the program where the given set of tests has not discovered errors) is totally error-free. The task of tester is to choose a finite set of tests which maximizes the probability to reveal majority of errors. A test is considered "good" if it does reveal yet unknown error. Any set of input data and user's interactions with the program together with expected output data and program responses may be considered as a "test".

The structure of domain of program testing is as follows:

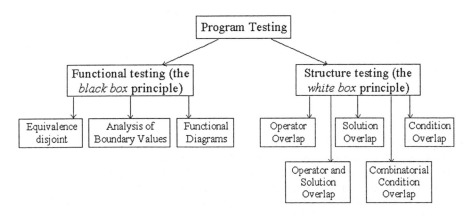

Fig. 2. The structure of program testing domain

The functional testing is aimed at revealing of contexts where the program does not correspond to its specification. The "white box" strategy supposes the structure of the program to be known and available for analysis.

6   Conclusion

The described project is in its early stage of development. Currently, different system components are studied separately. This is referred to, e.g., learner modeling, Internet programming, description of subject domain.

In this project much attention is paid to the learner modeling. This concept aggregates an abstraction of those user's features, which both may be measured automatically [5] and could be represented by a formal model. In contrast to [3] [6], where the maintaining of user model requires immediate participation of a human interface engineer, the use of the paradigm of multi-agent systems is expected to make the role of latter smoother and provide him/her more intelligent computer-aided support.

The results of the described investigations may help to elucidate some obscure questions related to different aspects of adaptability of software agents, both in the meaning of "agent"-"agent" and "agent"-"user" interaction.

7   Acknowledgements

Described project is partly supported by the Russian Foundation for Basic Studies (98-01-000081).

References

1. Brezillon, P.: Building Explanation During Expert-Computer Interaction. Proc. of East-West International Conference on Human-Computer Interaction EWHCI'92 (1992) 378–385
2. Cattell, R.G.G: Object Data Management. Addison-Wesley (1991)
3. Gavrilova, T., Averbukh, E., Voinov, A.: An Approach to Learning Software Based on Student Modeling. Lecture Notes in Computer Science N 1108: Third International Conference on Computer Aided Learning and Instruction in Science and Engineering. CALISCE'96, San-Sebastian (1996) 448–450
4. Gavrilova, T., Voinov, A.: Work in Progress: Visual Specification of Knowledge Bases. 11-th International Conf. On Industrial and Engineering Applications of Artificial Intelligence and Expert Systems IEA-98-AIE, Spain, Benicassim, Springer (1998) Vol. 2 717–726
5. Gavrilova, T., Chernigovskaya, T.,Voinov, A.,Udaltsov, S.: Intelligent Development Tool for Adaptive Courseware on WWW. 4-th International Conference on Computer Aided Learning and Instruction in Science and Engineering. June 15-17, Chalmers University of Technology Goteborg, Sweden (1998) 464–467
6. Gavrilova, T., Voinov, A.: Adaptive Interface Design and Scenario Control via User Modeling. In: Preprints of IFAC/IFIP/IFORS/IEA Symposium on Analysis, Design and Evaluation of Man-Machine Systems, MMS'95. MIT, Cambridge (1995) 535–540.

7. Goel, A. , Barber, K.S.: The Classification and Specification of a Domain Independent Agent Architecture. 11-th International Conf. On Industrial and Engineering Applications of Artificial Intelligence and Expert Systems IEA-98-AIE, Spain, Benicassim, Springer (1998) Vol. 1 568–576

8. van Harmelin: REFLECT project. ESPRIT Tech.Rep (1992)

9. Johanssen, G.: Design of Intelligent Human-Machine Interfaces. Proc. 3rd IEEE International Workshop on Robot and Human Communications, Nagoya (1994) 97–114

10. Kaas, R.: Student Modeling in Intelligent Tutoring Systems - Implications for User Modeling. In: User Models in Dialog Systems. USA (1989)

11. Myers, G.: The Art of Program Testing. Addison-Wesley Publishing Company (1980)

12. O'Hare, G., Jennings, N.: eds. Foundations of Distributed Artificial Intelligence. Sixth-Generation Computer Technology Series. Branko Soucek, Series Editor. John Wiley & Sons (1996)

13. Parodi, G., Ponta, D., Scapolla, A.M., Taini, M.: Cooperative and Distance Learning in Electronics Using Internet. In: Lecture Notes in Computer Science N 1108: Third International Conference on Computer Aided Learning and Instruction in Science and Engineering CALISCE'96, San-Sebastian, Spain, Springer-Verlag (1996) 213–219

14. Rich, E.: Users are Individuals: Individualizing User Models. Int. Journal of Man-Machine Studies, 3, No.18 (1983) 23–46

15. Shoham, Y.: Agent-oriented Programming. Artificial Intelligence, vol. 60 (1993) 51–92

16. Staniford, G., Paton, R.: Simulating Animal Societies with Adaptive Communicating Agents. In: Wooldridge M. Jennings N.R. (eds) Intelligent Agents - Theories, Architectures, and Languages. Lecture Notes in Artificial Intelligence Volume 890. Springer-Verlag (1995)

17. Takaoka, R., Okamoto, T.: An Intelligent Programming Supporting Environment based on Agent Model. IEICE Trans. INF. & SYST., vol. E80-D, No. 2 February (1997)

18. Wagner, E.: A System Ergonomics Design Methodology HCI Development. Proc. of East-West International Conference on Human-Computer Interaction EWHCI'92 (1992) 388–407

19. Weiss, G., Sen, S.: eds. Adaption And Learning In: Multi-Agent Systems. Springer-Verlag, Lecture Notes in Artificial Intelligence, Volume 1042. (1996) 238–245

A Multi-agent Solution for Advanced Call Centers

Bernhard Bauer

Siemens Corporate Technology, Munich, Germany

bernhard.bauer@mchp.siemens.de

Cornel Klein

Siemens I & C Networks, Munich, Germany

cornel.klein@icn.siemens.de

Abstract. In the past few years, call centers have been introduced with great success by many service-oriented enterprises such as banks and insurance companies. It is expected that this growth will continue in the future and that call centers will be improved by adding new functionality and by embedding call centers better into the workflow of a company. In this paper we show how agent technology can help to realize these goals. Agent-based approaches are becoming more and more mature for applications distributed over networks, supporting (dynamic) workflow and integrating systems and services of different vendors. We show by a typical example of a call center, the call center of a car rental agency, what the deficiencies of current call centers are and how agents can help to improve this situation.

1 Introduction

Over the last years, many customer-oriented enterprises (e.g. insurance companies, banks, mail order shops,...) have introduced call centers. Call centers are currently applied to domains like hot lines, tele-marketing, helpdesks, information services, mail order centers and advice services. The huge annual growth rates in this market field can only be continued if today's call centers are extended from simple transaction-oriented call handling systems to interactive, multi-medial customer oriented communication systems, embedded tightly into the workflows of enterprises.

During the seventies, structured programming was the dominating paradigm. The eighties were the decade of object orientation with data encapsulation and inheritance of behavior. Agent oriented software development is the paradigm of today. Compared to objects, agents are active by executing one or more internal threads. The activeness of agents is based on their internal states which include goals and conditions implying the execution of defined tasks. While objects need control from outside to execute their methods, agents know the conditions and intended effects of their actions by themselves and hence take responsibility of their needs. Furthermore, agents do not only act on their own but in cooperation with other agents. Multi-agent systems are a kind of social community of which the members depend on each other though acting individually on behalf of their users.

In this paper, we show how agents can be used to build advanced solutions for call centers. We will show how agents can solve the challenges of today's call centers, which among others are the integration of new multi-media communication techniques, the support of business processes, customer-specific services and the integration of legacy systems.

The remainder of this paper is structured as follows: First, agent technology is presented. Second, the car rental agency "Happy Driver" is introduced, representing a typical scenario for the application of call centers. For "Happy Driver", we sketch the existing call center solution and discuss its deficiencies. We design an agent based solution and use a simple scenario which shows how agents can solve the above mentioned challenges. Finally, we conclude and point out future work.

2 Agent Technology

Software agents are an innovative technology for the efficient realization of complex, distributed and highly interactive heterogeneous application systems. Software agents are software components which are characterized by **autonomy** (to act on their own), **re-activity** (to process external events), **pro-activity** (to reach goals), **cooperation** (to efficiently and effectively solve tasks), **adaptation** (to learn by experience) and **mobility** (migration to new places). On the one side agents must be specialized for the individual demands of their users to be used with minimal effort, on the other side they must communicate with other agents and external components in order to use or modify global relationships, see e.g. [3; 8; 11; 12].

Messages between agents must satisfy standardized communicative (speech) acts which define the type and the content of the messages (agent communication language (ACL) [3]) The order of exchanging messages of a certain type is fixed in protocols according to the relation of agents or the intention of the communication. For example, a PROPOSE message opens a negotiation process and ACCEPT or REJECT terminates it. A negotiation process is useful in equal opportunity scenarios like meeting scheduling [3].

Each agent maintains a model of its world, representing its current view of its environment. The goal of an agent is represented as a formula over the states of the world model, which should hold at some time in the future. The goal can either be described by a user, by an event of the surrounding or by another agent. Activating a new goal the agent starts a planning phase in order to calculate a plan from the actual state to the goal state within the states of the world model. Such a planning can be performed either by an agent itself or in combination with other agents. The deduction algorithm use heuristics. A planning component can be implemented e.g. using constraints and constraint handling rules (CHR), see e.g. [5; 6; 7], on topics of planning see [13] and the referred links. Such planning strategies can be used to schedule the dynamic workflow within a call center and the connected infrastructure of the company.

Multi-agent systems require an appropriate adaptation to changing environment and to dynamic demands of customers. Learning techniques are well suited for the adaptation of single agents and whole multi-agent systems, especially user profiles and preferences can be learned by an agent, see e.g. [4; 9; 10]. Learning the predilection of a customer leads to a content customer. The contentment of a customer is a main issue of call centers, since it is many times harder to acquire new customers than to hold existing customers.

3 Example: Car Rental Agency "Happy Driver"

From the point of view of an enterprise, the main motivation for introducing a call center is to provide an organizational unit which supports communication intensive business processes. As such, a call center is the externally visible "interface" to customers of modern service oriented companies, and hence has to be designed with great care.

The functions which are supported by today's call centers are: **Automatic call distribution (ACD)** - Incoming calls are distributed to human agents. **Automatic call handling** - Components such as automatic voice generation devices or voice recognition devices make it possible to handle some calls automatically. **Computer-Telephony Integration (CTI)** - provides additional support for the human call center agents. For instance, it is possible to dial automatically a phone number by clicking with the mouse on a telephone number on the screen.

As a typical example for the use of call centers, we introduce the example of a car rental agency. This example is used in the sequel to sketch an agent based call center solution and to highlight its benefits. Although our example is focused on a special case – the car rental agency "Happy Driver" – we believe that the presented scenario can easily be adapted to other applications of call centers.

3.1 Current situation

The car rental agency "Happy Driver" rents cars to private persons and to business travelers. It has branches in 50 locations in the nation. "Happy Driver" currently operates a centralized call center, which is accessible via the nationwide 0800-HAPPY-DRIVER number. In this centralized call center, 24 hours/day, incoming calls are automatically distributed up to 30 human agents which handle requests for information and for the reservation of cars. After the car has been picked up by the customer, the local branches are responsible for communicating with the customer, e.g. in case of a car accident or the modification of a rental contract. For this reason, the telephone number of the branch office is handed out to the customer when the car is delivered.

The current solution has some severe disadvantages concerning the functionality. We highlight the following:

Integration of new media types: The integration with new multi-media communication techniques such as the WWW, FAX and SMS is very weak. The operation of the web server is currently outsourced to an internet service provider (ISP). A reservation originating in the WWW currently generates a FAX message, which is sent to the call center, where the reservation request is handled manually. In particular, online confirmations of reservations are not possible.

Workflow support: As an additional service, "Happy Driver" delivers the rented cars directly to their customers home or office. Also, the cars can be returned this way. Field staff is used to drive the cars both between clients and branch offices as well as between branch offices and service centers (e.g. car wash, car maintenance, gas station etc.). The travel schedules and travel routes of the field staff are currently planned manually by a human dispatcher. As a result, it is not always possible to tell a calling customer immediately when and whether a car can be delivered to a certain place/time. Instead, the human call center agent has to call manually the dispatcher (who may be busy talking via mobile phone with the field staff) whether a certain delivery is possible. Conversely, in case the delivery is not possible due to some unforeseen condition (e.g. car accident), the customer is often neither informed about this problem nor are alternative solutions searched for pro-actively. In addition, field staff cannot always be reached by mobile phone due to a limited coverage of the mobile phone network.

Customer care: Customers calling the call center are treated the same way, independently whether they are first-time customers or whether they are regular business customers. However, the management of "Happy Driver" decided to pursue a more customer-oriented and aggressive marketing model. For instance, depending on customer profile, car availability and available reservations, regular business customers are always upgraded to a higher-class car when possible. Moreover, for these customers a specialized and simplified reservation process is envisaged, which can only be made by specially trained human call center agents. On the other hand, available low-end cars should be rented to private persons at discount rates in order to increase the utilization of the fleet.

Integration of legacy systems: Information systems of business partners are currently not integrated into the existing infrastructure. For instance, it is desirable that fees for gas refill can automatically be billed to the respective customer or that bills for car repair are automatically be forwarded to the customer or the insurance company.

Finding the right human call center agent: The current call center solution are mainly designed to deal with simple tasks such as reservations and requests for information. For that reason, one kind of human call center agent, which has been trained in a one week seminar, was sufficient. However, in order for improved customer satisfaction a more advanced scheme might be desirable. For instance, premium customers should be identified by their calling number and connected directly and immediately to a dedicated human call center agent. Moreover, qualified help in case of accidents should be available immediately, depending on the car type and the accident location.

Besides its limited functionality, the current call center solution also is not very economic:

Expensive centralized call center: The centralized call center requires a dedicated office. Moreover, in order to guarantee fast reaction time to incoming calls of customers, a sufficient amount of call center staff has to be available 24 hours/day. On the other hand, in the branch offices there is enough space for call center workplaces, and there is also available staff. Therefore a distributed call center is desirable.

Lack of automation: The increasing use of the WWW interface leads to a severe and unnecessary overload of the call center staff due to the manual handling of reservations from the WWW.

Insufficient support of business processes: The business process of one-way rents, where the car is returned at a different branch office, is not supported sufficiently. Since car reservation and car maintenance are made on a per-branch-office basis, cars have to be returned to their home branch office after a one-way rent by field staff. For this reason, one way rents can only be offered quite expensively to customers.

These disadvantages already indicate that software agents can be an ideal approach for an improved call center solution. In particular, we expect that an agent-based approach should be selected for developing a call center, since the following properties are satisfied by the call center scenario which can nicely be dealt with by a multi-agent system: distribution, co-ordination and communication, mobility, robustness and non-stop operation.

Viewing the above-mentioned general challenges of today's call centers and the particular disadvantages of the current solution of "Happy Driver", we expect that agent technology offers the following benefits:

Integration of new media types: As mentioned above future call centers have to support many new communication techniques (e.g. e-mail, FAX, SMS...) besides simple telephone calls. Nevertheless, independently of the used communication media, communicative acts may have the same meaning, and hence these communicative acts can be mapped to a internal SW agent language abstracting from the concrete syntax of the input. SW agent wrappers can be used for that purpose, mapping from the concrete input (e.g. an incoming e-mail message) to agent communication language and vice versa. In the case of "Happy Driver", a web interface can be an additional access mechanism to the call center.

Workflow support: Using the planning component of agents, dynamic workflow can easily be modeled by multi-agent system, see [1]. Distributed planning leads to a more sophisticated scheduling than the centralized approaches of today's workflow systems. Workflow can be optimized by negotiation between the SW agents about their capabilities and resources. Especially the data component can be taken into account. Moreover, learning techniques can be applied in order to derive estimations about the necessary time to perform certain tasks. This knowledge can be used as a working hypothesis for planning future workflows.

Integration of legacy software: Again the notion of agent wrapper helps to integrate components from business partners (e.g. from car maintenance companies) and of existing parts (e.g. the existing CTI system, the existing ACD system). More examples are facilities for speech recognition, e-mail gateways, web servers etc. Having components with the same supported functionality those SW agent wrappers can use the same communication language while talking to other call center agents.

Customer care: "know your customer" - learning techniques can be applied to learn the behavior of individual customers and of the customer behavior as a whole. The learned knowledge can be used to achieve more satisfied customers and to optimize the utilization of the car fleet.

In the following subsections we describe first of all the physical architecture, second the logical architecture, i.e. the agent architecture, and we finish this chapter with some scenarios of our agent based call center example.

3.2 Physical Architecture

The physical architecture (Fig.1) consists of the following components:

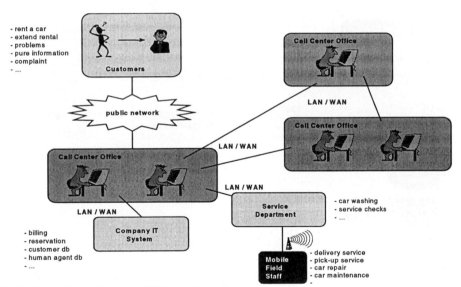

Fig. 1. Physical architecture of "Happy Driver"

- The customers of "Happy Driver";
- The field staff of "Happy Driver" and the associated human agents.
- Several physically distributed call center sites, involving both branch offices as well as teleworking sites;

- Additional sites, such as the field staff center, the car wash- and maintenance facilities etc.;
- Existing systems, such as the reservation & billing system as well as a web server.

The components belonging to "Happy Driver" are connected by LAN/WAN technology for data communication as well as by a fixed and wireless telephone network for voice communication. The data communication system and the telephone system are integrated with usual CTI technology, e.g. telephone calls can be initiated and terminated under control of the data system.

3.3 Logical Architecture

On top of the physical architecture of the previous section we can now define the logical architecture, i.e. the agent architecture, for the call center of "Happy Driver" (Fig. 2.)

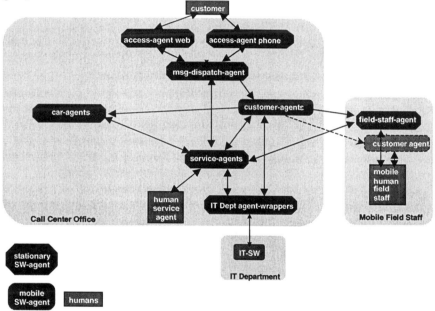

Fig. 2. Logical architecture of the call center of "Happy Driver"

The different kinds of software agents in this architecture have been introduced in order to solve the following tasks:

- A customer may contact the call center via different communication media. One example is a simple telephone call, where voice-announcements are used for output to the customer and touch tones or speech-recognition is used for customer input. Another example is world-wide web access. Access agents such as *access-agent-web* and *access-agent-phone* convert the incoming messages into a standardized internal format for inter-agent communication. This allows it to easily accommodate new communication media in the future. The other way round, these access agents provide facilities to send messages to customers. For instance, they may interface with an SMS gateway, an e-mail gateway or a FAX modem. Moreover, they may also establish telephone calls to customers.

- The *message-dispatch-agent* is responsible for dispatching incoming calls and web requests to dedicated service agents. For that reason, it first tries to identify the requested service and/or to identify the calling customer. After this, it forwards all incoming messages to the respective service agent. Moreover, it also creates a customer agent, based on the customer profile in the customer database. In the opposite direction, the *message-dispatch-agent* knows the communication channel which can be used to reach a particular customer, e.g. via telephone, wire-less communication or WWW.

- Depending on the incoming requests the best fitting *service agent* available at the moment for the problem is chosen. This is performed by negotiation of the SW agents depending on the service descriptions of the service agents, the capabilities of software agents, the availability of human service agents and the customers preferences. The service agents may be distributed over several call center offices and also include tele-workers at home, thus making a fully distributed call center possible. The goal of service agents is to make customers happy while at the same time maximizing the profit of the company.

- For each customer a *customer agent* exists knowing the data of a customer and especially his/her preferences to optimally process the belongings of him/her. Moreover, he maintains also data which is relevant for identifying customers, e.g. their telephone number or email address. Customer agents are responsible for pursuing the interest of customers. For instance, they proactively interact with field-staff agents in order to plan for a timely delivery of cars. Moreover, the customer agent maintains all states belonging to a particular communication session, e.g. during a telephone call. This allows to pass information already communicated by the customer to the call center between the different service agents. Note that customer agents are mobile. They are sent to hand-held mobile devices of field staff in order to support their work in order to allow field staff to work even if no online connection to the agent system is available.

- For each car a *car agent* exists, maintaining all knowledge about the car and pursuing the interests of the car. These include reservations for the car and other constraints (e.g. necessary maintenance times etc.).

- Several *agent-wrappers* are used to interface with legacy IT systems and to support the call center application, e.g. by providing persistent storage.

3.3 Example Scenario

We now demonstrate the benefits of our solution by means of a simple scenario:

- Customer Joe calls the 0800-HAPPY-DRIVER number. The call setup message is forwarded via a legacy wrapper from the existing CTI system to *access-agent-*

phone, which passes it to a *message-dispatch-agent.* Based on the phone number of the calling party, it is determined that Joe is a premium business customer. *Message-dispatch-agent* generates a new customer agent with the information stored in the database being accessible via IT-system-wrapper and searches for a suitable service-agent for premium customers. The appropriate human agent "Jacky" is found and the message-dispatch-agent requests the CTI system to forward the call to Jacky.

- Joe tells Jacky that he wants to make a one-way rent over the weekend from (A-City, December 1st) to (B-City, December 3th) with a low-budget car. Jacky enters the data into the according *service-agent*, which in turn sends the reservation to the appropriate *customer-agent.*

- The *customer-agent* is responsible for pursuing the goals of the customer. As such, it immediately makes a reservation at the *car-agent* of Car A. After the reservation has been confirmed, the information is propagated to the corresponding *service-agent*, which passes it to Joe. The telephone call ends.

- Another customer, Paul, wants to make a reservation from (B-City, December 4th) to (C-City, December 24th). Since it is planned that Car A is at B-City on December 4th, Car A is reserved.

- Since Joe is a premium business customer, some *service-agent* proactively checks whether a higher-class car can be delivered to Joe as well. Because there are sufficient higher-class cars in A-City for the weekend, it seems to be possible to give Joe SportsCar B. However, the re-negotiation with Car A fails, because this way Car A would not be available in B-City on December 4th.

- For that reason, *customer-agent* Paul is asked to take another car or to take SportsCar B. *Customer-agent* Paul negotiates with service agents and other car-agents. Finally, some other car is found which suites Paul's needs. Paul is not allowed to take SportsCar B, because the business policy does only allow premium customers to get such cars for three weeks.

- The pro-active planning of the field staff activities allows to deliver SportsCar B to Joe's office on the evening of November 30th, directly from the maintenance center. Joe is happy, because he gets a better car as he had paid for.

4 Conclusions and Prospective

We have shown that agents are a promising technology for the implementation of advanced call center solutions. It will be applied to the software within a call center, to connect existing SW of the IT department, to schedule the dynamic workflow of the call center and its surrounding, to adapt the customers predilections and to coordinate the mobile service staff. This guarantees customers to receive high quality support for their requests. The call centers provides all customers, human agents and mobile staff with sufficient information at any place and any time. This call center scenario will be able to reduce the waiting times and wrong processing of customer requests by optimal scheduling and consideration of customers' preferences and staffs' behavior. Call center providers get new means to advertise their services and to catch customers.

We are well aware that we have presented a high-level view of an advanced call center. For a concrete implementation, the given model has to be refined. An example where certain aspects of an agent-based call center are described in more detail can be found in [2]. It is planned to prototypically implement such a call center with different access-media and mobile devices. Especially with learning customer preferences and with dynamic workflow scheduling the tasks of the (human) service agents can be efficiently supported.

Acknowledgment The authors would like to thank their colleagues within Siemens I & C Networks for the information on call centers, their colleagues D. Steiner, G. Völksen and M. Schneider from Siemens Corporate Technology for fruitful input on this paper, as well as the unknown referees of the paper.

REFERENCES
[1] B.Bauer, M.Berger: *Using Agent Technology for Dynamic Workflow.* to be published 1999
[2] Brazier, F.M.T.; Jonker, C.M.; Jungen, F.J.; Treur, J., *Distributed Scheduling to Support a Call Centre: a Co-operative Multi-Agent Approach.* in: H.S. Nwana and D.T. Ndumu (eds.), Proceedings of the Third International Conference on the Practical Application of Intelligent Agents and Multi-Agent Technology (PAAM '98), The Practical Application Company Ltd, pp. 555-576, 1998.
[3] Foundation for Intelligent Physical Agents; Specifications 1997/1998: http://www.fipa.org/
[4] Forsyth, R.; Rada, R.: *Machine Learning*, 1986, Ellis Horwood Limited, Chichester, England, 1986.
[5] Frühwirth, T.; Abdennadher, S.; Meuss, H.: *Implementing Constraint Solvers: Theory and Practice.* in: Forum de la Recherche en Informatique'96, Tunis, Tunesia, July 1996.
[6] Frühwirth, T.; Abdennadher, S.: *Constraint-Programmierung: Grundlagen und Anwendungen*, Springer, September 1997.
[7] Java constraint Kit (JACK). http://www.fast.de/~mandel/JACK
[8] Maes, P.: *Modeling Adaptive Autonomous Agents*, in: Langton, C. (ed.): Artificial Life Journal, Vol. 1, No. 182, MIT Press, pp. 135-162,1994.
[9] Michalski, R.: *Understanding the Nature of Learning*, in: R. S. Michalski, J. G. Carbonell and T. M. Mitchell (eds.), Machine Learning - An Artificial Intelligence Approach, Morgan Kaufman, Los Altos, CA, 1986.
[10] Mitchell, T.M.: *Machine Learning*, McGraw Hill, 1997.
[11] Müller, J. (ed.): *Verteilte Künstliche Intelligenz: Methoden und Anwendungen*, BI Wissenschaftsverlag Mannheim, Leipzig, Wien, Zürich, 1993.
[12] O'Hare, G.; Jennings, N. (eds.): *Foundations of Distributed Artificial Intelligence*, John Wiley & Sons, Inc. New York, 1996.
[13] Carnegie Mellon University, PRODIGY Project Home Page, http://www-cgi.cs.cmu.edu/afs/cs/project/prodigy/Web/prodigy-home.html.

A Multiagent System for Emergency Management in Floods[1]

José Cuena, Martín Molina

Department of Artificial Intelligence
Technical University of Madrid
Campus de Montegancedo s/n
Boadilla del Monte, 28660-Madrid (SPAIN)
Tel: (34) 91 352 48 03; Fax: (34) 91 352 48 19; E-mail: {jcuena,mmolina}@dia.fi.upm.es

Abstract. This paper presents the architecture of a multiagent system for the problem of emergency management in floods. The complexity and distributed nature of the knowledge involved in this problem makes very appropriate the use of the multiagent technology to achieve adequate levels of robustness and flexibility for maintenance. The paper describes the types of agents defined to reason about the river basin behavior and the types of agents defined to make decisions about control and protection. According to this organization, three methods have been defined for social interaction: (1) an interaction method for prediction based on the water propagation, (2) a second method for a distributed decision about hydraulic control actions on reservoirs based on the principle of homogeneity of risk levels and (3) a last method for decision about civil protection based on the concept of social revenue.

1 Introduction

The possibilities of the new technologies for data sensoring and communications makes feasible to get on real time the information on rainfall, water levels and flows in river channels to monitor a flood emergency situation. In Spain there is an investment program aiming to install this type of equipment in the main river basins (SAIH, Spanish acronym for Automatic System of Hydrology Information). However, receiving a large amount of detailed data flow (every 5 minutes data of 150 sensors in a typical watershed) requires an intelligent interface able to translate the data flow to a conceptual framework close enough to the professional ideas required by the responsible persons.

To meet this objective, a system is now on course of development capable, first, to identify *what is happening* in terms of the relevant problematic events to be detected and their corresponding diagnosis, second, to predict *what may happen in the short term* assuming that the current state of control is maintained, third, *to recommend possible plans* of decisions acting on the causes of the detected problems and, fourth, to predict *what will happen if* the recommended plans or some variants for them are applied.

---

[1] This research is sponsored by the Dirección General de Obras Hidráulicas, Ministry of Environment, Spain.

Given that the SAIH program will develop several of such systems it is important that reusability of software be considered so an effort is required in designing a model system where modules responsible of solving general problems such as event detection, behavior simulation, etc. be adaptive and, hence, reused.

Also, given the extensions of the river basins it is impossible to install initially definitive versions of the systems because data available about the physical features of the river and reservoir systems are very unequally distributed (there are areas very well known and others not). This leads to use an open structure such as the knowledge based one where the results of experimentation may allow to modify the knowledge contents accordingly.

Autonomy of the models is required to ensure a good maintenance policy for extension of the system because if an additional part of the model is to be introduced a module responsible for this part may be created and integrated without need of modification of the rest of the architecture.

All these circumstances lead to an intelligent, knowledge based agent architecture where the main functions of problem detection, reservoir management, water resources behavior and civil protection resources management are concentrated in specialized agents integrated by relations of physical behavior and multiplan generation for flood problems management.

This paper presents an overview of the system architecture and the interaction model where two specific utility function laws for emergence management, the risk level of reservoirs and the social revenue for civil protection, are used as a basis for the model of interaction.

2 The User-System Interaction for Flood Management

An intelligent interface has been designed where the user may ask basic questions such as:
- *what is happening* to identify which are the active events in every moment,

Figure 1: Screen example of the user interface presented by the SAIDA system.

- *what may happen if* no decisions are taken modifying the current state of control to know the possible evolution of the problems detected. *what to do* to get recommendations of plans of actions on the causes of the problems detected.
- *what may happen if* the proposed plans are put in operation with additional variants proposed by the operator.

Currently, it is being developing a software system, called SAIDA, that implements this type of interaction. Figure 1 shows an example of the user interface presented by this system, where, at the top, the main set of basic types of questions are presented to the operator.

3 The Multiagent Architecture

To obtain the answers defined by the previous user-system interaction, an architecture of multiagents specialized in the different problems to solve will be implemented.

Two main types of agents have been designed:

* *Hydraulic agents* are responsible to give answers about the behavior of the following physical phenomena: the rainfall, the runoff produced by the rainfall incidence in the land, the concentration of the runoff in the main river channel, the reservoir operation receiving input flows from the different tributary channels and the flow in the lower levels on the river where overflow is produced creating the flooding problems is to be simulated.

The knowledge of the hydraulic agents is defined at two levels of abstraction:

- There is a synthetic level where the simulation function is described by a bayesian network relating in a cause effect structure probabilities of input states, internal states and output states representing the behavior of a component along the time lag required by the internal process.
- There is a detailed level where the behavior is modeled by a simulator generating for a case of the input flow distribution a time series of internal states and output states with a greater detail than the one presented by the network.

The synthetic level is useful to infer probabilities of effects values based on probabilities of cause values. However, if the user requires to understand a detailed simulator of a sample case compatible with the cause values assumed in the bayesian inference the deep level model will provide a detailed simulation.

* *Management agents*, responsible of the decisions for the river basin management. This type of agents include:

- *Problem detection agents*, responsible to detect possible patterns of problems on course and to ask the reservoir agents to modify their discharge policy by guaranteeing limits of flow upstream the potentially damaged areas. When the flow control in reservoirs makes unfeasible to suppress the flood risk, this type of agent asks for resources of civil protection to the corresponding agents.
- *Reservoir management agents*, which embody criteria for exploitation strategy as will be detailed later with capacity for autonomous reaction to the

reservoir situation (i.e. if too much inflow is produced the agent may get conscious of the need to open the spill for increasing discharge values according with its knowledge).

- *Civil protection resources agents,* responsible to provide with resources of different types according to the demands of the problem detection agents.

To structure the communication between agents, a model is built based on two layers:

- *Hydraulic layer.* Including dependence influences between hydraulic agents corresponding to the water flow relationships (for instance, there is a dependence in a reservoir of the tributary rivers providing input flows).
- *Decision layer.* Including dependence influence between decision agents corresponding to water resources risk management. For instance, a problematic area depends on the reservoirs for upstream water control and on the protection civil centers providing general services such as manpower and transport for population evacuation and protection.

Between both layers, agents can communicate according to their local needs. Thus, when decision agents (such as problem detection agents or reservoir management agents) need to know an estimation about the future behaviour of the river at a certain location, taking into account specific tentative hypotheses of external actions, they send the corresponding messages to the hydraulic agents that are responsible of simulating the physical behaviour.

The general features of the model are summarized in the figure 2.

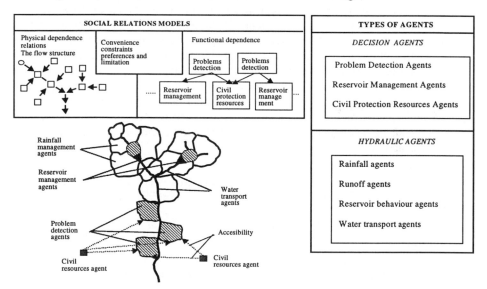

Figure 2: General structure of the agent based model

4 Interaction Methods

According to the previous multiagent architecture, the following interaction methods have been designed:

- Problem prediction. The objective of this interaction is to detect and predict the presence of problems.
- Reservoir management. This model must produce downstream limits in the discharge of every reservoir to contribute to avoid or to alleviate flood problems.
- Civil protection resources management (transport units, manpower for defense works, hospital places, ...) to the different problem areas.

4.1 Social Interaction for Problem Prediction

Local problem detection agents are responsible of detecting and predicting potential problems at specific locations of the river basin. For this purpose they receive input data from sensors, analyses them and, when a particular agent identifies a potential dangerous scenario, it asks for a prediction of behaviour to the corresponding hydraulic agents. Once, the problem detection agent receives the prediction from hydraulic agents, it interprets the information to conclude the level of severity of the future problem.

In order to predict the behaviour of the river, each type of hydraulic agent is specialized in simulating a local physical phenomena (rainfall, runoff, etc.). A particular model includes several instances of hydraulic agents. The social interaction between hydraulic agents is necessary to propagate downstream the prediction. This process starts with a particular hydraulic agent (e.g., an agent whose goal is to predict the water level of the river at a certain location), motivated by an external stimulus (for example, the operator or another agent). Then, this agent tries to apply its specific knowledge for prediction. As presented, this prediction can be performed at two levels of abstraction: synthetic or detailed. If this agent does not have information about the upstream river, then it looks for the immediate neighbour agents that can provide this information. Then, the agent sends messages to those upstream agents, that are further propagated until terminal agents. Finally, the agent waits for the upstream information and, then, simulates it local physical phenomena to produce as a result the required prediction. Figure 3 summarizes the social interaction method for problem prediction.

1. All **problem detection agents** evaluate the current situation. Agents that do not detect the presence of problems are deactivated.

2. **Problem detection agents** that identify a dangerous potential situation send messages to the corresponding **hydraulic agents** asking for a prediction of behaviour.

3. **Hydraulic agents** send messages recursively to upstream **hydraulic agents** asking for their local predictions. Answers are generated recursively to downstream agents.

4. **Hydraulic agents** respond to **problem detection agents** with the future behaviour.

5. **Problem detection agents** evaluate the future behaviour to conclude the level of severity of future problems.

Figure 3: Social interaction method for problem prediction.

4.2 Social Interaction for Reservoir Management

Once local problem detection agents are conscious of the existence of future problems they ask for limiting the water flow upstream their areas location. In general, different problem detection agents may ask for a limitation to the same reservoir, so the reservoirs must adapt their discharge policy to try not to exceed several limits.

The reservoir management is stated as follows. First, an ordered set of limit values established by each problem detection agent is associated to the different points upstream the problematic areas in the river channel network. Then, the problem to be solved for reservoir management is to keep to some extent the resulting discharge at these points below the given limits. Thus, the reservoir discharge must be below the maximum limit but, if this is not possible, below the lower limits of the given set. So the final answer will be which limits are guaranteed and which ones are impossible to maintain.

The problem may require to plan the discharge policy of a single agent (case 1) or a set of reservoir single agents converging in the point where the limits are proposed (case 2). It may happen that a chain of reservoir agents be conditioned by the limit (case 3) of, finally, it may happen that several chains converge in L (case 4).

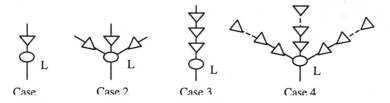

Case   Case 2.   Case 3   Case 4

Obviously, this is the most general case for local relations between reservoirs, connected in parallel or series, converging on a limit point. It exists the more general one where different branches converge (case 5) but this case is to be considered when limits are imposed at the end of the converging branches (L1, L2, ...).

Case 5

To describe the general method, first the isolated agent strategy criteria is presented and, later, its integration to deal with the more complex cases is analyzed. It is assumed that the discharge decision in an agent is based on the evaluation of its risk level R, which is represented with a numerical value between 0 and 1. The risk level is defined as a function of the ratio between the decrease of safety volume I and the safety volume S.

$$R = f(I/S)$$

The decrease of safety volume I is defined as:

$$I = 0 \qquad\qquad \text{when } U - D + C < V$$
$$I = U - D + C - V \qquad\qquad \text{when } U - D + C > V$$

where U is the upstream in flow volume, D the discharge, C is the initial volume of the reservoir and V is the objective volume[2]. The safety volume S, then, is defined as difference between the total volume T of the reservoir and the objective volume V, $S = T - V$. For instance, consider an example where the upstream volume U and the discharge D for the next four hours are respectively $U = 3.5$ Hm^3 and $D = 1.2$ Hm^3, the current volume of the reservoir is $C = 13.2$ Hm^3, the total volume $T = 17$ Hm^3 and the objective volume is $V = 14$ Hm^3. Then, the decrease of safety volume is $I = 3.5 - 1.2 + 13.2 - 14 = 1.5$ Hm3 and the safety volume is $S = 17 - 14 = 3$ Hm3. Therefore, the ratio $I/S = 1.5/3 = 0.5$.

Figure 4 shows two possible displays of the function f. Both have risk level values 1 for I values greater or equal to S but one is more optimistic than the other in the sense that in one of the versions, the risk R grows below the relation I / S and, in the other one, grows above this relation. Besides this mathematical definition of these concepts, a realistic model can include a symbolic knowledge base relating the risk level values with the I / S values, where it may be possible to introduce also the influence of the time of the year and other qualitative aspects.

The social law regulating reservoir management is that cooperative reservoirs must operate at similar risk levels so a model to infer the agent acceptable discharge for a single reservoir management agent may be built based on the acceptable range of risk levels: a default value R_{def}, assumed when no conditions are imposed, and an upper value R_{max} maximum acceptable value. This maximum value may be either a prefixed constant value or, in order to balance the situation of the reservoir and the downstream problem, it may be deduced according to the severity of that problem by using a specific knowledge base.

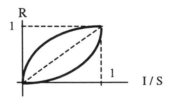

Figure 4: Two possible functions for the evaluation of the risk level R.

Given the safety volume S value and known the upstream volume U, it is possible to relate the risk level R_i for every discharge D_i as:

$$R_i = f((U - D_i) / S)$$

According to this relation, the minimum value of D_i, D_{min}, corresponds to $R_i = R_{def}$ and the maximum value of D_i, D_{max}, corresponds to $R_i = R_{max}$. A request for a

---

[2] that may be defined as the maximum acceptable volume.

limitation of discharge from a downstream problem can be satisfied when the required limit L is $D_{min} \le L \le D_{max}$.

To deal with the case of multiple reservoirs, as stated in the general law of cooperation, it is required to keep an homogeneous risk level among all the reservoir management agents so an individual agent must accept to increase its risk level if the other "cooperating colleagues" accept also to increase. If there is a request of limitation L from a downstream problem, the objective is to determine a set of discharge values D_k along every converging river such that:

> **Condition 1:** *The sum of the discharges D_k for all the reservoirs immediately upstream the problem must be less than the limit L (unless all the risk levels of the upstream reservoirs are in the upper limit risk level):*
>
> $$\sum_k D_k \le L$$

> **Condition 2:** *The sum of the discharges D_k of the reservoirs immediately upstream the location of the problem, to avoid unnecessary discharges, must be close enough to the limit L (where α is a predefined constant between 0 and 1 close to 1, for instance $\alpha=0.85$):*
>
> $$\sum_k D_k > \alpha.L$$

The procedure to attain this solution is based on a method that increases or decreases step by step the risk level of reservoirs, following the social rule that all the reservoirs must have similar risk levels.

1. **Problem detection agents** send to all **reservoir management agents** upstream the location of the problem a message asking for a limitation of discharge L. This messages are sent by turn following the social norm that upstream agents start first.

2. Each **reservoir management agent** computes the current discharge *Dmin* according the default risk level *Rdef* or the current risk level *R*.

3. If *condition 1* is not satisfied, one of the **reservoir management agents** increases its risk level $R := R + \Delta R$, and computes its new discharge value *Dk*. To select the agent to increase the risk level, the following social norms are applied: (1) the next reservoir to increase the risk level is the reservoir with *lowest* risk level, and (2) if all reservoirs have the same risk level, they increase their risk all together. If *condition 2* is not satisfied, the agent searches for an appropriate value of *R* following a dicotomic strategy (dividing ΔR by 2, decreasing the risk levels by ΔR and restarting the process).

4. Step 3 is repeated until *condition 1* is satisfied.

5. Step 2-3-4 are repeated until all problems have been considered.

Figure 5: Social interaction method for reservoir management.

The case where several limits are imposed along the axis of the river as a result of the demand generated by local detection agents may be managed by integration of this case of single limit by considering the limit satisfaction from upstream to downstream so once the problem is solved by a set of discharges in a set of reservoirs for a limit upstream these reservoirs enter to contribute to the solution on a position downstream with the initial values of discharge and risk level that solve the problems upstream. The current discharge values may be decreased, if the risk level increase is acceptable, to meet the conditions in the points downstream. Figure 5 summarizes the method for reservoir management.

4.3 Social Interaction for Civil Protection

If the social law for reservoir agents was the delivery of the required services with a mostly homogeneous minimum risk, *the social law for resource allocation in civil protection, is to maximize the social revenue in terms of an uniform social risk unit concept integrating the perspective of the local agent suffering the damage and the resource owner aiming to an adequate balance of the resources allocation.*

The following definitions allow to obtain a reasonable solution. A local detection agent requires civil protection resources to help in a damage situation where risk may be evaluated using a qualitative scale, for instance, $<low_1, low_2, average, high_1, high_2>$ and where the size may be evaluated in terms of the units requiring the use of resources (for instance, if a local dam is to be done if a number of persons is to be evaluated, the number of persons, if an area is to be isolated the square meters, measure, etc.). A composed measure can be considered summarizing both the size and risk level in an agent:

risk level of damage, number of units => local social revenue units

A simple example of this, could be to apply to the number of units a factor associated to the risk level, so for instance if the factors depending on risk level are <1, 1.50, 1.75, 2, 2.5> if there are 20 persons in high risk level 50 social revenue units are to be computed, 2.5 x 20.

Every agent for civil protection has a priority list regarding the areas of problems in such a way that a damage in city A is more relevant that a damage in an agricultural area B. According to this model a hierarchy of factors is associated to the levels of priority of resource allocation agent in such a way that the local detection agent A has a factor of 1.5 and local detection agent B has a factor of 1.8.

According to the previous concepts the resource allocation is performed by predefined slices of the available stock of resources. Every slice is assigned by computing for every demanding agent a definitive social value resulting from combining:

local social revenue, resource social value => definitive social value

For instance, if the available resources are 60 units and the slices are 3 at the current moment 20 should be assigned, if three agents ask for 12, 8, 10 units with respective local social revenues 20, 24, 12 and the resource social values 1.5, 1.20, 1, the definitive social values (assuming factors 30/12, 28.8/8, 12/10) are 2.5, 3,6, 1.2.

Then, 8 units will be allocated to the second agent, 12 units to the first agent and no resources will be allocated to the third agent.

The assignment via slices of total resources is conceived to ensure adaptivity to the evolution of the emergency situation. For instance, it may happen that the third agent damage situation in the next hour evolves in the bad sense so the local social value of intervention grows to 30 which if the other values remain the same will produce a new assignment of the second slice with 8 units to the second agent, 10 units to the third agent and 2 units to the first agent.

1. **Problem detection agents** send to the corresponding **civil protection agents** a message asking for certain resources.

2. Every **civil protection agent** selects the next **problem detection agent** to be considered and, then, assigns the corresponding resources according to (1) availability of resources and (2) the social value. The selection process is based on a local priority model.

3. Step 2 is repeated until all problems have been considered.

4. Steps 1-2-3 are repeated for each time slice.

Figure 5: Social interaction method for civil protection.

5 Conclusions

A brief description of an agent based architecture for the problem of flood emergency management where improvements have been produced with respect to previous approaches, [1], [2], [3], [4], in the aspects of a model design, reuse and maintenance. Also two social interaction models have been designed based in two general social utility criteria, the risk level in reservoirs and social revenue for resource allocation, which may generate a distributed model of intelligent operation acceptable enough for emergency managers which, given the knowledge based internal architecture of agents, may get explanations of the system proposals.

6 References

1. Cuena J.: "The Use of Simulation Models and Human Advice to Build an Expert System for the Defense and Control of River Floods". (IJCAI-83). Karlsruhe. Kaufmann, 1983.
2. Alonso M., Cuena J., Molina M.: "SIRAH: An Architecture for a Professional Intelligence". Proc.9th European Conference on Artificial Intelligence (ECAI'90). Pitman, 1990.
3. Cuena J., Molina M., Garrote L.: "An Architecture for Cooperation of Knowledge Bases and Quantitative Models: The CYRAH Environment". XI International Workshop on Expert Systems. Special conference on Second Generation Expert Systems. Avignon'91. EC2, 1991.
4. Cuena J., Ossowski S.: "Distributed Models for Decision Support" in "Multiagent Systems - A Modern Approach to Distributed Artificial Intelligence" Sen, Weiss (eds), AAAI/MIT Press. (1999).

Problem-Solving Frameworks for Sensible Agents in an Electronic Market

K. S. Barber, A. Goel, D. Han, J. Kim, T. H. Liu, C. E. Martin, R. McKay

The Laboratory for Intelligent Processes and Systems
Department of Electrical and Computer Engineering, ENS 240
University of Texas at Austin
Austin, TX 78712-1084
barber@mail.utexas.edu

Abstract.   The need for responsive, flexible agents is pervasive in the electronic commerce environment due to its complex, dynamic nature. Two critical aspects of agent capabilities are the ability to (1) classify agent behaviors according to autonomy level, and (2) adapt problem-solving roles to various situations during system operation.   Sensible Agents, capable of Dynamic Adaptive Autonomy, have been developed to address these issues. A Sensible Agent's "autonomy level" constitutes a description of the agent's problem-solving role with respect to a particular goal. Problem-solving roles are defined along a spectrum of autonomy ranging from command-driven, to consensus, to locally autonomous/master. Dynamic Adaptive Autonomy allows Sensible Agents to change autonomy levels during system operation to meet the needs of a particular problem-solving situation.   This paper provides an overview of the Sensible Agent Testbed and provides examples showing how this testbed can be used to simulate agent-based problem solving in electronic-commerce environments.

1 Introduction

Electronic markets are inherently complex and dynamic. These characteristics create many challenges for the automation of electronic-market tasks such as matchmaking, product aggregation, and price discovery [1]. The use of agent-based systems offers significant benefits to an electronic market through adaptable automated or semi-automated problem-solving and distribution of control and processing [13]. However, simply applying the agent-based paradigm to electronic market problems may not be enough to address the real-time demands of these systems. Agent-based systems operating in the electronic market domain are subject to dynamic situational changes across many dimensions: (1) *certainty of information* held by an agent or acquired by an agent (e.g. speculation about future trends in supply and demand), (2)

This research was supported in part by the Texas Higher Education Coordinating Board (#003658452) and a National Science Foundation Graduate Research Fellowship *resource accessibility* for a particular agent (e.g. money, products), (3) *goal constraints* for multiple goals (e.g. deadlines for goal completion, goal priorities), and (4) *environmental states* (e.g. cpu cycles, communication bandwidth).

As a result, electronic markets require agent-based problem solving to be flexible and tolerant of faulty information, equipment, and communication links. This research uses Sensible Agent-based systems to extend agent capabilities in dynamic and complex environments [2]. Sensible Agents achieve these qualities by representing and manipulating the interaction frameworks in which they plan.

Agent interactions for planning can be defined along a spectrum of agent autonomy as shown in Fig. 1. An agent's *level of autonomy* for a goal specifies the interaction framework in which that goal is planned. Although autonomy is traditionally interpreted as an agent's freedom from human intervention, this extended concept of autonomy refers to an agent's degree of freedom with respect to other agents.

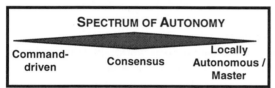

Fig. 1. The Autonomy Spectrum

An agent's autonomy increases from left to right. Agents may, in theory, operate at any point along the spectrum. Current research on Sensible Agents focuses on the three discrete, labeled autonomy level categories that define the endpoints and midpoint: (1) **Command-driven**— the agent does not plan and must obey orders given by a master agent, (2) **Consensus**— the agent works as a team member, sharing planning decisions equally with other agents, and (3) **Locally Autonomous/Master**— the agent plans alone and may or may not give orders to other agents.

Agents can be designed to operate at a single level of autonomy if (1) the application is simple, (2) the designer correctly predicts the types of problems agents will face, and (3) the environmental context and problem types remain constant. However, for complex applications in dynamic environments, the appropriate level of autonomy may depend on the agent's current situation. A Sensible Agent maintains solution quality in dynamic environments by using a technique called *Dynamic Adaptive Autonomy* (DAA). DAA allows a Sensible Agent to modify a goal's autonomy level during system operation. The process through which an agent chooses the most appropriate autonomy level is called *autonomy reasoning*.

2 Related Work

The organizational structure of agent-based systems, which is the coordination framework in which agents work to achieve system goals, has been the subject of much research over the past few decades [11;15;18]. One overall goal of multi-agent-systems research is adaptive self-configuration: allowing agents to reason about and change their organizational structure. Most self-organizing systems rely on a fixed number of explicit, predefined agent behaviors [6;10]. Others are based on adapting application-specific roles that agents can play during problem solving [7]. Sensible Agents use DAA, which allows run-time definition and adaptation of application-independent problem-solving roles.

Multi-agent researchers use simulation environments to test algorithms and representations. Existing simulation environments include: DVMT (now DRESUN) [12], and MACE [5]. Unfortunately, most of these testbeds do not support distributed heterogeneous computing environments. Additionally, these simulation environments are not designed to support third-party researcher usage.

3 Sensible Agent Architecture

As defined for this research, an "agent" is a system component that works with other system components to perform some function. Generally, agents have the ability to act and perceive at some level; they communicate with one another; they attempt to achieve particular goals and/or perform particular tasks; and they maintain an implicit or explicit model of their own state and the state of their world. In addition, Sensible Agents have many other capabilities supported by the Sensible Agent architecture shown in Fig. 2. Each Sensible Agent consists of four major modules:

- The **Perspective Modeler** (PM) contains the agent's explicit model of its local (subjective) viewpoint of the world. The overall model includes the behavioral, declarative, and intentional models of the self-agent (the agent who's perspective is being used), other agents, and the environment. The PM interprets internal and external events and changes its models accordingly. The degree of uncertainty is modeled for each piece of information. Other modules within the self-agent can access the PM for necessary information.

- The **Autonomy Reasoner** (AR) determines the appropriate autonomy level for each of the self agent's goals, assigns an autonomy level to each goal, and reports autonomy-level constraints to other modules in the self-agent. The AR handles all autonomy level transitions and requests for transition made by other agents.

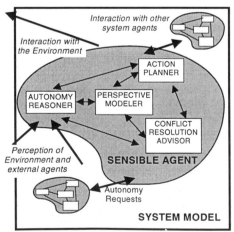

- The **Action Planner** (AP) interprets domain-specific goals, plans to achieve these goals, and executes the generated plans. Domain-specific problem solving information, strategies, and heuristics are contained inside this module. The AP interacts with the environment and other agents, and draws information from all other modules in the self-agent.

Fig. 2. The Sensible Agent Architecture.

Fig. 3. Functional View of Communication

- The *Conflict Resolution Advisor* (CRA) identifies, classifies, and generates possible solutions for conflicts occurring between the self-agent and other agents. The CRA monitors the AP and PM to identify conflicts. Once a conflict is detected, it classifies the conflict and offers resolution suggestions to the AP.

Since the AP has the only domain-specific implementation requirements, the other modules and Sensible Agent technology can be applied to many different problems.

4 Testbed Design

Accurate simulation of complex domains requires an environment that handles complex modeling issues and produces reasonable visual and numerical output. The Sensible Agent Testbed provides an environment for running repeatable Sensible Agent experiments. The wide range of functionality and implementation languages for each module requires a multi-platform and multi-language testbed environment. The current implementation allows integration of C++, Java, Lisp, and ModSIM implementations on Solaris, WindowsNT, and Linux platforms. Figure 3 illustrates how the different modules and the environment simulator are implemented as CORBA® objects that communicate through the Xerox Inter-Language Unification (ILU) distributed object environment [19].

The Common Object Request Broker Architecture (CORBA®) standard allows the use of the Interface Definition Language (IDL™) to formally define the interactions among agents and their modules. IDL™ permits Sensible Agent development in an language-independent manner and facilitates parallel research initiatives within this framework. The Internet Inter-Orb Protocol (IIOP™) standard makes it possible to connect the simulation to external ORBs™ [14] to further enhance extensibility.

The far right of Fig. 3 shows how intra-agent communication occurs without loading the inter-agent communication medium. Fig 4 shows a logical view of these communication pathways. The Sensible Agent System Interface (SASI) provides a single handle that other Sensible Agents and the environment can use for messaging and naming. The separation of the user interface from other agent functionality (the far left of Fig. 3) allows data collection and

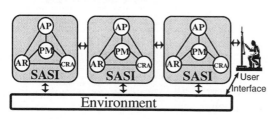

Fig. 4. Logical View of Communication

simulation viewing without local access to the greater computing power needed by the individual agents and modules.

Packages / Bundles		Items
Pkg.A	Pkg.B	CPU
		Motherboard
		Hard drive
		Memory
	Pkg.C	Case
		Monitor
		Keyboard
		Mouse

Fig. 5. Component Package Names

5 Example and Scenario

The following section describes an example of how a Sensible Agent-based electronic market system could operate. This example is intended only to demonstrate the capabilities of Sensible Agents in a simple electronic market. Agents are assumed to have a mechanism for posting and reviewing trades. Human buyers and sellers are assumed to register their goals with agents. The two primary objectives are:

- **BUY {product specification} for {monetary specification}**
- **SELL {product specification} for {monetary specification}**

The term "{product specification}" represents a variable that allows buyers to describe their needs and sellers to describe their offerings. Product specifications in this example consist of conjunctions and disjunctions of functional descriptions. Functional descriptions have been shortened to item names, as shown in Fig. 5. The term "{monetary specification}" describes how much a buyer will pay, or how much a seller will sell for.

The following is based on a sample Sensible Agent scenario in this domain. Buyers and sellers are interested in personal computer components. Fig. 5 shows the various package and item names that all the agents in the system understand. This scenario focuses on buyers: Agent 1 and Agent 2, and seller: Agent 3. Agent 1 wants to buy Pkg.B for $500, and Agent 2 wants Pkg.C for $500. Agent 3 offers Pkg.A for $1100, Pkg.B for $600, and Pkg.C for $600.

Agent 1 and Agent 2 form a consensus group to work on their goals. They decide to buy Pkg.A and split it. However, their combined purchasing power of $1000 conflicts with Agent 3's offering price of $1100. After some negotiation, Agent 3 is convinced to lower its price for Pkg.A to $1000. Pkg.A is purchased and split between Agent 1 and Agent 2. The details of this scenario are given below.

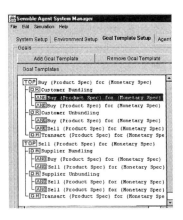

5.1 Goal Trees

Agents plan from AND/OR goal trees, which represent the agents' alternatives. The goal/subgoal relationship reflects a task-reduction planning paradigm. The goal tree displayed in Fig. 6 describes the options available to the agent in buying a product. The

Fig. 6. Sensible Agent Goal Tree

goal templates in this goal tree are instantiated by agents in this system. The items in curly braces, "{ }", identify those variables which must be instantiated in order to begin planning for a goal.

In this electronic market the goal to buy a product can be split up through bundling or unbundling by the customer. Customer bundling aggregates the purchase of two products into a single product that satisfies the higher level "buy" goal. Customer unbundling extracts a portion of a purchased product to satisfy the higher level goal and adds a sell goal to remove the excess components of the purchased product. The goal tree for the "sell" top-level goal exactly mirrors that for the "buy" top-level goal. Suppliers can aggregate products to sell as a package or disaggregate a product to sell as component pieces. These goal trees are recursive, because lower level goals are additional instances of top-level goals. The "transact" goal allows agents to terminate the recursion and carry out the trade. After performing this action, the owners of the respective objects are switched.

5.2   Autonomy Reasoner

In this example domain, Sensible Agents have multiple factors to consider when selecting an autonomy level for a particular goal. For example, a buyer agent may find a seller whose product is just what it was looking for. In this case, it may choose locally autonomous for its "buy" goal. On the other hand, it may not be able to find the exact item it is looking for. However, some sellers may offer a bundle that includes the desired item. This situation may prompt a buyer agent to enter an autonomy agreement with some other buyers who want the other parts of the bundle. Similarly, seller agents may form an autonomy group to sell their products as a bundle. Since money is involved, it is important for agents to be in the planning group for any multi-agent agreements they form. Therefore, in this domain, master/command-driven autonomy groups would not be seen as frequently consensus autonomy groups. However, master/command-driven agreements may still be more attractive under some circumstances. For example, communication bandwidth may be insufficient for negotiation, a mechanism often used in consensus groups to reach agreement.

Sensible Agents use their communication abilities for allocation of goals, negotiating Autonomy Level Agreements (ALA), and general information passing. Sensible Agents currently use a protocol based on the Contract Net Protocol [16] for allocating goals and planning responsibility through ALAs. A contract negotiation begins with a contract announcement, followed by bids from interested agents. The announcing agent makes a decision and awards the contract to one of the bidders.

5.3   Perspective Modeler

An agent's behaviors are determined by its internal states and external events as well as system and local goals. An agent's understanding of the external world is built from its interpretations of the states, events, and goals of other agents and the environment. A Sensible Agent must be able to perceive, appreciate, understand, and demonstrate cognizant behavior. Both behavioral and declarative knowledge are used by an agent to understand itself, other agents, and the environment. The declarative knowledge in the Perspective Modeler (PM) consists of a set of attributes for the self-agent and for other agents. Behavioral knowledge is modeled by Extended Statecharts (ESCs) [17] a derivative of Statecharts [9] with respect to temporal and hierarchical extensions.

A Sensible Agent represents the goals it has chosen to pursue in an Intended Goals Structure (IGS). The IGS differs from a goal tree. The goal tree contains templates for candidate goals, which have not been accepted by any agent. On the other hand, the IGS is an agent's representation of the actual goals it intends to achieve as well as any additional goals for which it must plan (external goals, intended by other agents). For a formal definition of agent intention, see [4]. The IGS contains AND-only compositions of these goals and therefore does not represent alternative solutions or future planning strategies.

The example IGS Viewer in Fig. 7 shows that autonomy levels are assigned to each goal in an agent's IGS. These autonomy assignments are classified in each goal's label as locally autonomous (LA), consensus (CN), command-driven (CD), or master (M). Fig. 7 shows Agent 2's IGS as it appears at the end of the trading scenario. It's own top-level goal, "Buy Pkg.C for 500," has been combined with Agent 1's goal, "Buy Pkg.B for 500," in a consensus agreement. This *goal element* (an individual entry in an agent's IGS) has been highlighted, and details about its autonomy assignment appear in the right-hand panel. Agent 2 uses the identifying number "0" for this goal element and has allocated 10 planning resource units toward achieving the goals. Both Agents 1 and 2 are planning (in consensus). Agent 2 identifies its intended goal, "Buy Pkg.C for 500," using the numerical identifier "1". Agent 2 identifies the goal element in this local model that contains Agent 1's intended goal, "Buy Pkg.B for 500," using the identifier "1-0", where "1" is Agent 1's id number and "0" identifies which goal element in the local model of Agent 1's IGS contains that intended goal. The autonomy assignment information also indicates that the planning group has authority to allocate sub-goals to either Agent 1 or Agent 2. Agent 2 has "COMPLETE" commitment to its goal of "Buy Pkg.C for 500," which means it cannot give up the goal. Agent 2 has "LOW" commitment to the consensus autonomy agreement with Agent 1, which means Agent 2 can dissolve this agreement without penalty. Finally, the independence level assigned for this planning group is "NONE," which means that Agents 1 and 2 are planning with the most restrictive set of social laws available.

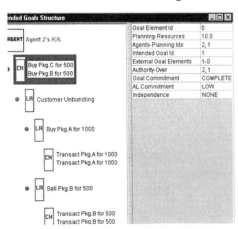

Fig. 7. Intended Goals Structure for Agent 2

Agents 1 and 2 determine that Agent 2 should perform "Customer Unbundling" in order to achieve the top-level consensus goal. Agent 2 is locally autonomous for this goal, and decides it should purchase Pkg.A for 1000 and then sell Pkg.B to Agent 1 (keeping Pkg.C for itself). Agent 2 and Agent 3 agree to perform the trade for Pkg.A, and then Agent 2 and Agent 1 agree on the Pkg.B transaction.

5.4   Action Planner

After the AR has assigned an autonomy level for a particular goal, the Action Planner (AP) performs the problem solving calculations. The AP is designed to hold multiple problem solving strategies and select the most appropriate for each goal. Currently, the AP uses a two staged planning methodology to (1) improve coordination among agents through the use of autonomy levels, and (2) leverage previous work in the planning field as applied to various domains [3].

The first stage consists of a hierarchical planner, which differentiates between those activities that are critical to success and those that are replaceable details. This stage builds the IGS through the selection of subgoals. The second stage provides an interface to plug in a domain specific planner. For each goal in the IGS that can be solved directly, this stage creates a plan by selecting the appropriate actions to execute. Autonomy levels are used as a selection aid for both stages [3].

In the electronic market, an agent may hold a goal element containing goals to buy Pkg.B for itself and goals from other agents to buy Pkg.C. The planning agent may then choose to instantiate a consumer unbundling goal. Underneath the consumer unbundling goal the AP could choose to Buy Pkg.A for the combined money from the parent goals. The components of Pkg.A are redistributed among the agents using the sell goal template. An instantiation of the transact goal is used to purchase Pkg.A. The second stage planner searches for a supplier that sells Pkg.A and performs the transaction. Planning by suppliers will operate in a similar manner. The AR may implement different ALAs for any of the instantiated goals, leading to more complex behavior. The Conflict Resolution Advisor (CRA) is used to check for inconsistencies in the IGS or plan.

5.5   Conflict Resolution Advisor

A Sensible Agent can dynamically select a suitable conflict resolution strategy according to (1) the nature of conflict (i.e. goal, plan or belief conflict), (2) the agent's social roles (represented by its autonomy levels), and (3) its solution preferences. Utility is used for decision making, which is the total weighted utility value of a specific solution for its attributes (see [8] for details of definition and calculation of utility based on autonomy level). In addition to evaluating potential solutions, agents should also use certain indices of utility (e.g. effectiveness, performance, and agent and system properties) to evaluate available CR strategies.

There are three kinds of decision-making styles that agents can make by tuning the weight factors: (1) CR strategies are selected after preferred potential solutions --

preferred solutions are the most important targets, (2) CR strategies are selected before preferred potential solutions -- the solutions will be the natural results of applying specific CR strategies, and (3) balanced consideration between preferred solutions and CR strategies.

6 Conclusions and Future Work

Electronic commerce environments are complex and dynamic in nature. Because system designers cannot predict every possible run-time situation, approaches to automating electronic market systems must be adaptive and flexible. Dynamic Adaptive Autonomy (DAA) is a technique that allows software agents to modify their planning interaction frameworks at run-time. Sensible Agents, capable of DAA, provide agent-based systems that are both adaptive and flexible. This paper has demonstrated one possible application of Sensible Agents to electronic markets through an example system description and scenario.

The Sensible Agent Testbed currently provides a set of functionality that greatly facilitates further work on each of the Sensible Agents modules. Future work for the Action Planner module includes enhancing existing planners to take advantage of DAA. The Autonomy Reasoner module's task of selecting the optimal planning framework in which to plan is critical to the future development of Sensible Agent-based systems. Currently, we are investigating techniques of reinforcement learning and case-based reasoning for this task. Future work for the PM module includes incorporating more advanced reasoning capabilities and continuation of work on ESC knowledge representation to model uncertainty and time. The communication capabilities of Sensible Agents will soon be extended to include a variety of negotiation protocols, and the ability to reason about which communication protocol to use in various situations.

References

[1] Bailey, J. P. and Bakos, Y.: An Exploratory Study of the Emerging Role of Electronic Intermediaries. International Journal of Electronic Commerce, 1, 3 (1997) 7-20.
[2] Barber, K. S.: The Architecture for Sensible Agents. In Proceedings of International Multidisciplinary Conference, Intelligent Systems: A Semiotic Perspective (Gaithersburg, MD, 1996) National Institute of Standards and Technology, 49-54.
[3] Barber, K. S. and Han, D. C.: Multi-Agent Planning under Dynamic Adaptive Autonomy. In Proceedings of IEEE International Conference on Systems, Man, and Cybernetics (San Diego, CA, USA, 1998) IEEE, 399-404.
[4] Cohen, P. R. and Levesque, H. J.: Intention is Choice with Commitment. Artificial Intelligence, 42 (1990) 213-261.
[5] Gasser, L., Rouquette, N. F., Hill, R. W., and Lieb, J. Representing and Using Organizational Knowledge in DAI Systems. In Distributed Artificial Intelligence, vol. 2, Gasser, L. and Huhns, M. N., (eds.). Pitman/Morgan Kaufman, London, (1989) 55-78.

[6] Glance, N. S. and Huberman, B. A.: Organizational Fluidity and Sustainable Cooperation. In Proceedings of 5th Annual Workshop on Modelling Autonomous Agents in a Multi-Agents World (Neuchatel, Switzerland, 1993) 89-103.

[7] Glaser, N. and Morignot, P. The Reorganization of Societies of Autonomous Agents. In Multi-Agents Rationality: Proceedings of the Eighth European Workshop on Modeling Autonomous Agents in a Multi-Agents World, Boman, M. and van de Velde, W., (eds.). Springer-Verlag, New York, (1997) 98-111.

[8] Goel, A., Liu, T. H., and Barber, K. S.: Conflict Resolution in Sensible Agents. In Proceedings of International Multidisciplinary Conference on Intelligent Systems: A Semiotic Perspective (Gaithersburg, MD, 1996) 80-85.

[9] Harel, D.: Statecharts: A Visual Formalism for Complex Systems. Science of Computer Programming, 8 (1987) 231-274.

[10] Ishida, T., Gasser, L., and Yokoo, M.: Organization Self-Design of Distributed Production Systems. IEEE Transactions on Knowledge and Data Engineering, 4, 2 (1992) 123-134.

[11] Kirn, S. Organizational Intelligence and Distributed Artificial Intelligence. In Foundations of Distributed Artificial Intelligence. Sixth-Generation Computer Technology Series, O'Hare, G. M. P. and Jennings, N. R., (eds.). John Wiley & Sons, Inc., New York, (1996) 505-526.

[12] Lesser, V. R.: A Retrospective View of FA/C Distributed Problem Solving. IEEE Transactions on Systems, Man, and Cybernetics, 21, 6 (1991) 1347-1362.

[13] Nwana, H. S., Rosenschein, J. S., Sandholm, T., Sierra, C., Maes, P., and Guttmann, R.: Agents-Mediated Electronic Commerce: Issues, Challenges and some Viewpoints. In Proceedings of Second International Conference on Autonomous Agents (Minneapolis/St. Paul, MN, 1998) ACM Press, 189-196.

[14] OMG, OMG Home Page. http://www.omg.org/ Current as of February 10, 1999.

[15] Singh, M. P.: Group Ability and Structure. In Proceedings of Decentralized A.I. 2, Proceedings of the 2nd European Workshop on Modelling Autonomous Agents in a Multi-Agents World (Saint-Quentin en Yvelines, France, 1990) Elsevier Science, 127-145.

[16] Smith, R. G. and Randall, D. Frameworks for Cooperation in Distributed Problem Solving. In Readings in Distributed Artificial Intelligence, Bond, A. H., (ed.). Morgan Kaufmann Publishers, Inc., San Mateo, CA, (1988) 61-70.

[17] Suraj, A., Ramaswamy, S., and Barber, K. S.: Extended State Charts for the Modeling and Specification of Manufacturing Control Software. International Journal of Computer Integrated Manufacturing, Special Issue on Design and Implementation of Computer-Integrated Manufacturing Systems: Integration and Adaptability Issues, 10, 1-4 (1997) 160-171.

[18] Werner, E. and Demazeau, Y.: The Design of Multi-Agents Systems. In Proceedings of Proceedings of the 3rd European Workshop on Modelling Autonomous Agents in a Multi-Agents World (Kaiserslautern, Germany, 1991) Elsevier Science Publishers, 3-28.

[19] Xerox, Inter-Language Unification -- ILU. ftp://ftp.parc.xerox.com/pub/ilu/ilu.html Current as of February 10, 1999.

A Model for Distributed Multi-agent Traffic Control

Christine Bel, Wim van Stokkum

Kenniscentrum CIBIT
P.O. Box 19210, 3501 DE Utrecht, The Netherlands
Tel: +31 (0)30 - 230 8900 Fax: +31 (0)30 - 230 8999
URL: http://www.cibit.nl Email: cbel@cibit.nl, wvstokkum@cibit.nl

Abstract. Railway traffic is not always being controlled optimally. Using the current approach the Dutch Railways can only use up to 70% of the capacity of the infrastructure and adjustments often lead to delays. It has been investigated whether the use of distributed multi-agent traffic control can enlarge the use of the capacity of the infrastructure, improve the ability to make adjustments when disturbances occur and reduce delays. It seems that the approach is successful in reducing delays. More important, however, is that when delays occur, travelers are less bothered by the delays, because their interests are taken into account when adjustments are made.

1 Introduction

The Railned BV company is responsible for the capacity management and the railway safety within the Dutch Railways. For environmental and economical reasons, railway transport, for both passengers and cargo, is stimulated by the government. Increasing numbers of passengers and the growing need for cargo transport force the Railways to investigate possibilities to make more efficient use of the infrastructure, while maintaining safety and high quality. These last two requirements are particularly difficult to satisfy. An increasing amount of traffic using the same infrastructure will require a more advanced way of ensuring the safety, because trains will have to be allowed to get closer to each other than they are allowed now (in the current situation, only one train is allowed at a time in a certain, relatively large, area). More traffic will also require a better anticipation on disturbances. In the current situation only a percentage of the maximum capacity of the infrastructure is used and the rest is used to anticipate on (minor) disturbances. When a higher percentage of the capacity is used, there is a need for a 'smarter' way to anticipate on the disturbances. In order to be able to consider alternative approaches Centre of Excellence CIBIT has been asked to explore the possibilities to optimize both railway traffic control and railway capacity, using innovative techniques and to investigate in what way information technology (IT) can contribute to optimization of the current process of traffic control and the technologies in which to invest for the next millennium.

2 Problem Description

In the current situation traffic control is determined way ahead. Of course time-tables should be available to the public, but times of arrival at points *between* the starting point of a train and its destination(s) as well as the train's speed aren't relevant to the public. Nevertheless, even these details are scheduled in advance. As long as everything goes as planned this system works perfectly. But, of course, in real life not everything goes as planned. There may be technical problems, more travelers than estimated, causing a train to make longer stops, drivers may be late, accidents may happen etc. All these disturbances together cause delays and make travelers miss their connecting trains. To avoid this as much as possible, adjustments to the initial traffic plans have to be made. This is very difficult, because adjustments can influence other plans, which will then have to be adjusted as well etc. To make a necessary amount of adjustments possible, today only 70% of the capacity of the infrastructure is used when making the initial plans. Apart from this, for security reasons, only one train at a time is allowed in a certain (predefined) area. This is also one of the causes of suboptimal use of the infrastructure. Because of the earlier mentioned increasing number of travelers (people are encouraged to travel by train instead of causing traffic jams going by car) and the growing amount of railway-cargo, there is a need to use the infrastructure more efficiently. If a better approach to anticipate on disturbances would be used, delays will be reduced, offering better service to the public. At the same time, using this approach, it may be possible to use a higher percentage of the infrastructure's capacity.

3 The multi-agent perspective

Centre of Excellence CIBIT has developed a model and a prototype for distributed real-time traffic control, based on multi-agent concepts. In this model centralized traffic control is minimized. Instead, most traffic control is carried out by the trains and elements of the infrastructure, such as platforms, crossings and points. The trains and elements of the infrastructure are autonomous agents, which communicate with each other, and try to reach their own goals. Only the different runs to make are planned in advance. The details of the traffic control within each run aren't fixed, but are determined by negotiation between the agents.

3.1 The agents

The model for distributed real-time traffic is focussed on the infrastructure level. On the infrastructure level two concepts play an important role:

- Infrastructure points (elementary points in the railway infrastructure: crossings, platforms and sections)
- Trains

These concepts are chosen to act as agents, because they are the concepts on the lowest possible level to which responsibilities can be allocated. By choosing this low level, the system works as distributed as possible, where agents only have to pursue their own goals, using their own information or communicate with others to obtain information about them. The information needed to make decisions and to keep the other agents informed is situated at agent level. For example, trains know best about their situation, location and speed and therefore are best suited to make an estimated time of arrival. Platforms, however, have all the information about which trains should pass or stop. Therefore, they are held responsible for the detection of conflicts. So, in the model developed, each of these concepts is implemented an autonomous agent, each having its own primary goals to achieve and its own knowledge to achieve these goals. In the following sections each agent is described in more detail.

3.2 Perspective of a train agent

The goal of the trains is to minimize delays for its passengers both in terms of time as well as in terms of experienced annoyance. In trying to arrive on time, it will close contracts with the infrastructure points about the time it is allowed to pass the infrastructure point. When a train starts a particular run it receives a list of the infra points it will have to pass. For each infra point the train will pass, it makes an estimate of its time of arrival. The train calculates how much time it will need to get somewhere by its:

- maximum speed
- ability to accelerate
- length and load of the train (especially important for cargo trains)
- fitness (age and state of maintenance)

The train will then communicate its estimated time of arrival to the infra points with a certain amount of uncertainty to compensate for possible disturbances. The distribution communicated to the points includes an estimated time of arrival as well as an earliest and latest time of arrival. After each interval (6 sec.) the train checks whether its predictions are still valid or new estimates have to be made, because of:

- good luck (e.g. 'wind in the back')
- bad luck (disturbances occurred)
- severe incidents (cow on rails)
- a more accurate estimate can be given (the train is closer to the point, and therefore can give a more precise estimation about its arrival time)

In this way, the infra points are always informed of up-to-date information about the times of arrival of the trains that will be passing the point in the future. All contracts together form a plan for the train to travel to its destination point. The train will adhere to this plan, unless it receives a message from an infra point with an alternative plan. Such messages are sent when a problem has to be solved. Solving problems will be covered in more detail in section 4.

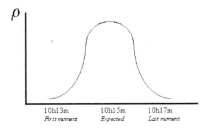

Fig. 1. Estimate of time of arrival a train sends to an infra point.

3.3 Perspective of an infrastructure point agent

The primary goal of the infrastructure elements is, of course, to ensure safety. A secondary goal, however, is to make optimal use of the infrastructure capacity. When an infra point receives the estimate from a train it compares this with the (earlier received) estimates of the other trains that are going to pass. It checks whether safety requirements are met. Whenever two or more trains estimate to arrive at approximately the same time a conflict is detected that must be solved in order to guarantee safety.

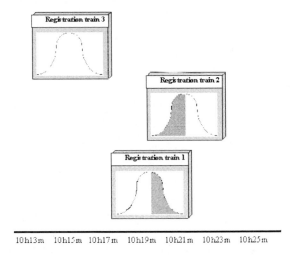

Fig. 2. Estimates of times of arrival from three trains; two overlapping

Detection of conflict To predict the change of a conflict to occur, the infra point calculates the amount of overlap between the estimates. The chance is calculated by determining the area of the intersection between the two (normally distributed) predictions. In this process safe margins between trains are taken into account. For example: there should be a certain amount of time between two trains passing a crossing or a platform. This amount of time is different for different infra points and differs also for passing trains, arriving trains and leaving trains.

Acceptable versus non acceptable risks Whenever the infra point detects a conflict it will decide whether the risks are acceptable if the conflict is not solved. Two factors are taken into account making this decision: *a predefined threshold* and *the amount of uncertainty involved*. The threshold value is introduced to exclude risks that are very unlikely to occur. In the simulation model a threshold value of 35% has proved to be acceptable. Whenever the chance of an accident to occur happens to be smaller than this threshold, the conflict will not be solved (unless the probability of a collision increases later on). The *uncertainty value* is introduced to exclude the solving of conflicts detected on base of very uncertain data. When the trains involved in the conflict at infra point X have to travel a relative long distance to point X, their estimates will have a large standard deviation, indicating the amount of uncertainty. On their way to the point a lot of events can occur. The standard deviations are taken into account when detecting the conflict. Whenever the conflict is detected based on large standard deviations, the conflict will not be solved immediately. The assumption is that the conflict will probably be solved later on by infra points earlier on track. If these points will not solve the conflict, the conflict will eventually be solved by point X in the future, because this point will receive (more accurate) updates of the estimates whenever the trains are closer to the point.

4   Solving a problem

As explained earlier, infra points are aware of the trains on a track. Each infra point receives estimates from each train passing the point. These estimates are always up-to-date and valid for a period of 6 seconds. Based on this information, the infra points will detect possible conflicts. Whenever a infra point receives new or updated estimates, it will detect possible conflicts and will decide if these conflicts are acceptable (for the time being) or not. When more than one non acceptable conflict is detected, it will rank those conflicts by probability of occurrence. Every 6 seconds each infra point has the chance to solve the most severe conflict.

Infra point proposes a plan to the trains involved in the conflict A conflict detected at an infra point always involves two or more trains. These trains have predicted to arrive at the same time (or at least: too close to eachother). For a selected non-acceptable conflict, the infra point will make up one or more plans to solve the problem. These alternative plans imply modified times of arrival for involved trains. For each plan the trains involved will receive a notification of the infra point of the plan, asking for the consequences of arriving earlier or later at a time specified in the plan.

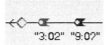

Fig. 3. Two trains on one track arriving too close together at a point

It is possible that a train receives more than one plan in the context of the same conflict (each having a different proposal for the arriving time). In figure 3 two trains ("3.02", "9.02") are involved and the point can not change their order of arrival. In figure 4 the point can change the order of the trains. In this case there will be two plans. Plan 1 is asking "3.02" to arrive later at the point than "9.02", plan 2 is asking "9.02" to arrive later than "3.02". Of course more than two trains can be involved, resulting in combinations of these plans made by the infra point.

Fig. 4. Two trains on different tracks arriving too close together at a point

Train determines consequences of a proposed plan Whenever a train receives a plan from some infra point, it will propagate the effect of the plan and calculate the costs. A plan includes a proposal for a new arriving time at a point. If this new arriving time is later than the time the train had planned itself, this implies that the train has to reduce speed at the track leading to the point. This mean it will also arrive later at the points following the particular point. When evaluating a plan it will estimate the new time of arrival at each platform following the point in case the plan is accepted.

Virtual costs for delays at a platform The costs of a plan (for a particular train) are determined by adding the cost of each delay at a platform following the point. In calculating the cost not only the time of delay is involved, but also the impact that delay will have on the level of quality experienced by the customers (passengers or cargo-clients). In determining the level of quality, the following aspects are taken into account:

- priority of trains (InterCity, EuroCity, Cargo,...)
- number of passengers getting of the train (known statistics)
- number of passengers that will miss their connection (and the extra time they'll have to wait until the next train)
- kind of passengers (number of regular customers with reduction cards and number of incidental customers with full-price tickets)

- in case of cargo transport: price one is willing to pay for the transport (similar with EMS services: paying for guaranteed arrival times)
- annoyance factor (how many times has this train lost negotiations)

The cost of a plan is the cumulated cost of all trains involved in the plan The infra point initiating a plan will receive from each train involved the cost of the plan for that particular train. The cumulation of the costs of the individual trains will result in the total cost of the entire plan. The infra point will then choose the plan with the lowest cost. This will be communicated to the trains, making this plan the new plan. Every 6 seconds the infrastructure commits to follow the current plan.

A plan can cause follow up conflicts Solving conflicts like those described before seems to be a straight-forward procedure. In practice however initiating a plan may cause other conflicts at other points. In the evaluation of the plan these follow-up conflicts are also taken into account. This consequence means that each agent must be able to build up and maintain it's own belief system in which hypothetical plans can be evaluated and (partially) be redrawn.

Handling exponential complexity Plans causing other conflicts may cause a situation in which problems can not be solved due to the exponential complexity of the problem solving method. In the model this problem has been tackled in the following ways:

- By taking uncertainty into account by solving follow-up conflicts, problem solving can be reduced to a local area.
- Trains are given autonomy to decide to commit to a plan without always evaluating them. Statistically, a train knows the margin it has in terms of time to give away at the beginning of its route.

5   Simulation evaluation

The model has been tested by building a simulator. The section Arnhem-Utrecht of the Dutch rail net has been implemented in a prototype. Figure 5 shows the situation at Arnhem. The simulator supports extension of the model (to cover a larger area or to add functionality) and serves as an instrument to evaluate this alternative means of traffic control. To be able to compare the alternative approach with the current approach, the current situation has been implemented as well. Therefore, it wasn't only possible to test the alternative approach and to see what happened in different circumstances, but also made it possible to get statistical results from the simulation.

As can be seen in figure 6, compared to the current approach, delays are reduced using the alternative approach. The alternative approach, however, has limited means to make adjustments in trying to solve very long delays (e.g. it has no ability to make diversions). The improvements made by the alternative approach are even more clear in figure 7, which shows the differences of 'costs', this is amount of

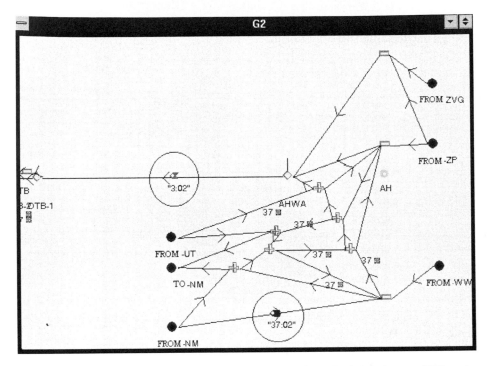

Fig. 5. The interface of the simulator: the situation at Arnhem. In the circles two (different kind of) trains are depicted. The color indicates what kind of train it represents.

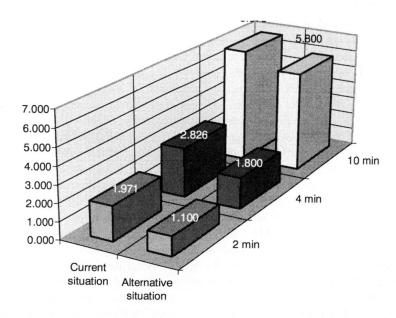

Fig. 6. Difference in mean delays between current and alternative situation

annoyance to the customer, by the different delays. These 'costs' are significantly lower using the alternative approach.

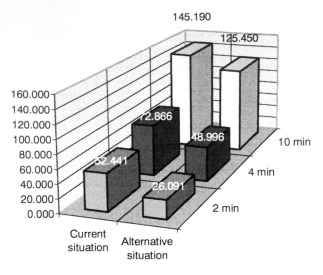

Fig. 7. Difference in costs of the delays between current and alternative situation

6 Discussion

The alternative model of traffic control has a some advantages. Though sometimes central control is still needed (e.g. to decide when a certain train service has to be canceled), most decisions will be made at a lower level: solutions can be offered and decisions can be made in a distributed manner. The results of the simulation show that the mean duration of short delays can be reduced by 25%. Although long delays are less easy to solve, the travelers are less bothered by these delays, because connections with other trains are being preserved by attaching priorities. The average virtual costs (annoyance experienced by customers) is reduced by 40%, independent of the average delay. In this way, the quality of the service experienced by the passengers will increase dramatically. Additional functionality can make the approach even more powerful. If the system of using strict time tables would be exchanged for an 'metro like' system, where there is no strict time table, but a guarantee that every few minutes a train comes by, more advantage can be taken of the strength of this approach. Also, when the model would be extended and agents would be allowed to make diversions, the approach may prove to be even more efficient than it is right now. To make the model useful for a larger area and to be able to make diversions which have an impact on a large area, it might be necessary to create some kind of 'middle management' agents which represents junctions or railway centres. Compare this 'middle management' with the collaborating agents forming 'holonic agents' which act in a corporate way [Bürckert et al.]. As stated in the introduction, growing traffic asks for alternative ways to optimize transportation. In [Antoniotti et al] semi-automated vehicles driving on a multiple merge junction highway, where merging and yielding is controlled in a distributed way. Although the intelligence of these vehicles is concentrated in their ability to calculate the best way to merge, rather than in communication and collaboration, it illustrates that also in other kinds of transport, distributed control can be used for optimisation.

References

Dunin-Keplicz, B., Treur, J.: Compositional formal specification of multi-agent systems. In: M. Wooldridge, N. Jennings (eds.): Intelligent Agents, Proc. of the First International Workshop on Agent Theories, Architectures and Languages, ATAL'94, Lecture Notes in AI, vol. 890, Springer Verlag, 1995, pp. 102-117.

Bürckert, H-J., Fischer, K., Vierke, G.: Transportation Scheduling with Holonic MAS The TeleTruck approach. In: Nwana, H. S., Ndumu, D. T. (eds.): Proc. of the Third International Conference and Exhibition on The Practical Application of Intelligent Agents and Multi-Agent Technology, PAAM'98, The Practical Application Company Ltd, 1998, pp. 577-590.

Antoniotti, M., Deshpande, A., Girault., A., Microsimulation Analysis of a Hybrid System Model of Multiple Merge Junction Highways and Semi-Automated Vehicles. Proceedings of the IEEE Conference on Systems, Man and Cybernetics, Orlando, FL, U.S.A. October 1997

Smart Equipment Creating New Electronic Business Opportunities

Rune Gustavsson

University of Karlskrona/Ronneby
Ronneby, Sweden
Rune.Gustavsson@ipd.hk-r.se
URL: www.sikt.hk-r.se/~soc/

Abstract. Ongoing deregulation of energy markets together with emergent Information and Communication Technologies totally changes the business processes for utilities. They have to change from selling a commodity such as kWh to build up new business relations based on user defined added value services. These topics have been in focus of an international project ISES. We report on some lessons learned from ISES and future work to guide utilities in the ongoing transformation. We claim that smart communicating equipment on the electric grid pro-vides important building blocks in creation of future value added services and products. To emphasise the importance of web-based technologies, we denote these services and products as eServices and eProducts.

1 Introduction

The three year long *ISES* project was completed at the end of 1998. ISES stand for Information/Society/ Energy/Systems and was co-ordinated by the R&D company EnerSearch AB. The success of the ISES project has been followed up by another international R&D effort in the form of an international *Research Academy of IT in Energy* (www.enersearch.se/).

The key objectives of the ISES project were to investigate the impact on utilities due to two major forces. Firstly the deregulation of the energy sector, secondly emergent IT solutions. The deregulation of the energy sector could mean that the utilities face a transition from having a monopoly of selling a commodity, kWh, to being players in a highly competitive market with very low profits on the same commodity. Emergent technologies such as using the electric grid as a communication channel, gave on the one hand, the utilities a communication network for free. On the other hand, customers of the utilities could easily go for other suppliers of the commodity unless they felt that the relation with the old supplier also gave some added value. In short, utili--ties should aim at getting out of the *"kWh trap"* by substituting selling a commodity with providing added-value services to the customer. Of course, this change of business strategies is much easier to say than to do.

The paper is organized as follows. First we paint a broad picture of a current set of application areas in the next section, Section 2. The following two sections, Section 3 and Section 4, give a review and summary of the ISES project backing up our statements in Section 2. In Section 5 we return to the issues put forward in Section 2 and describe a common framework for these applications. The paper ends with a section on some conclusions.

2 The Big Picture: Challenges and common infrastructures

A modern post industrial society, such as Sweden, faces at least the following three challenges:

- Efficient and environmental friendly creation and use of energy.
- Making more efficient and better products by adding suitable functionalities.
- Create safe, secure and trusted care at homes for elderly and disabled citizens.

The first topic was a focus in the ISES project and we will report some results in subsequent sections. The second challenge is sometimes phrased as *"From kWh to services"* as in the ISES project or as *"From refrigerators to selling freezing services"* or *"From heat exchangers to selling heating services"* as in a couple of ongoing projects. In an aging society we can not longer afford to support an increasing number of senior citizens in traditional institutions. Furthermore there are social benefits if elderly and handicapped people can remain as long as possible in their properly equipped homes. In the ISES project we have done a few studies and implementations of *"Smart homes"*.

We claim that although these applications are quite different applications they have very much in common with respect to supporting infrastructure. Basically, this is the motivation behind the Framework as described in Section 5. Our approach is to implement selected emerging technologies supporting suitable eServices and eProducts in the application areas above. As a research group, we are particularly focusing on agent technologies in these settings.

3 New business opportunities: Two-way communication between utilities and their customers

The Information Age is inconceivable without electricity. But, the way in which electricity is produced and distributed to its billions of customers, will radically change as a result of the Information Age. Global networks such as the Internet connect people on a worldwide scale. Local access networks - assuming different forms, from telephone copper, cable TV, wireless, satellite - serve to interconnect people at the level of local buildings, offices and homes. Today, even the power line can be used to this end. Emergent radio based access technologies supplement the power line in a transparent net-worked world. So to speak, "smart" industrial and household equipment, interconnected by networked microprocessors, is itself becoming part of the Information Society. Wall sockets and home electrical meters will be contact points not only for electricity, but also for information and communication.

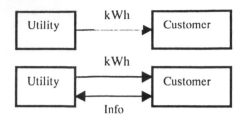

Fig. 1. Through advances in information and communication technology, two-way communication between the utility and its customers has become possible. As a result, the strategic position of energy businesses is shifting from pure product delivery (top) to customer-oriented service offerings (bottom).

These advances in information and communication technology (ICT) coincide with an on-going process of liberalization and deregulation of the energy and telecom industry sectors, in Europe and elsewhere, enhancing competition and customer pressure, and increasing societal demands, such as the concern for our environment. The combination of these technological, economical and societal factors has a profound influence on the strategic position of energy utilities. This strategic change, from one-way energy production orientation to two-way customer oriented service offerings, is visualized in Figure 1.

This transformation poses many challenges and opportunities for new services and applications, but also raises new questions. What kind of new applications are enabled by advances in ICT technology? How should the underlying business strategies and concepts be shaped? How will customers react upon new offerings? Here, we will discuss a specific type of applications of innovative information technology, namely, how to build new business processes based on smart equipment connected to the electric grid involving thousands (and more) of industrial and household devices and appliances.

4 Some Lessons learned from the ISES project

The basic model of interaction with a customer in a monopoly situation and in the case of a commodity is basically as follows. The customer buys the product or service without any further interest from the supplier. As an example: A customer decides that he/she wants to buy electricity from a utility. The utility responds by setting up a meter at the customer site and delivers the needed power. The customer then gets a bill referring to the amount of consumed kWh as well as tariffs and other rather technical information. Furthermore the bill is often based on predictions of consumption which makes it difficult to have a clear picture of the potentials of for instance energy saving or avoidance of expensive peaks in the consumption.

Having this in mind, it is quite clear that a first mean to create a better interaction with the customer is to design and implement a suitable form of an *Interactive Bill*. As a matter of fact this was also a first attempt in the ISES project to investigate interaction between customers and utilities, [4]. However, the attempt did not become a fullfledged prototype mainly for three reasons. Firstly, it turned out that the infrastructure, that is a sufficient smart meter at the customer site allowing a bi-directional communication was not in place in time to set up a real pilot study. However, the meter functioned excellent as a remote metering device meaning that the local power distributor had real-time access to a database containing a complete and automatic updated history of power consumption at individual meters in the test. The second reason, or obstacle, was at that time we did not have a clear picture of the factual content or of the presentation style of the Interactive Bill. Thirdly, it turned out that the local power distributor did not have a clear view of how to handle the billing in the new situation. Issues such as security and trust in electronic information exchange were not properly understood at that time. However, the lessons learned highly influence our mindset of future models of interactions between customers and service providers, Section 5.

4.1 Load balancing as a new business opportunity: A Society of HomeBots

As an attempt to create a first win-win business opportunity between utilities and their customers we focused on *Load balancing* on the electric grid. The challenge of efficiently managing the total energy system is easy to state. At each and every point in time and space, energy delivery through the production and distribution network must exactly match the wide variety of energy consumption demands stemming from extremely large numbers of devices and processes of all kinds at the customer side.

One possibility to increase the efficiency of the energy system is to try to manage customer equipment so as to reduce the temporal fluctuations in demand (Figure 2, below). As energy consuming devices are known as loads in this context, this is called load management. The goal of load management is to move demand from expensive hours to cheaper hours. This reduces costs, curtails energy system over- or undercapacity, and enhances the utilization degree of investments in existing energy network assets.

As illustrated in Figure 2, energy demand varies depending on the nature of the customer premises, and it shows big fluctuations over time.

The number of loads is several orders of magnitude larger than in current forms of load management but as we have demonstrated, it is nevertheless possible to implement efficient load management of this scale, [1] and [8]. The trick is to let smart equipment manage itself, programmed with the joint goal to find the optimum situation of energy use.

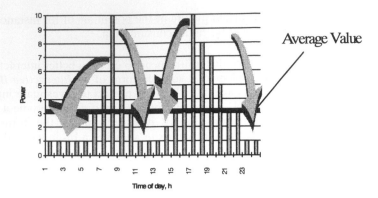

Fig. 2. The characteristic daily energy load curve of a residential household. The goal of energy load management is to move demand from expensive hours to cheaper hours. This reduces costs and improves the utilization of existing energy production and distribution network assets.

It is now technologically possible that software-equipped communicating devices *"talk to"*, *"negotiate"*, *"make decisions"* and *"co-operate with"* one another, over the low-voltage grid and other media. We use this concept to achieve distributed load management in a novel fashion: by a co-operating *"society of intelligent devices"* (Figure 3). These software agents we call *HomeBots*, [1].

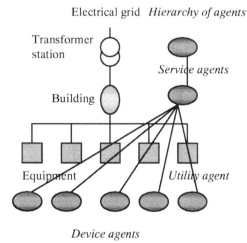

Fig. 3. Devices are equipped with smart small software programs called agents. These "Home-Bot" software agents communicate, act and co-operate as representatives assisting the customer, to achieve goals such as energy load management.

It is the responsibility of each HomeBot agent to make the best use of electricity at the lowest possible cost. Hence, HomeBot equipment agents must communicate and collaborate in order to achieve the desired overall optimum load situation. Equipment only becomes smart through co-operation. This is achieved by their working together on an electronic market. We have by now successfully demonstrated the technical feasibility of load management implemented as a society of agents using a computational market as a problem solving method [1]and [8].

4.2 Implementation: Hardware and Software

In order to set up experiments with high-level customer services we have built a test bed for *Smart homes*, [3] (Figure 4). Our system is a collection of software agents that monitor and control an office building in order to provide value added services. The system uses the existing power lines for communication between the agents and the electrical devices of the building, i.e., sensors and actuators for lights, heating, ventilation, etc. (ARIGO Switch Stations, see www.arigo.de). At our Villa Wega test site, the interaction with the devices at the hardware level is facilitated by an infrastructure based on LonWorks technology (www.echelon.com/).

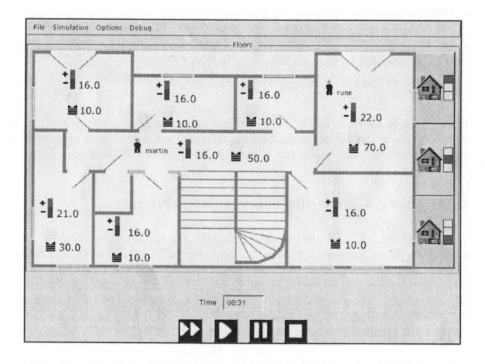

Fig. 4. A snapshot of the GUI that visualizes the state of the Villa Wega building in terms of temperature, light intensity of the rooms and the persons present in the rooms.

The objectives are both energy saving, and increasing customer satisfaction through value added services. Energy saving is realized, e.g., by lights being automatically switched off, and room temperature being lowered in empty rooms. Increased cus-

tomer satisfaction is realized, e.g., by adapting temperature and light intensity according to each person's personal preferences. Initial results from simulation experiments
of an office and its staff indicate that significant savings, up to 40 per cent, can be
achieved.

Currently, a simulation of the building environment is provided including a simulation of the control panel functionality (Figure 4). Our multi agent system and the interface communicate with the simulated building through the control panel. This design will simplify the integration of the multi agent system with the actual LonWorks
system of Villa Wega used for load balancing. The only modification necessary concerns the part in the interface that communicates with the control panel.

4.3 Outcomes of the ISES project

The most important, in this setting, conclusions of the ISES project are:

- Utilities are in general unprepared in how to get out of the 'kWh trap' and into
  customized added-value services.
- We have demonstrated the technical feasibility of load balancing on the electric
  grid utilizing multi agent technologies in the form of smart equipment.
- We have demonstrated the technical feasibility of high-level customer services
  based on smart equipment.

These findings are basic blocks in our present and planned activities within the Research Academy on IT in Energy (www.enersearch.se), and related projects as indicated in Section 2 above. The rest of the paper is an outline of those activities.

5 A framework supporting eServices and eProducts

We use the terms eService and eProducts to denote the fact that they are web-based in
a sense we will make more precise in the following paragraphs. The shift of business
focus from kWh to value-added customized services is but one example of a trend
which have accelerated due to the evolution of eCommerce. We can now for instance
order customized computers (Dell), dolls (Barbie), cars (BMW) or CDs (cductive) on
the net. Most of these success stories are either due to that there is a well-defined
added value (amazon.com) or a well-known product (barbie.com) on which to build
up the customer value and customer relation. In the case of utilities and kWh the story
is more complicated. The product offered by utilities, kWh, is a commodity with very
low profit in a deregulated market. Furthermore the customers of utilities do not want
to buy kWh per se but want proper functionality of their electric based appliances and
equipment. Lastly, utilities are used to mainly see their customers as 'loads' or 'two
holes in the wall'. These facts are well recognized by the utilities themselves and are
in focus of several R&D activities.

We strongly believe that a way to build up a customer relation for utilities is by offering services via smart equipment connected to the electric grid. Having said that, it is also quite obvious how the customers, such as manufactures, can use this equipment to improve their productivity. The equipment can also form a base for installing health care services supporting elderly and handicapped people in their homes.

The current evolution in communication and interconnection capabilities point out that we very soon will have technical possibilities to interconnect smart equipment seamlessly in networks. Besides efforts by consortia investigating power line communication (www.enersearch.se/), consortia such as Bluetooth (www.bluetooth.com/) develop mobile radio transmitters and Internet access points to wireless nets. Simultaneously companies such as Echelon (www.echelon.com/) develop internet capabilities in their next generation of devices. At the same time Microsoft (Windows CE) and Symbian Ltd. (www.symbian.com/) are developing appropriate operating system for embedded systems. In short, there soon will be technological possibilities to interconnect embedded systems into networks such as Internet. Commercial forces drive these efforts but they also have great implications on R&D efforts on, for instance, agent technologies. The open question is whether, and how, it is possible to design suitable information system connecting users and service providers to these nets of embedded smart equipment and thus enabling value added eServices and eProducts.

5.1 A Conceptual View

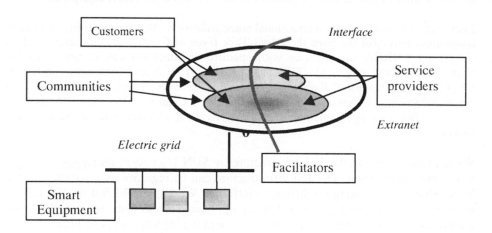

Fig. 5. Basic components of a service oriented electronic market place for eServices and eProducts.

Smart equipment connected to the electric grid is, as stated above, a potentially very profitable way to create and build up customer relations and hence new added value eServices and eProducts.

From Figure 5 we can identify at least three new models for business processes emanating from smart equipment. Firstly utilities, or vendors, can install such equipment and sell services such as energy management to customers. Secondly users can use the installed equipment to enhance customer site processes and businesses. Thirdly, utilities and customers can jointly develop new business areas by upgrading and extend the capabilities of the equipment.

A bare-bone technical infrastructure would not suffice as a platform for new business processes as indicated above. What will be needed is a closed infrastructure including customers and service providers, i.e., an *extranet*. The degree of closure depends, among other things, on the trust and integrity needed in the processes. In order to add support for service identification as well as service delivery and creating value for the customer we also need an explicit *context* and *context based services*. As an example, the concepts of *offer*, *bid*, *auction*, and *bill* have to have a common meaning or *semantics* (depending on the context) in order to become a basis for electronic supported business processes. A common name for this kind of context is a *community*. In summary, the kind of infrastructure we are aiming at can be described as follows. An extranet supporting several communities, which each is an environment enabling user-centric services. Of specific interest to us are *communities with smart equipment*. That is communities of users and service providers based on equipment and services in smart homes, e.g., Homebot based services, [1], [2], [5], and [6].

5.2 Implementations of eServices and eProducts based on smart equipment

There are at present several international standardization efforts concerning electronic commerce. Foremost we have the OMG Task force of Electronic commerce (TFEC) combining CORBA technologies with specific architectures and technologies for eCommerce (www.omg.org/). TFEC has also a RFI on Agent Technology in OMA. The European Network of Excellence AgentLink have also several SIGs focusing on electronic markets (www.agentlink.org/). Finally, we can mention the FIPA standardization efforts on agent mediated applications such as electronic markets (www. fipa.org/).

We are however using the Jini infrasructure by SUN Microsystems (www.sun.com/) in our present platform. The Jini infrastructure can be regarded as a society of Java Virtual Machines running on different platforms. We implement Jini on NT and Linux servers. Since both Echelon and Bluetooth have plans to make their products Jini compatible it is fair to say that we can implement the infrastructures of Figure 5 up to the community level for the set of applications we have indicated above. A strength of Jini is that we automatically will have a common semantic infrastructure (due to the common JVMs) as a basis for the communities. Our efforts will thus be to design and develop a suitable set of components, agents and ontology facilitators (basically mobile Java programs) to support new eServices and eProducts.

Some examples of a current set of eServices and eProducts are:

- eServices based on smart equipment where the smartness is based on co-operation between distributed equipment rather than on stand-alone systems. Equipment mentioned above in the applications of *Load balancing* and *Smart offices*. Extended by smart *water meters* to prevent flooding in houses.

- Higher order added-value eServices. *Integration* of Load balancing and Smart office services as mentioned above and eServices based on *Active recipe boxes*.

The concept of a Smart recipe box takes a user-centric stance on the ideas behind smart kitchens [7].

- eProducts. Districting heating based on *Smart heat exchangers*. The eProduct is initially based on the load management concept.

6 Conclusions

In the ISES project we have identified some basic ingredients which will enable utilities to change their kWh based business model to providing added-value services to customers. An important driving force in this transition is smart equipment connected to the electric grid and co-operating in providing these kinds of new services and products. Communities of customers and service providers on top of extranets provide the necessary infrastructure for those eServices and eProducts.

References

1. Akkermans, J.M., Ygge, F., and Gustavsson, R.. HOMEBOTS: Intelligent Decentralized Services for Energy Management. In J.F. Schreinemakers (Ed.) *Knowledge Management: Organization, Competence and Methodology*, Ergon Verlag, Wuerzburg, D, 1996.
2. Akkermans, J.M., Gustavsson, R., and Ygge, F.. An Integrated Structured Analysis Approach to intelligent Agent Communication. In *Proceedings of IFIP'98 World Computer Congress*, IT and Knowledge Conference, Vienna/Budapest, 1998.
3. Boman, M., Davidsson, P., Skarmeas, N., Clark, K., and Gustavsson, R.. Energy Saving and Added Customer Value in Intelligent Buildings. In *Proceedings of the Third International Conference on the Practical Application of Intelligent Agents and Multi-Agent Technology* (PAAM'98), pages 505-517, 1998.
4. van Dijk, E., Raven, R., and Ygge, F.. SmartHome User Interface: Controlling your Home through the Internet. In *Proceedings of DA/DSM Europe 1996*, Distribution Automation & Demand Side Management, Volume III, pp: 675 - 686, PennWell, 1996.
5. Gustavsson, R.. Multi Agent Systems as Open Societies - A design framework. In *Proceedings of ATAL-97, Intelligent Agents IV, Agent Theories, Architectures, and Languages*, LNAI 1365, pp. 329-337, Springer Verlag, 1998.
6. Gustavsson, R.. Requirements on Information Systems as Business Enablers. Invited paper, in *Proceedings of DA/DSM DistribuTECH'97*, PennWell, 1997.
7. Norman, D.A.. *The Invisible Computer*. The MIT Press, 1998.
8. Ygge, F.. *Market-Oriented Programming and its Application to Power Load Management*. Ph.D. thesis, ISBN 91-628-3055-4, CODEN LUNFD6/(NFCS-1012)/1-224/(1998). Lund University, 1998.

Agent Oriented Conceptual Modeling of Parallel Workflow Systems

Samir Aknine   **Suzanne Pinson**

LAMSADE, Université Paris Dauphine
Place du Maréchal De Lattre de Tassigny
75775 Paris Cedex 16
France
Aknine@lamsade.dauphine.fr

Abstract This research shows how to use ontology and conceptual models for the design of parallel workflow systems in a multi-agent perspective. To illustrate this work, we propose the example of cooperative writing of technical specifications in the telecommunications domain.
Keywords: Parallel workflow, ontology, conceptual models, design methodology.

1 Introduction

New systems are appearing to address the management of semi-structured information such as text, image, mail, bulletin-boards and the flow of work. These new client/server systems place people in direct contact with other people. Researchers and industrial engineers are actively developing workflow systems that assist users in achieving their tasks and enhance collaboration among actors in workflow activity processes[1]. So developers are facing the problem of developing large complex applications. More trouble is created by the lack of methodologies which offer an analysis approach to workflow applications, and a modeling approach to behavioral and structural knowledge, making it possible to define the interactions between the actors of a workflow system.

In [Siau, et al, 96], Siau, Wand and Benbassat wrote: Information modeling can be defined as the activity of formally describing some aspects of the physical and social world around us for the purpose of understanding and communication [Mylopoulos, 92]. Information modeling requires an investigation of the problems and requirements of the user community, and to do so, building a requirement specification for the expected system [Rolland et al, 92]. The product of an information modeling process is an information model which serves as a link between requirement specification and system development. The information model allows thinking and communicating about information systems during the analysis phase [Willuamsen, 93].

---

[1] In this article, we refer to human actors of the workflow system as actors.

A workflow model is a model which must offer the advantages of an information system model but moreover intervene in control, execution and validation of the modeled workflow processes, i.e., it must model the dynamic of the workflow. The current information system models do not offer all concepts that include the coordination and cooperation constraints between the workflow system users. So new conceptual models for these systems are essential.

In this article, we offer a new representation ontology of parallel workflow processes in a multi-agent perspective and we define a conceptual model of workflow processes adjusted to the development of reactive coordination functionalities and the parallelization of workflow activity processes. Our task representation ontology is generic, it has been influenced by the work on KADS methodology (De Hoog et al, 93; Schreiber et al, 93; Post et al, 97).

This is the structure of our article. In section 2, we detail, through a part of the application studied, the new form of parallelization of workflow activities offered. In section 3, we present our representation ontology of workflow activities and the conceptual models offered for the design of parallel workflow systems. Finally, we conclude this work in section 4.

2 A new reactive coordination mechanism for cooperative parallel workflow activity processes

Before describing the proposed ontology and conceptual model, we will briefly present the main characteristics of the cooperative application we studied to make this model valid.

2.1 Application context

The application consists in organizing activities of actors and in checking them during the process of cooperative writing of technical specifications in the telecommunications domain. In the cooperative writing process, the production of technical specifications requires four operations (writing, reading, reviewing and checking) on the three sub-specifications (the functional specification, the detailed specification and the referential specification).

We face several problems particularly due to the parallel execution of the tasks. The cooperative system offered [Aknine, 98a] is composed of several actors who take part in a technical specification writing process. To release actors from some coordination tasks, we have designed a cooperative system composed of software agents. For each actor, a software agent is associated. The actor executes the tasks of the workflow process (writing, reading, reviewing and coherence control); the agent assists and represents the actor in the system. He checks the task activation conditions which will be announced for other agents in the system when these conditions are satisfied. As the agent knows the actor's profile which corresponds to the tasks that the actor can do, the agent offers the actor to perform the announced tasks [Aknine, 98c]. Whenever a task is assigned to the competent actor, a cooperative problem solving process is then engaged between the agent and the actor the task is assigned to. For example, when the actor writes his detailed specification, the software agent analyses a set of telecommunication net components from different points of view and offers them to the actor who can select the most advantageous one for his purpose.

2.2 Parallel workflow activity processes

Figure 1 illustrates the execution plan of the workflow process according to the traditional approach. The functional specification (FS) writing activity is activated at instant t_0. According to the traditional definition of dependence between activities, the reviewing, the reading of the FS and the detailed specification writing activity are sequential activities. We can notice in this figure that a document which is not modified by any activity appears in input and output of an activity with the same version. A consulted document during the writing process appears as a resource of the activity.

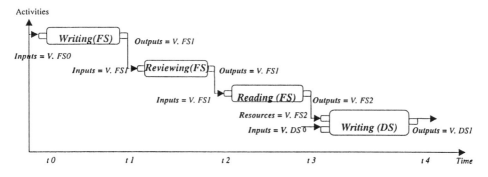

FS: Functional specification; DS: Detailed specification; V. ξ j: Version 'j' of the document ξ.
The first version of the functional specification is V.FS1, which has not been modified by the activity of reviewing, appears in input and output of this activity. On the other hand V.FS1 is modified by the activity of reading which gives a second version V.FS2. This version is a resource of the detailed specification writing activity.

Figure 1. Workflow process execution according to traditional approach.

Due to the fact that activities of a workflow process may sometimes be independent, they may be either fully or partially parallelized. This form of parallel execution is interesting because it can reduce the execution time of the activity process. Once the FS writing activity finishes, the activities of reviewing and reading of the FS as well as the activity of writing the detailed specification (DS) are immediately started.

Having a representation ontology where there is a relation of partial independence between activities and a reactive control mechanism which immediately warns the reviewer of the (FS), its reader and the writer of the detailed specification of each modification done to the (FS), enables us to take a risk on the parallel execution of independent activities and partially independent ones.

This figure shows part of the cooperative process and illustrates the favorable and the unfavorable cases of the execution of the partial independent tasks. In the first case (cf. figure 2.a), the parallelization is favorable as far as the execution time is

503

concerned because the execution of the reviewing activity of the functional specification, using version 1 of the functional specification (called V. FS1), does not generate a new version of the functional specification. Therefore, the partial independence relationship between the reviewing, the reading activities of the FS and the writing activity of the detailed specification are transformed into relations of independence between these three activities. In the second case (cf. figure 2.b), a new version of the functional specification (called V. FS2) generated by the reviewing activity lengthens the execution time of the parallelized writing process with [t₃ , t'₃] for the reading activity of the FS and [t₄ , t'₄] for the writing activity of the DS (cf. figure 2).

The tasks of reading the FS and writing the DS are stated again with the new version V.FS2 of the functional specification. The relation of partial independence is transformed into a relation of dependence between the parallelized activities.

We consider the modifications of the version of a document on which several activities are carried out in parallel, as a case of inconsistency. It creates conflict situations between the software agents of the system which control this document. In the following, we detail how the conflicts between the agents of the system are solved.

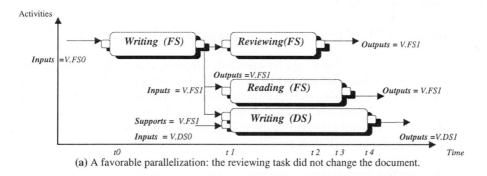

(a) A favorable parallelization: the reviewing task did not change the document.

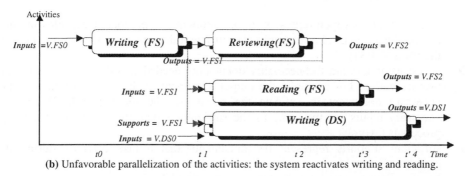

(b) Unfavorable parallelization of the activities: the system reactivates writing and reading.

FS: Functional specification; DS: Detailed specification; V. ξ ⱼ: Version 'j' of the document ξ.
Figure 2 Illustration of the parallelization of the possible independent activities on a part of a cooperative writing process.

3 Ontology and conceptual models for parallel workflow representation

A certain amount of work on multi-agent systems exploited ontology for designing information agents in various fields [Takeda et al, 96][Cutkosky et al, 93]. The basic concepts of these ontologies are objects, actions and space. The modeling of the three aspects characterizing a multi-agent system is not sufficient to represent parallel workflow activity processes and to solve possible conflicts between the agents. To complete this multi-agent modeling of the distributed phenomenon, we introduce the time factor and offer a modeling of possible conflicts between agents. Modeling the conflicts is significant because the agents of the system share knowledge which must be constantly coherent.

So we introduce in this article four levels of multi-agent modeling of parallel workflow processes: a temporal object model, an actor model, an activities model and a conflict model. These various models provide the agents of the system with the information which allows them to solve a problem collectively, detect and the solve possible conflicts between agents. Each one of these models is illustrated by the example of the process of cooperative writing of technical specifications in the telecommunications domain.

3.1 Parallel workflow activity processes representation ontology

Our purpose is to define an ontology for knowledge representation adjusted to parallel workflow systems at knowledge level [Newell, 82]. At a conceptual level, a task is a description of actions allowing the achievement of a goal [Ferber, 95]. We propose a script of a task combining a declarative description, a behavioral description and an extended control structure extended of the task description proposed by [Schreiber et al, 93] [De Hoog et al, 93] [Post et al, 97].

- **The declarative description** is composed of the following elements:

  ➢ sufficient inputs which represent, in our application, the documents allowing the completion of a task. For example, the sufficient document for the reading task of the FS is V. FS1;
  ➢ necessary inputs are the necessary documents needed to validate the results of task execution. For example, the output document of the reviewing task is the necessary document to validate the results of the reading task. When the necessary inputs are the same as the sufficient inputs the parallelized activity process is favorable;
  ➢ sufficient supports represent the documents that a task consults during its execution;
  ➢ necessary supports validate the final results of a task;
  ➢ task preconditions are the sufficient conditions for the execution of a task. These conditions control the existence of the sufficient inputs and the sufficient supports of the task;
  ➢ failure conditions are constraints on the necessary inputs and the necessary supports which invalidate the task execution;

➢ success states and failure states define the states of the necessary inputs and the necessary supports with which a task has been considered as valid or invalid. In our case, they represent the different document versions;

➢ priority is a numerical value attributed to each task. It allows conflict solving between partial independent tasks. The priority allows the separation between correct parallelized tasks and incorrect tasks;

➢ status defines the state of a task: active (i.e. the preconditions are satisfied), inactive (i.e. the preconditions are not satisfied), executed or inhibited. This state corresponds to a temporary state of a task during the occurrence of an incoherence in the system.

• **The behavioral description** indicates the necessary operational elements for task execution. Concerning our application of cooperative writing, the tasks are performed by a human actor in cooperation with its software agent, the part of a task reserved for an agent is limited to actions of answering queries sent by the actor. This description contains inference rules and procedures.

• **The control structure** allows software agents to be totally independent of a domain problem. It contains five types of method: (1) methods of precondition control for task execution; (2) activation methods which activate a task once the preconditions are verified; (3) execution methods of the behavioral part of a task and task execution control methods; (4) task inhibition methods activated when a conflict appears between partial independent tasks and (5) methods for task re-execution.

In the following part, we present the definitions which subtend our ontology for knowledge managing.

Definition 1: Dependence relation

Let T_i and T_j be two domain tasks. Let $\mathfrak{R}_D$ be a dependence relation between tasks.
$T_i \; \mathfrak{R}_D \; T_j$ means that T_i is functionally dependent on T_j if the outputs of T_j are the necessary inputs or resources for task T_i.

Definition 2: Independence relation

Let T_i and T_j be two domain tasks. Let $\mathfrak{R}_P$ be an independence relation between tasks.
$T_i \; \mathfrak{R}_P \; T_j$ means that T_i is functionally independent of T_j if the outputs of T_j are different from the inputs and resources for task T_i.

Definition 3: Partial Independence relation

Let T_i and T_j be two domain tasks. Let $\mathfrak{R}_I$ be a partial independence relation between tasks.
$T_i \; \mathfrak{R}_I \; T_j$ means that T_i is partially independent of T_j if the outputs of T_j can be the inputs or resources for task T_i.
The "can be" quantifier between the outputs of T_j and the inputs and supports of T_i shows the existence of some situations in which the execution of the task T_i is independent of the execution of the task T_j.

3.2 Workflow activity process conceptual models

The introduction of the three relations of dependence, independence and partial independence leads us to define four models which correspond to the different points of view we construct starting from our task representation ontology: (1) a domain object model; (2) an actor model; (3) a task functional dependence model and (4) a task conflict model.

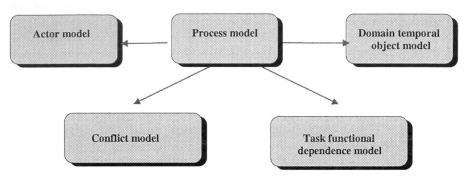

Figure 3 Parallel workflow modeling process

- **Domain temporal object model**

This describes the objects on which an actor can carry out some operations in order to perform the activities of the domain. We include in these objects a temporal attribute allowing the agents to recognize the successive transition states of an object. In the context of our application, the domain objects are the various documents presented (functional specification, detailed specification and reference specification). They represent the sufficient and necessary inputs, the sufficient and necessary resources and the outputs of a task.

- **Task dependence model**

The task dependence model is a representation used by the system to describe the various activities of a workflow process and the relations of precedence (cf. definition 4 and 5) between these activities. This model is illustrated by the example of the technical specification writing process described in figure 4. The task dependence model of our application is made up of the different specific activities deduced from the four generic activities (writing, reviewing, reading and checking) applied to the three sub-specifications (functional, detailed and reference) which form a technical specification. In this example, the arrow which connects the (FS) reviewing activity to the (FS) writing activity means that the activity of reviewing the FS is activated once the task of writing the (FS) finishes. From this model, the agents of the system complete the internal domain task model (cf. section 3.1). Indeed, the dependence relations between the activities expressed in this model arise in the task model as task activation conditions.

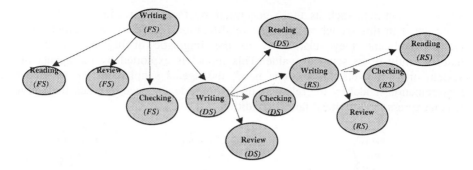

FS Functional Specification; DS Detailed Specification; RS Reference Specification;
---►*Task dependence relation.*

Figure 4 Task dependence model

- **Actor model**

In the task dependence model, the designer of the workflow system can notice that the execution of some activities depends on the one hand on a preliminary position of the activities of the system and on the other hand on the internal actors in the organizational system.
This model considers the actors with whom the agents cooperate. It describes the way in which the agent represents the profile of the actor, i.e., the activities which an actor can perform and its current activity. Since the agents are not able to execute the domain activities, they must know the actors able to execute them. This model is particularly significant because it makes possible to the agents to trace for the actors their activity plans according to their profiles.

- **Conflict model**

This model is a representation used by the system to describe the conditions of uncertainty in which the activities of a workflow process are executed in parallel. The connections between the nodes of the graph indicate the possible conflicts between the activities. A conflict is detected by an agent of the system when the sufficient inputs or resources of a task being executed are modified by the outputs of another parallelized task. Each arrow in this model starts at the task which generates the conflict and ends at the task affected by the inconsistency.
In this model, we represent only high intensity conflicts. For example, in our application, the modification of the version of the FS document by the (FS) reviewing activity carried out in parallel with the (FS) reading, (FS) checking and (DS) writing activities creates a situation of conflict among the four activities. This conflict is intense because the activity of reviewing the (FS) modifies the semantic contents of the FS. In the conflict model (figure 5), this situation is represented by arrows which start from the (FS) reviewing activity and go towards the (FS) reading, (FS) checking and (DS) writing activities.

However, conflicts such as those generated by the (FS) reading activity are not represented in this model because the modifications made by this activity are not really significant, they relate only to the linguistic quality of the document independently of its semantic value. This model is exploited by the agents of the system on two levels. On the one hand, the agents supplement the internal task representation model by the relations of partial independence between the activities which express the activities' failure conditions (cf. section 3.1).

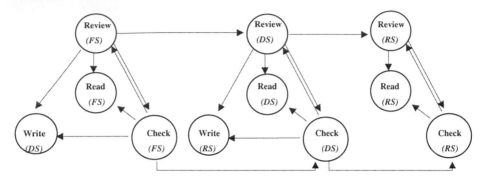

*FS functional Specification; **DS** Detailed Specification; **RS** Reference Specification;*
----► *Conflict.*

Figure 5 Task conflict model

On the other hand, the agent refers to the model of conflicts between the activities to assign priority to the tasks. We have chosen to give the greater weight to the tasks having maximal semi-independence value (the more a task has relationships of semi-dependence with other tasks, the more the task generates conflicts). The conflicts identified by the agents between the activities of the workflow are paired with the conflict model to define a resolution strategy. In our application, a conflict can be directly solved by the agents of the system when the priorities between the activities make it possible to separate the tasks whose execution is coherent from the tasks whose execution is incoherent. The conflict solving process can thus finish without intervention of the actors when the agents of the system reach an agreement. In other cases, i.e., when the tasks have the same priorities, the agents delegate the problem of conflict solving to the actors that they assist [Pinson & Moraitis, 96].

4 Conclusion

In this article, we have proposed a new approach for designing parallel workflow systems in a multi-agent perspective. These systems are able to adjust themselves in a reactive way to the evolution of the activity of a collective work group. The new form of parallel execution of activity processes suggested exploits the possible independence relation between the activities within a workflow process. We have illustrated it by the example of technical specifications writing in the telecommunications domain. To include the partial-independence relations in the representation of the activity, we took as a starting point the formalism suggested in KADS methodology to define a new ontology adapted to the representation of workflow activity processes.

From our activity representation ontology, we have developed four models for the representation of activity processes. The two main advantages of our approach are: the adaptability and the representation of conflict situations allowing their resolution. In this article, we particularly stressed on the modeling aspects. In [Aknine, 98b], we discussed in more detail the architecture of our software agent and the technical aspects of the operating of these agents in our system.

References

Aknine, S. "Exploiting Knowledge Ontology for Managing Parallel WorkFlow Systems", 18th International Conference on Expert Systems and Applied Artificial Intelligence, Springer Verlag, Cambridge, 1998a.

Aknine, S. "A Reflexive Agent Architecture applied to Parallel WorkFlow Systems", CE'98, 5th ISPE International Conference on Concurrent Engineering Research and Applications, Tokyo, Japan, July 15-17, 1998b.

Aknine, S. "Issues in Cooperative Systems. Extending the Contract Net Protocol", IEEE ISIC/CIRA/ISAS'98, A Joint Conference on the Science and Technology of Intelligent Systems, Maryland USA, September 14-17, 1998c.

Cutkosky, M., N., Engelmore, R., S., Fikes, R., E., Genesereth, M., R., Gruber, T., R., Mark, W., S., Tenenbaum, J., M. and Weber, J., C. "PACT: An Experiment in Integrating Concurrent Engineering Systems", IEEE Computer, January, 1993.

De Hoog, R., Martil, R. and Wielinga, B., J. "The CommonKads Model Set", Research Report, Esprit Project, KADS-II Consortium, 1993.

Miller, G., A. "WordNet: An On-Line Lexical DataBase", International Journal of Lexicography, 3-4, 1990.

Mylopoulos, J. "Conceptual Modeling and Telos", Loucopoulos, P. and Zicari, R. (eds.), Conceptual Modeling, Databases and Case, New York, 1992.

Newell, A. "The knowledge level", Artificial Intelligence,18, 1982.

Pinson, S. et Moraitis, P. "An Intelligent Distributed System for Strategic Decision Making", Group Decision and Negotiation, N° 6, Kluwer Academic Publishers, 1996.

Post, W., Wiellinga, B., J., De Hoog, R. and Schreiber, G. "Organizational Modeling in CommonKADS: The Emergency Medical Service", IEEE Expert, November-December, 1997.

Rolland, C. and Cauvet, C. "Trends and Prespectives in Conceptual Modeling", Loucopoulos, P. and Zicari, R. (eds.), Conceptual Modeling, Databases and Case, New York, 1992.

Schreiber, G., Breuker, J., Biedeweg, B. et Wiellinga, B., J. "Kads A Principled Approach to Knowledge Based Systems Development" Academic Press, 1993.

Siau, K., Wand, Y. and Benbassat, I. "When Parents Need Not Have Children: Cognitive Biases in Information Modeling", Constintopoulos, P., Mylopoulos, J. and Vassiliau, Y (eds.), Advanced Information Systems Engineering, 8th International Conference CAISE'96, May, 1996.

Takeda, H., Iwata, K., Sawada, A. and Nishida, T. "An Ontology-based Cooperative Environment for Real-world Agents", International Conference on Multi-Agent Systems, ICMAS'96, 1996.

Willuamsen, G. "Conceptual Modeling in IS Engineering", Executable Conceptual Models in Information Systems Engineering, 1993.

A Distributed Algorithm as Mobile Agents

Dwight Deugo

School of Computer Science, Carleton University
Ottawa, Ontario, Canada, K1S 5B6
deugo@scs.carleton.ca
http://www.scs.carleon.ca/~deugo

Abstract. We present a technique for mapping a distributed algorithm to a set of homogeneous mobile agents that solves the distributed control problem known as election. Our solution is as efficient as the corresponding distributed algorithm, but does not rely on message passing between custom protocols located on communicating nodes. Rather, the proposed solution relies on mobile agents. We also present an environment model that supports not only the mapping to election agents but other agent-based solutions to distributed control problems based on distributed algorithms.

1   Overview

Autonomous decentralized systems (ADSs) involve the cooperation and interoperability of large numbers – potentially millions - of systems distributed over complex computer and communication networks. These systems are preferred over sequential ones, or are simply unavoidable, for various reasons, some of which include: information exchange, resource sharing, reliability through replication and performance.

As computing power and cost continue in opposite directions, a world in which everyone has a computing device at home, at the office and on their person is not far from becoming a reality. Although computers are becoming cheaper, it is often not possible, or at least unadvisable, to equip everyone with their own printer, server, tape drive, database, and other computing resources. For this reason, resources must be shared. Moreover, they must also be reliable. Some parts of a system may fail, but others must be ready to take over when the situation occurs. As the complexity of the resulting applications increases, developers face the challenge to decrease the response times of their applications. ADSs can meet these requirements by splitting tasks over several systems. However, this choice forces the ADSs to solve another type of problem.

ADSs provide services to support both individual and collaborative tasks. In order to provide these services, there must be cooperation and interoperability among the systems. This type of cooperation is in response to user defined problems, generated externally from a set of entry points – systems – initiating the cooperative behavior. For these types of problems, the entry points are determined and controlled a priori.

The servicing of these types of problems creates another set of internal problems for the ADSs. ADSs are no longer reasoning about an externally generated problem, they are reasoning about a problem resulting from their organization as a distributed network. For this class of problem, assumptions about the ADSs' global state and time and ones about which system starts or solves the problem are not permitted.

Examples of these types of problems include election [4] and termination detection [2]. Well studied by the distributed algorithm community, we find it possible to transform – not directly – the algorithm into a collection of mobile agents that cooperatively negotiates a solution. The benefits of this approach include performance, pluggability, and reusability. In this paper, we describe one such example, providing a mobile agent solution to the distributed control problem known as election.

1.1 Contributions and Organization

This paper presents a technique for mapping and reusing a distributed algorithm to a construct a set of homogeneous mobile agents that solves the general distributed control problem know as election [12]. Specific contributions include the following:

- *We present mobile agents that solve the election distributed control problem.* Election is a cooperative information task where one system or node is elected the leader, and that it, and it alone, will initiate a task.
- *Our solution is as efficient as the distributed algorithm.* The solution, which does not rely on message passing between custom protocols on communicating nodes, relies on mobile agents transferring themselves between information systems and negotiating a leader. We show that the number of agent transfers is the same as the number of message sent in the original algorithm.
- *Our solution does not rely on the installation of custom agents on all information system.* This is the case for the distributed algorithm, where each communicating node must preinstall the identical election protocol.
- *We present an environment model for distributed agents that supports the mapping to not only election agents but other agent-based solutions to distributed control problems based on distributed algorithms.* Our solution relies on the installation of the same environment in each system in order for agents to meet and exchange information. However, once installed, it allows all types of agents to operate without any new installation.

It is imperative that information systems cooperate with one another. However, in certain situations, such as election, cooperation involves all information systems not just one or two. In these situations, simple server-centric techniques are inefficient and will not scale. What is required is a network-centric solution. The distributed algorithm community has already solved, at least theoretically, many of these types of problems. The problem with these solutions is that they rely on custom protocols and installations. This is impracticable in situations where the number of cooperating systems and nodes can reach the millions. One must 'install once, but run everything', which our solution does. Our solution clearly demonstrates the practicality of using network-centric agents for solving problems that occur in cooperative information systems, such as election. Our solution bridges the ADSs and distributed algorithm communities, and provides a working agent-based communication infrastructure.

We first begin with an analysis of the requirements of the election problem, restating them for an agent-based solution. Section 3 describes our agent-based infrastructure for distributed control problems and a solution to the problem of election in a unidirectional ring. In section 4, we discuss how our approach relates to others and in section 5 we summarize.

2   Requirement Analysis

Election, also called leader finding, was first proposed by LeLann [9]. The original problem, for a network of linked nodes each running a common protocol, was to start from a configuration where all processes are in the same state, and arrive at a configuration where exactly one process is in the 'leader' state and all other processes are in the state 'lost'.

An election protocol is an algorithm that satisfies the following properties: 1) each process has the same algorithm, 2) the algorithm is decentralized, and 3) the algorithm reaches termination.

For our agent-based solution, we propose a variant of the problem. A network of linked information systems, each starting from an initial configuration, arrive at a configuration where exactly one agent in the network is elected the leader and all other agents know about this and terminate themselves.

An agent system solving this problem satisfies the following properties: 1) each process, node or information system has a similar environment, 2) each process, node or information system initiates no more than one agent, 3) agents do not have to run common implementations (in our case they do), 4) agents are decentralized, and 5) one agent is elected as leader, and all others are terminated.

The move to an agent-based problem definition is motivated by the demands from real-world computing. It is too expensive to install new software every time one requires a different or new algorithm. Rather, we must permit software to move autonomously onto a system. We permit one, initial installation. However after that, the new mantra is 'Install once, run everything – not to be confused with Sun's mantra of 'write-once, run anywhere'. This implies that our solution must be light-weight in order to scale.

3   An Agent Infrastructure for Distributed Control Problems

Our proposed infrastructure for distributed control problems contains two essential objects: an environment and an agent. The environment is responsible for the movement of agents to other environments, the safe execution of an agent while in its domain and the posting of information that agents leave behind for others to read. The agent realizes the distributed algorithm, which in our case is election. It is important to note that the mapping from a distributed algorithm to an agent is not one-to-one.

3.1 Environment

An environment must be active on every information system in the network involved in the distributed control problem. Since the environment is a common component of the infrastructure, we must take care in developing its interface so that it will meet the requirements of other agent-based solutions to distributed control problems.

Agents must be able to move from the current environment to other environments located on different information systems. Therefore, an environment must support a general move operation that transports an agent from the current environment to another, given the new environment's universal resource locator. As information systems are often arranged in well-defined topologies, such as rings and trees, to simplify agent transport, we also propose specialized move operations for a given topology. However, agents that use these specialized operations must realize that they are explicitly making a topological assumption that will limit their use in other topologies.

Agents must be able to post briefings to any environment for other agents to read. Moreover, the environment may also need to post briefings for agents to read. Therefore, we propose environment posting operations. These operations require two parameters: a briefing identifier and text. The briefing identifiers are known a priori to both reading and writing agents and the environment. Therefore, the agent or environment writing the briefing knows how to identify the briefing and the agent wanting to read the briefing knows what identifier to use to retrieve it. It is possible for an agent to attempt to read a briefing that has not been posted. In this case, the operation returns nil, identifying to the agent that the briefing does not exist. Finally, two different briefings written with the same identifier results in the overwriting of the first briefing's text.

The environment must allow agents to execute their behaviors once they arrive. However, an environment makes no assumptions about what an agent will do. The only knowledge the environment has about an agent is that it responds to the message `actInEnvironment`. After the environment sends this message to the agent, passing a reference to the environment as a parameter, the agent may do any combination of the following actions: post briefings to the environment, read a briefing from the environment, transfer to another environment, update its internal state, simply decide to remain in the environment, or remove itself from the environment without further transport - termination.

Internally, an environment keeps track of agents that have moved to it and the posted briefings. Once started, the environment runs in a loop waiting for new agents to arrive, activating the agents that are present, and transporting the agents that want to leave.

3.2 Mobile Agent

A mobile agent is an autonomously entity that can move from one environment to another, read or post briefings, remain in its current environment, or terminate itself. An agent does not openly communicate with other agents using messages. Instead, agents exchange information by posting briefings to the information systems' environments.

Agents can make assumptions about the environments they execute in. The first assumption is called topological awareness. Making this assumption, an agent is aware of the network topology. Another assumption is called sense of direction. Making this assumption, an agent knows the "direction" to which it is heading in the network, e.g., to a child or parent system. Another assumption is called briefing knowledge. Making this assumption, an agent relies on other agents and the environments to post specific briefings. A final assumption is called process identity. Making this assumption, an agent relies on every information system having a unique identifier and that this information is posted to the corresponding environments as a briefing.

Although, not required, agents can be best understood as finite state machines. The activation point for an agent is its behavior `actInEnvironment`. If implement as a FSM, an agent's behavior resembles the following pattern:

```
actInEnvironment(env) {
   switch (currentState) {
      case STATE1: methodForState1(env);
      case STATE2: methodForState2(env);
        . .
      case STATEN: methodForStateN(env);
      default: errorProcedure("error");}}
```

3.3 The Ring Election Mobile Agent

Our proposed agent performs election in a ring topology. It assumes that each environment posts a briefing indicating the unique identifier of the information system. It assumes that initially no more than one agent exists in any environment. However, anywhere from 1 to n environments can start an agent. Finally, all agents move in a clockwise direction, which for this paper is to the left neighbor of the current information system.

We base our agent on the election distributed algorithm by Chang-Roberts [4], modified by Franklin [7] to consider a unidirectional network. In the Chang-Roberts algorithm, initially each node is active, but in each subsequent round some nodes become passive. In a round, an active node compares itself with its two neighboring active nodes in clockwise and anti-clockwise directions. If the node is a local minimum it survives the round, otherwise it becomes passive. Because nodes must have unique identities, it implies that half of them do not survive a round. Therefore, after $log\ n$ rounds only one node remains active, which is the leader.

In Franklin's algorithm, messages and the corresponding node identities are forwarded to the next clockwise neighbor and to that node's following neighbor. This final node is responsible for determining whether it becomes active or passive using the identities of its previous two active neighbors and its own.

The transformation from a distributed election algorithm, which passes messages between nodes, to a homogeneous set of agents, moving from one system to another leaving information for one another in the form of briefings, is a task of mapping the algorithm's protocols to an agent's internal states and then mapping the algorithm's data to a combination of internal agent state variables and environmental briefings.

In our approach, the environment is responsible for setting the *IDENTITY* briefing when it initializes. This briefing contains the unique identity of the information systems. Agents will only read this briefing. Agents also read and write the following briefings: *ROUND* and *WORKINGIDENTITY*. The ROUND briefing indicates the current working round of the environment. The WORKINGIDENTITY briefing is the identity that an active environment is assuming.

Internally, each agent keeps track of the following information:

- ElectedleaderIdentity: The identity of the information system elected as the leader
- OriginalIdentity: The identity of the active information system the agent began in at the beginning of the current round
- FirstNeighborIdentity: The identity of the first active information system the agent encountered in the current round
- Round: The current round the agent is participating in

The agents FSM is as follows:

```
actInEnvironment(env) {
   switch (currentState) {
      case InitiateState: initiate(env);
      case LocateFirstNeighbor: firstNeighbor(env);
      case LocateSecondNeighbor: secondNeighbor(env);
      default: errorProcedure("error");}}
```

When in the initiate state, an agent initializes its round and originalIdentity variables and the environment's WORKINGIDENTITY briefing. Since it is possible for this agent to be slow to activate, another agent that moved in and out of the environment may have activated the election computation. In this case, the current agent must terminate itself since it is no longer required. Note, in the following segment, **this** refers to the agent, since this is an agent's behavior.

```
initate(env){
   round = 1;
   if (env.getBriefing(ROUND) == NOBRIEFING){
      originalIdentity = env.getBriefing(IDENTITY);
      env.postBriefing (ROUND, 1);
      env.postBriefing(WORKINGIDENTITY,originalIdentity);
      setState(LocateFirstNeighbor);
      env.moveToLeftNeighbor(this);}
   else env.kill(this);} //another agent gone by
```

Once in agent is in the state LocateFirstNeighbor, three situations can occur. 1) It visits a passive environment, one previously removed from consideration for election. In this case, that agent simply moves to the next environment. 2) It reaches the environment it began in at the start of the round. In this case, all other environments must be passive. Therefore, the environment with the identity WORKINGIDENTITY and the agent, must be the only ones involved with election, so a leader can be elected. 3) The most common situation is that the agent finds its next active environment for its current round. In this situation, the agent records the environment's, WORKINGIDENTITY and moves on to find its second active environment.

```
firstNeighbor(env) {
    // Passive case, environment not involved
    if (env.getBriefing(ROUND) == NOBRIEFING){
        env.moveToLeftNeighbor(this);
        return;}
    // Agent returned to start environment
    if ((env.getBriefing(ROUND) <= round) &&
        (originalIdentity == env.getBriefing(WORKINGIDENTITY))){
        //back to original node, found leader
        electedleaderIdentity = originalIdentity;
        annouceLeaderFound(this);
        return;}
    // Agent found first active neighbor
    if (env.getBriefing(ROUND) <= round){
        firstNeighborIdentity = env.getBriefing(WORKINGIDENTITY);
        setState(LocateSecondNeighbor);
        env.moveToLeftNeighbor(this);
        return;}}
```

When that agent is in the state LocateSecondNeighbor, four situations can occur. 1) It visits a passive environment. In this case, that agent moves to the next environment. 2) It reaches the environment it began in at the start of the round. In this case, all other environments are passive, and the environment with smallest identity WORKINGIDENTITY is elected the leader. 3) The most common situation is that the agent finds the second active environment for its current round. In this situation, the agent decides whether to move on to the next round, or 4) terminate itself. It moves to the next round if the environment's WORKINGIDENTITY is less than the identities of the environments the agent passed through.

We summarize or agent-based to approach to election in a uni-directional ring as follows. 1) Initially one or more environments become active, each instantiating one agent. All other environments are passive. 2) In a round, an agent records the working identity of it current environment and then locates the next two active environments, recording their working identities while moving in a clockwise direction. 3) After finding the second active environment, the agent compares the working identities of all three environments. The agent remains active, as does the current environment – the environment it has just reach - and continues to the next round if the identity of its first visited environment is less than the original and the current one. In this case the agent's current environment assumes the new working identity and the agent repeats the steps.

Having an agent compare the identities of three environments (the one it started the round in, the active one it passed by, and the active one it is now stopped in) before considering to continue to the next round implies that at least half of the agents and environments do not survive a round. Therefore, after *log n* rounds – *n* the number of environments - only one agent remains active, which can elect any information system it wishes as the leader.

```
secondNeighbor(env) {
    /// Passive case, environment not involved in election
    if (env.getBriefing(ROUND)== NOBRIEFING){
        env.moveToLeftNeighbor(this);
        return;}
    // Agent returned to start environment
    if ((env.getBriefing(ROUND) <= round) &&
        (originalIdentity == env.getBriefing(WORKINGIDENTITY))){
        // there are two environments left
        if (originalIdentity < firstNeighborIdentity){
            electedleaderIdentity = originalIdentity;
            annouceLeaderFound(this);}
        env.kill(this);
        return;}
    // Found second neighbor
    if (env.getBriefing(ROUND) <= round){
        if ((firstNeighborIdentity < originalIdentity) &&
            (firstNeighborIdentity <
                env.getBriefing(WORKINGIDENTITY))){
        round = round + 1;
        originalIdentity = firstNeighborIdentity;
        firstNeighborIdentity = NOIDENTITY;
        env.postBriefing(ROUND, round);
        env.postBriefing(WORKINGIDENTITY,originalIdentity);
        setState(LocateFirstNeighbor);
        env.moveToLeftNeighbor(this);
        return;}
    else { //Droping out of election
        env.postBriefing(ROUND, NOBRIEFING);
        env.kill(this);
        return;}}
```

4   Related Work

Frameworks are reusable designs and compositions of different software components, with well-defined interfaces for using and extending them. [10]. The essential notion is that frameworks help decrease the effort in developing similar applications. Our proposed framework has two essential components: an environment and an agent, and has the underlying goal of enabling pluggable distributed computation without forcing the new installation of software on the information systems involved. Therefore, our approach is comparable with the other metacomputing frameworks, distributed object computing, and on distributed algorithms because of our reliance on them.

If we suppose that information systems are connected via the Internet, then our approach is similar to ones in metacomputing. Globus [5] is an infrastructure for running applications on networked virtual supercomputers. NPAC [6] attempts to perform similar computations using a Web-enabled concurrent virtual machines. Javelin [1] is a Java-based architecture for writing parallel programs, implemented over the Internet. All of these systems attempt to provide seamless parallelism for developing high-performance applications. Although our approach could also be used for this purpose, the focus of our approach is to support decision problems that involve the communication and cooperation of all nodes or information systems in a network. Election, for example, is a problem whose solution does not necessarily benefit from parallelism. In fact it is the parallelism, or at least the distributed nature of the information systems, that causes the need for fast, efficient, decentralized solutions to the election problem. We do not take a problem and attempt to solve it faster on a network of computers. Rather the network of computers is at the heart of the problem we are trying to solve.

Other approaches create frameworks for manipulating a distributed system of objects. One of the most well know C++ frameworks, recently converted to Java, is the Adaptive Communication Environment (ACE) [11]. ACE provides wrappers and components that perform common communication tasks. Aglets [8] is a Java-based framework that supports mobile agents. However, neither of these approaches attempts to perform global reasoning about the distributed system. This activity is left for the programmer. One framework that does reason about the global composition of a distributed system is the framework proposed by [3]. However, as its name suggests, the framework does not provide you with off-the-shelf solutions to distributed computing problems. Rather it focuses on the services required to solve these types of problems.

Many distributed algorithms [13] exist for handling generic distributed computing problems such as election [9] and broadcasting [12]. However, it has been our experience that few people, especially in industry, are aware of such algorithms. One of the contributing factors to this lack of knowledge is that, although companies know about CORBA, Java's Remote Method Invocation (RMI), Sockets, or agent technologies, they are not willing to use these facilities to build infrastructures supporting distributed algorithms. A second factor is that no agent infrastructure including the above mentioned algorithms exists. Our agent-based approach is an attempt to fill this gap by providing 'off-the-shelf' agents.

5   Summary

We have tested our approach on various network sizes and agent startup configurations. When found it was very important to remember not to make any assumptions about the time taken for agents to transfer themselves from one environment to another, the order of transfer and about how many agents were active in an environment at any one time.

Interoperability and cooperation between ADSs is not only required because of the needs and problems generated by external users of these systems. The fact that they are distributed generates problems that rely on these capabilities for the solutions.

Work has been done in the area of distributed algorithms on providing theoretical solutions to this class of problems. However, many of the existing frameworks have not supported their use, rather focusing on the services required by users of the frameworks. Our work bridges these two communities, by describing an agent-based infrastructure that provides the necessary framework and approach for using mobile agents, based on distributed algorithms, to negotiated solutions to problems that ADSs generate for themselves.

References

1. Cappello, P., Christiansen, B., Ionescu, M. F., Neary, M.O., Schauser, K.E., Wu, D., Javelin: Internet-Based Parallel Computing Using Java, Sixth ACM SIGPLAN Symposium on Principles and Practice of Parallel Programming (1997)
2. Chandrasekaran, S., Venkatesan, S., A Message-Optimal Algorithm for Distributed Termination Detection, Journal Parallel and Distributed Computing, 8, 3 (1990) 245-252
3. Chandy, K.M., Kiniry, J., Rifkin, A., Zimmerman, D., A Framework for Structured Distributed Object Computing, http://www.infospheres.caltech.edu/papers/framework/framework.ps, (1999)
4. Chang, E., Roberts, R., An Improved Algorithm for Decentralized Extrema Finding in Circular Arrangements of Processes, Communications of the ACM 22, (1979) 281-283
5. Foster, I., Kesselman, C., Globus: A Metacomputing Infrastructure Toolkit, Proceedings of the Workshop on Environments and Tools for Parallel Scientific Computing, SIAM, Lyon, France, August (1996).
6. Fox, Furmanski, W., Towards Web/Java based High Performance Distributed Computing – An Evolving Virtual Machine, Proceedings of the Fifth IEEE International Symposium on High Performance Distributed Computing, Syracuse, New York, August (1996).
7. Franklin, W. R., On an Improved Algorithm for Decentralized Extrema Finding in Circular Configurations and Processors, Communications of the ACM 25, 5 (1982) 336-337
8. Lange, D.B., Oshima, M., Programming Mobile Agents in Java – With the Java Aglet API, IBM Research, http://www.trl.ibm.co.jp/aglets/ (1997)
9. LeLann, G., Distributed Systems: Towards a Formal Approach, Proceedings of Information Processing '77, B. Gilchrist (ed.), North-Holland (1977) 155-160
10. Roberts, D., Johnson, R., Evolving Frameworks: A Pattern Language for Developing Object-Oriented Frameworks, Proceedings of Pattern Languages of Programs, Allerton Part, Illinois, September (1996)
11. Schmidt, D.C., ACE: An Object-Oriented Framework for Developing Distributed Applications, Proceedings of the Sixth USENIX C++ Technical Conference, Massachusetts, April (1994)
12. Tel, G., Topics in Distributed Algorithms, vol. 1 of Cambridge Int. Series on Parallel Computation, Cambridge University Press (1991)
13. Tel, G., Introduction to Distributed Algorithms, Cambridge University Press (1994)

Why Ontologies Are Not Enough for Knowledge Sharing*

Flávio S. Corrêa da Silva**
Inst. de Matemática e Estatística
Univ. de São Paulo, São Paulo, Brazil
fcs@ime.usp.br

Wamberto Weber Vasconcelos* * *
Depto. de Estatística e Computação
Univ. Estadual do Ceará, Ceará, Brazil
wvasconcelos@acm.org

Jaume Agustí
Inst. d'Investigació en Intel.ligencia Artificial
Campus UAB 08193, Barcelona, Spain
agusti@iiia.csic.es

David Robertson
Dept. of Artificial Intelligence
Edinburgh Univ., Scotland, Great Britain
dr@dai.ed.ac.uk

Ana Cristina V. de Melo
Inst. de Matemática e Estatística
Univ. de São Paulo, São Paulo, Brazil
acvm@ime.usp.br

Abstract. Knowledge sharing is difficult. One reason is that it is hard to decide how to describe a domain in a way which suits everyone interested in the knowledge. Tackling this problem has been a central theme of the surge in ontological research over recent years. Unfortunately, getting an agreed ontology is not the end of our problems, since the way we represent knowledge is intimately linked to the inferences we expect to perform with it. We introduce three inference systems and discuss the problems of having knowledge passing through them, which are representative of complex problems we may need to solve for knowledge sharing.

1 Introduction

Mainstream research in knowledge sharing among knowledge-based systems concentrates on the mappings between different domain-specific notations while making the assumption that the inference mechanisms of each system are compatible. Sometimes this assumption is explicit: to share knowledge, each system must translate its knowledge into a standard system of inference [17,13,10]. At other times the assumption is implicit: a standard knowledge representation language is provided but it is (mostly) left to users of the notation to choose compatible forms of inference [11,18].

There is good reason for making this assumption because it is difficult without it to guarantee that the meaning of knowledge expressed in one system is preserved when used by another system. In theory, this guarantee would require us to demonstrate that the models of the world permitted by a system supplying its knowledge include all the models of the world permitted by the system receiving that knowledge. If we allow donors of knowledge also to be recipients then the theoretical constraint becomes even stronger: the models permitted by

* Work sponsored by the Consortium British Council/CAPES (Brazil), Grant no. 070/98, and the Science Foundation of the State of Ceará (FUNCAP), Brazil.
** Partially sponsored by FAPESP, Grant no. 93/0603-01, and CNPq, Grant no. 300041/93-4.
* * * Partially sponsored by CNPq, Grant no. 352850/96-5.

all systems must coincide precisely. This raises a major practical problem because differences in models of the world often show up in the differing styles of inference which we use to derive consequences from the knowledge we represent. For instance, in an assembly system we might express rules of assembly using the notation of classical logic but our inferences might be resource-bounded. If the rules for this system were used within a "classical" deduction system then the resource limitation would be removed and we might be able to derive solutions which were unobtainable from the donor system.

There is no definitive solution for this problem, other than the impractically restrictive rule that all knowledge based systems must share the same models of the world. Alternatively, we have looked for localised solutions where the relationship between inference systems is close enough to allow sharing of knowledge in limited forms [8]. This sometimes involves loss of information but in many cases this loss may be detected, assessed and made tolerable. The purpose of this paper is to present some problems which occur with this approach and suggest ways of solving them. We do this by describing a high-level method for assessing how knowledge of particular kinds may be shared (Section 4). To make our discussion concrete, we use as examples three simple inference systems (introduced in Section 3). Our analysis is based on logic because this allows us to maintain a clear separation between logical axioms, describing domain knowledge, and the inference rules used to manipulate that knowledge. We can also be sure that if problems can be found at the idealised level of logical analysis then they will occur at least as strongly in actual applications.

2 A Simple Knowledge Sharing Scenario

Suppose that we have the task of combining knowledge-based systems involved in ordering appropriate personal computers. Three systems are involved:

- Our business is international so different computers may be assembled in different countries. We want each computer built in each country to be assembled from parts available in that country. Therefore we have a system giving a catalogue of the types of components which are available in each country and telling us how these components may be combined to produce product specifications for different countries. This is based on a classical logic and is described in Section 3.1.
- Our customers sometimes impose acceptability requirements on their orders – they may want, for example, to trade off cheapness against quality. As a consequence of a separate business analysis effort we have a system which partially describes these requirements. Since our requirements are imprecise, this is based on a fuzzy inference system described in Section 3.2.
- To assist in the control of our production processes we have a system for suggesting how the components we have in stock at each of our factories may be assembled to form the larger scale components of our computers. Since this form of advice is based on the assumption that each individual component is used once, and cannot then be used again, we use a resource-bounded inference system as described in Section 3.3.

We would like to allow these three independently constructed inference systems to be able to share their knowledge.

3   Three Simple Inference Systems

To avoid becoming lost in details it is necessary to keep the examples of this section simple. This is a limitation but not a serious one – the point of our paper is to show how some key problems may be anticipated when attempting to share knowledge so if we find problems in these simple systems we can expect at least these same problems in more complex systems.

3.1   A Classical Inference System

Our product specification system contains definitions of the role of different catalogue entries (*e.g.* $disk(dtype_1)$ says that the catalogue part $dtype_1$ is a disk):

$$disk(dtype_1) \quad disk(dtype_2) \quad memory(mtype_1) \quad screen(stype_1)$$
$$keyboard(ktype_1) \quad mouse(xtype_1) \quad button(btype_1)$$

Our product specification for a computer combines the appropriate components:

$$computer(D, M, S, K, X) \leftarrow$$
$$disk(D) \wedge memory(M) \wedge screen(S) \wedge keyboard(K) \wedge mouse(X)$$

Goals are satisfied in this system in a standard backward-chaining style based on classical logic, in which we attempt to find an instance satisfying the goal and can search for further instances on backtracking.

3.2   A Fuzzy Inference System

Our system for acceptability requirements uses a fuzzy logic which allows the degree of acceptability of any component type to be related to its properties [2, 6, 5, 7, 12, 16] (in our example these are its cheapness and the degree to which it is considered medium or top quality).

$$acpt(X) \leftarrow cheap(X) \wedge med\_qlty(X) \qquad acpt(X) \leftarrow top\_qlty(X)$$

We then associate the acceptability properties with fuzzy values:

$$cheap(dtype_1) : 0.8 \quad med\_qlty(dtype_1) : 0.7 \quad top\_qlty(dtype_1) : 0.1 \quad top\_qlty(dtype_2) : 0.6$$

Our fuzzy inference mechanism will operate by finding the maximum obtainable truth value for a given goal, with the truth value for any conjunction of goals taken as that of the least preferred conjunct. For instance, the proof for goal $acpt(dtype_1)$ is shown in Figure 1. The fuzzy values obtained are shown as labels

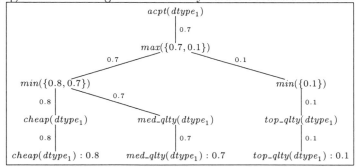

Fig. 1: Fuzzy Proof-Tree for $acpt(dtype_1)$

associated to branches of the tree, conveying the fact that that value was passed on to the parent node as part of the inference mechanism. For instance, fuzzy value 0.8 obtained in the left-most leaf node $cheap(dtype_1)$ is passed on to its parent node which, on its turn, passes it on to node *min*; the *min* node then calculates the minimum among its nodes and passes the result, 0.7, to the *max* node; the *max* node performs a similar calculation, but chooses the maximum of the values associated to its branches, passing this value to its parent node, the initial query. Deducted information (unifications or fuzzy values) is presented as labels of branches

3.3   A Resource-Bounded Inference System

Our assembly system contains definitions which explain how to construct components from subcomponents. For example, a two-button mouse (where m_2 denotes the type of mouse) needs to have two button subcomponents:

$$mouse\_comp(m_2, B_1, B_2) \leftarrow button\_comp(B_1) \land button\_comp(B_2)$$

The assembly system also records which individual components are in stock. For instance, we might have only two buttons, $button\_comp(b_1)$ and $button\_comp(b_2)$. The inference mechanism for this system is the same as for the specification system of Section 3.1 with the additional constraints that each fact can be used no more than once, and in the proposed order as occurring in the clauses. Thus, the goal $mouse\_comp(X, B_1, B_2)$ would succeed, with the instance $mouse\_comp(m_2, b_1, b_2)$ but there would be no solution for the goal $(mouse\_comp(X_1, B_1, B_2) \land mouse\_comp(X_2, B_3, B_4))$ because only two buttons are available in our database. This characterises a resource-bounded logic, similar to those found, for instance, in [15], [4], [3] and [9]

4   Method of Analysis

Our aim is to use examples of knowledge sharing between simple representatives of common forms of inference system to suggest problems which may occur. Our representative systems are those of Section 3. We first consider (in Section 4.1) whether knowledge sharing is needed at all. We then examine solutions to knowledge sharing which rely on systems who donate knowledge being viewed as oracles (Section 4.2). Finally, we explore solutions which rely on the knowledge of one system playing the role of a surrogate knowledge base for another (Section 4.3). In both these sections we consider sharing in each direction between classical and fuzzy systems and between classical and resource bounded systems. Each example of knowledge sharing raises a general question which is summarised in the sentences labelled **Question 1** to **9**. Although derived from logic, these questions apply to any sort of system with the properties described.

4.1   How Much Can Be Done Locally?

The first question we need to ask is whether we need to have three different communicating systems at all. Perhaps it would be easier simply to merge the systems into one large, heterogeneous system. The easy part of this would be merging the knowledge bases, since their syntax is similar and there is no obvious ontological conflict (when we use different names we have different concepts in mind). The difficulty comes in merging the inference mechanisms. We cannot simply combine them because it is not clear what it would mean to have an inference mechanism which was *simultaneously* classical, fuzzy and resource bounded. On the other hand, using each inference mechanism independently does not allow us to share knowledge between systems any more than when they were physically separate. To do that, we need to agree on the correspondences between the terms which we wish to share and assess the extent to which the conclusions derived by one inference system can be trusted by appropriate recipient systems.

Question 1 *Can the inference systems and knowledge bases be merged cost-effectively, avoiding the need for knowledge sharing?*

4.2 Can Donor Systems be Oracles?

Perhaps the simplest way in which systems can share knowledge is for the recipient to view the donor as an oracle, which always provides all the right answers. Whether we can maintain this level of trust depends on which sorts of system need to communicate.

Oracle: Classical Specification Receives Fuzzy Requirements If we wish to have our model of acceptability influence the specification of our computers then we could do this by defining new axioms which relate the fuzzy acceptability constraints to the specification. One axiom we might add to Section 3.1 is:

$$acpt\_comp(D, M, S, K, X) \leftarrow computer(D, M, S, K, X) \wedge econ\_to\_acpt(D) \wedge econ\_to\_acpt(M) \wedge$$
$$econ\_to\_acpt(S) \wedge econ\_to\_acpt(K) \wedge econ\_to\_acpt(X)$$

The $econ\_to\_acpt$ conditions are not defined within the specification system. Instead, we link them to the fuzzy definition of $acpt$ in the requirements system of Section 3.2. This requires us to provide translations [8] between the predicate names and also from the fuzzy values of the requirement system to the boolean truth values of the specification system [2, 19, 14]. We lose information this way because we are compressing the shades of meaning available in the fuzzy system into a true or false decision. The most obvious way to do this is by agreeing that whenever a fuzzy value exceeds a given threshold value it is translated into "true" in our classical system (shown as T^C below). Our translations are therefore:

Expression in donor	Expression in recipient
$acpt(X)$	$econ\_to\_acpt(X)$
Fuzzy value (μ), where $\mu \geq$ threshold value	Boolean truth (T^C)

This narrow channel of communication allows us to bring the acceptability requirements within the specification system but the results we obtain are only as reliable as the assumption we made in choosing the threshold value.

Question 2 *What information may be lost if we use a system with complex truth values as an oracle for a system with simpler truth values?*

Oracle: Fuzzy Requirements System Receives Classical Specification All the requirements of Section 3.2 refer to subcomponents but not to assembled computers. We might like to include information from the specification system of Section 3.1 in order to talk about requirements on possible computer systems. For instance, a computer system might be considered acceptable if its disk, memory and screen are acceptable, which we might express by:

$$acpt(computer(D, M, S, K, X)) \leftarrow avlble\_comp(D, M, S, K, X) \wedge acpt(D) \wedge acpt(M) \wedge acpt(S)$$

The need for translation here is similar to the previous example but in the opposite direction. Again, we must decide on the mapping between boolean and fuzzy values, this time associating classical truth with our fuzzy maximum of 1.

Expression in donor	Expression in recipient
$computer(D, M, S, K, X)$	$avlble\_comp(D, M, S, K, X)$
Boolean truth ($\top^C$)	Fuzzy value (1)

The assumptions made in sharing this knowledge are similar to the previous case. Although in our example the mapping of truth values seems obvious, since we either have a specification or not, we could imagine circumstances when we might set a lower fuzzy value (for example, if we had reason to believe that specifications were not always reliable). The mapping above between the classical truth value $\top^C$ and the fuzzy value 1 is correct [2, 1, 14], since we use an appropriate minimum combinator. If different combinators had been used, the morphism between these logics might have been destroyed .

Question 3 *How do we approximate when using a system with simple truth values as an oracle for a system with complex truth values?*

Oracle: Resource-Bounded Assembler Receives Classical Specification

Our resource-bounded system is only allowed to use the supply of buttons which it has in stock but on some occasions we might want to guess which components we are able to assemble if we ordered sub-components of the appropriate type from our catalogue of part types. This catalogue resides in the specification system of Section 3.1. To connect to it, we must translate the appropriate component names. Translating the truth values of both logics, which lost information in our previous example, is safe here because the notion of truth in both logics is similar (although the consequences obtainable from the resource bounded logic are a subset of those for the classical system):

Expression in donor	Expression in recipient
$button(B)$	$button\_comp(B)$
Boolean truth ($\top^C$)	Boolean truth ($\top^R$)

This allows the assembly system to satisfy goals which it was previously unable to solve. For instance, in Section 3.3 we were unable to find a solution for $mouse\_comp(X_1, B_1, B_2) \wedge mouse\_comp(X_2, B_3, B_4)$ but now we can use the specification system to obtain the solution $mouse\_comp(m_2, b_1, b_2) \wedge mouse\_comp(m_2, btype_1, btype_1)$. The two instances of $btype_1$ come from our specification system and we must understand these not as instances of particular buttons (like b_1 and b_2) but as "promises" that we could obtain buttons of the type $btype1$ from our catalogue. Consequently, we are changing the interpretation of $mouse\_comp$ in the resource-bounded system - making it no longer resource bounded (since we can get as many instances of $btype_1$ from the classical system as we need) and having it generate both objects in stock and markers from the catalogue.

Question 4 *If we use as an oracle a system with fewer constraints on acceptable solutions than the system receiving its advice, can we check that the additional solutions it may generate will be helpful?*

Oracle: Classical Specification Receives Resource Bounded Assembly

We might decide that for some components our specification system should, instead of looking up the appropriate part type in the catalogue, use the assembly system to generate a description of it. For instance, we might remove the information about *mouse* from the axioms of Section 3.1 and connect to the resource-bounded system of Section 3.3 by allowing *mouse* to succeed in

the specification system if *mouse_comp* succeeds in our assembly system. Our translations in this case are similar to those of the previous example but the direction is reversed:

Expression in donor	Expression in recipient
$mouse\_comp(M, B_1, B_2)$	$mouse(M)$
Boolean truth ($\top^R$)	Boolean truth ($\top^C$)

As in the previous example, we must be aware when choosing this sort of connection that we are also changing the meaning of predicates in the classical system. In particular, we are limiting the number of solutions we can find for $mouse(M)$. Previously we would have obtained a solution to the goal $(mouse(M_1) \wedge mouse(M_2))$ with both M_1 and M_2 bound to $xtype_1$. Now this goal would fail because our resource-bounded system (which now supplies the definition of *mouse*) uses up its supply of buttons when providing an instance for M_1 and is incapable of supplying an instance for M_2.

Question 5 *If we use as an oracle a system with more constraints on acceptable solutions than the system receiving its advice, are these all the solutions we need?*

4.3 Can a System be a Surrogate for Another?

In Section 4.2 we used the donor system as an oracle and therefore we had no opportunity to ensure that the inference it performed was consistent with the inference patterns known to the recipient. This meant that we had to tolerate various forms of potential discrepancy – if this is not possible then we may consider whether we can continue to use the inference method of the recipient system but with the donor supplying only a surrogate knowledge base, replacing the knowledge of the recipient at appropriate points during inference.

Surrogacy: Classical Specification Receives Fuzzy Requirements Recall that in Section 4.2 we used our fuzzy system as an oracle to tell us whether components used in the specification system were acceptable. This meant that information coming from the fuzzy system was determined according to fuzzy inference rules, rather than the classical inference rules of the specification system. An alternative is to use the classical inference method alone but use appropriate fuzzy knowledge as a surrogate for missing classical knowledge. The translation required to do this is identical to that of Section 4.2 but the results we obtain are different because the fuzzy inference method requires exhaustive search of appropriate parts of its knowledge base while the classical inference method may only search part of it. For instance, Figure 2 shows two proofs we would get

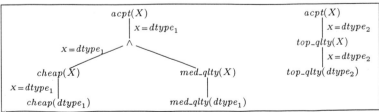

Fig. 2: Classical Inferences using Knowledge Base of Section 3.2

if we set the threshold for μ as 0.5 and use the classical inference mechanism on the knowledge base of Section 3.2 to satisfy $acpt(X)$. Notice that these no

longer contain fuzzy values because our translation mechanism converts these to a boolean truth value for the classical system. By contrast, the fuzzy logic inference mechanism applied to the same goal gives us the single proof shown in Section 3.2. The advantage of using the classical inference method is that each individual proof may be shorter because it does not require exploration of all relevant knowledge. The disadvantage is that we no longer produce a single "most likely" result – instead we get all results which exceed the 0.5 translation threshold.

Question 6 *Will the amount of search performed by your inference strategy be sufficient when applied to a surrogate knowledge base designed for more extensive search?*

Surrogacy: Fuzzy Requirements System Receives Classical Specification Using a classical inference method in the previous example allowed us to generate solutions based on partial searches of the available fuzzy knowledge. An alternative strategy if we wish to preserve the fuzzy system's exhaustive search is to use the fuzzy inference method with the classical knowledge. For example, we may wish to define acceptability requirements which are not fuzzy and are therefore more naturally expressed in a classical style. An example might be that we will accept a component which is similar to another one which we know is acceptable. Formally, we might represent this within the classical system as

$$econ\_to\_acpt(X) \leftarrow econ\_to\_acpt(Y) \wedge similar(X, Y)$$

To allow this to be used by our fuzzy inference method we connect the *econ_to_acpt* predicate to the *acpt* predicate of the fuzzy system and translate from boolean to fuzzy truth values as before. Unfortunately, this approach can create problems, as we can demonstrate by attempting a proof of *acpt(X)*. This is similar, at the beginning, to the proof in the stand-alone fuzzy system of Section 3.2. The important difference is when the connection between *acpt* (in the fuzzy system) and *econ_to_acpt* (in the classical system) is used to include the requirement given above in the fuzzy logic search. The problem with the new definition is that it is recursive, so our fuzzy logic inference mechanism will get stuck in an infinite loop when using it.

Question 7 *Will your inference strategy explore uncharted areas of the search space if applied to a surrogate knowledge base designed for less exhaustive search?*

Surrogacy: Resource-Bounded Assembler Receives Classical Specification When using our classical specification as an oracle for the resource bounded system (see Section 4.2) we had the difficulty that the classical system could generate from its catalogue a limitless supply of (identically named) buttons. It could do this because its inference is not resource constrained. If we wish to impose resource constraints uniformly in this situation we can use the classical knowledge base as a surrogate for the resource-bounded inference mechanism. In this case, the translations would be identical to those of Section 4.2. However, the results of inference are different because uniform application of the resource constraint would prevent goals which succeed in the oracle version, such as $(mouse\_comp(m_2, b_1, b_2) \wedge mouse\_comp(m_2, btype_1, btype_1))$, from succeeding now because two uses of $button(btype_1)$ are no longer allowed.

Although easy to describe, this sort of application of proof constraints is not always easy to do. Three possible strategies are:

- We could simply import the rules from the donor system, translating them into the relevant format for the recipient's inference mechanisms. This is straightforward for our example because the notations of each system are similar. It is more difficult for dissimilar logics and, even if the translation is straightforward, may require us to import larger amounts of knowledge than we would like.
- We could be strict about how often we allow the recipient inference mechanism to access the donor knowledge base. For instance, in our example we could impose the resource constraint by never attempting the same goal more than once using the classical system. The problem with this strategy is that it loses information.
- We could require that each donor system returns the proof tree describing how it found the conclusion to each given goal. This proof tree could then be checked retrospectively for compatability with the recipient inference strategy. This approach is more expensive than the previous one because the donor system must record the proof and may have gone to a great deal of effort only to be rejected. However, it gives more information about the capabilities of the donor system, making it more likely that the recipient could make an informed choice.

Question 8 *If your inference mechanism places resource limits on proofs, how will these be applied to its surrogate knowledge base?*

Surrogacy: Classical Specification Receives Resource Bounded Assembly In Section 4.2 we saw that using the resource-bounded system as an oracle supplying information to the classical specification system gave an unsatisfactory mixture of metaphors when constructing possible computer specifications. All components were treated as if they were constantly available except for the mouse which (because it was obtained from the resource-bounded system) was in limited supply. One way of lifting this restriction is by using the classical inference mechanism with the knowledge base of the resource-bounded system. The translations between notations remain the same as in Section 4.2 but we can now obtain as many copies of mouse components of type m_2 as we need - so the stock system is behaving like the catalogue system in this respect.

Question 9 *If your inference mechanism is not resource limited but its surrogate knowledge base is, does it matter that this restriction has been lifted?*

5   Conclusions

Often when we write a knowledge base we have in mind some inference strategy which will harness that knowledge to solve problems. This influences the way we describe the knowledge, so that our representation is not entirely neutral with respect to inference. Consequently, if we share knowledge bases without considering the interaction between the inference methods of the donor and recipient systems then we may produce results which are unexpected. This paper examines this sort of interaction with the aid of three simple logical systems and uses these to generate a series of generic questions which aspiring knowledge sharers might ask themselves if they plan to share knowledge between systems with differing inference methods.

Our paper raises questions but does not provide a computational architecture for solving them. One promising avenue of investigation, which we are beginning to study, is that the full semantics of systems of inference can often be divided into two components: the part which is responsible for connecting to the syntax of the knowledge base and the part which determines the control strategy used in problem solving. Preliminary implementation of a knowledge sharing architecture which uses meta-interpretation to automate this method is underway.

References

1. J. Agusti, F. Esteva, P. Garcia, L. Godo, R. Lopez de Mantaras, and C. Sierra. Local Multi-Valued Logics in Modular Expert Systems. *Journal of Experimental & Theoretical Artificial Intelligence*, 6(3):303–321, 1994.
2. J. Agusti, F. Esteva, P. Garcia, L. Godo, and C. Sierra. Combining Multiple-valued Logics in Modular Expert Systems. In *VII Conf. on Uncertainty in A. I.*, 1991.
3. J. Andrews, V. Dahl, and F. Popowich. Characterizing Logic Grammars: a Substructural Logic Approach. *Journal of Logic Programming*, 26:235–283, 1996.
4. A. J. Bonner and M. Kifer. Transaction Logic Programming. Technical Report CSRI 270, University of Toronto, 1992.
5. F. S. Correa da Silva. Automated Reasoning About an Uncertain Domain. Technical report, Edinburgh Univ., Dept. of A. I., 1991.
6. F. S. Correa da Silva and D. V. Carbogim. A System for Reasoning with Fuzzy Predicates. Technical Report RT-MAC-9413, IME, Univ. of São Paulo, 1994.
7. F. S. Correa da Silva, D. S. Robertson, and J. Hesketh. Automated Reasoning with Uncertainties. In *Knowledge Representation and Uncertainty*, chapter 5. Springer Verlag LNAI 808, 1994.
8. F. S. Correa da Silva, W. W. Vasconcelos, and D. S. Robertson. Cooperation Between Knowledge-Based Systems. In *Proc. IV World Congress on Expert Systems*, pages 819–825, Mexico City, Mexico, 1998.
9. J. Y. Girard. Linear Logic. *Theoretical Computer Science*, 50:1–102, 1987.
10. P. p. i. Gray. KRAFT – Knowledge Reuse and Fusion/Transformation. http:// www.csd.abdn.ac.uk/ apreece/ Research/ KRAFT/ KRAFTinfo.html.
11. N. Guarino, editor. *Formal Ontology in Information Systems*. IOS Press, 1998.
12. R. C. T. Lee. Fuzzy Logic and the Resolution Principle. *Journal of the ACM*, 19:109–119, 1972.
13. R. Neches and D. Gunning. The Knowledge Sharing Effort. http://www-ksl.stanford.edu/knowledge-sharing/papers/kse-overview.html.
14. J. A. Reyes, F. Esteva, and J. Puyol-Gruart. Defining and Combining Multiple-Valued Logics for Knowl. Based Syst. In *Proc. 1st Catalan A. I. Congr.*, 1998.
15. M. V. T. Santos, P. E. Santos, F. S. Correa da Silva, and M. Rillo. Actions as PROLOG Programs. In *IEEE Proc. of Joint Symposia on Intell. & Systems*, 1996.
16. E. Y. Shapiro. Logic Programming with Uncertainties - a Tool for Implementing Rule-based Systems. In *Proc. 8th IJCAI*, 1983.
17. V. S. p. d. Subrahmanian. Hermes – a Heterogeneous Reasoning and Mediator System. http:// www.cs.umd.edu/ projects/ hermes/index.html.
18. M. Uschold and M. Gruninger. Ontologies: Principles, Methods and Applications. *Knowledge Engineering Review*, 11(2):93–136, 1996.
19. C. Zhang and M. Orlowska. Homomorphic Transformation Among Inexact Reasoning Models in Distributed Expert Systems. Technical Report 135, Dept. of Computer Science, University of Queensland, 1989.

A Novel Approach for Detecting and Correcting Segmentation and Recognition Errors in Arabic OCR Systems

Khaled Mostafa[1] Samir I. Shaheen[2] Ahmed M. Darwish[2] Ibrahim Farag[3]

[1]Information Technology Department, Faculty of Computers and Information
[2]Computer Engineering Department
{shaheen, darwish} @frcu.eun.eg
[3]Institute of Statistical Studies and Research
[1, 2 & 3]Cairo University, Giza 12613 EGYPT

Abstract. In this paper, we propose a new approach for detecting and correcting segmentation and recognition errors in Arabic OCR systems. The approach is suitable for both typewritten and handwritten script recognition systems. Error detection is based on rules of the Arabic language and a morphology analyzer. This type of analysis has the advantage of limiting the size of the dictionary to a practical size. Thus, a complete dictionary for roots, which does not exceed 5641 roots, the morphological rules and all valid patterns can be kept in a moderate size file. Recognition channel characteristics are modeled using a set of probabilistic finite state machines. Contextual information is utilized in the form of transitional probabilities between letters of previously defined vocabulary (finite lexicon) and transitional probabilities of garbled text. The developed detection and correction modules have been incorporated as a post-processing phase in an Arabic handwritten cursive script recognition system. Experimental results show a considerable enhancement in performance.

1 Introduction

Humans utilize context in reading extensively that quiet often they do not even realize they have read a misspelled word correctly. Thus, it appears that contextual techniques might have a great potential for detecting and correcting errors resulting from early stages in an OCR system. Examples of these errors that may occur during the segmentation include splitting or merging. In the splitting case, the segmentor introduces an extra break point where a character is split into two, thus introducing an extra symbol that creates confusion later. In the merging case, the segmentor misses the correct break point, thus, two symbols become single symbol, leading to a recognition error later. During the recognition phase, the popular error is substituting where a character is incorrectly identified as a different one.

The solution to such errors through contextual techniques is achieved through string correction algorithms. Basically, there are two approaches to solve the string correction problem: the metric space approach [1], and the probabilistic approach [2].

The metric space approach looks for the Minimum distance d($\mathbf{X}$,$\mathbf{Y}$) overall $\mathbf{X}$ where X is the correct word and Y is the observed word. The distance between two strings that allows insertion and/or deletion is also referred to as the Levenshtien metric [3]. A commonly used generalization of the Levenshtien distance is the minimum cost of transforming string $\mathbf{X}$ into string $\mathbf{Y}$ when the allowable operations are character insertion, deletion, and substitution [4]. By using different weights for the three different edit operations, we obtain the weighted Levenshtien distance [2]. A recent review of these techniques is presented in [5] and speeding up techniques for approximate string matching is presented in [6].

The probabilistic approach is based on Bayes theory and looks for Maximum probability P($\mathbf{X}$,$\mathbf{Y}$) overall $\mathbf{X}$ [2]. In this approach, the problem is tackled as an optimization problem given certain statistics. These statistics are letter transition probabilities in the source text and the letter confusion probabilities that characterize the noisy channel in the OCR.

A number of algorithms that utilize both contextual constraints of the text and peculiarities of the recognition channel have been proposed [7], [8]. They assume a text recognition model as shown in figure 1. The noisy channel is a process that introduces errors into the correct text. The text enhancer utilizes contextual constraints, known to be present in the correct text, as well as channel characteristics, and attempts to determine the correct text from the garbled text. The contextual knowledge used may be in the form of probabilities of letters, letter pairs, and letter triplets, probabilities of words, a list of words acceptable in global or local context (i.e. a lexicon), a grammar describing the syntax of the language, or a semantic network representation of the embodied concept.

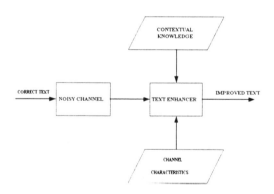

Fig. 1. Model of text garbling and enhancement.

Bozinovic et al [8] presented a string correction algorithm for the output of a cursive script recognition channel under the assumption that the string is a garbled version of a word in a finite dictionary. The algorithms are based on Bayes theory which utilizes a channel model in terms of a probabilistic finite state machine. A tree structure representation of the dictionary is used to improve the performance of the search process. A recent version of this representation is presented in [9]. The algorithm proposed and used here is of the same nature as [8] but adjusted to work with Arabic character primitives which may represent part of a character [10]. A major contribution is that the algorithm is also used to solve the problem of binding and dot assignment of the recognized primitives in addition to the error correction. Also contextual knowledge, based on the characteristics of the Arabic language, is combined with the algorithm to improve its performance.

Aside from this introductory section, the paper is composed of seven more sections. Section 2 introduces the letter sequence validity check. Section 3 explains the error detection using lexical and morphological validity check. The channel modeling using the probabilistic finite state machine is presented in section 4. Section 5 gives the theoretical details of the likelihood computation. Details of the implementation and initialization are presented in section 6. Experimental results are discussed in section 7. We finally conclude in section 8 with observations and recommendations.

2 Letter Sequence Validity

Letter sequence validity considers the combination of letters in the word. Due to the fact that certain letter sequences are not allowed in Arabic, any word containing such a sequence is immediately flagged as an error. Some of these rules follow directly from the difficulty of pronouncing two consecutive letters of similar phonation. Based on studies from [11,12 &13] the following pairs of letters cannot be present together in Arabic word:

- "ج" and "ص" thus, "صولجان" is not an Arabic word.   – "ذ" and "س".
- "ج" and "ق" thus, "منجنيق" is not an Arabic word.   – "ز" and "س".
- "ط" and "ج" thus, "طاجن" is not an Arabic word.

A full statistical study of letter sequences in Arabic roots was conducted by Ali Mousa [14] and following is a sample of invalid sequences presented in that study.

- "ث" can not be preceded by"ظ ض ص ش س ز ذ ت" or followed by "ص س ز ذ ت ظ".
- "ح" can not be preceded by "هـ غ ع خ" or followed by "غ ع ظ خ ث".
- "غ" can not be preceded by "ك ع ظ خ ح ج" or followed by "ق ع ظ ذ خ ح ج".

3 Lexical and Morphological Validity

Arabic linguists classify words according to many properties [15]. One of these properties is the ability to apply morphological rules to the word. According to this classification, the word is either declinable or indeclinable. Declinability means the ability to connect to relational letters, dualistic letters, ...etc. The different indeclinable nouns in Arabic are:

أنواع الأسماء غير المتصرفة(تلازم حالة واحدة):
الضمير ، اسم الإشارة، الاسم الموصول، اسم الشرط، اسم الاستفهام، الكناية، الظرف، اسم العدد.

Declinable words may also be classified according to their ability of derivation.

According to this classification, the word is either derivable or non-derivable (fixed). It is important to note that the Arabic language is an algorithmic language, i.e. most of its words follow the morphology rule. The words that are fixed represent a very small fraction of the Arabic words. A list of the different fixed nouns is given below:

أنواع الأسماء الجامدة (غير مأخوذة من الفعل):

اسم العلم، اسم الجنس، بعض أسماء المكان، بعض أسماء الآلة، المصدر المجرد، بعض المصادر الميمية.

Thus, before the morphology analyzer is put into operation, the input word is checked against a dictionary containing all indeclinable words in the Arabic language. If the input word is not one of the indeclinable word, then it is analyzed by the morphology analyzer. Our assumption is that the general form of any derivational word is as follows:

Word == Prefix (es) + Core + Suffix (es)

The plural indicates that the same word may contain more than one prefix and/or more than one suffix.

The morphology analyzer works in four steps namely, prefix processing, suffix processing, core processing, and compatibility of parts of the word processing.

3.1 Prefix Processing

The Arabic language has many prefixes, examples of which are: present letters e.g. "أ ن ت ى", future letters e.g. "س ل ف", determination article "ال", question letter "أ", in addition to many other articles affecting the word. Some prefixes are used only with nouns while others are used with verbs. The determination prefix, for example, distinguishes between a verb and a noun and it also makes that noun determinate. A table for prefixes of verbs and nouns has been devised.

3.2 Suffix Processing

Suffixes are common in Arabic words and they have a very important effect on both the syntactic and semantic phases. There are five types of suffixes in Arabic:

- Subject pronouns: These may take different forms according to the tense of the verb. Moreover, for each tense, the forms of subject pronoun differ according to the order of the subject (first person "متكلم", second person "مخاطب", or third person "غائب") and the state of the subject (single "مفرد", dual "مثنى", and plural "جمع"). Subject pronouns connected to present tense verbs differ according to the state of the verbs in the present tense (Indicative "مرفوع", apocopate "مجزوم", and Subjunctive "منصوب").

- Object pronouns: These may take different forms according to the order of the object (first person "متكلم", second person "مخاطب", or third person "غائب") and the state of the object (single "مفرد", dual "مثنى", and plural "جمع").

- Guard 'Noon' " نون الوقاية" is added to the verbs just before 'Yaa' of first person "ياء المتكلم" (e.g. أعطانى أخرجنى).

- The plural indicator "واو الجمع" is added to the verbs just before the plural indicator "ميم الجمع" (e.g. سألتمونى أعطيتموه).

- The emphasizing Noon "نون التوكيد" is connected to present and imperative tenses verbs only. The verb must not be connected to subject pronoun to accept the addition of Noon Al-tawkeed (e.g. ليسجننّ ليكونـن).

3.3 Core Processing

The most difficult part of an Arabic word analyzer lies in the core processing. The augmented three-letter verbs may take several morphological patterns as shown below:

تفعّل تفاعل إنفعل إفتعل إفعلّ إستفعل إفعوعل إفعولّ فاعل أفعل الصيغ المختلفة للفعل الثلاثي المزيد: فعّل

The four-letter verbs may take several other morphological patterns. The main problem of core processing is the irregularity of verb roots that have vowel letters "حروف علّه". These verbs are classified into five categories:

- METHAL which starts with a vowel (e.g. وعد - وصم).
- AGOUAF which has a vowel in the middle (e.g. باع - خاف).
- NAQUES which has a vowel at the end (e.g. رمى - سعى).
- LAFEEF MAFROUCK which has one at the beginning and another at the end (e.g. وفى - ولى).
- LAFEEF MAKROON which has either first and middle or middle and last letters (e.g. روى - حيى).

The verbs that belong to these five types of roots cause most of the problems in morphology processing. The reason for this is that roots always need a modification while applying the morphology rules.

4 The Probabilistic Finite State Machine

Assuming the segmentation step breaks each connected part of a word into several primitives [16]. Common segmentor errors are either missing a primitive, thus leading to merging two symbols together, or splitting a primitive into two subprimitives, thus generating an extra symbol. The segmentor can be modeled by a channel consisting of a number probabilistic finite states machines (PFSM) equal to the number of primitives. Each PFSM consists of three states: S1 which is the initial state, S2 which is an intermediate state, and S3 which is the final state. Transition from S1 to S3 through S2 occurs in case of missing primitive and transition from S1 to S3 directly occurs in case of splitting a primitive.

In the following, we will adopt the following definitions

b_i The symbols of the output Alphabet (I=1.2....., 68). where 68 is the number of primitives.

$q(b_j)$ The probability that the PFSM will produce symbol b_i during the transition from state S1 to state S3.

$q'(b_j)$ The probability that the PFSM will produce symbol b_i during the transition from state S1 to state S2.

$q''(b_j)$ The probability that the PFSM will produce symbol b_i during the transition from state S2 to state S3.

P_o The probability of transition from state S1 to state S3 without producing any output.

P_1 The probability of transition from state S1 to state S3 and producing an output symbol.

P_2 The probability of transition from state S1 to state S2

P_3 The probability of transition from state S2 to state S3. (always = 1).

Each PFSM must satisfy the following constraints:

$$P_o + P_1 + P_2 = 1,$$
$$q(b_1) + q(b_2) + ... + q(b_{68}) = 1,$$
$$q'(b_1) + q'(b_2) + ... + q'(b_{68}) = 1, \text{ and} \tag{1}$$
$$q''(b_1) + q''(b_2) + ... + q''(b_{68}) = 1.$$

When a correct input string passes through the channel, the output of the channel is obtained by applying the corresponding channel model (PFSM) to each symbol of the input string and concatenating the output string for each PFSM. Errors due to splitting into two symbols are accurately modeled in the PFSM by the P_2 transition. Additional states can be introduced to take into account splitting into more than two symbols. Merging errors are only approximated by the P_o transition since they do not reflect what symbol the "deleted" one was merged with.

5 Likelihood Computation

Given that $Y = y_1, y_2, ..., y_N$ is an observed word of N characters (i.e. output of the channel and input to the string correction algorithm) and X is one of a set of possible legal strings.

According to Bayes rule, the probability of X being the correct one is:

$$P(X|Y) = P(X,Y) / P(Y) \tag{2}$$

The correct word X' is the one that maximizes $P(X|Y)$. In other words X' is the one that satisfies the condition: $P(X'|Y) > P(X|Y)$ for all Xs where (X' not equal to X).

Maximizing the conditional probability $P(X|Y)$ over all legal X is equivalent to maximizing the joint probability $P(X,Y)$ since $P(Y)$ is independent of X.

Let $X = x_1, x_2, ..., x_m$ be a string of length m of the word that produce Y, $Y_{1,n}$ be the prefix of length n of word Y, $L(X_{1,m},Y)$ be the probability that $X_{1,m}$ produces Y and $P(X_{1,m},Y_{1,n},Y_{n+1,N})$ be the probability that $X_{1,m}$ produces $Y_{1,n}$ where $Y_{n+1,N}$ is the remainder of Y, then

$$L(X_{1,m}, Y) = \sum_{n=0}^{N} P(X_{1,m}, Y_{1,n}, Y_{n+1,N})$$

$$= \sum_{n=0}^{N} P(X_{1,m}) * P(Y_{1,n}, Y_{n+1,N} | X_{1,m}) \tag{3}$$

$$= P(X_{1,m}) * \sum_{n=0}^{N} P(Y_{1,n} | X_{1,m}) * P(Y_{n+1,N} | X_{1,m}, Y_{1,n})$$

Assuming that $Y_{n+1,N}$ does not depend significantly on $X_{1,m}$, then $P(Y_{n+1,n}|X_{1,m},Y_{1,n})$ can be approximated by $P(Y_{n+1,n}|Y_{1,n})$.

Let $Q(Y|X_{1,m})$ be { $P(F|X_{1,m})$, $P(Y_{1,1}|X_{1,m})$, ..., $P(Y_{1,N}|X_{1,m})$ }, $R(Y)$ be { $P(Y_{1,N}|F)$, $P(Y_{2,N}|Y_{1,1})$, ..., $P(F|Y_{1,N})$ }, then $L(X_{1,m},Y) = P(X_{1,m})$ * | $Q(Y|X_{1,m})$. $R(Y)$ |
where $|Q.R|$ is the inner product between the two vectors, F is an empty string.

$R(Y)$ can be computed using the first order transitional probabilities for the garbled text according to the relation:

$$P(Y_{1,N}|Y_{1,i-1}) = P(y_i|y_{i-1}) * ... * P(y_N|y_{N-1}) \qquad (4)$$

where $i = 1,2, ..., N+1$, $Y_{1,N} = y_1, y_2, ..., y_N$, $y_o = F$ and $Y_{k,j} = F$ if $j < k$

$P(X_{1,m})$ can be computed using the first order transitional probabilities for the correct text according to the relation:

$$P(X_{1,m}) = P(X_1|F) * P(X_2|X_1) * ... * P(X_m|X_{m-1}) \qquad (5)$$

$Q(Y|X_{1,m})$ can be computed using the fact that $X_{1,m}$ can account for/produce $Y_{1,n}$ in at most three ways:

1- $X_{1,m-1}$ produces $Y_{1,n}$ and x_m was deleted (due to merge).
2- $X_{1,m-1}$ produces $Y_{1,n-1}$ and x_m produces y_n (substitution)
3- $X_{1,m-1}$ produces $Y_{1,n-2}$ and xm produces y_{n-1} y_n (splitting)

Since these cases partition the space of possibilities of $X_{1,m}$ producing $Y_{1,n}$ we have:

$$Q_i(Y|X_{1,m}) = Q_i(Y|X_{1,m-1}) * P_o(x_m) + Q_{i-1}(Y|X_{1,m-1}) * P_1(x_m) * q(x_m,y_n)$$
$$+ Q_{i-2}(Y|X_{1,m-1}) * P_2(x_m) * q'(x_m,y_{n-1}) * q''(x_m,y_n) \qquad (6)$$

where $i=0, 1, ..., N$.
The string correction algorithm could be summarized as follows:

```
Set X=F;        push (X);
repeat          if (empty stack) return Y      //no correction occurs
                pop(X)
                for x= first_Arabic_character to last_Arabic_character
                        begin
                        Z = concatenation (X,x);
                        if Z is a prefix of a valid dictionary word and L(Z,Y) >0
                                then push(Z)
                        end;
                Sort stack in descending order of L values;
                Z = top element of the stack;
until (Z is a valid dictionary word) and  (Q(Y|Z) > 0)
return Z
```

6 Implementation and Initialization Issues

To implement the morphology analyzer, a set of rules is built one for each morphology pattern. There are some morphology patterns that need more than one rule for each root type. The root and morphology pattern extraction are achieved using simple template matching; i.e. every pattern has a template, and the system puts the input word on this template and if the word does not fit, another template is chosen and the process is repeated until a successful match is found or an error message is returned. The output of the morphological analyzer is a list of prefixes, the root, the morphology pattern, suffixes, and the word type (verb or noun). These word components must be compatible with each other. i.e. the prefixes must be compatible with the morphology pattern, and the suffixes must be compatible with the word type, ...etc. For example, the emphasizing 'ن' cannot be connected to a noun. If this occurs, this means that the whole word is morphologically wrong although each component is individually correct. Hence, the morphological analyzer can check whether each input word is a valid Arabic word or not. If it is not a valid Arabic word then the error correction algorithm is used to correct the recognition error.

Two steps are required to initialize the correction algorithm: transitional probabilities and PFSM probabilities.

- Transitional probabilities of the correct text, $P(X_{1,m})$, are easily computed using the dictionary. Transitional probabilities of the garbled text, $R(Y)$, are easily computed using the garbled text.
- PFSM probabilities depend on the characteristics of the recognition system. In other words, they depend on the merging and splitting probabilities for each primitive. These characteristics can be obtained by testing the recognition system using real input data and comparing the input and the output of the system. The following list illustrates PFSM probabilities for character "ـضـ" for the segmentor and the system reported in [10] and [16].
  probability of deletion (merging) $P_o = 0.0$,
  probability of substitution $P_1 = 0.95$,
  probability of splitting $P_2 = 0.05$,
  probability of substitution q"(ـضـ) = ٠,٩٤,
  probability of substitution q"(ـخـ) = 0.03,
  probability of splitting q"(ـنـ) = 1,
  probability of substitution q"(ـفـ) = 0.03,
  probability of splitting q"(ـصـ) = 1.

For each word four inputs are fed into the detection and correction algorithms: the number of parts (isolated non-connected segments of the word known as Part of the Arabic Word (PAW)), the recognized primitives, recognized diacritics, and the distribution of diacritics. The output is the word with the maximum likelihood probability $L(X_{1,m},Y)$ such that the number of PAWs of the reconstructed word = number of PAWs in the input word. Words having the same diacritic types and vertical position associations are favoured; if none is found, this condition is relaxed.

7 Experimental Results

To determine the performance and efficiency of the proposed error detection approach and error correction algorithm, a database was established and experiments were conducted. The database was constructed using a set of 30 randomly selected documents containing 1843 words. A dictionary of about 800 words was extracted from the previously constructed database. Transitional probabilities were computed from the dictionary words. Three attributes were associated with each word in the dictionary: ordered primitives labels of the word, number of parts of the word (PAW) and types and positions of all diacritics associated with the word. A simulation for the recognition channel errors was applied on the dictionary words in order to obtain the garbled text. Transitional probabilities for the garbled text were computed using the garbled version of the dictionary text. All distinct words in the test database were fed into the simulator to randomly introduce one to three errors on each word. Error detection algorithms were applied on the simulated output of the recognition system. There are three possible outputs of the morphology analyzer:

1. Valid root, prefixes and suffixes are compatible to that root.
2. Invalid root. The extracted root is not in the roots dictionary associated with the analyzer.
3. Incompatibility between the root and the prefixes and suffixes of the word.

Table 1 shows an example of the output of the morphology analyzer. If the output is "1" then the recognized word is assumed to be correct and is written in the recognition output file. If the output is "2" or "3" then the recognized word is assumed to be wrong.

A word that has been determined to be erroneous, by either letter sequence test or morphological test is passed to correction algorithm. The percentage of the corrected words was 95.3%. Improper corrections are not all due to algorithm malfunction. Some improper corrections are due to the fact that the correct word is not present in the limited dictionary in the first place.

Table 1. An example of error detection using morphology analyzer.

Output code	word under test
1	يذهبان
1	أعطيناكموه
3	يذهبتان
2	رمينه
1	أكلنا

8 Conclusion

In this paper we have presented a new morphological based approach for detecting errors originated from segmentation and recognition phases in Arabic OCR systems and a PFSM based algorithm for correcting these errors. Morphological analysis is performed to verify that the recognized word is a valid root of the Arabic language. This is done by reducing the word to its canonical form, i.e. the root, verifying that the removed letters are valid augments. Then, the root itself is checked against a dictionary, which contains all the valid roots of the Arabic language, to verify the validity of the root itself. Also, word components (prefixes, the root, the morphology pattern, and suffixes) are checked for the compatibility with each other. The main advantage of this approach is the usage of a limited size dictionary to check the validity of unlimited number of words in Arabic language. The structure of the correction algorithm is based on Bayes theory which utilizes channel model in terms of probabilistic finite state machines and contextual information in the form of transitional probabilities between letters. The output of the algorithm is a valid dictionary word that maximizes the likelihood probability between the input word and the output word. Experiments conducted using a dictionary of 800 words indicate that the presented approach reflects the most that can be gained from the utilization of both contextual information of the text and characteristics of the recognition channel of the system.

References

1. Galil, A.A.Z., *Pattern Matching Algorithms*, Oxford University Press, 1997.
2. Srihari, S.N., *Computer Text Recognition And Error Correction*, IEEE Computer Society Press, 1984.
3. Levenshtien, V.I., "Binary Codes Capable of Correcting Deletions, Insertions, and Reversals", *Cybernetics and Control Theory*, vol. 10, 1966, pp. 707-710.
4. Masek, W.J., "A Faster Algorithm Computing String Edit Distances," *Journal of Computer and Systems Sciences*, vol. 20, Feb. 1980, pp. 18-31.
5. Stephen, G.A., *String Searching Algorithms*, Lecture Notes on Computing, World Scientific Pub. 1994.
6. Du, M.W. and Chang, S.C., "An Approach to Designing Very Fast Approximate String Matching Algorithms", *IEEE Trans. on Knowledge and Data Engineering*, vol. 6, no. 4, Aug. 1994, pp. 620 - 633.
7. Hull, J.and Srihari, S., "Experiments in Text Recognition with Binary n-Grams and Viterbi Algorithm", *IEEE Trans. Pattern Anal. and Machine Intell.*, vol. PAMI-4., no. 9, Sept. 1982, pp. 520-530.
8. Bozinovic, R. and Srihari, S., "A String Correction Algorithm for Cursive Script Recognition", *IEEE Trans. Pattern Anal. and Machine Intell.*, vol. PAMI-4. no. 6, Nov. 1982.
9. Shang, H. and Merrettal, T.H., " Tries for Approximate String Matching," *IEEE Trans. on Knowledge and Data Engineering*, vol. 8, no. 4, Aug. 1996, pp. 540 - 547.
10. Mostafa, K. and Darwish, A.M., "A Novel Approach for Arabic Handwritten Cursive Script Recognition", *Journal of Eng. & Applied Sc.*, (to Appear, Feb. 1999).
١١. الخليل ابن أحمد الفراهيدي،*كتاب العين*، تحقيق د/مهدى المخزومى، د/ابراهيم السامرائى، طبعه دار و مكتبه الهلال.
١٢. عبد الرحمن جلال الدين السيوطى، *المزهر فى علوم اللغة*، تحقيق محمد أحمد جاد المولى، على البجاوى، محمد أبو الفضل إبراهيم، طبعه دار الجيل، بيروت.
١٣. ابن منظور المصرى، *لسان العرب*، تحقيق عبدالله على الكبير، محمد أحمد حسب الله، هاشم محمد الشاذلى، طبعه دار المعارف، القاهرة.
١٤. د. على حلمى موسى، *دراسه إحصائية لجذور مفردات اللغة العربية*، مطبوعات جامعه الكويت، ١٩٧١.
١٥. أنطوان الدحداح، *معجم تصريف الأفعال العربية*، مكتبه لبنان، الطبعة الثالثة، ١٩٩٦.
16.Mostafa, K. and Darwish, A.M., "Robust Baseline Independent Algorithms for Segmentation and Reconstruction of Arabic Handwritten Cursive Script" *Proc. SPIE Document Recognition and Retrieval VI*, vol. 3651, San Jose, Jan 1999.

Face Recognition Using Principal Component Analysis Applied to an Egyptian Face Database

Mohammad E. Ragab[1] Ahmed M. Darwish[2] Ehsan M. Abed[1] Samir I. Shaheen[2]

[1]Electronics Research Institute, Tahrir St., Dokki, Giza, 12411 EGYPT
[2]Computer Engineering Department, Cairo University, Giza, 12613 EGYPT
{darwish, shaheen @frcu.eun.eg}

Abstract. Although face recognition is highly race-oriented, to-date there is no Egyptian database of face images for research purposes. This paper serves two purposes. First we present the efforts undertaken to build the first Egyptian face database (over 1100 images). Second we present a variant algorithm based on principal component analysis (PCA) but adjusted to Egyptian environment. In order to conduct face recognition research under realistic circumstances, no restrictions have been imposed on the volunteers (eyeglasses, moustaches, beards, and veils (hijab)). Furthermore, photos, for each volunteer, were taken during two sessions that are two months apart (March and May). Meanwhile, multiple light sources have been used. More than 1000 experiments have been carried out to evaluate the approach under different conditions. A new pentagon-shaped mask has been devised, which has proven suitable to enhance the recognition rate.

1 Introduction

Computer face recognition can open the door to numerous applications specifically in security. The face recognition problem has been tackled using different methodologies. One of the first approaches used geometrical features such as singular points [1] or distance measurements [2]. Recently, the use of the Karhunen-Loeve (KL) expansion for the representation of faces has generated renewed interest [3]-[6]. Another example is the use of profile images such as the work presented in [7]-[11]. Some very recent approaches include thermal image processing techniques [12], and using combined biometric features [13]. Neural Networks has been also used to address several problems but these are out of the scope of this paper.

This paper describes a three-stage project that is thought to be the first serious effort in Egyptian face recognition. The first stage involved building an Egyptian face database. This is important because system performance highly depends on the training database. The second stage involved the development of a recognition algorithm that is based on principal component analysis (PCA) with several distinct modifications. Once all the tools were in place, the third stage involved the massive experimentation to study the effect of every parameter in the system from an imaging point of view and from an algorithmic point of view. Over 1000 experiments have been conducted to test the effect of training set size, number of eigenfaces, different masks, rotations, lighting conditions, moustaches, eyeglasses, veils (hijab) and others.

Aside from this introductory section, this paper is composed of four more sections. Section 2 introduces the PCA approach. Setup and statistics for the face database are described in section 3. Summary of results is presented in section 4. We finally conclude in section 5 with observations and recommendations for future work.

2 Proposed Approach

The image space is highly redundant when used to describe faces. Principal Component Analysis (PCA) can be used to reduce the dimensionality of the face image by representing it using a small number of eigenvalues. PCA derives its basis from the KL transform or the Hotelling transform [14].

The aim is to construct a face space with each component not correlated with any other. This means that the covariance matrix of the new components should be diagonal. In addition, the new space should maximize the correlation of each component with itself.

Assume x_i and y_i are the vectors describing the face in the image space and the new space respectively, $\mathbf{X}$ and $\mathbf{Y}$ are the matrices containing several vectors representing N faces in the image space and the new space respectively. Let $\mathbf{P}$ be the transformation matrix, then

$$\mathbf{Y} = \mathbf{P}^T \mathbf{X} \tag{1}$$

$$\mathbf{X} = \mathbf{P}\mathbf{Y} \tag{2}$$

From (1) and (2), we deduce that $\mathbf{P}\mathbf{P}^T = \mathbf{I}$, i.e. $\mathbf{P}'s$ columns are orthonormal to each other. The relation between the covariance matrices, $\mathbf{S}_x$ and $\mathbf{S}_y$, becomes as follows:

$$\mathbf{S}_y = \mathbf{Y}\mathbf{Y}^T = \mathbf{P}^T \mathbf{X}\mathbf{X}^T \mathbf{P} = \mathbf{P}^T \mathbf{S}_x \mathbf{P} \tag{3}$$

If we choose $\mathbf{P}$ to be the matrix containing the eigenvectors of $\mathbf{S}_x$, then:

$$\mathbf{S}_x \mathbf{P} = \Lambda \mathbf{P} \tag{4}$$

where Λ is the diagonal matrix containing the eigenvalues of $\mathbf{S}_x$. Substituting from (3) we get,

$$\mathbf{S}_y = \mathbf{P}^T \Lambda \mathbf{P} = \Lambda \mathbf{P}^T \mathbf{P} = \Lambda \tag{5}$$

Thus, $\mathbf{S}_y$ will be a diagonal matrix containing the eigenvalues of $\mathbf{S}_x$.

Hence, we can now represent the image of a face with a limited number of coefficients equal to, k, where k is the $rank(\mathbf{X}*\mathbf{X}^T)$. We can further reduce the dimensionality to n ($n<k$) by omitting some principal components. This is done by dropping the smallest $k-n$ eigenvalues and their associated eigenvectors.

The proposed approach, thus, proceeds in two steps, an offline one-time step which is eigenface generation and an online repetitive step with every query which is the transformation of the input face image to the face space.

2.1 Eigenface Generation

PCA computes the basis of a space represented by the training vectors (faces). The basis vectors are eigenvectors which are defined in the image space and can be viewed as faces, thus they are known as the eigenfaces. Depending on the required accuracy and computational complexity of the following steps, the M eigenfaces corresponding to the largest M eigenvalues are retained and the rest are skipped.

2.2 Transformation to Face Space

The transformation of any image of a face from the image space to the face space is very simple using the following:

$$f_F = \mathbf{E}^T f_I \tag{6}$$

where, $\mathbf{E}$ is the matrix of the eigenfaces, f_I is a face in the image space, and f_F is the same face in face space. Thus, it is basically a projection operation in which the components of the face are calculated in directions of the most important eigenfaces.

Once this is done, then, the recognition process is reduced to finding the best match between the coefficients of the given query face and the coefficients of faces stored in the database.

2.3 Center of Attention Enhancement

Better results could be obtained by removing the background from the image. This increases the effect of the face data (pixels) on the computed eigenfaces. One contribution of the work presented here is the experimentation with different masks and the development two masks, a pentagon mask for general purposes, and an oval mask for females with veils. Samples of both masks are included in figures 3-5.

3 Building an Egyptian Face Database

Several databases with various sizes and diverse circumstances have been used in face recognition research at universities and research institutes. Sample databases include 2 frontal views of 20 Japanese used in [2], 40 faces used in [15], the MIT database of 27 views for 16 persons used in [16], the Olivetti database of 10 views for 40 persons, Weizmann database consisting of 30 views for 28 persons [16] and the Bern database of 10 views for 30 persons, and the FERET database, the largest database so far, consisting of 7562 images for 3000 persons [6]. Thus, except for the FERET most databases consist of a humble number of persons with very limited number of views (usually two).

Most existing databases have imposed several restrictions on volunteers, such as excluding people with eyeglasses or moustaches. To target real life applications, we did not impose any restrictions on the volunteers and made sure to include all distinct features from the race and traditional point of view such as females wearing veils (hijab). The database contains a fairly large size of images consisting of 10 views with different angles of rotation for 117 persons. Three views were taken in March 1998, and the rest were taken two months later in May to allow realistic changes in appearance to take place (e.g. growing a beard and changing hairstyle). This is as opposed to cosmetic changes usually done by waiting few minutes and retaking the pictures. Meanwhile, multiple light sources have been used and varied. In the following, we present the setup, statistics and light correction procedures.

3.1 Setup

Figure 1, shows the plan of the location and person seating used during photograph shooting. We would like to point out the following:
- We had multiple light sources: fluorescent light, sunlight (coming through the window, which differs drastically during the day and also differs from March to May), and flash. This resembles real life applications.
- The normal lighting conditions were realized by having the window-curtain half-open and the fluorescent light on.
- Darker lighting conditions were realized by closing the curtain completely, the fluorescent light, however, was left on. Thus, it was dark but not dim.
- During flash photography, curtains were half-open and fluorescent light was on.
- Volunteers were photographed in March 1998 then in May of the same year.
- All the photographs were taken between 10:00 am and 4:00 p.m.
- We used Casio QV-700 digital camera, mounted on a tripod to avoid blurs.

3.2 Statistics

For each volunteer, we took the following photographs:
- In March 3 views in normal lighting conditions with the camera in the first position. The 3 views are: Frontal view, volunteer's seat has the direction (f), 45° view (direction m), and 90° side view (direction s).
- In May 7 views as follows: 3 at the same lighting conditions, angles, and distance from camera as those in March, 1 at the same lighting conditions, but with seat in the direction (r) 22.5° (to study the effect of changing pose), 1 frontal view with flash while the camera is at the first position, 1 frontal view in darker lighting conditions while the camera is at the first position (to study the effect of changing lighting conditions negatively, and positively) and 1 frontal view at the normal lighting conditions while the camera is at the second position (to study the effect of changing scale.

The only restriction we have imposed on the volunteers was to ask them to avoid any emotional expressions such as showing sadness, happiness or anger.
Volunteer profile and database statistics could be summarized as follows:
- Volunteers' ages range from mid-twenties (23) to mid-fifties (56).
- In March 117 volunteers were photographed (3 views).

544

- In May 111 volunteers were photographed (7 views).
- Volunteers were 82 males and 35 females.
- 20 males were wearing eyeglasses in March, increased to 21 in May.
- 4 females were wearing eyeglasses.
- 32 males had a moustache only. 8 had a beard and a moustache in March increased to 10 in May. No one had a beard without a moustache.
- 28 females were wearing a veil (hijab), 7 females were not wearing a veil.
- In the frontal views of March 4 males and 2 females blinked.
- In the frontal views of March 2 males and 1 female had their lips slightly parted showing sort of an intention to smile.

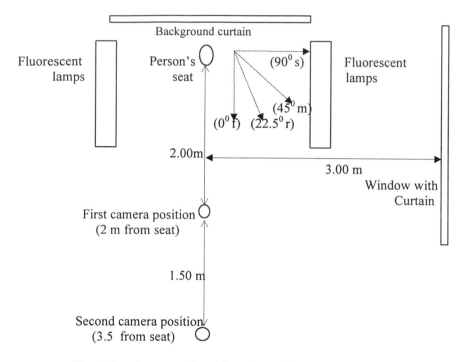

Fig. 1. Imaging setup (dotted lines show different seating angles).

3.3 Fluorescent Correction

Fluorescent light causes flickering that cannot be noticed by the human eye. However, it is sensed by the digital camera and leads to changes in brightness from an image to another. A correction procedure was carried out as follows:

We collected all images photographed under the same lighting conditions, and for each image of them we calculated the following gray levels: Minimum, maximum and average gray level for the background curtain, b_m, b_x and b_a, Minimum gray level in the whole image, g_m, Maximum gray level within face in the image, g_x.

We then computed the average of these values for the whole set of images taken under the same lighting conditions.

$$b_{sm} = \frac{1}{N} \sum_i^N b_m(i)$$ (7)

where, $b_m(i) = b_m$ for image i and thus, b_{sm} becomes the average minimum value for the whole set.

Correction then, proceeded by fixing 7 gray levels on each image and using a spline interpolation to figure out the correction mapping between input gray levels and output (corrected) gray levels. A sample of such mapping is shown in figure 2a. The seven gray levels chosen were as follows: b_m, b_x, b_a, and g_m were forced to the average of the whole set b_{sm}, b_{sx}, b_{sa} and g_{sm}, respectively, g_x was forced to the global maximum for this face in the whole set, to increase the dynamic range of gray levels for each face, and 0 (darkest black) and 255 (lightest white) were kept as is. Figure 2b shows samples of images before and after correction.

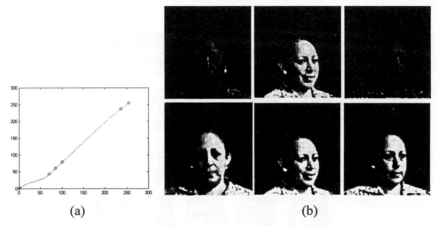

(a)   (b)

Fig. 2. Fluorescent light correction. (a) mapping (input (horizontal) vs. output (vertical) gray levels). (b) Sample images before (upper) and after (lower) correction.

4 Experimental Results

We formed two sets of the database of faces. Each one of them contains of the same number of males, females, males with moustaches and beards, females with veils (hijab) and persons with eyeglasses. One set was used for training to obtain the eigenvectors in which we code other faces. The second set was used for queries.

We have used two distance measures to compare faces, the Euclidean distance, d_e, and the Mahalanobis distance, d_m, defined respectively as follows:

$$d_e(x,y)^2 = \sum (x_i - y_i)^2$$ (8)

$$D_m(x,y)^2 = \sum \lambda_i^{-1} (x_i - y_i)^2$$ (9)

where x_i and y_i are the component of the two faces in direction of the ith eigenface and λ_i is the ith eigenvalue. Thus, Mahalanobis distance pays equal attention to all components in direction of all eigenfaces [17].

We have carried out the numerous experiments totaling to over 1000 queries. Table 1 lists samples of the results obtained from these experiments. Figures 3-5 show samples of the training sets, eigenfaces and the oval mask used for females with veil. A complete set of tables and plots for the results can be found in [18]. For most of the experiments, we have gathered statistics for:

- Varying the number of training faces from 10 to 57 and for each training set of faces varying the number of eigenvectors from 10 to 57.
- Using the Euclidean and Mahalanobis distance.
- Recording recognition percentage of perfect hit (correct face is the top ranked by the query result), best three matches (correct face is within the highest three ranked faces) and best five matches (correct face within the highest five ranked faces).

Frontal views recognition rate was higher than 68% without co-locating eyes of matched faces. If eyes were to be co-located we would expect a higher recognition rate.

Fig. 3. A sample training set of frontal views without background or hair

Fig. 4. Some eigenfaces extracted from a training set of frontal views.

Table 1. Sample results for different queries. R: recognition rate, NTF: number of training faces, NEV: number of eigenvalues.

Case	R %	NTF	NEV
1. Frontal views			
a. normal lighting, with background, perfect hit.	21.8	57	25
b. a with removing background and most of hair.	30.9	50	45
c. previous + histogram equalized, perfect hit.	32.7	40	35
d. c with best three matches.	40.0	40	35
e. c with best five matches.	41.8	50	30
f. e, Mahalanobis distance	45.4	57	30
g. f + removing males with moustaches.	48.6	40	20
h. g + removing females with veils.	68.2	30	25
i. Females with veils (hijab) using a special oval mask.	61.5	40	30
2. Rotated views			
a. 45° given same angle two months earlier, no histogram equalization, perfect hit	27.3	57	50
b. Side views given side views two months earlier, no histogram equalization, perfect hit	27.3	57	20
c. 22.5° given frontal views two months earlier, no equalization, perfect hit	14.6	40	15
d. a with histogram equalization, Mahalanobis distance, best 5, excluding females with veils.	63.6	50	40
e. b with histogram equalization, Mahalanobis distance, best 5, excluding females with veils.	52.4	57	30
f. c with histogram equalization, Mahalanobis distance, best 5, excluding females with veils.	36.4	15	11
g. Different angles given 2 months earlier all-together, histogram equalization, Mahalanobis distance, best 5, excluding females with veils	50.0	30	30
3. Scaled views			
a. No histogram equalization, perfect hit	5.5	40	10
b. Histogram equalization, Mahalanobis distance, best 5matches, excluding females with veils.	18.2	50	20
4. Different lighting conditions			
a. darker lighting conditions given normal 2 months earlier, no equalization, perfect hit	18.2	30	30
b. flash given normal 2 months earlier, no equalization, perfect hit	10.9	10	10
c. a but with histogram equalization	27.3	20	20
d. b but with histogram equalization	9.1	57	20
e. c + Mahalanobis distance, best 5matches, excluding females with veils.	50.0	20	20
f. d + Mahalanobis distance, best 5matches, excluding females with veils.	28.6	50	40

From table 1 and rest of results in [18] we notice that:

- Removing the background and most of the hair gives a considerable enhancement in the recognition rate.
- Histogram equalization has very little effect under normal lighting conditions, considerably enhances the recognition rate under dark lighting conditions and has a negative effect under flashlight.
- Recognition given a different view is almost half the case of the same view.
- Mahalanobis distance gives better results than Euclidean distance.
- Considering the best five matches is slightly better than the best three matches.
- It seems that moustaches result somehow in a dominant eigenvector, that once present causes any face with a moustache to be considered a hit.
- Concentrating on the inner oval area of the face considerably enhanced the recognition rate for females (figure 5). This is expected because it avoids features from veil and hair style to dominate the recognition process, thus, reducing the effect of such changes from March to May (a normal case for females)
- The system still holds well even with a mixture of views present at the same time altogether.
- While the system performance was quite acceptable for rotations, scaled cases presented a real problem where the recognition rate was at its minimum.

Fig. 5. Oval mask used to deal with veils.

5 Conclusion

In this paper, we have presented a complete study on using a PCA based approach to recognition of Egyptian faces. The new set of masks and implementation specifics introduced, proved to be effective leading to satisfactory results of over 68% correct recognition rate for frontal views. Another major contribution of the work presented here is efforts exerted in creating the first Egyptian face database. Unlike previous databases used by other researches, it contains a fairly large number of images (over 1100 images) taken under realistic multiple source lighting circumstances with no restrictions imposed on the volunteers (eyeglasses, moustaches, beards, and veils). Furthermore, images were taken two months apart allowing changes in features from the first to the second shooting (e.g. growing a beard).

We intend to carry out a study to find out if we can correlate specific eigenfaces with specific features of the face (like gender or race). We also hope we can keep track of the volunteers so that we can take several more sets of photographs periodically in the next 10 years to study the effect of aging.

549

Acknowledgement   The authors would like to thank Dr. Esmat Abdel-Fattah for supporting the project and all the volunteers who participated to make this database possible.

References

1. Kaya, Y. and Kobayashi, K., "A Basic Study on Human Face Recognition," in *Frontiers of Pattern Recognition* (S. Watanabe, ed.), pp. 265-289, New York: Academic Press, 1972.
2. Kanade, T., *Computer Recognition of Human Faces*, Basel and Stuttgart: Birkhauser, 1977.
3. Sirovich, L. and Kirby, M., " Low-dimensional Procedure for the characterization of Human Face," *Journal of the Optical Society of America*, vol. 4, pp. 519-524, 1987.
4. Kirby, M. and Sirovich, L., "Application of the Karhunen-Loeve Procedure for the Characterization of Human Faces," *IEEE Trans. Pattern Analysis and Machine Intelligence*, vol.12, 1990.
5. Turk, M.A. and Pentland, A.P., "Face Recognition Using Eigenfaces," *Proceedings IEEE Inter. Conf. on Computer Vision and Pattern Recognition*, pp. 586-591, 1991.
6. Pentland, A., Moghaddam, B., Starner, T. And Turk, M., "View-based and Modular Eigenspaces for Face Recognition," *Proceedings IEEE Inter. Conf. on Computer Vision and Pattern Recognition*, 1994.
7. Kaufman, G.J. And Breeding, K.J., "The Automatic Recognition of Human Faces from Profile Silhouettes," *IEEE Trans. System, Man, and Cybernetics*, vol. 6, pp. 113-121, 1976.
8. Harmon, L. and Hunt, W., "Automatic Recognition of Human Face Profiles," *Computer Graphics and Image Processing*, vol. 6,pp. 135-156, 1977
9. Harmon, L., Kuo, S. and Ramig, P., "Identification of Human Face Profiles by Computer," *Pattern Recognition*, vol. 10, pp. 301-312, 1978.
10. Harmon, L., Khan, M., Lasch, R. and Ramig, P., "Machine Identification of Human Faces," *Pattern Recognition*, vol. 13, pp. 97-110, 1981.
11. Wu, C. and Huang, J., "Human Face Profile Recognition by Computer," *Pattern Recognition*, vol. 23, pp. 255-259, 1990.
12. Yoshitomi, Y., Miyaura, T., Tomita S. and Kimura, S., "Face Identification Using Thermal Image Processing," *Proceedings. 6th IEEE International Workshop on Robot and Human Communication*, pp. 374-379, 1997.
13. Dieckmann, U., Plankensteiner, P. and Wagner, T., "SESAM: A Biometric Person Identification System Using Sensor Fusion," *Pattern Recognition Letters*, vol.18, no.9, pp. 827-33, 1997.
14. Jain, A.K., *Fundamentals of Digital Image Processing*, Prentice Hall, 1989.
15. J. Buhmann, M. Lades, and C.V.D. Malsburg, "Size and Distortion Invariant Object Recognition by Hierarchical Graph Matching," *Proceedings, Inter. Joint Conf. on Neural Networks*, pp. 411-416, 1990
16. Zhang, J., Yan, Y. and Lades, M., "Face Recognition: Eigenface, Elastic Matching, and Neural Nets," *Proceedings of the IEEE*, vol. 85, pp. 1423-1435, 1997.
17. Costen, N. and Craw, I., " Automatic Face Recognition: What Representation?," *University of Aberdeen,* Scotland, 1996.
18. Ragab, M.E., *Face Recognition Using Principal Component Analysis Applied to an Egyptian Face Database*, M.SC. Thesis, Cairo University, Feb 99.

Using Labelled and Unlabelled Data to Train a Multilayer Perceptron for Colour Classification in Graphic Arts

Antanas Verikas[1], Adas Gelzinis[2], and Kerstin Malmqvist[1]

[1] Centre for Imaging Science and Technologies,
Halmstad University, Box 823, Halmstad, S 30118, Sweden,
antanas.verikas@cist.hh.se,
[2] Department of Applied Electronics,
Kaunas University of Technology, Studentu 50, 3031, Kaunas, Lithuania

Abstract. This paper presents an approach to using both labelled and unlabelled data to train a multi-layer perceptron. The unlabelled data are iteratively pre-processed by a perceptron being trained to obtain the soft class label estimates. It is demonstrated that substantial gains in classification performance may be achieved from the use of the approach when the labelled data do not adequately represent the entire class distributions. The experimental investigations performed have shown that the approach proposed may be successfully used to train networks for colour classification in graphic arts.

1 Introduction

Numerous classifiers and associated learning algorithms have been developed. A common feature of nearly all the approaches is the assumption that class labels are known for each input data vector used for training. This is true for neural networks such as multilayer perceptron and radial basis functions. These neural networks are usually trained to minimize the squared distance to the target class value. Knowledge of class labels is also required for parametric classifiers such as a mixture of Gaussian experts. The training for such classifiers typically involves dividing the training data into subsets by class and then using the maximum likelihood estimation to separately learn each class density.

Labelled data can be plentiful for some applications. For others, such as medical imaging, quality control in a halftone multi-coloured printing the correct class labels can not be easily obtained for a significant part of the vast amount of training data available. The difficulty in obtaining class labels may arise due to incomplete knowledge or limited resources. An expensive expertise is often required to derive class labels. Besides, labelling of the data is often a very tedious and time-consuming procedure. A significant part of the data often remains unlabelled.

The practical significance of training with labelled and unlabelled data was recognised in [2]. The theoretical results obtained in [1] are less optimistic concerning the value of unlabelled data for learning classification problems. By contrast, Towell [7], Shashahani and Landgrebe [6], Miller and Uyar [3],[4] have obtained substantial gains in classification performance when using both labelled and unlabelled data. However, despite the promising results, there has been little work done on using together labelled and unlabelled data. One reason is that conventional supervised learning approaches such as the error back propagation have no direct way to incorporate unlabelled data and, therefore, discard them. In this paper, we propose an approach to using both labelled and unlabelled data to train a multilayer perceptron. The unlabelled data are iteratively preprocessed by a perceptron being trained to obtain the soft class label estimates.

The remainder of the paper is organised as follows. In the next section, we briefly describe the related work. The learning approach proposed is presented in section three. The fourth section presents results of the experimental investigations. Section five gives conclusions of the work.

2 Related work

The learning algorithm proposed by Towell uses conventional neural network supervised training techniques except that it occasionally replaces a labelled sample with a synthetic one [7]. The synthetic sample is the centroid of labelled and unlabelled samples in the neighbourhood of the labelled sample. Therefore, the algorithm uses both labelled and unlabelled samples to make local variances estimates.

Another approach to using unlabelled data for learning classification problems relies on probability mixture models. A mixture-based probability model chosen by Shashahani and Landgrebe [6], and Miller and Uyar [3],[4] is the key to incorporate unlabelled data in the learning process. In [6], the conditional likelihood is maximised, while Miller and Uyar [3],[4] maximise the joint data likelihood given by

$$\log L = \sum_{\mathbf{x}_i \in \mathbf{X}^u} \log \sum_{l=1}^{L} \alpha_l f(\mathbf{x}_i/\theta_l) + \sum_{\mathbf{x}_i \in \mathbf{X}^l} \log \sum_{l=1}^{L} \alpha_l P[c_i/\mathbf{x}_i, m_i = l] f(\mathbf{x}_i/\theta_l) \quad (1)$$

where $f(\mathbf{x}_i/\theta_l)$ is one of L component densities, with non-negative mixing parameters α_l, such that $\sum_{l=1}^{L} \alpha_l = 1$, θ_l is the set of parameters of the component density, $\mathbf{X}^u$ is the unlabelled data set and $\mathbf{X}^l$ is the labelled data set. The class labels are also assumed to be random quantities and are chosen according to the probabilities $P[c_i/\mathbf{x}_i, m_i]$, i.e. conditioned on the selected mixture component $m_i \in \{1, 2, ..., L\}$ and on the feature values. The optimal classification rule for this model is given by the following selector function with range in the class label set:

$$S(\mathbf{x}) = \text{argmax}_k \sum_{j} P[c_i = k/m_i = j, \mathbf{x}_i] P[m_i = j/\mathbf{x}_i] \quad (2)$$

where

$$P[m_i = j/\mathbf{x}_i] = \frac{\alpha_j f(\mathbf{x}_i/\theta_j)}{\sum_{l=1}^{L} \alpha_l f(\mathbf{x}_i/\theta_l)} \quad (3)$$

The EM algorithm is used to maximise the likelihood. The EM algorithm iteratively guesses the value of missing information. The algorithm uses global information in this process and, therefore, it may not perform well on some problems. We use the approach chosen by Miller and Uyar [4] for our comparisons.

3 Training the network

3.1 The network

The network used is a multilayer perceptron. Let $o_j^{(q)}$ denote the output signal of the jth neuron in the qth layer induced by presentation of an input pattern $\mathbf{x}$, and $w_{ij}^{(q)}$ the connection weight coming from the ith neuron in the $(q-1)$ layer to the jth neuron in the qth layer. Assume that $\mathbf{x}$ is an augmented vector, i.e. $x_0 = 1$. Then $o_j^{(0)} = x_j$, $o_j^{(q)} = f(net_j^q)$,

$$net_j^{(q)} = \sum_{i=0}^{n_{q-1}} w_{ij}^{(q)} o_i^{(q-1)} \tag{4}$$

where $net_j^{(q)}$ stands for the activation level of the *neuron*, n_{q-1} is the number of neurons in the $q-1$ layer and $f(net)$ is a sigmoid activation function.

3.2 Learning set

We assume that the learning set $\mathbf{X}$ consists of two subsets $\mathbf{X} = \{\mathbf{X}_l, \mathbf{X}_u\}$, where $\mathbf{X}^l = \{(\mathbf{x}^1, c_1), (\mathbf{x}^2, c^2)...(\mathbf{x}^{N_l}, c^{N_l})\}$ is the labelled data subset and $\mathbf{X}^u = \{\mathbf{x}^{N_l+1}, ..., \mathbf{x}^{N_l+N_u}\}$ is the unlabelled data subset, $\mathbf{x}^n \in R^K$ is the nth data vector, $\mathbf{c}^n \in I = \{1, 2, ..., Q\}$ is the class label, Q is the number of classes, N_l and N_u are the number of labelled and unlabelled data points, respectively; $N = N_l + N_u$.

The target values for the labelled data subset $\mathbf{t}^1, ..., \mathbf{t}^{N_l}$ are encoded according to the scheme $1 - of - Q$, i.e. $t_k^n = 1$, if $c^n = k$ and $t_k^n = 0$, otherwise.

Target values for unlabelled data It has been shown that the MLP trained by minimising the mean squared error can be viewed as a tool to estimate the posteriori class probability from the set of input data [5], i.e.

$$P(c_j/\mathbf{x}) = \frac{\exp(o_j^L(\mathbf{x}, \mathbf{w}))}{\sum_{i=1}^{Q} \exp(o_i^L(\mathbf{x}, \mathbf{w}))} \tag{5}$$

where L is the number of layers in the network. We use this estimate of the posteriori class probability to obtain target values for unlabelled data.

Let M^n be the set of indices of the k_{nn} nearest neighbours of the unlabelled data point $\mathbf{x}^n$, i.e.

$$\| \mathbf{x}^n - \mathbf{x}^k \| < \| \mathbf{x}^n - \mathbf{x}^i \|, \forall k \in M^n, \forall i \notin M^n; N_l + 1 \leq n \leq N; 1 \leq k, i \leq N \tag{6}$$

Then, the target vector $\mathbf{t}^n$ for the unlabelled data point $\mathbf{x}^n$ is given by

$$\mathbf{t}^n = \frac{\sum_{l \in M^n} \mathbf{t}^l}{k_{nn}} \tag{7}$$

where

$$t_j^l = \frac{\exp(o_j^L(\mathbf{x}^l, \mathbf{w}))}{\sum_{i=1}^Q \exp(o_i^L(\mathbf{x}^l, \mathbf{w}))}, \forall j = 1, 2, ..., Q, \forall n = N_{l+1}, ..., N \tag{8}$$

if $N_l + 1 \leq l \leq N$ and

$$t_j^l = 1 \text{ or } 0 \tag{9}$$

if $1 \leq l \leq N_l$.

3.3   Learning algorithm

We assume that regions of lower pattern density usually separate data classes. Therefore, decision boundaries between the classes should be located in such low pattern density regions. First, the network is trained using labelled data only. Then, the target values for unlabelled data are estimated using (7). Next, the network is retrained using both the labelled and unlabelled data and the target values for the unlabelled data are re-estimated. In the following, the training and re-estimation steps are iterated until the classification results obtained from the network in a predetermined number of subsequent iterations stop changing or the number of iterations exceeds some given number. We use the error back propagation algorithm to train the network. The network is trained by minimising the sum squared error augmented with the additional regularisation term

$$E(\mathbf{w}) = \frac{1}{2} \sum_{n=1}^N \sum_{j=1}^Q (o_j^{(L)n}(\mathbf{w}) - t_j^n)^2 + \beta \sum_{i=1}^{N_W} w_i^2 \tag{10}$$

where N_W is the number of weights in the network and β is the regularisation coefficient. The learning algorithm is encapsulated in the following five steps.

1. Train the network using labelled data only.
2. Evaluate the posteriori class probabilities for unlabelled data using (8).
3. Calculate target values for the unlabelled data using (7).
4. Train the network using both the labelled and unlabelled data.
5. Stop, if the classification results obtained from the network in a predetermined number of subsequent iterations stop changing or the number of iterations exceeds some given number; otherwise go to Step 2.

4   Experimental testing

The learning approach developed was compared with the EM algorithm proposed in [4] and the conventional error back propagation learning when only labelled data are exploited. We used data of two types to test the approach: the 2D artificial data, and the data from real application aiming to separate colours of a halftone multi-coloured picture.

4.1 Tests for artificial data

We performed two series of experiments with artificial data. In the first series, a two-class separation problem that requires linear decision boundary was considered. In the second series, the network had to develop highly non-linear decision boundaries for classifying two-dimensional data into three classes.

A two-class problem The data set is shown in Fig. 1. The data are Gaussian with the same covariance matrix. The optimal decision boundary for the data is linear. There are 2000 data points in the class 'o' and 400 in the class '+'.

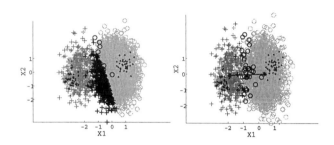

Fig. 1. A two-class ('+' and 'o') classification problem. Classification result obtained from the MLP trained on labelled data only (left). Classification result obtained from the MLP trained on both labelled and unlabelled data (right).

The black dots illustrate the labelled data. The unlabelled data points shown in grey are correctly classified. The classification errors are shown in black '+' and 'o'. Only 40 data points from each class are labelled. As it often happens in practice, the labelled data do not adequately represent the class distributions. The one hidden layer perceptron used in the experiment contained four nodes in the hidden layer.

Fig. 1 (left) illustrates the typical classification result obtained from the MLP trained on the labelled data only. As can be seen from Fig. 1 (left), the labelled data are classified correctly. However, the average classification error for the unlabelled data is 8.98%.

The classification result obtained using the proposed training approach is shown in Fig. 1 (right). The number of nearest neighbours used in the experiment is $k_{nn} = 8$. The improvement obtained from the use of both the labelled and unlabelled data in the learning process should be obvious. The EM algorithm yielded a similar error rate for this particular data set: 1.89% for the EM algorithm with $L = 2$ mixture components and 1.77% for the approach proposed.

The next experiment aims to justify that the decision boundary found is situated in a place sparse in data points. First, we uniformly distribute some number of data points, let say 100, on the line connecting centres of the classes.

The line is shown in Fig. 1 (right). For each data point on the line we then evaluate the average distance to the k_{nn} nearest neighbours in the data set. To evaluate the distance we always used the same number of the nearest neighbours as for the target value estimates when training the network. The average distance evaluated for all the data points on the line reflects the density of the data points in the neighbourhood of the line. Fig. 2 (left) illustrates the distribution of the average distance obtained using $k_{nn} = 8$. The characters '+' and 'o' show to which class the points of the line are assigned. As can be seen from Fig. 2 (left), the class label changes in the place of the lowest pattern density. A more accurate estimate of the average distance (the generalised average distance) is obtained by averaging the estimates from several lines drawn in parallel to the line connecting the centres of the classes. The end points of the ith line are given by

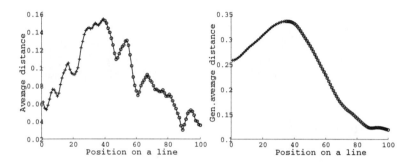

Fig. 2. Distribution of the average distance when $k_{nn} = 8$ (left). Distribution of the generalised average distance when $k_{nn} = 50$ (right).

$$\mathbf{p}_c^i = \mathbf{p}_c^* + (0, (((N_{lin} + 1)/2) - i)\Delta h), c = 1, 2; i = 1, 2, ..., N_{lin} \qquad (11)$$

where c is the class index, $\mathbf{p}_c^*$ is the centre of the class c, N_{lin} stands for the number of the lines, and Δh is a constant. The generalised average distance from the jth point on the generalised line to the k_{nn} nearest neighbours is given by

$$d_j = \frac{1}{N_{lin}} \sum_{i=1}^{N_{lin}} d_j^i \qquad (12)$$

where d_j^i is the average distance from the jth point on the ith line.

We use local information to obtain the target value estimates when training the network. We use also the same degree of locality (the same number of nearest neighbours) when estimating the generalised average distance distributions. An increase in the number of nearest neighbours used smoothens the estimates. Fig. 2 (right) illustrates the case. The value of $k_{nn} = 50$ has been used in this experiment. Due to the over-smoothing, the peak of the distribution is moved towards the less populated class. The figure shows that the class label, again, changes at the top of the distribution. Therefore, the decision boundary is also moved towards the less populated class. However, from the point of view of the estimate obtained, the decision boundary is located in the place of the lowest pattern density.

A three-class problem The data set used in this experiment is shown in Fig. 3. There are three classes in the set: 'o', '+', and '□' containing 1683, 1739, and 1735 data points, respectively. All the data points, except those shown as black dots, are unlabelled. There are only 40 labelled data points from each class. The labelled data extremely badly represent the class distributions. The MLP used in this experiment contained ten nodes in the hidden layer. Fig. 3 (left) shows a

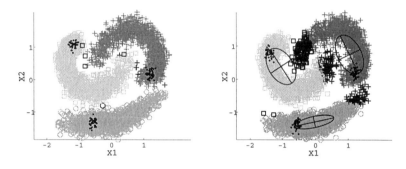

Fig. 3. A three-class ('o', '+', and '□') classification problem. Classification result obtained from the network trained on both labelled and unlabelled data (left). Classification result obtained from the EM algorithm for $L = 3$ mixture components (right).

typical classification result obtained from the network trained according to the approach proposed. The data points shown in grey as well as the labelled data are correctly classified. The classification errors are shown in black 'o', '+', and '□'.

We have experimented with the EM algorithm using different number of mixture components. In all the trials the obtained classification accuracy was far bellow the accuracy illustrated in Fig. 3 (left). Fig. 3 (right) displays one outcome from the EM algorithm for $L = 3$ mixture components. It is obvious that it is not enough to use three mixture components for modelling the class distributions. Using more mixture components, however, causes errors in the estimate of the probabilities $P[c_i/\mathbf{x}_i, m_i]$. The performance of the EM algorithm improves considerably if labelled data are spread over the entire class distributions. For such a case, the classification error rate obtained from the EM algorithm with $L = 8$, and the network trained according to approach proposed was 0.31% and 0.19%, respectively.

4.2   Experiments with real data

Nowadays, multi-coloured pictures in newspapers, books, journals and many other places are most often created by printing dots of cyan (**c**), magenta (**m**), yellow (**y**), and black (**k**) inks upon each other through screens having different raster angles. Fig. 4 (above-left) illustrates an example of an enlarged grey scale view of a small area of a newspaper picture that contains dots of all the four inks. There are many factors contributing to the quality of such colour prints. One factor that influences the colour impression of the picture is the size and shape of the areas covered by different inks. Different printing conditions can yield a different distribution of size of the printed dots as well as a different degree of shape irregularities of the dots. Therefore, quality inspection and control of the printing process implies evaluation of the percentage of area covered by inks of different colours and estimation of the actual size and shape of the printed dots. The measurements can be done automatically if we know the colour of every pixel of the picture.

Solving the colour classification task in a supervised way raises a problem of labelling the training data [9]. For example, it is not a trivial problem to assign labels (one of **c**, **m**, **y**, **w** (white paper), **cy**, **cm**, **my**, **cmy**, and **k**) for each pixel of the image shown in Fig. 4 (above-left). It is not clear what labels pixels located on the borders of the dots should acquire. Besides, manual labelling is a very tedious procedure. In the remainder of the paper, we show that the proposed learning technique can be successfully used to perform colour separation in images taken from halftone multi-coloured pictures.

In all the experiments presented here, every pixel was described by the normalised variables $i = R + G + B$, $j = R - B$, and $k = R - 2G + B$. These variables are obtained by performing a linear transformation of the R, G, B vector by eigenvectors of the covariance matrix of the R, G and B variables (under the assumption that the variables are of equal variances and covariances) [8].

In the first experiment, pictures created by printing dots of **c** and **y** inks have been used. Images recorded from such pictures contain four colour classes, namely, **c**, **y**, **w**, and **cy**. The task was to extract "yellow dots" by classifying pixels into **c**, **y**, **w**, and **cy** colour classes. Pixels of the **y**, and **cy** colour classes then represent the "yellow dots". The distribution of the learning data set used in this experiment is shown in Fig. 5 (left). The labelled data and the corresponding class labels are shown in black in the figure. To make the labelling process less tedious, labels were assigned only for pixels from some central parts of the printing dots and the **cy** and **w** areas. As can be seen, the labelled data do not adequately represent the entire class distributions. There are 14770 pixels in total. Among them 200 pixels from each class are labelled. The class **w** is most populated, and the class **cy** is the one least populated. The network we used to solve the problem contained 5 hidden nodes. The number of the nearest neighbours has been chosen to be $k_{nn} = 25$. Fig. 4 (above) gives a possibility to compare two solutions to the task: the results obtained from the EM algorithm (above-middle) and the approach proposed (above-right). The results reveal the superiority of the approach proposed over the EM algorithm in this particular

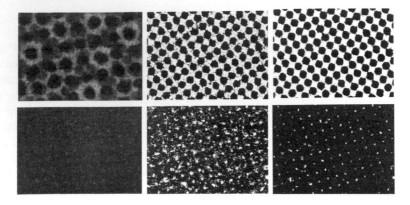

Fig. 4. An enlarged grey scale view of a part of a newspaper picture that was created by printing dots of cyan, magenta, yellow, and black inks (above-left). The "Yellow dots" found by the EM algorithm (above-middle) and the approach proposed (above-right). An image containing pixels of two colour classes: **m** and **w** (below-left). Classification result obtained from the MLP trained on: labelled data only (below-middle), both labelled and unlabelled data (below-right).

application. The EM algorithm has created noise by assigning some pixels to class **y** when the pixels actually belong to class **w**. The classification result obtained from the EM algorithm is visualised in Fig. 5 in the ijk colour space. The figure clearly shows that the class **y** was favoured in this experiment.

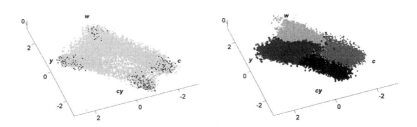

Fig. 5. Distribution of the learning set in the ijk colour space (left). Classification result in the ijk colour space for the EM algorithm (right).

In graphic arts, it is important to accurately measure the size of halftone printing dots. Such a need arises when studying interaction between different types of ink, paper and printing devices. It is not an easy task if we have to deal with very small or very large tonal values. Fig. 4 (below-left) presents an example of such a task. Approximately 97% of the area of the picture the image was taken from is covered by a magenta ink. The task is to determine the percentage of the "white" areas. The middle and the right images of Fig. 4 (below) illustrate two solutions to the problem, namely, the result obtained from the MLP trained on labelled data only and the result obtained from the approach proposed, respectively. Comparing the results, we find that an obvious improvement in classification accuracy has been obtained when using the proposed training approach.

5 Conclusions

We have presented an approach to using both labelled and unlabelled data to train a multi-layer perceptron. The approach banks on the assumption that regions of low pattern density usually separate data classes. Decision boundaries developed during training according to the approach proposed are positioned in such low pattern density regions.

We have demonstrated experimentally that substantial gains in classification performance may be achieved from the use of the approach when the labelled data do not adequately represent the entire class distributions. In most of the tests performed, we found superiority of the proposed training approach over the EM algorithm and the MLP trained on labelled data only. Encouraging experimental results have been obtained when using the approach proposed for colour classification in graphic arts.

References

1. Castelli, V., Cover, T. M.: On the exponential value of labeled samples. Pattern Recognition Letters **16** (1995) 105-111
2. Lippman, R.P.: Pattern Classification Using Neural Networks. IEEE Communications Magazine **27** (1989) 47-64
3. Miller, D.J., Uyar, H.S.: Combined learning and use for a mixture model equivalent to the RBF classifier. Neural Computation **10** (1998) 281-293
4. Miller, D.J., Uyar, H.S.: A mixture of experts classifier with learning based on both labelled and unlabelled data. Mozer, M.C., Jordan, M.I., Petsche, T. (eds). Advances in Neural Information Processing Systems **9** MIT Press (1997) 571-577
5. Ruck, D.W., Rogers, S.K., Kabrisky, M., Oxley, M.E., Suter, B.W.: The multilayer perceptron as an approximation to a Bayes optimal discriminant function. IEEE Trans. Neural Networks **1** (1990) 296-298
6. Shashahani, B., Landgrebe, D.: The Effect on Unlabelled Samples in Reducing the Small Sample Size Problem and Mitigating the Huges Phenomenon. IEEE Transactions on Geoscience and Remote Sensing **32** (1994) 1087-1095
7. Towell, G.: Using unlabeled data for supervised learning. Mozer, M.C., Jordan, M.I., Petsche, T. (eds). Advances in Neural Information Processing Systems **9** MIT Press (1997) 647-653
8. Verikas,A., Malmqvist, K., Bergman, L.: Colour Image Segmentation by Modular Neural Network. Pattern Recognition Letters **18** (1997) 173-185
9. Verikas, A., Malmqvist, K., Bergman, L., Signahl, M.: Colour Classification by Neural Networks in Graphic Arts. Neural Computing & Applications **7** (1998) 52-64

A Novel 3D-2D Computer Vision Algorithm for Automatic Inspection of Filter Components

MA Rodrigues and Y Liu

AI and Pattern Recognition Research Group
Department of Computer Science
The University of Hull, Hull, HU6 7RX, UK
http://www2.dcs.hull.ac.uk/aise/aise.html

Abstract. This paper describes investigation results on the design of a real time, computer vision based system for automatic inspection of filter components in a manufacturing line. The problem involves reasoning about an object's 3D structure from 2D images. In computer vision, this is normally referred to as a 3D-2D problem. In this paper, we first present a geometrical analysis of image correspondence vectors synthesised into a single coordinate frame. The analysis is based on geometrical considerations that are fundamentally different from analytical, perspective, or epipolar geometries. The camera setup stems from the geometrical implications of such analysis and from the given background knowledge of the task within the context of the production line. We then describe a novel geometrical algorithm to estimate parameters of interest that include depth estimation and the position and orientation of the camera in world coordinate frame. The algorithm provides the closed form solution to all estimated parameters making full use of distance between feature vectors and angle information. For a comparative study of algorithm performance, we also developed an algorithm based on epipolar geometry. Experimental results show that the geometrical algorithm performs significantly better than the algorithm based on epipolar geometry.

1 Introduction

We are investigating the design of a real time, automatic vision system for inspection of filter components in a manufacturing line. The main parameters of interest include physical dimensions such as width, height, depth, diameter, and also texture, such as typical of aluminium meshes or polystyrene coatings. The essence of the problem is to reason about the 3D structure of an object from sets of 2D images. In computer vision, this is called a 3D-2D correspondence problem. Many algorithms mainly based on epipolar geometry have been proposed to solve 3D-2D problems. Examples include techniques based on conservation of distance [1], triangular geometry [6], [9], and iterative methods [5], [7], [8], [10], [11], [12]. However, such algorithms do not explicitly use distance between feature points and angular information as constraints to calibrate the orientation and position parameters of the camera even though it seems that such information would increase the performance of the algorithms. Second, these algorithms are normally based on iterative methods which have a disadvantage of depending on the initial value of unknowns and that the best solution cannot be guaranteed to be found. Despite such shortcomings, it must be pointed out that algorithms based on epipolar geometry have the advantage of simplicity of implementation while representing the only common linear method to solve all 3D-2D and 2D-2D problems. This is particularly relevant considering that much research effort has gone into the 2D-2D problem [2], [3], [4], and significantly less into the 3D-2D problem.

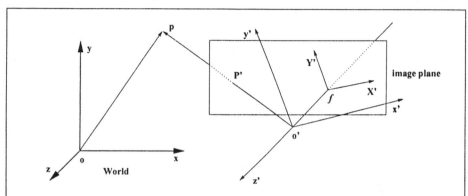

Fig. 1. 3D-2D problem: given all information about a scene (**p**) in **oxyz** coordinate frame and an image (**P′**) of this scene taken from another viewpoint **o′**, how to estimate the orientation and position of the camera at **o′** in **oxyz** coordinate frame and the structural information of this scene in camera centred coordinate frame **o′x′y′z′**? In the figure, f represents the focal length of the camera, and (**p**, **P′**) is called a 3D-2D correspondence.

In the 3D-2D problem of filter components several parameters are known and controllable, such as the position and orientation of the camera. We are thus, able to incorporate these into the setup so that implementation is more effective and reliable. Figure 1 highlights the problem and the setup. We align the optical axis of the camera so that it is parallel to the **z** axis of the reference coordinate frame and coincides with the focal point of the camera in the **xy** plane of reference. As a result, the image plane will be parallel to the **xy** plane of the reference. In order to estimate the parameters of interest including the orientation $\hat{\mathbf{R}}$ and position $\hat{\mathbf{t}}$ of the camera, and the filter structural data $\hat{\mathbf{z}}'_i$ describing the depth of each visible point in camera centred coordinate frame, we developed a novel algorithm called Simplified Geometrical Algorithm (SGA), providing the closed form solutions to all calibrated parameters making full use of distance and angle information. For a comparative study of algorithm performance, we also developed an algorithm based on the epipolar geometry called Simplified Epipolar Geometry Algorithm (SEA). The algorithms operate on the images to calibrate the transformation parameters $(\hat{\mathbf{R}}, \hat{\mathbf{t}}, \hat{\mathbf{z}}'_i)$, and relative calibration errors e_R, e_t, and $e_{z'}$ of these parameters can be estimated and used as the main criteria to decide whether to accept or reject a component.

The rest of this paper is organised as follows. The theoretical foundations for the camera setup are provided in Section 2, the novel SGA and SEA algorithms are described in Section 3 and validation of algorithms is described in Section 4. Finally, some conclusions are drawn in Section 5.

2 Theoretical Foundations for the Camera Setup

Given two sets of 2D or 3D correspondences, many methods to calibrate rigid body transformation parameters have been proposed such as the ones described in Section 1. Such methods mainly consider the relationships between object feature points and their corresponding image points based on analytic, perspective, or epipolar geometries. The geometrical considerations of our method are fundamentally different from such geometries, as we mainly consider the relationships between correspondences that have been synthesised into a single coordinate frame. The relationship between a point $\mathbf{p}$ described in one coordinate frame and its corresponding point $\mathbf{p}'$ described in another coordinate frame after a rigid body transformation can be represented as follows:

$$\mathbf{p}' = \mathbf{R}(\mathbf{p} - \mathbf{t}) \tag{1}$$

where $\mathbf{R}$ represents the rigid body rotation matrix (corresponding to an orthonormal matrix with determinant equal to 1), $\mathbf{t}$ represents the translation vector of the rigid body transformation and $(\mathbf{p}, \mathbf{p}')$ is called a correspondence. Our method is based on gometrical analysis of correspondence points with the following general assumptions and constraints: (i.) all transformations must be rigid body transformations; (ii.) all correspondences must undergo the same rigid body transformation; (iii.) all correspondences must be synthesised into a single coordinate frame; (iv.) the rigid body transformation must include a rotational part, in other words, the rotation angle θ of the rigid body transformation must be defined as: $0 < \theta < \pi$.

The geometrical considerations and formalisation of our method in 2D and in 3D are described as follows. In 2D, feature points and their correspondences $(\mathbf{p}, \mathbf{p}')$ are plotted into a single coordinate frame. It is verified that the perpendicular bisectors of correspondence vectors $\mathbf{CV} = \mathbf{p_i} - \mathbf{p'_i}$ ($i = 1, 2, \cdots$) all intercept at a fixed point $\mathbf{c}$. Plotting the translation vector $\mathbf{t}$ at the origin of the coordinate frame, it is verified that its perpendicular bisector also intercepts at the same fixed point. It is also verified that the including angle between the lines passing through any correspondence $(\mathbf{p}, \mathbf{p}')$ and the fixed point is equal to the rotation angle θ of the transformation. Formalising the above geometrical properties regarding 2D rigid body transformations we have:

Property 1 *There is one and only one point $\mathbf{c}$ in 2D which is uniquely determined by the rigid body transformation equidistant to any correspondence $(\mathbf{p}, \mathbf{p}')$ subject to the same rigid body transformation and the including angle between $\mathbf{p} - \mathbf{c}$ and $\mathbf{p}' - \mathbf{c}$ is equal to the rotation angle θ of the transformation.*

Because we are interested in analysing geometrical properties from sets of image correspondences $(\mathbf{p}, \mathbf{p}')$, we must look at the problem from such perspective. From image correspondence, the critical point $\mathbf{c}$ can be determined which is then used to calibrate the rotation angle θ of the transformation, the rotation matrix $\mathbf{R}$, and the translation vector $\mathbf{t}$. Thus, given two non-parallel correspondence vectors $\mathbf{CV_1}$ and $\mathbf{CV_2}$, their perpendicular bisectors will intersect at the point $\mathbf{c}$ which can be estimated by:

$$\mathbf{c} = \left(\begin{matrix} (\mathbf{p_1} - \mathbf{p_1'})^{\mathrm{T}} \\ (\mathbf{p_2} - \mathbf{p_2'})^{\mathrm{T}} \end{matrix} \right)^{-1} \left(\begin{matrix} \frac{\mathbf{p_1^T p_1} - \mathbf{p_1'^T p_1'}}{2} \\ \frac{\mathbf{p_2^T p_2} - \mathbf{p_2'^T p_2'}}{2} \end{matrix} \right) \tag{2}$$

If the two correspondence vectors are parallel then the two perpendicular bisectors of $\mathbf{CV_1}$ and $\mathbf{CV_2}$ will coincide and the critical point $\mathbf{c}$ is undetermined. Once the critical point is known, the rotation angle θ can be estimated by:

$$\cos \theta = \frac{(\mathbf{p} - \mathbf{c})^{\mathrm{T}} (\mathbf{p'} - \mathbf{c})}{(\mathbf{p} - \mathbf{c})^{\mathrm{T}} (\mathbf{p} - \mathbf{c})} \tag{3}$$

As a result, the rotation matrix $\mathbf{R}$ of the transformation is uniquely determined by:

$$\mathbf{R} = \left(\begin{matrix} \cos \theta & \sin \theta \\ -\sin \theta & \cos \theta \end{matrix} \right) \tag{4}$$

Finally, the translation vector can be represented as a correspondence vector $\mathbf{CV_0} = \mathbf{t} - \mathbf{0}$, which can be estimated by:

$$\mathbf{t} = (\mathbf{I} - \mathbf{R}^{\mathrm{T}})\mathbf{c} \tag{5}$$

This completes the analysis in 2D. In 3D, given a set of image correspondences $(\mathbf{p_i}, \mathbf{p_i'})$, find the vector difference $\mathbf{CV_i} - \mathbf{CV_j}$ ($i \neq j$). The perpendicular bisector planes of such vectors are found to intercept at a fixed line. We call this fixed line the rotation axis $\mathbf{h}$. Projecting all correspondences $(\mathbf{p_i}, \mathbf{p_i'})$ on a plane perpendicular to the rotation axis $\mathbf{h}$ yields a set of projected points $(\underline{\mathbf{p_i}}, \underline{\mathbf{p_i'}})$. Looking at this plane from a 2D perspective, we found that the perpendicular bisector of projected correspondence vectors $\mathbf{CV_i}$ also intercept at a fixed point. We call this fixed point the critical point $\mathbf{c}$ in 3D. Moreover, the perpendicular bisectors of the projected translation vector $\underline{\mathbf{t}}$ on this plane also intercepts at the same point, and the including angle measured at the interception point between projected correspondences $(\underline{\mathbf{p_i}}, \underline{\mathbf{p_i'}})$ is equal to the rotation angle θ of the transformation. Formalising the above, we have the following property as depicted in Figure 2.

Property 2 *There is one and only one invariant point $\mathbf{c}$ in 3D which is equidistant to the projected correspondences $(\underline{\mathbf{p}}, \underline{\mathbf{p'}})$ on the plane perpendicular to the rotation axis $\mathbf{h}$ and the including angle between vectors $\underline{\mathbf{p}} - \mathbf{c}$ and $\underline{\mathbf{p'}} - \mathbf{c}$ is equal to the rotation angle θ of the transformation.*

564

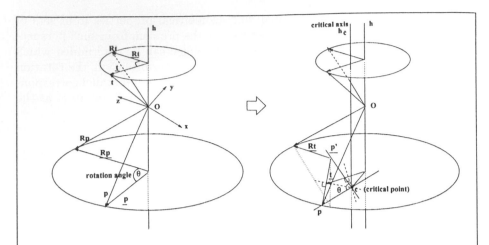

Fig. 2. The relationships among correspondence $(\mathbf{p}, \mathbf{p}')$, critical point $\mathbf{c}$, rotation axis $\mathbf{h}$, and translation vector $\mathbf{t}$ in 3D as defined by Property 2.

If we change our viewpoint and look carefully at such interesting geometric properties, we can determine expressions to estimate the rotation axis $\mathbf{h}$, the critical point $\mathbf{c}$ in 3D, and the transformation parameters rotation angle, rotation matrix, and translation vector $(\theta, \mathbf{R}, \mathbf{t})$ described as follows. Given three non-collinear correspondence vectors $\mathbf{CV_i}$ $(i = 1, 2, 3)$, the rotation axis $\mathbf{h} = (h_1, h_2, h_3)^T$ can be uniquely determined as:

$$\mathbf{h} = \frac{(\mathbf{CV_2} - \mathbf{CV_1}) \times (\mathbf{CV_3} - \mathbf{CV_1})}{\|(\mathbf{CV_2} - \mathbf{CV_1}) \times (\mathbf{CV_3} - \mathbf{CV_1})\|} \tag{6}$$

If the correspondence vectors $\mathbf{CV_i}$ are collinear, then the rotation axis $\mathbf{h}$ can also be uniquely determined as $\mathbf{h} = \mathbf{0}$. However, a nil rotation axis is not useful because correspondences cannot be projected on a plane perpendicular to such an axis. Once a valid rotation axis is known, correspondences $(\mathbf{p}, \mathbf{p}')$ are projected on a plane perpendicular to $\mathbf{h}$:

$$\underline{\mathbf{p}} = (\mathbf{I} - \mathbf{h}\mathbf{h}^T)\mathbf{p}, \quad \underline{\mathbf{p}}' = (\mathbf{I} - \mathbf{h}\mathbf{h}^T)\mathbf{p}' \tag{7}$$

Given two non-parallel projected correspondence vectors $\underline{\mathbf{CV_1}}$ and $\underline{\mathbf{CV_2}}$, the critical point $\mathbf{c}$ in 3D can be uniquely estimated by:

$$\mathbf{c} = \frac{\underline{\mathbf{p_2}}^T\underline{\mathbf{p_2}} - \underline{\mathbf{p_2}'}^T\underline{\mathbf{p_2}'}}{2(\underline{\mathbf{CV_2}})^T\mathbf{H}(\underline{\mathbf{CV_1}})}\mathbf{H}(\underline{\mathbf{CV_1}}) + \frac{\underline{\mathbf{p_1}}^T\underline{\mathbf{p_1}} - \underline{\mathbf{p_1}'}^T\underline{\mathbf{p_1}'}}{2(\underline{\mathbf{CV_1}})^T\mathbf{H}(\underline{\mathbf{CV_2}})}\mathbf{H}(\underline{\mathbf{CV_2}}) \tag{8}$$

where

$$\mathbf{H} = \begin{pmatrix} 0 & -h_3 & h_2 \\ h_3 & 0 & -h_1 \\ -h_2 & h_1 & 0 \end{pmatrix} \tag{9}$$

If the projected correspondence vectors are parallel, then their perpendicular bisectors will coincide and the critical point $\mathbf{c}$ in 3D cannot be uniquely estimated as there are infinite solutions for it. When a projected correspondence $(\underline{\mathbf{p}}, \underline{\mathbf{p}'})$ and the critical point $\mathbf{c}$ are known, the rotation angle θ can be estimated as:

$$\cos\theta = \frac{(\underline{\mathbf{p}} - \mathbf{c})^{\mathbf{T}}(\underline{\mathbf{p}'} - \mathbf{c})}{(\underline{\mathbf{p}} - \mathbf{c})^{\mathbf{T}}(\underline{\mathbf{p}} - \mathbf{c})} \tag{10}$$

and the rotation matrix $\mathbf{R}$ is estimated by:

$$\mathbf{R} = \mathbf{I} - \mathbf{H}\sin\theta + (1 - \cos\theta)\mathbf{H^2} \tag{11}$$

where $\mathbf{H}$ is defined by Equation 9. From the described geometry, it is known that the projected translation vector on the plane perpendicular to the rotation axis $\mathbf{h}$ is equal to $(\mathbf{I} - \mathbf{R^T})\mathbf{c}$. Thus, given a correspondence $(\mathbf{p}, \mathbf{p}')$, then the projected translation vector along the rotation axis $\mathbf{h}$ can be estimated as $\mathbf{hh^T}(\mathbf{p} - \mathbf{p}')$ resulting in an estimation of the translation vector as:

$$\mathbf{t} = (\mathbf{I} - \mathbf{R^T})\mathbf{c} + \mathbf{hh^T}(\mathbf{p} - \mathbf{p}') \tag{12}$$

This completes the analysis in 3D. From image correspondence $(\mathbf{p}, \mathbf{p}')$ in 2D and 3D we have developed a set of explicit expressions making full use of feature vectors and angular information. These equations are useful to the estimation of rigid body transformation parameters (e.g. $\mathbf{R}, \mathbf{h}, \mathbf{t}, \theta$) and to the design and analysis of real world applications.

3   Algorithm Description

A special camera setup for 2D image acquisition and structural analysis of 3D data has been designed as highlighted in Figure 1. First, we use a reference coordinate frame $\mathbf{oxyz}$ to describe the ideal position $(\mathbf{p})$ of the 3D filter at a fixed place in the manufacturing line. Then perspective images $(\mathbf{P}')$ of the filter are acquired from a suitable place with the optical axis of the camera parallel to the $\mathbf{z}$ axis of the reference coordinate system and the focal point of the camera on the $\mathbf{xy}$ plane. As a result, the image plane will be parallel to the $\mathbf{xy}$ plane of the reference coordinate frame. From this setup, it is known that the camera can only move in the $\mathbf{xy}$ plane of the reference coordinate frame $\mathbf{oxyz}$ and rotate around the optical axis $\mathbf{h} = (0, 0, 1)^{\mathbf{T}}$. Then the 3D orientation $\mathbf{R}$ and the 3D position $\mathbf{t}$ of the camera in the reference coordinate frame $\mathbf{oxyz}$ can be represented as:

$$\mathbf{R} = \begin{pmatrix} \cos\theta & \sin\theta & 0 \\ -\sin\theta & \cos\theta & 0 \\ 0 & 0 & 1 \end{pmatrix}, \quad \mathbf{t} = \begin{pmatrix} t_x \\ t_y \\ 0 \end{pmatrix} \tag{13}$$

The relationship between the point $\mathbf{p} = (\mathbf{x}, \mathbf{y}, \mathbf{z})^{\mathbf{T}}$ in the $\mathbf{oxyz}$ coordinate frame and the image point $\mathbf{P}' = (\mathbf{X}', \mathbf{Y}')$ can be represented as:

$$z'\begin{pmatrix} P' \\ 1 \end{pmatrix} = \mathbf{R}(\mathbf{p} - \mathbf{t}) \tag{14}$$

where z' represents the depth of point $\mathbf{p}$ in camera centred coordinate frame, $\begin{pmatrix} P' \\ 1 \end{pmatrix}$ represent the homogeneous coordinate of image point $\mathbf{P}'$, and $\mathbf{p}, \mathbf{R}, \mathbf{t}$ are as defined in Equation 1. Assuming a unit focal length of the camera ($f = 1$) for convenience of computation, Equation 14 is equivalent to:

$$ z\mathbf{P}' = \mathbf{R}(\begin{pmatrix} x \\ y \end{pmatrix} - \mathbf{t}) \tag{15} $$

where $\mathbf{R}$ is a 2D rotation matrix and $\mathbf{t}$ is a 2D translation vector. Obviously, Equation 15 represents a 2D rigid body transformation where the 2D correspondence $(\mathbf{p}, \mathbf{p}') = ((\mathbf{x}, \mathbf{y})^{\mathbf{T}}, z\mathbf{P}')$. Our aim is to estimate the transformation parameters $(\hat{\theta}, \hat{\mathbf{R}}, \hat{\mathbf{t}})$ and the filter 3D structure data $(\hat{z}')$ given a number n (usually, $n \geq 3$) of 3D-2D correspondences. The following 2 algorithms are thus proposed:

Algorithm 1: Simplified Geometrical Algorithm (SGA)

1. Using Equation 2, estimate a set of critical points $\mathbf{c_i}$ in 2D corresponding to $(\mathbf{p_i}, \mathbf{p_i'})$ and $(\mathbf{p_{i+1}}, \mathbf{p_{i+1}'})$. The finally calibrated critical point $\hat{\mathbf{c}}$ in 2D is estimated by median filtering $\mathbf{c_i}$.
2. Using Equation 3, estimate a set of the cosines a_i of rotation angles corresponding to $(\mathbf{p_i}, \mathbf{p_i'})$. The finally calibrated rotation angle θ is estimated as: $\hat{\theta} = \arccos\{average\ of\ band\ pass\ filtered\ a_i\}$.
3. The 2D rotation matrix $\hat{\mathbf{R}}$ can be estimated by Equation 4.
4. The 2D translation vector $\hat{\mathbf{t}}$ of the camera can be estimated by Equation 5.
5. The depth $\hat{z}_i'$ of each visible point on the filter in camera centred coordinate frame can be estimated by:

$$ \hat{z}_i' = \frac{\frac{\mathbf{R_1}(\mathbf{p_i}-\hat{\mathbf{t}})}{\mathbf{X_i'}} + \frac{\mathbf{R_2}(\mathbf{p_i}-\hat{\mathbf{t}})}{\mathbf{Y_i'}}}{2} \tag{16} $$

where $\mathbf{R_1}$ and $\mathbf{R_2}$ are the row vectors of rotation matrix $\hat{\mathbf{R}}$.

For a comparative study of algorithm performance, a second algorithm called Simplified Epipolar Geometry Algorithm (SEA) has been developed given the following definitions. From [2], [3], it is known that the essential matrix $\mathbf{E}$ can be estimated by $\mathbf{E} = \mathbf{TR}$ where $\mathbf{T} = \begin{pmatrix} 0 & 0 & t_y \\ 0 & 0 & -t_x \\ -t_y & t_x & 0 \end{pmatrix}$. Therefore, $\mathbf{E} = \mathbf{TR} =$

$$ \begin{pmatrix} 0 & 0 & t_y \\ 0 & 0 & -t_x \\ -t_y & t_x & 0 \end{pmatrix} \begin{pmatrix} \cos\theta & \sin\theta & 0 \\ -\sin\theta & \cos\theta & 0 \\ 0 & 0 & 1 \end{pmatrix} = \begin{pmatrix} 0 & 0 & t_y \\ 0 & 0 & -t_x \\ -t_y\cos\theta - t_x\sin\theta & -t_y\cos\theta + t_x\sin\theta & 0 \end{pmatrix} $$

Thus, $e = (e_1, e_2, e_3)$ can be estimated by: $\mathbf{e} = (\mathbf{A^TA})^{-1}\mathbf{A^Tb}$, where

$$ \mathbf{A} = \begin{pmatrix} x_1 & X_1'z_1 & Y_1'z_1 \\ x_2 & X_2'z_2 & Y_2'z_2 \\ \vdots & \vdots & \vdots \\ x_n & X_n'z_n & Y_n'z_n \end{pmatrix} \text{ and } \mathbf{b} = (-y_1, -y_2, \cdots, -y_n)^T. $$

Algorithm 2: Simplified Epipolar Algorithm (SEA)

1. The sine and cosine of the rotation angle can be estimated by: $\begin{pmatrix} \hat{\sin} \theta \\ \hat{\cos} \theta \end{pmatrix} =$
   $\begin{pmatrix} e_3 & -e_2 \\ -e_2 & -e_3 \end{pmatrix}^{-1} \begin{pmatrix} e_1 \\ 1 \end{pmatrix}$

2. The finally calibrated rotation angle $\hat{\theta}$ can be estimated by: $\hat{\theta} = (\arcsin \hat{\sin} \theta + \arccos \hat{\cos} \theta)/2$

3. The 2D rotation matrix $\hat{\mathbf{R}}$ up to a sign can be estimated using Equation 4.

4. The 2D translation vector up to a sign can be estimated by: $\hat{\mathbf{t}} = \mathbf{R}^{\mathbf{T}} \begin{pmatrix} -e_3 \\ e_2 \end{pmatrix}$

5. The depth of each visible point $\hat{z}_i'$ on the filter up to a sign can be estimated using Equation 16. Because the filter is in front of the camera, the depth of each point should be positive. Therefore, the finally calibrated 2D $(\hat{\mathbf{R}}, \hat{\mathbf{t}})$ and depth $\hat{z}_i'$ is given by:

   if $\sum_{i=1}^{k} sign(z_i') \leq 0$, then $\hat{\mathbf{R}} = -\hat{\mathbf{R}}$, $\hat{\mathbf{t}} = -\hat{\mathbf{t}}$, and $\hat{z}_i' = -\hat{z}_i'$ where
   $sign(x) = \begin{cases} 1 & \text{if } x \geq 0 \\ -1 & \text{otherwise} \end{cases}$, and $k \geq 3$ is application-dependent.

4 Experimental Results

Both SGA and SEA algorithms were implemented as follows. From a 3D image database, we selected 30 points $\mathbf{p_i} = (\mathbf{x_i}, \mathbf{y_i}, \mathbf{z_i})^{\mathbf{T}}$ $(i = 1, 2, \cdots, 30)$ on a 3D object described in a reference coordinate frame. These points were subject to controlled rotations of 20, 30, and 50 degrees around a rotation axis $\mathbf{h} = (0, 0, 1)^{\mathbf{T}}$ with a constant translation vector $\mathbf{t} = (3, 4, 0)^{\mathbf{T}}$ yielding their correspondence points $\mathbf{p_i'} = (\mathbf{x_i'}, \mathbf{y_i'}, \mathbf{z_i'})^{\mathbf{T}}$ in camera centred coordinate frame. Finally, correspondence points were projected on the image plane $z' = 1$ assuming a focal length f of the camera equal to 1. We thus, have one set of 3D object points in the reference coordinate frame and their corresponding 2D perspective image points $\mathbf{P_i'} = (\mathbf{X_i'}, \mathbf{Y_i'})^{\mathbf{T}}$ $(i = 1, 2, \cdots, 30)$ on the image plane. These are the control sets of points to serve as reference for error estimation.

We then added Gaussian noise to the coordinates of object points and their image points and used the proposed SGA and SEA algorithms to estimate the orientation and position parameters represented as the 3D rotation matrix $\hat{\mathbf{R}}$, the position parameter represented as the 3D translation vector $\hat{\mathbf{t}}$ and the structural data represented as depth $\hat{z}_i'$. We defined the relative calibration error of 3D rotation matrix $\hat{\mathbf{R}}$ as $e_R = ||\hat{\mathbf{R}} - \mathbf{R}||/||\mathbf{R}||$, the relative calibration error of 3D translation vector as $e_t = ||\hat{\mathbf{t}} - \mathbf{t}||/||\mathbf{t}||$, and the relative calibration error of the structure as $e_{z'} = ||\hat{\mathbf{z}}_i' - \mathbf{z}_i'||/||\mathbf{z}_i'||$. Experimental results are summarised in Tables 1 and 2.

Rot. angle (deg)	20				30				50			
Calibration methods	c_R (%)	e_t (%)	$e_{z'}$ (%)	time (sec.)	e_R (%)	e_t (%)	$e_{z'}$ (%)	time (sec.)	e_R (%)	e_t (%)	$e_{z'}$ (%)	time (sec.)
SGA	3.23	14.08	3.31	0.01	3.17	12.71	-0.76	0.005	2.42	9.99	10.44	0.01
SEA	6.29	24.84	3.07	0.02	7.76	30.11	-3.34	0.01	10.32	40.16	14.28	0.02

Table 1. Relative average calibration errors and calibration time of different algorithms for data corrupted by Gaussian noise with mean zero and deviation 0.05.

Rot. angle (deg)	20				30				50			
Calibration methods	e_R (%)	e_t (%)	$e_{z'}$ (%)	time (sec.)	e_R (%)	e_t (%)	$e_{z'}$ (%)	time (sec.)	e_R (%)	e_t (%)	$e_{z'}$ (%)	time (sec.)
SGA	10.13	45.53	-5.44	0.005	6.36	25.06	13.89	0.01	10.25	42.94	14.41	0.01
SEA	17.33	55.18	-6.89	0.01	19.74	66.45	20.23	0.02	21.28	85.14	-12.16	0.02

Table 2. Relative average calibration errors and calibration time of different algorithms for data corrupted by Gaussian noise with mean zero and deviation 0.1.

An overall analysis of the tables reveals that the SGA algorithm is more accurate and displays a better performance than the SEA algorithm. The main reason is that the SEA algorithm does not take into consideration the rigid constraints of distance between feature points and does not use angular information to estimate transformation parameters. Experiments have shown that the SEA algorithm is not only sensitive to noise, but also sensitive to rounding errors. Unless accurate enough data can be obtained, satisfactory performance of the algorithm cannot be guaranteed.

Moreover, the SGA algorithm has still room for further performance improvements when compared with the SEA method, as the latter makes full use of all points information while the SGA algorithm uses only partial points information, such as those far from the critical point. Thus, by using fewer feature points, it is possible to vastly reduce the computation time of the algorithm without loss of accuracy and this is critical for real time applications. The tables also show that the accuracy of the algorithms tends to decrease when data are corrupted by heavier noise. This is not a surprising result and, relatively, while the SGA's accuracy is more tolerant to noise, the accuracy of the SEA algorithm rapidly deteriorates with noise.

5 Conclusions

A special system setup for a real time automatic inspection of filter components has been designed based on the analysis of geometrical properties of image correspondence vectors synthesised into a single coordinate frame. The setup design is likely to render automatic inspection feasible and reliable within given performance constraints. The proposed SGA algorithm based on this special setup provides the closed form solutions to all estimated parameters necessary to decide whether a filter is to be rejected or not. While our approach is substantially different from epipolar geometry, we pointed out that linear algorithms based on epipolar geometry represent a common linear approach to solve 3D-2D and 2D-2D problems. For a comparative study with the SGA algorithm, we also developed an algorithm, SEA, based on epipolar geometry. Experimental results have demonstrated that the overall performance of the SGA algorithm is superior to SEA's performance.

The experimental results have shown that the SGA algorithm represents a practical proposition for the task at hand. The algorithm will be tested and the analysis method further refined in the actual production line and further results will be reported in the near future.

References

1. A. Mitiche and J. K. Aggarwal. A computational analysis of time-varying images. *Handbook of Pattern Recognition and Image Processing*, T. Y. Yuong, K. S. Fu, Academic Press, 1986.
2. J. Weng, T. S. Huang, and N. Ahuja. Motion and structure from two projective views: algorithms, error analysis, and error estimation. *IEEE Trans on Pattern Analysis and Mach Intell*, Vol. 11, No. 5, pp. 451–476, 1989.
3. R. Y. Tsai and T. S. Huang. Uniqueness and estimation of three-dimensional motion parameters of rigid objects with curved surfaces. *IEEE Trans on Pattern Analysis and Mach Intell*, Vol. 6, No. 1, pp. 13–27, 1984.
4. T. S. Huang and O. D. Faugeras. Some properties of the E matrix in two-view motion estimation. *IEEE Trans on Pattern Analysis and Mach Intell*, Vol. 11, No. 12, pp. 1310–1312, 1990.
5. R. M. Haralick, H. J. Chung-Nan Lee, X. Zhuang, V. G. Vaidya, and M. A. Kim. Pose estimation from corresponding point data. *IEEE Trans on Pattern Analysis and Mach Intell*, Vol. 19, No. 6, pp. 1426–14461, 1989.
6. S. Linnainmaa, D. Harwood, and L. S. Davis. Pose determination of a three-dimensional object using triangular pairs. *IEEE Trans on Pattern Analysis and Mach Intell*, Vol. 10, No. 5, pp. 634–647, 1987.
7. P. R. Wolf. *Elements of photogrammetry*. New York: McGraw-Hill, 1974.
8. S. Ganapathy. Decomposition of transformation matrices for robot vision. *Pattern Recognition Letters*, Vol. 2, pp. 401–412, 1989.
9. M. Fischler and R. C. Bolles. Random sample consensus: a paradigm for model fitting with applications to image analysis and automated cartography. *Communications of ACM*, Vol. 24, no. 6, 99. 381–395, 1981.
10. S.H. Joseph. Optimal pose estimation in two and three dimensions. *Comp Vision and Image Understanding*, Vol. 73, no. 2, 1999, pp 215–231.
11. D. Oberkampf, D.F. DeMenthon, and L.S. Davis. Iterative pose estimation using coplanar feature points. *Comp Vision and Image Understanding*, Vol 63, no.3, 1996, pp 495–511.
12. H. Araujo, R.L. Carceroni, and C.M. Brown. A fully projective formulation to improve the accuracy of Lowe's pose estimation algorithm. *Comp Vision and Image Understanding*, Vol. 71, no. 2, 1998, pp 227–238.

A Robust and Unified Algorithm for Indoor and Outdoor Scenes Based on Region Segmentation

E. Zagrouba[1], T. Hedidar[2] and A. Jaoua[2]

1 : LARIMA, FSM, Route de l'environnement 5000 Monastir, Tunisia
e-mail : Ezzeddine.Zagrouba@fsm.rnu.tn
2 : ERPAH, FST, Campus Universitaire Ras Tabia 1005 Tunis, Tunisia

Abstract. Segmentation is one of the most important steps leading to the analysis of processed image data. This paper describes a segmentation system based on the application of *Total Gradient Histogram Method* (*TGH*) and *Relative Contrast Method* (*RC*). The implementation is processed in two strategies. In the first strategy, we implement monothresholding. The second strategy based on multithresholding is processed through three steps : highest thresholds, highest thresholds by range of grey level, and hierarchical thresholding segmentation. Through all implementations approaches on stereoscopic scenes, an other process using a rule based system is done in order to improve segmentation results. Finally, we apply our algorithms of segmentation to images from many other domains.
Keywords : Connexity, Multithresholding, Hierarchical segmentation.

1. Introduction

A classic stereoscopic system is decomposed sequentially into five steps : cameras calibration, image acquisition, segmentation, matching and 3D reconstruction. The segmentation step is the basis of all recognition and interpretation process, its main goal is to extract entities that have a strong correlation with objects or parts of objects of the real world contained in the observed scene. The big influence of segmentation results on ulterior process and the difficulty to maximise useful information and to minimise unsuitable information explain the great interest for authors in the area of image processing. Schematically, we can distinguish two possible approaches : *edge-based* segmentation and *region-based* segmentation. The detection of discontinuity in the first approach is made with known operators like : ROBERTS, SOBEL, CANNY,... [1], [4], [5]. It is followed by others processes allowing to link edge elements : edge following using heuristic search, edge following as dynamic programming, [3], [6], [8]. The aim of region-based segmentation is to describe the image into related parts by being based on region intrinsic criteria. We can classify region-based segmentation methods into three categories : spatial measures methods [12], pixel grouping methods and region growing methods [17]. In the first category, the image is fragmented into parts according to global properties measured on the whole image or on zones of the image. The knowledge is usually represented by a histogram of image features. In the second category, the grouping in regions can be redirected by the similarity of the pixel with its neighbours as it can be made by taking into account the relationship pixel-region. Finally, methods of the third category can be divided into three classes : Region merging , region splitting and split and merge.

The region merging (bottom-up) is the natural method of region growing [9]. Initially, each pixel represent a region and regions satisfying merging criteria are merged. The region splitting method (top-down) begins with the whole image represented as a single region which does not usually satisfy homogeneity criterion. Therefore the existing image regions are sequentially split into a set of related and disjoint regions. The split and merge method is a combination of splitting and merging methods in view to pull profit simultaneously from their advantages [10]. The homogeneity criterion plays a major role in split-and-merge algorithms, just as it do in other region growing methods [2]. The two above briefly discussed approaches, (edge-based and region-based) present insufficiencies. Indeed, the most common problems of edge-based approach are edge presence in locations where there is no border and no edge presence where a real border exists [13]. Also, the region-based segmentation strongly depends on lighting conditions and cameras position which generate badly localised boundaries. These insufficiencies create a major handicaps for next placement steps in matching and 3D reconstruction. This brings us to seek an hybrid approach [15], exploiting the duality of the edge and region approaches (edges delimit regions and vice versa) and pulling profit simultaneously from their advantages. Our work has for principal objective the realisation of a segmentation system that seeks to improve the quality of the segmentation by the application of intelligible and effective algorithms so as to insure a reliable results and to avoid errors propagation. The study of the different methods presented by KOHLER in [7], has oriented us to choose spatial measuring methods using an heterogeneity function. These methods constitute a combination of edge-based and region-based segmentation. They can take profit simultaneously from their advantages by integration of the gradient notion in the calculation of histogram. Indeed, these methods are not very sensitive to lighting and allow a good location of boundaries. We have retained after several tests : a *Total Gradient Histogram Method* (TGH) and a *Relative Contrast Method* (RC). The different processes constituting our segmentation system are illustrated in the following figure.

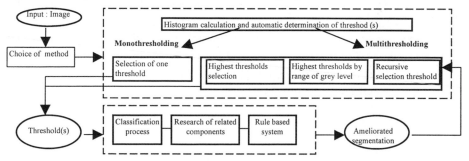

Fig.1. *Description of the segmentation system*

In a first section, we present the bimodal thresholding. The second section deals about multithresholding according to three approaches. The third section is devoted to the improvement of results. Finally, the fourth section is reserved to the experimentation results.

2. bimodal Thresholding

2.1 Definitions

In the bimodal thresholding, we use a single threshold so as to determine the class associated to each pixel of the image. We present here respectively functions of histogram calculation according to TGH and RC methods respectively :

- for TGH, the threshold s is determined by maximising the gradient histogram :

$$H(s) = \underset{n \in [0,255]}{Max} (H(n)).\qquad(1)$$

- for RC :
$$H(s) = \begin{cases} C(s)/N(s) & \text{if } N(s) \neq 0 \\ 0 & \text{otherwise} \end{cases}\qquad(2)$$

Where : s is a grey level in the internal [0, 255], C(s) is the cumulated contrast observed on edge pixels having a grey level equal to s and N(s) is the number of edge pixels detected by threshold s.

N(s) Calculation : Let pa, pb two neighbour pixels having respectively, Ia is the grey level of pa and Ib is the grey level of pb such that Ia < s < Ib

$$N(s) = \sum_{pa,pb,neigbors} p(Ia, Ib, s) \text{ and } p(Ia, Ibs) = \begin{cases} 1 & \text{if } Ia < s < Ib \\ 0 & \text{otherwise} \end{cases}\qquad(3)$$

C(s) Calculation

$$C(s) = \sum_{pa, pb \text{ neigbors}} cr(Ia, Ib, s) \text{ and } cr(Ia, Pb, s) \begin{cases} \min(s-Ia, Ib-s), & \text{if } Ia \leq s < Ib \\ 0, & \text{otherwise} \end{cases}\qquad(4)$$

2.2 Algorithms

With TGH method calculation histogram, we choose SOBEL mask which enable the horizontal and vertical edge detection. In practices gradient calculation is undertaken with norm 1 and only pixels having a norm greater than 2 are taken into account. The figure 2 represents the histogram of the "room.im" image (fig.3) obtained with TGH and RC methods. For the criterion C used by the RC method (neighbouring pixels selection), several choices are possible [11]. In our case, we have choose the pixel *pv* having the greatest gradient norm among all adjacent.

After labelling, we sweep the image from left to right and from up to bottom by applying the following algorithm (*ph* : high pixel, *pl* : left pixel and *pc* : current pixel).

```
if          class (ph) ≠ Class (pc) and class (pl) ≠ class (pc)
then        create a new region R (pc) ←{pc}
else if     ∃! neighbour pixel of pc (noted pv)
            such that class(pc)=class (pv)
then        R (pv) ← R (pv) ∪{pc}
else if     R (pl) =R (ph)
```

```
then        R (pl) ← R (pl)∪{pc}
else        R (pl) ←R (pl) ∪R (ph) ∪ {pc}
```

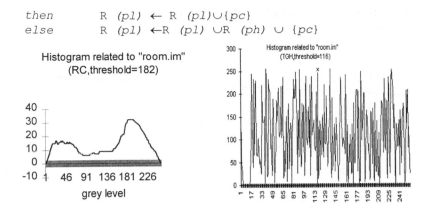

Fig.2. *Histograms (RC,TGH)*

2.3 Experimentation and conclusion

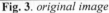

Fig. 3. *original image*

Fig. 4. *final monothresholding (TGH)*

The TGH method shows far more regions than the RC method (fig.4). The majority of these regions are generally qualified as "noise regions". These regions are du to the quantification and discretisation problems of the CCD collector and specially to the inherent disadvantages of this method. This method does not show luminous sources in the different observed scenes despite the big contrast they mark with neighbours (fig. 3 and 4). This leads us to conclude that the TGH method selects thresholds that do not necessary appear regions having most contrasted borders. On the other hand, the RC method generates less regions, but these regions mark a big contrast (luminous sources). This method selects thresholds of manner to generate the most well contrasted edges and the less weak contrasted edges. However, we can obtain undesirable subsections of realistic regions (the notice in the middle of the scene is subdivided into regions witch do not describe perfectly the scene). On images constituted of a clear background and dark objects or conversely, the selection of only one threshold could give good results. Nevertheless, with images presenting several objects having no well distinction between background and objects like the room image (fig.3), the selection of one threshold does not suffice. Indeed, we can't release all objects or facets of objects contained in the scene. This has brought us to select several thresholds in the histogram. Different alternatives of multithresholding are discussed in the next section.

3. Multimodal Thresholding

To be able to solve problems caused by monothresholding, we select several thresholds according to three following approaches.

3.1 Selection of highest thresholds

The different thresholds chosen in this strategy correspond to greatest peaks released by the histogram. A peak p in the histogram is defined by:

$$H \ (p) > = H \ (p - 1) \text{ and } H \ (p) > H \ (p+1). \tag{5}$$

With the selection of greatest thresholds, we notice that the number of regions has considerably increased for realists as well as for superfluous ones (appearance of chair, luminous sources, ...fig.5). The results can be far more satisfying than that corresponding to the first strategy. It remains to solve some questions that are in narrow relationship with this strategy such, the automatic determination of the thresholds number witch will be able to maximise the number of realistic regions.

3.2 Selection of the two highest thresholds released by TGH and RC

Tests undertaken with TGH and RC methods show a complementarily results as we come to raise it in (§2.1). So as to better exploit this complementarily, we have decided to select highest thresholds released by the two methods. This approach shows more realistic regions and less hummed ones than previously (fig.5). However, the selection of a great number of highest thresholds, even if it realises a more important number of realistic regions, it will be always more sensitive to the appearance of hummed regions. In the approach of highest selection thresholds we have omitted tests undertaken with the RC method. Indeed, we see that these last are without interest because highest thresholds will be in the most share of cases situated to straight of the first threshold detected in the histogram of the method, it is to tell correspondent to greatest grey levels. However, the selection of thresholds corresponding to highest peaks by range of grey level will give again more interested results. It is what we present in the next section.

3.3. Selection of highest thresholds by range of grey level

In this approach, we select highest thresholds by range of grey level. The number of thresholds chosen in the different tests has been determined experimentally. Results of the different relative tests are far more satisfying than those obtained in the monothresholding strategy (fig.7 and 9). Indeed, they show more realistic regions. Nevertheless, they release as many hums regions as realistic ones. In similar images that we have tested, we notice that the selection of single threshold does not suffice to release most present objects in the scene. Furthermore, the selection of several thresholds applying the second or the third strategy give a very important number of hums and non realistic regions. This explains that global calculations on the image comprise a risk of superposition of information. We have therefore opted for a method of segmentation by recursive divisions on the image, where calculations become gradually local. This choice is presented in the next section.

3.4 Hierarchical segmentation

In this approach, calculations are localised by restriction of divisions to a given zone of the image in order to control effects of segmentation and to overcome the global and local dissimilarities on images. The recursive algorithm for this approach is based on a criterion of region divisibility (CRD). In the calculation of the histogram of the TGH and the RC methods, we don't take into account pixels belonging to the borders of the asked region. Indeed, this can influence the determination of the threshold because these pixels mark a big contrast with the interior of the region to be divided. For the region divisibility criterion (CRD), we impose that the surface will be superior to a certain threshold $(s1)$, in order to not divide the small regions that will falsify the threshold histogram calculation. Also, we impose that the difference Max-Min level grey of the asked region will be greater than a threshold s2 [14]. Experimentally, we have fixed $s1$ and $s2$ respectively to 100 and 10. (fig.8 and 10).

4. Segmentation Improvement

In order to succeed matching process, it is imperative to increase the similarity between segmentations obtained on the same scene. Frequently, the similarity is found generally wrong because of next discrepancies [15], [16] : massive presence of small regions, excrescence of regions and labelling errors. These discrepancies (defects) drive to connect unduly objects facets of the scene. These regions present several problems indeed, they have little chance to exist on both images, therefore to be matched and exploited for 3D construction. We classify these last in four groups that we review sequentially in next sections : very small regions (noise); threadlike regions, non realistic regions (with irregular frontiers); and non significant regions.

4.1 Noise elimination

A very small region is found generally inside a real region or well on borders of a region that has no physical existence. In the first case, we merge the hums region to their mothers. In the second case, we merge the hums region to a neighbour that has no physical reality and that will undergo even a particular improvement process. Generally, the number of hums regions exceeds 80% from total regions released by TGH method. The processing of these regions has clearly improved segmentation results, as we can notice it on the final results.

4.2 Threadlike region elimination

A region is considered threadlike if its length or its width is inferior to 2 pixels. Threadlike regions are generally found in transition zones. These regions will be merged with their mothers.

4.3. Weak carves regions elimination

Our algorithm produces regions of small carve, but greater than those of the first category. These regions present an enough raised gradient and they are not eliminated by the algorithm presented in 4.1. It is for this reason that we consider them belonging to an independent category. The TGH method is more sensitive to this kind of problem. These regions are du to abnormal subsections caused by false selection of the adequate threshold. To eliminate a region R having a small carves, we merge it with the neighbour region R' having the most common long frontier.

4.4. Non realistic regions elimination

Non realistic regions often have irregular frontiers fathered by classification errors, or by other phenomena such that : presence of the shade, objects homogeneous, ... [16]. To be able to eliminate these regions, we have developed a based rule-system whose each insures the fusion of a particular region category.

® **Rule 1.** It insures the fusion of regions having a common weak gradient frontier

```
For each region R' adjacent to R
   If (average grey level of R∪R') < T and
      (average gradient of the common frontier) < TG
   Then  merge R and R'
```

® **Rule 2.** It merges adjacent regions having a weak gradient of the common frontier. The common frontier must be long enough.

```
For  each region R' adjacent to R
   If  (average of grey level of R∪R') < T and
       (gradient average of the common frontier) < TG'
       and (Frontier Length) > TL
   Then    merge R and R'
```

® **Rule 3.**

```
For  each region R' adjacent to R
   If  (average of grey level of R∪R') < T and
       (gradient average of the common frontier) < TG'
       and(Frontier Length) > TL'
   Then    merge R and R'
```

The length threshold TL' fixed in a static manner in rule2 does not allow to eliminate some non realistic regions. We have to make dynamic this variable in rule 3 according to the following principe : let R and R' two neighbour regions if the length of the cord between A and B is superior to (1+D).AB and the (R∪R') average grey level is inferior to T then we merge R and R'. In this section, we have proposed algorithms of elimination of superfluous information so as to render intelligible the segmentation and as a result to facilitate the matching process. Nevertheless, it remains difficult sometimes to master non realistic regions. Indeed, their elimination can father the elimination also of realistic ones (window in fig.4). We can not therefore insure the quality of the segmentation only during direct alteration between segmentation and matching, thing allowed by hierarchical segmentation.

577

These different rules of improvement are not inevitably applicable on outdoor scenes. Indeed, these last can present particularities that we will evoke in the next section.

Fig. 5. *highest thresholds (TGH)*

Fig. 6. *highest thresholds (TGH and RC)*

Fig.7. *highest thresholds by range*

Fig. 8. *Hierarchical strategy level 4 (TGH)*

Fig. 9. *highest thresholds by range (RC)*

Fig. 10. *Hierarchical strategy level 4 (RC)*

5. Application of algorithms on outdoor scenes

In the opposition of interior scenes, outdoor scenes admit some particularities. Indeed, objects of these last can have frontiers that are not inevitably vertical or horizontal; as they can present irregularities had to their natural forms. We try in this section, to put in obviousness results of application of our algorithms on this type of scenes. The first image (fig.11) represents a general view of « St-Michel ». We have applied on this image different tests. Good results are obtained by recursive application of RC method four times (Fig.12). Analogous results have been obtained by the selection in a first stage of the two highest thresholds released by TGH and RC methods, then the application of the recursive hierarchical segmentation of the RC method.

Fig. 11. *original image*

Fig. 12. *Hierarchical strategy level 4 (RC)*

6. Conclusion

Image segmentation is one of the most important steps leading to the analysis of processed image data. It's main goal is to divide an image into parts having a strong correlation with objects or areas of the real world contained in the image. First, segmentation process is discussed through edges and region approaches. Then, our system based on the application of *Total Gradient Histogram Method* (*TGH*) and *Relative Contrast Method* (*RC*) is presented. The implementation of the system is processed according to two strategies. In the first strategy, we implement monothresholding. The results given by the last one don't detect most real regions existing in the observed scene. The second strategy based on multithresholding, is processed through three steps : in the first step, we choose the highest thresholds detected in the histogram. In the second step, we choose the highest thresholds by range of grey. In the third one, we apply hierarchical segmentation. Through all implementations approaches on stereoscopic scenes, an other process using a rule based system is done in order to make better the segmentation results. Finally, we apply our algorithms of segmentation to images from many other domains. Our system give good results according to the main goal presented below. We think that the hierarchical segmentation approach constitute a fundamental interest. Indeed, it allows directly alteration of division and matching. In that, it does not only allow validation of released regions but also construct a principal convergence criterion. This method could have be adopted the long of ulterior works by integrating it in a global vision system. The application of rule-based system has considerably ameliorated the quality of process segmentation. We are convinced that multithresholding process segmentation specially by highest thresholds strategy could have be coupled with the hierarchical segmentation process. In first place we select an optimal number thresholds that make release more realistic regions. Then, we can refine this initial segmentation by an hierarchical one. This will be able to enlarge the chance of matching specially in these firsts steps.

References

1. CANNY J.F.: A computational approach to edge detection, PAMI n°6 Vol8, November 1986.
2. Chen S Y, W C Lin, and CT Chen : Split-and-merge image segmentation based on localised feature analysis and statistical tests. CVGIP Graphical Models and Image Processing, 1991.
3. Matthieu Cord, Florence Huet & Sylvie Philipp : Optimal adjusting of edge detectors to extract close contours, Proc. 10th Scandinavian Conference on Image Analysis, Lappeenranta, Finlande, Jun 1997.
4. Didier Demigny & Tawfik Kamleh : A Discrete expression of Canny's Criteria for Step Edge Detection Performances Evaluation, IEEE Pattern Analysis and Machine Intelligence, 1997.
5. DERICHE R. : Optimal Edge Detection Using Recursive filtering, First International Conference on Computer Vision, London, 1987].
6. Florence Huet & Sylvie Philipp : High-scale edge study for segmentation and contour closing in textured or noisy images., IEEE SSIAI'98,Tucson, Arizona, USA, Avril 1998.
7. R.KOHLER. : a segmentation system based on thresholding, In Computer Graphics and Image Processing, volume 4, n°3, pages.
8. A., A MARTELLI., « *Edge detection using heuristic search methods* », CGIP n°1, August 1972.
9. NJ Nilsson. : Principals of Artificial Intelligence, Springer Verlag, Berlin, 1982.
10. T. PAVLIDIS, H. HOROWITZ « Picture Segmentation by a Directed Split-and-merge Procedure » Proc 2nd IJCPR, Août 1974.
11. S. RANDRIAMASY. : Segmentation descendante coopérative en régions de pairesd'images stéréoscopiques, PhD thesis, University Paris IX-Dauphine.
12. D.R REDDY, R. OHLANDER, K PRICE. : picture segmentation using a recursive region splitting method, In computer graphics and Image processing, volume 8, 1978.
13. M. SONKA, V. HLAVAC, R. BOYLE : Image Processing, Analysis and Machine Vision, Chapmen & Hall Computing, London 1993.
14. E. ZAGROUBA, C, KREY : A rule-based system for region segmentation improvement in sterevision, IS&T-SPIE International Technical Conference of San Jose, Volume 21-82 Image and Video Processing II, pp, 357-367, California-USA, 7-9 Février 1994.
15. E. ZAGROUBA : Construction de Facettes Tridimensionnelles par Mise en Correspondance de Régions en Stéréovision, PhD Thesis INP Toulouse, France, Septembre 1994.
16. E. ZAGROUBA : 3-D facets constructor for stereovision, 5th International Conference on Artificial Intelligence. ROME, Italie, September 1997.
17. W. Zucker. : Region growing Childhood and adolescence, Computer Graphics and Image Processing, 382- 1976

SAPIA: A Model Based Satellite Image Interpretation System

F. Ramparany and F. Sandt

Cap Gemini France - ITMI Div. , BP 87, 38244 Meylan, France

SAPIA is a satellite image interpretation assistant system. SAPIA is based on a model of the scene which incorporates knowledge about the shape of objects to recognized, geometrical and structural relationships among objects and operative knowledge concerning which image processing operator is appropriate for identifying and extracting objects. It salient features include:
- its ability to work cooperatively with photo-interprets, without intruding his or her current flow of work.
- the facilities it provides to photo-interprets for smoothly integrating new information in the model, and for improving the performance of the system.

In this paper, we give an overview of the system, enlight its main characteristic features and discuss them with respect to work done elsewhere.

1 Introduction

Image interpretation consists of deriving an abstract sensible description from an image. This description should be built in such a way as to be used in a report or to support a further decision making process ([Pau92]). Typically, in the domain of aerial or satellite image interpretation, specific objects (such as "runway", or "taxiway") and their characteristics (such as spatial location and orientation) should be identified and precisely reported.

When drawing maps from a raw satellite image, photo-interpreters basically perform three kinds of activities:
- identify relevant domain objects on the plan. (e.g., runways in the case of an airport).
- draw their geometrical shape (e.g., runways are depicted as linear ribbons).
- annotate the drawing with specific information. (e.g., buildings are labeled with reference to their function: airport, fuel tank, parking hall,...).

One difficulty is to extract of only the pertinent objects, which are strongly context dependent (e.g., the distinction between a forest and a field is meaningless for an airport but is essential in an agricultural context). Another difficulty is to identify the objects whose shape and appearance are variable ([Gri97], [Cas97]).

For these reasons photo interpretation still remains the task of a rare handful of human experts (i.e. photo-interpreters). From another perspective, the joint development of observation satellites and digital processing technology has increased both the quality of satellite images and the quantity of the available information, shifting the bottleneck of information processing from data acquisition to data interpretation. Therefore much effort has been devoted to entirely or partially automating satellite or aerial image interpretation ([KPC96],[ZJ93], [GGM91]).

Although most of these efforts yield interesting results, the progress achieved is far from initial expectations. The main shortcomings of existing image interpretation systems include:
- the instability to noise in the image ([Bev89])
- the difficulty to update the system when new objects need to be detected ([HS93])
- the difficulty to easily adapt the system to new domains ([SCG94]).

We have developed SAPIA *Système d'Aide à la Photo-Interprétation Avancée* a satellite image interpretation system that addresses some of these shortcomings by:
- integrating the photo-interpreter within the process loop,
- explicitly modeling the objects to be recognized in the image, as well as the knowledge necessary for their recognition,
- using the expressiveness of constraint specification as a modeling paradigm,
- using constraint propagation for pruning the interpretation search tree, and speeding up the efficiency of the interpretation process.

In the following sections, we first provide a detailed description of SAPIA modeling language and its constraint specification constructs. We then introduce the interpretation process and show how it controls the mechanisms of constraint propagation and solution generation. Lastly, we describe our interpretation problem in terms of constraint management problems and discuss the way we have addressed them.

2 Modeling using constraint specification

SAPIA exploits a generic model of the scene, which describes its physical components and the relationships each component may share with another. Component properties and relationships are essentially geometrical. They impose constraints on measures such as size, shape, and relative position of one component with respect to another component.

As we will see later, the interpretation process controls a constraint propagation module which dynamically propagates information between the various elements of interpretation as soon as some progress has been made.

2.1 Modeling objects

As mentioned previously, domain objects are central to the process of image interpretation. In the SAPIA system the image interpretation process focuses on objects, and primitive tasks of this process amount to identifying and locating domain objects from the raw image.

In the SAPIA system, objects are described from three viewpoints:
- geometrical and photometrical properties, which describe how the objects look in the image,

- relationships with other objects,
- image interpretation operative knowledge, which states how this object should be detected in the image (i.e. what kind of image processing operators should be used).

2.2 Modeling geometrical and photometric properties

The relevant objects are to be searched for in the images as specific shapes with photometric properties. For example, runways are searched for as straight ribbons, taxiways are searched for as smoothly curved lines or ribbons. Current shapes in SAPIA are essentially bi-dimensionnal and include rectangles, circular arcs, circles, straight ribbons, curved ribbons, small regions and polygonal lines.

Object shapes are parametric shapes. For example, a ribbon is characterized by its length, width and greylevel. The SAPIA modeling language enables objects to be described through constraints on those parameters. For instance a constraint such as *"length * width ≤ 1000"* states that the surface of a ribbon should not exceed $1000m^2$.

2.3 Modeling relations

An object in the model can share relations with other objects. Representing relations for image understanding achieves two goals. One is to capture structuring information that allows the scene description to prune the interpretation search space. For example beacons are usually aligned with runways; if all runways have been identified, it does not make sense to search for beacons anywhere but in the alignment of runways. The other advantage is to guide the interpretation process flow. The use of relations for controling the interpretation process is discussed in section 3.

More specifically, decomposition relationships and geometrical constraints provide structure to the model, making the modeling language more expressive, and the interpretation process more efficient. The figure 1 illustrates the type of relations that can be represented in SAPIA models.

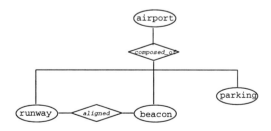

Fig. 1. examples of relation in SAPIA models

Decomposition relations Decomposition relationships enable complex objets to be split into simpler ones. For example Airports can be described as a complex object consisting of runways, taxiways, parkings, etc... Some of these sub-objects are themselves complex objects.

Geometrical relations Geometrical constraints are used to specify layout or orientation relationship among different objects. More generally, constraint based modeling has been widely used for the following reasons:

1. most concepts in the real world can be conceptualized in terms of objects properties and relationships between objects. Constraint based languages are close to end-users terminology.
2. specifying a constraint between two variables does not make any assumption about its use, or about the direction along which this constraint (considered as a relation) will be exploited.

More generally, from the stand point of constraint representation and expressivness, the requirements of our application can be stated as follows:

Non linear constraints : e.g., to state that the surface of a rectangular object is at most $100m^2$, we set the constraint *"width * length* ≤ 100"

Nary constraints : A example of geometrical constraints is the inclusion of a point in a rectangle. This constraint involves at least 7 variables: the coordinates of the point and the coordinates of the center of the rectangles, its length, width and orientation.

Continuous constraints : Although our interpretation system operates on digital, sampled images, our scene model is expressed in absolute coordinates and units. For example length values are expressed in meters, not in pixels.

2.4 Image interpretation operative knowledge

In order to search for an object in the image, one looks for other objects as clues, or for some low level features which are characteristic of the objects. For example, runway can be located by first looking for beacons (this is the way pilots drive their plane in foggy weather conditions) or by finding their borders as long linear edges. This information is made explicit in the SAPIA model as it is useful for controlling the flow of construction of the interpretation map.

Objects are identified and extracted from the image using image analysis and pattern recognition techniques (called in SAPIA a "specialist"). For each domain object, for a given weather, operative sensor conditions, we usually can find one specific image processing algorithm which is appropriate for identifying the specific shape corresponds to the domain object in the image.

Thus, the word specialist should implicitly stand for a technique specialised for retrieving "this" object under "these" conditions. These conditions are embodied in our system in a structure called "context of interpretation".

In the case of a complex object, the interpretation strategy consists of specifying the order in which its components will be searched for.

3   Using the model

Process control strategy is a feature that distinguishes image interpretation systems from one another. Process control strategy plays a salient role in the efficiency of the process and determines the global flow of interpretation.

Basically, the control strategy of a process defines the planning and the scheduling of activities that will achieve the process. In the context of image interpretation, it includes the selection of objects which are to be found in the image, the search order, as well as the way (selection techniques to be used) to search for them.

3.1   SAPIA control architecture

The global control mechanism of the SAPIA interpretation process is similar to that of a blackboard architecture. The main components of the control architecture are displayed in figure 2.

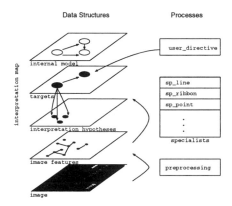

Fig. 2. SAPIA control architecture

The interpretation map is a shared structure which registers the objects which can potentially be found in the image and their probable occurrences in the image. The objects stored in the interpretation map are called targets.

The internal model is an explicit generic description of the objects that can be found in the image. Its content is expressed using the approach described in section 2. The probable occurrences are called interpretation hypotheses; they are represented as instances of the objects of the internal model, and caracterized with a confidence factor. Interpretation hypotheses also contain the shape information of the objects.

The interpretation map content is accessible by the image processing specialists and through user intervention.

Interpretation tasks or activities are triggered in response to changes in the interpretation map status and result in new modifications. Interpretation tasks are stored in an agenda structure and scheduled according to a priority level which is determined on-line on the basis of their relevance with respect to other objects currently searched for, or already found. For example, if a runway has been identified, it is quite meaningful to look for a beacon.

Interpretation task execution results either in the creation and scheduling of new interpretation tasks corresponding to searching for sub-objects, when the current task corresponds to a composite object, or in the triggering of a specialist when the current task corresponds to a simple object. A specialist is a SAPIA image processing module that implements a pattern recognition algorithm which extracts image features such as edges, regions, lines, junctions, apar, ribbons, or rectangles from raw images.

Our system integrates a constraint satisfaction algorithm and user inter-action. the integration scheme specifies user interaction as either a constraint posting for pointing search actions, or an hypothesis generation for focusing the search

The basic intent underlying the flow of control is to instantiate an object of the internal model (e.g. a runway) and install it as a target in the interpretation map. Once labelled as a target, an object is destined to be searched for as a real physical instance. However its description is still generic, as it involves parameters (e.g. length, width,...) which might not be instantiated.

Searching for a target triggers the corresponding specialist, which implements an image processing and pattern recognition algorithm. This algorithm will extract image features (lines, points, areas, polygons...) from the rough image. Those features will provide potential instantiations of the target, which are called interpretation hypotheses.

3.2   System User cooperation

User interaction is handled as a modification of the interpretation map. There are two possible modifications:

- add a new target to the interpretation map. This will add in the agenda a new search task with the highest priority
- the creation of a new constraint on one object of the internal model. If a point or a direction (two points) have been entered through a mouse click, a constraint is posted stating that the point(s) should be located inside the geometrical shape of the target object.

Whenever a change has been made to the interpretation map (e.g. target creation, target complete or partial instantiation,...) the constraint progagation engine is invoked. Because partial interpretation instantiates some variables, the constraint propagation system will exploit these instantiations by propagating their effects on other variables which share constraints with the instantiated variable.

3.3 Constraint propagation mechanism

In terms of constraint processing, SAPIA rely on the functionality of a commercially available constraint solving library. We simply highlight here the features we found necessary for our system.

Dynamically changing problem : new constraints are added while the interpretation process is progressing, such as the user pointing on the image to focus the search or the user requesting a new object to be detected.

Global optimization : The most plausible interpretation should be presented while synthezising the current map.

Handling of heterogeneous constraints : Such as in the previous example, boolean, integer and continuous constraint.

More generally, we believe these features should hold for any model-based interpretation problem solving method (data analysis, classification).

4 Illustrative scenario

We illustrate here the functions of SAPIA when used on a typical SPOT2 [1]satellite image.

The photo-interpreter can either take the initiative, by selecting which object to look for (eventually by pointing on the image where to look for it), or let SAPIA control the flow of interpretation. In the later case, all objects that compose the site of interest (an airport) will be searched automatically, in an order specified by the model.

At some point during the work session, whatever mode has been adopted (user vs. system initiative), a runway will be searched in the image.

SAPIA will invoke the runway specialist in order to find regions that satisfy the geometrical properties of runways. The runway specialist implements an image processing algorithm that extracts linear ribbons. The minimum length and maximum width of the ribbon are parameters of the algorithm. For instance when looking for runways, the minimum length of 50m and maximum width of 10m are set as parameters of the algorithm. These values are made explicit in the model of the runway.

Figure 3-(a) displays all runway hypotheses that have been detected by SAPIA.

Thus 24 linear ribbons satisfying runway geometrical constraints have been detected in the image. For the sake of better visualisation of the results, the hypotheses will be displayed in black, while the original image will be displayed in inverse video in the background.

SAPIA displays the highest probability hypothesis in yellow, whereas other hypotheses are displayed in green. In figure 3-(a), *hyp*1 is the highest probability hypothesis. Once this hypothesis is validated (either manually or by default),

---

[1] CNES ©1996 - Distribution Spot Image - processed by Cap Gemini

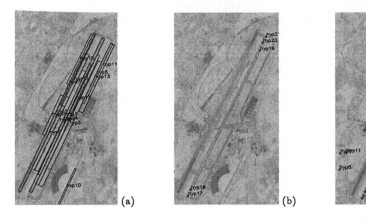

(a)   (b)   (c)

Fig. 3. Runways and Beacons interpretation

neighbouring objects will be searched. Geometrical relations are used as pointers to neighbouring objects. A beacon is one runway neighbouring object. When beacons are searched they will be searched in the alignment of this runway. The beacon specialist implements an image processing algorithm that extracts highly contrasted objects with small size. In addition to the minimum contrast and the maximum size, the region area of search (minimum and maximum bounds for the coordinates) constitutes the call parameters of the algorithm.

Beacons satisfying the geometrical alignment relation constraint are displayed in figure 3-(b). There are only 5 such beacons. Without the propagation of the geometrical relation constraint about one hundred beacons would have been extracted as shown in figure 3-(c). Thus constraint management enables to dramatically reduce the size of the interpretation tree in the solution space. More generally, we discuss the gain of using constraint management systems in image interpretation tasks, in the following section.

5   Discussion and Conclusion

The use of constraint in image processing dates from 1982 with the work of David Waltz ([Wal75]), with his seminal work on 2D line labelling. Two limitations on using constraints in visual task modeling have been:

- the implicit direction of constraint processing. While posting a constraint, the way the constraint will be processed and exploited during inference should be kept in mind, as it should be expressed in a certain way to produce a certain effect.
- the linearity of the constraint specification. Constraints among variables should remain linear otherwise it would not be processed by the constraint handling system.

Integrating constraint specification and processing in model based vision has been successfully experimented in SAPIA, specifically with respect to the previously mentioned points.

Our approach was to rely on a generic and low level constraint solving engine, and integrate it with our interpretation algorithm.

An alternative integration approach exists: Instead of using a generic constraint solving engine we could have used a geometric constraint solver ([WFH+93])

drawbacks :

generic constraint algorithms are not complete. They do not necessarily detect the global inconsistency of the set of constraints (no solution exists). They usually do not eliminate all inadequate candidates of the solution space.

This drawback can be fixed in some cases by extending the constraint propagation engine with specific constraint solving techniques. for example (orthogonal relationship).

This limitation is not dramatic for interpretation task, as candidate solutions are generated by the sensors (oneline data filtering).

advantages :

flexibility: for example tangent, concentric distances only are provided with the geometric constraint library. with a generic constraint handling library one may introduce new geometric constraints, or construct elaborate constraints from combining more primitive ones (ex. logical combination of existing constraints) or even combine numerical constraints with symbolic constraints.

In general, a generic constraint solving engine supports the integration of heterogeneous constraints in a sound uniform framework.

example:

```
size < 30m and 2 beacons or size > 30m and 4 beacons
```

The constraint solver library handles the solution bookeeping overhead.

Using a declarative model for image interpretation permits reusability and improves maintainability of the system. It also sets the interface between the image interpretation system and the administrator (the person in charge of adapting the system to the domain - usually a photo-interpreter) at the appropriate level.

Constraints are used as an information communication medium among image features which could be located at disjoint locations in the image.

Specialists do not need to know each other to cooperate. Communication is done indirectly, by

- interpretation task decomposition, which transmits control directives and schedules activities among specialists
- parameters domain reductions, resulting from partial interpretation and constraint propagation

With respect to other systems integrating constraint processing engines, our approach is unique because constraints are handled in a dynamic fashion. Constraints are not used only as a filtering or consistency checking mechanism, but also as a guiding mechanism to focus the search performed by the specialists through propagation.

Acknowledgements

The work presented in this paper was supported by DRET/ETCA under contract No. 94 01 048. The authors wish to thank Florence Germain, Bruno Bouyssounouse, Olivier Martinet, Christian de Sainte Marie, Renaud Zigmann and Yannick Hervouet from Cap Gemini Group who have contributed to the work presented in this paper, Jean-Claude Bauer for reviewing early drafts of the papers, and Christophe Thomas and Christian Millour from DRET/ETCA for fruitful discussions on the specification of SAPIA.

References

[Bev89]   J.R. Beveridge. Segmenting images using localised histograms and region merging. *International Journal of Computer Vision*, 2:311–347, 1989.

[Cas97]   T.A. Cass. Polynomial-time geometric matching for object recognition. volume 21, pages 36–61, Netherlands, 1997. Kluwer Academic Publishers.

[GGM91]   G. Giraudon, P. Garnesson, and P. Montesinos. MESSIE : un systeme multi specialistes en vision. application a l'interpretation en imagerie aerienne. *Traitement du Signal*, 9(5):403–419, 1991.

[Gri97]   W.E.L. Grimson. Object recognition research at MIT. volume 21, pages 36–61,5–8, Netherlands, 1997. Kluwer Academic Publishers.

[HS93]   R.M. Haralick and L.G. Shapiro. Computer and robot vision. In *Computer and Robot Vision*, volume 2. Addison-Wesley Publishing Company, Inc., 1993.

[KPC96]   T.H. Kolbe, L. Plümer, and A.B. Cremers. Using constraints for the identification of buildings in aerial images. In *Proceedings of PACT'96*, pages 143–154. Practical Applications, 1996.

[Pau92]   L.F. Pau. Context related issues in image understanding. In *Handbook of Pattern Recognition and Computer Vision*, pages 741–768. World Scientific Publishing Company, 1992.

[SCG94]   F. Sandakly, V. Clement, and G. Giraudon. Un systeme multi-specialiste pour l'analyse de scene 3d. In *Proceedings de la conference Reconnaissance des Formes et Intelligence Artificielle*, volume 1, pages 611–616. AFCET, jan. 1994.

[Wal75]   D. Waltz. Understanding line drawings of scenes with shadows. In *Psy-chCV75*, pages 19–91, 1975.

[WFH+93]   W.Bouma, I. Fudos, C. Hoffmann, Jiazhen Cai, and R. Paige. A geometric constraint solver. *Technical Report*, CSD-TR-93-054, 1993.

[ZJ93]   A. Zlotnick and P.D. Carnine Jr. Finding road seeds in aerial images. *CVGIP Image Understanding*, 57(2):243–260, 1993.

Hierarchical Multifeature Integration for Automatic Object Recognition in Forward Looking Infrared Images

Shishir Shah[1] and J. K. Aggarwal[2]

[1] Wayne State University, Laboratory for Visual Computing, Dept. of Electrical & Computer Engineering, Detroit, MI 48202, U.S.A.
[2] The University of Texas at Austin, Computer & Vision Research Center, Dept. of Electrical & Computer Engineering, Austin, TX 78712, U.S.A.

Abstract. This paper presents a methodology for object recognition in complex scenes by learning multiple feature object representations in second generation Forward Looking InfraRed (FLIR) images. A hierarchical recognition framework is developed which solves the recognition task by performing classification using decisions at the lower levels and the input features. The system uses new algorithms for detection and segmentation of objects and a Bayesian formulation for combining multiple object features for improved discrimination. Experimental results on a large database of FLIR images is presented to validate the robustness of the system, and its applicability to FLIR imagery obtained from real scenes.

1 Introduction

This paper addresses the problem of object recognition in complex scenes imaged by a single sensor or registered multiple sensors. Object recognition comprises of identifying the location of the object of interest and recognizing its identity and pose. It is perceived as a high-level task in computer vision, relating semantic knowledge in terms of configuration of known objects [1]. The overall process includes preprocessing for image enhancement and is followed by the detection stage, which localizes the regions of interest in the image with the assistance of *a priori* known or an auxiliary source of information. The subsequent processing uses this information to segment the object and extract features to obtain a complete representation. Object recognition is then achieved by comparing descriptions of *a priori* known object models, which are generalized descriptors defining object classes.

A central problem in computer (machine) vision, and automatic object recognition (AOR) in particular, is obtaining robust descriptions of the objects of interest in an image. In most complex scenes, and especially so in AOR applications, the images vary with environmental conditions and natural lighting, which

---
* This work was supported by the Army Research Office Contracts DAAH-94-G-0417 and DAAH 049510494.

increases the complexity of the recognition problem. The object or the region of interest may also be "intentionally" camouflaged, occluded, or surrounded by clutter. In order to build a system that can succeed in a realistic environment, certain simplifications and assumptions about the environment and the problem domain are generally made. This simplification process introduces uncertainties into a problem that may create inaccuracies or difficulties in the reasoning abilities of a system unless these uncertainties are represented and handled in a suitable manner. In this paper, we present a methodology that uses multiple

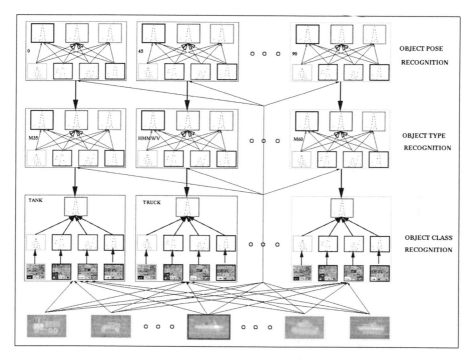

Fig. 1. The general flow of the object recognition paradigm.

object features along with a cooperative decision integration for robust recognition. The proposed system uses a hierarchical scheme where the lowest level recognizes the class of the object, while the higher levels use the information from the lower levels, as well as the features from the original image, for classification within each class. A schematic of this structure is shown in figure 1. No information about the 3D structure (such as a 3D model [2] of the object is available; the only *a priori* information is a database containing images of the object viewed at different angles (aspects) and distances. The methodology is applied to recognition of objects in second generation Forward Looking Infrared (FLIR) images. Multiple images from a single sensor or multiple sensors could be used as input data to the system. Multiple object features from the input image(s) are computed and a Bayesian implementation of the methodology is presented, in which each mapping is characterized by the probability density function of the feature. Thus, object modeling is used to design the classifiers and the distribution of object features is modeled as a mixture of Gaussians. An adaptive Expectation-Maximization (EM) algorithm is used to find the parameters of the normal distributions and a supra-Bayesian scheme is used for decision integration at the highest level.

The rest of the paper is organized as follows: Section 2 details the multifeature hierarchical architecture based on Bayesian statistics for recognition of the object class, type, and pose. The object feature representations used and the computation of conditional probability density, along with the final decision integration is also presented in this section. The algorithms for detection and segmentation of the objects in FLIR images are briefly presented in section 3. Section 4 presents the experimental results using the developed recognition system. Finally, conclusions and a summary of this study are presented in section 5.

2 Multifeature Object Recognition

Object recognition is the classification of objects into one of many *a priori* known object classes. In addition, it may involve the estimation of the pose of the object and/or the track of the object in a sequence of images. Humans recognize objects and understand complex scenes with multiple objects, noise, clutter, occlusion, and camouflage with great ease. Humans are able to recognize as many as 10,000 distinct objects [3] under varying viewing conditions, while a state-of-the-art object recognition system can recognize relatively few objects.

Multisensor/multifeature fusion is now widely accepted as being indispensable in complex recognition applications due to the limitations of discriminatory information obtained from a single feature or sensor, poor imaging conditions, or the effect of counter-measures or camouflage [4]. The motivation behind the design of multifeature integration system stems from the realization that single feature measurements inherently incorporate varying degrees of uncertainty and are occasionally spurious and incorrect. Inspired by biological organisms, which are essentially multifeature perception systems, the development of intelligent systems that use multiple sources of information to extract knowledge about the sensed environment seems a natural step forward. The shortcomings of single sensors can be overcome by employing redundancy and diversity [5].

2.1 Probabilistic Formulation

Recognition involves deciding the class of the object in the image from a set of known classes (C_i, $i = 1, \ldots, K$), its type from a set of known objects (T_j, $j = 1, \ldots, L$), and its pose or aspect β_n, $n = 1, \ldots, M$ using information from the image ($\mathbf{X}$). In Bayesian decision theory, this information can be determined by finding the C_i, T_j, and β_n that maximize the posterior probability $P(\beta_n, T_j, C_i | \mathbf{X})$. The *a posteriori* probability can be derived from Bayes theorem. More precisely (dropping the subscripts for clarity),

$$P(\beta, T, C, \mathbf{X}) = P(\beta | T, C, \mathbf{X}) P(T | C, \mathbf{X}) P(C | \mathbf{X}) P(\mathbf{X}). \tag{1}$$

We also have that

$$P(\beta, T, C, \mathbf{X}) = P(\beta, T, C|\mathbf{X})P(\mathbf{X}). \tag{2}$$

Therefore, from equations (1) and (2), we have

$$P(\beta, T, C|\mathbf{X}) = P(\beta|T, C, \mathbf{X})P(T|C, \mathbf{X})P(C|\mathbf{X}). \tag{3}$$

The maximum a posteriori probability (MAP) estimates of the object class, type, and pose $(\widehat{C, T}, \beta)$ is obtained as:

$$
\begin{aligned}
\widehat{C, T}, \beta &= \arg\max_{C,T,\beta} P(\beta, T, C|\mathbf{X}) \\
&\equiv \arg\max_{C,T,\beta} \ln P(\beta, T, C|\mathbf{X}) \\
&\equiv \arg\max_C [(\arg\max_T [(\arg\max_\beta P(\beta|T, C, \mathbf{X})) + P(T|C, \mathbf{X})) + P(C|\mathbf{X}))]. \quad (4)
\end{aligned}
$$

The last simplification in the above analysis is obtained because the logarithm is a monotonically increasing function. Since $P(\beta|T, C, \mathbf{X})$, $P(T|C, \mathbf{X})$, and $P(C|\mathbf{X})$ are always positive, from a classification point of view, the MAP estimates are obtained by finding C_i, T_j, β_n such that

$$
\begin{aligned}
(P(\beta_n|T_j, C_i, \mathbf{X}) &+ P(T_j|C_i, \mathbf{X}) + P(C_i|\mathbf{X})) > \\
(P(\beta_l|T_k, C_o, \mathbf{X})] &+ P(T_k|C_o, \mathbf{X}) + P(C_o|\mathbf{X})), \quad (5)
\end{aligned}
$$

2.2 Multifeature Integration

For robust recognition, it is important to characterize the input-output relationship to understand the behavior of the system under a wide variety of inputs. Such a measure, in turn, also helps to improve the prediction capability, an important feature that is essential for practical use of the system. In the proposed hierarchical multifeature integration framework, the lowest level in the hierarchy identifies the class of the object (e.g., tank vs. truck), while the higher levels use the information from the lower levels, as well as features extracted from the original object, for classification within each class to determine object type and pose. As shown in figure 1, each level consists of classifier modules, each of which is an expert for a single object representation. The modular classifier at the lowest level is trained to recognize the object class under different object types and aspects. When presented with an input, each classifier provides a measure of confidence of that object belonging to a represented class. A similar recognition architecture has been proposed in [6], where the recognition is based on object parts. Input objects are segmented into parts, and each module in the hierarchy is an expert on recognizing a single part. Recognition is perfomred by accumulating probability estimates from each expert using the recursive Bayes' update rule as parts are presented in a sequential manner. This makes the process extremely sensitive to object segmentation and determination of parts. In contrast, the architecture proposed here allows for intermediate levels of recognition (e.g. Truck class recognized before M35 truck) and accounts for variability in object appearance and faults in segmentation through multifeature representations.

A number of features and representations for object recognition have been proposed in the past [7, 8]. We propose a novel approach to object representation that does not use a single feature, instead uses a feature bank, whose contribution for final recognition are learned from examples. The response of such a multifeature bank contains much more information than does a single feature. Each feature is modeled using a mixture of Gaussians to represent the conditional probability distribution. To keep the theory simple, we will only consider the first layer of hierarchy here. The objective is to recognize different object classes based on class experts represented by a multifeature bank. Consider the general case where there are K possible classes of objects $C_i, i = 1, \ldots, K$. Let each class be represented by m different features. Each classifier within a module at this level evaluates the *a posteriori* probability for each input. Let X_m represent the input feature vector for the mth feature for the input under consideration and $\mathbf{X} = \{X_1, X_2, \ldots, X_m\}$ represent the entire feature bank. Given the *a priori* probability of each class $P(C_i)$ and the conditional density for each feature $p(X_m|C_i)$, we can compute the posterior probability of the observed feature being object C_i using the Bayes Rule:

$$P(C_i|X_m) = \frac{p(X_m|C_i)P(C_i)}{p(X_m)} \tag{6}$$

For each class module, we are interested in a two class discrimination, where the object is recognized by the expert module or is rejected. Thus, in a two case discrimination the posterior probability given by the classifier is:

$$P(C_i|X_m) = \frac{p(X_m|C_i)P(C_i)}{p(X_m|C_i)P(C_i) + p(X_m|\bar{C}_i)P(\bar{C}_i)} \tag{7}$$

where, $p(X_m|\bar{C}_i)$ is the conditional density function of the other object classes and $P(\bar{C}_i)$ is the prior probability of observing the remaining classes. Given the posterior estimates from the individual classifiers, the goal of the combining stage is to produce a single estimate that maximizes the probability for object class recognition while reducing clutter and false alarms. Various integration methods have been proposed in the past [9]. We formulate a supra-Bayesian integration in which the posterior estimates from each classifier are assumed to have a probability distribution and, based on the means and variances of the outputs, we can formulate an optimal decision scheme. Strictly speaking, Bayesian theory holds true only for individual decision makers, but if the group decision is viewed as a collaborative effort, the effect is externally Bayesian. The integration module, just as individual classifiers, is interested in estimating the probability of observing an object. So, given m individual classifiers, each providing a measure of its subjective probability of observing an object class, $P(C_i|X_m)$, the integration has to result in the combined probability P_I. Considering that P_I has to obey the laws of probability if an integrated measure of the posterior is to be estimated, the integration may be written as:

$$P_I(C_i|X_1, X_2, \ldots, X_m) = \prod_{j=1}^{m} P(C_i|X_j) \tag{8}$$

Assuming that the individual classifier posteriors are Gaussian distributed, then the integrated posterior decision simplifies to:

$$P(C_i|\mathbf{X}) = \frac{[\prod_{j=1}^{m}(\frac{P(C_i|X_j)}{P(C_i)})^{w_j}]P(C_i)}{[\prod_{j=1}^{m}(\frac{P(C_i|X_j)}{P(C_i)})^{w_j}]P(C_i) + [\prod_{j=1}^{m}(\frac{P(\bar{C}_i|X_j)}{P(\bar{C}_i)})^{w_j}]P(\bar{C}_i)} \tag{9}$$

where w_j is the contribution of each of the distributions. This can be considered as the confidence measure for individual classifiers. As each of the classifiers is designed to identify a single object class, we know that there is sufficient diversity and complementarity within the estimates. Thus the weights associated with each of the classifiers plays an important role in deciding the contribution from each estimate. This is mainly due to the fact that the integrator module does not have the same information that is seen by each of the classifiers. Evaluating the log likelihood of equation 9 and assuming that the combined probability ratios provide the final probability as

$$P(\ln(\frac{P(C_i|X_1)}{1 - P(C_i|X_1)}),\ldots,\ln(\frac{P(C_i|X_m)}{1 - P(C_i|X_m)})|C_i) \tag{10}$$

and

$$P(\ln(\frac{P(\bar{C}_i|X_1)}{1 - P(\bar{C}_i|X_1)}),\ldots,\ln(\frac{P(\bar{C}_i|X_m)}{1 - P(\bar{C}_i|X_m)})|\bar{C}_i) \tag{11}$$

and if the joint distributions are multivariate normal densities with mean μ_{C_i} and $\mu_{\bar{C}_i}$ and covariance $\Sigma_{C_i \bar{C}_i}$, then the weights for the individual classifiers can be computed by:

$$\mathbf{w} = \Sigma_{C_i \bar{C}_i}^{-1}(\mu_{C_i} - \mu_{\bar{C}_i}) \tag{12}$$

This result provides an intuitive insight to the integration of decisions. In general, when all the classifiers provide similar estimates, the combining results in the peaking of that estimate. On the other hand, and more importantly, when the classifiers do not agree on an estimate, their reliability has to be considered. According to the weight assignment in equation 12, the reliability associated with each of the classifiers will depend on how different its estimate is from rest of the classifiers, and how much diversity exists within the estimates. Having computed the posterior probability, they can be used to determine the final classification of the input, that is, the input is classified as belonging to object class i if $P(C_i|\mathbf{X}) > P(C_j|\mathbf{X}) \, \forall j \neq i$.

2.3   Feature Modeling and Object Parameters

In this section, we describe how the conditional density functions and the model parameters are estimated for each expert module. Due to the complex and non-Gaussian distribution of the object features, we model the data using a mixture of Gaussians. Modeling of data is an important consideration in designing statistical classifiers. The simplest way to model non-Gaussian data is to use the histograms of the training data. However, classification based on this method does not generalize well from the training data to the test data. Maximum likelihood estimators [10] compute piecewise estimates of one-dimensional density functions. This approach can be regularized by introducing a penalty term. Such methods are attractive, but rely on a predefined model of the density function. They also do not generalize well in the case of mixture models unless coupled with other optimization techniques.

We use the Expectation-Maximization (EM) algorithm to determine the parameters for the mixture of Gaussians model to estimate the density function [11, 12]. A stagewise k-means procedure is used, where the initial guess for the cluster centroids is obtained by splitting the centroids resulting from the previous stage. Given a set of features and the number of kernels, a basic procedure is repeated. Initially, the number of kernels or components is set to one. The centroid of all training points is computed and a measure of the mean and in-class deviation is computed. The component weights is computed as a ratio of the number of training points in the corresponding component and the total training points. A new estimate of the means, variances, and the distortion are computed iteratively until convergence.

2.4 Model Prior Estimation

For the lowest level in the hierarchy, the model prior is based on the importance in differentiating one object class from the other. In such a case, given no *a priori* information, equal prior probabilities, $P(C_i)$ are assumed and fixed at $1/K$. However for multiple feature recognition, some representation features/models may be more significant than the others. For example, in trying to distinguish between an M60 tank and a T72 tank, the Zernike moment representation may be more salient. This can be expressed as the prior probability for that individual classifier. The relative importance of each model can be determined by examining the model representation for a given class and the remaining models.

The prior probability for each classifier is then computed as:

$$P(C_i) = \frac{\sum_{k=1, X \in C_i}^{K} X_k}{\sum_{j=1, X \in C_i, j \neq i}^{M} X_j} \tag{13}$$

The analysis for the higher levels of hierarchy are easily extended. The outputs of the uppermost level can be inputs to the next higher level. For each of the models, the probability of the object belonging to the given class is computed and integrated for final recognition.

3 Detection, Segmentation, and Models

In this section, we briefly describe the algorithm that was used to detect and segment the object from the image. The models used for object representation are also briefly discussed.

Robust detection of manmade objects in complex scenes is obtained by learning multiple feature models in images [13]. The methodology is based on a modular structure consisting of multiple classifiers, each of which solve the problem independently based on its input observations. Each classifier module is trained to detect manmade object regions and a higher order decision integrator collects evidence from each of the modules to delineate a final region of interest. A reliability measure for each classifier module is computed to determine its contribution to the final decision.

The detected regions were individually analyzed by performing a connected component analysis on the output of pixel classification and regions consisting of less than 100 pixels were removed. A region growing procedure with compactness and edge linearity constraints [12] was used to isolate final regions. An initial pixel connectivity analysis is performed to identify grouped regions and, based on the *a priori* known object size, the region is grown to incorporate neighboring pixels without violating the compactness and regularity of the object. Compactness and regularity measures are defined based on boundary segments and corner points.

For the recognition stage, each module uses four features, each represented by a conditional probability density function. To capture the variation in surface intensity, we use the eigen image representation [14] up to the 10th order. Object boundaries are extracted from the image and their eigen representation is also used. After segmenting the images, the objects are thresholded to get binary images. The shape of the object is represented using Zernike moments [15] up to order 8, and the Distance Transform [16].

4   Experimental Results

This section presents results of using the proposed methodology for object recognition in second generation FLIR images. For the results presented here, five objects from two classes were considered. Figure 2 shows example images of these objects. Figure 3 shows an example of a typical input image and the result of detection and segmentation. The segmented images are used for the recognition process. For each object, a total of 144 images were considered (2 sets of $0 - 360°$ at $5°$ intervals) for training. Since the two sets of 72 images were imaged from roughly the same elevation and acquired in varying environmental conditions, the training set captured a wide range of variations in the objects.

Fig. 2. Typical FLIR images of objects used in recognition experiments.

To test the system, five object types under different viewing conditions and varying segmentation outputs were presented to the fully trained recognition system. For each object, 144 images were tested, giving a total of 720. Out of the 720 images, 450 were segmented well, 120 were segmented poorly and included parts of the background or had part of the object missing, while the remaining 150 were considered bad as they included objects that were occluded and thus included part of background objects cluttering the object of interest. Table 1 gives the recognition results obtained from these experiments. Overall the class recognition rate was 94.1%, object type rate was 92.2%, and pose estimation rate was 83.7%.

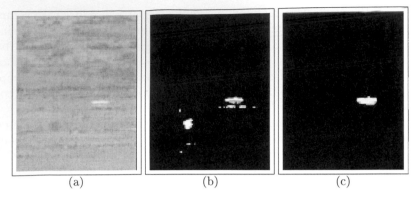

(a)   (b)   (c)

Fig. 3. Preprocessing steps applied to typical FLIR image (a), the results obtained after initial detection (b), and the result of final segmentation (c).

Object Seg.	Class Recog.	Type Recog.	Pose Recog.	Mis-Classified
Good	442/450 = 98.2%	438/450 = 97.3%	402/450 89.3%	8/450 = 1.7%
Poor	112/120 = 93.3%	109/120 = 90.8%	99/120 82.5%	8/120 = 6.6%
Bad	124/150 = 82.6%	117/150 = 78.0%	102/150 68.0%	26/150 = 17.3%

Table 1. Recognition results obtained using the Bayesian HMS. Refer to text for details.

5 Summary and Conclusion

In this paper we have presented a hierarchical methodology and results for object recognition based on multifeature integration. The task of recognizing object characteristics such as object class, type, and pose are achieved as the input traverses upward in the hierarchy. The classification at higher levels is achieved by considering the decision at the lower levels as well as considering the input features directly. At each level of hierarchy, multiple features are computed for representing the objects and modular computational structures consisting of multiple classifiers are used to perform recognition. Each classifier within a module is trained to recognize a single feature and tries to solve the problem of recognition based on its input observation. A higher level decision integrator oversees and collects evidence from each classifier within a module and combines it to provide a final decision while considering the redundancy and diversity of individual classifiers. A Bayesian realization of the methodology is presented. Each classifier module models the object signature probability density function based on the computed features and the final integration is achieved in a supra-

Bayesian scheme. Results are presented for recognition of five different objects belonging in two classes with varying pose. The methodology can easily be extended to registered multisensor data as input.

References

1. A. Rosenfeld, "Image analysis: Problems, progress and prospects," *PR*, vol. 17, pp. 3–12, January 1984.
2. F. Arman and J. K. Aggarwal, "Model-based object recognition in dense depth images - a review," *ACM Computing Surveys*, vol. 25, no. 1, pp. 5–43, 1993.
3. I. Biederman, "Human image understanding: Recent research and a theory," *Computer Vision, Graphics and Image Processing*, vol. 32, pp. 29–73, 1985.
4. C. Chu and J. K. Aggarwal, "The integration of image segmentation maps using region and edge information," *IEEE Transactions on Pattern Analysis and Machine Intelligence*, vol. 15, no. 12, pp. 1241–1252, 1993.
5. J. Manyika and H. F. Durrant-Whyte, *Data Fusion and Sensor Management: A decentralized information-theoretic approach*. Ellis Horwood, 1994.
6. D. Nair and J. Aggarwal, "Robust automatic target recognition in 2nd generation flir images," in *Proceedings of 3rd IEEE Workshop on Applications of Computer Vision*, (Sarasota, Florida), pp. 311–317, December 1996.
7. D. Ballard and C. Brown, *Computer Vision*. Prentice-Hall, Inc., 1982.
8. W. E. L. Grimson, *Object Recognition by Computer: The role of geometric constraints*. MIT Press, Cambridge, 1990.
9. J. Kittler, M. Hatef, and R. P. W. Duin, "Combining classifiers," in *International Conference on Pattern Recognition*, pp. 897–901, 1996.
10. B. W. Silverman, *Density Estimation for Statistics and Data Analysis*. London: Chapman and Hall, 1986.
11. A. P. Dempster, N. M. Laird, and D. B. Rubin, "Maximum likelihood from incomplete data via the EM algorithm.," *Journal of the Royal Statistical Society*, vol. 39-B, pp. 1–38, 1977.
12. S. Shah and J. K. Aggarwal, "A Bayesian segmentation framework for textured visual images," in *Proc. of Computer Vision and Pattern Recognition*, pp. 1014–1020, 1997.
13. S. Shah and J. K. Aggarwal, "Multiple feature integration for robust object localization," in *Proc. Computer Vision and Pattern Recognition (to appear)*, (Santa Barbara), 1998.
14. A. Pentland, B. Moghaddam, and T. Starner, "View-based and modular eigenspaces for face recognition," *Proc. of Computer Vision and Pattern Recognition*, pp. 84–91, 1994.
15. A. Khotanzad and Y. Hong, "Invariant image recognition by Zernike moments," *IEEE Transactions on Pattern Analysis and Machine Intelligence*, vol. 12, pp. 489–497, 1990.
16. D. Paglieroni, "Distance transforms: Properties and machine vision applications," *Graphical Models and Image Processing*, vol. 54, pp. 56–74, January 1992.

Air-Crew Scheduling through Abduction

A.C. Kakas and A. Michael

Department of Computer Science, University of Cyprus,
P.O.Box 537, CY-1678 Nicosia, CYPRUS
Email: {antonis, antonia}@cs.ucy.ac.cy

Abstract. This paper presents the design and implementation of an air-crew assignment system based on the Artificial Intelligence principles and techniques of abductive reasoning as captured by the framework of Abductive Logic Programming (ALP). The aim of this work was to produce a system for Cyprus Airways that can be used to provide a solution to the airline's crew scheduling problem whose quality was comparable with the manual solutions generated by human experts on this particular problem. In addition to this the system should also constitute a tool with which its operators could continually customize the solutions to new needs and preferences of the company and the crew. The abductive approach (using ALP) adopted in our work offers a flexible modeling environment in which both the problem and its constraints can be easily represented directly from their high-level natural specification. This high-level representation offers two main advantages in the development of an application: (i) *modularity* with a clear separation of the two main issues of validity and quality (optimality) of the solution and (ii) *flexibility* under changes of the requirements of the problem.

1. Introduction

Abduction is the process of reasoning to explanations for a given observation according to a general theory that describes the domain of observation. In Artificial Intelligence abduction manifests itself as an important form of inference for addressing a variety of problems (eg. [14, 16, 10, 5, 2]). These problems include, amongst others, reasoning with incomplete information, updating a database or belief revision and formalizing and addressing application problems like that of planning, diagnosis, natural language understanding and user modeling. Several applications of abduction exist (eg. [10, 4, 3, 13, 17]) and we are now beginning to see the emergence of the first "engineering principles" for the development of abductive applications.

In this paper we use the framework of Abductive Logic Programming (ALP) [10, 5] to develop an abductive-based system for the application of air-crew scheduling. The problem of air-crew scheduling (e.g. [1, 9]) is concerned with the assignment of air-crews to each of the flights that an airline company has to cover over some specific time period. This allocation of crew to flights has to respect all the necessary constraints (validity) and also minimize the crew operating cost (quality or optimality). The validity of a solution is defined by a large number of complex constraints, which express governmental and international regulations, union rules, company restrictions etc. The quality of the schedule is specified by its cost, but also by the needs and preferences of the particular company or crew.

We present an air-crew scheduling system based on abduction, developed for the application of Cyprus Airways Airlines. The aim of this work was to produce a system that can be used to provide a solution to the airline's crew scheduling problem whose quality was comparable with the manual solutions generated by human experts. In addition to this the system should also constitute a tool with which it will be possible for its operators to continually customize the solutions to new unexpected requirements while the solution (crew-roster) is in operation or additional foreseen needs and preferences of the company and the crew. The system therefore can be used, not only for the production of a valid and good quality initial schedule, but also as a rescheduling tool that can help human operators to adjust an existing solution to the necessary changes.

Various approaches, within OR and AI (e.g. [8, 15, 1, 9]), have been adopted for solving the crew scheduling problem. Although these methods have succeeded in solving the problem efficiently, for specific problem cases, they lack a general and simple way of specifying the problem its constraints. Moreover, these methods lack flexibility and their adaptation to changes in airline company rules is not very practical. Recent Constraint Logic Programming based approaches (e.g. [8, 15, 7]) have succeeded in providing a more declarative way of describing scheduling problems, but these still have to be represented directly in terms of the special domain constraints which can be handled by the constraint solver. Thus again the representation of the problem looses some of its flexibility.

The rest of the paper is organized as follows. Section 2 introduces the problem of air-crew scheduling and rescheduling. In section 3, we give a brief overview of the ALP framework and how this can be used in applications. Section 4 presents the architecture of the air-crew scheduling system. In section 5, we present some of the experiments carried out to evaluate this system. The paper concludes, in section 6. An extended version of this paper [12] can be obtained from the authors.

2. Air-Crew Scheduling

Air-Crew Scheduling is the problem of assigning cockpit and cabin crews to the flights that a company has to carry out over some predefined period of time. The flight schedule of a company consists of flight legs, which are non-stop flights between two airports, for a predefined period (usually a month). Given the flight schedule, the crew scheduling task is concerned with the allocation of crew personnel to each flight leg, in a way that minimizes the crew operating cost.

The crew scheduling problem is extremely difficult and combinatorial in nature due to the large number and the complexity of the constraints involved, e.g. governmental and international regulations, union rules, company restrictions etc.
An abductive approach (e.g. using Abductive Logic Programming - ALP) offers a flexible modeling environment in which both the basic problem and its constraints can be easily represented directly from their high-level natural specification. It then provides an effective automatic translation (reduction) of this high-level representation to the lower-level computational goals (constraints) which need to be solved. In this way ALP offers the possibility to combine the advantages of modularity and flexibility with efficient constraint solving.

2.1. The Problem Domain

Let us first introduce some of the basic terms used in air-crew scheduling.

flight leg : The smallest planning unit and it's a non-stop flight (link) between two airports. *pairing or rotation :* A sequence of flight legs which could be taken by a single crew member.

flight duty time (FDT) : The sum of the flight time of a rotation, the time spent on the ground in between two succeeding legs and the time required for briefing and debriefing of the crew.

deadhead (DH) : The positioning (by another plane) of crew, as a passenger, to another airport in order to take a flight assignment.

transfer : The transferring of crew via road (by bus or car) from one airport to another in order to take an assignment.

day-off : A full non-working day, with some additional requirements on the time which the duty of the previous day ends or on the beginning of the next day.

rest period : A continuous period of time between two duties when the crew has no assignments.

stay-over (S/O): A rest period that a crew spends away from base. Stay-Overs are necessary because of the upper bounds which exist on the duty period of a crew.

stand-by : A period of time when the crew has to be near the base and ready to take an assignment, if necessary, within one hour's notice.

available (AV) : A crew member is said to be available on a day that is not off and on which the crew has no duty.

The input to the crew scheduling problem is the flight schedule of an airline company for a specific period (usually a month). The flight schedule consists of all the flight legs that the company has to cover in that period. Given the flight schedule, together with the necessary information about each available crew member (e.g. their name and rank), the required output is a set of valid and good quality, if possible optimal, (monthly) rosters for each member.

The **quality** (or optimality) of the solution depends on three major issues: cost, fairness and preferences. A good solution should minimize the crew operating cost. Among the main factors which can result in high-cost schedules are: the number of DHs, transfers, S/Os and the total number of duty hours, over some specific period, of a crew (overtime rate is paid if this number exceeds a certain limit). A second factor for quality is the fairness of the solution on the distribution of the flying hours, type of allocated flights, DHs, S/Os, stand-bys, availability, day-offs etc. Another important issue for the quality of a solution is the extent in which the (crew or company) preferences are satisfied. This issue can sometimes also determine the validity of a solution.

The **validity** of the rosters is defined through a number of complex rules and regulations which can be grouped in classes according to the type of information that they constrain. Some of the most common groups of constraints in air-crew scheduling are the *temporal* and *location* constraints, the *FDT* constraints and the constraints on the *rest* and *working times* of the crew. A number of other constraints which express specific requirements, or preferences, of the particular application, or rules of the specific airline company or union may also be necessary for valid or better quality solutions.

2.2. Rescheduling - Changes

In the domain of air-crew scheduling changes occur very frequently either due to flight delays or cancellations, new flight additions, or because of crew unavailability. It is common that such changes happen on the day of operation, in which case they have to be dealt with immediately. Also apart from these changes which happen unwillingly, a number of other changes may often be voluntarily made by the crew administration on the solution found initially, in order to reflect special preferences and thus make the final solution more desirable. For these reasons the automation of the rescheduling process is crucial in the crew scheduling problem.

In general, the changes which are required fall in one of the following categories.
- Change the crew of a flight.
- Reschedule a flight, when the times of a flight are changed (e.g. delayed).
- Cancel a flight.
- Add a new flight.

These changes may violate the validity of the old solution or make the old schedule of poorer quality. We want to reestablish validity or quality without having to discard the old solution and recompute from the beginning, but rather by keeping the unaffected part unchanged and by rescheduling only those assignments which may be affected. In air-crew scheduling it is also necessary that any change can be accommodated by changing the old schedule within 48 hours with the fewest possible changes on the old assignments of crew members.

3. The ALP Approach to Applications

Abduction is the process of reasoning to explanations for a given goal (or observation) according to a general theory that describes the problem domain of the application. The problem is represented by an abductive theory. In Abductive Logic Programming (ALP) an abductive theory is defined as follows.

Definition 3.1 (Abductive Theory)

An abductive theory in ALP is a triple <P,A,IC> where P is a logic (or constraint logic) program, A is a set of predicate symbols, called abducibles, which are not defined (or are partially defined) in P, and IC is a set of first order closed formulae, called integrity constraints.

In an abductive theory <P,A,IC>, the program P models the basic structure of the problem, the abducibles play the role of the answer-holders, for the solutions to particular tasks (goals) in the problem, and the integrity constraints IC represent the validity requirements that any solution must respect. A goal G is a logic (or constraint logic) programming goal. A solution to a goal G is an abductive explanation of G defined as follows.

Definition 3.2 (Abductive Solution)

An abductive explanation or solution for a goal G is a set Δ of ground abducible formulae which when added to the program P imply the goal G and satisfy the integrity constraints in IC, i.e.

$$P \cup \Delta \models_{lp} G \qquad \text{and} \qquad P \cup \Delta \models_{lp} IC$$

where $\models_{lp}$ is the underlying semantics of Logic Programming or Constraint Logic Programming. More generally, the solution set Δ may also contain existentially quantified abducibles together with some constraints on these variables (see [11]).

The computational process for deriving the abductive solution (explanation) consists of two interleaving phases, called the abductive and the consistency phases. In an *abductive phase*, hypotheses on the abducible predicates are generated, by reducing the goals through the model of the problem in P, thus forming a possible solution set. A *consistency phase* checks whether these hypotheses are consistent with respect to the integrity constraints. During a consistency phase it is possible for new goals to be generated, if these are needed in order to ensure that the hypotheses so far can satisfy the integrity constraints. In turn these new goals can generate further abducible assumptions, to be added to the solution set. It is also possible that the consistency phase refines the solution set of assumptions generated originally - by setting constraints on the existential variables involved in the abducible assumptions - when this restriction can help ensure the satisfaction of (some of) the integrity constraints.

4. The Air-Crew Scheduling System

4.1. The System Architecture
A crew scheduling system has been developed using the ALP approach and applied to the real-life problem of the airline of Cyprus Airways. The general structure of the system is shown in Figure 4.1.

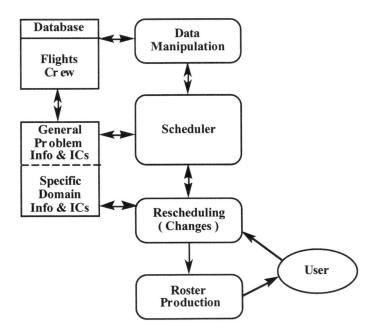

Figure 4.1 : Air-Crew Scheduling System Architecture

The Database holds information about all available crew members and the flight schedule of the company, which consists of all the flight legs that have to be covered for a predefined period (usually a month). Given the flight schedule of an airline, a preprocessing phase ("Data Manipulation") constructs possible pairings of flight legs which can be assigned to a single crew without exceeding the given maximum flight duty time (FDT). In FDT we include 1hr for briefing and 1/2hr for debriefing of crew after each leg, plus the time spent on the ground between the succeeding legs. One such pairing, in the Cyprus Airways problem, might for example be the "Larnaca→Athens→Larnaca" rotation. Other flight legs which cannot be paired in one duty remain as "single" flights.

In the main "computational" part of the system, the "Scheduler", flights (pairings or single) are assigned, via the process of abductive reasoning, to the selected crew members. The output of this phase is a consistent set of assignments of flights, deadhead flights, transfers, stay-overs and day-offs to crew members. The required solution should cover all flight legs exactly once and it should be of good quality (e.g. with low cost, balanced etc.). This set of assignments will go through the "Roster Production" phase which will then extract all the information needed, like for example the (monthly) roster for each crew member, the total flying and duty hours of each crew for that period, the crew available on some specific day etc.

4.2. The Scheduler
Given the set of flights (pairings or single), generated in the preprocessing ("Data Manipulation") phase, the scheduler is concerned with the assignment of crew members to each of these flights and to other related duties. Figure 4.2 shows a graphical representation of the model of this problem, captured by the program P of the overall ALP theory <P, A, IC> which represents the whole problem. With this modeling of the problem there is a need for only one abducible in A, which *assigns* crew members to different types of duty tasks (e.g. flights, stand-bys, day-offs etc.). Therefore the solution (schedule) will consist of a set of assignments of crew to tasks, in a way that satisfies the integrity constraints IC on the problem.

One natural feature, which is represented in the basic model of this problem in the program P, is that the flights on consecutive days are more related among them than with flights on other days, in the sense that many constraints involve flights which are temporally near to each other. Furthermore, an even stronger dependency exists between flights which are on the same day. This observation suggests that a better (computationally) modeling of the problem would be one where the flights are scheduled in increasing date order and, within the same day, in increasing order of departure time.

Scheduling the flights per day and in chronological order can also facilitate the derivation of better quality schedules, in the sense that certain operations ("Day Operations"), which improve the solution, can be performed at the end of each day. For example, consider the case of deadhead crew which has to return to base after some number of stay-overs (depending on his/her location). This crew has to be assigned as a passenger on a return (deadhead) flight on some specific day. Other additional assignments which are required, on a daily basis, are assignments of crew to transfers, stand-bys and day-offs.

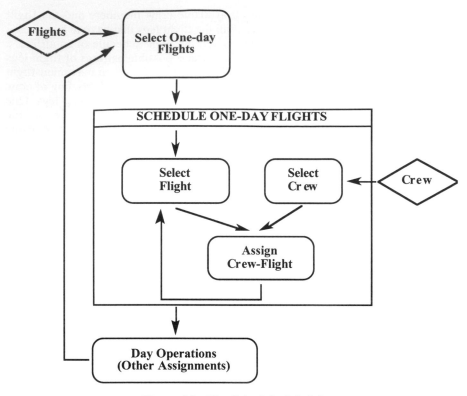

Figure 4.2 : The Scheduler Module

Another important quality of solution issue is that of balancing. A good schedule should be fair, or well balanced, at several levels. One such level is the balancing of the flying and duty hours for each crew member, another would be the balancing on the stand-bys, balancing on the stay-overs, or even fairness on the flight destinations of each crew. These issues can be treated incrementally in the model by adopting different strategies (e.g. load balancing algorithms) in the selection of the crew members.

The Constraints
One of the characteristics of the crew scheduling problem is the large number and the variety of the constraints to be considered. We present here a small representative sample of these, as they are included in the integrity constraints IC of the ALP formulation of the problem, that demonstrate the concepts proposed by this approach.

Temporal constraints
One hard temporal constraint necessary for valid solutions is that no crew member can be assigned two flights at the same time. This is represented as an integrity constraint, by the logical formula in IC,

$\neg$ (assign(Crew,FlightA), assign(Crew,FlightB),
         FlightA $\neq$ FlightB, overlap(FlightA,FlightB)   ).

where "overlap" is defined in the program P by

overlap(FlightA,FlightB) $\leftarrow$
     flight_details(FlightA,DeptDateA,DeptTimeA,ArrDateA,ArrTimeA),
     departure(FlightB,DDB,DTB),
     between(DeptDateA,DeptTimeA,DDB,DTB,ArrDateA,ArrTimeA).

The predicate "between" holds when the departure date and time of one flight is in the flight period of the other flight. This will generate, when this integrity constraint is reduced in the consistency phase of the computation, equality and inequality constraints on the values of the two variables "FlightA" and "FlightB".

Duty Period
The duty period is a continuous block of time when a crew member is on duty. The maximum duty period of a crew cannot exceed a certain amount (MaxDuty), which is (usually) less than or equal to 12 hours and it is calculated according to this crew's previous assignments. This is represented by the following integrity constraint in IC.

$\neg$ (assign(Crew,Flight),
     flight_period(Flight,Period), departure(Flight,DeptDate,DeptTime),
     MaxDuty(Crew,MD), DutyPeriod is MD-Period,
     subtract(DeptDate,DeptTime,DutyPeriod,NewDate,NewTime),
     on_duty(Crew,NewDate,NewTime)   ).

The predicate "on_duty" is defined in the model of the problem in the program P and decides if a crew is on duty on a particular date and time. Note that again the reduction of the predicates "on_duty", "between" and "between_flights" will generate a number of arithmetic constraints on the date and time of the flight that can be assigned to the selected crew.

Rest Period
There must be at least MinRest hours rest period between any two consecutive duties. MinRest is greater than or equal to 12 and it is calculated according to the previous assignments of the crew. This integrity constraint is given by the following logical formula in IC :

$\neg$ (assign(Crew,Flight),
     departure(Flight,DeptDate,DeptTime),
     on_new_duty(Flight,DeptDate,DeptTime),
     end_prev_duty(Crew,DeptDate,DeptTime,EndOfDuty),
     time_difference(EndOfDuty,DeptDate,DeptTime,RestPeriod),
     MinRest(Crew,MR), RestPeriod < MR   ).

where "on_new_duty" defines whether a flight is the beginning of a new duty period and "end_prev_duty" specifies the end of the duty which is immediately before some given date and time.

In addition, to general constraints as above there are often other requirements on the problem stemming from particular policies of the specific company and crew preferences. The abductive formulation with its modular representation of the problem, facilitates a direct representation of these either with additional integrity constraints in IC or through the further refinement of the model in P.

4.3. The Rescheduling Module

The problem of rescheduling is an important task in air-crew scheduling due to the frequent changes that occur in this environment, which often make the existing schedule inconsistent or of poorer quality. In the ALP framework we can develop a rescheduling system from the initial representation in <P,A,IC>. The sets of abducibles A and integrity constraints IC remain the same with the possibility of new constraints to be added, whereas P is extended to model the rescheduling process. The process of rescheduling, in ALP, is facilitated by the (natural) ability of abduction to reason with an existing set of hypotheses, in this case the old solution. Thus abduction can help to restore the consistency of the old solution with minimal changes on the affected assumptions and others depending on them.

The air-crew scheduling system presented above has been extended to model rescheduling. The rescheduling module was implemented to handle four general types of changes, flight cancellation, new flight addition, the rescheduling (e.g. change on the date or time) of an existing flight, or the selection of an alternative crew for a given flight. This module is interactive in the sense that the user can either suggest a crew for a particular flight, in which case the system will check if this choice is valid and to what extend it will affect the rest of the old schedule, or decide whether to accept a system proposed selection of crew. This rescheduling facility allows the system to be used (by an expert) both as a tool for improving the quality of an existing solution as well as a rescheduling tool that will adapt, on the day of operation, an old schedule to the changes that may occur. More details can be found in the extended version of this paper [12].

5. Evaluation of the System

An abduction-based system for air-crew scheduling has been implemented for the problem of Cyprus Airways Airlines in ALP on top of ECLiPSe [6]. Several experiments have been carried out in order to evaluate the system along the following three dimensions :
- The quality of the solution, compared to the solution of an expert.
- The flexibility of the system, and its ability to be specialized to additional or changing needs and preferences of the problem.
- The ability of the system to help the process of rescheduling.

The Cyprus Airways airline company, although small in size, contains the full complexity of the air-crew scheduling problem. The problem can be divided into a number of independent phases each of which is concerned with the scheduling of one class of crew members. For the purposes of these experiments we focus only on one type of crew, namely the captains. Cyprus Airways has 28 captains of several ranks (e.g. captains, training captains, technical pilots and flight managers). The system has been tested for busy, with a heavy flight schedule, and for less busy months of the year. The generated schedules were of good quality on the cost, as compared with an expert's solution (e.g. comparable number of deadheads, stay-overs etc.). Furthermore, our solutions were better in the balancing of the duty and flying hours assigned to each crew member over the month. This also affects (favourably) the total cost of the solution.

Other experiments were carried out with this system to show that equally good solutions can be found with less number of pilots. Below is an indicative table on some of these experiments (on a SUN Ultra 64Mb RAM), carried out for busy and less busy (ordinary) months of the year.

	Busy		Ordinary		
	28 plts	26 plts	28 plts	24 plts	21 plts
Time	560 sec	950 sec	145 sec	150 sec	152 sec
Deadheads	23	25	10	10	10
Stay-Overs	9	8	16	16	16
Transfers	83	85	39	39	39

The flexibility of the system is demonstrated through the addition of new integrity constraints which express requirements or preferences of the company or crew. These constraints are divided in three categories, namely the **local**, **group** or **global**, according to the number of resources (pilots) they involve.

Local integrity constraints refer to individual crew members and have no significant effect on the computational time. **Group** constraints involve a number of pilots which have something in common (e.g. of the same rank) and have a minor effect on the efficiency of the system. The third type of constraints, the **global** constraints, affect all crew members and can affect the computation by increasing the time from 10% up to 86%. In some cases, when these constraints make the problem very tight, the system is unable (within a reasonable time) to find a solution. Experiments were again carried out for busy and ordinary months of the year. The results are displayed in the table below.

	Local ICs	Group ICs	Global ICs
Ordinary	146 sec	160 sec	170 sec
Busy	560 sec	768 sec	1047 sec

It is important to note that, due to the modularity of the abductive system (as discussed in section 3.2 previously), such additional requirements are accommodated into the system with no extra (programming) effort. We simply express these in logical form as integrity constraints and add them to the system.

Rescheduling
Using the solutions generated for a busy and an ordinary month, a number of rescheduling experiments were carried out for each class of changes, as presented in section 2.2. The rescheduling module of the system is interactive and can be used as

an advisory to the operator, offering incrementally choices that would correct the old schedule. In each case, the operator is informed of other changes that would be needed if s/he accepts this choice.

On average for changes of the schedule (apart from cancelations which as expected take very little time) on an ordinary month the system suggests new assignments within 2-3 secs. For busy months the systems needs 3-6 secs on average. Again more details can be found in [12].

6. Conclusions and Future Work

In this paper we have presented an air-crew scheduling system based on Abduction. The system was developed and tested for a medium-sized application, that of Cyprus Airways Airlines, which, although average in size, it still contains the full complexity of the crew scheduling problem. Several experiments were carried out, using this system, with promising results.

The paper has explored the advantages of using an abductive approach for solving scheduling problems. It has demonstrated how the high-level modeling environment, offered by abduction, helped to develop a *modular and flexible* scheduling system: *modularity* with a clear separation of the two main issues of validity and quality (optimality) of the solution and *flexibility* under changes of the requirements of the problem. The modularity of the system allows the adoption of different strategies or heuristics which can lead to better quality of the solutions. The flexibility of the approach can facilitate the adaptation (or specialization) of the system to different airline companies.

We have also studied how abduction can help us to tackle the problem of adjusting or correcting existing solutions in order to accommodate changes in the application environment, namely the problem of rescheduling. An interactive rescheduling module has been integrated in the developed system that allows the operator to adjust or perfect a schedule to new (unexpected) requirements or additional specific needs of the company or crew. Thus the utility of the system can be summarized as follows:

- Generate a good first solution to the problem.
- Adjust a generated solution and improve its quality.
- Adapt an existing solution to unexpected changes occurring before or on the day of operation.

In the future we want to investigate how the AI techniques, that we have adopted in this work, and the system developed based on these can be applied for the solution of other air-crew scheduling problems in different airline companies. Future work can also study improvements on the computational efficiency of the system. One such interesting direction would be to integrate in the abductive process a specialized constraint solver, for the effective handling of the lower-level goals which are generated through the abductive reductions.

Acknowledgments

We would like to thank Cyprus Airways Airlines for their collaboration in the development of this application. Part of this work has been carried out with financial help from the EU, KIT program.

References

1.   L.Bianco, M Bielli, A.Mingozzi, S.Ricciardelli, M.Spadoni. A heuristic procedure for the crew rostering problem. European Journal of Operational Research, 58:272-283, 1992.
2.   G.Brewka. Principles of Knowledge Representation. CSLI Publications, 1996.
3.   L. Console, L. Portinale and D. Dupré. Using Compiled knowledge to guide and focus abductive diagnosis. Journal of IEEE Transactions on Knowledge and Data Engineering, 8(5):690-706, 1996.
4.   L.Console, M.L.Sapino, D.Theseider Dupre. The role of abduction in database view updating. Journal of Intelligent Systems, 1994.
5.   M.Denecker and D.De Schreye. SLDNFA: an abductive procedure for abductive logic programs. Journal of Logic Programming, 1997.
6.   ECLiPSe User Manual. ECRC Munich, Germany, 1993.
7.   F.Focacci, E.Lamma, P.Mello and M.Milano. Constraint Logic Programming for the Crew Rostering Problem. In Proceedings of the Third International Conference on the Practical Applications of Constraint Technology, PACT'97, pp. 151-164, 1997.
8.   N.Guerinik, M.Van Caneghem. Solving Crew Scheduling Problems by Constraint Programming. Proc. 1st International Conference on Principles and Practice of Constraint Programming, 1995.
9.   M.Kress, B.Golany. Optimizing the assignment of aircrews to aircraft in an airlift operation. European Journal of Operational Research, 77:475-485, 1994.
10. A.C.Kakas, R.A.Kowalski, F.Toni. Abductive Logic Programming. Journal of Logic and Computation 2(6), 1993.
11. A.C.Kakas, A.Michael. Integrating Abductive and Constraint Logic Programming. Proc. 12th International Conference on Logic Programming, MIT Press, pgs. 399-415, 1995.
12. A.C.Kakas, A.Michael. An abductive-Based Scheduler for Air-Crew Assignment. Department of Computer Science, University of Cyprus, Technical Report TR-98-17, 1998.
13. T.Menzies. Applications of Abduction: Knowledge-Level Modeling. International Journal of Human Computer Studies, August, 1996.
14. D.Poole, R.G.Goebel, Aleliunas, Theorist: a logical reasoning system for default and diagnosis. The Knowledge Fronteer: Essays in the Representation of Knowledge, (Cercone and McCalla eds), Springer Verlag Lecture Notes in Computer Science, 1987.
15. C.Pavlopoulou, A.P.Gionis, P.Stamatopoulos, C.Halatsis. Crew Pairing Optimization Based on CLP. Proc. 2nd International Conference on the Practical Application of Constraint Technology, London, UK, 1996.
16. D.Poole. A logical framework for default reasoning. Artificial Intelligence vol.36, 1988.
17. M.P.Shanahan. Event Calculus Planning Revisited. Proceedings of the 1997 European Conference on Planning (ECP 97), 1997.

Unfold/Fold Inductive Proof: An Extension of a Folding Technique

Rym Salem and Khaled Bsaïes

Faculté des Sciences de Tunis Campus Universitaire,
1060, Le Belvédère, Tunis Tunisia
email : Khaled.Bsaies@fst.rnu.tn

1   Introduction

The folding transformation plays an important role as well in the process of improving programs by transformations [TS84, BD77] as in performing proofs [SM88, BM79]. The success of the folding of an initial member (conjunctions of atoms) in an other obtained from it after applying some unfoldings is in general not decidable [PP89b]. This problem was studied in the context of transforming logic programs for quite restrictive specifications (members and programs) [PP89b, Ale92]. We propose in this paper a comparison between the different approaches for resolving the folding problem, and we propose an extension in order to deal with more general specifications.

2   Preliminaries

In the following, we assume that the reader is familiar with the basic terminology of first order logic such as *term, atom, formula, substitution, matching, most general unifier* (mgu) and so on. An *expression* is either a term, atom, conjunction of atoms. We assume knowledge of the semantics of Prolog, minimum Herbrand model...[Llo88], [AVE82].

Definition 1 Implicative formula, member. An implicative formula is a first order formula of the form:

(1) $\forall X (\exists Y \Delta \leftarrow \exists Z \Lambda)$ where: Δ and Λ are conjunctions of atoms.
$X = Var(\Delta) \cap Var(\Lambda)$, $Y = Var(\Delta) \setminus Var(\Lambda)$ and $Z = Var(\Lambda) \setminus Var(\Delta)$. A such formula is abbreviated by $\Delta \Leftarrow \Lambda$. Δ and Λ are called members. Δ is the left member and Λ s the right member.

Definition 2 Validity of formula. The logic formula (1) is valid in $\mathcal{M}(P)$ (least Herbrand model), if the following condition holds:
for all ground substitution σ of the variables of X if there exists a ground substitution θ of the variables of Z such that $(\Lambda\sigma)\theta$ is valid in $\mathcal{M}(P)$ then there exists a ground substitution δ of the variables of Y such that $(\Delta\sigma)\delta$ is valid in $\mathcal{M}(P)$

3 The folding problem

3.1 Formalization of the problem

Let Δ_1 be a member. Δ_1 is said to be foldable if there exists a sequence of members $\Delta_2, \ldots, \Delta_n$ such that:
- for i=2...n, Δ_i is obtained by unfolding (The unfolding used here follows the definition of the unfolding transformation proposed in [TS84]) an atom among Δ_{i-1},
- $\Delta_n = \Gamma(\Delta_1 \sigma) \Pi$, such that Γ and Π are conjunctions of atoms and σ is a substitution called the folding substitution of Δ_1.

3.2 Criterion of the folding success

The fact that Δ_1 appears in Δ_n has as consequence that the unfolded atoms are defined by recursive clauses. The existence of the folding substitution σ depends on:
- the inductive relation defined in these clauses,
- the instantiation of terms in the unfolded atoms,
- the relationship between the variables in the different atoms of Δ_1.

The crucial problem is the choice, at each step, of the atom to be unfolded. Such a choice needs to define a selection rule.

We note that the considered members Δ_i are those obtained by using recursive clauses of the definitions of predicate symbols of the unfolded atoms.

4 Selection rules

In the literature, two selection rules were defined for a member representing the body of a clause. We recall in this section these two approches and we present a comparison between them.

4.1 The results of Pettorossi and Proieitti

A selection rule called SDR for Synchronized Descent Rule, was introduced by Pettorossi and Proietti [PP89a]. This rule allows to guide the unfolding process. It is in some cases a decision procedure for the folding problem.

Given an ordering on atoms, the SDR rule selects the greater atom with respect to this ordering. One of the problems with this method is the definition of a suitable ordering of atoms.

4.2 The results of Alexandre

A schema based approach was proposed by Alexandre [Ale92] in order to characterize the clauses of a program.

Thanks to these schemas, a static analysis of the program permits to define a selection rule for always choosing an atom among the initial member to be folded Δ_1.

Let $\Delta_1 : p_1(u_{11}, u_{12}, \ldots, u_{1n_1}) \ldots p_k(u_{k1}, u_{k2}, \ldots, u_{kn_k})$ be the member to be folded such that, all the predicate symbols p_i are defined by linear recursive clauses having the following form $p_i(T_{i1}, T_{i2}, \ldots, T_{in_i}) \leftarrow \Delta\ p_i(t_{i1}, t_{i2}, \ldots, t_{in_i})\ \Theta$. The defined schemas characterize the couples (T_{ij}, t_{ij}). The schema of the considered recursive clause corresponds to the set of couples (T_{ij}, t_{ij}). Conditions regarding the folding are determined in terms of the numbers n_i of unfoldings of the atoms $p_i(u_{i1}, u_{i2}, \ldots, u_{in_i})$ of Δ_1. From the study of the elementary members constructed of two atoms projected according to only one argument, the global conditions of the folding of Δ_1 are determined by using the projections of the atoms of Δ_1 according to the arguments containing the variables having more then one occurrences.

Program	Member to fold	Conditions
$p_1(T_1) \leftarrow p_1(t_1)$ $p_2(T_2) \leftarrow p_2(t_2)$	$p_1(u_1(x))p_2(u_2(x))$	n_1, n_2

The unfoldings to be done in order to succeed the folding (when it is possible), are determined by resolving the following equation system having the k unknown $n_1, \ldots, n_k$:

$$\mathcal{S} \begin{cases} condition(n_i, n_j) \\ n_1 + \ldots + n_k \geq 1 \end{cases}$$

4.3 Comparison and choice of an approach

The SDR rule is a general rule which can be applied to all type of programs. However, it presents the following limitations:
• the choice criterion of the atom to be unfolded seems to be arbitrary with respect to the criterion of the success of the folding: it is rather a heuristic,
• the choice of a suitable atom ordering is not obvious ,
• the SDR function is not totally defined when the atom ordering is partial,
• the atoms to be unfolded are known by comparing atoms in the current member, thus it is not possible to know these atoms from the starting point (initial member).

The schema based approach presents essentially the following advantages:
• the choice criterion of the atoms to be unfolded is objective with respect to the criterion of the success of the folding,
• the schemas are a compilation of the pertinent informations of the program,
• a static study of the program is possible,
• the approach is incremental, new schemas can be added without influencing the old ones.

Our goal is to resolve the folding problem for members with respect to more general programs, the schema based approach proposed by Alexandre suits our purpose.

In the following, we propose an extension of the schema based approach for a more general class of members and programs.

5 Proposition of an extension

5.1 Restrictions done by Alexandre

Let $\Delta_1 : p_1(u_{11}, u_{12}, \ldots, u_{1n_1}) \ldots p_k(u_{k1}, u_{k2}, \ldots, u_{kn_k})$ be the member to be folded, such that all the predicate symbols p_i are defined by linear recursive clauses having the following form: $p(T_1, T_2, \ldots, T_n) \leftarrow \Delta\ p(t_1, t_2, \ldots, t_n)\ \Theta$. Alexandre restricted his study to the following clauses and members:

- Inductive schemas: $t_i \ll T_i$, $t_i = T_i$, $t_i \notin \mathcal{V}ar(T_i)$. These cases correspond respectively to the three arguments of the definition of the *times* predicate:
$times(s(x), y, z) \leftarrow times(x, y, t)plus(y, t, z)$,
- $t_i \ll T_i$, otherwise t_i is a variable,
- u_{ij} is a variable,
- $p_i(u_{i1}, u_{i2}, \ldots, u_{in_i})$ is linear,
- if x is a variable having two occurrences in two decreasing positions, then they decrease in the same manner.

5.2 The proposed extension

We note that the considered schemas list is not exhaustive, however a large number of common specifications are written following these schemas. We then consider only these inductive schemas.

The assumption, that terms used in the recursive atoms of the bodies of clauses are all variables, is quite restrictive. We propose to extend the method by considering any terms. In order to manipulate these terms, we propose to explicit their form. The determination of the form of terms in the general case is quite difficult, we restrict our study to terms representing the natural numbers and lists.

Moreover, the used clauses have any decrease, and the atoms of the member to be folded are linear.

5.3 Representation of terms

The idea we propose to extend the results of Alexandre consists on representing terms by the minimum information, sufficiently enough to study the behavior of the unfolding and the success of the folding of members containing these terms. Thus we explicit the form of terms in terms of their type constructors:

- a term t having as type the *natural number* type is written $s(s(...(s(x))...)$,
- a term t having as type the *list* type is written $cons(a_1, cons(a_2, ..., cons(a_n, l)...)$.

Note that two informations are necessary for representing terms:
- the inductive variable of the term (x and l), having the same type of the term,
- the size of the term (number of constructions) with respect to the inductive variable.

For the natural number these informations are sufficient. For the list, these two informations could be sufficient by considering other conditions concerning Δ_1 and P. We give here not exhaustively some necessary conditions:

- the atoms of Δ_1 are linear
- if $t = cons(a_1, cons(a_2, ..., cons(a_n, l)...)$ is a term of the member Δ_1, the a_i's are variables having only one occurrence in Δ_1

We consider the case where these informations are sufficient, therefore we denote terms by

- $s(s(...s(x)...))$ is represented by $s^n(x)$
- $cons(a_1, cons(a_2, ..., cons(a_n, l)...)$ is represented by $cons^n(l)$

In the general case, a term t is denoted $c^n(x)$, where x is a variable (of natural number type or list type) and c is the constructor of x.

This abstraction can be extended to members and programs, we then consider $\widetilde{P}$ and $\widetilde{\Delta_1}$ as the program and the member obtained from P and Δ_1 by applying the proposed term representation. We propose to study the conditions of the folding of $\widetilde{\Delta_1}$ with respect to $\widetilde{P}$. The results concerning the folding of $\widetilde{\Delta_1}$ with respect to $\widetilde{P}$ are transposed for the folding of Δ_1 with respect to P, since we are in the case where $c^n(x)$ represents sufficiently the initial term.

5.4 Conditions of the folding

According to the representation of the terms, a term $t = c^n(x)$ is denoted, following its inductive schema where it appears in Δ_1, as
- $\tau_{k,r}(x^n) : (c^{k+r}(x),\ c^k(x))$,
- $\iota_k(x^n) : (c^k(x),\ c^k(x))$,
- $\omega_k(x^n) : (c^k(x),\ z)$.

with x^n is denoted x if n=0.

The results we have proposed here do not concern only variables having more then one occurrence in a member, but also variables having only one occurrence. In fact, this limitation has been done by Alexandre, since the terms he consider are reduced to variables, a matching is always possible from variable to a term. The variables to be studied are those having several occurrences and those having one simple occurrence appearing in constructed terms. The results we have found are summarized in the tableau 1.

Example 1. The predicate *plus* expressing the natural number addition is specified by the following program P:

Schéma	Condition
$\tau_{k,r}(x^a)$	$(a \leq k) \vee (n = 0)$
$\iota_k(x^a)$	none
$\omega_k(x^a)$	$(k = 0) \vee (n = 0)$
$\{\ \tau_{k_1,r_1}(x^a),\ \tau_{k_2,r_2}(x^b)\ \}$	$(a > k_1) \wedge (b > k_2)$ $n_1 = n_2 = 0$ $(a \leq k_1) \vee (b \leq k_2)$ $n_1 * r_1 = n_2 * r_2$
$\{\ \iota_{k_1}(x^a),\ \tau_{k_2,r_2}(x^b)\ \}$	$n_2 = 0$
$\{\ \omega_{k_1}(x^a),\ \tau_{k_2,r_2}(x^b)\ \}$	$n_1 = n_2 = 0$
$\{\ \iota_{k_1}(x^a),\ \omega_{k_2}(x^b)\ \}$	$n_2 = 0$

Table 1. Elementary schemas of members having linear atom and constructed terms

$$P \begin{cases} plus(0, x, x) & \leftarrow \\ plus(s(x), y, s(z)) & \leftarrow plus(x, y, z) \end{cases}$$

The schema of the recursive clause of the definition of P is $(\tau_{0,1}, \iota_0, \tau_{0,1})$. Let $\Delta_1 : plus(s(y), z, t)plus(x, t, w)$ be the member to be folded. The variables to be considered are y and t. These variables are in the schemas $\tau_{0,1}(y^1)$ and $\{\tau_{0,1}(t), \iota_0(t)\}$. The associated conditions for the folding are respectively $n_1 = 0$ (first line of the tableau 1) and $n_1 = 0$ (fifth line of the tableau 1). The system of the conditions of the folding is then:

$$\mathcal{S} \begin{cases} n_1 = 0 \\ n_1 + n_2 \geq 1 \end{cases}$$

the minimal solution is $(n_1, n_2)=(0, 1)$. By unfolding one time the second atom of Δ_1 we obtain the member $\Delta_2 : plus(s(y), z, t)plus(x, t, w)$, Δ_1 is foldable in Δ_2 with the folding substitution ε

6 Application

We are interested in applying the folding results for proving implicative formulas $\varphi : \Gamma \Leftarrow \Delta$, such that Δ and Γ are conjunctions of atoms defined by clauses of a logic program P. The used transformations are the unfolding left, the unfolding right, the folding left, the folding right and the simplification defined in [SM88].

The proof is based on an implicit induction, the inductive relations used are defined by the clauses of the program P. The proof process is guided by reducing differences between the two member of the implication. Thus, when we obtain the same member in both the two parts of the implication, a simplification is then possible. Obtaining a formula having same members (left and right), is allowed by the folding of the current member in the initial formula.

We consider the proof of the associativity property of the predicate *plus*. This property can be expressed by the following formula:

$$\varphi_1 : plus(x, y, t)plus(t, z, w) \Leftarrow plus(y, z, u)plus(x, u, w)$$

The program P defining the clauses specifying the addition is that of the example 1. We look for folding the right member Δ. the terms of this member are variables, then we only consider the variable u (having two occurrences in Δ) in order to determine the conditions of the folding of Δ. The schemas of occurrences of u are $(\tau_{0,1}(u), \iota_0(u))$. The condition of the folding associated to this schema is $n_1=0$ (fifth line of the table 1). The system of the folding conditions is then:

$$\mathcal{S} \begin{cases} n_1 = 0 \\ n_1 + n_2 \geq 1 \end{cases}$$

its minimal solution is $(n_1, n_2)=(0, 1)$. By unfolding one time the second atom of Δ we obtain the formula

$$\varphi_2 : plus(s(x), y, t)plus(t, z, s(w)) \Leftarrow plus(y, z, u)plus(x, u, w).$$

The member Δ of φ_1 is foldable in the right member of φ_2 with the folding substitution ε. A right folding of φ_1 in φ_2 gives the formula

$$\varphi_3 : plus(s(x), y, t)plus(t, z, s(w)) \Leftarrow plus(x, y, k)plus(k, z, w).$$

We note that the members of φ_3 resemble each other. If we unfold one time each atom of the left member of φ_3, we obtain the formula

$$\varphi_5 : plus(x, y, t)plus(t, z, w) \Leftarrow plus(x, y, k)plus(k, z, w).$$

The simplification of the members of φ_5 is possible thanks to the substitution $\theta=\{t|k\}$ and gives an empty clause. This is the proof in th inductive case of the associativity of the predicate *plus*.

7   The theorem prover

The theorem prover used follows from the work of Kanamori and Seki for the verification of Prolog programs using an extension of execution [KS86]. This prover is able to prove by induction some first order formulas of the form $\Delta \Leftarrow \Lambda$, where Δ and Λ are conjunctions of atoms.

The general idea of the strategies is to apply some unfolding and a folding on the two parts of the implicative formula in order to obtain formulas in the form $\Gamma \Leftarrow \Gamma$ or $\Gamma \Leftarrow false$ or $true \Leftarrow \Gamma$ (i.e. the formula true). We describe the deduction rules in the form of inference rule $\frac{e}{f}(NAME)$, in the following e and f are implicative formulas or a set of implicative formulas, $Name$ is the name of the rule, this means that in the proof process; to prove e it is sufficient to prove f. Moreover for every rule we give the conditions of application.

Definition 3. Simplification

$$\frac{\Delta,\ A\ \Leftarrow\ \Gamma,\ B}{(\Delta\ \Leftarrow\ \Gamma)\theta}\ (SIMP)$$

θ is an existential-mgu (i.e. the domain of θ is a subset of the existential variables set of π) of A and B.

Definition 4 Unfolding right.

$$\frac{\Gamma\ \Leftarrow\ \Delta,A}{\{\,(\Gamma\ \Leftarrow\ \Delta,\Delta_i)\theta_i\,,\,i\in[1,..k]\}}\ (UNF\_R)$$

$E=\{c_1,...c_k\}$ is the set of clauses of the program P such that $c_i\ :\ B_i\leftarrow\Delta_i$ and there exists $\theta_i=mgu(B_i,A)$.

If $E=\emptyset$, then we generate the formula $\Gamma\Leftarrow false$, that can be reduced to the formula $true$.

Definition 5. Unfolding left

$$\frac{\Gamma,A\ \Leftarrow\ \Delta}{\{(\Gamma,\ \Delta_i)\theta_i\ \Leftarrow\ \Delta,\ i\in[1,..k]\}}\ (UNF\_L)$$

$E=\{c_1,...c_k\}$ is the set of the clauses of the program P such that $c_i\ :\ B_i\leftarrow\Delta_i$ and such that there exists $\theta_i=existential\text{-}mgu(B_i,A)$.

Definition 6. Folding right

$$\begin{array}{c}\Lambda\Leftarrow\Pi\\[4pt]\dfrac{\Gamma\ \Leftarrow\ \Delta_1,\Delta_2}{(\Gamma\Leftarrow\Lambda\theta,\Delta_2)}\ (FOL\_R)\end{array}$$

• θ is a substitution such that $\Pi\theta=\Delta_1$
• the formula is $\Lambda\Leftarrow\Pi$ is an ancestor of the formula $\Gamma\Leftarrow\Delta_1,\Delta_2$ in the proof tree.

Definition 7. Folding left

$$\begin{array}{c}\Lambda\Leftarrow\Pi\\[4pt]\dfrac{\Gamma_1\Gamma_2\ \Leftarrow\ \Delta}{(\Pi\theta,\Gamma_2\Leftarrow\Delta)}\ (FOL\_L)\end{array}$$

• θ is a substitution such that $\Lambda\theta=\Gamma_1$
• the formula $\Lambda\Leftarrow\Pi$ is an ancestor of the formula $\Gamma_1\Gamma_2\Leftarrow\Delta$ in the proof tree.

8 A complete example

This example is intended to illustrate the proposed algorithm. Show the binomial theorem:

$$(x+1)^n = \sum_{k=0}^{n} C_n^k x^k, \quad n \geq 0$$

The definite program under consideration, specifies predicates to express the binomial theorem.

The program specifying the binomial example.

(1) $plus(0, x, x)$ $\leftarrow$
(2) $plus(s(x), y, s(z)))$ $\leftarrow plus(x, y, z)$

(3) $mul(0, x, 0)$ $\leftarrow$
(4) $mul(s(x), y, z)$ $\leftarrow mul(x, y, t)plus(y, t, z)$

(5) $power(0, x, s(0))$ $\leftarrow$
(6) $power(s(x), y, z)$ $\leftarrow power(x, y, t)mul(y, t, z)$

(7) $ps([\,], z, z)$ $\leftarrow$
(8) $ps(z, [\,], z)$ $\leftarrow$
(9) $ps([x_1|z_1], [x_2|z_2], [x|z]) \leftarrow ps(z_1, z_2, z)plus(x_1, x_2, x)$

(10) $bin(0, [s(0)])$ $\leftarrow$
(11) $bin(s(x), z)$ $\leftarrow bin(x, y)ps([0|y], y, z)$

(12) $seq(x, [\,], 0)$ $\leftarrow$
(13) $seq(x, [y|z], w)$ $\leftarrow seq(x, z, t)mul(x, t, a)plus(y, a, w)$

The formula to be proved is the following:

$$\varphi \ ps(k, l, t), seq(x, t, w) \Leftarrow seq(x, k, a), seq(x, l, b), plus(a, b, w)$$

9 Conclusion

Our investigation has its original motivation in producing efficient logic programs by way of transformations and proofs [AB97]. In unfold/fold transformation, efficiency is often improved by performing folding step. This paper addresses aspects of these related problem.

The interesting feature of the work presented here is an extension of Alexandre results concerning a schema based technique for studying the folding problem. This extension will permit to pursue this work for mechanizing the proof strategies.

References

[AB97]   F. Alexandre and K. Bsaïes. A Methodology for constructing Logic Programs. *Fundamenta Informaticae*, 29(3):203–224, 1997.

[Ale92] F. Alexandre. A Technique for Transforming Logic Programs by Fold-Unfold Transformations. In M. Bruynooghe and M. Wirsing, editors, *PLILP92*, volume 631 of *Lecture Notes in Computer Science*, pages 202–216. Springer-Verlag, August 26 - 28 1992. Leuven, Belgium.

[AVE82] K.R. Apt and M.H Van Emden. Contribution to the Theory of Logic Programming. *Journal of the Association for Computing Machinery*, 29(3):841–862, 1982.

[BD77]   R.M Burstall and J.A Darlington. Transformation System for Developing Recursive Programs. *Journal of the Association for Computing Machinery*, 24(1):44–67, 1977.

[BM79]   R.S. Boyer and J.S. Moore. *Computational Logic*. Academic Press, 1979.

[KS86]   T. Kanamori and H. Seki. Verification of Prolog Programs Using an Extension of Execution. In *3rd International Conference on Logic Programming*, pages 475–489. LNCS 225, 1986.

[Llo88] J.W. Lloyd. *Foundations of Logic Programming*. Springer-Verlag, 1988.

[PP89a] A. Pettorossi and M. Proietti. The Automatic Construction of Logic Programs. In *IFIP WG2.1 Meeting*, January 1989. Preliminary Version.

[PP89b] A. Pettorossi and M. Proietti. Decidability Results and Characterization of Strategies for the Development of Logic Programs. In G. Levi and M. Martelli, editors, *6th International Conference on Logic Programming*, Lisbon (Portugal), 1989. MIT Press.

[SM88]   A. Sakuraï and H. Motoda. Proving Definite Clauses without Explicit Use of Inductions. In K. Furukawa, H. Tanaka, and T Fujisaki, editors, *Proceedings of the 7th Conference, Logic Programming '88*, Tokyo, Japan, April 1988. LNAI 383, Springer-Verlag.

[TS84]   H. Tamaki and T. Sato. Unfold/Fold Transformation of Logic Programs. In Sten-Åke Tärnlund, editor, *Proceedings of the Second International Logic Programming Conference*, pages 127–138, Uppsala, 1984.

On Solving the Capacity Assignment Problem Using Continuous Learning Automata*

B. John Oommen and T. Dale Roberts

School of Computer Science
Carleton University
Ottawa ; Canada : K1S 5B6.

Abstract. The Capacity Assignment problem focuses on finding the best possible set of capacities for the links that satisfy the traffic requirements in a prioritized network while minimizing the cost. Apart from the traditional methods for solving this NP-Hard problem, one new method that uses Learning Automata (LA) strategies has been recently reported. This method uses discretized learning automata [1]. The present paper shows how the problem can be solved using continuous learning automata. The paper considers the realistic scenario when different classes of packets with different packet lengths and priorities are transmitted over the networks. After presenting the two well-known solutions to the problem (due to Marayuma and Tang [2], and Levi and Ersoy [3]) we introduce our new method that uses continuous LA, which is comparable to the discretized version, and is probably the fastest and most accurate scheme available.

I. INTRODUCTION

In this paper we study the Capacity Assignment problem which focuses on finding the best possible set of capacities for the links that satisfies the traffic requirements in a prioritized network while minimizing the cost. Unlike most approaches, which consider a single class of packets flowing through the network, we base our study on the more realistic assumption that different classes of packets with different packet lengths and priorities are transmitted over the networks. Apart from giving a brief overview of the problem and a report of the existing schemes, we present a new continuous learning automata strategy for solving the problem. This strategy is analogous to the discretized version already reported [1] except that it operates in a continuous probability space, and is thus easier to comprehend since it is more akin to the well-studied families of learning automata.

Data networks are divided into three main groups which are characterized by their size, these are Local Area Networks (LANs), Metropolitan Area Networks (MANs) and Wide Area Networks (WANs). An Internetwork is comprised of several of these networks linked together, such as the Internet. Most applications of computer networks deal with the transmission of logical units of information or messages, which

---

* Partially supported by the Natural Sciences and Engineering Research Council of Canada. Contact e-mail address : oommen@scs.carleton.ca.

are sequences of data items of arbitrary length. However, before a message can be transmitted it must be subdivided into packets. The simplest form of a packet is a sequence of binary data elements of restricted length, together with addressing information sufficient to identify the sending and receiving computers and an error correcting code.

There are several tradeoffs to be considered when designing a network system. Some of these are difficult to quantify since they are criteria used to decide whether the overall network design is satisfactory. This decision is based on the designer's experience and familiarity with the requirements of the individual system. As there are several components to this area, a detailed examination of the pertinent factors, which are primarily cost and performance, can be found in [4] and [5].

In the process of designing computer networks the designer is confronted with a trade-off between costs and performance. Some of the parameters effecting the cost and performance parameters used in a general design process are listed above, but, in practice, only a subset of these factors are considered in the actual design. In this paper we study scenarios in which the factors considered include the location of the nodes and potential links, as well as possible routing strategies and link capacities.

The **Capacity Assignment (CA) Problem** specifically addresses the need for a method of determining a network configuration that minimizes the total cost while satisfying traffic requirements across all links. This is accomplished by selecting the capacity of each link from a discrete set of candidate capacities that have individual associated cost and performance attributes. Although problems of this type occur in all networks, in this paper, we will only examine the capacity assignment for prioritized networks. In prioritized networks, packets are assigned to a specific priority class which indicates the level of importance of their delivery. Packets of lower priority will be given preference and separate queues will be maintained for each class.

The currently acclaimed solutions to the problem are primarily based on heuristics that attempt to determine the lowest cost configuration once the set of requirements are specified. These requirements include the topology, the average packet rate, or the routing, for each link, as well as the priorities and the delay bounds for each class of packets. The result obtained is a capacity assignment vector for the network, which satisfies the delay constraints of each packet class at the lowest cost.

The primary contribution of this paper is to present a continuous Learning Automaton (LA) solution to the CA problem. Apart from this fundamental contribution of the paper, the essential idea of using LA which have actions in a "meta-space" (i.e., the automata decide on a *strategy* which in turn determines the physical action to be taken in the real-life problem) is novel to this paper and its earlier counterpart [1]. This will be clarified in Section IV.

I.1 ASSUMPTIONS AND DELAY FORMULAE

The model used for all the solutions presented have the following features [3] :
1. Standard Assumptions : (a) The message arrival pattern is Poissonly distributed, and (b) The message lengths are exponentially distributed.
2. Packets : There are multiple classes of packets, each packet with its own (a) Average packet length, (b) Maximum allowable delay and (c) Unique priority level, where a lower priority takes precedence.

3. Link capacities are chosen from a finite set of predefined capacities with an associated fixed setup cost, and variable cost/km.
4. Given as input to the system are the (a) Flow on each link for each message class, (b) Average packet length measured in bits, (c) Maximum allowable delay for each packet class measured in seconds, (d) Priority of each packet class, (e) Link lengths measured in kilometers, and (f) Candidate capacities and their associated cost factors measured in bps and dollars respectively.
5. A non-preemptive FIFO queuing system [6] is used to calculate the average link delay and the average network delay for each class of packet.
6. Propagation and nodal processing delays are assumed to be zero.

Based on the standard network delay expressions [3], [6], [7], all the researchers in the field have used the following formulae for the network delay cost :

$$T_{jk} = \frac{\eta_j \cdot \left(\sum_l \frac{\lambda_{jl} \cdot m_l}{\eta_j \cdot C_j} \right)^2}{(1 - U_{r-1})(1 - U_r)} + \frac{m_k}{C_j} \tag{1.1}$$

$$U_r = \sum_{l \in V_r} \frac{\lambda_{jl} \cdot m_l}{C_j} \tag{1.2}$$

$$Z_k = \frac{\sum_j T_{jk} \cdot \lambda_{jk}}{\gamma_k}. \tag{1.3}$$

In the above, T_{jk} is the Average Link Delay for packet class k on link j, U_r is the Utilization due to the packets of priority 1 through r (inclusive), V_r is the set of classes whose priority level is in between 1 and r (inclusive), Z_k is the Average Delay for packet class k, $\eta_j = \sum_l \lambda_{jl}$ is the Total Packet Rate on link j, $\gamma_k = \sum_j \lambda_{jk}$ is the Total Rate of packet class k entering the network, λ_{jk} is the Average Packet Rate for class k on link j, m_k is the Average Bit Length of class k packets, and C_j is the Capacity of link j. As a result of the above, it can be shown that the problem reduces to an integer programming problem, the details of which can be found in [4].

II. PREVIOUS SOLUTIONS

II.1 THE MARAYUMA -TANG SOLUTION

The Marayuma/Tang (MT-CA) solution to the Capacity Assignment (CA) problem [2] is based on several low level heuristic routines adapted for total network cost optimization. Each routine accomplishes a specific task designed for the various phases of the cost optimization process. These heuristics are then combined, based on the results of several experiments, to give a composite algorithm. We briefly describe each of them below but the details of the pseudocode can be found in [2], [3] and [4].

There are two initial capacity assignment heuristics, SetHigh and SetLow:

(a) **SetHigh**: In this procedure each link is assigned the maximum available capacity.

(b) **SetLow**: On invocation each link is assigned the minimum available capacity.

The actual cost optimization heuristics, where the motivating concept is to decide on increasing or decreasing the capacities using various cost/delay trade-offs, are:

(a) **Procedure AddFast**: This procedure is invoked when all of the packet delay requirements are not being satisfied and it is necessary to raise the link capacities while simultaneously raising the network cost, until the bounds are satisfied.

(b) **Procedure DropFast**: This procedure is invoked when all of the packet delay requirements are being satisfied but it is necessary to lower the link capacities, and thus lower the network cost, while simultaneously satisfying the delay bounds.

(c) **Procedure Exc**: This procedure attempts to improve the network cost by pairwise link capacity perturbations.

To allow the concatenation of these heuristics, the algorithm provides two interfaces, ResetHigh, used by DropFast, and ResetLow, used by AddFast below :

(a) **ResetHigh**: Here the capacity of each link is increased to the next higher one.

(b) **ResetLow**: Here the capacity of each link is decreased to the next lower one.

After performing several experiments using these heuristics on a number of different problems, Marayuma/Tang determined that a solution given by one heuristic can often be improved by running other heuristics consecutively. The MT-CA algorithm is the best such composite algorithm (see [2], [3] and [4]).

II.2 THE LEVI/ERSOY SOLUTION

To our knowledge the faster and more accurate scheme is the Levi/Ersoy solution to the CA problem (LE-CA) [3], based on simulated annealing. The process begins with an initial random, feasible solution and creates neighbor solutions at each iteration. If the value of the objective function of the neighbor is better than that of the previous solution, the neighbor solution is accepted unconditionally. If, however, the value of the objective function of the neighbor solution is worse than the previous solution it is accepted with a certain probability. This probability is the **Acceptance Probability** and is lowered according to a distribution called the **Cooling Schedule**.

Since the simulated annealing process is a multi-purpose method, its basic properties must be adopted for the CA problem. In this case, the solution will be a **Capacity Assignment Vector**, C, for the links of the network. Therefore, $C = (C_1, C_2, C_3, ..., C_i, ..., C_m)$ where m is the total number of links and C_i takes a value from the set of possible link types/capacities. The objective function is the minimization of the total cost of the links. Neighbor solutions, or assignment vectors, are found by first selecting a random link and randomly increasing or decreasing its capacity by one step. Feasibility is constantly monitored and non-feasible solutions are never accepted. The pseudocode for the actual algorithm is given in [3] and [4].

III. LEARNING AUTOMATA

Learning Automata have been used to model biological learning systems and to find the optimal action which is offered by a random environment. The learning is accomplished by actually interacting with the environment and processing its responses to the actions that are chosen, while gradually converging toward an ultimate goal. A complete study of the theory and applications of this subject can be found in [8], [9].

The learning loop involves two entities, the **Random Environment** (RE) and a **Learning Automaton** (LA). Learning is achieved by the automaton interacting with the environment by processing responses to various actions and the intention is that the LA learns the optimal action offered by the environment.

The actual process of learning is represented as a set of interactions between the RE and the LA. The LA is offered a set of actions $\{\alpha_1, ..., \alpha_r\}$ by the RE it interacts with, and is limited to choosing only one of these actions at any given time. Once the LA decides on an action α_i, this action will serve as input to the RE. The RE will then respond to the input by either giving a **reward**, ('0'), or a **penalty**, ('1'), based on the **penalty probability** c_i associated with α_i. Based upon the response from the RE and the current information it has accumulated so far, the LA decides on its next action and the process repeats. The intention is that the LA learns the **optimal action** (that is, the action which has the minimum **penalty probability**), and eventually chooses this action more frequently than any other action.

Variable Structure Stochastic Automata (VSSA) can be described in terms of time-varying transition and output matrices. However, they are usually completely defined in terms of **action probability updating schemes** which are either **continuous** (operate in the continuous space [0, 1]) or **discrete** (operate in steps in the [0, 1] space). The action probability vector $P(n)$ of an r-action LA is $[p_1(n), ..., p_r(n)]^T$ where, $p_i(n)$ is the probability of choosing action α_i at time 'n', and satisfies $0 \le p_i(n) \le 1$, and the sum of $p_i(n)$ is unity.

A VSSA can be formally defined as a quadruple (α, P, β, T), where α, P, β, are described above, and T is the updating scheme. It is a map from $P \times \beta$ to P, and defines the method of updating the action probabilities on receiving an input from the RE. Also they can either be ergodic or absorbing in their Markovian behavior. Since we require an absorbing strategy, the updating rule we shall use is for the Linear Reward-Inaction (L_{RI}) scheme. The updating rules for the L_{RI} scheme are as follows, and its ε-optimal and absorbing properties can be found in [8], [9].

$$p_i(n+1) = 1 - \sum_{j \ne i} \lambda_r p_j(n) \qquad \text{if} \qquad \alpha_i \text{ is chosen and } \beta=0$$
$$p_j(n+1) = \lambda_r p_j(n) \qquad \text{if} \qquad \alpha_i \text{ is chosen and } \beta=0$$
$$p_j(n+1) = p_j(n) \qquad \text{if} \qquad \alpha_i, \alpha_j \text{ chosen, and } \beta=1,$$

where $\lambda_r (0 < \lambda_r < 1)$ is the parameter of the scheme. Typically, λ_r is close to unity.

IV. THE CONTINUOUS AUTOMATA SOLUTION TO CA

We now propose a continuous LA which can be used to solve the CA problem. The Continuous Automata Solution to CA (CASCA) algorithm is faster than the MT and LE algorithms and also produces superior cost results.

This solution to the CA problem utilizes the capacity assignment vector nomenclature discussed for the Levi/Ersoy solution [3]. The capacities of the links are represented by a vector of the form $(C_1, C_2, ..., C_i, ..., C_n)$, where C_i is chosen from a finite set of capacities (e.g. 1200, 2400, ..., etc.), and n is the maximum number of links.

In this solution each of the possible link capacities of the capacity assignment vector has an associated **probability vector** of the form (I_{ij}, S_{ij}, D_{ij}), where

I_{ij} is the probability that the current capacity j of link i is *increased*,

S_{ij} is the probability that the current capacity j of link i is *unchanged*, and

D_{ij} is the probability that the current capacity j of link i is *decreased*.

The final solution vector will be comprised of the capacities, C_i, that exhibit S_{ij} probability values that are closest to the converging value of unity. In a practical implementation this value is specified by the user and is reasonably close to unity. Indeed, the closer this value is to unity, the higher the level of accuracy.

We now present the various initial settings for the probability vector. By virtue of the various values for C_i, there are three possible settings for the initial probability vector (I_{ij}, S_{ij}, D_{ij}) given below as Init1, Init2 and Init3 respectively. To explain how this is done, we shall refer to the index of a capacity as its "capacity-index". Thus, if the set of possible capacities for a link are {1200, 2400, 3600, 4800, 9600} the corresponding capacity-indices are {0, 1, 2, 3, 4} respectively. Using this terminology we shall explain the initial settings and the updating strategies for our scheme.

Init 1: This is the scenario when the capacity-index of the link is at the *lowest* possible value, 0, called the left boundary state. This means that the capacity cannot be lowered further. In such a case,

$$I_{i0} = 1/2, \ S_{i0} = 1/2, \ D_{i0} = 0,$$

because the value can be increased or stay the same, but cannot be decreased.

Init 2: This is the scenario where the capacity-index of the link is at the *highest* possible value, n, called the right boundary state. This means that the capacity cannot be raised further. Thus,

$$I_{in} = 0, \ S_{in} = 1/2, \ D_{in} = 1/2,$$

because the value can be decreased or stay the same, but cannot be increased.

Init 3: This is the scenario where the capacity-index of the link is at one of the interior values, referred to as the interior state. This means that the capacity can be raised or lowered or maintained the same, and hence,

$$I_{ij} = 1/3, \ S_{ij} = 1/3, \ D_{ij} = 1/3 \qquad \text{for } 0 < j < n.$$

The next problem that arises is that of determining when, and how, to modify the probability values for a given link/capacity combination. Initially, a random feasible capacity assignment vector is chosen and assumed to be the current best solution. After this step, the algorithm enters the learning phase which attempts to find a superior cost by raising/lowering the capacities of the links using the L_{RI} strategy. At each step of this process the capacity of *every single link* is raised, lowered or kept the same based on the current action probability vector associated with the link. Based on the properties of the new capacity vector, the associated probability vector for this assignment is modified in two cases to yield the updated solution and the new capacity probability vector. We consider each of these cases individually.

Case 1 : The new capacity assignment is feasible. Since this means that no delay constraints are violated, the probability vector is modified in the following manner :

(a) If the capacity was increased we raise D_{ij}, the Decrease probability of the link,

(b) If the capacity stayed the same we raise S_{ij}, the Stay probability of the link, and,

(c) If the capacity was decreased we raise D_{ij}, the Decrease probability of the link.

Case 2 : The new capacity assignment is feasible *and* the cost of the network has been reduced. Since this means that the new assignment results in a lower cost than the previous best solution the probability vector is modified as :

(a) If the capacity was increased we raise D_{ij}, the Decrease probability of the link,

(b) If the capacity stayed the same we raise S_{ij}, the Stay probability of the link, and,

(c) If the capacity was decreased we raise S_{ij}, the Stay probability of the link.

It is important to remember that we are always trying to minimize cost, we thus never attempt to reward an increase in cost, by raising the increase probability, I_{ij}.

The next question we encounter is one of determining the degree by which the probability vectors are modified. These are done in terms of two user defined quantities - the first, λ_{R1}, is the reward parameter to be used when a feasible solution is reached, and the second, λ_{R2}, is used when the solution *also* has a lower cost. As in learning theory, the closer these values are to unity, the more accurate the solution. It should also be noted that the rate of convergence to the optimal probability vector decreases as the parameters increase.

As stated previously, this paper not only presents a new solution for the CA problem, but also introduces a new way of implementing LA. Unlike traditional LA, in our current philosophy we proceed from a "meta-level" whereby each LA chooses the *strategy* (increase the capacity, decrease the capacity, or let the capacity remain unchanged) which is then invoked on the selected link to set the new capacity. In this way the LA always selects its choice from a different action space rather than from a vector consisting of all the available capacities. The actual algorithm is in [4], [5].

V. EXPERIMENTAL RESULTS

In order to evaluate the quality of potential solutions to the CA problem an experimental test bench [5] must be established. This mechanism will establish a base from which the results of the algorithms can be assessed in terms of the comparison criteria. In this case the comparison criteria is the cost of the solution and the execution time. The test bench consists of the two main components described below.

First, the potential link capacities and their associated cost factors are specified as inputs. Each link capacity has two cost entries - the initial setup cost of establishing the link, and a cost per kilometer of the length of the link. Each of these cost factors increases as the capacity of the link increases.

The next step is to establish a set of sample networks that can be used to test the various solution algorithms. Each network will possess certain characteristics that remain the same for each algorithm, and therefore allow the results of the solutions to be compared fairly. The set of networks that will be used in this paper are shown in Table 5.1 below. Each network has a unique I.D. number given in column 1 and is composed of a number of nodes connected by a number of links, given in column 2, with the average length of the links given in column 3. Each network will carry multiple classes of packets with unique priority levels. The classes of packets which the network carries is given in column 4 while the average packet rate requirements, for each class over the entire network, is given in column 5.

In the suite of networks used in the test bench the network I.D. indicates the average size and complexity of the network. This means that network 4 is substantially more complex when compared with network 1 in terms of the number of links and the type and quantity of packet traffic carried.

NET I.D.	# LINK	AVE: LINK LEN:	PACKET CLASSES	AVE: PACKET RATES	PACKET PRIORITY	DELAY BOUND	PACKET LEN
1	6	54.67	1	13	3	0.013146	160
			2	13.5	2	0.051933	560
			3	14.5	1	0.914357	400
2	8	58.75	1	14.375	3	0.013146	160
			2	15.625	2	0.051933	560
			3	15.125	1	0.914357	400
			4	15.5	4	0.009845	322
3	12	58.08	1	15.417	3	0.053146	160
			2	15	2	0.151933	560
			3	15.5	1	0.914357	400
			4	17.083	4	0.029845	322
4	12	58.08	1	15.417	3	0.053146	160
			2	17	2	0.151933	560
			3	15.5	1	0.914357	400
			4	17.083	4	0.029845	322
			5	17.33	5	0.000984	12
5	48	54.67	1	13	3	0.013146	160
			2	13.5	2	0.051933	560
			3	14.5	1	0.914357	400

Table 5.1 : Characteristic values of the networks used in the test-bench.

Each of the sample networks that is used to test the algorithms carry a distinct type of packet traffic, and these are catalogued in Table 5.1 above. Each network, given by the network I.D. in column 1, carries a number of different packet classes, given in column 4. Each packet class has its own distinct priority, given in column 6, delay bound, given in column 7, and length, given in column 8. The delay bound indicates the maximum amount of time that the packet can stay undelivered in the network. This type of network carries packets of three different types:
1. Packet class one has a priority level of three. Each packet of this class has an average length of 160 bits with a maximum allowable delay of 0.013146 seconds.
2. Packet class two has a priority level of two. Each packet of this class has an average length of 560 bits with a maximum allowable delay of 0.051933 seconds.
3. Packet class one has a priority level of one. Each packet of this class has an average length of 400 bits with a maximum allowable delay of 0.914357 seconds.
   A sample network similar to Network Type 1 has nodes {1,2,3,4,5}, and edges
   {(1,2), (1,4), (3,4), (3,5), (4,5) and (2,5)}
with edge lengths L1, L2, L3, L4, L5 and L6 respectively. Each of the six links, L1 - L6, can be assigned a single capacity value from Table 5.1 and the average of the lengths will be specified by the quantity "average length" of Network Type 1. Additional details of the setup/variable costs are included in [4], [5].

In order to demonstrate that the new algorithm achieved a level of performance that surpassed both the MT and LE algorithms, an extensive range of tests were performed. The best results obtained are given in the table below. The result of each test is measured in terms of two parameters, Cost and Time which are used for comparison with the previous algorithms. In the interest of time we have only considered one large network in this series of tests, namely Network #5 which consists of 48 links, since the execution times of the previous solutions, especially the MT algorithm, take a considerable time to produce results for larger networks.

Scheme	Category	Net 1	Net 2	Net 3	Net 4	Net 5
MT-CA	Cost ($)	5735.04	12686.10	11669.30	53765.90	43341.90
	Time (sec)	0.22	0.77	1.86	2.08	67.45
LE-CA	Cost ($)	5735.04	7214.22	10295.70	45709.60	39838.40
	Time (sec)	0.22	1.21	1.10	1.93	4.01
CASCA	Cost ($)	4907.68	6937.90	9909.11	40348.90	37307.40
	Time (sec)	0.11	0.39	0.70	1.33	4.12

Table 5.2 : Best results for all algorithms.

The results displayed in the tables demonstrates that the LA solution to the CA problem produces superior results when compared with both the MT and LE solutions. The new algorithm also has consistently lower execution times in most cases when compared to either of the previous solutions. For example, we consider the tests for Network #4 [5, 4]. The MT algorithm finds its best cost as $53,765.90 and takes 2.08 seconds while the LE algorithm finds a best cost of $45,709.60 and takes 1.93 seconds. The CASCA algorithm finds a best cost of $40348.90 and takes only 1.33 seconds which is superior to either of the previous best costs. Hundreds of other experiments have been carried out which demonstrate identical properties. More detailed results can be found in [4] and [5].

It is obvious that as the reward values get closer to unity the accuracy of each cost value improves but the execution times also increases. This means that the algorithm can be optimized for speed (decrease the parameters λ_{R1}, λ_{R2}), accuracy (increase the parameters λ_{R1}, λ_{R2}) or some combination of the two that the user finds appropriate for his particular requirements. Also, the value of λ_{R2} should always be set lower than, or equal to, λ_{R1} since this is only invoked when a lower cost solution is found and is not used as much as λ_{R1} invoked when any feasible solution is found.

VI. CONCLUSIONS

In this paper we have studied the Capacity Assignment (CA) problem. This problem focuses on finding the lowest cost link capacity assignments that satisfy certain delay constraints for several distinct classes of packets that traverse the network. Our fundamental contribution has been to design the Continuous Automata Solution to CA (CASCA) algorithm which is the LA solution to the problem. This algorithm generally produces superior low cost capacity assignment when compared with the MT-CA and LE-CA algorithms and also proves to be substantially faster. Indeed, to the best of our knowledge, the LA automata solutions are the fastest and most accurate schemes available.

The problem of incorporating topology and routing considerations in the network design remains open. Also, the problem of relaxing the Kleinrock assumption is still a challenge.

REFERENCES

[1]   Oommen, B. J. and Roberts, T. D., "A Fast and Efficient Solution to the Capacity Assignment Problem using Discretized Learning Automata, *Proceedings of IEA/AIE-98, the Eleventh International Conference on Industrial and Engineering Applications of Artificial Intelligence and Expert Systems*, Benicassim, Spain, June 1998, Vol. II, pp. 56-65.

[2]   Maruyama, K., and Tang, D. T., "Discrete Link Capacity and Priority Assignments in Communication Networks", *IBM J. Res. Develop.*, May 1977, pp. 254-263.

[3]   Levi, A., and Ersoy, C., "Discrete Link Capacity Assignment in Prioritized Computer Networks: Two Approaches", *Proceedings of the Ninth International Symposium on Computer and Information Services*, November 7- 9, 1994, Antalya, Turkey, pp. 408-415.

[4]   Oommen, B. J. and Roberts, T. D., "Continuous Learning Automata Solutions to the Capacity Assignment Problem". Unabridged version of this paper. Submitted for Publication. Also available as a technical report from the School of Computer Science, Carleton University, Ottawa ; Canada : K1S 5B6.

[5]   Roberts, T. D., *Learning Automata Solutions to the Capacity Assignment Problem*, M.C.S. Thesis, School of Computer Science, Carleton University, Ottawa, Canada : K1S 5B6.

[6]   Bertsekas, D. and Gallager, R., *Data Networks* Second Edition, Prentice-Hall, New Jersey, 1992.

[7]   Kleinrock, L., *Communication Nets: Stochastic Message Flow and Delay*, McGraw-Hill Book Co., Inc., New York, 1964.

[8]   Narendra, K. S., and Thathachar, M. A. L., *Learning Automata*, Prentice-Hall, 1989.

[9]   Lakshmivarahan, S., *Learning Algorithms Theory and Applications*, Springer-Verlag, New York, 1981.

[10]   Oommen, B. J., and Ma, D. C. Y., "Deterministic Learning Automata Solutions to the Equi-Partitioning Problem", *IEEE Trans. Comput.*, Vol. 37, pp 2-14, Jan. 1988.

[11]   Ellis, Robert L., *Designing Data Networks*, Prentice Hall, New Jersey, 1986, pp. 99-114.

[12]   Etheridge, D., Simon, E., *Information Networks Planning and Design*, Prentice Hall, New Jersey, 1992, pp. 263-272.

[13]   Gerla, M., and Kleinrock, L., *On the Topological Design of Distributed Computer Networks*, IEEE Trans. on Comm., Vol. 25 No. 1, 1977, pp. 48-60.

Supervised Parameter Optimization of a Modular Machine Learning System

Manuel LINK , Masuhiro ISHITOBI

Mitsubishi Materials Corporation, Knowledge Industry Dept., Advanced Systems Center
Koishikawa 1-3-25, Bunkyo-ku, Tokyo 112, Japan
Phone: +81-3-5800-9322, FAX: +81-3-5800-9376
manuel@mmc.co.jp, tobi@mmc.co.jp

Abstract. To model a complex manufacturing process effectively with commonly used machine learning methods (Rule Induction, MLP, ANFIS) further preprocessing steps of feature selection, feature value characterization and feature value smoothing are required. The model prediction error serves as the fitness measure. It depends strongly on the parameter settings of the processing modules. Qualitatively better processing parameter settings are found by iterative training process where these settings are modified based on the quality of the previous learning process, controlled by downhill algorithms (Evolutionary Strategy, Downhill Simplex). This enables us to find a suitable model describing a process with less effort. The traditional optimizing process of determining signal processing parameters with heuristics may be standardized and preserved through this mechanism. **Keywords.** Fuzzy Logic, Machine Learning, Neural Networks, System Identification, Rule Induction, Evolutionary Strategy, copper plant control system

1. Introduction

Increasing demands of product quality, economization and savings on resources have to work alongside rising standards of industrial control. In spite of growing availability of process signals, due to advancements of science and technology, it is rather difficult to develop multi-dimensional controllers analytically. Frequently, uncertain system parameters or imperfect operator knowhow make a proper control-design impossible. If the dominant process signals of a complex manufacturing process like e.g. temperature, flow rate and product quality are simply fed to common machine learning algorithms, accuracy and generalization may remain unsatisfactory because of uncertain signal observations, such as noise, time delay, etc.. Empirical evidence suggests that having an appropriate input space characterization set is more effective for solving a real problem than developing and tuning machine learning algorithms themselves.

From the physical point of view, there are usually some **known** aspects of a process, for instance a time lag between controller and the sensor. Performing the identification of the input space characterization by a separate module to machine learning reduces the computational burden of the machine learning module, and usually improves the modeles generalization (Fig. 1).

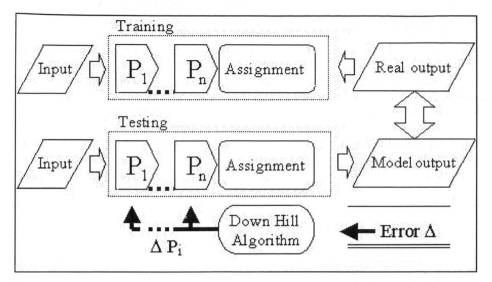

Fig. 1. The concept

'Sourcing out' the identification of the input space characteristics allows separate tuning of extracted parameters for each signal and each physical property, prior to machine learning. Input space characterization can be appropriately quantified according to the predictions that the better the characterization the better the prediction accuracy. Trial, and parameter-dependent model quality, lead to a parameter set with, at least local, minimal, remaining, prediction error. If the assumed, extracted physical phenomena apply, and the assigned signals are sufficiently representative, the processing parameter set with the smallest model error can be regarded as the identified parameters of the real, physical system.

2. Modular model structure

In our system, every necessary function for machine learning is identified and functionally realized as a modular component. Therefore a machine learning system is composed of several, sequentially connected, modular, components. Combination of several components allows a wide range of machine learning implementation.

2.1. Processing Components

Some necessary signal processing methods are categorized and realized as modular component.

- **Feature-Selection** Component: Large databases with several features and lots of samples should be stripped down to simplify analysis and assignment. Feature selection of actually-related signals is used to reduce data volume.

- **Clustering** Component: If the sample concentration varies in different areas of the feature parameter space, the position of a given number of cluster centers can be calculated, e.g. by the k-Means algorithm, minimizing the Euclidean distance sum between all training samples and assigned centers. Although this method is accompanied by loss of information, it may increase the generalization ability of some machine learning methods and therefore improve the prediction capability.
- **Smoothing** Component: As measured data is usually noisy, all numerical data is usually smoothed. The optimal degree of smoothing is frequently uncertain. The stronger the smoothing intensity, the lesser the remaining noise. However, data inherent information about certain system properties gets lost if smoothing gets too strong. Thus, optimal intensity of smoothing may be an uncertain model parameter.
- **Time Lag** Component: Several physical processes contain a timelag among simultaneously measured signals. Although the existence of a time delay is expected, it might be rather difficult to quantify the exact length of interval between cause and effect, which is regarded as another uncertain parameter.
- **Range Filter** Component: Valid signals usually fall into a certain range. Signal data out of this range should be removed to reduce noisy data. This can be done by a range-filter module.
- **Classification** Component: This component is used when discrete values are necessary for certain types of machine learning algorithms like rule induction or when simplification by grouping is necessary. Training samples are grouped into several different classes. One is the fixed-width method where the ranges of all samples are devided into a certain number of sub ranges having equal width. Another one is the fixed-number method where the total number of samples are distributed constantly in that way that each class contains the same number of samples.

2.2. Fuzzy Components

Particularly in the case of operator-controlled processes, a rough model of some inherent system relations may be extracted in the form of a fuzzy system (Zadeh, 1965; Zimmermann, 1988). Fuzzification, inference and defuzzification are realized as three distinct modular components.

- **Fuzzification** and **Defuzzification** Components: Fuzzification and defuzzification require the definition of some membership functions (MF) for each signal, that describe certain signal qualities. The problem is the determination of ideal MF parameters that best represent the physical system, which is another type of uncertain parameter.
- **Inference** Component: Rule based inference between in- and output MFs is generally user-defined and fixed, because common inference rules (MIN, MAX, Average or Product operators) are not continuously tunable and therefore not optimizable by downhill methods.

Because of the modular nature of fuzzy systems, it is also only possible to use **either** fuzzification **or** inference. In the first case, the fuzzified data can be input to a machine learning module. If machine learning is done by a neural network, a neural-fuzzy system arises.

2.3. Statistical Analysis Component (Feature Selection)

If several input signals are available, it may be difficult to select the most qualified ones to achieve a satisfactory model. Feature selection is achieved by statistical discrimination analysis. A common algorithm to do this is the discrete Mutual Information Weight (Shannon, 1948; McGill, 1955) between all classes / clusters of the input features and all output classes. The mutual information between two variables is the average reduction in uncertainty about the second variable, given a value of the first (Dietterich, 1997; Wettschereck et al., 1997).

We assume that the better the processing parameters apply, the easier it is to assign corresponding in- and output classes, and consequently the bigger the Mutual Information Weight becomes. Thus, beside using the Mutual Information Weight for feature selection, it can be regarded as the quality of the processing parameters, which can be optimized by use of a downhill algorithm that changes the processing parameters appropriately; described below.

2.4. Machine Learning Components

In principle any machine learning algorithm will be adequate for processing parameter optimization by downhill algorithms, if a remaining model error (respectively the assignment quality) is returned after training. The following machine learning algorithms have been implemented as modular components:

- MLP Component: A common Multi Layer Perceptron (Rummelhart, 1986; Rosenblatt, 1988), consisting of m layers with n nodes each, sigmoid transfer-functions and back-propagation ('SNNS' , Zell et al., 1995),
- Induction Component: Rulebased Induction (Quinlan, 1986, 1993) is a decision tree generator to classify input data to a number of output classes,
- ANFIS Component: ANFIS (Jang, 1992) is a network-based fuzzy inference system with back propagated membership function parameter optimization.

2.5. Parameter Optimization Components

Several of the above mentioned processing modules include uncertain parameters that have a strong impact on the machine learning result. Generally one processing parameter set exists which will result in the smallest model error after machine learning. Downhill algorithms can be used to optimize these processing parameters by keeping track of the training results, changing them for better or worse. Implemented components use the following algorithms:

- Evolutionary Strategy Component (Rechenberg, 1973): Based on random parameter variation in the hope of obtaining higher scores on the fitness function, which in this case is represented by the prediction results of the learnt model, and
- Downhill Simplex Component (Nelder,. Mead, 1965): Downhill Simplex is an algorithm to minimize an n-dimensional function by the use of n+1 parameter sets that define 'corners' of an n-dimensional simplex. In each iteration, the corner with the biggest error is reflected and expanded away from the high point (Press et al., 1996).

3. Application Example

3.1. The Process

This example describes a model of a large scale manufacturing process of a copper casting plant. Copper is molten in a furnace and temporarily stored in a buffer before casting (fig. 2).

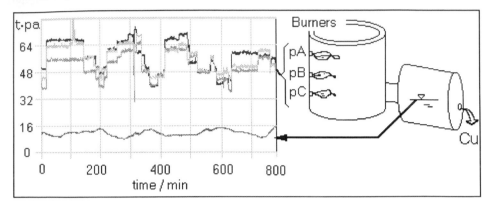

Fig. 2. The process

Due to erratic temperature fluctuations and discontinuous feed of copper plate, the melting process is highly complex and non-linear and it is difficult to describe it analytically. To guarantee a continuous flow of the molten copper at the best possible rate, it is necessary to control the copper level inside the buffer as precisely as possible. The only available signals are the gas pressures of three furnace burners (which can be regarded to be approximately proportional to the furnace temperatures) at three different furnace locations, and the weight of the buffer (which is directly related to the copper level inside). The feeding rate of copper plate and the flowing rate of the molten copper are unknown.

3.2. Parameter Optimization by MLP Assignment

Feature selection by training of a Multi Layer Perceptron with two hidden layers (three and four nodes) shows that the neural network performs best, i.e. the training error becomes smallest, when all three pressure signals are used as inputs. The result of direct assignment between channel-wise normalized pressure signals and corresponding charge signals, trained with standard back propagation, is shown in figure 3. Extending this structure by a smoothing module, the calculation structure as shown at the left side of figure 4 comes into being. In this case, three smoothing parameters for each separate input signal are regarded as uncertain. Here, the reciprocal of the training error of the neural network is used as fitness measure which is to be maximized. Because of the small parameter space of three dimensions, an evolutionary strategy is used, because former experiments have shown that the simplex algorithm tends to oscillate when use with only a few dimensions.

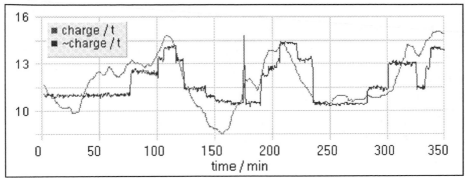

Fig. 3. Assignment of unprocessed signals via MLP

The Graph in figure 4 shows the optimization results like the development of the smoothing intensities, the mutation step size of the evolutionary strategy algorithm, and the success referring to the quality of the first evaluated parameter set.

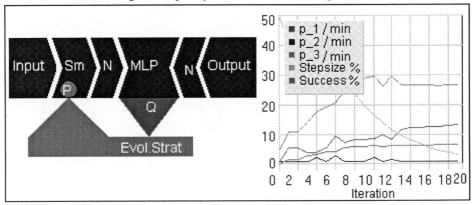

Fig. 4. Calculation structure; progress during optimization with Evolutionary Strategy

Figure 5 shows the application results. On top the optimized system was applied to the training data set. Below it was applied to unseen test samples.

3.3. Parameter Optimization by Rule Induction Assignment

Rule induction requires discrete target values. So the charge signal is assigned to 10 k-Means-clusters. Comparison of the classification scores for each mutual signal combination reveals a best classification by rule induction if, again, all three pressure signals are used. Assignment of unprocessed signals to the clusters results in figure 6.

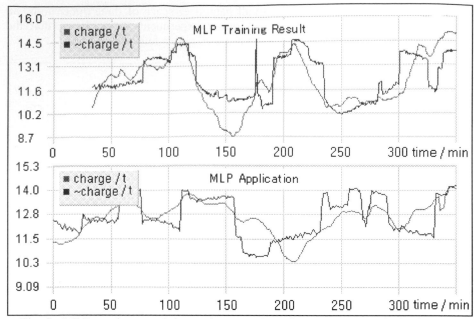

Fig. 5. MLP results on training and unseen data after parameter optimization

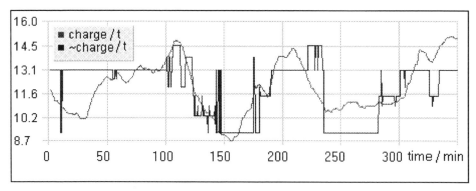

Fig. 6. Assignment of unprocessed signals using Rule Induction

The structure is extended by a smoothing module, as described for the neural network above. Thus the complete model looks like the block diagram in figure 7. The fitness measure used for downhill optimization is arbitrarily defined as the number of misclassifications, weighted by the distance between assigned and correct output class. Again, an evolutionary strategy is used because of the relatively low dimensional parameter space. The development of smoothing parameters, step size and optimization success are shown in the graph of figure 7.

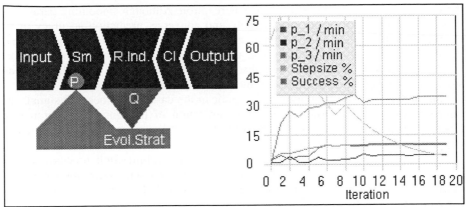

Fig. 7. Processing optimization for Rule Induction via Evolutionary Strategy

Application of the optimized system to the training data is shown on top of figure 8; application to unseen test samples is shown below.

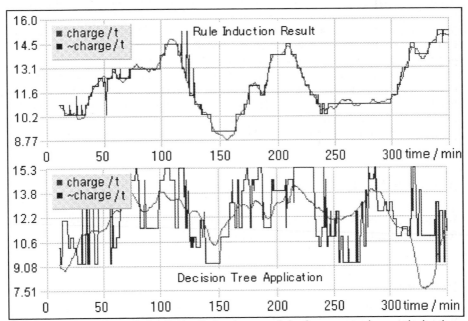

Fig. 8. Application of the Rule Induction model after processing optimization: training samples (up) and application to unseen cases

4. Results

Downhill optimization of the smoothing intensity has only little effect in the case of the prediction accuracy of a Multi Layer Perceptron. Although smoothing parameters of two input signals increase considerably and converge, prediction accuracy of the training samples remain nearly unchanged. Application to untrained samples reveals marginal improvements of prediction accuracy. Training and testing results show that, either the available input signals, or the pre- processing by only smoothing are not sufficient to obtain a proper prediction by use of a Multi Layer Perceptron.

Rule Induction classification is rather poor in the case of unprocessed signals and improves enormously during downhill optimization of the smoothing parameters. However, application to unknown data reveals that there is hardly any improvement in generalization of the rule-based model.

It is interesting to notice, that the results of the optimization process depends strongly on the classification method used, i.e. obtained smoothing intensities after optimization are quite different for each input signal depending on whether the assignment is done by rule induction or by MLP.

5. Conclusion

It has been shown, that the sourcing out of certain system properties into the pre-processing part of a machine learning structure is an effective measure to improve the performance of some machine learning systems. Along these lines it became obvious that it is reasonable to use the prediction error as a measure of fitness of the extracted parameters for parameter optimization by down hill algorithms.

Because of different results in the optimization of processing parameters in the entire example (depending on the assignment method) it is not possible to interpret results as real-system parameters. However, in principle, the described optimization method can be expected to be useful for machine modelling as well as for system parameter identification in the field of cybernetics.

Acknowledgement

We thank Dr. James G. Shanahan (A.I. Group, University of Bristol, UK) for his comments and for his advice on English.

References

Buckley, J.J. and Hayashi, Y.: Fuzzy neural networks: A survey, Fuzzy sets and systems, vol. 66, pp. 1-13, 1994
Dietterich, Thomas G.: Machine Learning Research - Four current Directions, AI Magazine, p. 108, AAAI, Winter 1997

Jang, J.-S. R.: ANFIS: Adaptive Network-based Fuzzy Inference System, IEEE Transactions on Systems, Man and Cybernetics, vol. 23, no. 3, May 1993

McGill, W.: Multivariate Information Transmission, IEEE Transactions on Information Theory, vol. 1, pp. 93-111, 1955

Nelder, J.A. and Mead, R.: Computer Journal, vol. 7, pp. 308-313, 1965

Press, William et al.: Downhill Simplex Method in Multidimensions, Numerical Recipes in C, Cambridge University Press , 1996

Quinlan, J.R.: Induction of Decision Trees, Machine Learning 1, pp. 81-106, 1986; C4.5: Programs for Machine Learning, Morgan Kaufmann, 1993

Rechenberg, I.: Optimierung technischer Systeme nach Prinzipien der biologischen Evolution, Frommann Holzboog, 1973

Rosenblatt, F.: The perceptron: a probabilistic model for information storage and Organization in the Brain, Psychological Preview 65: 386-408, 1988

Rummelhart, D. E. and McClelland, J. L.: Parallel distributed Processing, MIT Press, Ch.8, pp. 318-362, 1986

Shannon, C. E.: A mathematical Theory of Communication, Bell Systems Technology Journal, 27, pp. 379-423, 1948

Weiss, S. M. and Indurkhya, N.: Reduced Complexity Rule Induction, Learn and Knowledge Acquisition, pp. 678-684, 1997

Wettschereck, D., Aha, D. W., Mohri, T.: Technical Report AIC-95-012, Washington, D.C.: Navel Research Laboratory, Navy Center for Applied Research in Artificial Intelligence, 1997

Zadeh, L.: Fuzzy Sets, Information and Control, 1965

Zell, A. et al.: SNNS User Manual 4.0, ftp://ftp.informatik.uni-stuttgart.de, 1995

Zimmermann, H.-J.: Fuzzy Set Theory and its Applications, Kluwer-Nijhoff Publishing, Boston, 1988

Non-supervised Rectangular Classification of Binary Data

[1] Hedia Mhiri Sellami , [2] Ali Jaoua

[1] Equipe ERPAH: Faculté des sciences de Tunis.
Institut Superieur de Gestion, 41; Av de la liberté. Bouchoucha.
Tunis
Email: Hedia.Mhiri@isg.rnu.tn

[2] King Fahd University of Petroleum and minerals,
Information and Computer Science Department
P.O. Box: 126, Dhahran 31261,
Email: ajaoua@ccse.kfupm.edu.sa

Abstract: Rectangular decomposition [6] has proved to be useful for supervised data classification, for learning, or data organisation, and information extraction [7]. In this paper, we propose an adaptation of rectangular decomposition to non supervised data classification. Initial experiments and comparison with main other classification methods, have given us promising results. The proposed approach is based on successive optimal rectangle selection, from which we extract different classes that give a partition.

Keyword: Rectangular Decomposition. Non supervised classification. Partition

In this paper we propose a classification strategy which is inspired from the rectangular decomposition method [1]. As rectangular decomposition uses binary data, our classification strategy also concerns binary tables. The important difference is that the first method generates not disjoint classes while our strategy gives disjoint classes. In the first section of this paper we present some theoretical concepts we will need in the following sections. In the second section we present the principles of our strategy. In the third section we compare the results generated by our strategy with those generated by some classification methods. In the last section, we present the perspectives of our work.

I Preliminary concepts

The purpose of the rectangular decomposition is to find an economical binary relation coverage [5, 9].

1-1 Definition

A binary relation R between E and F is a subset of the «cartesian» product ExF [1]. An element of R is denoted (x,y).
For any binary relation we associate the subsets given as follows :
* The set of images of e defined by
e.R = { e' / (e,e') $\in$ R }.
* The set of antecedents of e' defined by
R.e' = { e / (e,e') $\in$ R }.
* The domain of R defined by
Dom = {e/ $\exists$ e' : (e,e') $\in$ R }.
* The range of R defined by
Cod = { e' / $\exists$ e (e,e') $\in$ R }.
* The cardinality of R defined by
Card = number of pairs in R.
* The inverse relation of R is given by
R^{-1} = { (e,e') / (e',e) $\in$ R }.
* The relation I, Identity, of a set A is given by
I(A) = { (e,e) / e$\in$ A }.
We define the relative product of R and R' by the relation RoR' = { (e,e') / $\exists$ t : (e,t) $\in$ R and (t,e') $\in$ R' } where the symbol " o " represents the relative product operator.

1-2 Definition

Let R be a binary relation defined between E and F. A rectangle of R is the product of two sets (A,B) such that A $\subseteq$ E, B $\subseteq$ F, and AxB $\subseteq$ R. A is the domain of the rectangle and B its range.

1-3 Proposition

Let R be a binary relation defined between E and F and (a,b) an element of R. The union of rectangles containing (a,b) is:

$$\Phi R(a,b) = I(b.R^{-1}) \text{ o } R \text{ o } I(a.R)$$

$\Phi R(a,b)$ is called the elementary relation containing (a,b) [1].

1-4 Definition

Let R be a binary relation defined between E and F. A rectangle (A,B) of R is "maximal" when AxB $\subseteq$ A'xB' then A=A' and B=B'.

1-5 Definition

A rectangle containing (a,b) of relation R is "optimal " if it achieves a maximal gain among all the maximal rectangles containing (a,b). The gain in storage space realized by a rectangle RE=(A,B) is measured by

$$g(RE) = [card(A) \times card(B)] - [card(A) + card(B)] [6].$$

In their paper dealing with rectangular decomposition, the authors [1] propose an heuristic based on the " branch and bound method [2]. The idea is to decompose the relation R in elementary relations PR' and select the one which might give the optimal rectangle. The authors use the following gain formula:

$$g(PR') = (r/d*c)(r - (d+c))$$

where
r = cardinal(PR') (i.e. Number of pairs of relation PR');
d=cardinal(dom(PR')) (i.e. Numbers of elements of the domain of PR');
c= cardinal(cod(PR')) (i.e. Numbers of elements of the range of PR').

II The principles of decomposition

Let R be a binary relation, the first step is to determine for each pair (a,b) of R the elementary relation R(a,b). Then, for each elementary relation, we select the domain of R(a,b) and suppose its cardinal equal to d. The range of R(a,b) is also determined and we suppose its cardinal equal to c. Gain g is given by the last formula
g= (r/d*c)(r - (d+c)). The selected elementary relation is the one giving the maximum value of the gain g. To illustrate this, we use the following binary relation having ten observations corresponding to ten types of computers. This binary relation has also ten variables each one corresponds to a property that a computer may have. The value of " 1 " means that the corresponding computer has a specific property [3].

```
    1 2 3 4 5 6 7 8 9 10
a   1 0 1 0 1 0 0 1 0 1
b   0 1 0 1 0 1 1 0 1 0
c   1 0 0 0 0 0 0 1 1 0
d   1 0 1 0 0 0 0 1 0 0
e   0 1 0 1 0 1 1 0 1 0
f   0 1 0 0 0 1 1 0 1 0
g   0 1 0 0 0 0 0 1 0 1
h   1 0 1 0 1 1 0 1 1 1
i   1 0 0 1 0 0 0 0 0 1
j   0 1 0 1 0 0 1 0 0 0.
```

The principle of our algorithm is to begin by selecting an observation, for example a, and for each couple (a,x) ∈ R we determine ΦR(a,x) and calculate the corresponding value of the gain (g) as defined before (above). Each value of (g) and the corresponding dom(ΦR(a,x)) are stored in a file which will be used to select the maximum value of the gain (g). We consider that the dom(ΦR(a,x)) associated to the maximum gain is a class. The observations belonging to this class will be retrieved from the initial data set. As $\Phi R(a,x) = I(x.R^{-1})oRoI(a.R)$ and if we want to determine $\Phi R(a,x)$ we should at the begiring determine $I(a.R)$ which is illustrated by this schema:

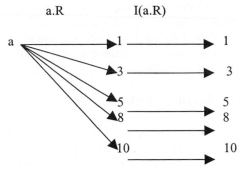

Fig. 1 Representation of $I(a.R)$

After this we should determine $I(1.R^{-1})$ which is illustrated by the schema:

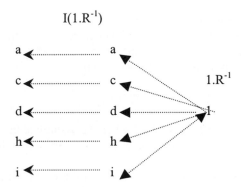

Fig. 2 Representation of $I(1.R^{-1})$

The following schema represents $\Phi R(a,1)$:

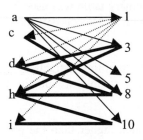

Fig. 3 Representation of $\Phi R(a,1)$

We also have dom($\Phi R(a,1)$) = { a, c, d, h, i }, its cardinal is 5 and cod($\Phi R(a,1)$) = { 1, 3, 5, 8, 10} and its cardinal is 5. The value of r is 17 and we obtain a gain equal to 4 .

Before repeating this process on the other remaining observations, we remove the set of observations which belong to the selected class, from the initial binary data. So at each step the cardinal of our sample is reduced. Our process is repeated until we have no more observations. This program gives us a partition of our initial data and each class of this partition corresponds to a rectangle.

III Comparison with classification methods

To evaluate our strategy, we compare our partition to those generated by other classification methods. The criteria selected to make a comparison between partitions is inertia [3]. We calculate the value of **R= B/T** where B is the inertia between classes and T is the total inertia of the partition [3,4]. The general formula of total inertia is:

$$T= \Sigma^n_{i=1} \; p_i d_M^2 (x_i,g)$$

Where g is the global center of gravity,
d_M^2 is the simple Euclidean metric, and p is the weight relevant to the observation x_i.
When we have our population in a partition and when we associate a centre of gravity g_l to each one the general definition of *B* (Between) is:

$$B= \Sigma^k_{l=1} \; \mu_l d_M^2 (g_l ,g).$$

R represents the percentage of inertia retained when we assimilate the observations to the centre of gravity corresponding to each class. The partition is better when the centre of gravity is greater.
We test our program on some samples and we present the results of three examples. The first is with 10 observations, the second is with 40 observations and the last is with 59 observations.

First test:
The first sample corresponds to the above binary data. The binary table was used by Celleux [3] who executes a program of crossed classification on it. This classification method gives the following three classes :
{ {a,d,h} ; {b,e,f, j} ; {c,g, i} }.
The value of R corresponding to this partition is 0.61731. This partition extracts 61% of the total inertia. When we use our strategy on the same sample we obtain a partition composed of the following four classes :
{ {a,d,h} ; {b,e,f, j} ; {c, i} ; {g} }.
The value of the associated R is equal to 0.703147; it is much higher than the one generated by the crossed classification.

Second test:

This one is composed of 40 observations and twelve variables. An observation corresponds to an elementary function which may be present or not in a management file system [4]. When using crossed classification on this table the author recuperates four classes. The value of the corresponding R is equal to 0.472020. This partition extracts 47% of the total inertia. When we use our program on the same sample we obtain a partition containing seven classes and the corresponding R is 0.475639. Our R is greater (48%) than the one generated by the crossed classification method (47%).

Third test:

The third sample is composed of 59 observations and 26 variables. Each observation corresponds to a "plate-buckle" of belt relevant to the eighth century and found in the north-east of France. Each " plate-buckle " is described by 26 criteria corresponding to the technique of fabrication, shape, etc... [8]. The archaeologists need to establish a typology between these plates and to show the evolution of the fabrication process. For this purpose, they apply several classification methods on this data. At the first step they apply two hierarchical classifications on the table, using the observations and the other using the variables. These two hierarchical classifications generate a partition of 6 classes and the value of the corresponding R is 0.764042. This partition extracts about 76% of the total inertia. At the second step the archaeologists made permutations on colons and on arrows to obtain a certain partition [8]. The R associated to it is equal to 0.723848. This partition extracts about 72% of the total inertia.

At the third step the archaeologist used correspondence analysis. From the projection of the observations on the first axes they obtain some partition of the initial data. This method generates 9 classes and the corresponding R has a value of 0.762482. This partition extracts about 76% of the total inertia. At the forth step the archaeologists operate two automatic clustering on the data. They obtain 11 classes with an R equal to 0.811843. This partition extracts about 81% of the total inertia.

To compare our strategy with these four methods we apply our program on the 59 observations. Our strategy generates 7 classes with an R equal to 0.760584. Our partition extracts about 76% of the total inertia. With these results we can say that our strategy gives better result than the permutation strategy. Our results are similar to those generated by the two hierarchical classifications and by the factorial analysis. Finally the application of the automatic classification gives better percentage of inertia than our strategy, but it is applied two times. Our approach make a common classification for both objects and their properties.

The following table can summurize the value of R relevant to the three samples:

Method	Sample 1 10 observations	Sample2 40 observations	Sample3 59 observations
Our strategy	70%	48%	76%
Crossed-classification	61%	-	-
Crossed-classification	-	47%	-
Hierarchical-classification	-	-	76%

Graphical-treatment	-	-	72%
Factorial-analysis	-	-	76%
Automatic-classification	-	-	81%

IV Conclusion

Rectangular decomposition can generate coverage from binary tables. These coverage are not disjoint and as our aim is to generate partition we propose some modifications on the rectangular decomposition method to obtain disjoint classes. The results generated by our strategy are promising. As a matter of fact, partition generated by our program extracts a percentage of inertia which is better than the one extracted by some classification methods. We can also mention that our strategy has a complexity equal to $O(n^2)$, and can give results since its first application while some classification methods must be applied two or three times to give similar results. Despite this good result our next step is to test our method on samples with bigger size. In the second step, we will compare between the variables contribution generated by our strategy and the others.

References

1. N. Belkhiter, C. Bourhfir, M.M. Gammoudi, A. Jaoua, Le Thanh, M. Reguig (1994). Décomposition rectangulaire optimale d'une relation binaire : Application aux bases de données documentaires. Revue INFOR vol 32 Feb 1994.

2. G.Brassard, P. Bratley (1987). Algorithmique, conception et analyse. Masson, Les Presses de l'Université de Montréal, pp. 197-200.

3. G.Celleux, E. Diday ; G. Govaert ; Y. Lechevallier H. Ralambondrainy. Classification automatique des données. Dunod informatique 1989.

4. E. Diday, J. Lemaire, J. Pouget, F. Testu. Eléments d'analyse de données. Dunod 82.

5. M.R. Garey et D.S. Johnson (1979). Computers and Intractability: A guide to the Theory of NP-Completeness. W.H. Freeman, 1979.

6. R.Khcherif , A. Jaoua. Rectangular Decomposition Heuristics for Documentary Databases. Information Science Journal, Intelligent Systems, Applications. 1997

7. M.Maddouri, A. Jaoua. Incremental rule production : toward a uniform approach for knowledgr organisation. . In Proceeding of the ninth international conference on Industrial and engineering applications of artificial intelligence and expert systems. Fukuota, Japan June 1996.

8. P. Perin; H. Leredde. Les plaques-boucles Merovingiennes. Les dossiers de l'archéologie. N° 42, Mars-Avril 1980.

9. J. Riguet (1948). Relations binaires, fermetures et correspondances de Galois. Bulletin de la Société mathématiques de France 76, pp. 114-155.

Synthesizing Intelligent Behavior: A Learning Paradigm

Mohamed Salah Hamdi and Karl Kaiser

Fachbereich Informatik, Universitaet Hamburg
D-22527 Hamburg, Germany.
e-mail: hamdi/kaiser@informatik.uni-hamburg.de

Abstract. This paper presents a self-improving reactive control system for autonomous agents. It relies on the emergence of more global behavior from the interaction of smaller behavioral units. To simplify and automate the design process techniques of learning and adaptivity are introduced at three stages: first, improving the robustness of the system in order to deal with noisy, inaccurate, or inconsistent sensor data, second, improving the performance of the agent in the context of different goals (behaviors), and third, extending the capabilities of the agent by coordinating behaviors it is already able to deal with to solve more general and complex tasks.

1   Introduction

The autonomy of an agent is defined as the ability to operate independently in a dynamically changing and complex environment. The classical AI approach towards autonomous agents is based on logical foundations, as exemplified by Shakey [18]. More recently, an alternative approach, known as the bottom-up or behavior-based approach, has been explored by several researchers, e.g. Brooks [3]. Some important limitations of this new approach are its lack of goal-directedness (it provides no guarantee to fulfill the goal) and flexibility (the control is entirely wired and the robot is limited to the behaviors implemented by its designer). In recent years much effort has been invested to improve the original proposal. Representative examples of these new approaches include the work described in [2, 6, 7] concerned with hybridizing reactive and classical systems and in [16, 17] concerned with introducing effective learning capabilities.

Our work is focused on improving the above mentioned shortcomings of reactive systems by combining run-time arbitration, goals and learning. The approach we propose for designing autonomous agents relies on the emergence of more global behavior from the interaction of smaller behavioral units. The basic idea consists of achieving goals by switching basic-behaviors on and off. To do this, priorities for the basic-behaviors are computed at each time-step using the current sensor data and the knowledge of the designer. Because it may be very difficult, if not impossible, for a human designer to incorporate enough world knowledge into an agent from the very beginning, machine learning techniques may play a central role in the development of intelligent autonomous agents. We introduced learning and adaptivity techniques at three different stages of the design process: first, improving the robustness of the system in order to deal with noisy, inaccurate, or inconsistent sensor data. This is achieved by using a self-organizing map and by integrating the knowledge of the designer into this map. Second, improving the performance of the agent with regard to the individual goals (behaviors) separately. This is achieved by using exploration algorithms combined with dynamic programming techniques. And third, coordinating the individual goals to solve more general and complex tasks and get an optimal overall behavior of the system. This is achieved by combining a dynamic self-organizing network with reinforcement learning.

In this paper we give a brief summary of the control architecture. Section 2 describes the design methodology and the basic idea (see [8] for details). Section 3 gives an overview of the part concerning the improvement of the robustness of the system (see [9, 10] for details). Section 4 describes the part concerned with improving the performance of the system in the context of different goals (see [11] for details). Section 5 describes the method for coordinating goals (see [12] for details).

2 Control architecture

We investigate a methodology using concurrent processes (in the following referred to as behaviors) and a priority-based arbitration scheme. The behaviors are organized in different levels (see figure 1). The lowest level consists of a set of

Fig. 1. *Organization of the behaviors (goals) in different complexity levels.*

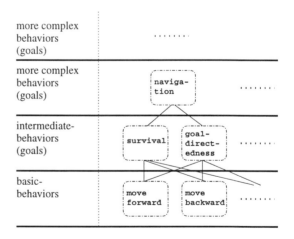

basic-behaviors which define the basic abilities of the agent. They interact in order to fulfill the tasks the agent should deal with. They can be implemented using any appropriate method. The next higher level consists of a set of intermediate-behaviors or goals that can be specified by switching basic-behaviors on and off. To achieve a certain goal, each of the basic-behaviors is assigned at each time-step a priority which is a value that reflects how good is the basic-behavior for fulfilling the required goal. To compute the priorities of the basic-behaviors for a given goal, the designer is asked first to use all its knowledge about the system to model the dependency of the basic-behaviors on the sensors (see figure 2). The dependency of a basic-behavior on the sensors is specified by several functions that state the effects of perceptual values on the activity of the basic-behavior (with respect to the given goal). The priority of the basic-behavior is then calculated by summing up all these effects.

All subsequent levels consist of more and more complex goals (behaviors). The higher is the level the higher is its competence. Complex goals are achieved by coordinating less complex goals. Coordination consists of computing at each time-step for each basic-behavior its priority with regard to the complex goal. The priority is obtained by combining the priorities with regard to the individual less complex goals.

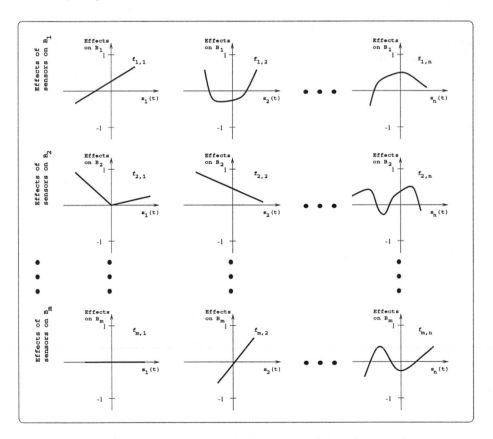

Fig. 2. *Designing a goal (intermediate-behavior) by specifying the dependency between sensors and basic-behaviors. To achieve a specific goal basic-behaviors are selected according to priorities that are computed at each time-step by summing up the effects of all perceptual values on each basic-behavior. $s_1(t)$ is the perceptual value delivered by sensor number 1, $s_2(t)$ that by sensor number 2 and so on. $B_1, B_2, \ldots, B_m$ are the basic-behaviors.*

3   Improving the robustness of the system

The component of a behavior dedicated to the computation of its priority with respect to a certain goal (the priority-component) may be thought of as the formulation of an input-output mapping, the input being some pattern of raw sensor data and the output being a value expressing the impact of the current environmental situation on the activation of the behavior with regard to that goal. This mapping, initially specified by the designer, may be adjusted and tuned using a neural network. Artificial neural networks are able to learn on-line and also have useful generalization characteristics. Noisy and incomplete input patterns can still trigger a specific response that was originally associated with more complete data.

The network algorithm that we use consists of the combination of two parts. The first part is a Kohonen self-organizing feature map [15] that is common for all behaviors. The second part is distributed over all behaviors. Figure 3 shows the integration of the association network in the original model for computing priorities. Two types of learning are being simultaneously applied: unsupervised learning concerned with building the self-organizing map and supervised learning in each behavior concerned with integrating the knowledge of the designer (priority–component) into the system by associating each region of the map (each neural unit) with priority values for the different behaviors(see [9, 10] for details).

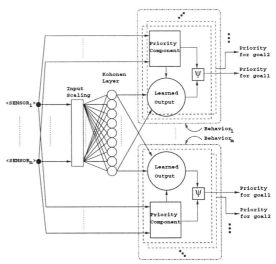

Fig. 3. *Addition of an association network to the original model for computing priorities. Input values (sensor data) are routed both to the Kohonen layer and to the priority–component of each behavior. The output of a priority–component is used to train the corresponding network part in order to integrate the available knowledge into the neural control system.*

Figure 4 shows for example how the robustness of an agent acting in a simulated dynamic environment is improved using such a neural structure. The dynamic environment involves the agent, enemies, food, and obstacles. The enemies chase the agent, food provides the agent with additional energy, and moving costs the agent energy. At each time-step the agent has four basic-behaviors to choose from: "move forward", "move backward", "move left", and "move right".

The agent is allowed to see only the area surrounding it. It has four sensors that are used to sense objects in the directions forward, backward, left and right. Each of the sensors delivers a real perceptual value which depends on the kind of the detected object and its distance from the agent. A play ends when the agent collides with an enemy or an obstacle or runs out of energy. The agent should try to survive as long as possible.

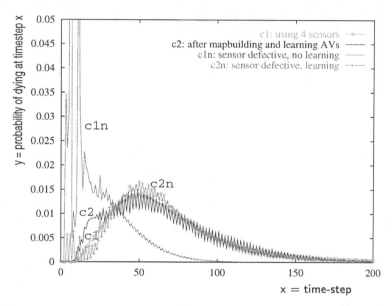

Fig. 4. *Improving the robustness of the agent in the context of the goal "survival". c1: Performance of the agent when the dependencies between sensors and basic-behaviors (priority-components) are specified by the designer. c2: The performance of the agent after map-building and integrating the knowledge of the designer into the neural structure is almost the same as that of the teacher (the priority-components are used during learning as a teacher). c1n: The agent performs very bad when a sensor is defective and no neural structure is used. c2n: With the neural structure there is only a small performance deviation when a sensor is defective.*

4 Improving the performance of the system with regard to specific goals

The computations presented above assume that the designer of the agent is able to model the dependency of behaviors on sensors in the context of all goals and specify the necessary functions correctly. Unfortunately, this is usually a difficult task. The designer has usually only little knowledge about the system. Even when the effects of sensors on behaviors are known, this knowledge is often only immediate, i.e. the delayed consequences of actions are unknown.

Reinforcement learning (RL) provides an appropriate framework for solving these problems. It aims to adapt an agent to an unknown environment according to rewards. Many RL methods exist that can handle delayed reward and uncertainty, e.g., Q-learning. The goals these systems learns, however, are implicit, i.e., what is learned in the context of one goal, cannot be applied in the context of another goal (e.g., all of the Q-values are goal dependent). Furthermore, most of these systems converge slowly.

The method we use achieves the purpose of RL and at the same time deals with many of these problems. It consists of identifying the environment using exploration and then determining an optimal policy using dynamic programming.

The proposed improvement method was shown to be very effective especially when the designer has only little knowledge about the system. Figure 5 shows for example how the agent acting in the simulated environment described in the previous section (with slightly changed parameter settings) is able to learn the goal "survival" starting from a control structure for which only the effects of one of the four sensors are known. The method has also the advantage to be able to learn in the context of many different goals simultaneously. Another advantage is the principled exploration method which seems to make the learning system faster then other RL paradigms. See [11] for details.

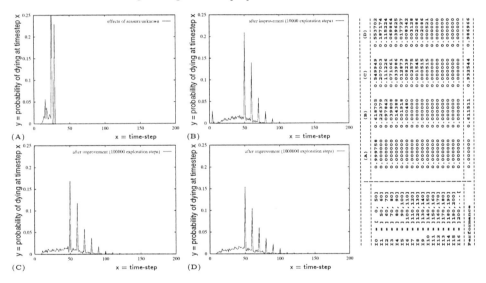

Fig. 5. *Learning the goal "survival" starting from a control structure for which the effects of many sensors are unknown. The table summarizes the probabilities of dying in the intervals $I_0, .., I_{16}$ for the cases (A): at the beginning, (B): after 10000 exploration steps, (C): after 100000 exploration steps, and (D): after 1000000 exploration steps and gives the corresponding performance.*

5 Extending the capabilities of the agent

Extending the capabilities of an agent is done by combining behaviors it is already able to cope with to produce more complex emergent ones. The method consists of a combination of map-building using a dynamic self-organizing feature map with output and reinforcement learning. The main role of the map-building strategy is to remember previous successful control parameters (coordination parameters) and use them when the agent faces similar circumstances. The main role of the reinforcement learning strategy is to reinforce associations (mappings between situations and control parameters) that tend to produce useful results. The basic learning scheme consists of the following steps: first, using the feature map to generate control parameters for the current situation, second, observe the agent operating with these parameters, and third, update the feature map according to the outcome.

This behavior coordination method was evaluated through many simulation studies and was shown to be very effective (see [12]). Figure 6 shows for example how an agent (an extended version of the agent used in the previous sections) that learns to coordinate the goals "survival" and "goal-directedness" to achieve the more complex goal "navigation" (see figure 1) incrementally increases its performance. The learning system (the system that uses this coordination method) was compared with a random system that changes the coordination parameters randomly at each time-step (initial configuration of learning system) and with a static system that uses fixed coordination parameters which were determined manually and found to be quite effective (a desirable configuration for the learning system).

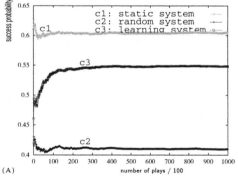

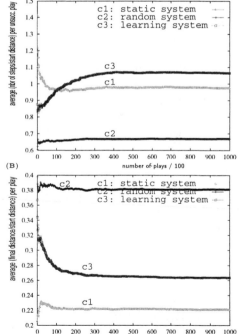

Fig. 6. *Learning to coordinate goals to deal with more complex ones. Comparison of the performance of the learning system with a random system and with a static system.*

6   Related work

Our work situates itself in the recent line of research that concentrates on the realization of complete artificial agents strongly coupled with the physical world and usually called "embedded" or "situated" agents. Examples of this trend include [1],[14], [4],[5], and [23].

While there are important differences among the various approaches, some common points seem to be well established. A first, fundamental requirement is that agents must be able to carry on their activity in the real world and in real time. Another important point is that adaptive behavior cannot be considered as a product of an agent in isolation from the world, but can only emerge from a strong coupling of the agent and its environment.

Our research has concentrated on a particular way to obtain such a coupling: the use of self-organization and reinforcement learning to dynamically develop a robot controller through interaction of the robot with the world.

Self-organization is a type of behavior exhibited by dynamical systems, and examples of this behavior are abundant in many areas of science (see, e.g., [13]). Self-organization refers to the ability of an unsupervised learning system to develop specific detectors of different signal patterns. Unsupervised learning (e.g. [15]) tasks differ from supervised learning (e.g. [19]) tasks in the type of information supplied to the system. In supervised learning, a "teacher" must supply the class to which each training example belongs, whereas in unsupervised learning the goal is to look for regularities in the training examples. Self-organization in learning adaptive controllers is useful because it is unlikely that designers would have detailed knowledge of how to characterize the environment in terms of the agent's sensors. Thus, it is not easy to select the appropriate control parameters values the controller must use when facing specific environments. A system that is able to characterize environmental situations autonomously can associate useful control parameters to environmental situations and adapt the controller successfully.

Reinforcement learning (see, e.g., [20]) has recently been studied in many different algorithmic frameworks. In [21], for example, Sutton, Barto and Williams suggest the use of reinforcement learning for direct adaptive control. Mahadevan and Connell ([16]) have implemented a subsumption architecture ([3]) in which the behavioral modules learn by extended versions of the Q-learning algorithm ([22]). Reinforcement learning methods combine methods for adjusting action-selections with methods for estimating the long-term consequences of actions. The basic idea in reinforcement learning algorithms such as Q-learning ([22]) is to estimate a real-valued function, $Q(.,.)$, of states and actions, where $Q(x,a)$ is the expected discounted sum of future rewards for performing action a in state x and performing optimally thereafter.

7   Conclusions

This paper has introduced a new Method for the design of intelligent behavior in a behavior-based architecture using priorities and learning techniques. The method relies on the emergence of more global behavior from the interaction of smaller behavioral units.

Fairly complex interactions can be developed even with simple basic-behaviors. In particular, behavior patterns that appear to follow a sequential plan can be realized by the agents when there is enough information in the environment to determine the right sequencing of actions. The addition of elements such as a memory of past perceptions and reactions improves the level of adaptation to the dynamics of the environment.

The agents are strongly coupled with their environment through their sensorimotor apparatus, and are endowed with an initial control structure that has the ability to adapt itself to the environment and to learn from experience. To develop an agent, both explicit design and machine learning have an important role. Although, no much initial knowledge is required (all dependency functions can be chosen arbitrarily, e.g., all equal to zero), it is sensible that the designer implements all its knowledge about the system into the initial control structure. This will avoid long learning periods and allow the system to become able to deal with its task in a satisfactory way early enough.

For our experiments, we used simulated agents. The proposed learning methods were shown to be very effective especially when the designer has only a little knowledge about the system. Simulation appears at the first moment to be inadequate for real world robots. However, simulated environments have proved very useful to test design options, and the results of simulations seem to be robust enough to carry over to the real world without major problems, provided the sensory and motor capacities of the robots are similar to those of their simulated counterparts. Furthermore, since the use of real robots is too time-consuming, training in a simulated environment and then transferring the resulting controller to a real robot can be a viable alternative. At least in some cases it is possible to use behavioral modules learned in simulated environments as a starting point for real robot training.

The results obtained are encouraging and suggest that the whole architecture is a promising approach to building complete autonomous learning systems. The architecture is not limited to robotic agents. It can also be applied to design goal-directed behavior for other kinds of agents such as software agents.

References

1. P. Agre and D. Chapman. Pengi: An implementation of a theory of activity. In *Proceedings of the Sixth National Conference on Artificial Intelligence, AAAI-87*. Morgan Kaufmann, Los Altos, CA, 1987.
2. Ronald C. Arkin. Integrating behavioral, perceptual, and world knowledge in reactive navigation. *Robotics and Autonomous Systems*, 6:105–122, 1990.
3. R. A. Brooks. A robust layered control system for a mobile robot. *IEEE Journal of Robotics and Automation*, RA-2(1):14–23, April 1986.
4. R. A. Brooks. Elephants don't play chess. *Robotics and Autonomous Systems*, 6(1-2):3–16, 1990. Special issue on designing autonomous agents.

5. R. A. Brooks. Intelligence without reason. In *Proceedings of the Twelfth International Joint Conference on Artificial Intelligence, IJCAI-91*, pages 569–595, 1991.

6. R.J. Firby. Adaptive execution in complex dynamic worlds. Ph.D. Dissertation YALEU/CSD/RR#672, Yale University, Jan. 1989.

7. E. Gat. Integrating planning and reacting in a heterogeneous asynchronous architecture for controlling real-world mobile robots. In *Proceedings of AAAI-92*, pages 809–815, San Jose, CA, July 1992.

8. M. S. Hamdi. A goal-oriented behavior-based control architecture for autonomous mobile robots allowing learning. In M. Kaiser, editor, *Proceedings of the Fourth European Workshop on Learning Robots*, Karlsruhe, Germany, December 1995.

9. M. S. Hamdi and K. Kaiser. Adaptable local level arbitration of behaviors. In *Proceedings of The First International Conference on Autonomous Agents, Agents'97*, Marina del Rey, CA, USA, February 1997.

10. M. S. Hamdi and K. Kaiser. Adaptable arbitration of behaviors: Some simulation results. *Journal of Intelligent Manufacturing*, 9(2):161–166, April 1998.

11. M. S. Hamdi and K. Kaiser. Improving behavior arbitration using exploration and dynamic programming. In *Proceedings of the 11th International Conference on Industrial and Engineering Applications of Artificial Intelligence and Expert Systems*, Benicàssim, Castellón, Spain, June 1998.

12. M. S. Hamdi and K. Kaiser. Learning to coordinate behaviors. In *Proceedings of the 13th biennial European Conference on Artificial Intelligence (ECAI-98)*, Brighton, UK, August 1998.

13. E. Jantsch. *The Self-Organizing Universe*. Pergamon Press, Oxford, New York, 1980.

14. L.P. Kaelbling. An architecture for intelligent reactive systems. In *Proceedings of the 1986 Workshop Reasoning about Actions and Plans*, pages 395–410. Morgan Kaufmann, San Mateo, CA, 1986.

15. T. Kohonen. *Self-Organization and Associative Memory*. Springer Series in Information Sciences 8, Heidelberg. Springer Verlag, 1984.

16. S. Mahadevan and J. Connell. Automatic programming of behavior-based robots using reinforcement learning. *Artificial Intelligence*, 55(2):311–365, 1992.

17. J.d.R. Millàn and C. Torras. Efficient reinforcement learning of navigation strategies in an autonomous robot. In *Proceedings of The International Conference on Intelligent Robots and Systems, IROS'94*, 1994.

18. Nils J. Nilsson. Shakey the robot. Technical Note 323, SRI AI center, 1984.

19. D. E. Rumelhart, G. E. Hinton, and R. J. Williams. Learning representations by back-propagating errors. *Letters to Nature*, 323:533–535, 1986.

20. R. S. Sutton. Reinforcement learning architectures for animats. In Jean-Arcady Meyer and Stewart W. Wilson, editors, *Proceedings of the First International Conference on Simulation of Adaptive Behavior*, Cambridge, MA, 1990. MIT Press.

21. R. S. Sutton, A. G. Barto, and R. J. Williams. Reinforcement learning in direct adaptive optimal control. In *Proceedings of the American Control Conference*, Boston, MA, 1991.

22. C. J. C. H. Watkins. *Learning from Delayed Rewards*. University of Cambridge, England, 1989. Ph.D. Thesis.

23. S. D. Whitehead and D. H. Ballard. Learning to perceive and act by trial and error. *Machine Learning*, 7(1):45–83, 1991.

Automatic Input-Output Configuration and Generation of ANN-based Process Models and Their Application in Machining

Zs. J. VIHAROS; L. MONOSTORI

Computer and Automation Research Institute, Hungarian Academy of Sciences
Kende u. 13-17,H-1111 Budapest, Hungary, Tel.: (+36 1) 4665644, Fax: (+36 1) 4667503
viharos@sztaki.hu, monostor@sztaki.hu

Abstract

Reliable process models are extremely important in different fields of computer integrated manufacturing. They are required e.g. for selecting optimal parameters during process planning, for designing and implementing adaptive control systems or model based monitoring algorithms. Because of their model free estimation, uncertainty handling and learning abilities, artificial neural networks (ANNs) are frequently used for modelling of machining processes. Outlying the multidimensional and non-linear nature of the problem and the fact that closely related assignments require different model settings, the paper addresses the problem of automatic input-output configuration and generation of ANN-based process models with special emphasis on modelling of production chains. Combined use of sequential forward search, ANN learning and simulated annealing is proposed for determination and application of general process models which are expected to comply with the accuracy requirements of different assignments. The applicability of the elaborated techniques is illustrated through results of experiments.

Introduction

Modelling methods can be used in several fields of production e.g. in planning, optimisation or control. The production in our days incorporates several stages, the workpiece goes through a number of operations (Fig. 1.).

The output of one operation is the input of an another one or it is a feature of the end product. To build a model for a production chain, models have to be ordered to every stage of production. A chain of operations connected by their input-output parameters can model the sequence of production operations.

Operations have several input- and output parameters and dependencies among them are usually non-linear, consequently, the related model has to handle multidimensionality and non-linearity.

Artificial neural networks (ANNs) can be used as operation models because they can handle strong non-linearites, large number of parameters, missing information. Based on their inherent learning capabilities, ANNs can adapt themselves to changes in the production environment and can be used also in case there is no exact knowledge about the relationships among the various parameters of manufacturing.

Some error is usually incorporated into modelling of real processes, the model estimates its output variables only with a limited accuracy. The error by the output side of an operation model in the production chain depends on its own error and the error incorporated into the input variables of the model. These input variables are usually output parameters from previous operations. Consequently, model errors can be summed up and, therefore, the accuracy of the individual models is a crucial factor in the modelling of production chains.

A lot of effort has been made to apply ANNs for modelling manufacturing operations 00. The assignments to be performed determined the input-output configurations of the models, i.e. the parameters to be considered as inputs and the ones as outputs.

Considering the input and output variables of a given task together as a set of parameters, the ANN model estimates a part of this parameter set based on the remaining part. This partitioning strongly influences the accuracy of the developed model especially if dependencies between parameters are non-invertable. In different stages of production (e.g. in planning, optimisation or control) tasks are different, consequently, the estimation capabilities of the related applied models are different even if the same set of parameters is used.

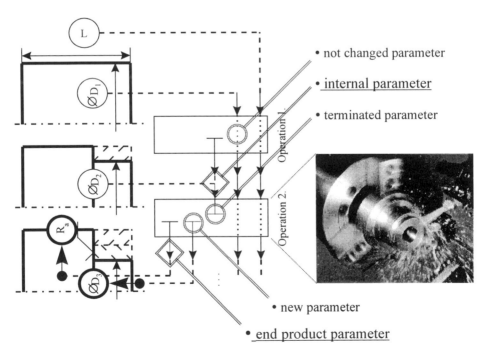

Connections of the operations of the production chain through workpiece parameters. Stages of material removal from an axle, the related operations in the middle and the related parameter stream of the workpiece along the production chain are illustrated from left to right 0. **Fig. 1.**

One of the main goals of the research to be reported here was to find a general model for a set of assignments, which can satisfy the accuracy requirements. Research was also focused on how to apply the general model for various tasks.

Accordingly, the structure of the paper is as follows:

- Section 2 gives a short survey of approaches to modelling and monitoring of machining processes.
- In Section 3 the proposed method for automatic generation of ANN-based process models is described which are expected to be applicable for different assignments.
- The application phase of the general process model is detailed in Section 4. A novel technique based on simulated annealing search is introduced to find the unknown parameters of the model in given situations. The results of experimental runs justify the approach.
- Conclusion and further research issues are presented in paragraph 5.

ANN based approaches to modelling and monitoring of machining processes

Several approaches can be found in the literature to represent the knowledge of manufacturing operations 0000. The aim of this paragraph is to show the large variety of tasks and related input-output configurations of ANNs.

An interesting example is presented by Knapp & Wong 0 who used ANNs in planning.

ANN is also used for ordering of resources to workcenters 0.

Cutting tool selection is realised by Dini 0.

To generate an optimum set of process parameters at the design state of injection molding, Choi *et al.* use an ANN model 0.

The compensation of thermal distortion was the goal of Hatamura *et al* 0.

A fuzzy neural network is used for cutting tool monitoring by Li & Elbestawi 0. Optimisation and search for input variables are presented in the work by Rangwala & Dornfeld 0.

Monostori described models to estimate and classify tool wear 0. The paper presents different input-output configurations of ANN models according to various tasks.

A model building for creep feed grinding of aluminium with diamond wheels is presented by Liao & Chen 0. The paper also calls the attention to the problem that the measurement could not been satisfactory handled by the chosen ANN model realizing one to one mapping.

Automatic generation of ANN-based process models

The automatic generation of appropriate process models, i.e. models, which are expected to work with the required accuracy in different assignments, consists of the following steps:

- Determination of the (maximum) number of output parameters (No) from the available N parameters which can be estimated using the remaining Ni = N - No input parameters within the prescribed accuracy.
- Ordering of the available parameters into input and output parameter sets having Ni and No elements, respectively.
- Training the network whose input-output configuration has been determined in the preceding steps.

These steps can be formulated as follows. A search algorithm is needed to select all the possible outputs from the given set of parameters with regard to the accuracy demands. This algorithm results in a general ANN model, which realises mapping between the parameters of the given parameter set. The largest number of outputs can be found, the accuracy demands are satisfied and the ANN model is built up 0.

A general ANN model for drilling is shown in figure 2.

In the developed method the estimation error is used to evaluate an ANN configuration. This error assures the user that all of the outputs can be estimated within an average error given in advance.

Experimental results

To test the behaviour of the developed algorithm the case of non-invertable dependencies were investigated first ($x_2=x_1^2$, $x_3=x_1^2+x_2^2$, $x_4=x_1^2+x_2^2+x_3^2$, $\sin(x)$). Favourable results of these investigations promised real world applicability, too.

In the following space, results are presented with four engineering assignments where the required models work on the same parameter set but the feasible input-output configurations of these models are different.

1. The first task is planning. A surface has to be machined by turning to achieve roughness (parameter: R_a[mm]) demands of the customer. The engineer has to determine the tool (parameters: cutting edge angle: χ[rad], corner radius: r_ε[mm]), the cutting parameters (parameters: feed: f[mm/rev], depth of cut: a[mm], speed: v[m/min]) and predict phenomenon during cutting (parameters: force: F_c[N], power: P[kW] and tool life: T[min]) consequently a model is needed where R_a serves as input and other parameters as outputs. Usually, the customer gives only an upper limit for the roughness, in contrast to other parameters.

2. The second task is to satisfy the roughness demands of the customer but with a given tool. In this case the R_a, χ, r_ε are inputs and f, a, v, F_c, P, T are outputs.

3. The third task is to control the running cutting process with measured monitoring parameters such as force and power. Measured values of these parameters can be used as information about the current state of the cutting process. In this case R_a, χ, r_ε, F_c, P serve as input and f, a, v, T as outputs. The CNC controller has to select the appropriate cutting parameters to produce the requested surface.

4. The fourth task is the same as the third one but the CNC controller can change only the 'f 'and 'a' parameters because v is prescribed. This case needs a model with inputs R_a, χ, r_ε, F_c, P, v and with outputs f, a, v, T.

These assignments show several input-output configurations for modelling dependencies between the different elements of a parameter set. The question arises: which model describes the cutting process in the best way, i.e. with the highest accuracy? The heuristic search algorithm can answer this question.

In practical implementation sensors, machine controllers and computers would provide a part of parameters of an ANN operation model. For simulating the machining process in the investigations to be reported here all information were generated via theoretical models, which are functions of several input variables. It should be stressed that in a practical implementation theoretical models are not necessary. The validity of the equations is determined by the minimum and maximum boundaries of the parameters. Four equations are used in this paper for the above engineering tasks (force, power, tool life and roughness) 0.

$$F_c = 1560 \cdot f^{0.76} \cdot a^{0.98} \cdot (\sin(\kappa))^{-0.22}, P = 0.039 \cdot f^{0.79} \cdot a \cdot v \tag{1}$$

$$T = 1.85 \cdot 10^{10} \cdot f^{-0.7} \cdot a^{-0.7} \cdot v^{-3.85}, R_a = 8.5 \cdot f^{1.8} \cdot a^{0.08} \cdot v^{-0.9} \cdot r_\varepsilon^{-0.5},$$

where the boundaries of the equations are as follows:

$$f : 0.1 \cdots 0.4[mm/rev], a : 1 \cdots 4[mm], \kappa : 1.3 \cdots 1.66[rad], v : 75 \cdots 200[m/\min], \tag{2}$$

$$r_\varepsilon : 0.4 \cdots 1.2[mm], T : 5 \cdots 60[\min], \text{consequently}, Fc \approx : 800 \cdots 3000[N],$$

$$P \approx : 3.8 \cdots 13.5[kW], R_a \approx : 0.0015 \cdots 0.023[mm]$$

To create learning and testing parameter sets random values were determined in the allowed range of f, a, χ, v, r_ε considering also the boundaries of T and R_a, Fc, P, T while calculating their values using the above equations. The dependencies between parameters f, a, χ, v, r_ε, Fc, P, T, R_a were experienced as invertable in the given parameter range only the variable χ is the exception, consequently, to get an accurate ANN model the variable χ has to be always input. A hundred data vectors were created as stated above. To test this type of problems the described input-output configuration and model building approach were repeated a hundred times. Several variations of input-output configurations were generated. The allowed average estimation error was given as ±2.5%. Fifteen different ANN configurations were generated as results 0. The variable χ is always on the input size of the ANN model as expected.

The test all of the configurations shows that there are no significant differences among their estimation capabilities.

The results indicate that the developed technique is able to generate process models with the required accuracy, moreover, under given circumstances a result is a set of applicable models each guaranteeing the required accuracy performance.

Satisfying various assignments with the general model

Some parameters of a process are usually known by the user and modelling is expected to determine the other parameters while satisfying some constraints. In the previous paragraph a search method was introduced to select a general ANN model which is accurate enough and can be used for different assignments. Consequently, in almost every case a part of input and a part of output variables of the general model are known by the user and the task of the modelling is to search for the remaining, unknown input and output parameters like in the engineering tasks presented before (Fig. 2.). A search method can solve this task. The search space consists of unknown input parameters. The task for the search method can be formulated as follows: It has to find the unknown input parameters but at the same time it satisfy three conditions (Fig. 2.):

1. One point of the search space can be represented by one possible value set of the unknown input parameters. After placing these parameters together with the known input parameters to the input side of the given ANN an output vector can be calculated by the ANN estimation (forward calculation). The first condition assures that only that points of the search space can be accepted as result, which can adequately estimate the known output parameters by using forward calculation. To measure the deviation between estimated and known output parameters an error can be calculated. For the search algorithm the user can prescribe upper limit for this error.

2. The second condition for the unknown input parameters is determined by the validity of the ANN model. This validity is usually specified by the data set used for the training 0. Boundaries of the model can be handled by minimum and maximum values of the related parameters like in the engineering tasks presented above. This means that the search algorithm can take values for the unknown input parameters only from the related allowed intervals.

3. The third condition relates also to the validity of the ANN. These boundaries come forward by that part of the estimated output vector, which is unknown by the user. Because of the limited validity of the ANN model there are also boundaries for parameters of this part of the estimated output vector. Values of the unknown input parameters are only acceptable if the estimated values of the unknown output parameters are within their allowed range. To measure this condition an error can be calculated for that unknown output parameters which estimated values are out of their boundaries. For the search algorithm the user can prescribe an upper limit also for this type of error.

The search algorithm is terminated if all of the three conditions above are met. Simulated annealing has been selected as search method 0. In the simulated annealing search an error value is ordered to all points of the search space. In the developed algorithm this value is the maximum of error1 and error2 presented above. The algorithm searches for the minimum error point.

The simulated annealing technique has a special parameter, the temperature, which decreases during the search algorithm. The algorithm discovers the search space by repeated change from the current point into a neighbour point. A probability value is used to evaluate a neighbour incorporating information about the error difference between the neighbour and the current point and about the current temperature. The algorithm stops if no neighbour can be selected and the current error value is below the prescribed error limit. This simulated annealing algorithm works on the discrete points of the search space. To realise this, the parameters of unknown part of the input vector consist of the discrete points of the related intervals. The distance between two points of an interval is chosen to satisfy the accuracy requirements of the estimation prescribed by the user.

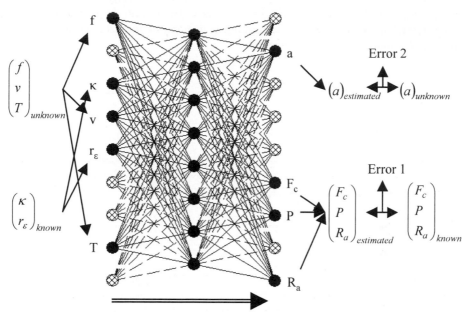

The simulated annealing search is used to satisfy various tasks of the user without regard to the given ANN configuration. Error 2 is used to hold the search between the boundaries of the ANN model, while error 1 measures the distance between estimated and known outputs. The search space consists of unknown input parameters, the evaluation of one point based on the maximum of error 1 and error 2. The developed algorithm searches for the minimum error value. This picture shows the third engineering task presented above. **Fig. 2.**

As a result, this algorithm gives one solution for a given assignment of the user. To look for a larger number of solutions the search has to be repeated.

Results of the simulated annealing search

Tests of the non-invertable dependencies enumerated above show the applicability of this search algorithm.

The results of the four engineering assignments presented in the paper are also worth mentioning. There are a large number of solutions for each of the enumerated assignments. To represent the whole interval of solutions for each parameter the search algorithm was repeated a hundred times at each assignment. To get a simple view about the possible solution field the maximum and minimum values of the results were selected for all parameters, for each task. These parameter fields are listed in Figure 3. Results in this table show the descending interval of acceptable parameters from the planning phase to the CNC control. The requested value of parameter R_a is special because the user gives only upper limit for this parameter. In the assignments the allowed highest value for the roughness of the produced surface is 0.014 mm. The tool used for cutting is determined in the second task, values of related parameters are χ=1.549 rad, r_ε=0.7394 mm. In monitoring, measured values of force and power were Fc=2247N and P=8.69kW, respectively. In the fourth engineering task the prescribed speed value was v=161 m/min. In every case the task of the modelling was to satisfy the roughness demand of the user through choosing appropriate values of related parameters.

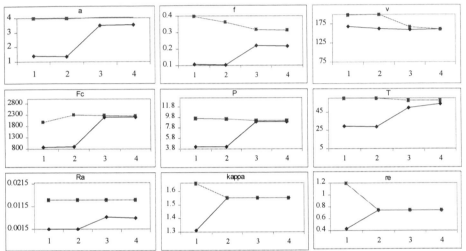

Descending intervals of allowed parameter fields in cutting in the four engineering tasks presented above. The horizontal axis represents the number of the given tasks.
Fig. 3.

By every case from planning to CNC control one or more new parameter(s) becomes to be restricted to one value.

Results show that by the first planning task a large field of parameters can be chosen to satisfy the user demands. Using the given tool in the second task possible fields of intervals are only a bit smaller. The intervals in the third task, in monitoring the cutting process with measured parameters, are much smaller. In the fourth task when the speed is prescribed allowed intervals become even more smaller. It should be stressed that these results were received with only one ANN model with the same input-output configuration and using the developed simulated annealing search method, indicating the acceptability of the techniques presented here. The developed sequential forward selection algorithm determined the appropriate input-output configuration automatically, showing that the realisation of the new concept works adequately.

Conclusions and further research issues

Outlying the importance of accurate process models in the control of production chains, generation and application of ANN-based process models were addressed in the paper with special emphasis on the automatic input-output configuration of general process models which are expected to satisfy the accuracy requirements of a set of related modelling assignments. Combined use of sequential forward search, ANN learning and simulated annealing is proposed. The applicability of the elaborated techniques was illustrated through results of experiments.

Several steps of the new model are to improve further. Some of them are:

- By searching the appropriate input-output configuration a method could be useful which prunes the variables which are not important for output estimations.
- By searching the unknown variables an optimisation is to be included in the method like in the paper of Rangwala & Dornfeld 0., e.g. cost minimisation, manufacturing time minimization, etc. This is important only if there are more solutions for the given task.

The above improvements will be subject of future publications.

Acknowledgement

This work was partially supported by *National Research Foundation,* Hungary, Grant No. F026326 and T026486. A part of the work was covered by the Nat. Comm. for Techn. Dev., Hungary Grants (EU-96-B4-025 and EU-97-A3-099) promoting Hungarian research activity related to the ESPRIT LTR Working Groups (IiMB 21108 and IMS 21995).

Literature

[1] Choi, G.H.; Lee, K.D.; Chang, N.; Kim, S.G., 1994, Optimization of the process parameters of injection molding with neural network application in a process simulation environment, CIRP Annals, Vol. 43/1, pp. 449-452.
[2] Chryssolouris, G.; Lee, M.; Pierce, J.; Domroese, M., 1990, Use of neural networks for the design of manufacturing systems, Manufacturing Review, Vol. 3, No. 3, pp. 57-63.
[3] Devijver, P. A.; Kittler, J.; Pattern recognition, a statistical approach. Book. Prentice-Hall International Inc., England, London, 1982,
[4] Dini, G., 1995, A neural approach to the automated selection of tools in turning, Proc. of 2nd AITEM Conf., Padova, Sept. 18-20, pp. 1-10.
[5] Hatamura, Y.; Nagao, T.; Mitsuishi, M.; Kato, K.I.; Taguchi, S.; Okumura, T.; Nakagawa, G.; Sugishita, H., 1993, Development of an intelligent machining center incorporating active compensation for thermal distortion, CIRP Annals, Vol. 42/1, pp. 549-552.
[6] Kis, T.; Introduction into artificial intelligence (in Hungarian). Book. AULA Press, Budapest, Edited by I. Futó, 1999.
[7] Knapp, G.M.; Wang, Hsu-Pin, 1992, Acquiring, storing and utilizing process planning knowledge using neural networks, J. of Intelligent Manufacturing, Vol. 3, pp. 333-344.

[8] Krupp, F. Gmbh; Widia-Richtwerte für das drehen von Eisenwerkstoffen. Book. Fried. Krupp Gmbh., Germany, Essen, 1985.

[9] Li, S.; Elbestawi, M.A., 1996, Fuzzy clustering for automated tool condition monitoring in machining, J. of Mechanical Systems and Signal Processing, pp. 321-335.

[10] Liao, T.W.; Chen, L.J., 1994, A neural network approach for grinding processes: modeling and optimization, Int. J. Mach. Tools Manufact., Vol. 34, No. 7, pp. 919-937.

[11] Markos, S.; Viharos, Zs. J.; Monostori, L.; Quality-oriented, comprehensive modelling of machining processes. Proc. of 6th ISMQC IMEKO Symposium on Metrology for Quality Control in Production, September 8-10, 1998, Vienna, Austria, pp. 67-74.

[12] Monostori, L., 1993, A step towards intelligent manufacturing: Modeling and monitoring of manufacturing processes through artificial neural networks, CIRP Annals, 42, No. 1, pp. 485-488.

[13] Monostori, L.; Barschdorff, D., 1992, Artificial neural networks in intelligent manufacturing, *Robotics and Computer-Integrated Manufacturing*, Vol. 9, No. 6, Pergamon Press, pp. 421-437.

[14] Monostori, L.; Egresits Cs.; Kádár B., 1996, Hybrid AI solutions and their application in manufacturing, Proc. of IEA/AIE-96, The Ninth Int. Conf. on Industrial & Engineering Applications of Artificial Intelligence & Expert Systems, June 4-7, 1996, Fukuoka, Japan, Gordon and Breach Publishers, pp. 469-478.

[15] Monostori, L.; Márkus, A.; Van Brussel, H.; Westkämper, E., 1996, Machine learning approaches to manufacturing, CIRP Annals, Vol. 45, No. 2, pp. 675-712.

[16] Monostori, L; Egresits, Cs; Hornyák, J; Viharos, Zs. J.; Soft computing and hybrid AI approaches to intelligent manufacturing. Proc. of 11th International Conference on Industrial & Engineering Applications of Artificial Intelligence & Expert Systems, Castellon, 1998, pp. 763-774

[17] Rangwala, S.S.; Dornfeld, D.A., 1989, Learning and optimization of machining operations using computing abilities of neural networks, IEEE Trans. on SMC, Vol. 19, No. 2, March/April, pp. 299-314.

[18] Salomon, R; Hemmen, L; Accelerating backpropagation through dynamic self-adaptation. Journal of Neural Networks, Great Britain, 1996. pp. 589-601.

[19] Tang, Z.; Koehler, G. J.; Deterministic global optimal FFN training algorithms. Journal of Neural Networks, Great Britain, 1994. pp. 301-311.

[20] Tarng, Y.S.; Ma, S.C.; Chung, L.K., 1995, Determination of optimal cutting parameters in wire electrical discharge machining, Int. J. Mach. Tools Manufact., Vol. 35, No.12, pp. 1693-1701.

[21] Tollerane, B.; SuperSAB: fast adaptive back propagation with good scaling properties, *Neural Networks*, Vol. 3, pp. 561-573.

[22] Viharos, Zs. J.; Monostori, L.; Markos, S; Selection of input and output variables of ANN based modeling of cutting processes. Proceedings of the X. Workshop on Supervising and Diagnostics of Machining Systems of CIRP, Poland, 1999 (under appear)

[23] Viharos, Zs. J.; Monostori, L.; Optimization of process chains by artificial neural networks and genetic algorithms using quality control charts. Proc. of Danube - Adria Association for Automation and Metrology, Dubrovnik,1997. pp. 353-354

[24] Volper, D. J.; Hampson, S. E.; Quadratic function nodes: use, structure and training. Journal of Neural Networks, Great Britain, 1990. pp. 93-107.

DCL: A Disjunctive Learning Algorithm for Rule Extraction

Saleh M. Abu-Soud[1], Mehmet R. Tolun[2]

[1] Department of Computer Science, Princess Sumaya
University College for Technology, Royal Scientific Society,
Amman 11941, Jordan
abu-soud@rss.gov.jo
[2] Dept. of Computer Engineering, Middle East Technical
University, Inonu Bulvari, Ankara, Turkey 06531
tolun@ceng.metu.edu.tr

Abstract. Most concept learning algorithms are conjunctive algorithms, i.e. generate production rules that include AND-operators only. This paper examines the induction of disjunctive concepts or descriptions. We present an algorithm, called DCL, for disjunctive concept learning that partitions the training data according to class descriptions. This algorithm is an improved version of our conjunctive learning algorithm, ILA. DCL generates production rules with AND/OR-operators from a set of training examples. This approach is particularly useful for creating multiple decision boundaries. We also describe application of DCL to a range of training sets with different number of attributes and classes. The results obtained show that DCL can produce fewer number of rules than most other algorithms used for inductive concept learning, and also can classify considerably more unseen examples than conjunctive algorithms.

1 Introduction

AI systems that learn by example can be viewed as searching a concept space that may include conjunctive and disjunctive concepts. Although most of the inductive algorithms developed so far can learn and thus generate IF-THEN type of rules using conjunction (i.e. AND-operator), the number of disjunctive concept learning systems have been quite limited. While each method has its own advantages and disadvantages the common tool they utilize are decision trees that are generated from a set of training examples. Decision tree-based approaches to inductive concept learning are typically preferred because they are efficient and, thus, can deal with large number of training examples. In addition, the final output (i.e. IF-THEN rule) produced is symbolic and, therefore, not difficult for domain experts to interpret.

In the 80's the best known algorithm which takes a set of examples as input and produces a decision tree which is consistent with examples was Quinlan's ID3 algorithm [1]. ID3 was derived from the *Concept Learning System (CLS)* algorithm described by [2]. ID3 had two new features that improved the algorithm. First, an information-theoretic splitting heuristic was used to enable small and efficient decision trees to be constructed. Second, the incorporation of windowing process that enabled the algorithm to cope with large training sets [3]. With these advantages, ID3 has achieved the status of being in the mainstream of symbolic learning approaches and a number of derivatives are proposed by many researchers. For example, ID4 by [4], ID5 by [5], GID3 by [6], and C4.5 by [7]. Other notable inductive learning algorithms include CN2 [8], BCT [9], AQ11 [10], OC1 [11], RULES [12], ASSISTANT 86 [13].

There are many concepts that cannot be described well in conjunctive terms only. One of the examples is the concept of *cousin,* since a cousin can be the son or daughter of an uncle or aunt (which are also disjunctive concepts).

On the other hand some concepts are inherently conjunctive, for example, a hierarchical representation such as [big brother] that corresponds to a conjunctive definition, namely: $X_1 =$ big and $X_2 =$ brother where X is the instance to be classified.

It is also not very difficult to produce further examples of disjunctive concepts. Let us suppose that our examples have attributes *color* and *shape*, such as [blue brick] [red sphere]

In this example {red, green, blue} are the possible colors and {brick, wedge, sphere, pillar} are the possible shapes. Assuming that the final concept covers [green sphere], [red pillar], [red wedge], and [green pillar] but none of the other attribute-value pairs. This concept cannot be represented in terms of a conjunctive concept because there is no attribute that is *shared* between all the positives.

Given a set of training instances, each with associated class labels, the aim is to find a disjunctive description that correctly classifies these instances to the extent possible. In general methods that address this task accept as input a set of classified instances, and generate an expression in *disjunctive normal form (DNF)*, a decision list, or competitive disjunction for use in classifying future instances. In other words, the difference between conjunctive and disjunctive concepts learners lie in that the latter approach lets each class to be associated with more than one attribute.

Most of the research on the induction of disjunctive concepts and has built upon variants of the HSG (Heuristic Specific-to-General) and HGS (Heuristic General-to-Specific) algorithms

In this paper we present a disjunctive concept learning algorithm called DCL. This algorithm produces IF-THEN rules that have the OR-operator from a set of training examples in addition to the AND-operator, which most inductive algorithms use.

DCL is an improved version of a conjunctive learning algorithm called ILA [14], [15], [16], and [17]. It extracts the same number of rules as ILA generates, but adds disjuncts to the L.H.S. of the produced rules. This results in producing a set of general rules, which can classify more unseen examples than ILA and most other inductive algorithms. We also describe the application of DCL to a set of problems demonstrating the performance of the algorithm.

2 DCL: The Disjunctive Concept Learning Algorithm

DCL is a kind of a sequential covering algorithm as it is based on the strategy of generating one rule, removing the data it covers and then repeating this process. However, DCL performs this on a set of positive and negative training examples separately. Usually the generated rule has a high coverage of the training data.

The sequential covering algorithm is one of the most widespread approaches to learning disjunctive sets of rules. It reduces the problem of learning disjunctive set of rules to a sequence of simpler problems, each requiring that a single conjunctive rule be learned. Because it performs a greedy search, formulating a sequence of rules without backtracking, it is not guaranteed to find the smallest or best set of rules that cover the training examples. Decision tree algorithms such as ID3 on the other hand, learns the entire set of disjuncts simultaneously.

Many variations of the sequential covering approach have been explored, for example AQ family that predates the CN2 algorithm. Like CN2, AQ learns a disjunctive set of rules that together cover the target function.

Our algorithm searches a space of disjunctive descriptions rather than a smaller space of conjunctive descriptions. This task is often more difficult than conjunctive ones. DCL is nonincremental although there are some incremental algorithms, for example ISC (Incremental Separate and Conquer) algorithm. Most disjunctive methods place no restriction on the number of terms in each disjunction. The NSC (Nonincremental Separate and Conquer), NEX (Nonincremental Induction Using Exceptions), and NCD (Nonincremental Induction of Competitive Disjuncts) algorithms introduce new terms only when they find training instances that are misclassified by existing ones, as do their incremental counterparts.

DCL is a new algorithm for generating a set of classification rules for a collection of training examples. The algorithm to be described in section 2.2 starts processing the training data by sorting the example set according to class (decision) attribute values. It then constructs sub-tables for each different class attribute value. Afterwards it makes comparisons between attribute values of an attribute among all sub-tables and counts their number of occurrences. Starting off with the maximum number of occurrences it then immediately begins generating rules until it marks all rows of a sub-table classified. DCL then repeats this process for each sub-table and attribute values of each attribute. Finally, all possible IF-THEN rules are derived when there are no unmarked rows left for processing.

2.1 General Requirements

1. The examples are to be listed in a table where each row corresponds to an example and each column contains attribute values.
2. A set of m training examples, each example composed of k attributes, and a class attribute with n possible decisions.
3. A rule set, R, with an initial value of ϕ.
4. All rows in the tables are initially unmarked.

2.2 The DCL Algorithm

Step1 Partition the table, which contains m examples into n sub-tables. One table for each possible value that the class attribute can take.

(* Steps 2 through 9 are repeated for each sub-table *)

Step2 Initialize attribute combination count j as j=1.

Step3 For the sub-table under consideration, divide the attribute list into distinct combinations, each combination with j distinct attributes.

Step4 For each combination, count the number of occurrences of attribute values that appear under the same combination of attributes in unmarked rows of the sub-table under consideration but not under the same combination of attributes of other sub-tables. Call the first combination with the maximum number of occurrences as max-combination.

Step5 If max-combination = ϕ, increase j by 1 and go to Step 3.

Step6 Mark all rows of the sub-table under consideration, in which the values of max-combination appear, as classified.

Step7 Add a rule to R whose LHS comprise attribute names of max-combination with their values separated by AND operator(s) and its RHS contains the decision attribute value of the sub-table.

Step8 If there is more than one combination having the same number of occurrences as max-combination, then classify only the rows last marked as classified in step 6. For such a combination, add it to the LHS of the current rule with an AND-operator separating the attributes within the combination and with an OR-operator separating the combinations.

Step9 If there are still unmarked rows go to Step 4. Otherwise move on to process another sub-table and go to Step 2. If no sub-tables are available, exit with the set of rules obtained so far.□

3. A Description of the Disjunctive Concept Learning Algorithm

DCL is a disjunctive concept learning algorithm for extracting IF-THEN rules from a history of previously stored database of examples. An example in the database is described in terms of a fixed set of attributes, each with its own set of possible values.

As an illustration of the operation of DCL, let us consider the Object Classification training set given in Table 1. This set consists of seven examples (i.e. m=7) with three attributes (k=3) and one decision (class) attribute with two possible values (n=2).

In this example, "size", "color", and "shape" are attributes with sets of possible values {small, medium, large}, {red, blue, green, yellow}, and {brick, wedge, sphere, pillar} respectively.

Table 1. Object Classification Training Set [3]

Instance	Size	Color	Shape	Decision
1	medium	blue	brick	yes
2	small	red	wedge	no
3	small	red	sphere	yes
4	large	red	wedge	no
5	large	green	pillar	yes
6	large	red	pillar	no
7	large	green	sphere	yes

Since *n* is two, the first step of the algorithm generates two subtables (Table 2 and Table 3).

Table 2. Training Set Partitioned According to Decision 'Yes'

Example old	new	Size	Color	Shape	Decision
1	1	medium	blue	brick	yes
3	2	small	red	sphere	yes
5	3	large	green	pillar	yes
7	4	Large	Green	Sphere	yes

Table 3. Training Set Partitioned According to Decision 'No'

Example old	new	Size	Color	Shape	Decision
2	1	Small	red	wedge	no
4	2	Large	red	wedge	no
6	3	Large	red	pillar	no

Applying the second step of the algorithm, we consider first the sub-table in Table 2:
For j=1, the list of attribute combinations is: {size}, {color}, {shape}.

Starting off with one attribute combination and counting the number of occurrences of attributes under each combination that appear in Table 2 but not in Table 3, the following number of occurrences are obtained:

1. For the combination {size} the value "medium" appeared once.
2. For the combination {color} the value "blue" appeared once and "green" appeared twice.
3. For the combination {shape} the value "brick" appeared once and "sphere" appeared twice.

The maximum number of occurrences is therefore, two for color attribute-value "green" and shape attribute-value "sphere". Thus, considering any one of these values will not affect the final outcome. So, let us consider the first maximum value, i.e. twice-occurring color attribute-value of "green". According to this selection, rows 3 and 4 of Table 2 will be marked as classified, and a rule will be extracted. Its LHS will be <IF color is green...>. Now let us see if other conditions will be added to this rule's premises. According to step 6 of DCL algorithm, there is another attribute value {shape : sphere} that also appeared twice. Since this attribute-value appears in the second and fourth rows in Table 2, while {color : green} appears in the third and fourth rows, it will not be considered. The rule is then completed as follows:

Rule1: IF color is green THEN class is yes

Now, the same steps will be repeated for the unmarked examples in Table 2(i.e. rows 1 and 2). By applying these steps again, we obtain:

1. "medium" attribute value of {size} appeared once.
2. "blue" attribute value of {color} appeared once.
3. "brick" and "sphere" attribute values of {shape} occurring once.

Considering the "medium" attribute value of {size}, we obtain the LHS of the next rule as <IF size is medium...>. Now since {color : blue} and {shape : brick} occurred once and all of them also belong to the first row in Table 2, we can add them to the conditions on the LHS of this rule with an OR-operator. The LHS of the second rule becomes <IF size is medium *OR* color is blue *OR* shape is brick...>. {shape : sphere} will not be considered, since it does not belong to row 1 in Table 2. This rule will be completed as follows:

Rule2: IF size is medium *OR* color is blue *OR* shape is brick THEN class is yes.

After this step, there is only one unmarked row (i.e. the second row) in Table 2 with one value that satisfies the condition in step 3 that is "sphere" attribute value of {shape}. Since we have only this value, it will constitute a separate rule:

Rule3: IF shape is sphere THEN class is yes.

Row 2 in Table 2 is now marked as classified and we proceed on to Table 3. Applying the same steps on Table 3, we give an initial value of *1* to *j*, and then divide the attributes into three distinct combinations, i.e. {size}, {color}, and {shape}. In these combinations we have {shape : wedge} which appear in Table 3 and at the same does not appear in Table 2. This attribute value will again constitute a separate rule:

Rule4: IF shape is wedge THEN class is no.

Still we have one row unmarked in Table 3(i.e. the third row) and no value satisfied the condition of step 3 for combinations {size}, {color}, and {shape}. This will force the algorithm to increase the value of *j* to *2* (i.e. 2-attribute combinations {size and color}, {size and shape}, and {color and shape}) and go to step 3.

Only two of the attribute values satisfy the condition in step 4, namely {size : large & color : red} and {color : red & shape : pillar}. Since both of them appear in the same row in Table 3, both will be included in the LHS of the rule that will be extracted in this step. AND-operator separates the attributes of each combination whereas OR-operator separates the combinations themselves. This rule will be as follows:

Rule5: IF (size is large *AND* color is red)*OR* (shape is pillar *AND* color is red)
     THEN class is no.

Finally, row 3 in Table 3 is marked as classified. Now, since all of the rows in Table 3 are marked as classified and no other sub-table is available, the algorithm halts.

4. Evaluation of Disjunctive Concept Learning Algorithm

4.1 Evaluation Parameters

The evaluation of inductive learning algorithms is not an easy process, because the wide range of tasks on which these algorithms can be applied. However, such evaluation process must include at least the following important parameters: (1) number of rules generated, (2) average number of conditions in the generated rules, and (3) the ability of classifying unseen examples.

The original ILA algorithm generates less number of rules than some well-known inductive algorithms such as ID3 and AQ family algorithms as discussed in [14-17]. DCL extracts the same number of rules that ILA generates. For this reason, we will not concentrate on the first parameter, i.e. number of rules generated.

For the second parameter, that is, average number of conditions in the generated rules, it is quite important for the rules produced by conjunctive concept learning algorithms to have the minimum number of conditions in their LHS as possible. This is because our aim in such algorithms is to generate the most simple and general rules as possible, which can classify the examples listed in the table correctly.

However, classifying the known examples correctly is not enough to consider the classifying algorithm successful. For learning algorithms to have the ability of classifying as much of unseen examples as possible is crucial. One way to achieve that, is to construct disjunctive algorithms, which produce rules with disjuncts or OR-operators. This is of some help in classifying more unseen examples. Of course, as the number of disjuncts gets larger, the ability of classifying unseen examples becomes higher. So, during the evaluation process of DCL, we will concentrate on number of disjuncts available in the generated rules of this algorithm.

4.2 Training Sets

To exploit the strength aspects of DCL and to fully evaluate its performance, we used seven training sets in the experiments performed on our algorithm. The characteristics of these training sets are summarized in Table 4.

These training sets are automatically generated realistic data sets, using the Synthetic Classification Data Set Generator (SCDS) version 2. *

Using synthesized data sets is another important way to assess inductive learning algorithms. It is stated by Melli G. [*], that *"Learning-From-Example (LFE) algorithms are commonly tested against real world data sets, particularly the ones stored at Irvin University Database Repository. Another important way to test learning-from-example algorithms is to use well understood synthetic data sets. With synthetic data sets classification accuracy can be more accurately assessed..."*

Table 4. Characteristics of Training Sets.

	Number Of Examples	Number Of Attributes	Average Values Per Attribute	Number Of Class Values
Data Set 1	100	3+1	20	3
Data Set 2	300	5+1	5	2
Data Set 3	500	5+1	3	4
Data Set 4	500	19+1	6	4
Data Set 5	1000	12+1	6	3
Data Set 6	5000	15+1	10	2
Data Set 7	10000	25+1	100	6

4.3 Experiments

4.3.1 Number of Generated Rules

DCL is exposed to seven data sets as listed in Table 4. The two well-known algorithms ID3 and AQ, in addition to our original algorithm ILA, have also been chosen for comparing their results with DCL on the same data sets.

Table 5 summarizes the number of rules generated by each algorithm on the seven data sets.

Table 5. Number of Rules Generated.

	DCL	ILA	AQ	ID3
Data Set 1	17	17	21	37
Data Set 2	51	51	84	129
Data Set 3	64	64	173	201
Data Set 4	111	111	201	268
Data Set 5	203	203	409	620
Data Set 6	627	627	995	1357
Data Set 7	892	892	1448	2690

As stated earlier in this section, it is noted in Table 5 that DCL produces the same number of rules that ILA was producing. It is noted also that both DCL and ILA produced less number of rules among other algorithms.

4.3.2 Number of Disjuncts

When an inductive learning algorithm generates general rules, that is, rules with less number of conditions in their LHS, the ability to classify unseen examples gets higher. This statement is mainly true when dealing with conjunctive algorithms.

In the case of disjunctive learning algorithms, the situation is somewhat different. In addition to the number of conditions in the generated rules, the rules with more disjuncts the rules with more ability to classify more unseen examples.

As in the case of ILA, DCL extracts general rules with minimum number of conditions. We observe the number of disjuncts in the generated rules, and see the relationship between this and the classification ratio of unseen examples. Table 6 shows the average number of disjuncts in the rules generated after applying DCL on the seven training sets. Of course, this value is zero for other algorithms: ILA, AQ, and ID3, because they are all conjunctive algorithms.

Table 6. The Average Number of Disjuncts in DCL Generated Rules.

Data Set	Average number of disjuncts
Data Set 1	2.734
Data Set 2	2.901
Data Set 3	2.162
Data Set 4	4.861
Data Set 5	0.000
Data Set 6	3.630
Data Set 7	4.770

Since the number of disjuncts depends mainly on the data stored in the training set itself, this may cause some kind of irregularity in the relationship between number of attributes and average number of disjuncts in the seven data sets in Table 6. For example, even thought Data Set 5 has 12 attributes, it has zero disjuncts in its rules. While Data Set 1 has only 3 attributes and the average number of disjuncts in its generated rules is 2.734. Despite this irregularity, it is noted that there is a strong relationship between the number of attributes and the average number of disjuncts in each data set. It is clear that in general, as number of attributes gets larger, the average number of disjuncts also becomes larger.

4.3.3 Classification of Unseen Examples

In order to test the three algorithms for the ability of classifying unseen examples, each training set has been divided into two sets. The first set containing a sub set of the training examples on which the algorithms are run, while the second set contains rest of the examples which are selected randomly to form the unseen examples on which the generated rules from all algorithms are tested.

Tests are conducted on different sizes of data as follows:

Partition I: about 2/3 of the original set is kept as the training set and 1/3 as unseen examples.
Partition II: about 1/2 of the original set is kept as the training set and 1/2 as unseen examples.
Partition III: about 1/3 of the original set is kept as the training set and 2/3 as unseen examples.

To enhance the generality of the results, these tests have been conducted on the above cases for five times, each time with different (*randomly selected*) examples in both sets that contain the training examples and the unseen examples.

As noted earlier, the number of generated rules is the same for DCL and ILA. But the situation is different when considering classification of unseen examples as a factor. DCL proves significant improvements over the original algorithm ILA, and both have better results than AQ and ID3.

Table 7 demonstrates the average of error rates for the five tests of applying the four algorithms on the seven training sets for the same cases above.

It is clear from Table 11 that DCL has achieved a significant improvement on the original algorithm ILA. The average error rates of all algorithms show that DCL has got the lowest error rates for the data sets tested. In fact, in classifying unseen examples DCL is better than ILA, AQ, and ID3 by a factor 1.4, 1.8, and 2.6 respectively.

In Table 11 we also note that as the average number of disjuncts in the resulted set gets large, as the error rate of the same data set in classifying the unseen examples becomes smaller than that of other algorithms, as happened in data sets 2 and 3.

The worst case of DCL occurs when the generated rules do not have any disjuncts, as in data set 5. In this case, DCL behaves in the same way ILA algorithm does.

Table 7. The Average of Error Rates of Classifying Unseen Examples.

	Partition	DCL	ILA	AQ	ID3
Data Set 1	I	30.7%	42.9%	51.9%	57.3%
	II	27.9%	31.3%	42.3%	62.7%
	III	21.5%	26.8%	35.5%	49.9%
Data Set 2	I	22.9%	43.9%	51.4%	67.1%
	II	22.4%	44.1%	49.0%	58.2%
	III	16.7%	38.5%	46.3%	62.6%
Data Set 3	I	30.2%	35.1%	38.8%	66.1%
	II	31.5%	36.2%	43.4%	61.8%
	III	27.1%	30.5%	41.5%	69.2%
Data Set 4	I	4.2%	6.2%	12.0%	28.3%
	II	6.3%	9.6%	16.4%	34.9%
	III	4.7%	7.3%	14.3%	46.1%
Data Set 5	I	41.2%	41.2%	47.3%	53.8%
	II	32.7%	32.7%	38.2%	42.7%
	III	34.3%	34.3%	41.1%	47.2%
Data Set 6	I	15.2%	27.4%	31.7%	44.3%
	II	7.4%	12.8%	17.8%	28.9%
	III	7.0%	10.2%	17.2%	31.0%
Data Set 7	I	6.2%	18.6%	32.8%	37.7%
	II	5.6%	15.3%	30.4%	41.4%
	III	3.4%	11.1%	27.1%	40.9%
Average		19.01%	26.48%	34.59%	49.15%

5 Conclusions

A new disjunctive concept learning system, DCL, is presented. DCL achieves the lowest error rates amongst decision tree based ID3 and rule extraction algorithms AQ and ILA. Over seven synthetic data sets, DCL achieved an average generalization accuracy of 81%, whereas the performance of other systems ranged from 50.85% to 73.52%. DCL's accuracy in unseen test examples is clearly better than ID3, ILA, and AQ.

In all of the tested cases DCL produced the most general production rules which are also fewer in number.

Another conclusion of this work is that as the number of attributes increase the number of disjuncts (OR-operators) found in an output IF-THEN rule increases. However, this conclusion may not be regarded as definitive because this generally depends on the nature of the composed data set.

References

[1] Quinlan, J.R.(1983). "LearningEfficient Classification Procedures and their Application to Chess End Games". In R.S. Michalski, J.G. Carbonell & T.M. Mitchell, *Machine Learning, an Artificial Intelligence Approach,* Palo Alto, CA: Tioga, 463-482.

[2] Hunt, E.B., Marin J., & Stone, P.J. (1966). Experiments in Induction. New York, London : Academic Press.

[3] Thornton, C.J. (1992). Techniques in Computational Learning-An Introduction, London:Chapman & Hall.

[4] Schlimmer, J.C. & Fisher, D. (1986). "A Case Study of Incremental Concept Induction". *Proc. of the Fifth National Conference on Artificial Intelligence,* Philadelphia, PA: Morgan Kaufmann, 496-501.

[5] Utgoff, P.E. (1988). "ID5: An Incremental ID3", *Proc. of the Fifth National Conference on Machine Learning,* Ann Arbor, MI, University of Michigan, 107-120.

[6] Irani, Cheng, Fayyad, and Qian, (1993). "Applying Machine Learning to Semiconductor Manufacturing", IEEE Expert, 8(1), 41-47.

[7] Quinlan, J.R.(1993). C4.5: Programs for Machine Learning. Philadelphia, PA: Morgan Kaufmann.

[8] Clark, P. & Boswell, R.(1991). "Rule Induction with CN2:Some Recent Improvements". In J. Siekmann(ed.). Lecture Notes in Artificial Intelligence, Berlin, Springer-Verlag, 151-163.

[9] Chan, P.K.(1989). "Inductive Learning with BCT", Proc. Sixth International Workshop on Machine Learning. Cornell University, Ithaca, New York, 104-108.

[10] Michalski, R.S., & Larson, J.B. (1978). "Selection of most representative training examples and incremental generation of VL1 hypothesis: The underlying methodology and the descriptions of programs ESEL and AQ11 (Report No. 867)". Urbana, Illinois: Department of Computer Science, University of Illinois.

[11] Murthy, S.K., Kasif, S., & Salzberg, S. (1994). "A System for Induction of Oblique Decision Trees", Journal of Artificial Intelligence Research, 2, 1-32.

[12] Pham, D.T. & Aksoy, M.S.(1995). "RULES: A Simple Rule Extraction System", Expert Systems with Applications, 8(1), 59-65.

[13] Cestnik, B., Kononenko, I., & Bratko, I. (1987). "ASSISTANT 86:A Knowledge-Elicitation Tool for Sophisticated Users", in I. Bratko & N. Lavrac(eds.), Progress in Machine Learning, Wilmslow, UK: Sigma Press, 31-45.

[14] Abu-Soud S., "A Framework for Integrating Decision Support Systems and Expert Systems with Machine Learning", Proceeding of the 10th Intr. Conference on industrial and Engineering Applications of AI and ES, June 1997, Atlanta, USA.

[15] Tolun M. and Abu-Soud S. (1998). "ILA: An Inductive Learning Algorithm for Rule Discovery", The International journal of Expert Systems with Applications, 14(3), April 1998, 361-370.

[16] Tolun M., M. Uludag, H. Sever, and Abu-Soud S., "ILA-2 An Inductive Learning Algorithm for Knowledge Discovery", to appear in the intr. Journal of Cybernetics

[17] Abu-Soud S. and Tolun M. (1999), "A Disjunctive Concept Learning Algorithm for Rule Generation", The Proceedings of the 17th IASTED International Conference on Applied Informatics, Innsbruck, Austria.

Modeling and Simulating Breakdown Situations in Telecommunication Networks

Aomar Osmani

LIPN, Institut Galilée
Avenue J.-B. Clément, 93430 Villetaneuse, France
ao@lipn.univ-paris13.fr

Abstract. In this article we describe a framework for model-based diagnosis of telecommunication management networks.
Our approach consists in modeling the system using temporal graphs and extended transducers and simulating given breakdown situations. The simulation process associates to each breakdown situation a set of sequences of observations, building up a learning database. This database is used to classify breakdown situations in the space of the observations. Our application uses imprecise temporal information expressed by pointisable intervals and requires their integration. In this article a new temporal transducer dealing with such information is proposed.
Telecommunication management networks have a hierarchical structure; we propose a complete propagation algorithm of imprecise events in these systems.

Key words: modeling, simulating behavior, transducer, temporal graph, temporal reasoning, telecommunication networks.

1   Introduction

The management of breakdown situations is a crucial problem for telecommunication management networks. The network complexity requires artificial intelligence techniques to assist the operators in supervision tasks. Initially expert systems techniques were proposed [10], presently various techniques are used: neural networks [9], constraint satisfaction problems [15], Petri networks [4]. These methods are based on the analysis of the breakdown situations observed in the real system.
The continual change of the telecommunication networks reduces the effectiveness of these methods. The model based techniques [2, 3, 13] propose a framework based on the adaptive modeling of the system and the prediction of the behavior of the system from the simulation model. We propose a model-based technique to diagnose breakdown situations in telecommunication networks. This technique consists in modeling the network, simulating given breakdown situations and discriminating characteristics of simulated situations. Finally, the results of discrimination are used to recognize the learned situations in the real system.

In this article, we deal with the modeling and simulation tasks. A temporal graph and a new formalism of an extended transducer are used for the modeling. Telecommunication management networks have a hierarchical structure; we propose a complete propagation algorithm to generate all possible behaviors of the system when breakdown situations are simulated.

In section 2 we introduce a network telecommunication example motivating our work. Section 3 introduces the concepts used for modeling the system. Section 4 presents simulation algorithms and describes the mathematical module. Finally, Section 5 summarizes the paper.

2 Example and motivation

Let us consider the simple configuration of Asynchronous Transfer Mode (ATM) network described by the figure 1. The network (TN) made up of the supervision center, three components (switches) C_1, C_2 and C_3 and users connected to this network. Each component C_i contains a local supervisor Ls_i and a set of slots of the transmissions routing $c_{i1}, \ldots, c_{im_i}$. Our main interest is to diagnose breakdown situations in these components.

When a breakdown situation happens in the system, sets of events are emitted from local slots to the local supervisors, which process these events and send sets (possibly empty) of resulting events to the supervision center.

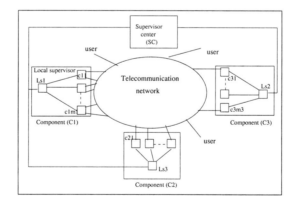

Fig. 1. Structure of simplified real network

To study the behavior of the telecommunication network when breakdown situations happens knowledge about the telecommunication network itself is not required [12]. In the rest of the paper a telecommunication network is viewed as a black box and our study is focused on the telecommunication management network (TMN) [8].

The structure of the TMN is hierarchical; the following figure (figure 2) gives the

TMN network associated to the TN of the figure 1. We refer to this application to illustrate our formalism of representation.

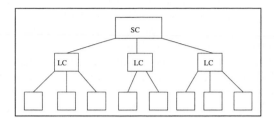

Fig. 2. View of the management network associated to the figure 1

3 Arithmetic of intervals

Reasoning about symbolic propagation of events generates for each simulated situation an exponential number of behaviors. To reduce this complexity we propose a new arithmetic of intervals. In this section, we introduce this arithmetic, which is used in the temporal graph and to extend the formalism of transducers (see section 4).

Let us consider the intervals $I_i = [I_i^-, I_i^+]$ $i = \overline{1..n}$. In classical arithmetic of intervals $I_i + I_j = [I_i^- + I_j^-, I_i^+ + I_j^+]$. This way of computing remains available if we need to generate all possible behaviors of the system, but in several cases we must reduce the number of behaviors. We define a p_interval I as a pair of two classical intervals $(\overline{I}, \underline{I})$. In this framework, a classical interval I is represented by (I, I). The addition operation is defined as follow:

$$I_i \oplus I_j = ([min(I_i^- + I_j^+, I_i^+ + I_j^-), max(I_i^- + I_j^+, I_i^+ + I_j^-)], I_i + I_j) = (\overline{I}, \underline{I})$$

Let us consider I and J two p_intervals. This definition is extended to p_intervals by:

$$(\overline{I}, \underline{I}) \oplus (\overline{J}, \underline{J}) = (\overline{\overline{I} \oplus \overline{J}}, \underline{\underline{I} \oplus \underline{J}})$$

The operation $\odot$ is defined as follows:

$$I \odot J = ([max(\overline{I}^-, \overline{J}^-), min(\overline{I}^+, \overline{J}^+)], [max(\underline{I}^-, \underline{J}^-), min(\underline{I}^+, \underline{J}^+)])$$

The operation $\odot$ is used to define the extended transducer (section 4.2). We note $t(I) = (\overline{t}(I), \underline{t}(I))$ the length of the interval I, where $\overline{t}(I) = t(\overline{I})$ and $\underline{t}(I) = t(\underline{I})$.

Remark 1. In our application only these operations are used. So, we limit the definition of this arithmetic to those operations.

4 Modeling

The first step in studying a system is to build a model of it. To model the system we use two abstraction models: structural and behavioral models.

4.1 Structural model

The telecommunication management network is made up of a set of components connected each other. When a component C receives an event (e, t) two cases are possible: the event changes the state of the component, in which case $(C, (e, t))$ defines a non-transmitter state or the event does not modify anything in the component, in which case $(C, (e, t))$ defines a transmitter state.

At the structural level, only the connections and the states (transmitter, non-transmitter) of the components matter, the behaviors of the components are not accessible.

The structure of the system is modeled by a temporal graph G(N(S),A(T)).

- $N = \{C_1, C_2, \ldots, C_n\}$, where the C_i's are the components of the system and $S_i \in \{transmitter,\ non-transmitter\}$ defines the state of the component C_i when it receives events. The parameter S is used and updated by the simulation module;
- $A = \{A_{ij}, i, j \in \{1, \ldots, n\}\}$, where $A_{ij} = (C_i, C_j)$ defines an oriented connection between the components C_i and C_j. T defines the time parameter. Time is modeled by convex intervals. Let us consider $t \in T, t = [t^-, t^+]$. $A_{ij}(t)$ means all events emitted from the component C_i to the component C_j at an arbitrary time-point k, arriving at C_j in the interval $[k + t^-, k + t^+]$. For our telecommunication application time intervals simulate transition duration of events trough telecommunication lines. Experts give temporal intervals T; they are domain-dependant knowledge.

4.2 Behavioral model

Each component of the system is modeled by a collection At of attributes, a mathematical module M_m (section 5.2), a set of input/output ports $\{Pr_1, \ldots, Pr_n\}$ and a set of extended transducers (section 4.2) $\{Tr_1, \ldots, Tr_m\}$, which defines the behavioral model of the component.

$(\forall C_i \in N), C_i = \{At, \{Pr_1, \ldots, Pr_n\}, Mm, \{Tr_1, \ldots, Tr_m\}\}$ (figure 3).

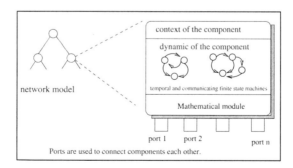

Fig. 3. Structure of the component

Extended transducers Finite state machines are usually used to model dynamic systems [13][6]. Particularly to model the behavior of telecommunication networks [14] [5].

To describe the component behaviors, we introduce a new formalism of an extended transducer (ET). ET is defined by an 9-uplet: $< X_i, X_o, Q, I, F, \delta, At, A, T >$.

- Q is the set of states of the transducer. Each state is labeled by a time interval. $I \in Q$, $F \in Q$ are, respectively, the initial state and the set of final states of the transducer;
- X_i, X_o define input and output alphabets, respectively. Let consider the set of events e_k $k = \overline{1..n}$ received (resp. emitted) by a component C from the ports Pr_j $j = \overline{1..m}$ at the time interval t_{ij}. $\forall x_k \in X_i$ (resp. $\forall x_k \in X_o$)), $x_k = (e_k, Pr_j, t_{kj})$;
- At: to model the system, we use non-deterministic transducers. The context (At) represents the information about the environment. These informations transform the non-deterministic transducers into deterministic ones;
- A: actions computed by the transducer in its environment when transitions occur;
- The transition function: $\delta : Q \times \{X_i \cup \{\epsilon\}\} \times At \times T_I \times T_E \longrightarrow Q \times X_o^* \times T \times A.$

$$(q_1, t_1) \times (e_i, Pr_i, t_i) \times At \times t_I \times t_E \rightarrow (q_2, t_2) \times (e_o, Pr_o, t_o)^* \times a^*$$

$$with \ t_2 = (t_1 \odot t_i) \oplus t_I \ and \ t_o = (t_1 \odot t_i) \oplus t_E$$

- T: simple or pair of intervals models temporal informations. According to the transition function defined above the interpretation of temporal information are: t_I models the imprecision of internal transitions of the component, it represents the time needed by the component to switch from the state q_1 to the state q_2. t_E models the imprecision of external propagation of the events. t_i and t_o model respectively the imprecision of received and emitted events. t_1 and t_2 model the time lower bounds needed by the components to be able to change their states. t_i, t_o, t_1 and t_2 are consistently maintained by our simulating module.

Example The figure 4 gives an example of extended transducer.

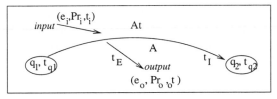

Fig. 4. *Example of an extended transducer*

This figure means that when the transducer is in the state q_1 at time t_1, if it receives event e_i on the port Pr_i at time t_i and context At is satisfied then it switches to state q_2 after time t_I and updates t_2. It sends event e_o on the port Pr_o at time t_o, and computes actions a^* in the context of the transducer.

5 Simulation

When the model is built, given breakdown situations $\{SIC_1, \ldots, SIC_p\}$ are simulated. For each SIC_i $i = \overline{1..p}$ a new kind of component (*simulator component*) is added to the model. SIC_i is a non-deterministic component, having a cause ports that can receive messages from the simulator and effect ports, which are connected to the physical components of the model. It simulates one breakdown situation (see figure 5).

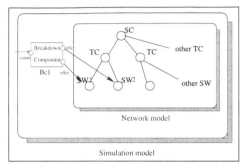

Fig. 5. Example of simulator component

The graph $G_i = (N_i, A_i(T)$ composed by the temporal graph $G=(N,A(T))$ and a breakdown situation SIC_i defines the simulation model of the situation SIC_i. To simulate this situation, a cause port associated to SIC_i is spontaneously triggered and events are propagated in the model of the network to the supervision center (SC). The SC is the alone observable component in the real system. Events, which arrived to this component, are called alarms.

To generate the representative learning database two kinds of simulation algorithms are used: a propagation algorithm, which ensures the consistency and the completeness of the propagation of events between components, and a deduction algorithm, which describes the dynamic of each component. The deduction algorithm is involved by the propagation algorithm.

5.1 Propagation algorithm (PA1)

The principle of PA1 is to trigger the components SIC_i $i = \overline{1..p}$ one by one. Dealing with imprecise temporal information generates for each SIC_j $j = \overline{1..k}$ several possible behaviors of the system. For each behavior, a sequence E_{ji} $i = \overline{1..n}$ of alarms is collected: PA1 generates a learning database (SIC_j, E_{ij}) and a *learning window* (LW) [1].

Let consider the graph $G_{is} = (N_{is}, A_{is}(T))$ as the sub-graph of G_i, where $(\forall C_j \in N_i)$, $C_j \in N_{is}$ iff C_j belongs to a path $\{SIC_i, SC\}$. G_{is} is an oriented graph from SIC_i to SC. This orientation defines the order of processing

---

[1] Learning window LW is a pair (t_{LW}, nb_{LW}) where t_{LW} represents the maximal duration of all sequences of alarms and nb_{LW} represents the maximal cardinal of all sequences of alarms.

of components by the simulator module.

Let consider the relation $\preceq$ where $C_1 \preceq C_2$ means that the node C_1 of the temporal graph G_{is} must be processed before C_2. $(N_{is}, \preceq)$ defines the partial ordering of nodes processing. In hierarchical systems and particularly in telecommunication management networks, the reader may easily verify that $(N_{is}, \preceq)$ defines a lattice.

The following algorithm calculates the graph G_{is} from the graph G_i, computes the total order of processing of the components using lattices properties and builds the learning database (this algorithm computes also the learning windows used by the discrimination module) (see figure 6).
We note $E_i = (E_{ij})$ for $i = \overline{1..n}$. $E_{ij} = (e_{i1}, e_{i2}, \ldots, e_{ij})$ is a sequence of events.

```
Algorithm (PA1): this algorithm is computed for all SIC_i
   C_i = successor(C_j) in G_si means C_i ≼ C_j
   C_i = left_neighbor(C_j) if successor(C_i) = successor(C_j) and x(C_j) = x(C_i) + 1 (see figure 5).
   c(n) defines the component identified by the number n.
Input: SIC_i, G(N,A(T));
Output: Learning database {LW_i, (SIC_i, E_ij) : j = 1..n};
begin
   1. Selecting components implicated in the simulated situation.
      N_i = ∅;
      for all C_j if C_j is in a path (SIC_i, SC) then N_i = N_i ∪ C_j;
   2. Computing lattice of components:
   x(SC)=0; y(SC)=0; n(SC)=0;
   For each node c/ c ∈ N_i do
      x(c)=x(n(successor(c)))+1; y(c)=y(n(left_neighbor(c))+1;
      n(c)= (max( n(c)/ y(c)=x-1))+y;
   n(SIC)= max(n(successors(SIC)))+1;
      2. events propagation
   nn=n(SIC);
   While (nn!=0) do
      for each event e at input of c(nn) do
         if (e, t,Pr, c(nn)) is transmitter then
            Propagate e to the C_j = successor(c(nn)) and update t: t = t ⊕ t(c(nn),C_j)
      Computes deducting_algorithm: DA(c) / n(c)=nn;
      nn=nn-1;
   if (n(c)=0) then E_i = mathematical_module(SC);
   for all j do
      LW_i = (max(t_LW_i, t(E_ij)), max(nb_LW_i, nb_events_in_E_ij);
      save learning example (SIC,E_ij);
   save learning window LW_i;
      LW = (max(t_LW, t_LW_i), max(nb_LW, nb_i))
end.
```

Fig. 6. *Propagation algorithm*

5.2 Deduction algorithm

During the process of simulation, components receive imprecise sequences of events via their ports from the other components (see figure 7). The mathematical module generates all possible orderings of these events consistent with temporal information.

Given a set of sequences of events generated by the mathematical module, the deduction algorithm computes transition functions of the extended transducers describing the behavior of the component defined in the section 4.2 and updates the context of the component.

Mathematical module The following figure shows an example of situation computed by the mathematical module. The component C_i receives m sequences of events $E_1^{n_1}, \ldots E_i^{n_i 2}, \ldots, E_m^{n_m}$ from the components $C_1, C_2, \ldots, C_m$ respectively.

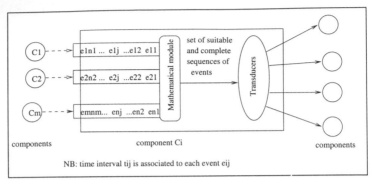

Fig. 7. *Mathematical module*

The main interest of the mathematical module is to generate complete sequences $E_i^n \; i = \overline{1..k}$ [3]. Temporal information are described by pointisable intervals.

In order to reason about these objects we use a subclass of Allen algebra[1]. This subclass called pointisable algebra was introduced by Vilain and Kautz [?]. This is done applying the metric relations of composition and intersection defined in Dechter and al [7]. In our application only two basic relations are needed to compare two events: $<, >$.

The goal is not to check for consistency (finding one solution), but to compute all consistent solutions in order to study all possible behaviors of the system. We have proposed [11] an algorithm (GS1) to compute all possible sequences and maintaining temporal information.

6 Conclusion

In this article, we have presented a model-based technique to simulate behaviors of dynamic systems with imprecise temporal information modeled by intervals. A structural model is based on a temporal graph. A new formalism of transducers is proposed to model the behavior of the components.

The goal of the simulation task is to build a representative learning database of simulated situations. To compute different behaviors of the system, we have developed two kinds of algorithms: a propagation algorithm (PA1) and a deduction algorithm. The deduction algorithm uses the mathematical module and computes the transition functions of the transducers. The propagation algorithm

---

[2] The sequence $E_i^{n_i}$ is defined as a sequence of n_i temporal events $((e_{i1}, t_{i1}), \ldots, (e_{in_i}, t_{in_i}))$.

[3] The sequences E_i^n are complete if and only if $n = \sum_1^m n_j$.

(PA1), proposed in this article, ensures the consistency and the completeness of behaviors associated to simulated situations particularly in TMN systems.

References

1. J. F. Allen. Maintaining knowledge about temporal intervals. *Communications of the ACM*, 26(11):832–843, 1983.
2. N. Azarmi et al. Model-Based Diagnosis in Maintenance of Telecommunication Networks. In *Proceedings of the International Workshop on Principles of Diagnosis(DX'93)*, pages 46–59, Aberyswyth(UK), 1993.
3. S. Bibas, M.O. Cordier, P. Dague, F. Lévy, and L. Rozé. Gaspar: a model-based system for monitoring telecommunication networks. In *Proceedings of the IEEE International Conference on Computational Engineering In Systems Applications (CESA'96)*, Lille(France), 1996.
4. R. Boubour and C. Jard. Fault detection in telecommunication networks based on petri net represnetation of alarm propagation. In *Proccedings of the 18th int. Conf. on Application and theory of Petri Nets*, ed. by Azéma P. and Balbo G.-Spring-Verlag, juin,1997.
5. A. Bouloutas. *Modeling Fault Management in Communication Networks*. PhD thesis, Columbia University, 1990.
6. A. Bouloutas, G. Hart, and M. Schwartz. Fault identification using a finite state machine model with unreliable partially obseved data sequences. *Network Management and Control*, pages 1074–1083, 1993.
7. R. Dechter, I. Meiri, and J. Pearl. Temporal constraint networks. In R. J. Brachman, H. J. Levesque, and R. Reiter, editors, *Knowledge Representation*, pages 61–95. MIT Press, London, 1991.
8. F. Krief and A. Osmani. Gestion de réseaux : Structure fonctionnelle d'un gestionnaire d'alarmes intelligent dans un réseau atm. *De Nouvelles Architectures pour les Communications (DNAC)*, Ministère des Télécoms, Paris (France), 1998.
9. P. Leray, P. Gallinari, and E. Didelet. Diagnosis tools for telecommunication network traffic management. pages 44–52, Bochum, Germany, 1996.
10. R. Molva, M. Diaz, and J.M. Ayache. Dantes: an expert system for real-time network troubleshooting. In *Proceedings of 8th IJCAI 1987*, pages 527–530. IJCAI, 1987.
11. A. Osmani. Simulating behavior of dynamic systems: Reasoning about uncertain propagation of events. In *DEXA'99*. Florence, Italy, submitted 1999.
12. A. Osmani, L. Rozé, M.-O. Cordier, P. Dague, F. Lévy, and E. Mayer. Supervision of telecommunication networks. In *European Control Conference*. Karlsruhe, Germany, to appear 1999.
13. M. Riese. Diagnosis of extended finite automata as a dynamic constraint satisfaction problem. In *Proceedings of the International Workshop on Principles of Diagnosis(DX'93)*, pages 60–73, Aberyswyth(UK), 1993.
14. M. Riese. Réalisation d'un système de diagnostic basé sur modèle pour les protocoles de communication. *CFIP*, 1993.
15. M. Sqalli and E.C. Freuder. A constraint satisfaction model for testing emulated lans in atm networks. In *Proceedings of the International Workshop on Principles of Diagnosis(DX'96)*, Val Morin, 1996.

Handling Context-Sensitive Temporal Knowledge from Multiple Differently Ranked Sources

Seppo Puuronen, Helen Kaikova

University of Jyvaskyla, P.O.Box 35, FIN-40351 Jyvaskyla, Finland
e-mail: {sepi,helen}@jytko.jyu.fi

Abstract. In this paper we develop one way to represent and reason with temporal relations in the context of multiple experts. Every relation between temporal intervals consists of four endpoints' relations. It is supposed that the context we know is the value of every expert competence concerning every endpoint relation. Thus the context for an interval temporal relation is one kind of compound expert's rank, which has four components appropriate to every interval endpoints' relation. Context is being updated after every new opinion is being added to the previous opinions about certain temporal relation. The context of a temporal relation collects all support given by different experts to all components of this relation. The main goal of this paper is to develop tools, which provide formal support for the following manipulations with temporal context: how to derive temporal relation interpreted in context of multiple knowledge sources and how to derive decontextualized value for every temporal relation. Decontextualization of a temporal relation in this paper means obtaining the most appropriate value from the set of Allen's interval relations, which in the best possible way describes the most supported opinion of all involved experts concerning this relation. We discuss two techniques to obtain the decontextualized value. First one (the *minimax* technique) takes into account only the worst divergences when calculates the distances between temporal relations. The second one (the *weighted mean* technique) takes into account all divergences. The modified technique is also considered that takes into account the probability distribution of point relations within the Allen's set and recalculates the appropriate expert support in accordance with «meta-weights» before making decontextualization.

1 Introduction

The problem of representation and reasoning with temporal knowledge arises in a wide range of disciplines, including computer science, philosophy, psychology, and linguistics. In computer science, it is a core problem of information systems, program verification, artificial intelligence, and other areas involving modelling process. Some types of information systems must represent and deal with temporal information. In some applications, such, for example, as keeping medical records, the time course of events becomes a critical part of the data [1]. Temporal reasoning has been applied in planning systems [2], natural language processing [5], knowledge acquisition [14] and other areas.

Reasoning with multiple experts' knowledge is one of central problems in expert systems design [9, 15, 11]. When the knowledge base is built using opinions taken from several sources then it usually includes incompleteness, incorrectness, uncertainty and other pathologies. The area of handling temporal knowledge is not an exception. One possible way to handle it is based on using context of expertise.

It is generally accepted that knowledge has a contextual component. Acquisition, representation, and exploitation of knowledge in context would have a major contribution in knowledge representation, knowledge acquisition, explanation [3] and also in temporal reasoning [6, 4, 12]. Contextual component of knowledge is closely connected with eliciting expertise from one or more experts in order to construct a single knowledge base.

In [13] a formal knowledge-based framework was developed and implemented for decomposing and solving that task that supports acquisition, maintenance, reuse, and sharing of temporal-abstraction knowledge. The logical model underlying the representation and runtime formation of interpretation contexts was presented. Interpretation contexts are relevant for abstraction of time-oriented data and are induced by input data, concluded abstractions, external events, goals of the temporal-abstraction process, and certain combinations of interpretation contexts. Knowledge about interpretation contexts is represented as context ontology and as a dynamic induction relation over interpretation contexts and other proposition types. Several advantages were discussed to the explicit separation of interpretation-context.

In this paper we develop one way to represent temporal relations in the context of multiple experts. Every relation between temporal intervals consists of four endpoints' relations. It is supposed that the context we know is the value of every expert competence concerning every endpoint relation. Thus the context for an interval temporal relation is one kind of compound expert's rank, which has four components appropriate to every interval endpoints' relation. Context is being updated after every new opinion is being added to the previous opinions about certain temporal relation. The context of a temporal relation collects all support given by different experts to all components of this relation. The main goal of this paper is to develop methods, which provide formal support for the following manipulations with temporal context:

- how to define *general representation* of temporal relation, which takes into account an appropriate context;
- how to derive temporal relation interpreted in the context of *multiple knowledge sources*;
- how to derive *decontextualized value* for every temporal relation.

Decontextualization of a temporal relation in this paper means obtaining the most appropriate value from the set of Allen's interval relations [1], which in some reasonable way describes the most supported opinion of all involved experts concerning this relation.

2 Basic Concepts

Temporal knowledge in this paper is represented by temporal relations between temporal intervals. It is assumed that temporal knowledge about some domain can be obtained from several different knowledge sources.

Temporal relation is considered as one of Allen's relations [1] between temporal intervals.

Temporal relation between two temporal intervals A and B is represented using Hirsch notation [7] as 2*2 matrix $[a_{ij}]_{AB}$, where a_{ij} defines one of $\{<,=,>\}$ relations between *i-th* end of interval A and *j-th* end of interval B. For example: relation

$\begin{bmatrix} < & < \\ > & < \end{bmatrix}_{AB}$ in Hirsch notation corresponds to Allen's *overlaps* relation (Fig. 1).

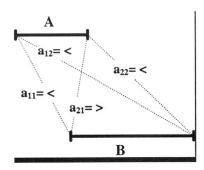

Fig. 1. Components of Hirsch matrix for *overlaps* relation

There are thirteen Allen's temporal relations. They are as follows in Hirsch notation with explicit indication of the relation's name:

$$\begin{bmatrix} < & < \\ < & < \end{bmatrix}_{AB}^{before} ; \begin{bmatrix} > & > \\ > & > \end{bmatrix}_{AB}^{after} ; \begin{bmatrix} < & < \\ = & < \end{bmatrix}_{AB}^{meets} ; \begin{bmatrix} > & = \\ > & > \end{bmatrix}_{AB}^{met-by} ; \begin{bmatrix} > & < \\ > & < \end{bmatrix}_{AB}^{during} ;$$

$$\begin{bmatrix} < & < \\ > & > \end{bmatrix}_{AB}^{includes} ; \begin{bmatrix} < & < \\ > & < \end{bmatrix}_{AB}^{overlaps} ; \begin{bmatrix} > & < \\ > & > \end{bmatrix}_{AB}^{overlapped-by} ; \begin{bmatrix} = & < \\ > & < \end{bmatrix}_{AB}^{starts} ; \quad (1)$$

$$\begin{bmatrix} = & < \\ > & > \end{bmatrix}_{AB}^{started-by} ; \begin{bmatrix} > & < \\ > & = \end{bmatrix}_{AB}^{finishes} ; \begin{bmatrix} < & < \\ > & = \end{bmatrix}_{AB}^{finished-by} ; \begin{bmatrix} = & < \\ > & = \end{bmatrix}_{AB}^{equals}$$

We consider *context* as some additional information about a source of temporal knowledge. Context in this model represents knowledge about the quality of expertise concerning each temporal relation. Context has the same structure as temporal relation and it is represented by a matrix $\left[q_{ij}\right]_S$ where q_{ij} $(0 < q_{ij} < n)$ is the value of the expertise quality of knowledge source S concerning relation of a type a_{ij}, n - number of knowledge sources. For example, if $n = 3$, then the following context of some knowledge source S:

$$\begin{bmatrix} 2.99 & 0.2 \\ 1.48 & 0.01 \end{bmatrix}_S \tag{2}$$

can be interpreted as follows. The knowledge source S has almost absolute experience when it claims something about a relation between the beginnings of any two possible intervals. In the same time it is not experienced to define a relation between ends of possible intervals. It is also weak to define a relation between the beginning of one of possible intervals and the end of another one. It is moderately well experienced to define a relation between the end of one of possible intervals and the beginning of another one.

The definition of a context in this model is analogous to the experts' ranking system represented in [8]. However an essential extension in this model is that an expert's rank (rank of a knowledge source) has its own structure. Thus it may occur that an expert can improve some parts of his compound rank and in the same time loose something within another part of the rank. The evaluation of an expert competence concerning each domain attribute (object or relation) is also presented in [10] without taking into account the structure of every relation. Here we share an expert competence among all structural components of any relation.

3 General Representation of a Temporal Relation

We define a *contextual support* c_{ij}^r given to ij-th component of the relation $[a_{ij}]_{AB}$ from the context $[q_{ij}]_S$ as follows:

$$c_{ij}^r = \begin{cases} q_{ij}, & \text{if } r = a_{ij}; \\ 0, & \text{otherwise.} \end{cases} \tag{3}$$

For example, if the relation is $\begin{bmatrix} < & < \\ > & < \end{bmatrix}_{AB}$ and the context is $\begin{bmatrix} 2.99 & 0.2 \\ 1.48 & 0.01 \end{bmatrix}_S$, then contextual support $c_{22}^< = 0.01$, $c_{21}^> = 1.48$, $c_{22}^= = 0$ and so on.

We define general representation of a temporal relation between two intervals A and B taken from source S as follows:

$$\begin{bmatrix} c_{11}^<, c_{11}^=, c_{11}^> & c_{12}^<, c_{12}^=, c_{12}^> \\ c_{21}^<, c_{21}^=, c_{21}^> & c_{22}^<, c_{22}^=, c_{22}^> \end{bmatrix}_{AB}^S, \tag{4}$$

where c_{ij}^r is a contextual support for the fact that i and j endpoints are connected by r relation.

4 Interpretation in the Context of Multiple Knowledge Sources

The context of multiple knowledge sources is considered when two or more knowledge sources give their opinions about the relation between the same temporal intervals. Such context can be derived from the contexts of each knowledge source by summing up the appropriate contextual support.

Let there be n knowledge sources S_1, S_2, …, S_n giving their opinions about relation between intervals A and B. The resulting relation interpreted in the context of multiple knowledge sources $S_1 \cap S_2 \cap ... \cap S_n$ should be derived as follows:

$$\left[c_{ij}^r \right]_{AB}^{S_1 \cap S_2 \cap ... \cap S_n} = \left[c_{ij}^r \right]_{AB}^{S_1} + \left[c_{ij}^r \right]_{AB}^{S_2} + ... + \left[c_{ij}^r \right]_{AB}^{S_n} = \left[\sum_{k=1, \; c_{ij}^r \in \left[c_{ij}^r \right]^{S_k}}^{n} c_{ij}^r \right]_{AB}^{S_1 \cap S_2 \cap ... \cap S_n} , \tag{5}$$

where «+» defines a *semantic sum* [16] for interpreted temporal relations.

Let there be two knowledge sources S_1, S_2, and their contexts, for example, are:

$$\begin{bmatrix} 2.99 & 0.2 \\ 1.48 & 0.01 \end{bmatrix}_{S_1} \text{ and } \begin{bmatrix} 0.14 & 0.02 \\ 2.7 & 0.8 \end{bmatrix}_{S_2} . \tag{6}$$

Let these two sources name the relation between intervals A and B respectively as: *A overlaps B* and *A starts B*. Thus we have two following opinions:

$$\begin{bmatrix} < & < \\ > & < \end{bmatrix}_{AB}^{\begin{bmatrix} 2.99 & 0.2 \\ 1.48 & 0.01 \end{bmatrix}_{S_1}} = \begin{bmatrix} 2.99, \; 0, \; 0 & 0.2, \; 0, \; 0 \\ 0, \; 0, \; 1.48 & 0.01, \; 0, \; 0 \end{bmatrix}_{AB}^{S_1} \text{ and}$$

$$\begin{bmatrix} = & < \\ > & < \end{bmatrix}_{AB}^{\begin{bmatrix} 0.14 & 0.02 \\ 2.7 & 0.8 \end{bmatrix}_{S_2}} = \begin{bmatrix} 0, \; 0.14, \; 0 & 0.02, \; 0, \; 0 \\ 0, \; 0, \; 2.7 & 0.8, \; 0, \; 0 \end{bmatrix}_{AB}^{S_2} .$$

According to the above definition, the relation between intervals A and B in the context of multiple sources S_1, S_2 is as follows: $\begin{bmatrix} 2.99, \; 0.14, \; 0 & 0.22, \; 0, \; 0 \\ 0, \; 0, \; 4.18 & 0.81, \; 0, \; 0 \end{bmatrix}_{AB}^{S_1 \cap S_2} .$

5 Deriving a Decontextualized Value for a Temporal Relation

Decontextualization of a temporal relation in this paper means obtaining the most appropriate value from the set of Allen's interval relations [1], which in some reasonable way describes the most supported opinion of all involved experts concerning this relation. We call this as the decontextualized value for a temporal relation and consider the decontextualization assuming that the probabilities for the interval temporal relations are equally distributed within Allen's set.

We define 2x2 *distance* *matrix* between two temporal relations

$$\begin{bmatrix} c_{11}^<,c_{11}^=,c_{11}^> & c_{12}^<,c_{12}^=,c_{12}^> \\ c_{21}^<,c_{21}^=,c_{21}^> & c_{22}^<,c_{22}^=,c_{22}^> \end{bmatrix}_{AB}^{S_1} \quad \text{and} \quad \begin{bmatrix} \tilde{c}_{11}^<,\tilde{c}_{11}^=,\tilde{c}_{11}^> & \tilde{c}_{12}^<,\tilde{c}_{12}^=,\tilde{c}_{12}^> \\ \tilde{c}_{21}^<,\tilde{c}_{21}^=,\tilde{c}_{21}^> & \tilde{c}_{22}^<,\tilde{c}_{22}^=,\tilde{c}_{22}^> \end{bmatrix}_{AB}^{S_2} \quad \text{as:} \quad \left[d_{ij} \right]_{AB}^{S_1,S_2}, \text{ where:}$$

$$d_{ij} = abs\left(\frac{c_{ij}^> + 0,5 \cdot c_{ij}^=}{c_{ij}^< + c_{ij}^= + c_{ij}^>} - \frac{\tilde{c}_{ij}^> + 0,5 \cdot \tilde{c}_{ij}^=}{\tilde{c}_{ij}^< + \tilde{c}_{ij}^= + \tilde{c}_{ij}^>} \right)$$

The physical interpretation of formula for calculating distance matrix is shown in Fig. 2. Every ij-th component of a temporal relation matrix has some amount of contextual support for each «<», «=» and «>» values. These supports for each component of the two compared relations are distributed as physical objects on the virtual lath, which is being marked as closed interval [0,1], as it is shown in Fig. 2: «<» on the left end of the interval, «>» on the right end of the interval, and «=» on the middle of the interval.

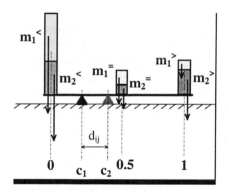

Fig. 2. Physical interpretation of distance measure

«Masses» of support are respectively: $m_1^< = c_{ij}^<$, $m_1^= = c_{ij}^=$, $m_1^> = c_{ij}^>$ and $m_2^< = \tilde{c}_{ij}^<$, $m_2^= = \tilde{c}_{ij}^=$, $m_2^> = \tilde{c}_{ij}^>$. We calculate the two points of balance separately for support distribution of the appropriate components of the two relations. The distance d_{ij} between the balance point for $m_1^<$, $m_1^=$, $m_1^>$ and the balance point for $m_2^<$, $m_2^=$, $m_2^>$ is taken as the ij-th component of the resulting distance matrix.

5.1 The *Minimax* Decontextualization Technique

The first *minimax* decontextualization technique, we consider, calculates the *distance* between two temporal relations as maximal value of distance matrix:

$$D_{AB}^{S_1,S_2} = \max_{ij} d_{ij}. \tag{7}$$

We define the *operation of contextual adaptation* of the temporal relation:

$$\begin{bmatrix} c_{11}^<,\overline{c_{11}},c_{11}^> & c_{12}^<,\overline{c_{12}},c_{12}^> \\ c_{21}^<,\overline{c_{21}},c_{21}^> & c_{22}^<,\overline{c_{22}},c_{22}^> \end{bmatrix}_{AB}^{S_1} \tag{8}$$

to the temporal relation:
$$\begin{bmatrix} \widetilde{c}_{11}^<,\widetilde{\overline{c}}_{11},\widetilde{c}_{11}^> & \widetilde{c}_{12}^<,\widetilde{\overline{c}}_{12},\widetilde{c}_{12}^> \\ \widetilde{c}_{21}^<,\widetilde{\overline{c}}_{21},\widetilde{c}_{21}^> & \widetilde{c}_{22}^<,\widetilde{\overline{c}}_{22},\widetilde{c}_{22}^> \end{bmatrix}_{AB}^{S_2} \quad \text{as:}$$

$$\begin{bmatrix} c_{11}^< \cdot \varphi_{11},\, \overline{c}_{11} \cdot \varphi_{11},\, c_{11}^> \cdot \varphi_{11} & c_{12}^< \cdot \varphi_{12},\, \overline{c}_{12} \cdot \varphi_{12},\, c_{12}^> \cdot \varphi_{12} \\ c_{21}^< \cdot \varphi_{21},\, \overline{c}_{21} \cdot \varphi_{21},\, c_{21}^> \cdot \varphi_{21} & c_{22}^< \cdot \varphi_{22},\, \overline{c}_{22} \cdot \varphi_{22},\, c_{22}^> \cdot \varphi_{22} \end{bmatrix}_{AB}^{S_1 \to S_2}, \tag{9}$$

where $\varphi_{ij} = \dfrac{\widetilde{c}_{ij}^< + \widetilde{\overline{c}}_{ij} + \widetilde{c}_{ij}^>}{c_{ij}^< + \overline{c}_{ij} + c_{ij}^>}$ are the *coefficients of adaptation.*

We define a *decontextualized value* for a temporal relation as such temporal relation from Allen's set adapted to the original one, which has lowest value of distance with the original relation. If there are more than one Allen's relations with the same distance concerning the original relation, then the next maximal elements of the appropriate distance matrixes are used to resolve the conflict.

Let us consider an example of deriving decontextualized value for the relation:

$$\begin{bmatrix} 5,\ 4,\ 3 & 0,\ 6,\ 7 \\ 3,\ 3,\ 3 & 9,\ 0,\ 3 \end{bmatrix}_{AB}^{S_1}. \tag{10}$$

First we make adaptation of all Allen's relations to the above relation. For example adapted relations *before* and *equals* look as follows:

$$\begin{bmatrix} 12,\ 0,\ 0 & 13,\ 0,\ 0 \\ 9,\ 0,\ 0 & 12,\ 0,\ 0 \end{bmatrix}_{AB}^{before \to S_1} ; \quad \begin{bmatrix} 0,\ 12,\ 0 & 13,\ 0,\ 0 \\ 0,\ 0,\ 9 & 0,\ 12,\ 0 \end{bmatrix}_{AB}^{equals \to S_1}. \tag{11}$$

After that, we calculate distance matrixes between the original relation and all adapted Allen's relations as follows:

$$\begin{bmatrix} 0.417 & 0.769 \\ 0.5 & 0.25 \end{bmatrix}_{AB}^{before,S_1} ; \begin{bmatrix} 0.583 & 0.231 \\ 0.5 & 0.75 \end{bmatrix}_{AB}^{after,S_1} ; \begin{bmatrix} 0.417 & 0.769 \\ 0 & 0.25 \end{bmatrix}_{AB}^{meets,S_1} ;$$

$$\begin{bmatrix} 0.583 & 0.269 \\ 0.5 & 0.75 \end{bmatrix}_{AB}^{\substack{met \\ -by,S_1}} ; \begin{bmatrix} 0.583 & 0.769 \\ 0.5 & 0.25 \end{bmatrix}_{AB}^{during,S_1} ; \begin{bmatrix} 0.417 & 0.769 \\ 0.5 & 0.75 \end{bmatrix}_{AB}^{\substack{inclu- \\ des,S_1}} ;$$

$$\begin{bmatrix} 0.417 & 0.769 \\ 0.5 & 0.25 \end{bmatrix}_{AB}^{\substack{over- \\ laps,S_1}} ; \begin{bmatrix} 0.583 & 0.769 \\ 0.5 & 0.75 \end{bmatrix}_{AB}^{\substack{over- \\ lapped \\ -by,S_1}} ; \begin{bmatrix} 0.083 & 0.769 \\ 0.5 & 0.25 \end{bmatrix}_{AB}^{starts,S_1} ; \tag{12}$$

$$\begin{bmatrix} 0.083 & 0.769 \\ 0.5 & 0.75 \end{bmatrix}^{\substack{started \\ -by,S_1}}_{AB} ; \begin{bmatrix} 0.583 & 0.769 \\ 0.5 & 0.25 \end{bmatrix}^{\substack{fini- \\ shes,S_1}}_{AB} ; \begin{bmatrix} 0.417 & 0.769 \\ 0.5 & 0.25 \end{bmatrix}^{\substack{fini- \\ shed \\ -by,S_1}}_{AB} ;$$

$$\begin{bmatrix} 0.083 & 0.769 \\ 0.5 & 0.25 \end{bmatrix}^{equals,S_1}_{AB} .$$

Now we can calculate the appropriate distances as follows:

$$D_{AB}^{before,S_1} = D_{AB}^{meets,S_1} = D_{AB}^{during,S_1} = D_{AB}^{includes,S_1} = D_{AB}^{overlaps,S_1} = D_{AB}^{\substack{overlapped \\ -by,S_1}} = \tag{13}$$

$$= D_{AB}^{starts,S_1} = D_{AB}^{\substack{started \\ -by,S_1}} = D_{AB}^{finishes,S_1} = D_{AB}^{\substack{finished \\ -by,S_1}} = D_{AB}^{equals,S_1} = 0.769;$$

$$D_{AB}^{after,S_1} = D_{AB}^{met-by,S_1} = 0.75. \tag{14}$$

There are two pretenders to be a decontextualized value for the relation we investigate. They are Allen's relations *after* and *met-by*. Comparing other components of the appropriate distance matrixes one can see that the relation *after* located more close to the relation under investigation. Thus the decontextualized value for the relation $\begin{bmatrix} 5,\ 4,\ 3 & 0,\ 6,\ 7 \\ 3,\ 3,\ 3 & 9,\ 0,\ 3 \end{bmatrix}^{S_1}_{AB}$ is temporal relation *after*: $\begin{bmatrix} > & > \\ > & > \end{bmatrix}^{after}_{AB}$.

The above decontextualization technique is based on a *minimax* principle. It means that we find minimal value among maximal divergences between the original relation and all Allen's relations.

5.2 The *Weighted Mean* Decontextualization Technique

Another possible decontextualization technique is based on a *weighted mean* principle. It means that the resulting distance between the original relation and every of Allen's relations is calculated as weighted mean between all components of the distance matrix. Weight for each component is defined by an appropriate contextual support value. We consider this second decontextualization technique more detail. Let it be the following temporal relation:

$$\begin{bmatrix} c_{11}^{<},c_{11}^{=},c_{11}^{>} & c_{12}^{<},c_{12}^{=},c_{12}^{>} \\ c_{21}^{<},c_{21}^{=},c_{21}^{>} & c_{22}^{<},c_{22}^{=},c_{22}^{>} \end{bmatrix}^{S_1}_{AB} . \tag{15}$$

We define the 2x2 *weight matrix* as: $\left[w_{ij} \right]^{S_1}_{AB}$, where: $w_{ij} = \dfrac{\sum\limits_{r} c_{ij}^{r}}{\sum\limits_{pqr} c_{pq}^{r}}$.

Let we also have another temporal relation $\begin{bmatrix} \tilde{c}_{11}^{<}, \tilde{c}_{11}^{=}, \tilde{c}_{11}^{>} & \tilde{c}_{12}^{<}, \tilde{c}_{12}^{=}, \tilde{c}_{12}^{>} \\ \tilde{c}_{21}^{<}, \tilde{c}_{21}^{=}, \tilde{c}_{21}^{>} & \tilde{c}_{22}^{<}, \tilde{c}_{22}^{=}, \tilde{c}_{22}^{>} \end{bmatrix}_{AB}^{S_2}$, which is

adapted to the first one. Let we have the following distance matrix for the two relations: $[d_{ij}]_{AB}^{S_1, S_2}$, which is calculated according to the definition.

We define distance between the two above relations according the weighted mean technique as follows:

$$D_{AB}^{S_1, S_2} = \sum_{ij} w_{ij} \cdot d_{ij} . \tag{16}$$

The decontextualized value for any relation according to the weighted mean technique is defined as that Allen's temporal relation, which has minimal distance to the decontextualized relation, after being adapted to it. If there are more than one Allen's relations have minimal distance, then we select one, which has minimal distance according to the minimax technique.

Let us use the weighted mean technique to decontextualize the relation of the above example. The weight matrix can be calculated as follows:

$$\begin{bmatrix} 0.261 & 0.283 \\ 0.195 & 0.261 \end{bmatrix}_{AB}^{S_1} . \tag{17}$$

Thus the decontextualized value for the relation (10) according to weighted mean technique is temporal relation *meets*, which is totally different than one obtained by the minimax technique.

6 Conclusion

In this paper we develop one way to represent and reason with temporal relations in the context of multiple experts. Context for an interval temporal relation is some kind of compound expert's rank, which has four components appropriate to every interval endpoints' relation. The context of a temporal relation collects all support given by different experts to all components of this relation. Decontextualization of a temporal relation in this paper means obtaining the most appropriate value from the set of Allen's interval relations, which in the best possible way describes the most supported opinion of all involved experts concerning this relation. We discuss two techniques to obtain the decontextualized value. First one (the minimax technique) takes into account only the worst divergences when calculates the distances between temporal relations. The second one (the weighted mean technique) takes into account all divergences. Both ones seem reasonable as well as their meta-weighted modification. However further research is needed to recognize features of the appropriate problem area that define right selection of decontextualization technique in every case.

Acknowledgment: This research is partly supported by the grant of the Academy of Finland.

References

1. Allen, J.: Maintaining knowledge about temporal intervals. Communications of the ACM 26 (1983) 832-843
2. Bacchus, F.: Using Temporal Logics for Planning and Control. In: TIME '96 – Proceedings of the International Workshop on Temporal Representation and Reasoning, Key West, Florida, USA (1996)
3. Brezillon, P., Cases, E.: Cooperating for Assisting Intelligently Operators. In: Proceedings of Actes International Workshop on the Design of Cooperative Systems (1995) 370-384
4. Chittaro, L., Montanari, A.: Efficient Handling of Context Dependency in the Cached Event Calculus. In: TIME '94 – Proceedings of the International Workshop on Temporal Representation and Reasoning, Pensacola, Florida, USA (1994)
5. Guillen, R., Wiebe, J.: Handling Temporal Relations in Scheduling Dialogues for An MT System. In: TIME '96 – Proceedings of the International Workshop on Temporal Representation and Reasoning, Key West, Florida, USA (1996)
6. Haddawy, P.: Temporal Reasoning with Context-Sensitive Probability Logic. In: TIME '95 – Proceedings of the International Workshop on Temporal Representation and Reasoning, Melbourne Beach, Florida, USA (1995)
7. Hirsch, R.: Relation algebra of intervals. Artificial Intelligence 83 (1996) 267-295
8. Kaikova, H., Terziyan, V.: Temporal Knowledge Acquisition From Multiple Experts. In: Shoval P., Silberschatz A. (eds.): Proceedings of NGITS'97 – The Third International Workshop on Next Generation Information Technologies and Systems, Neve Ilan, Israel (1997) 44-55
9. Mak, B., Bui, T., Blanning, R.: Aggregating and Updating Experts' Knowledge: An Experimental Evaluation of Five Classification Techniques. Expert Systems with Applications 10(2) (1996) 233-241
10. Puuronen, S., Terziyan, V.: Modeling Consensus Knowledge from Multiple Sources Based on Semantics of Concepts. In: Albrecht M., Thalheim B. (eds.): Proceedings of the Workshop Challenges of Design, 15-th International Conference on Conceptual Modeling ER'96, Cottbus (1996) 133-146
11. Puuronen, S., Terziyan, V.: Voting-Type Technique of the Multiple Expert Knowledge Refinement. In: Sprague, R. H. (ed.): Proceedings of the Thirtieth Hawaii International Conference on System Sciences, Vol. V. IEEE CS Press (1997) 287-296
12. Ryabov, V., Puuronen, S., Terziyan V.: Uncertain Relations between Temporal Points. In: STeP-98 - Human and Artificial Information Processing, Finnish Conference on Artificial Intelligence, Jyvaskyla, Publ. of the Finnish AI Society (1998) 114-123
13. Shahar, Y.: Dynamic Temporal Interpretation Contexts for Temporal Abstraction. In: Chittaro, L., Montanari, A. (eds.): Temporal Representation and Reasoning, Annals of Mathematics and Artificial Intelligence 22(1,2) (1998) 159-192
14. Shahar, Y., Cheng, C.: Model-Based Abstraction of Temporal Visualisations. In: TIME '98 – Proceedings of the International Workshop on Temporal Representation and Reasoning, Sanibel Island, Florida, USA (1998), to appear
15. Taylor, W., Weimann, D., Martin, P.: Knowledge Acquisition and Synthesis in a Multiple Source Multiple Domain Process Context. Expert Systems with Applications 8(2) (1995) 295-302
16. Terziyan, V., Puuronen, S.: Multilevel Context Representation Using Semantic Metanetwork. In: Context-97 – International and Interdisciplinary Conference on Modeling and Using Context, Rio de Janeiro (1997) 21-32

Introduction to Reasoning about Cyclic Intervals

Aomar Osmani

LIPN, Institut Galilée
Avenue J.-B. Clément, 93430 Villetaneuse, France
ao@lipn.univ-paris13.fr

Abstract. This paper introduces a new formalism for representation and reasoning about time and space. Allen's algebra of time intervals is well known within the constraint-based spatial and temporal reasoning community. The algebra assumes standard time, i.e., time is viewed as a linear order. Some real applications, however, such as reasoning about cyclic processes or cyclic events, need a representational framework based on (totally ordered) cyclic time. The paper describes a still-in-progress work on an algebra of cyclic time intervals, which can be looked at as the counterpart of Allen's linear time algebra for cyclic time.

Key words: Temporal reasoning, temporal constraints, cyclic intervals.

1   Introduction

The well-known temporal interval algebra that proposed by Allen in his 1983 paper [1] considers time as being a linear order. In brief, the element of the algebra are relations that may exist between intervals of time. It admits 2^{13} possible relations between intervals and all of these relations can be expressed as sets of definite simple relations, of which there are only thirteen: precedes, meets, overlaps, starts, during, finishes, equals relations and their converses.

This paper is about representing and reasoning in cyclic intervals. From now on, we shall refer to intervals of linear time as ℓ-intervals, and to intervals of totally ordered cyclic time as c-intervals; furthermore, all over the paper, cyclic time will refer to totally ordered cyclic time. We shall provide the counterpart for cyclic time of Allen's influential algebra of linear time intervals.

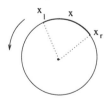

Fig. 1. Cyclic interval $x = (x_r, x_l)$.

We model cyclic time with the circle $\mathcal{C}_{O,1}$ centered at O and of unit radius. A c-interval x will be represented as the pair (x_r, x_l) of its two extremities; we do not allow empty intervals and the length of the c-intervals is less than 2π, which means that we assume $x_l \neq x_r$. We adopt the convention that cyclic time flows in trigonometric direction. Finally, an issue which has raised much attention in philosophy is whether an interval includes its extremities; whether or not this is the case is, however, irrelevant to this work.

2 The algebra of cyclic intervals

The algebra is indeed very similar to Allen's linear time algebra [1]; the difference, as already alluded to, is that we consider cyclic time.

The objects manipulated by the algebra are c-intervals; the atomic relations consist of all possible qualitative configurations of two c-intervals in cyclic time; in other words, all possible qualitative configurations of two c-intervals on circle $\mathcal{C}_{O,1}$. As it turns out, there are 16 such configurations, which are illustrated in Figure 2.

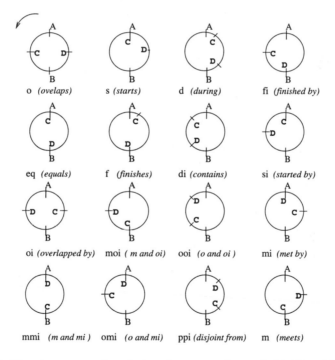

Fig. 2. The sixteen atomic relations between c-intervals [A,B] and [C,D].

Definition 1 (atomic relation). *An atomic relation is any of the 16 possible configurations a pair of c- intervals can stand in. These are s (starts), eq (equals), si (is-started-by), d (during), f (finishes), oi (is-overlapped-by),* *moi (is-meets-and-overlapped-by), ooi (is-overlapped-and-overlaps-by), mi (is-met-by), mmi (meets-and met- by (complements)), omi (overlaps-and-is-met-by), ppi (is-disconnected-from), m (meets), o (overlaps), fi (is-finished-by) and di (contains). We denote by $\mathcal{A}$ an atomic relation between two c-intervals and by A the set of all atomic relations.*

Remark 1

- *The relation ppi is similar to the DC relation (DisConnection) of the Region Connection Calculus (RCC) [10]. In Allen's algebra [1], because time is a linear order, disconnection splits into two relations, < (before) and > (after).*
- *Because we consider time as being a total cyclic order, an interval can meet and be overlapped by, overlap and be met by, overlap and be overlapped by, or complement another interval (see relations moi, omi, ooi, mmi).*

Definition 2 (general relation). *A general relation $\mathcal{R}$ is any subset of A. Such a relation is interpreted as follows:*

$$(\forall I, J)((I \; \mathcal{R} \; J) \Leftrightarrow \bigvee_{\mathcal{A} \in \mathcal{R}} (I \; \mathcal{A} \; J))$$

Definition 3 (converse). *The converse of a relation $\mathcal{R}$ is the relation $\mathcal{R}i$ such that*

$$(\forall I, J)((I \; \mathcal{R} \; J) \Leftrightarrow (J \; \mathcal{R}i \; I))$$

Remark 2 *The converse operation is such that*

$$(\forall \mathcal{R})((\mathcal{R}i)i = \mathcal{R})$$

Figure 3 provides the converse for each of the atomic relations. For a general relation R:

$$\mathcal{R}i = \bigcup_{\mathcal{A} \in \mathcal{R}} \{\mathcal{A}i\}$$

$\mathcal{A}$	o	f	d	s	eq	m	ppi	omi	ooi
$\mathcal{A}i$	oi	fi	di	si	eq	mi	ppi	moi	ooi

Fig. 3. The converse for of the atomic relations.

Definition 4 (intersection). *The intersection of two relations $\mathcal{R}_1$ and $\mathcal{R}_2$ is the relation $\mathcal{R}$ consisting of their set-theoretic intersection ($\mathcal{R} = \mathcal{R}_1 \cap \mathcal{R}_2$):*

$$(\forall I, J)((I\mathcal{R}J) \Leftrightarrow ((I\mathcal{R}_1 J) \wedge (I\mathcal{R}_2 J)))$$

For instance, if $\mathcal{R}_1 = \{eq, ooi, m\}$ and $\mathcal{R}_2 = \{eq, o, f\}$ then $\mathcal{R} = \{eq\}$.

Definition 5 (composition). *The composition, $\mathcal{R}_1 \otimes \mathcal{R}_2$, of two relations $\mathcal{R}_1$ and $\mathcal{R}_2$ is the strongest relation $\mathcal{R}$ such that:*

$$(\forall I, J, K)((I\mathcal{R}_1 K) \wedge (K\mathcal{R}_2 J) \Rightarrow (I\mathcal{R}J))$$

	o	s	d	fi	f	di	si	oi	moi	ooi	mi	mmi	omi	ppi	m
m	oi,mi ppi	ppi	ppi	moi mmi,m	m	ooi omi o,fi,di	o,fi,di	o,fi,di	di	di			s,eq,si	si	si
ppi	di,si,oi mi,ppi	ppi	ppi	di,si,oi mi,ppi	ppi	di,si,oi,moi ooi,mi,mmi omi,ppi m,o,fi	ppi,m,o fi,di	ppi,m,o fi,di	di	di	ppi,m,o fi,di	di	di	ppi,m,o s,d,fi,si,eq f,di,si,oi mi	di,si,oi mi,ppi
omi	oi moi ooi	mi,mmi omi	ooi	mi,mmi omi	omi	ooi	oi,moi ooi	ooi,omi o	ooi,omi o	ooi,omi	omi	o	fi,eq f	di,si,oi moi ooi	d
mmi	oi	mi	ppi	moi	m	ooi	oi	o	ooi,omi o	o	omi	o	fi	di	di
mi	di,si oi	mi	ppi	di,si oi	ppi	di,si oi moi ooi	si	mi,mmi omi	ppi m,o	ppi m,o s,d	mi,mmi omi	omi	ppi m,o s,d	di	di,si oi moi ooi
ooi	di,si,oi moi,ooi	moi	ooi	di,si oi ooi	ooi	di,si,oi,moi ooi,mi,mmi omi,ppi m,o,fi	ooi omi o,fi,di	ooi,omi o	ooi,omi o	ooi,omi,o s di,fi,eq,f di,si,oi,moi	ooi,omi o,s,d	di	di,si,oi moi ooi	di	di
moi	d,f oi	moi	ooi	di,si oi moi ooi	oi moi ooi	di,si,oi mi ppi	moi mmi,m	oi,moi,ooi mi,mmi,o omi,ppi,m	oi,moi,ooi mi,mmi,o omi,ppi,m	ooi,omi o,s,d	ppi m,o	moi mmi,m	di,si,oi moi ooi	di	di
oi	o,s,d,fi eq,f,di si,oi	d,f oi	d,f,oi moi,ooi	di,si oi	oi moi ooi	di,si,oi mi,ppi	oi mi ppi	oi,moi,ooi mi,mmi,o omi,ppi,m	ooi,omi o	ooi,omi o,s,d	ppi m,o	o	o,s,d	fi,di	o,fi,di
si	o,fi di	s,eq si	d,f,oi moi,ooi	fi,eq f	f	di	si	oi mi ppi	o,fi,di	o,fi,di	fi	eq	fi,eq f	di	si
di	ooi,omi o,fi,di	ooi omi,o fi,di	ooi,omi,o s,d,fi,eq f,di,si oi,moi	di	di,si oi moi ooi	di	di	di,si,oi moi,ooi	o,fi,di	ooi,omi o,s,d	di,si oi moi ooi	di	di,si,oi moi ooi	di,si,oi,moi ooi,mi,mmi omi,ppi m,o,fi	ooi omi o,fi,di
f	o,s,d	d	d	fi	fi,eq f	di,si oi moi ooi	si	oi moi ooi	moi mmi,m	ooi	di,si oi moi ooi	fi	omi	ppi	m
fi	ppi,m o	ppi m,o s,d	ooi,omi o,s,d	fi	fi,eq f	di	di,si oi moi ooi	di,si oi	di,si oi	di,si oi	mi	mi	mi,mmi omi	ppi	moi mmi,m
d	ooi,omi o,s,d	d	d	d	d	ooi,omi,o s,d,fi,eq f,di,si oi,moi	s,eq si	oi,moi,ooi mi,mmi,o omi,ppi,m,o,s	d,f oi	d,f,oi moi,ooi	mi,mmi omi	mi	oi moi ooi	d,f,oi,moi ooi,mi,mmi omi,ppi m,o,s	d,f,oi moi,ooi
s	ooi omi	o	d	ppi m,o s,d	d	ooi omi,o fi,di	s,eq si	oi,moi,ooi mi,mmi,o omi,ppi,m	ooi,omi o	ooi,omi o,s,d	mi	eq	mi,mmi omi	ppi	ppi
o	oi,moi,ooi mi,mmi,o omi,ppi,m	ooi omi,o	ooi,omi o,s,d	ppi,m o	o,s,d	ooi,omi o,fi,di	o,fi di	o,s,d,fi eq,f,di si,oi	d,f oi	di,si,oi moi,ooi	di,si oi	oi	oi moi ooi	di,si,oi mi,ppi	oi,mi ppi

Figure 4: The composition table of atomic relations.

If we know the composition for atomic relations, we can compute the composition of any two relations R_1 and R_2:

$$R_1 \otimes R_2 = \bigcup_{\mathcal{A}_1 \in R_1, \mathcal{A}_2 \in R_2} \mathcal{A}_1 \otimes \mathcal{A}_2$$

Stated otherwise, what we need is to provide a composition table for atomic relations . This is presented in the table below (figure 4).
For instance, if $\mathcal{R}_1 = \{eq,\ mi,\ d\}$ and $\mathcal{R}_2 = \{eq,\ s\}$ then $\mathcal{R} = \{eq, s, mi, oi, f, d\}$.

3 Networks of c-intervals

A c-interval network $\mathcal{N}$, which we shall also refer to as a CSP of c-intervals, is defined by:

(a) a finite number of variables X=$\{X_1, \ldots, X_n\}$ of c-intervals;
(b) a set of relations of the algebra of c-intervals $\mathcal{R}_{11}, \ldots, \mathcal{R}_{ij}, \ldots, \mathcal{R}_{nn}$ between a pairs $(X_i, X_j)_{i,j=\overline{1..n}}$ of these variables.

Remark 3 (normalized c-interval network) *We assume that for all i, j at most one constraint involving X_i and X_j is specified. The network representation of $\mathcal{N}$ is the labeled directed graph defined as follows:*
(1) the vertices are the variables of $\mathcal{N}$;
(2) there exists an edge (X_i, X_j), labeled with $\mathcal{R}_{ij}$, if and only if a constraint of the form $(X_i\mathcal{R}_{ij}X_j)$ is specified.

Definition 6 (matrix representation). *R is associated with an $n \times n$-matrix, which we shall refer to as R for simplicity, and whose elements will be referred to as $R_{ij}, i, j \in \{1, \ldots, n\}$. The matrix R is constructed as follows:*
(1) Initialise all entries of R to the universal relation $u\mathcal{R}$: $R_{ij} := u\mathcal{R}, \forall i, j$;
(2) $R_{ii} := eq, \forall i$;
(3) $\forall i, j$ such that R contains a constraint of the form $(X_i\mathcal{R}_{ij}X_j)$: $R_{ij} := R_{ij} \cap \mathcal{R}_{ij}; R_{ji} := R_{ij}$.

Definition 7. *[4] An instantiation of R is any n-tuple of $(X)^n$, representing an assignment of a c-interval value to each variable. A consistent instantiation, or solution, is an instantiation satisfying all the constraints. A sub-CSP of size k, $k \leq n$, is any restriction of R to k of its variables and the constraints on the k variables. R is k-consistent if any solution to any sub-CSP of size $k-1$ extends to any k-th variable; it is strongly k-consistent if it is j-consistent, for all $j \leq k$.*

1-, 2- and 3-consistency correspond to node-, arc- and path-consistency, respectively [8, 9]. Strong n-consistency of P corresponds to global consistency [2]. Global consistency facilitates the exhibition of a solution by backtrack-free search [4].

Remark 4 *A c-interval network is strongly 2-consistent.*

Definition 8 (refinement). *A refinement of R is any CSP R' on the same set of variables such that $R'_{ij} \subseteq R_{ij}, \forall i, j$. A scenario of R is any refinement R' such that R'_{ij} is an atomic relation, for all i, j.*

4 A constraint propagation algorithm

A constraint propagation procedure, $propagate(R)$, for c-interval networks is given below. The input is a c-interval network R on n variables, given by its $n \times n$-matrix. When the algorithm completes, R verifies the path consistency condition: $\forall i, j, k(R_{ij} \subseteq R_{ik} \otimes R_{kj})$.

The procedure is indeed Allen's propagation algorithm [1]; it makes use of a queue $Queue$. Initially, we can assume that all variable pairs (X_i, X_j) such that $1 \leq i < j \leq n$ are entered into $Queue$. The algorithm removes one variable pair from $Queue$ at a time. When a pair (X_i, X_j) is removed from $Queue$, the algorithm eventually updates the relations on the neighboring edges (edges sharing one variable with (X_i, X_j)). If such a relation is successfully updated, the corresponding pair is placed in $Queue$ (if it is not already there) since it may in turn constrain the relations on neighboring edges. The process terminates when $Queue$ becomes empty.

```
 1. procedure propagate(R)
 2. { repeat
 3.    { get next pair (Xᵢ, Xⱼ) from Queue;
 4.     for k := 1 to n
 5.     { Temp := Rᵢₖ ∩ (Rᵢⱼ ⊗ Rⱼₖ);
 6.       if Temp ≠ Rᵢₖ
 7.       { add-to-queue(Xᵢ, Xₖ); Rᵢₖ := Temp; Rₖᵢ := Tempi (converse of temp)}
 8.       Temp := Rⱼₖ ∩ (Rⱼᵢ ⊗ Rᵢₖ);
 9.       if Temp ≠ Rⱼₖ
10.       { add-to-queue(Xⱼ, Xₖ); Rⱼₖ := Temp; Rₖⱼ := Tempi;}
11.     }
12.   }
13.   until Queue is empty;
14. }
```

It is well-known in the literature that the constraint propagation algorithm runs into completion in $O(n^3)$ time [1, 11], where n is the size (number of variables) of the input network.

5 The neighborhood structure of the algebra

The idea of a conceptual neighborhood in spatial and temporal reasoning was first introduced by Christian Freksa [3]. The neighboring elements correspond to situations which can be transformed into one another with continuous change (two elements of the lattice which are immediate neighbors of each other correspond to situations which can be transformed into each other without going through any intermediate situation) [3]; the lattice in turn can be exploited for efficient reasoning.

In order to provide our conceptual neighborhood structure for the c-interval algebra, we proceed in a similar way as did Ligozat for the definition of his lattice for Allen's ℓ-interval algebra [6]. A c- interval $I = (I_r, I_l)$ (i.e., an interval of circle $C_{O,1}$) partitions $C_{O,1}$ into 4 regions, which we number $0, 1, 2, 3$, as illustrated in Figure 4.

The relation of another c-interval $J = (J_r, J_l)$ relative to I is entirely given by

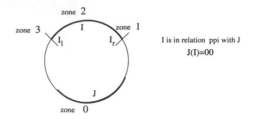

Fig. 4. The regions in the circle defined by a given c-interval $I = (I_r, I_l)$.

the ordered sequence of regions (regions determined by c-interval I) consisting of the region to which belongs the right endpoint of J and the regions we "traverse" when scanning the rest of c-interval J in trigonometric direction. For instance, the sequence $J(I) = 1230$ (the position of J compared to I) corresponds to the atomic relation s: "1" gives the region to which belongs the right endpoint of J, "23" say that regions number 2 and 3 are traversed twice when we scan the rest of c-interval J and "0" is the region of the left endpoint of J. Figure 5 provides for each of the atomic relations the corresponding sequence.

atomic relation	corresponding sequence
s	1230
eq	123
si	12
d	01230
f	0123
oi	012
moi	0123
ooi	23012
mi	01
mmi	301
omi	2301
ppi	00
m	30
o	230
fi	23
di	22

Fig. 5. The sequence corresponding to each of the atomic relations.

The figure 6 gives a complete conceptual neighborhood structure between all atomic relation between *c*-intervals. This structure will be used to define subclass of a convex and a preconvex relations between *c*- intervals. It opens the way to compute tractable class in this algebra.

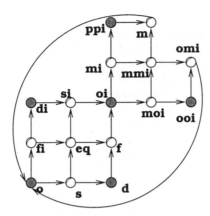

Fig. 6. The conceptual neighborhood structure.

6 Summary and future work

We have provided an interval algebra for (totally ordered) cyclic time, i.e., the counterpart for cyclic time of Allen's influential algebra of linear time intervals. This work is still in progress, and raises many questions, among which:

- Is the constraint propagation algorithm complete for atomic relations, or, more generally, for a subclass including all atomic relations? This question is important in the sense that a positive answer to it would mean that a general problem expressed in the algebra can be solved using a solution search algorithm à la Ladkin and Reinefeld [5].
- How can the conceptual neighborhood structure we have provided be exploited for efficient reasoning? Such a use of a conceptual neighborhood structure has been discussed for Allen's interval algebra [1] by Freksa [3] and Ligozat[7][6]. However, an important point should be taken into account when dealing with this question: for Allen's algebra, the idea of a conceptual neighborhood is intimately related to the idea of a convex [6]. the same structure is used by Ligozat to propose the preconvex subclass relations and to prove that this subclass is maximal tractable in Allen algebra. And this does seem to be the case for the algebra of cyclic time intervals.

References

1. J F Allen. Maintaining knowledge about temporal intervals. *Communications of the Association for Computing Machinery*, 26(11):832–843, 1983.
2. R Dechter. From local to global consistency. *Artificial Intelligence*, 55:87–107, 1992.
3. C Freksa. Temporal reasoning based on semi-intervals. *Artificial Intelligence*, 54:199–227, 1992.
4. E C Freuder. A sufficient condition for backtrack-free search. *Journal of the Association for Computing Machinery*, 29:24–32, 1982.
5. P Ladkin and A Reinefeld. Effective solution of qualitative constraint problems. *Artificial Intelligence*, 57:105–124, 1992.
6. G Ligozat. Towards a general characterization of conceptual neighbourhoods in temporal and spatial reasoning. In F D Anger and R Loganantharah, editors, *Proceedings AAAI-94 Workshop on Spatial and Temporal Reasoning*, 1994.
7. G Ligozat. Tractable relations in temporal reasoning: Pre-convex relations. In R Rodríguez, editor, *Proceedings of ECAI-94 Workshop on Spatial and Temporal Reasoning*, 1994.
8. A K Mackworth. Consistency in networks of relations. *Artificial Intelligence*, 8:99–118, 1977.
9. U Montanari. Networks of constraints: Fundamental properties and applications to picture processing. *Information Sciences*, 7:95–132, 1974.
10. D Randell, Z Cui, and A Cohn. A spatial logic based on regions and connection. In *Proceedings KR-92*, pages 165–176, San Mateo, 1992. Morgan Kaufmann.
11. M B Vilain and H Kautz. Constraint propagation algorithms for temporal reasoning. In *Proceedings AAAI-86*, Philadelphia, August 1986.

A Multiple-Platform Decentralized Route Finding System

Hans W. Guesgen[1] and Debasis Mitra[2]

[1] Computer Science Department, University of Auckland
Private Bag 92019, Auckland, New Zealand
hans@cs.auckland.ac.nz
[2] Department of Computer Science, Jackson State University
P.O. Box 18839, Jackson, MS 39217, USA
dmitra@stallion.jsums.edu

Abstract. This paper describes a simple route finding system that is not specific to any particular road map and is platform independent. The main features of the system are an easy-to-use graphical user interface and quick response times for finding optimal or near optimal routes. The system can be run on a desktop computer from home or the office, accessed through the internet, or even run on a palmtop computer sitting in a car.

To speed up the process of finding an optimal or near optimal route, we evaluated a number of heuristic search strategies. It turned out that heuristic search can give substantial efficiency improvements and that inadmissible heuristics can be appropriate if non-optimal solutions are permitted. Furthermore, it turned out that incorporating heuristic knowledge about commonly used intersections, through subgoal search, can dramatically increase search performance.

1 Route Finding and Search

With the increasing number of cars on our roads, it becomes more and more important to have an efficient means for finding a good route from some point A in a city to some point B. Printed road maps, as they have been used up to now, are not sufficient any more and will most likely be replaced with electronic systems in the near future.

The two main components of any route finding system are the route planning and route guidance systems. The route guidance component takes a given path through the roading network and provides the driver with a description or set of directions to accomplish the task of navigating. This may be a map display of the route or audio instructions together with a visual compass direction. The route planning system finds a path through the roading network which conforms to the users preferences about the required route.

This paper describes a route finding system that is not specific to any particular map (although we used Auckland as our test bed) and is platform independent. The main features of the system are a simple-to-use graphical user interface and quick response times for finding optimal or near optimal paths. The system can be run on a desktop computer from home or the office, accessed through the internet, or even run on a palmtop computer sitting in a car.

Platform independence is achieved through writing the route finding program entirely in Java. This however does not account for the speed and memory differences between, for example, a palmtop computer and a top of the range desktop. Because the system runs at a reasonable speed on a good desktop computer, it does not mean that it will run well enough to be useful on a smaller system. To account for this, two types of minimum path search algorithms have been used, one which is fast but uses a lot of memory, the other slower and using less memory. Both algorithms can be set to find optimal and non-optimal paths. A non-optimal search will be quicker than an optimal one, allowing the system to be run on low-end computers in reasonable time.

The rest of this paper is divided into two parts. The first part describes the main algorithms used in the system; the second part describes the user interface of the system.

2 Heuristic Search for Finding the Shortest Path

Most of the heuristic algorithm used for finding the shortest path from a point A in a city to a point B are based on the A* algorithm, which is a variant of uniform-cost search [3]. The A* algorithm attempts to find an optimal path from A to B in a network of nodes by choosing the next node in a search based on the total weighting of the path found so far. In addition to plain uniform-cost search, the A* algorithm uses heuristics to decrease the effective branching factor of the search, thereby trying to avoid combinatorial explosion as search spaces increase in size.

A number of heuristics are suggested in the literature which are applicable to the route finding. These heuristic evaluation functions include the air distance heuristic and the Manhattan metric heuristic [4]. The air distance heuristic is an admissible heuristic while the Manhattan heuristic is inadmissible, as the latter has the potential to overestimate the actual shortest distance. To test the pruning power of the heuristics on a real road network, we performed some experiments in the road network of Auckland, using the heuristic search as well as uninformed brute-force search. We found that uninformed brute-force search expands almost every node in the search area in finding the goal state. With the air distance heuristic, the search is more focussed and the node expansions describe a more elliptical path. The Manhattan heuristic is even more focussed than the air distance heuristic with a decrease in the size of the minor axis of the ellipse over the air distance estimate. Table 1 and Figure 1 summarize these results.

Standard heuristic search techniques use no other sources of knowledge than an evaluation function to guide search from a start to goal state. Of use to decrease the amount of search is the concept of intermediate states. This uses the idea that a search between the start and an intermediate state and the intermedi-

Table 1. Results of performing 1000 shortest path trials with no heuristic, air distance heuristic, and Manhattan heuristic. The average node reduction rate (compared to uninformed search) over the 1000 trials indicates that the air distance heuristic produces a significant gain in node expansion efficiency over uninformed search and that the Manhattan heuristic is significantly better than air distance.

	Air distance heuristic	Manhattan heuristic
Node reduction rate	$\mu = 21.8\%$, $\sigma = 10.0\%$	$\mu = 16.7\%$, $\sigma = 13.4\%$
Path inaccuracy rate	—	$\mu = 1.4\%$, $\sigma = 2.8\%$

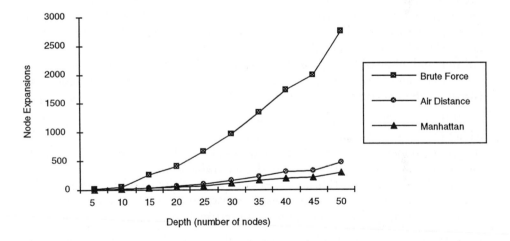

Fig. 1. Graph showing node expansions for the minimum distance problem. The two heuristically-guided searches produce significantly less node expansions than blind search especially for higher search depths.

ate state and the goal decreases search complexity by splitting the search into two smaller complexity searches. These ideas are embodied in two heuristic search concepts: bidirectional search and search using subgoals. Bidirectional searches approximate a subgoal lying in the conjectured middle of the search space and direct two opposing searches toward this point. Subgoal searches make use of explicitly defined subgoals and direct search through these nodes. If a series of subgoals can be identified before a search is begun, then this knowledge can be used to create an island set, which adds greater heuristic knowledge to the heuristic evaluation function.

If at least one of the islands lies on optimal path to the goal state, then directing a heuristic search through this island to the goal has the potential of dividing the exponent in the complexity term. Often, however, there is more than one island that may occur on the optimal path. Island sets which contain more than one island on the optimal path are called multiple level island sets [2]. In the domain of route finding systems, islands correspond to intersections that commonly appear among possible routes within the road network. These tend to be busy intersections through which most traffic passes or intersections which lead traffic to avoid obstacles.

Chakrabarti et al. [1] were the first to implement an algorithm which combined heuristic search with sub-goal knowledge. Their algorithm, Algorithm I, makes use of a heuristic that guides a search through a single island out of a larger island set. The heuristic used is an estimate of the shortest path from the current node to the goal node constrained to pass through the most favorable node in the island set. This has been proved by Chakrabarti et al. to be both optimal and admissible given that the nodes in the island set are on an optimal path from the start to the goal. Thus given a series of possible islands in an island set the heuristic will direct the search toward the most favorable island.

Dillenburg [2] has created an algorithm which makes use of multiple level island sets, Algorithm In. Given an island set and a parameter E that specifies the total number of island nodes on an optimal path between start and goal, the algorithm will provide an optimal path constrained to pass through the island nodes. Like Algorithm I, this is proved to be both optimal and admissible. Thus Algorithm In is potentially of more use in route finding applications as it is often known that more than one island is on an optimal path.

Although multiple island sets are usually more adequate for route finding than single island sets, they tend to increase the complexity of the heuristic computation. Since there usually is no particular order on the islands in a multiple island set, we have to consider each permutation of the islands to find an optimal path, which might nullify the advantage of using multiple islands. This problem can be avoided if an a-priori ordering of islands is known. We implemented a heuristic which works on the assumption that the islands in the island set have an ordering. Unlike the permuted heuristic, the ordering is already known and thus computation is limited to finding a choice of islands with the given ordering which minimizes the heuristic estimate.

To evaluate the effectiveness of heuristic search with islands, we ran Algorithm In on a series of test problems. First, the algorithm was applied to a problem where perfect knowledge of the existence of optimal islands was available. As Figure 2 shows, the number of node expansions decreases significantly

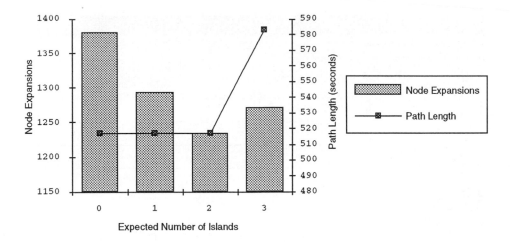

Fig. 2. Multilevel island search using a problem where perfect knowledge of the existence of optimal island was available.

as the number of island increases. The optimal path length (guaranteed when $E = 0$) is maintained when the correct number of islands is given (E between 0 and 2). However, if the expected number of islands is incorrect ($E = 3$), then the computed path length will be suboptimal.

In a second series of experiments, Algorithm In was applied to 700 problems of finding the optimal route between a source and a destination in different regions of the Auckland area. Islands for routes between the two regions were discovered by calculating every optimal route between the regions and then randomly choosing 5 nodes that were contained in over 95% of these routes [2]. In addition, another island node was included, which was known to occur in only 75% of all routes. The algorithm was run with different values for parameter E, using several island heuristics. The results of these experiments can be found elsewhere [5].

3 The User Interface

For the system to be accepted by a wide range of potential users, we added a graphical user interface to it, providing menus, a toolbar, entry and status panels, and a map view. It allows the user to perform a variety of operations, including the following:

- Entering the current position and the destination.
- Showing the road map at various resolutions.
- Displaying the proposed route.

Figure 3 shows a screen snapshot of the system after a route has been computed.

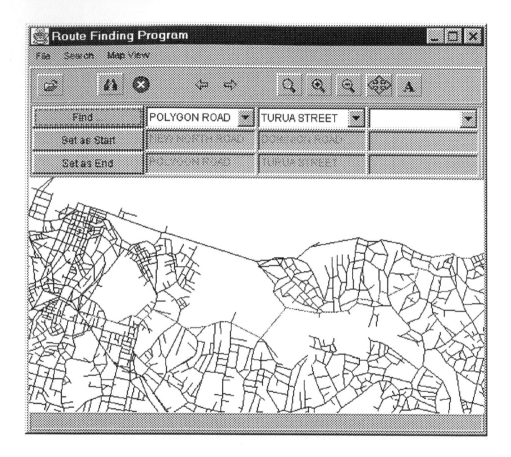

Fig. 3. Screen snapshot of the route finding system. The intersection between New North Road and Dominion Road has been selected as the current position and the intersection between Polygon Road and Turua Street as the destination. The road map is shown as set of connected line segments without street names, zoomed in on the part containing the current position and the destination.

The display of the map on the screen is a crucial aspect of the system. The map has to be computed automatically from the data given by the underlying database and therefore lacks the sophistication of manually designed maps. To compensate for this, the map can be displayed at various resolutions, with or without the street names displayed (see Figure 4 for an example of the map at its minimal resolution and maximal resolution, respectively).

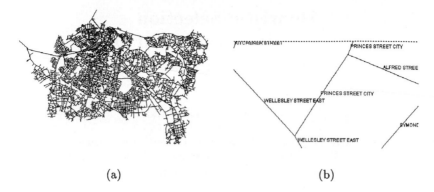

(a) (b)

Fig. 4. The map shown at different resolutions. In (a), the entire area of Auckland is displayed, at the cost of resolution and missing details such as street names. In (b), individual intersections can easily be located, but only a few blocks are shown.

4   Summary

In this paper, we introduced a platform-independent route finding system, its underlying algorithms, and its user interface. We showed that heuristic search can be used to reduce excessive node retrieval and memory use. Experimental comparisons were made between non-heuristic and heuristic approaches. It was shown that heuristics reduce considerably the combinatorial explosion evident in non-directed uniform-cost search, in particular when using the non-admissible Manhattan heuristic instead of the admissible air distance heuristic.

A further improvement was achieved by using islands, which can be discovered from the road network by identifying commonly used intersections. However, this discovery is sometimes difficult in the Auckland road map, as many regions are close together and a range of possible routes is possible.

References

[1] P.P. Chakrabarti, S. Acharaya, and S.C. de Sarkar. Heuristic search through islands. *Artificial Intelligence*, 29:339–348, 1986.

[2] J.F. Dillenburg and P.C. Nelson. Improving search efficiency using possible subgoals. *Mathematical and Computer Modelling*, 22(4–7):397–414, 1995.

[3] R.E. Korf. Artificial intelligence search algorithms. Technical Report TR 96-29, Computer Science Department, UCLA, Los Angeles, California, 1996.

[4] J. Pearl. *Heuristics: Intelligent Search Strategies for Computer Problem Solving*. Addison-Wesley, Reading, Massachusetts, 1984.

[5] J. Pearson. Heuristic search in route finding. Master's thesis, Computer Science Department, University of Auckland, Auckland, New Zealand, 1998.

Heuristic Selection
of Aggregated Temporal Data
for Knowledge Discovery

H.J. Hamilton and D.J. Randall

Department of Computer Science, University of Regina
Regina, Saskatchewan, Canada, S4S 0A2
e-mail: {*hamilton, randal*}@cs.uregina.ca

Abstract. We introduce techniques for heuristically ranking aggregations of data. We assume that the possible aggregations for each attribute are specified by a domain generalization graph. For temporal attributes containing dates and times, a calendar domain generalization graph is used. A generalization space is defined as the cross product of the domain generalization graphs for the attributes. Coverage filtering, direct-arc normalized correlation, and relative peak ranking are introduced for heuristically ranking the nodes in the generalization space, each of which corresponds to the original data aggregated to a specific level of granularity.

1 Introduction

We present a knowledge discovery technique for automatically aggregating temporal data, by using generalization relations defined by domain experts. In relational databases, temporal data are common. Temporal attributes are used to specify the beginning or ending of an event or the duration of an event. We refer to any attribute whose domain represents date and/or time values as a *calendar attribute*. Although recent advances have been made in the management of temporal data in relational databases [7], less research has focused on the development of techniques for generalizing and presenting temporal data. Let us look at generalizing and presenting in turn.

A salient problem when generalizing temporal data is that temporal values can be generalized in different ways and to different levels of granularity. Given a database containing a calendar attribute, we might want to aggregate the data by month, week, day of month, day of week, day, hour, hour of day, ten-minute interval, or many other levels of granularity. Statistics, such as the *number or instances* or the *percentage of instances*, can be recorded for each temporal value at a level of granularity. For example, if the calendar attribute specifies login times and the recorded statistic is the number of instances, a summary to *day of month* shows the number of logins for each day of the month (regardless of which month it occurred in), while a summary to *day* (denoted *YYYYMMDD*) shows the number of logins for each distinct day.

Our approach to this problem uses domain generalization graph to guide the generalization of calendar data extracted from a relational database. In previous work, we defined a *domain generalization graph* (or *DGG*) [4] for calendar attributes [9] and also one for duration attributes [8]. The calendar DGG explicitly identifies domains appropriate for relevant levels of temporal granularity and the mappings between the values in these domains. Generalization is performed by transforming values in one domain to a more general domain, according to directed arcs in the domain generalization graph.

Given the ability to aggregate temporal data to many levels of granularity, a second salient problem is to select appropriate aggregations for display. We address this problem using two fundamental ideas: (1) only one of a set of similar aggregations need be displayed, and (2) the user may provide criteria that allow rough ranking of the aggregations. We use heuristic techniques to filter out aggregations similar to others and rank the remainder according to interest.

A data mining system based on the approach described above can automatically identify temporal granularities relevant to a particular knowledge discovery task on a set of data. After generalization, an appropriate selection of data visualizations can be displayed to the user, prepatory to more detailed analysis.

The remainder of this paper is organized as follows. In Section 2, we discuss related work, giving an overview of domain generalization graphs and connections to recent research in temporal representation. In Section 3, we provide an example generalization space for a set of DGGs. In Section 4, we describe heuristic techniques for selecting interesting aggregations. We illustrate these techniques by applying them to a simple knowledge discovery task. We present conclusions in Section 5.

2 Generalization for Knowledge Discovery

A *domain generalization graph* (DGG) includes all generalization relations relevant to an attribute [4]. DGGs facilitate the specification and management of multiple possible paths of generalization.

A DGG is defined as follows [4]. Let $S = \{s_1, s_2, \ldots, s_n\}$ be the domain of an attribute, let D be the set of partitions of set S, and $\preceq$ be a binary relation (called a *generalization relation*) defined on D, such that $D_i \preceq D_j$ if for every $d_i \in D_i$ there exists $d_j \in D_j$ such that $d_i \subseteq d_j$. The generalization relation $\preceq$ is a partial order relation and $\langle D, \preceq \rangle$ defines a partial order set from which we can construct a *domain generalization graph* $\langle D, E \rangle$ as follows. First, the nodes of the graph are elements of D. Second, there is a directed arc from D_i to D_j (denoted by $E(D_i, D_j)$) iff $D_i \neq D_j$, $D_i \preceq D_j$, and there is no $D_k \in D$ such that $D_i \preceq D_k$ and $D_k \preceq D_j$. The partial order set $\langle D, \preceq \rangle$ is transitively closed and is a lattice. In practice, we give up the lattice property and use partial DGGs that include only the bottom, top, and the partitions selected by domain experts.

In relational databases, time is represented by database timestamps that include time and/or date and that correspond to global (absolute) timepoints in many temporal representation schemes. Thus, for knowledge discovery in relational databases that include timestamps, the most specific domain consists of

716

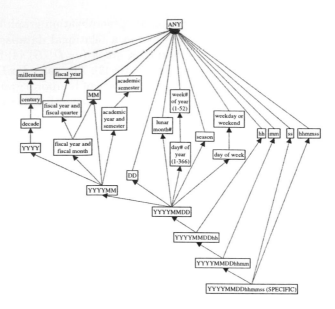

Fig. 1. Calendar DGG [9]

timestamps and less granular domains are defined as intervals. For our purposes, time is defined as linear, discrete, and bounded.

In [9], we proposed a default DGG for calendar attributes, and we developed techniques for adapting this DGG according to a particular data set or a user's interest. For example, if all data are from a single week within one month, then domains with only one value occurring in the data (such as the month domain) are pruned from the DGG. Higher-level nodes in the DGG represent generalizations of the most specific domain, which is time measured in seconds (represented by YYYYMMDDhhmmss timestamps). The arcs connecting the nodes represent generalization relations. To handle data containing calendar values specified to finer granularity, e.g., microseconds, more specific nodes could be added to the DGG.

Other research that has investigated the generalization of temporal data to multiple levels of granularity includes the work of Bettini *et al.* on discovering frequent event patterns with multiple granularities in time sequences [1] and Sharar's framework for knowledge-based temporal abstraction [10]. In the knowledge discovery literature, researchers have considered rule discovery from time series [3].

3   Example Generalization Space

In this section, we describe an application of knowledge discovery with two DGGs. The data to be summarized are 9273 (Username, LoginTime) pairs collected over a one week period in January 1998. Assuming additional information is available about the User Academic Year for each user, we explore the data

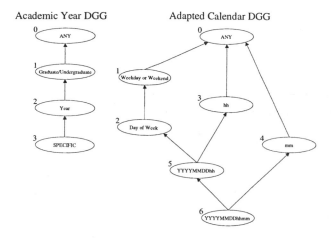

Fig. 2. User Academic Year DGG and Adapted Calendar DGG

for relationships involving users and login time, looking for distinct perspectives on this data. In [9], we explain how the calendar DGG is adapted for this application and how the data are generalized. Here, we look at the generalization space formed by the cross product of the DGGs for User Academic Year and LoginTime.

For the LoginTime attribute, we adapt the calendar DGG given in Figure 1 by combining nodes that do not correspond to distinctions in the data. For the usernames, we use a User Academic Year DGG based on a concept hierarchy that classifies usernames first into Year1, Year2, Year3, Year4, MSc, and PhD students, and then into Undergraduate and Graduate students. This DGG has nodes labelled SPECIFIC, Year, Graduate/Undergraduate, and ANY. Figure 2 shows the adapted calendar DGG alongside the User Academic Year DGG for usernames. We use the closed world assumption to constrain the cardinality of the SPECIFIC domain by assuming the only students in the domain are those who logged in at least once (by happenstance there are exactly 500) and the only time of interest is that from the first login to the last login (approximately $7 \times 24 \times 60 = 10080$ distinct minutes).

A generalization space is formed by taking the cross product of the relevant DGGs; Figure 3 shows the generalization space for DGGs in Figure 2. The aggregations corresponding to the nodes in the generalization space are obtained by mapping the original data to the most specific node (bottom left) and generalizing the data each time an arc is followed until all nodes have been reached (for an algorithm see [4]). Each node (i, j) in this generalization space contains the aggregation formed by generalizing usernames to node i in the User Academic Year DGG and the login times to node j in its DGG.

For small generalization spaces, such as that given in Figure 3, the user may choose to explore directly by selecting nodes and obtaining detailed summary information. In this case, the generalization space can be shown directly to the user (one tool for displaying and manipulating a generalization space is

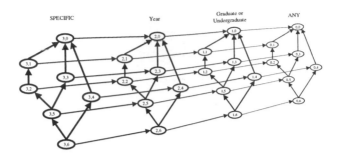

Fig. 3. The Generalization Space

GenSpace [6]). For larger generalization spaces, filtering and ranking functions can be applied before display, as described in the next section.

As an example of displaying aggregations, Figure 4 shows the rightmost seven nodes in the generalization space (where $i = 0$), which correspond to ignoring the username attribute and aggregating the data according to only the Login-Time attribute. For this display, we assume that nodes representing domains with between 2 and 100 distinct values are shown graphically and others are summarized in tabular form (using terminology explained in the next section).

4 Heuristic Selection Techniques

Filtering and ranking functions can be applied to identify promising nodes (domains) in large generalizaton spaces. A promising domain might provide an effective overview of all data or a means of identifying anomalous instances. We illustrate both possibilities by applying coverage filtering and relative peak ratio ranking to the generalization space given in Section 3. *Coverage filtering*, a simplified version of credibility filtering [5], removes all domains where the majority of domain values do not have any occurrences. The intuition is that in an appropriate aggregation, most entries are not zeros. *Relative peak ratio (RPR) ranking* ranks domains according to the ratio between the number of occurrences of the most frequently occurring domain value and the average number of occurrences. Relative peak ranking is a much simplified version of comparing the distance between the observed distribution of values and the expected distribution, such as is done in [2] using relative entropy. A user applies a variety of filtering and ranking techniques in an attempt to discover interesting knowledge.

4.1 Coverage Filtering

In Table 1, several statistics relevant to the 28 nodes (each representing a separate domain at a distinct level of generality) in the generalization space are given. In the table, i is the node index into the User Academic Year DGG, and $|i|$ is the size of the domain in this DGG with this index. Similarly, j is the node index in the Calendar DGG and $|j|$ is the size of its domain. For node $(1, 5)$, $i = 1$, $|i| = 2$, $j = 5$, and $|j| = 7 \times 24 = 168$ since values for Username have

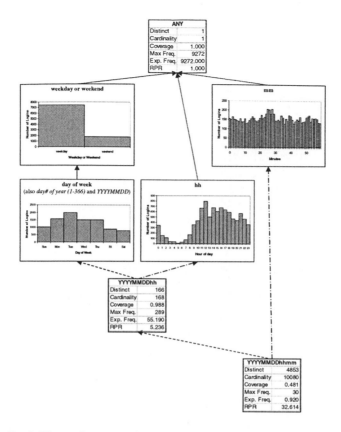

Fig. 4. LoginTimes Aggregated According to the Adapted Calendar DGG

been generalized to the *Graduate/Undergraduate* domain where there are two possible values and those for the LoginTime attribute have been generalized to *YYYYMMDDhh*, where there are 7 days with 24 hours each in the time interval of the data. Distinct gives the number of distinct values in the data at this level of generality, Cardinality is the cardinality of this domain $(|i||j|)$, and Coverage is the the ratio of occurring distinct values to all possible distinct values. The number of distinct instances in the data for node (1, 5) is 313, indicating that at least one graduate student and one undergraduate student logged in during almost every hour for the entire period. Of the 28 nodes in the generalization space, the coverage of 7 nodes is less than 0.5 (Prune = *), and the coverage of a further 3 nodes is less than 0.9 (Prune = +). The coverage of 15 nodes is 1.0.

4.2 Direct-arc normalized correlation

Direct-arc normalized correlation can be used to assess whether the nodes connected by arcs provide significantly different perspectives. The term "direct-arc" indicates only those nodes directly connected by an arc in the generalization space are examined. We apply it to a generalization space to which coverage

Node		Sizes		Domain Coverage			
i	j	$|i|$	$|j|$	Distinct	Cardinality	Coverage	Pruned
0	0	1	1	1	1	1.000	
0	1	1	2	2	2	1.000	
0	2	1	7	7	7	1.000	
0	3	1	24	24	24	1.000	
0	4	1	60	60	60	1.000	
0	5	1	168	166	168	0.988	
0	6	1	10080	4853	10080	0.481	*
1	0	2	1	2	2	1.000	
1	1	2	2	4	4	1.000	
1	2	2	7	14	14	1.000	
1	3	2	24	48	48	1.000	
1	4	2	60	120	120	1.000	
1	5	2	168	313	336	0.932	
1	6	2	10080	5916	20160	0.293	*
2	0	6	1	6	6	1.000	
2	1	6	2	12	12	1.000	
2	2	6	7	42	42	1.000	
2	3	6	24	142	144	0.986	
2	4	6	60	360	360	1.000	
2	5	6	168	834	1008	0.827	+
2	6	6	10080	7464	60480	0.123	*
3	0	500	1	500	500	1.000	
3	1	500	2	780	1000	0.780	+
3	2	500	7	2024	3500	0.578	+
3	3	500	24	3434	12000	0.286	*
3	4	500	60	6657	30000	0.222	*
3	5	500	168	5455	84000	0.065	*
3	6	500	10080	8585	5040000	0.002	*

Table 1. Coverage of Domains in the Generalization Space

filtering has first been applied. For example, the distribution of login frequencies among the hours of the day ($j = 3$) for each of Year1, Year2, Year3, and Year4 ($i = 2$) students is significantly correlated with the distribution of login frequencies for Undergraduate students ($i = 1$), and the distribution for each of M.Sc. and Ph.D. students ($j = 2$) is significantly correlated with those of Graduate students ($i = 1$). In this case, nodes (2,3) and (1,3) do not provide significantly different perspectives, as perhaps can seen from Figures 5 and 6.

We now explain the normalized correlation method. The distribution of the frequencies for each time period are normalized by computing the fraction of logins in each category. As implemented in Statistica, the underlying statistical engine for our analysis, the method requires, in the more specific domain, at least 2 distinct values for every attribute. Thus, the method cannot be applied to any arc beginning with a node (i, j) where either i or j is 0, since the ANY domains have only 1 possible value for every attribute. As well, at least 3 values are required to determine a distribution, so the number of distinct combinations of values for the attributes that are not generalized along an arc must be at least 3.

Table 2 summarizes the results of normalized correlation for arcs among nodes with coverage equal to at least 0.9 where the method can be applied. The first entry indicates that the distribution of students in Year1, Year2, Year3, Year4, MSc, and PhD for each minute after the hour (mm) correlates with that of all minutes, except among students logging in at 28 minutes after the hour (a

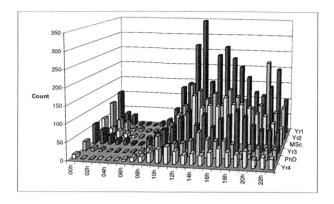

Fig. 5. Node (2,3): "Year" and "Hour of Day" (hh)

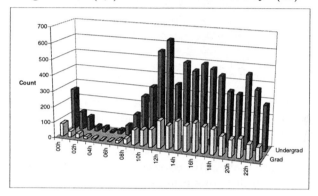

Fig. 6. Node (1,3): "Graduate/Undergraduate" and "Hour of Day" (hh)

partial cause is identified below by relative peak ranking). Most pairs of nodes given in Table 2 are correlated. Only one node from each such pair needs to be examined when searching for perspectives on the data. The table shows that nodes (2,4) and (2,3) may be worth examining with regard to specific exceptions.

4.3 Relative Peak Ranking

Table 3 provides information about the *peak*, i.e., the value in a domain that occurs most frequently. For each domain, the Peak Value column identifies the peak itself, the Freq column gives its observed frequency, the ExpFreq column gives the expected frequency for each value in this domain (assuming a uniform distribution), the RPR column gives the *relative peak ratio*, (i.e., the ratio between the observed frequency and the expected frequency of the peak), and the Rank column ranks the aggregations according to their RPR scores, with Rank 1 as the highest ranked. According to this heuristic ranking, the aggregation featuring the most interesting anomaly is that corresponding to node (3, 6). Frequency in this aggregation indicates the number of times that a specific user logged in a single minute. User276 logged in 28 times in the minute beginning

Source Node	Dest Node	Significant Correlations	Attempted Correlations	Exceptions
2,4	2,0	59	60	minute 28
2,4	1,4	5	6	Year4
2,3	2,0	17	24	hour 01, 03, 04, 05, 06, 07, 21
2,3	1,3	6	6	
2,2	2,1	7	7	
2,2	1,2	6	6	
2,1	2,0	2	2	
1,4	0,4	2	2	
1,3	0,3	2	2	
1,2	0,2	2	2	

Table 2. Correlations in Direct-Arc Generalizations ($p < 0.05000$)

at 21:28 on January 18, 1998. When our method identified this anomaly, people to whom we showed it commonly reacted with extreme surprise followed by disbelief. Further investigation showed that the result was correct; the user in question executed a script that repeatedly invoked the file transfer protocol (ftp) program to copy files, such that a separate login occurred for each file.

5 Conclusion

We have described heuristic techniques for selecting aggregations of data for display. The overall goal is to select aggregations that give different perspectives on the data. A generalization space of possible aggregations is defined by domain generalization graphs associated with attributes. Coverage filtering eliminates aggregations that have inadequate coverage of the generalized domain. Direct-arc normalized correlation allows pruning of aggregations that are not distinct from other aggregations. Future research could test the proposed approach on more complex knowledge discovery tasks involving larger databases. Additional heuristic measures could improve filtering and ranking.

Acknowledgement: We thank the Natural Sciences and Engineering Research Council of Canada and the Institute for Robotics and Intelligent Systems.

References

1. C. Bettini, X. S. Wang, S. Jajodia, and J.-L. Lin. Discovering frequent event patterns with multiple granularities in time sequences. *IEEE Transactions on Knowledge and Data Engineering*, 10(2):222-237, March/April 1998.
2. R. Feldman and I. Dagan. Knowledge discovery in textual databases (KDT). *Proceedings of the First International Conference on Knowledge Discovery and Data Mining (KDD'95)*, pp. 112-117, Montreal, August 1995.
3. G. Das *et al.* Rule discovery from Time Series. *Proceedings of the Fourth International Conference on Knowledge Discovery and Data Mining (KDD'98)*, pp. 16-22, New York, August 1998.
4. H.J. Hamilton, R.J. Hilderman, and N. Cercone. Attribute-oriented induction using domain generalization graphs. *Proceedings of the Eighth IEEE International Conference on Tools with Artificial Intelligence (ICTAI'96)*, pp. 246-253, Toulouse, France, November 1996.

723

Node		Peak				
UAY	Cal	Peak Value	Freq	ExpFreq	RPR	Rank
0	0	ANY-ANY	9272	9272.000	1.0	28
0	1	ANY-weekday	7473	4636.000	1.6	24
0	2	ANY-Tuesday	1989	1324.571	1.5	26
0	3	ANY-12h	802	386.333	2.1	22
0	4	ANY-28m	206	154.533	1.3	27
0	5	ANY-1998/01/20,12h	289	55.190	5.2	14
0	6	ANY-1998/01/20,21:28	30	0.920	32.6	9
1	0	ugrad-ANY	7046	4836.000	1.5	25
1	1	ugrad-weekday	5666	2318.000	2.4	19
1	2	ugrad-Tuesday	1535	662.286	2.3	20
1	3	ugrad-12h	628	193.167	3.3	17
1	4	ugrad-28m	168	77.267	2.2	21
1	5	ugrad-1998/01/20,12h	230	27.595	8.3	12
1	6	ugrad-1998/01/18,21:28	30	0.460	65.2	7
2	0	yr1-ANY	3107	1545.333	2.0	23
2	1	yr1-weekday	2616	772.667	3.4	16
2	2	yr1-Tuesday	784	220.762	3.6	15
2	3	yr1-12h	347	64.389	5.4	13
2	4	yr2-28m	74	25.756	2.9	18
2	5	yr2-1998/01/18,21h	157	9.198	17.1	11
2	6	yr2-1998/01/18,21:28	30	0.153	195.7	4
3	0	user231-ANY	599	18.544	32.3	10
3	1	user231-weekday	598	9.272	64.5	8
3	2	user231-Tuesday	257	2.649	97.0	6
3	3	user231-12h	156	0.773	201.9	3
3	4	user276-28m	30	0.309	97.1	5
3	5	user276-1998/01/18,21h	144	0.110	1304.6	2
3	6	user276-1998/01/18,21:28	29	0.002	15763.6	1

Table 3. Relative Peak Ratio Ranking of Domains in the Generalization Space

5. H.J. Hamilton, N. Shan, and W. Ziarko. Machine learning of credible classifications. In A. Sattar (ed.), *Advanced Topics in Artificial Intelligence, Tenth Australian Joint Conference on Artificial Intelligence (AI'97)*, pp. 330-339, Perth, Australia, November/December, 1997.
6. R.J. Hilderman, L. Li, and H.J. Hamilton. Data visualization in the DB-Discover system. In G. Grinstein, A. Wierse, U. Fayyad, A. Gee, and P. Hoffman (eds.), *Issues in the Integration of Data Mining and Data Visualization*, Springer-Verlag, Berlin, 1999. In press.
7. N.A. Lorentzos and Y.G. Mitsopoulos. SQL extension for interval data. *IEEE Transactions on Knowledge and Data Engineering*, 9(3):480-499, May/June 1997.
8. D.J. Randall, H.J. Hamilton, and R.J. Hilderman. A technique for generalizing temporal durations in relational databases. In *Eleventh International FLAIRS Conference (FLAIRS-98)*, AAAI Press, Sanibel, FL, pp. 193-197, May 1998.
9. D.J. Randall, H.J. Hamilton, and R.J. Hilderman. Generalization for calendar attributes using domain generalization graphs. *Proceedings of the Fifth International Workshop on Temporal Representation and Reasoning (TIME-98)*, IEEE CS Press, pp. 177-184, Sanibel, Florida, May 1998.
10. Y. Shahar. A framework for knowledge-based temporal abstraction. *Artificial Intelligence*, 90(1-2):79-133, 1997.

Architectural Knowledge Representation Using the Galois Lattice Technique

Dr. D. Chitchian[1], Prof. Dr. ir. S. Sariyildiz[1], Prof. Dr. H. Koppelaar[2]

[1] Delft University of Technology, Faculty of Architecture,
Technical Design and Information Group,
Berlageweg 1, 2628 CR Delft, The Netherlands
{D.Chitchian, S.Sariyildiz} @bk.tudelft.nl
[2] Delft University of Technology, Faculty of Information Technology and Systems,
Zuidplantsoen 4, 2628 BZ Delft, The Netherlands
{H.Koppelaar} @kbs.twi.tudelft.nl

Abstract. Representation of spatial information is very important in architecture. Information of a design may be organized as the objects and relations between them. Besides the limitation of those techniques, essentially they are problem specific, the main inefficiency of those methods is that the hierarchical nature of architectural information cannot be represented by them. To combat the inefficiency they have to handle information in a hierarchical manner. In such a representation system not only we can decompose information into smaller pieces also we can focus on the required parts of the problem without involving with many detailed unnecessary information. Also the implicit information in design specifications that is vital must be extracted. Without such information the generated solutions will not be so realistic. The extracted information is essential for hierarchical abstraction of the given design problem. A technique namely the Galois lattice, as a mathematical model, has been used to construct a hierarchical representation scheme using the design specification and to externalize the hidden information in that specification.

1   Introduction

Design representation is the key issue in solving design problems. There exist methods for representing design knowledge. The representation methods should be convenient and efficient for representing the required design information. Graph theory is a branch of mathematics. Its importance has been increased with the emergence of computer techniques. The use of graphs for modeling and solving architectural design problems can be seen in most researchers' works in this domain. Graphs as a means of representation can serve as tools for capturing the architectural design knowledge with respect to some relationships, i.e. adjacency, between the objects or elements of design.

In a rectangular representation, a floor plan is divided into smaller rectangles. It is a so called *dimensionless* method, because the rectangle dimensions are not specified precisely. The relationships between the elements in this method can be only topological such as adjacency. This characteristic of rectangular dissection makes it a good representation method compared to the adjacency graph. In rectangular representation, because of its limitation, certain arrangements of locations are not possible. The rectangular dissections have been used in many previous attempts in designing floor plans [1], [2].

The main limitation of those techniques are as follows. In a rectangular representation every location has four neighbor locations, so locations with more or less than four adjacent locations cannot be represented. In a graph representation the main problem is when we add a new node or link to the graph, we have to verify if the graph still is planar. So at every step, adding a new location to a plan, we must check the graph against this problem. Another main inefficiency of these methods is that the hierarchical nature of floor planning cannot be accommodated by them.

Consequently we need to resort to other means of representation techniques to cope with architectural design problems. They have to deal with hierarchical information. In such a representation system not only we can decompose the information into smaller pieces also we can focus on the required parts of the problem without involving with many detailed unnecessary information.

2 Hierarchical Representation

Those techniques discussed before may be helpful in some design problems for representation purposes, but definitely not in architectural design problems dealt with in this work. Thus we will find out a method that can represent for example a floor plan with various levels of detailed and abstracted representations. The appropriateness of such a method comes from its underlying, tree-based structure, that enables it to deal with spatial information with a hierarchical nature.

The key issue in such a representation method is that the whole information about the design problem is not kept together at one level. Rather only the information, i.e. related components and the relations among them, about any cluster[1] is represented together at one level. Therefore, information concerning a complicated system can be represented in a hierarchy. So a hierarchical abstraction technique decomposes design information into some simplified and manageable levels.

The components of the floor plan of a building are space blocks or locations such as rooms, corridors, hall-ways, and so on. The locations of a plan are connected to each other. A connection between two locations means there is a relationship between them. For instance, one location can be accessed via the other location. Sometimes the connection shows functional relationship, for example in a factory the production lines go through different locations to become assembled. It is very important that at every step of a design process we able to represent the components or elements related to that step plus the relationships among them. In this sense at every step of a design, we can focus on the part of the information relevant to it and detailed information will be left out at that level. The other good point in a hierarchical representation is that related clusters of information are brought together without considering details of each cluster.

---

[1] A cluster at one level contains one or more components of one level above or below that level.

A system consisting of many elements or objects and the interrelationship between them is a complex system. To study the complexity of such a system we need some analytical methods or tools. We simplify the given information into levels, then the common characteristics or interrelationship among the elements of the system become clear. For instance, in designing a floor plan with a set of locations the activities assigned to them can be used for identifying the clusters of locations that share activities. Without analytical means it is hard to cope with the problems of generating floor plans. An analytical method can be used for abstracting a floor plan with many locations in it into hierarchy of spatial clusters starting with detailed level and ending with a more abstracted one. Having such a hierarchy of clusters of locations makes the generation of floor plans an easy design task. This is in fact the main strategy in this work for designing floor plans of buildings.

Q-analysis has emerged over the past decades as a relatively new approach to data analysis in a wide range of fields as urban design, social networks, medical diagnostics, and many others. For a more comprehensive review of the practical aspect of this method see [3]. The original development of Q-analysis was the work of Atkin [4], [5]. Q-analysis is in fact only a method of cluster analysis that identifies connectivities of different dimensional strengths between elements of a given set. The deficiency of Q-analysis is this: the algorithm uses a chain of clusters of elements for identification of connectivities at successive dimensional levels among the elements. So the precise nature of the structure of correspondence between elements of the set cannot be identified in this framework. For detailed discussion of these points see [6], [7].

3 The Galois Lattice Technique

The history of *Galois lattice*, goes back many years [8] and recently re-emerged in the work of Ho and this approach appears in his series of remarkable papers [9], [10], [11], [12]. In this section the Galois lattice will be explained in detail also we will see how this technique can be used for clustering locations of a floor plan and how they will be represented as a hierarchical structure with some abstraction levels. Theoretical aspects of this lattice will not be discussed, rather we will concentrate only on practical aspects of this technique. Then an algorithm will be given for construction of the lattice representing the group of locations at various abstraction levels.

The Galois lattice interprets the given information about a system. The technique reveals and analyzes the association between a set of elements and their relations through some (common) features. The inputs for this technique are:

- A set of elements identifying the objects of the system.
- A set of features representing characteristics the objects of the system have.

To understand the technique we resort to an example. Assume the elements of the first set are some locations in a small school namely an office, classroom-1, classroom-2, classroom-3, laboratory-1, laboratory-2, a private-toilet, and a public-toilet. Also the second set consists a person or group of persons responsible for doing some activity in one or more locations. For this example, we assume that the second set contains: a principal, teacher-1, teacher-2, teacher-3, and three groups of students. The type of data for the lattice is a binary table in this example represents a relationship between a set, *M*, of locations (1, 2, 3, 4, 5, 6, 7, 8), and set, *N*, of persons (P, T1, T2, T3, S1, S2, S3). In the example given locations are referred by numbers also persons' name are abbreviated for simplicity. Note that *objects* and *features* in a lattice may cover a wide range of interpretations. For instance, they could be: people and their common characteristics or events they engaged, buildings and costs, locations and the service types running there, and so on. A relation between locations and assigned people for those locations is shown in the following Table 1(a). Assigning features to an object is called as a *mapping* $\lambda: M \rightarrow N$ such that, to each x in M,

$$\lambda(x) = \{\, y \mid y \text{ in } N, \text{ and } y \text{ is a feature of } x \,\}\,. \qquad \textbf{(1)}$$

Table 1. Arrays (a) $\lambda \subseteq M \times N$, (b) $\lambda^{-1} \subseteq N \times M$

(a)

M \ *N*	P	T1	T2	T3	S1	S2	S3
1	1	1	1	1	0	0	1
2	1	1	0	0	1	0	0
3	1	0	1	0	0	1	0
4	1	0	0	1	0	0	1
5	1	1	0	0	1	1	0
6	0	0	1	0	0	0	0
7	1	1	1	1	0	0	0
8	0	0	0	0	1	1	1

(b)

N \ *M*	1	2	3	4	5	6	7	8
P	1	1	1	1	1	0	1	0
T1	1	1	0	0	1	0	1	0
T2	1	0	1	0	0	1	1	0
T3	1	0	0	1	0	0	1	0
S1	0	1	0	0	1	0	0	1
S2	0	0	1	0	1	0	0	1
S3	1	0	0	1	0	0	0	1

The set of all ordered pairs (x, y), where $x \in M$, and $y \in N$, is denoted by $M \times N$, the Cartesian product of the sets M and N. Such a pair is termed the element of the lattice. We see that λ is a *subset* of the set $M \times N$, that is, $\lambda \subseteq M \times N$. The inverse of $\lambda: M \rightarrow N$ is $\lambda^{-1}: N \rightarrow M$, which maps each feature to a set of objects. This inverse relation $\lambda^{-1} \subseteq N \times M$, shown in Table 1(b), is obtained from that of $\lambda \subseteq M \times N$ by interchanging the rows and the columns of the Table 1(a). The relation between λ and λ^{-1} can be represented as:

$$y\, \lambda^{-1} x \Leftrightarrow x \lambda y\,. \qquad \textbf{(2)}$$

The main purpose behind interpreting data given in Table 1 is to identify the similarity and the distinction an object or group of objects have from others. The lattice is capable of finding precisely similarities and distinctive characteristics among the objects of a system. The similarities between objects are identified in terms of the common features they have.

As mentioned above relationships λ and λ^{-1} are mappings from $M \times N$ and $N \times M$

As mentioned above relationships λ and λ^{-1} are mappings from $M \times N$ and $N \times M$ respectively. The mapping can be composed into $\lambda^{-1} \cdot \lambda$ or $\lambda \cdot \lambda^{-1}$ where "$\cdot$" is the composition by *Boolean multiplication*. For brevity, these two relations can be written as: $\alpha = \lambda \cdot \lambda^{-1}$, and $\sigma = \lambda^{-1} \cdot \lambda$. Row i of Table 1(a) is denoted by λ_i, and column j by λ^j, also cell ij by λ^j_i. Therefore the arrays α and σ can be formed where

$$\sigma_i = \prod_{\substack{j=1 \\ \lambda^j_i \neq 0}}^{n} \lambda^j, \qquad \alpha_i = \prod_{\substack{j=1 \\ \lambda^j_i \neq 0}}^{m} \lambda_j. \tag{3}$$

where $\prod$ denotes term-wise Boolean multiplication on the vectors λ_j and λ^j. With respect to the Table 1(a), $\lambda^1 = [1, 1, 1, 1, 1, 0, 1, 0]$, and $\lambda^3 = [1, 0, 1, 0, 0, 1, 1, 0]$, then

$$\sigma_1 = \prod \lambda^j = \lambda^1 \cdot \lambda^3 = [1, 0, 1, 0, 0, 0, 1, 0]. \tag{4}$$

Repeating this calculation for all elements the results are shown in Table 2.

Table 2. Arrays (a) $\sigma = \lambda^{-1} \cdot \lambda$, (b) $\alpha = \lambda \cdot \lambda^{-1}$

		___	___	___	M	___	___	___				P	T1	T2	T3	S1	S2	S3
		1	2	3	4	5	6	7	8									
	1	1	0	0	0	0	0	0	0		P	1	0	0	0	0	0	0
	2	0	1	0	0	1	0	0	0		T1	1	1	0	0	0	0	0
	3	0	0	1	0	0	0	0	0		T2	0	0	1	0	0	0	0
M	4	1	0	0	1	0	0	0	0	N	T3	1	0	0	1	0	0	0
	5	0	0	0	0	1	0	0	0		S1	1	1	0	0	1	0	0
	6	1	0	1	0	0	1	1	0		S2	0	0	0	0	0	1	0
	7	1	0	0	0	0	0	1	0		S3	0	0	0	0	0	0	1
	8	0	0	0	0	0	0	0	1						(b)			

(a)

Each element of the lattice can be shown by $(A, \lambda(A))$, where A is a subset of objects and $\lambda(A)$ is a subset of common features that the elements of A have. The set A of elements represents the similarities between objects and features identifying the lattice. There exist operations that can be applied to any pair of elements (object and its features) of a lattice. Applying these operations on any pair yields another element of the lattice. Two of these operations are called the *meet* $\wedge$ (defining the greatest lower bound of the pair of elements) and the *join* $\vee$ (defining the least upper bound of the pair of elements).

If only some elements of a lattice are known, we can apply these two operations to find more elements of that lattice. Starting with the minimal elements of a lattice we can generate extra elements by applying the join operation. We proceed until there is no new element to be added to the set of elements. When all the elements are identified, we can classify and separate them with respect to the number of objects and common features they have. Relations between the elements at various levels can be identified through the common features they have.

4 An Algorithm for Constructing the Galois Lattice

In this section we provide an explanation of Ho's algorithm for making Galois lattices. The first step of the algorithm is the generation of the minimal[2] elements or terms of the lattice. Then other elements can be found applying the join and the meet operations on already known elements. The minimal elements can be found either by inspecting arrays λ and λ^{-1} depicted in Table 1, or by pairing the rows of arrays α and σ by the rows and columns of the array λ. Here we try both ways, note that anyhow the duplicated terms must be discarded. In the first way, the elements are found by distinguishing common features of every object also knowing each feature shared by some objects.

For the previously given example common features of any individual object can be found from Table 1(a) shown before. Also every individual feature shared by some objects is identified from Table 1(b). Note that objects here are locations and features are some people assigned to some locations for achieving some activities there. The minimal terms will be identified from the arrays shown in Table 1. First we find out the terms representing people responsible for doing activity in one location, they are:

For location 1 the term (1, P T1 T2 T3 S3), means that location 1 is assigned for persons P, T1, T2, T3, and S3 to do some activity there. For locations 2 and 5 the term (2 5, P T1 S1), means that persons P, T1, and S1 are responsible for some activity in location 2. Note that location 5 is also assigned for these persons probably for other activities. For location 3 the term (3, P T2 S2), the meaning should be clear. For location 4 the term (1 4, P T3 S3). For location 5 the term (5, P T1 S1 S2). For location 6 the term (1 3 6, T2). For location 7 the term (1 7, P T1 T2 T3). For location 8 the term (8, S1 S2 S3).

Now we figure out the terms identifying any individual person assigned to one or more locations for his or her activity. These terms are:

For person P the term (P, 1 2 3 4 5 7), means that person P achieves his or her activity in locations 1, 2, 3, 4, 5, and 7. For person T1 the term (P T1, 1 2 5 7), the meaning is clear the only note needed here is that these locations are already assigned for person P too. For person T2 the term (T2, 1 3 6 7). For person T3 the term (P T3, 1 4 7). For person S1 the term (S1, 2 5 8). For person S2 the term (S2, 3 5 8). For person S3 the term (S3, 1 4 8).

Note that in writing terms it is not important to specify first the objects or the features. We denote here objects before the comma sign and features after it. The minimal terms after discarding the duplicated terms are the following fourteen terms for this example.

(1, P T1 T2 T3 S3)	(1 4, P T3 S3)	(8, S1 S2 S3)	(1 3 6 7, T2)	(3 5 8, S2)
(2 5, P T1 S1)	(5, P T1 S1 S2)	(1 2 3 4 5 7, P)	(1 4 7, P T3)	(1 4 8, S3)
(3, P T2 S2)	(1 7, P T1 T2 T3)	(1 2 5 7, P T1)	(2 5 8, S1)	

---

[2] Minimal terms describe every object as a distinctive set of features it has, also they identify every feature possessed by a unique set of objects.

The second step after identifying the minimal terms of a lattice is to determine extra terms by applying the join operation to the above terms. This should be done within few rounds starting with the first round the terms having only one object. So in round 2 with terms with two objects, then in the next round with terms with three objects in them and so on.

The join, $\vee$, operation has been explained before. Assume two lattice terms are $(A, \lambda(A))$ and $(B, \lambda(B))$, the join operation is defined as:

$$(A, \lambda(A)) \vee (B, \lambda(B)) = (\lambda^{-1}[\lambda(A) \cap \lambda(B)], \lambda(A) \cap \lambda(B)) . \tag{5}$$

where $\cap$ is the usual set intersection operation. One point with respect to the above formula, the term $\lambda(A) \cap \lambda(B)$ represents the subset of features both objects in A and B have. So the term $\lambda^{-1}[\lambda(A) \cap \lambda(B)]$ identifies the subset of objects that each possesses this set of features. For the sake of completeness we add terms $(M, \varnothing)$ and $(\varnothing, N)$[3] to the minimal terms of the lattice. We proceed with the further steps of the algorithm.

Assume S to be the initial set of fourteen terms mentioned above plus the term $(M, \varnothing)$. The new terms of the lattice that are generated will be added to this set. Also assume L as a set that contains the terms that already linked for constructing the lattice. So at first $L = (\varnothing, N)$, then every term of set S that is considered while constructing the lattice will be added to set L. The input to the algorithm is:

$S = \{(\text{fourteen terms}) \cup (M, \varnothing)\}$
$L = \{(\varnothing, N)\}$
$n = 1$ where $S(n)$ denotes all the terms in S with n objects.

The steps of the algorithm for constructing the lattice are:

Step 1: If S is empty, go to step 3; else form
$\qquad S(n) = \{(A, B) \mid (A, B) \in S, A \text{ has } n \text{ objects}\}$,
Step 2: If $S(n)$ is empty, set $n = n + 1$, go to step 1; else do what follows.
$\qquad$ 2.1 Delete terms in $S(n)$ from S.
$\qquad$ 2.2 Form the join between each pair of terms in $S(n)$ and if the generated term is not already a term in S so add the term to S.
$\qquad\qquad$ Form the join between each term in $S(n)$ and each term in L and if the generated term is not already a term in S so add the term to S.
$\qquad$ 2.3 Terms in $S(n)$ will be the terms of the lattice[4] at level n, so each term of $S(n)$ should be linked by an edge to any terms of the lattice at one lower

---

[3] Terms $(M, \varnothing)$ and $(\varnothing, N)$ in fact represent terms $(1\ 2\ 3\ 4\ 5\ 6\ 7\ 8, 0)$ and $(0, P\ T1\ T2\ T3\ S1\ S2\ S3)$ respectively. Meaning that there exist no person assigned into all locations for doing his/her activity, also there is no any single location shared by all person in this example.

[4] Elements of the lattice, in graphical representation, are linked by an edge. Also the linked elements of each lattice element can be shown as a set.

level if the term of $S(n)$ *covers*[5] any term at lower level.

2.4 Form the meet between each pair of terms in $S(n)$ and if the generated term is already in L delete this term from L.
Form the meet between each term in $S(n)$ and each term in L and if the generated term is already in L delete this term from L.

2.5 Add the terms of $S(n)$ to L, set $n = n + 1$, go to step 1.

Step 3: Stop.

Applying this algorithm to the example given in this section, results the lattice shown graphically in Figure 1. We can easily identify how particular locations of the plan aggregated into a larger cluster if they are shared by some people. For instance at the first level, the level with single location, persons P, T1, T2, T3, and S3 are assigned to location 1 also location 2 is considered for persons P, T1, and S1 and so on. At the next level, we can identify persons who are assigned to pairs of locations, namely 1, 4 and 1, 7. For example persons P, T3, and S3 use the locations 1 and 4. Therefore, the relationships between persons with respect to sharing the same locations for some purposes now become clear.

Considering the result of the Galois lattice, depicted in Figure 1, we can cluster the given locations at various levels. Clusters at different abstraction levels are shown in Table 3. Note that in each abstraction level there exist two series of numbers separated by a hyphen with the exception at the first abstraction level. The first series shows the location(s) related to that abstraction level and the last number (after the hyphen sign) indicates the number of persons sharing those locations at that level. For instance, at level-2 the second row represents locations 1 and 7 as a single cluster which is shared by four persons, also at level-3 the same row shows locations 1, 4 and 8 are used only by one person and so on.

At the last abstraction level, sixth in this example, all locations which merged into a single cluster are not shared by anybody. This means that there is no one person or more who share all locations of the building, of course this is almost true. The number indicating how many persons share the same locations at any abstraction level in fact represents the association among locations corresponding to that level. So the higher a number is the more relationships exist among those locations. These numbers help us to create one or more clusters at every abstraction level.

The other point we should consider for clustering locations at different abstraction levels is how other locations are appeared at a higher abstraction level. For instance, in the result of the Galois lattice for the given example (shown in the above table) at level-2 locations 1 and 4 belong to a cluster that is shared by three persons. Also at this level locations 1 and 7 are related to each other because they are shared by four persons. At the next higher level location 7 is somehow related to this cluster, by means of sharing with the same persons. With respect to the degree of association between locations at different abstraction levels also the appearance of other

---

[5] $(A, \lambda(A))$ is said to cover $(B, \lambda(B))$, if $(B, \lambda(B)) < (A, \lambda(A))$ and there exists no other element $(X, \lambda(X))$ such that $(B, \lambda(B)) < (X, \lambda(X)) < (A, \lambda(A))$.

732

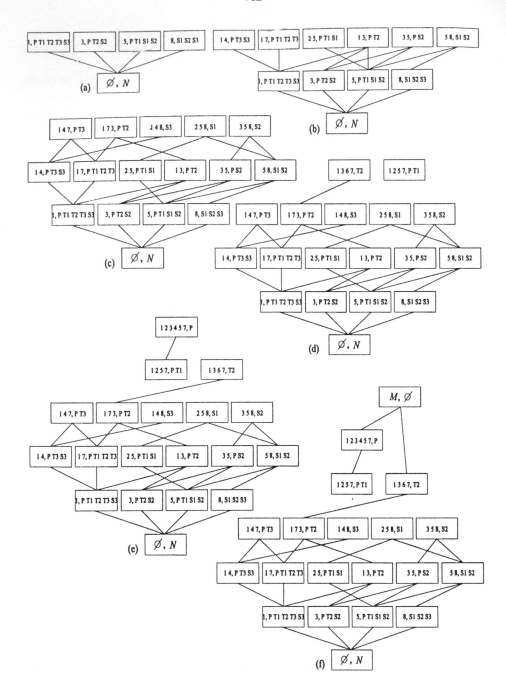

Figure 1. (a), (b), (c), (d), (e), and (f) are lattices derived from the given data

Table 3. Clusters at different abstraction levels

Level-1	Level-2	Level-3	Level-4	Level-5	Level-6
1-5	1,4-3	1,4,7-2	1,2,5,7-2	1,2,3,4,5,7-1	1,2,3,4,5,6,7,8,9-0
3-3	1,7-4	1,4,8-1	1,3,6,7-1		
5-4	1,3-2	1,7,3-2			
8-3	2,5-3	3,5,8-1			
	5,3-2	2,5,7-2			
	8,5-2				

locations at higher abstraction levels, we can collapse the related locations into one cluster. In fact, this is how we cluster locations of a building and allow them to be expanded until we come to a single cluster at the last level of abstraction that includes all locations of that building. Other floor plan design information, such as the adjacency requirements among locations, can help us for collapsing some locations together or separating them into different clusters.

5 Conclusions

Floor plan generation is viewed as a twofold activity that involves spatial representation of plans and a modification process. In the first task, a hierarchical representation scheme is constructed using the given requirements and specifications of the design problem. In the next task, the requested floor plan is generated using the hierarchical representation in a stepwise manner such that some conditions are fulfilled, i.e. the required adjacency locations. Building such hierarchy is very important in solving floor plans design problems. Because it clusters the complicated information of the given problem into some chunk of associated small information. Having such hierarchy of information at our disposal eases the generation process of the plans.

We have seen how a floor plan can be clustered or abstracted using the Galois technique. The technique enables us to create a hierarchy of locations of a floor plan in question. The hierarchy starts with a floor plan considering the given locations in a separate cluster. At every abstraction step, the system puts one or more locations into a single cluster. The abstraction process ends when the system provides a single cluster covers all locations or with few clusters where the system cannot proceed any more. Usually vital information is missing in design specifications. In most cases it, however, is stated implicitly in the specifications. Therefore we have to make them explicit in order to manipulate them while generating floor plans. The implicitly stated information in the floor plan design specifications may be the association between locations of the plan with respect to people sharing some locations, for example who shares one or more specific locations, which location(s) are shared by most people, and which locations are related and what are the association degrees between them. Rather the specification only shows which locations are assigned to which persons. The Galois lattice also has been used for externalizing the hidden information.

The author has been applied this technique for automating the process of floor plan generation [13]. Note that the explained technique can be used in other engineering and industrial design fields to automate the associated design processes. Therefore this technique deserves more attention in solving design problems.

References

1. Steadman, J. P.: Graph theoretic representation of architectural arrangement. Architectural research and teaching, Vol. 2. No 3. (1973)
2. Galle, P.: Abstraction as a tool of automated floor-plan design. Environment and Planning B, Vol. 13. (1986) 21 - 46
3. Macgill, S. M.: Structural analysis of social data: a guide to Ho's Galois lattice approach and a partial re-specification of Q-analysis. Environment and Planning B, Vol. 17. (1985) 1089 - 1109
4. Atkin, R. H.: Mathematical Structure in Human Affairs. Crane, Russak & Company Inc. New York (1974)
5. Atkin, R. H.: Combinatorial Connectivities in Social Systems; An Application of Simplicial Complex Structures to the Study of Large Organizations. Birkhauser Verlag Stuttgart (1977)
6. Macgill, S. M.: Cluster analysis and Q-analysis. International Journal of Man-Machine studies, Vol. 20. (1984)
7. Macgill, S. M., Springer, T.: An alternative algorithm for Q-analysis. Environment and Planning B, Vol. 14. (1987) 39 - 52
8. Ore, O.: Mathematical relations and structures. Bulletin of the American Mathematical Society, Vol. 48. (1942) 169 - 182
9. Ho, Y-S.: The planning process: mappings between state and decision spaces. Environment and Planning B, Vol. 9. (1982a) 153 -162
10. Ho, Y-S.: The planning process: a formal model. Environment and Planning B, Vol. 9. (1982b) 377 - 386
11. Ho, Y-S.: The planning process: fundamental issues. Environment and Planning B, Vol. 9. (1982c) 387 - 395
12. Ho, Y-S.: The planning process: structure of verbal descriptions. Environment and Planning B, Vol. 9. (1982d) 397 - 420
13. Chitchian, D.: Artificial Intelligence for Automated Floor Plan Generation. PhD Thesis. Technology University of Delft, The Netherlands (1997)

Automated Solving of the DEDS Control Problems

František Čapkovič

Institute of Control Theory and Robotics, Slovak Academy of Sciences
Dúbravská cesta 9, 842 37 Bratislava, Slovak Republic
utrrcapk@nic.savba.sk,
WWW home page: http://www.savba.sk/~utrrcapk/capkhome.htm

Abstract. Petri nets (PN) and oriented graphs (OG) are used for modelling of discrete event dynamic systems (DEDS). Model-based analysing the behaviour of the system to be controlled with respect to knowledge about the control task specifications (criteria, constraints, etc.) yields the automated solving of the control synthesis problem.

1 Introduction

DEDS consist of many cooperating subsystems. Their behaviour is influenced by occurring discrete events that start or stop activities of the subsystems. In the other words, DEDS are asynchronouos systems with concurrency or/and parallelism among the activities of their subsystems. Usually, they are large-scale or/and complex. Typical representants of DEDS are flexible manufacturing systems (FMS), transport systems, different kinds of communication systems, etc. Because DEDS are very important in human practice, the demand of the successful and efficient control of them is very actual. The control task specifications (constraints, criteria, etc.) are usualy given verbally or in another form of non-analytical terms. The main problem of the DEDS control synthesis is to express them in a suitable form in order to satisfy them properly.

2 The DEDS modelling

PN are frequently used for DEDS modelling. They can be understood to be the bipartite oriented graphs, i.e. the graphs with two kinds of nodes (positions and transitions) and two kinds of edges (oriented arcs emerging from the positions and entering the transitions on one hand, and oriented arcs emerging from the transitions and entering the positions on the other hand). Formally, the PN structure is

$$\langle P, T, F, G \rangle \quad ; \quad P \cap T = \emptyset \quad ; \quad F \cap G = \emptyset \tag{1}$$

where

$P = \{p_1, ..., p_n\}$ is a finite set of the PN positions with p_i, $i = 1, n$, being the elementary positions. The positions represent states of the DEDS subsystems activities.

$T = \{t_1, ..., t_m\}$ is a finite set of the PN transitions with t_j, $j = 1, m$, being the elementary transitions. The transitions express the DEDS discrete events.

$F \subseteq P \times T$ is a set of the oriented arcs entering the transitions. It can be expressed by the arcs incidence matrix $\mathbf{F} = \{f_{ij}\}$, $f_{ij} \in \{0, M_{f_{ij}}\}$, $i = 1, n$; $j = 1, m$. Its element f_{ij} represents the absence (when 0) or presence and muliplicity (when $M_{f_{ij}} > 0$) of the arc oriented from the position p_i to its output transition t_j. Hence, the oriented arcs represent the causal relations between the DEDS subsystems activities and their discrete events.

$G \subseteq T \times P$ is a set of the oriented arcs emerging from the transitions. The arcs incidence matrix is $\mathbf{G} = \{g_{ij}\}$, $g_{ij} \in \{0, M_{g_{ij}}\}$, $i = 1, m$; $j = 1, n$. Its element g_{ij} expresses analogically the absence or presence and multiplicity of the arc oriented from the transition t_i to its output position p_j. In such a way the oriented arcs express the causal relations between the DEDS discrete events and their subsystems activities.

The PN dynamics can be formally expressed as follows

$$\langle X, U, \delta, x_0 \rangle \quad ; \quad X \cap U = \emptyset \tag{2}$$

where

$X = \{\mathbf{x}_0, \mathbf{x}_1, ..., \mathbf{x}_N\}$ is a finite set of the state vectors of the PN positions in different situations with $\mathbf{x}_k = (\sigma_{p_1}^k, ..., \sigma_{p_n}^k)^T$, $k = 0, N$, being the n-dimensional state vector of the PN in the step k. Here, $\sigma_{p_i}^k \in \{0, c_{p_i}\}$, $i = 1, n$ is the state of the elementary position p_i in the step k - passivity (when 0) or activity (when $0 < \sigma_{p_i}^k \leq c_{p_i}$); c_{p_i} is the capacity of the position p_i, i.e. the maximal number of tokens that can be placed into the position; k is the discrete step of the PN dynamics development; T symbolizes the matrix or vector transposition. In the PN theory the state vector is named to be PN marking or the vector of PN marking.

$U = \{\mathbf{u}_0, \mathbf{u}_1, ..., \mathbf{u}_N\}$ is a finite set of the state vectors of the PN transitions in different situations with $\mathbf{u}_k = (\gamma_{t_1}^k, ..., \gamma_{t_m}^k)^T$, $k = 0, N$ being the m-dimensional control vector of the PN in the step k. Here, $\gamma_{t_j}^k$, $j = 1, m$ is the state of the elementary transition t_j in the step k - enabled (when 1), i.e. able to be fired, or disabled (when 0), i.e. not able to be fired.

$\delta : X \times U \longmapsto X$ is a transition function of the PN marking.

$\mathbf{x}_0$ is the initial state vector of the PN.

The simplest form of the linear discrete dynamic k-invariant model of the DEDS, based on an analogy with the ordinary PN (OPN), can be written - see e.g. [1] - as follows

$$\mathbf{x}_{k+1} = \mathbf{x}_k + \mathbf{B}.\mathbf{u}_k \quad , \quad k = 0, N \tag{3}$$

$$\mathbf{B} = \mathbf{G}^T - \mathbf{F} \tag{4}$$

$$\mathbf{F}.\mathbf{u}_k \leq \mathbf{x}_k \tag{5}$$

where

k is the discrete step of the DEDS dynamics development.

$\mathbf{x}_k = (\sigma_{p_1}^k, ..., \sigma_{p_n}^k)^T$ is the n-dimensional state vector of the system in the step k. Its components $\sigma_{p_i}^k \in \{0, c_{p_i}\}$, $i = 1, n$ express the states of the DEDS elementary subprocesses or operations - 0 (passivity) or $0 < \sigma_{p_i} \leq c_{p_i}$ (activity); c_{p_i} is the capacity of the DEDS subprocess p_i as to its activities.

$\mathbf{u}_k = (\gamma_{t_1}^k, ..., \gamma_{t_m}^k)^T$ is the m-dimensional control vector of the system in the step k. Its components $\gamma_{t_j}^k \in \{0, 1\}$, $j = 1, m$ represent occurring of the DEDS elementary discrete events (e.g. starting or ending the elementary subprocesses or their activities, failures, etc. - 1 (presence) or 0 (absence) of the corresponding discrete event.

$\mathbf{B}$, $\mathbf{F}$, $\mathbf{G}$ are, respectively, $(n \times m)$, $(n \times m)$ and $(m \times n)$- dimensional structural matrices of constant elements. The matrix $\mathbf{F} = \{f_{ij}\}$; $i = 1, n$, $j = 1, m$; $f_{ij} \in \{0, M_{f_{ij}}\}$ expresses the causal relations among the states of the DEDS and the discrete events occuring during the DEDS operation, where the states are the causes and the events are the consequences - 0 (nonexistence), $M_{f_{ij}} > 0$ (existence and multiplicity) of the corresponding causal relations. The matrix $\mathbf{G} = \{g_{ij}\}$; $i = 1, m$, $j = 1, n$; $g_{ij} \in \{0, M_{g_{ij}}\}$ expresses very analogically the causal relation among the discrete events (the causes) and the DEDS states (the consequences). Both of these matrices are the arcs incidence matrices. The matrix $\mathbf{B}$ is given by them.

$(.)^T$ symbolizes the matrix or vector transposition.

When the PN transitions are fixed on the corresponding oriented arcs among the PN positions - see Fig. 1 - we have a structure that can be understood to

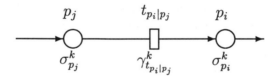

Fig. 1. An example of the placement of a transition on the oriented arc between two positions p_i and p_j

be the ordinary OG

$$\langle P, \Delta \rangle \tag{6}$$

where

$P = \{p_1, ..., p_n\}$ is a finite set of the OG nodes with p_i, $i = 1, n$, being the elementary nodes. They are the PN positions.

$\Delta \subseteq P \times P$ is a set of the OG edges i.e. the oriented arcs among the nodes. The functions expressing the occurrency of the discrete events (represented above by the PN transitions) represent its elements. The set can be expressed in the form of the incidence matrix $\Delta = \{\delta_{ij}\}$, $\delta_{ij} \in \{0,1\}$, $i = 1, n$; $j = 1, n$. Its element δ_{ij} represents the absence (when 0) or presence (when 1) of the edge oriented from the node p_i to the node p_j containing the PN transition.

It can be said that there is only one difference between the PN-based model and the OG-based one. Namely, the PN transitions are fixed on the oriented edges between corresponding nodes (the PN positions) in the OG structure. However, as a matter of fact, the elements of the set Δ are functions. Namely, the set $\Delta = \Delta_k \subseteq (P \times T) \times (T \times P)$. To introduce exactly the weights δ_{ij}, $i = 1, n$; $j = 1, n$, the OG dynamics can be formally expressed (in analogy with the above PN-based approach) as follows

$$\langle X, \delta_1, \mathbf{x}_0 \rangle \tag{7}$$

where

$X = \{\mathbf{x}(0), \mathbf{x}(1), ..., \mathbf{x}(N)\}$ is a finite set of the state vectors of the graph nodes in different situations with $\mathbf{x}(k) = (\sigma_{p_1}^k(\gamma), ..., \sigma_{p_n}^k(\gamma))^T$, $k = 0, N$, being the n-dimensional state vector of the graph nodes in the step k; $\sigma_{p_i}^k(\gamma) \in \mathbf{x}(k)$, $i = 1, n$ is the functional state of the elementary node p_i in the step k (such a state depends in general on the corresponding input transitions of the position p_i and its numerical value depends on the fact whether the transitions are enabled or disabled in the step k); k is the discrete step of the graph dynamics development.

$\delta_1 : (X \times U) \times (U \times X) \longmapsto X$ is the transition function of the graph dynamics. It contains implicitly the states of the transitions (the set U is the same like before) situated on the OG edges.

$\mathbf{x}(0)$ is the initial state vector of the graph dynamics.

Consequently, the k-variant OG-based linear discrete dynamic model of the DEDS can be written as follows

$$\mathbf{x}(k+1) = \Delta_k . \mathbf{x}(k) \quad , \quad k = 0, N \tag{8}$$

where

k is the discrete step of the DEDS dynamics development.

$\mathbf{x}(k) = (\sigma_{p_1}^k(\gamma), ..., \sigma_{p_n}^k(\gamma))^T$; $k = 0, N$ is the n- dimensional state vector of the DEDS in the step k; $\sigma_{p_i}^k(\gamma)$, $i = 1, n$ is the state of the elementary subprocess p_i in the step k. Its activity depends on the actually enabled input transitions. The variable γ formally expresses such a dependency.

$\Delta_k = \{\delta_{ij}^k\}$, $\delta_{ij}^k = \gamma_{t_{p_i|p_j}}^k \in \{0, 1\}$, $i = 1, n$; $j = 1, n$, because the set Δ_k can be understood to be in the form $\Delta_k \subseteq (X \times U) \times (U \times X)$. This matrix expresses the causal relations between the subprocesses depending on the occurrence of the discrete events. The element $\delta_{ij}^k = \gamma_{t_{p_i|p_j}}^k \in \{0, 1\}$ expresses the actual value of the transition function of the PN transition fixed on the OG edge oriented from the node p_j to the node p_i.

3 The system dynamics development

The development of the PN-based k-invariant model is the following

$$\mathbf{x}_1 = \mathbf{x}_0 + \mathbf{B}.\mathbf{u}_0 \tag{9}$$

$$\mathbf{x}_2 = \mathbf{x}_1 + \mathbf{B}.\mathbf{u}_1 = \mathbf{x}_0 + \mathbf{B}.\mathbf{u}_0 + \mathbf{B}.\mathbf{u}_1 \tag{10}$$

$$\vdots \qquad \vdots \qquad \vdots$$

$$\mathbf{x}_k = \mathbf{x}_{k-1} + \mathbf{B}.\mathbf{u}_{k-1} = \mathbf{x}_0 + \mathbf{B}.\mathbf{u}_0 + \mathbf{B}.\mathbf{u}_1 + \ldots + \mathbf{B}.\mathbf{u}_{k-2} + \mathbf{B}.\mathbf{u}_{k-1} \tag{11}$$

$$\mathbf{x}_k = \mathbf{x}_0 + \mathbf{B}.\sum_{i=0}^{k-1} \mathbf{u}_i \tag{12}$$

$$\mathbf{x}_k = \mathbf{x}_0 + \mathbf{W}_k.\mathbf{U}_k \tag{13}$$

$$\mathbf{W}_B = \underbrace{[\mathbf{B},\ \mathbf{B},\ \ldots,\mathbf{B},\ \mathbf{B}]}_{k-times} \tag{14}$$

$$\mathbf{U}_k = (\mathbf{u}_{k-1}^T, \mathbf{u}_{k-2}^T, \ldots, \mathbf{u}_1^T, \mathbf{u}_0^T)^T \equiv (\mathbf{u}_0^T, \mathbf{u}_1^T, \ldots, \mathbf{u}_{k-2}^T, \mathbf{u}_{k-1}^T)^T \tag{15}$$

Hence, the matrix $\mathbf{W}_B$ can be understood to be an analogy with the controllability or/and reachability matrix $\mathbf{W}_{A,B} = [\mathbf{B},\ \mathbf{A}.\mathbf{B},\ \ldots, \mathbf{A}^{k-2}.\mathbf{B},\ \mathbf{A}^{k-1}.\mathbf{B}]$ of the linear systems in the classical system or/and control theory, however with $\mathbf{A} = \mathbf{I}_n$. Consequently, the criterion of the controllability or/and reachability (concerning the $rank(\mathbf{W}_{A,B})$) can be utilized.

The dynamical development of the k-variant model is the following

$$\mathbf{x}(1) = \boldsymbol{\Delta}_0.\mathbf{x}(0) \tag{16}$$

$$\mathbf{x}(2) = \boldsymbol{\Delta}_1.\mathbf{x}(1) = \boldsymbol{\Delta}_1.\boldsymbol{\Delta}_0.\mathbf{x}(0) \tag{17}$$

$$\vdots \qquad \vdots \qquad \vdots$$

$$\mathbf{x}(k) = \boldsymbol{\Delta}_{k-1}.\mathbf{x}(k-1) = \boldsymbol{\Delta}_{k-1}.\boldsymbol{\Delta}_{k-2}.\ldots.\boldsymbol{\Delta}_1.\boldsymbol{\Delta}_0.\mathbf{x}(0) \tag{18}$$

$$\mathbf{x}(k) = \boldsymbol{\Phi}_{k,0}.\mathbf{x}(0) \tag{19}$$

$$\boldsymbol{\Phi}_{k,j} = \prod_{i=j}^{k-1} \boldsymbol{\Delta}_i \quad ; \quad j = 0, k-1 \tag{20}$$

The multiplying is made from the left. Meaning of the multiplying and additioning operators in the development of the k-variant model have symbolic interpretation. An element $\phi_{i,j}^{k,0}$, $i = 1, n$; $j = 1, n$ of the transition matrix $\boldsymbol{\Phi}_{k,0}$ is either a product of k functional elements (such transition functions express the "trajectory" containing the sequence of elementary transitions that must be fired in order to reach the final state x_i^k from the initial state x_j^0) or a sum of several such products (when there exist two or more "trajectories" from the initial state to final one).

Consequently, any nonzero element δ_{ij}^k of the matrix $\boldsymbol{\Delta}_k$ gives us information about reachability of the state $\sigma_{p_i}^{k+1}$ from the state $\sigma_{p_j}^k$. Hence, any element $\phi_{i,j}^{k2,k1}$ of the transition matrix $\boldsymbol{\Phi}_{k2,k1}$ gives us information about the reachability of the state $\sigma_{p_i}^{k2}$ from the state $\sigma_{p_j}^{k1}$.

4   The control problem solving

Symbolically, the process of the control problem solving can be expressed by means of the following procedure:

START

- input of both the initial state $\mathbf{x}_0$ and the terminal one $\mathbf{x}_t$
- solving the system of diophantine equations $\mathbf{B}.\mathbf{U}_? = \mathbf{x}_t - \mathbf{x}_0$
- *if* (the nonnegative integer solution $\mathbf{U}_?$ does not exist) *then goto* END (because this fact means that the terminal state $\mathbf{x}_t$ is not reachable)
- the solution $\mathbf{U}_?$ exists, but there is no information about the number N of the steps that are necessary to reach the terminal state $\mathbf{x}_t$ and it is necessary to find N. Therefore,
- $K = 0$

LABEL 1:

- $\mathbf{x}(K+1) = \mathbf{\Delta}.\mathbf{x}(K)$
- *if* ($\mathbf{x}_t \subseteq \mathbf{x}(K+1)$) *then* (*begin* $N = K$; *goto* LABEL 2; *end*) *else* (*begin* $K = K+1$; *goto* LABEL 1; *end*)

LABEL 2:

- $k = 0$; $\mathbf{x}_k = \mathbf{x}_0$
- $\mathbf{x}(0) = \mathbf{x}_0$; $\mathbf{x}(k) = \mathbf{x}(0)$

LABEL 3:

- doing the step of the k-variant model: $\mathbf{x}(k+1) = \mathbf{\Delta}_k.\mathbf{x}_k$
- generation of the possible reachable state vectors $\mathbf{x}_{k+1} \in \mathbf{x}(k+1)$ of the k-invariant model by means of enabling the corresponding transitions being expressed by means of their transition functions - the elements of the vector $\mathbf{x}(k+1)$; it means that something like the reachability tree is generated
- *if* ($k < N$) *then* (*begin* $k = k+1$; *goto* LABEL 3; *end*)
- consideration of the control task specifications (criteria, constraints, etc.)
- choice of the most suitable (optimal) sequence of both the state vectors $\mathbf{x}_k$; $k = 0, N$ of the k-invariant model and the corresponding control vectors $\mathbf{u}_k$; $k = 0, N$ (the final result of the control synthesis)

END

The details of the solving procedure are strongly dependent on the actual system to be controlled as well as on the actual control task specifications of the control synthesis problem to be solved.

5 An example of DEDS modelling and their control problem solving

Consider the maze problem introduced by Ramadge and Wonham in [8]. Two "participants" - in [8] a cat and a mouse - can be as well e.g. two mobile robots or two automatically guided vehicles (AGVs) of the FMS, two cars on a complicated crossroad, two trains in a railway network, etc. They are placed in the maze (however, it can also be e.g. the complicated crossroad, etc.) given on Figure 2 consisting of five rooms denoted by numbers 1, 2,..., 5 connecting by the doorways

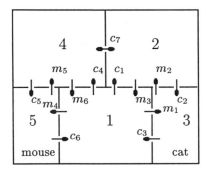

Fig. 2. The maze structure.

exclusively for the cat denoted by $c_i, i = 1, 7$ and the doorways exclusively for the mouse denoted by $m_j, j = 1, 6$. The cat is initially in the room 3 and the mouse in the room 5. Each doorway can be traversed only in the direction indicated. Each door (with the exception of the door c_7) can be opened or closed by means of control actions. The door c_7 is uncontrollable (or better, it is continuously open in both directions). The controller to be synthetized observes only discrete events generated by sensors in the doors. They indicate that a participant ist just running through. The control problem is to find a feedback controller (e.g. an automatic pointsman or switchman in railways) such that the following control task specifications - three criteria or/and constraints will be satisfied:

1. The participants never occupy the same room simultaneously.
2. It is always possible for both of them to return to their initial positions (the first one to the room 3 and the second one to the room 5).
3. The controller should enable the participants to behave as freely as possible with respect to the constraints imposed.

At the construction of the PN-based model of the system the rooms 1 - 5 of the maze will be represented by the PN positions p_1 - p_5 and the doorways will be represented by the PN transitions. The permanently open door c_7 is replaced by means of two PN transitions t_7 and t_8 symbolically denoted as c_7^k and c_8^k.

The PN-based representation of the maze is given on Figure 3. The initial state vectors of the cat and the mouse are

$$^c\mathbf{x}_0 = (0\,0\,1\,0\,0)\,, \quad ^m\mathbf{x}_0 = (0\,0\,0\,0\,1)^T \tag{21}$$

The structure of the cat and mouse control vectors is

$$^c\mathbf{u}_k = (c_1^k,\ c_2^k,\ c_3^k,\ c_4^k,\ c_5^k,\ c_6^k,\ c_7^k,\ c_8^k)^T;\ c_i^k \in \{0,1\},\ i = 1,8$$
$$^m\mathbf{u}_k = (m_1^k,\ m_2^k,\ m_3^k,\ m_4^k,\ m_5^k,\ m_6^k)^T;\ m_i^k \in \{0,1\},\ i = 1,6$$

The parameters of the cat model are
$$n = 5 \qquad m_c = 8$$

$$\mathbf{F}_c = \begin{pmatrix} 1 & 0 & 0 & 1 & 0 & 0 & 0 & 0 \\ 0 & 1 & 0 & 0 & 0 & 0 & 1 & 0 \\ 0 & 0 & 1 & 0 & 0 & 0 & 0 & 0 \\ 0 & 0 & 0 & 0 & 1 & 0 & 0 & 1 \\ 0 & 0 & 0 & 0 & 0 & 1 & 0 & 0 \end{pmatrix} \quad \mathbf{G}_c = \begin{pmatrix} 0 & 1 & 0 & 0 & 0 \\ 0 & 0 & 1 & 0 & 0 \\ 1 & 0 & 0 & 0 & 0 \\ 0 & 0 & 0 & 1 & 0 \\ 0 & 0 & 0 & 0 & 1 \\ 1 & 0 & 0 & 0 & 0 \\ 0 & 0 & 0 & 1 & 0 \\ 0 & 1 & 0 & 0 & 0 \end{pmatrix}$$

and the parameters of the mouse model are
$$n = 5 \qquad m_m = 6$$

$$\mathbf{F}_m = \begin{pmatrix} 1 & 0 & 0 & 1 & 0 & 0 \\ 0 & 0 & 1 & 0 & 0 & 0 \\ 0 & 1 & 0 & 0 & 0 & 0 \\ 0 & 0 & 0 & 0 & 0 & 1 \\ 0 & 0 & 0 & 0 & 1 & 0 \end{pmatrix} \quad \mathbf{G}_m^T = \begin{pmatrix} 0 & 0 & 1 & 0 & 0 & 1 \\ 0 & 1 & 0 & 0 & 0 & 0 \\ 1 & 0 & 0 & 0 & 0 & 0 \\ 0 & 0 & 0 & 0 & 1 & 0 \\ 0 & 0 & 0 & 1 & 0 & 0 \end{pmatrix}$$

This model was used in the approaches presented in [2], [3] and the general knowledge-based approach to the control synthesis of DEDS was presented in [5]. Knowledge representation was described in [4], [6], [7]. At the construction of the OG-based model the matrices $^c\mathbf{\Delta}_k$ and $^m\mathbf{\Delta}_k$ of the system parameters are the following

$$^c\mathbf{\Delta}_k = \begin{pmatrix} 0 & 0 & c_3^k & 0 & c_6^k \\ c_1^k & 0 & 0 & c_8^k & 0 \\ 0 & c_2^k & 0 & 0 & 0 \\ c_4^k & c_7^k & 0 & 0 & 0 \\ 0 & 0 & 0 & c_5^k & 0 \end{pmatrix} = \begin{pmatrix} 0 & 0 & ^c\delta_{13}^k & 0 & ^c\delta_{15}^k \\ ^c\delta_{21}^k & 0 & 0 & ^c\delta_{24}^k & 0 \\ 0 & ^c\delta_{32}^k & 0 & 0 & 0 \\ ^c\delta_{41}^k & ^c\delta_{42}^k & 0 & 0 & 0 \\ 0 & 0 & 0 & ^c\delta_{54}^k & 0 \end{pmatrix}$$

$$^m\mathbf{\Delta}_k = \begin{pmatrix} 0 & m_3^k & 0 & m_6^k & 0 \\ 0 & 0 & m_2^k & 0 & 0 \\ m_1^k & 0 & 0 & 0 & 0 \\ 0 & 0 & 0 & 0 & m_5^k \\ m_4^k & 0 & 0 & 0 & 0 \end{pmatrix} = \begin{pmatrix} 0 & ^m\delta_{12}^k & 0 & ^m\delta_{14}^k & 0 \\ 0 & 0 & ^m\delta_{23}^k & 0 & 0 \\ ^m\delta_{31}^k & 0 & 0 & 0 & 0 \\ 0 & 0 & 0 & 0 & ^m\delta_{45}^k \\ ^m\delta_{51}^k & 0 & 0 & 0 & 0 \end{pmatrix}$$

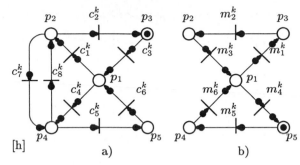

Fig. 3. The PN-based representation of the maze. a) possible behaviour of the cat; b) possible behaviour of the mouse

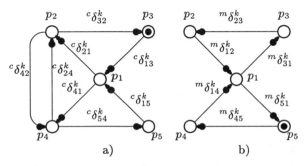

Fig. 4. The OG-based model of the maze. a) possible behaviour of the cat; b) possible behaviour of the mouse

The transitions matrices for the cat and mouse are the following

$$
{}^{c}\mathbf{\Phi}_{k+2,k} = {}^{c}\mathbf{\Delta}_{k+1}.{}^{c}\mathbf{\Delta}_{k} =
$$

$$
= \begin{pmatrix}
0 & c_3^{k+1}.c_2^k & 0 & c_6^{k+1}.c_5^k & 0 \\
c_8^{k+1}.c_4^k & c_8^{k+1}.c_7^k & c_1^{k+1}.c_3^k & 0 & c_1^{k+1}.c_6^k \\
c_2^{k+1}.c_1^k & 0 & 0 & c_2^{k+1}.c_8^k & 0 \\
c_7^{k+1}.c_1^k & 0 & c_4^{k+1}.c_3^k & c_7^{k+1}.c_8^k & c_4^{k+1}.c_6^k \\
c_5^{k+1}.c_4^k & c_5^{k+1}.c_7^k & 0 & 0 & 0
\end{pmatrix}
$$

$$
{}^{c}\mathbf{\Phi}_{k+3,k} = {}^{c}\mathbf{\Delta}_{k+2}.{}^{c}\mathbf{\Delta}_{k+1}.{}^{c}\mathbf{\Delta}_{k} =
$$

$$
= \begin{pmatrix}
c_3^{k+2}.c_2^{k+1}.c_1^k + c_6^{k+2}.c_5^{k+1}.c_4^k & c_6^{k+2}.c_5^{k+1}.c_7^k & \vdots \\
c_8^{k+2}.c_7^{k+1}.c_1^k & c_1^{k+2}.c_3^{k+1}.c_2^k & \vdots \\
c_2^{k+2}.c_8^{k+1}.c_4^k & c_2^{k+2}.c_8^{k+1}.c_7^k & \vdots \\
c_7^{k+2}.c_8^{k+1}.c_4^k & c_4^{k+2}.c_3^{k+1}.c_2^k + c_7^{k+2}.c_8^{k+1}.c_7^k & \vdots \\
c_5^{k+2}.c_7^{k+1}.c_1^k & 0 & \vdots
\end{pmatrix}
$$

744

$$\left. \begin{matrix} \vdots & 0 & c_3^{k+2}.c_2^{k+1}.c_8^k & 0 \\ \vdots & c_8^{k+2}.c_4^{k+1}.c_3^k & c_1^{k+2}.c_6^{k+1}.c_5^k + c_8^{k+2}.c_2^{k+1}.c_8^k & c_8^{k+2}.c_4^{k+1}.c_6^k \\ \vdots & \boxed{c_2^{k+2}.c_1^{k+1}.c_3^k} & 0 & c_2^{k+2}.c_1^{k+1}.c_6^k \\ \vdots & c_7^{k+2}.c_1^{k+1}.c_3^k & c_4^{k+2}.c_6^{k+1}.c_5^k & c_7^{k+2}.c_1^{k+1}.c_6^k \\ \vdots & c_5^{k+2}.c_4^{k+1}.c_3^k & c_5^{k+2}.c_7^{k+1}.c_8^k & \boxed{c_5^{k+2}.c_4^{k+1}.c_6^k} \end{matrix} \right)$$

$${}^m\boldsymbol{\Phi}_{k+2,k} = {}^m\boldsymbol{\Delta}_{k+1}.{}^m\boldsymbol{\Delta}_k =$$

$$= \begin{pmatrix} 0 & 0 & m_3^{k+1}.m_2^k & & m_6^{k+1}.m_5^k \\ m_2^{k+1}.m_1^k & 0 & 0 & 0 & 0 \\ 0 & m_1^{k+1}.m_3^k & 0 & m_1^{k+1}.m_6^k & 0 \\ m_5^{k+1}.m_4^k & 0 & 0 & 0 & 0 \\ 0 & m_4^{k+1}.m_3^k & 0 & m_4^{k+1}.m_6^k & 0 \end{pmatrix}$$

$${}^m\boldsymbol{\Phi}_{k+3,k} = {}^m\boldsymbol{\Delta}_{k+2}.{}^m\boldsymbol{\Delta}_{k+1}.{}^m\boldsymbol{\Delta}_k =$$

$$= \left(\begin{matrix} m_3^{k+2}.m_2^{k+1}.m_1^k + m_6^{k+2}.m_5^{k+1}.m_4^k & 0 & \vdots \\ 0 & m_2^{k+2}.m_1^{k+1}.m_3^k & \vdots \\ 0 & 0 & \vdots \\ 0 & m_5^{k+2}.m_4^{k+1}m_3^k & \vdots \\ 0 & 0 & \vdots \end{matrix} \right.$$

$$\left. \begin{matrix} \vdots & 0 & 0 & 0 \\ \vdots & 0 & m_2^{k+2}.m_1^{k+1}.m_6^k & 0 \\ \vdots & \boxed{m_1^{k+2}.m_3^{k+1}.m_2^k} & 0 & m_1^{k+2}.m_6^{k+1}.m_5^k \\ \vdots & 0 & m_5^{k+2}.m_4^{k+1}.m_6^k & 0 \\ \vdots & m_4^{k+2}.m_3^{k+1}.m_2^k & 0 & \boxed{m_4^{k+2}.m_6^{k+1}.m_5^k} \end{matrix} \right)$$

The states reachability trees are given on Fig. 5 and Fig. 6. It can be seen that in order to fulfille the prescribed control task specifications introduced above in the part 5, the comparison of the transition matrices of both animals in any step of their dynamics development is sufficient. Because the animals start from the defined rooms given by their initial states, it is sufficient to compare the columns 3 and 5. Consequently,

1. the corresponding (as to indices) elements of the transition matrices in these columns have to be mutually disjuct in any step of the dynamics development

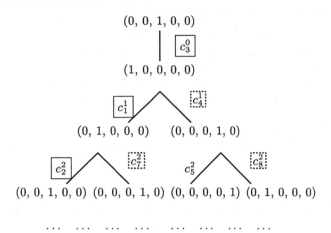

Fig. 5. The fragment of the reachability tree of the cat

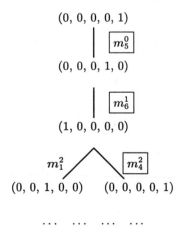

Fig. 6. The reachability tree of the mouse

in order to avoid encounter of the anomals on the coresponding "trajectories".

2. if they are not disjunct they must be removed. Only elements with indices [3,3] and [5,5] of the matrices $\Phi_{k+3,0}$ represent the exception. Namely, they express the trajectories making the return of the animals to their initial states possible. In case of the elements with indices [3,3] the element of the matrix $^c\Phi_{k+3,0}$ should be chosen. It represents the trajectory of the cat making their come back possible. In case of the elements with indices [5,5] the element of the matrix $^m\Phi_{k+3,0}$ should be chosen. It represents the trajectory of the mouse making their come back possible.

3. in the matrix $^c\Phi_{k+3,k}$ two elements in the column 3 (with the indices [2,3] and [4,3]) stay unremoved, because of the permanently open door. It can be seen that also the elements of the column 5 of this matrix (with indices [2,5] and [4,5]) stay removed. It corresponds to the prescribed condition that othervise the movement of the animals in the maze should be free.

6   Conclusions

The advantage of the presented approach to the automated DEDS control problem solving is that it automatically yields the complete solution of the problem in the elegant form. Even in analytical terms, if it is necessary.

References

1. Čapkovič, F.: Modelling and justifying discrete production processes by Petri nets. *Computer Integrated Manufacturing Systems*. Vol. **6**, No 1, February 1993, pp. 27-35.
2. Čapkovič, F.: A Petri nets-based approach to the maze problem solving. In: Balemi, S., Kozák, P. and Smedinga, R. (Eds.): *Discrete Event Systems: Modelling and Control*. Birkhäuser Verlag, Basel - Boston - Berlin, 1993, pp.173-179.
3. Čapkovič, F.: Computer-aided design of intelligent control systems for discrete event dynamic systems. In: Mattson, S.E., Gray, J.O. and Cellier, F. (Eds.): *Proc. of the IEEE/IFAC Joint Symposium on Computer-Aided Control System Design CACSD'94*, Tucson, Arizona, USA, March 7-9, 1994. Printed in USA, IEEE Catalogue #94TH0619-7, 1994, pp. 55-60.
4. Čapkovič, F.: Using Fuzzy Logic for Knowledge Representation at Control Synthesis. *BUSEFAL*, **63**, 1995, pp. 4-9.
5. Čapkovic, F.: Knowledge-Based Control of DEDS. In: Gertler, J.J., Cruz, J.B. and Peshkin, M. (Eds.): *Proc. of the 13th IFAC World Congress 1996*, San Francisco, USA, June 30-July 5, 1996, Vol. J, (also Compact Disc, Elsevier Science Ltd., Pergamon, 1996, paper J-3c-02.6), pp. 347-352.
6. Čapkovič, F.: Petri nets and oriented graphs in fuzzy knowledge representation for DEDS control purposes. *BUSEFAL*, **69**, 1997, pp. 21-30.
7. Čapkovič F.: Fuzzy knowledge in DEDS control synthesis. In: Mareš, M., Mesiar, R., Novák, V., Ramík, J. Stupňanová, A. (Eds.): *Proc. of the 7th International Fuzzy Systems Association World Congress - IFSA'97*, Prague, Czech Republic, June 25-29, 1997, Academia, Prague, Czech Republic, 1997, pp. 550-554.
8. Ramadge, P.J.G., Wonham, W.M.: The control of discrete event systems. In: Ho, Y.CH. (Ed.): *Proceedings of the IEEE*, Vol. **77**, No 1, 1989, pp. 81-98.

A More Efficient Knowledge Representation for Allen's Algebra and Point Algebra

Jörg Kahl, Lothar Hotz, Heiko Milde, and Stephanie Wessel

Laboratory for Artificial Intelligence, University of Hamburg
Vogt-Koelln-Str. 30, 22527 Hamburg, Germany
kahl@kogs.informatik.uni-hamburg.de

Abstract. In many AI applications, one has incomplete qualitative knowledge about the order of occurring events. A common way to express knowledge about this temporal reasoning problem is Allen's interval algebra. Unfortunately, its main interesting reasoning tasks, consistency check and minimal labeling, are intractable (assuming $P \neq NP$). Mostly, reasoning tasks in tractable subclasses of Allen's algebra are performed with constraint propagation techniques. This paper presents a new reasoning approach that performs the main reasoning tasks much more efficient than traditional constraint propagation methods. In particular, we present a sound and complete $O(n^2)$-time algorithm for minimal labeling computation that can be used for the pointisable subclass of Allen's algebra.

1 Introduction

In many AI applications, one has incomplete qualitative knowledge about the order of occurring events. It is a temporal reasoning task to complete the event order as far as possible. A common way to express knowledge about this task is Allen's interval algebra $\mathcal{A}$ [1]. The algebra can express any possibly indefinite relationship between two intervals. Complete knowledge about their temporal relationship is expressible with one of the thirteen mutually exclusive basic relations depicted in Figure 1.

Relation	Symbol	Inverse	Meaning
A before B	b	bi	
A meets B	m	mi	
A overlaps B	o	oi	
A starts B	s	si	
A during B	d	di	
A ends B	f	fi	
A equals B	e	e	

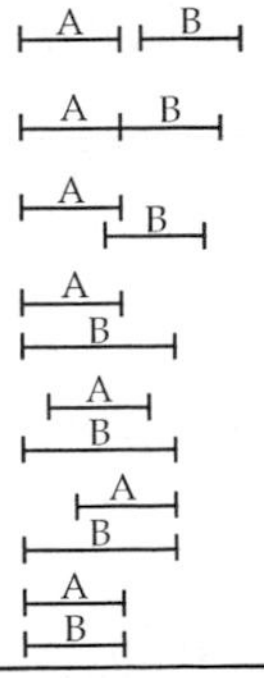

Figure 1. Basic interval relations

Mainly, there are two reasoning tasks arising in $\mathcal{A}$:

- Consistency maintenance decides if new temporal knowledge incorporated into the actual knowledge base is consistent. In terms of constraint networks one has to check if there is a consistent scenario among the alternatively defined basic relations. In the following, we will call this problem ISAT.
- Question answering consists of providing answers to queries to the possible relative order between time relations. The main problem in terms of constraint networks is to determine for every network edge the subset of basic relations that is part of a consistent scenario. This task is called the minimal labeling or strongest implied relation problem ISI.

Unfortunately, [6] proof that ISAT($\mathcal{A}$) and ISI($\mathcal{A}$) are NP-complete. An alternative representation form is the less expressive point algebra $\mathcal{TP}$ [6]. This algebra has time points instead of intervals as its primitives and therefore contains only three basic relations (for two time points P_1, P_2 the possible relations are $P_1 < P_2, P_1 = P_2$, and $P_1 > P_2$). Like in $\mathcal{A}$, any disjunction of basic relations is allowed resulting in $2^3 = 8$ elements.

The restricted expressiveness of $\mathcal{TP}$ is rewarded with the tractability of ISAT and ISI, which are defined as in $\mathcal{A}$. Interestingly, [6] show that a subclass of $\mathcal{A}$, the pointisable algebra $\mathcal{P}$, can be expressed within $\mathcal{TP}$ (see [5] for an enumeration).

For ISAT($\mathcal{P}$), a $O(n^2)$-time algorithm (w.r.t. the number of time points) can be found in [4]. Additionally, [4] presents the so far best ISI($\mathcal{P}$) algorithm for minimal labeling computation which is $O(n^4)$-time in worst case.

This paper presents an alternative reasoning approach that solves ISI($\mathcal{P}$) in $O(n^2)$-time, too. In the remainder, we present the data structure called 'ordered time line' on which our reasoning takes place and outline the algorithm inserting time point algebra constraints into ordered time line. We conclude with an outlook to further research.

2 Instantiation intervals

Our reasoning approach is influenced by van Beek's instantiation algorithm for ISAT($\mathcal{P}$) [4]. Its first step is the transformation of all time interval constraints into constraints relating pairs of interval endpoints. These constraints can be expressed within $\mathcal{TP}$. Afterwards, van Beek finds a consistent instantiation of all constraint variables. The basic relations between them finally give an ISAT solution.

Unlike van Beek, we represent time points by intervals qualitatively constraining the time period in which time points can be instantiated. As an example, Figure 2 depicts the transformation and instantiation of $A\{b, m\}B \in \mathcal{P}$ into instantiation intervals. The built up total order of instantiation interval endpoints on an imaginary time line will be called 'ordered time line' OTL. Note that instantiation interval lengths and their position within OTL do not have any specific values. In Figure 2, they are chosen arbitrarily.

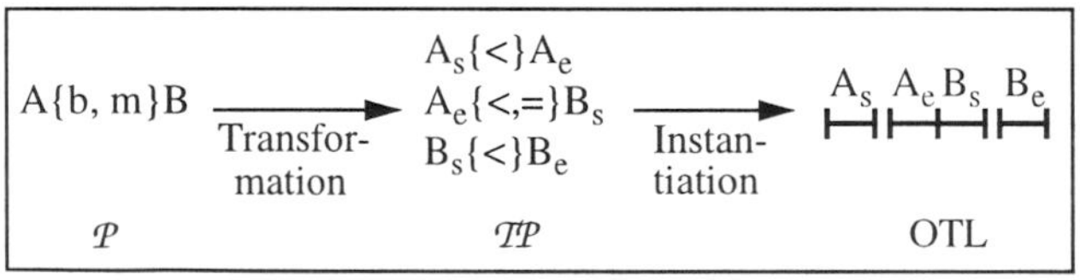

Figure 2. Transformation of $A\{before, meets\}B$ into instantiation intervals.

Table 1 presents six relations on our time primitives (a, b, and c are instantiation interval start- or endpoints of arbitrary time point variables). With them, we accomplish the expressiveness of $\mathcal{TP}$ respectively $\mathcal{P}$.

Table 1. Time primitive relations and their semantics

Type	1	2	3	4	5	6
Relation	$a \cdot b$	$\begin{vmatrix} a \\ b \end{vmatrix}$	$\begin{matrix} a \\ b \end{matrix}$	$\begin{matrix} a \cdot b \\ c \end{matrix}$	$\begin{vmatrix} a \\ b \end{vmatrix} \cdot \begin{vmatrix} a \\ b \end{vmatrix}$	$a(\neg b)$
Semantics	$a < b$	$a = b$	$(a < b) \vee (b < a)$	$(a < b) \wedge ((c < a \wedge c < b) \vee (a < c \wedge c < b) \vee (a < c \wedge b < c))$	$(a < a) \wedge (b < b) \wedge (a = b)$	$a \neq b$

In the following, we briefly elucidate the six relation types. Totally ordered time primitives are the simplest relation form in OTL (1). They express a strict time hierarchy from left to right. Some of our time primitives may occur at the same time (2), while some may occur in every possible permutation (3). Within a permutation of primitives, time primitives are allowed to have a local order (4). Besides local type 1 relations depicted in Table 2, type 2 relations may also occur in a time primitive permutation. A set of time primitives in type 2 relation can be defined over a time period (5). This means, some sets of instantiation interval start or end points are represented as intervals. These so called 'split time primitives' are also allowed within type 2 relations. For some pairs of time primitives, we just know that they do not occur at the same time (6).

Some OTL syntax examples are in order. We confront the examples in Table 2 covering all six relation types with their semantically equivalent constraint sets.

Table 2. OTL syntax examples with semantically equivalent constraint sets

OTL syntax example	Constraint set
$\langle T_1 \mid \cdot \mid T_1 \rangle \cdot \langle T_2 \mid \cdot \mid T_2 \rangle$	$\{T_1 < T_2\}$
$\langle T_1 \mid \cdot \begin{vmatrix} \mid T_1 \rangle \\ \langle T_2 \mid \end{vmatrix} \cdot \mid T_2 \rangle$	$\{T_1 \leq T_2\}$
$\begin{matrix} \langle T_1 \mid & \mid T_1 \rangle \\ \langle T_2 \mid & \mid T_2 \rangle \end{matrix} \cdot \langle T_3 \mid \cdot \mid T_3 \rangle$	$\{T_1 < T_3, T_2 < T_3\}$
$\begin{matrix} \langle T_1 \mid & \mid T_1 \rangle \cdot \langle T_4 \mid & \mid T_3 \rangle \\ \langle T_2 \mid & \mid T_2 \rangle & \cdot \langle T_3 \mid \cdot & \mid T_4 \rangle \end{matrix}$	$\{T_1 < T_3, T_2 < T_3, T_1 < T_4\}$

Table 2. OTL syntax examples with semantically equivalent constraint sets

OTL syntax example	Constraint set
$\left\langle \begin{matrix} \langle T_1\| \\ \left\langle\begin{matrix}\|T_1\rangle \\ \langle T_2\|\end{matrix}\right\| \\ \langle T_3\| \end{matrix} \right\| \cdot \|T_3\rangle \cdot \langle T_4\| \cdot \left\| \begin{matrix} \|T_1\rangle \\ \left\|\begin{matrix}\|T_1\rangle \\ \langle T_2\|\end{matrix}\right\| \\ \|T_4\rangle \end{matrix} \right\rangle$	$\{T_1 \le T_2, T_3 < T_4\}$
$\langle T_1(\neg T_2)\|\ \|T_1(\neg T_2)\rangle$ $\langle T_2(\neg T_1)\|\ \|T_2(\neg T_1)\rangle$	$\{T_1 \ne T_2\}$

The most important feature of our instantiation representation is its totally ordered set of time primitives. The modification of an element within this set updates the relation to all other concerned time primitives. Thus, to obtain an ISAT solution, all constraints have to be processed only once.

Furthermore, since instantiation intervals comprise all possible instantiations of the represented time points, they also comprise all possible basic relations to other time points. Thus, with an interval representation, we obtain an ISI solution without further computations.

3 The OTL inserting algorithm

From the eight relations comprised by $\mathcal{TP}$, we can exclude six of them from our inserting algorithm:
- $A\ ?\ B$: Since these 'constraints' contain no constraining information about A or B, we can simply omit them.
- $A\ \varnothing\ B$: If constraints of this form occur, no consistent scenario is possible within the given constraint set. Thus, the return value for ISAT($\mathcal{P}$) and ISI($\mathcal{P}$) is **false**.
- $A = B$: In this case, like in [4], we can condense the two constraint variables A, B into a new constraint variable AB.
- $A \ge B$: We transform these constraints into $B \le A$.
- $A < B$, $A > B$: We transform both constraints into two relations $A \le B$, $A \ne B$ and $B \le A$, $A \ne B$, respectively.

Thus, our OTL inserting algorithm only has to consider time point constraints of the form $A \le B$ and $A \ne B$. The algorithm has to comprise case discriminations about:
- the two possible relations,
- how many of the two variables of the constraint to insert are already instantiated,
- if both variables are already instantiated, the basic relation holding between them (in addition to the thirteen basic Allen relations, point-interval and point-point relations must be considered).

Summing up all possibilities, our OTL inserting algorithm has to consider about 300 different syntax cases. Fortunately, most syntax cases can be treated equally resulting in 18 cases. Due to space limitation, we cannot present the algotithm in this paper (see [2] for a presentation and a proof of its soundness and compelteness). Instead, we subsequently present an application example. Figure 3 shows a constraint set in form of a network and its corresponding strongest implied relations which the algorithm has to compute.

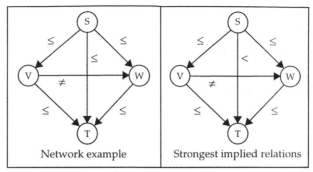

Figure 3. Constraint network example and its strongest implied relations

In Table 3, we show the successive constraint transformations and the corresponding reasoning results in OTL.

Table 3. Inserting an example set of time point constraints into OTL

Constraint to insert	$S \leq W$	$S \leq V$	$S \leq T$
Ordered time line	$\langle S\rvert \cdot \begin{vmatrix}\lvert S\rangle \\ \langle W\rvert\end{vmatrix} \cdot \lvert W\rangle$	$\langle S\rvert \cdot \begin{vmatrix}\lvert S\rangle \\ \langle W\rvert \\ \langle V\rvert\end{vmatrix} \cdot \begin{matrix}\lvert W\rangle \\ \lvert V\rangle\end{matrix}$	$\langle S\rvert \cdot \begin{vmatrix}\lvert S\rangle \\ \langle W\rvert \\ \langle V\rvert \\ \langle T\rvert\end{vmatrix} \cdot \begin{matrix}\lvert W\rangle \\ \lvert V\rangle \\ \lvert T\rangle\end{matrix}$
Constraint to insert	$V \leq T$	$W \leq T$	$V \neq W$
Ordered time line	$\langle S\rvert \cdot \begin{vmatrix}\lvert S\rangle \\ \langle W\rvert \\ \langle V\rvert \\ \langle\lvert\lvert V\rangle\rvert \\ \lvert\langle T\rvert\rvert\end{vmatrix} \cdot \begin{matrix}\lvert W\rangle \\ \lvert T\rangle \\ \lvert\lvert V\rangle\rvert \\ \lvert\langle T\rvert\rvert\end{matrix}$	$\langle S\rvert \cdot \begin{vmatrix}\lvert S\rangle \\ \langle W\rvert \\ \langle V\rvert \\ \lvert V\rangle \\ \langle\lvert W\rangle\rvert \\ \lvert\langle T\rvert\rvert\end{vmatrix} \cdot \begin{matrix}\lvert T\rangle \\ \lvert V\rangle \\ \lvert\lvert W\rangle\rangle \\ \lvert\langle T\rvert\rvert\end{matrix}$	$\langle S\rvert \cdot \begin{vmatrix}\lvert S\rangle \\ \langle W(\neg V)\rvert \\ \langle V(\neg W)\rvert \\ \lvert V(\neg W)\rangle \\ \langle\lvert W(\neg V)\rangle\rvert\rvert \\ \langle T\rvert\end{vmatrix} \cdot \begin{matrix}\lvert T\rangle \\ \lvert\lvert V(\neg W)\rangle\rvert \\ \lvert\lvert W(\neg V)\rangle\rangle \\ \langle T\rvert\end{matrix}$

The first step, inserting $S \leq W$ into OTL, is obvious due to the instantiation interval semantics (cf. Section 2). The following two constraints are inserted in the same way. The next constraint $V \leq T$ is more difficult to integrate, because both constraint variables are already instantiated into OTL. Both variables are integrated into a split structure (cf. Section 2). The constraint $W \leq T$ is integrated in the same way. The insertion of the last constraint $V \neq W$ is also obvious due to the instantiation interval semantics.

The most interesting query in our example is the temporal order between S and T. We can compute (in linear time) that S can be instantiated before T but not after T. Additionally, we compute that S and T cannot be instantiated at the same time. This relation prevent V and W because the split time primitive consisting out of V, W, and T cannot shrink into a time point (because of $V \neq W$).

4 Outlook to further research

In the near future, we want to develop and implement an algorithm based on the OTL approach for ISAT($\mathcal{A}$) and ISI($\mathcal{A}$). For this, we have to develop strategies for finding an inserting order of constraints into OTL with minimal instantiation interval disjointedness.

Furthermore, It seems promising to investigate more expressive knowledge representations than Allen's approach. Especially, its inability of stating time order information including more than two variables and of describing cyclic behavior are serious expressiveness restrictions.

Because of their similarity to quantitative intervals, instantiation intervals can be easily combined with quantitative information. Thus, our approach seems far more suitable for a combination of quantitative and qualitative constraints than approaches using the path consistency algorithm like [3].

References

1. Allen, J.F.: Towards a general theory of action and time. Artificial Intelligence 23 (1984)
2. Kahl, J., Hotz, L., Milde, H., Wessel, S. E.: Efficient temporal reasoning with instantiation intervals. LKI-report LKI-M-99/1. University of Hamburg (1999)
3. Kautz, H. A., Ladkin, P. B.: Integrating metric and qualitative temporal reasoning. In: Proceedings AAAI-91 (1991)
4. Van Beek, P.: Reasoning about qualitative temporal information. In: Artificial Intelligence 58 (1992)
5. Van Beek, P., Cohen, R.: Exact and approximate reasoning about temporal relations. In: Computational Intelligence 6 (1990)
6. Vilain, M., Kautz, H.: Constraint propagation algorithms for temporal reasoning. In: Proceedings AAAI-86 (1986)

Unified Chromosome Representation for Large Scale Problems

Hoda A. Baraka*, Saad Eid*, Hanan Kamal* and Ashraf H. Abdel Wahab**

* Faculty of Engineering, Cairo University, **hbaraka@idsc1.gov.eg**
** Computers and Systems Dept., Electronics Research Institute, Dokki, Giza,
ashraf@eri.sci.eg

Abstract. Genetic Algorithms have been successfully applied to the function optimization problem. However, the main disadvantage of this technique is its large chromosome length and hence long conversion time specially when applied to functions with a large number of parameters. In this paper, a new chromosome representation scheme that reduces the chromosome length is proposed. The scheme is also domain independent and may be used with any function. Results and a comparison between the conventional chromosome representation and the proposed one are presented.

1. Introduction

Originally, GAs have been used in function optimization problems [1]. Although their success, GAs had a considerable disadvantage, namely, the large chromosome size. In binary representation, a chromosome represents the number of parameters as encoded in binary format. Thus, the chromosome length is proportional to the number of parameters and the required accuracy. In large optimization problems (i.e. problems involving large number of variables), this representation leaDS to a long chromosome structure and hence a long convergence time. Slow convergence time is basically due to the extended chromosome length, the large size of the population and finally the types of operators used by the algorithm.

Since the representation scheme (the representation language and the encoding of the chromosome) is a crucial factor for the success of GAs, many research work has been performed in discovering the best representation scheme suitable for a class of problems [2]. The selection of a "good" representation scheme helps reducing the convergence time and obtaining successful results but may affect the structure of the GA operators [3]. To help with the slow convergence problem, Georges Harik et al [4] suggested a compact form of GA in which the population is represented as a probability distribution over the set of solutions. This paper presents a solution to the slow convergence time of GAs when applied to large scale optimization problems. In this paper, a unified chromosome representation that is independent of the problem domain is presented. This representation results in reducing the chromosome length, and consequently enhancing the performance of GA. The paper is organized as follows : section II presents the unified chromosome representation, section III describes the experiments done, section IV presents the results obtained and finally the conclusion is presented in section V.

2. The unified chromosome representation

In conventional representation, each parameter is encoded (in binary format) and the concatenation of the encoded parameters forms the chromosome. In the proposed scheme, the chromosome length is defined as the total number of bits sufficient to represent all the permutations of the encoded parameters. This definition not only limits the chromosome length but also is domain independent. In real world, the domain of search for any problem having n unknowns can be defined by the following tuple :

$$O\_DS = (O\_n, O\_D_i, i=1,2,....,n \quad O\_L_i) \ i=1,2,....,n \qquad (1)$$

Where
  O_DS = Original domain of search
  O_n = Original number of unknowns or parameters
  O_D_i = Original Domain of values defined for parameter I
  O_L_i = Original Length of domain for parameter I

A point in O_DS is identified by the set $(O\_P_1, O\_P_2, ...,O\_P_n)$ where $O\_P_i \in O\_D_i$.

Thus, the number of points in the search domain is equal to :

$$N = \text{Permutations } (O\_L_i) \qquad (2)$$

For example, a system with 5 parameters and six values associated with each parameter has a domain of search of 15625 points. If n becomes 25 parameters, then the domain of search will increase to 244140625 points. This simple example reflects the effect of large number of parameters on the size of the search space. Fig. 1. shows the relation between the search space size and the number of parameters for different domain sizes.

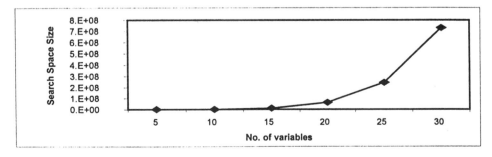

Fig. 1. The relation between the number of variables and the search sapce size. Each variable has 6 values associated with it.

It is required to map the original search domain into another less complex domain. In this paper, a proposed unified domain of search is characterized by the following tuple:

$$U\_DS = (U\_n, U\_D, U\_L) \qquad (3)$$

Where :
  U_DS = unified domain of search
  U_n = unified number of genes, where $U\_n < O\_n$
  U_D = unified domain for all parameters
  U_L = length of unified domain

The unified domain U_D is defined to be the set of integer values ranging from zero to 10.
  U_D= {0,1,2,3,4,...,10} and U_L = 11 values.

Note that in the proposed unified domain, all parameters have the same domain of values (set to the range of integer values from 0 to 10). The main difference between the proposed and conventional schemes is that while the concatenation of the parameters forms the chromosome in the conventional representation, in the proposed representation the chromosome is formed by selecting the minimum set of bits required to represent all the permutations of the parameters, i.e. parameters may share more than one bit with each other. In the next section, the mapping algorithm from the original search space to the unified one is presented.

2.1. Original domain to unified domain mapping algorithm

In order to present the problem using the unified domain representation defined in (3), a mapping algorithm is used to convert the O_DS to the U_DS. The steps of the algorithm are described below :

1. Calculate the minimum chromosome length. $Min.L = \log_2 O\_n$
2. Calculate $U\_n = Rnd(Min.L, O\_n)$
3. For every parameter O_P_i in the original domain do the following :
   a. Calculate the number of unified parameters used in the representation of O_P_i
      $$N_{i\,O\,P_i} = Rnd(1, U\_n)$$
   b. Determine the set of unified parameters used to represent O_P_i as follows :
      b.1. for I=1 to $N_{O\,P_i}$
         $U\_set_{i\,O\,P_i} = Rnd(1, U\_n)$
      b.2. Check that set_U_P is unique, i.e. the set is not used by another original parameter, otherwise the same parameter will be referenced again.
   c. Calculate the value of O_P_i based upon the unified parameters :
      c.1. Choose a random value for every unified parameter in the set of
         $O\_P_i(U\_set\_U)_i$
         value = $Rnd(0,10)$ as the unified domain includes only integers from 0 to 10 .
      c.2. $O\_P_i = Map.Fn(U\_set\_V)_I$

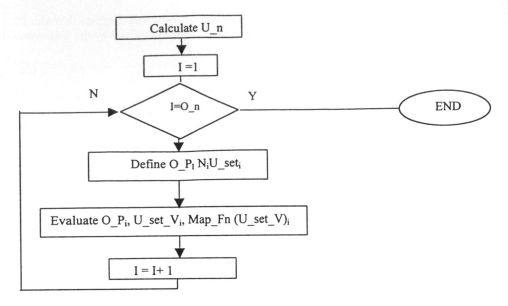

Fig. 2. Flowchart of the proposed algorithm

2.2 Mapping functions

The module **Evaluate O_Pi** defines the relation between the original parameter and the selected unified parameters existing in the U_set$_i$. A number of mapping functions was tested to define this relation, and to analyze the effect of different mapping functions on the performance of the genetic algorithms. The mapping functions used are :

1. Sum function :

$$O\_P_i = \sum_{j=1}^{U\_n} U\_Pj \quad \text{for U\_Pj} \in \text{U\_set}_I \tag{4}$$

2. Product function :

$$O\_P_i = \prod_{j=1}^{U\_n} U\_Pj \quad \text{for U\_Pj} \in \text{U\_set}_I \tag{5}$$

3. Exponential summation

$$O\_P_i = e^{\Sigma U\_P}{}_j \tag{6}$$

3. Negative exponential summation :

$$O\_P_i = e^{-\Sigma U\_P_j} \tag{7}$$

5. Root Mean Square :

$$O\_P_i = RMS (\Sigma U\_Pj /N) \tag{8}$$

If we have *n* variables then we need to represent these variables as a binary string using the optimum number of genes z. The problem is how to select the best number of genes r that gives the minimum number of repetition. Note that the value of r that gives the minimum number of repetition is equal to the value that gives maximum number of combinations.

$$\text{Number of combinations (NCOM)} = n!/(n-r)! \; r! \tag{9}$$

Thus in order to obtain the maximum number of combinations, we evaluate the maximum value for equation (9) using different values for r. This point will define the best number of genes that should be used to represent the n variables.

2.3. Modified Genetic Algorithms M_GA

The resulting M_GA can now be stated as follows :

    1. Initialization module (Define the O_DS)
    2. Mapping module.
    3. SGA Repeat :
        Select
        Crossover
        Mutate
        Modified fitness module.

As it is seen, two points have been changed in the SGA, the first point is the addition of the mapping module and this is executed once in the algorithm. The second point is the modified fitness module. This module evaluates the original parameters of the system based on the results obtained by the GA operators on the unified chromosome and using the mapping function. This means that the modified fitness module consists of two parts : the first part evaluates the value of the original parameters of the system, and the second part calculates the fitness function from the original parameters as was done classically. It is worth noting that in the classical SGA a mapping function was also used to reflect the domain range used. So, basically, what has been added to the SGA is a more complex mapping function module.

3. The Domain of the Experiments

The M_GA has been applied to following three integer programming problems :

   1 - Max Σ x(i) where n = number of variables and x is an integer number that takes the value of one or zero.
   2 - Max Σ (x(i)-1)^2 where n = number of variables and x is an integer number that takes the value of one or zero.
   3 - Max Σ (x(I))^2 where n = number of variables and x is an integer number that takes the value of one or zero.

Also, five mapping functions have been used and compared with the SGA. These mapping functions are :

   1. $x = INT ((z_1+z_2+ \;+z_n)/N)$
   2. $x = INT (z_1*z_2* \;*z_N)$
   3. $x = INT [exp (- (z_1+z_2+ \; + z_n)/N))]$
   4. $x = INT [exp (-z_1*z_2* \;*z_N)]$
   5. $x = INT [RMS ((z_1+z_2+ \;+z_n)/N)]$

The performance of the proposed representation was assessed in the three mentioned integer programming problems using all mapping functions. The number of variables used in each case ranged from 20 to 1000 variables. The results of applying these mapping functions are presented and discussed in the next section.

4. Results

The results obtained from applying the proposed representation is classified into two categories. In both categories, the experiments were conducted on the three simple problems drawn from the domain of integer programming and discussed above. The first category presents a comparison between the five mapping functions defined in the proposed representation and the conventional SGA (the GENESIS simulator was used for this purpose) with the number of variables ranging from 20 to 100. The second set of results represent a comparison between the defined mapping functions when applied to the same problems with number of variables ranging from 200 to 1000. Table 1 summarizes the results of the first category for the three problems. Each row specifies the experiments conducted for a specific number of variables. The values in each row in the table indicate the iteration number at which a solution was found for every method. The results showed faster convergence for all mapping functions as compared to the SGA. The reduction in chromosome size achieved is equal to 0.1 i.e. the chromosome length is ten times shorter than the one used in the SGA. Figures 3 and 4 show the results of the best mapping function (M3) observed so far versus the conventional SGA for the second and third problems and number of variables equal to 50 and 100 respectively.

759

For the second set of experiments, table 2 shows the iterations at which the mapping functions have discovered a solution. It should be noted that, for this set of experiments, only the mapping functions are compared together since the GENESIS simulator can represent more than 100 parameters. Figures 5 and 6 show the results of the second category for 500 and 1000 variables respectively. The obtained results showed that the third mapping function (M3) is the best one in terms of the convergence time. Also, the second mapping function (M2) did not perform quite well in most of the cases. This is mainly due to the nature of the function itself as it performs a multiplication operation between all variables. The chromosome size was also kept at 0.1 of the size of a similar chromosome if SGA was to be used.

For these simple integer programming problems, the results showed that the proposed unified representation outperformed the conventional parameter encoding and representation -using discrete locations for each encoded variable- in terms of required storage and hence the convergence time has been considerably reduced.

Table 1. Comparison between mapping functions and SGA for problems 1,2 and 3.

# of variables	Problem 1						Problem 2						Problem 3					
	M1	M2	M3	M4	M5	SGA	M1	M2	M3	M4	M5	SGA	M1	M2	M3	M4	M5	SGA
30	10	10	10	10	10	1000	10	10	10	10	10	1300	10	10	10	10	10	1200
40	10	10	10	10	10	1700	10	10	10	10	10	2600	10	10	10	10	10	2000
50	10	10	10	10	10	2700	10	10	20	10	10	2200	10	40	10	10	10	3800
60	10	50	10	10	10	4700	10	10	10	10	10	4200	10	40	10	10	10	5000
70	10	100	10	10	10	5300	10	10	56	10	10	4100	10	200	10	10	10	900
80	10	100	10	10	10	8000	10	10	30	100	10	7000	10	200	10	10	10	5700
90	10	200	10	10	10	4600	10	10	100	20	10	9100	10	100	10	10	10	6600
100	10	400	10	10	10	6000	50	10	30	50	10	9300	10	400	10	10	10	6000

Table 2. Comparison between mapping functions for problems 1 and 3

# of variables	Problem 1					Problem 3				
	M1	M2	M3	M4	M5	M1	M2	M3	M4	M5
200	50	-	50	50	50	100	-	100	100	100
300	1000	-	50	50	200	400	-	100	100	100
400	600	1100	50	50	300	400	800	100	100	500
500	800	1400	50	200	800	1200	1800	100	100	500
600	1000	7000	50	200	800	1500	3000	100	100	1000
700	1800	5000	200	300	1300	2200	-	300	300	1500
800	2100	-	400	900	1900	2300	4000	600	1000	2500
900	3000	-	900	1000	2800	3300	-	800	1300	3000
1000	3000	-	1000	1400	2200	3600	-	1000	1100	2800

*The (-) mark means that the solution was not found in an acceptable time limit.

5. Conclusion

This paper presented a new representation scheme for binary chromosomes that reduces the chromosome size and hence accelerates the convergence of the population. The proposed representation was tested and compared with SGA using three simple problems from the domain of integer programming. The representation is domain independent and its use for other, non binary representation schemes is under implementation.

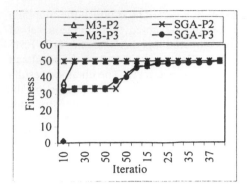

Fig. 3. M3 and SGA. Variables = 50.   Fig. 4. M3 and SGA. Variables =100.

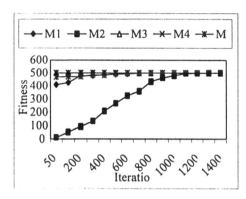
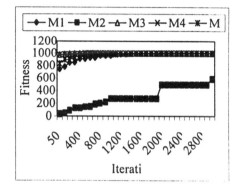

Fig. 5. Mapping functions comparison. Variables=500   Fig. 6. Mapping functions.
Variables=1000

References

1. Goldberg D. (1989) 'Genetic Algorithms in Search, Optimization and Machine Learning' Addison-Wesley, 1989.
2. Manderick, B., de Weger, M., & Spiessens, P. (1991) 'The genetic algorithm and the structure of the fitness lanDScape' Proceedings of the Fourth International Conference on Genetic Algorithms, La Jolla, CA, Morgan Kaufmann
3. De Jong K., Spears W., 'On the state of Evolutionary Computation'. In Proceedings of the 1993 International Conference on Genetic Algorithms, Urbana-Champaign, IL, pages 618-623
4. Harik G., Lobo F. & Goldberg D. (1997) 'The Compact Genetic Algorithm', IlliGAL Report No. 97006 August 1997.
5. Holland J. H., (1975) 'Adaptation in natural and artificial systems', Cambridge CA MIT press, first edition 1975.
6. Shaefer, C. G., (1987) 'The ARGOT strategy : adaptive representation genetic optimizer technique' proceedings of the Second International Conference on Genetic Algorithms, Cambridge, MA: Lawrence Erlbaum.

A Novel Framework for Hybrid Intelligent Systems

Hesham A. Hefny[1], Ashraf H. Abdel Wahab[2] , Ahmed A. Bahnasawi[3] & Samir I. Shaheen[4]

1 Arab Organization For Industrialization, Cairo, E-mail: hehefny@hotmail.com
2 Computer & System Dept., Electronics Research Institute, Cairo, E-mail: ashraf@eri.sci.eg
3 Electronics & Communication Dept., Faculty of Engineering, Cairo Univ., E-mail: Bahnsawi@alpha1-eng.cairo-eun.eg
4 Computer Engineering Dept., Faculty of Engineering, Cairo Univ., E-mail: sshaheen@frcu.eun.eg.

Abstract.   The aim of this paper is to generalize the approaches used for developing hybrid intelligent systems based on integration of neural networks, fuzzy logic and   genetic algorithms. The paper introduces the concepts of intelligent artificial life system (IALS) space as a generalized conceptual space which represents a framework for all   possible integration schemes of such intelligence technologies. Concepts like order of intelligence, degree of intelligence and   types of hybrid coupling are also illustrated. Based on such concepts, a proposed philosophy for hybrid integration schemes in the IALS space is presented and discussed.

1. Introduction

Artificial Neural Networks (ANNs), Fuzzy Logic (FL) and Genetic algorithms (GAs) have received much interest recently as the most popular intelligence technologies. Their great success in various fields of real world applications convinced many of the AI researchers of the value of developing computational techniques based on an analogy with nature. Since the main target of the field of AI is to build artificial systems that exhibits a similar behavior to that of human, and due to the fact that each of such computational techniques simulates a single aspect of natural intelligence, it is believed that combining such techniques together into hybrid models is the next step towards achieving the target of building intelligent systems. The fact that the above three intelligence technologies can complement   each other makes the idea of combining   them in hybrid models widely acceptable [1]. However, it is still unclear for many researchers what is the best approach to develop a hybrid model? In otherwords, what are the major rules that should be considered during the development of such hybrid models?. In this paper, a generalized framework for developing such hybrid models will be introduced and discussed. Such a framework is introduced as a 3-dimensional conceptual space called "Intelligent Artificial Life System" (IALS) space. In the next section, the classification of hybrid intelligent systems is presented together with the concepts of order and degree of intelligence. Section 3 presents the concept of the IALS space. The motivations for developing hybrid models are discussed in section 4. First, second and third order intelligent models in the IALS space are given in sections 5, 6 and 7 respectively. Section 8 illustrates the major drawbacks that may be found in the hybrid intelligent models. General rules for developing new approaches for hybrid modelling are given in section 9. Section 10 introduces two approaches for developing third order hybrid models. The philosophy behind such approaches are covered in section 11. Finally, section 12 concludes the paper.

2. Hybrid Intelligent Systems

Classifications of hybrid intelligent systems have been studied extensively in the literature [2]. However, a simplified classification approach will be adopted in this paper. According to this approach, any hybrid intelligent system can be characterized by two features:

Order of Intelligence: which is the number of different intelligence technologies combined in the hybrid.
Degree of Intelligence: which is the degree of sophistication of the used intelligence technologies. This directly affects the performance of the hybrid models, for example:

i- increasing the degree of intelligence for first order intelligent systems may be performed by nested application of the same intelligence technology, e.g. building networks of many neural networks.

ii- second order intelligent models may be obtained simply by sequential application of two different intelligence technologies. However, the degree of intelligence of the hybrid model may be increased considerably if strong coupling of the two techniques is established.

Hybrid intelligent models based on different integrations of ANNs, FL and GAs are of special significance as they combine the most important aspects of intelligence of living beings, together, namely: learning, decision making and living behaviors.

3. Intelligent Artificial Life Systems (IALS)

Intelligent Artificial Life System (IALS) is a new concept used to refer to those hybrid systems which integrate both living and intelligent behaviors. This may be stated in the following relation [3]:

Living behavior + Learning mechanism + Decision making ability = IALS

Considering the IALS as a result of combining three different and independent intelligence technologies, one can imagine the IALS as a point in a 3-dimensional space in which each of the adopted technologies represents a certain axis of reference. This space may be called "Intelligent Artificial Life Space".

4. Motivations For Developing Hybrid Intelligent Models

The purpose of combining single intelligence technologies together in one hybrid model should be one of the following [3]:
1 - To retain features while removing drawbacks of such intelligence techniques.
2 - To improve the performance of the intelligent model. This results in better error level, faster training process, less structure complexity, and/or higher reasoning capabilities.

5. First Order Intelligent Models In The IALS Space

Each axis in the IALS space, corresponds to intelligent models based on a single intelligence technology. Each of such technologies has its own advantages and drawbacks. For example, the major advantages of ANNs are their abilities to adapt (or learn), to generalize and to make highly nonlinear mapping among different domains. The highly parallel processing nature of ANNs provide them also with the fault tolerance properties. However, the main drawbacks are their poor reasoning capabilities, and the minima problem. On the other hand, FIS (Fuzzy-Logic Inference Systems) have the ability to make decisions based on inexact knowledge. They follow the same way that human beings use to make decisions in real life, namely, they use linguistic models. Such models have high reasoning and nonlinear mapping capabilities. However, the main drawback of FIS is that, they lack the adaptation property of the ANNs. Both ANNs and FIS can be considered as two different "information processing systems", or equivalently: "modelling techniques", however this is not the case for GAs. GAs are highly parallel searching techniques that can search multi-peak spaces without suffering the local minima problem associated with the hillclimbing methods. Also, it is not dependent on continuity of the parameter space. Although GA can not be used alone as a modeling technique , it is quite efficient when used with other modeling techniques.

6. Second Order Intelligent Models In The IALS Space

Hybrid models based on integration of GAs and FL results in adaptive fuzzy systems that use global search of GAs to modify: fuzzy rule sets, fuzzy relational matrices, or fuzzy membership functions [4]-[5]. Due to the global optimization nature of GAs, hybrid combinations of GAs and ANNs have been suggested to avoid local minima problems that may occur during the gradient-descent based learning phase or even to select the best topology of the network. It has also been found that such an integration increases the fault-tolerance characteristic of ANNs, see [6]-[9]. Integration of ANNs and FL has also been studied extensively and examples are found in [10]-[15].

7. Third Order Intelligent Models In The IALS Space

The success of second order intelligent models encourages the research to step forward towards third order modeld. However, the work done in this area is still little when compared with second order intelligent models [16]-[17].

8. Drawbacks Of Hybrid Intelligent Models

The main drawbacks that may arise when developing second or third order intelligent models are[3]:

Weak Coupling: This may arise when the adopted approach to develop the hybrid model is based on sequential application of different techniques without being able to merge their characteristics together.
High Structure Complexity: For some hybrid models, (e.g. FL + ANNs), large number of layers and nodes per layer may be needed.
Complicated Learning Algorithms: For some hybrid models with large structure and different forms of nodes equations, the learning process becomes complicated.
High Computational Complexity: This is expected for most of the developed hybrid models, as it represents the cost paid for combining different computational techniques to get better behavior.

9. Developing New Approaches For Hybrid Models

Since there are enormous number of possible hybrid models, there is no single methodology that one can follow to develop a hybrid intelligent system. However, for developing a hybrid intelligent system based on integration of the above intelligence technologies, the designer needs the following [3]:

1 - A solid background of the three intelligence technologies together with the features and drawbacks of each one.
2 - A motivation or a target that he needs to achieve using such a hybrid model.
3 - An evaluation methodology based on features and drawbacks of the obtained hybrid model.

If this is clear in the designer's mind, he will be able to put a philosophy for integrating the individual techniques together to develop the required hybrid model. Such a philosophy means determining the role of each intelligence technology in the proposed hybrid model. In the following section, an approach to integrate ANNs, FL & GAs into a one third order hybrid intelligent model will be presented.

10. The Proposed IALS Models

Two approaches are introduced in this section. Both approaches are of the form of rule-based expert network, in which the opaque structure of the classical neural networks is replaced by a transparent one with fuzzy-neurons acting as fuzzy IF-THEN rules. The first approach is considered to be a generalized framework for many possible approaches. The second approach is derived from the first one.

10.1 The First Approach

This approach can be considered as a generalized form of adaptive multilevel fuzzy inference systems. The architecture of the model is basically a multilayer network of fuzzy-neurons which act as fuzzy IF-THEN rules. Different forms of Fuzzy rules are introduced through different realization techniques, i.e. AND, OR, NAND & NOR fuzzy neurons [3],[18]. GAs have been adopted as the learning mechanism to preserve generality of the model. Fig. 1 shows the generalized structure of the proposed fuzzy-neuron model.

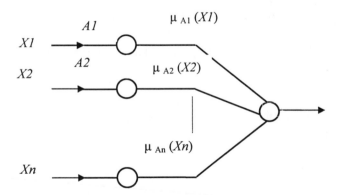

Fig. 1. The basic structure of the proposed fuzzy neuron model

Considering a multilayer network structure of such fuzzy-neuron nodes, then the operation of node j in layer S of the network can be described by any of the following If-Then rules:

$$\text{IF } \{ \overset{n_{s-1}}{\underset{i}{AND}} \, (\, O_i^{s-1} \text{ is } A_{ij}^s \,) \} \text{ THEN } (\, y_j^s \text{ is } B_j^s \,)$$

$$\text{IF } \{ \overset{n_{s-1}}{\underset{i}{OR}} \, (\, O_i^{s-1} \text{ is } A_{ij}^s \,) \} \text{ THEN } (\, y_j^s \text{ is } B_j^s \,)$$

$$\text{IF } \{ \overset{n_{s-1}}{\underset{i}{NAND}} \, (\, O_i^{s-1} \text{ is } A_{ij}^s \,) \} \text{ THEN } (y_j^s \text{ is } B_j^s \,)$$

$$\text{IF } \{ \overset{n_{s-1}}{\underset{i}{NOR}} \, (\, O_i^{s-1} \text{ is } A_{ij}^s \,) \} \text{ THEN } (\, y_j^s \text{ is } B_j^s \,)$$

$$j = 0,1,\ldots\ldots, n_s \,, \quad s = 1,2,\ldots \tag{1}$$

where n_s is the number of nodes (rules) in layer S, i and j are indexes for inputs and outputs of the layer S. The *firing degree* of the node is computed by *aggregating* all input membership functions through an appropriate T-norm or S-norm operation according to the required logical operation of each node:

$$net_j^s = \underset{i}{T} \{ \mu_{A_{ji}^s} (O_i^{s-1}) \} \qquad \text{(Fuzzy-AND node)}$$

$$net_j^s = \underset{i}{S} \{ \mu_{A_{ji}^s} (O_i^{s-1}) \} \qquad \text{(Fuzzy-OR node)}$$

$$net_j^s = 1 - \underset{i}{T} \{ \mu_{A_{ji}^s} (O_i^{s-1}) \} \qquad \text{(Fuzzy-NAND node)}$$

$$net_j^s = 1 - \underset{i}{S} \{ \mu_{A_{ji}^s} (O_i^{s-1}) \} \qquad \text{(Fuzzy-NOR node)} \tag{2}$$

where O_i^{s-1} is the output of node i in layer $s\text{-}1$. The crisp output of each node is computed by a defuzzification operation for the output fuzzy set:

$$O_j^s = \text{DFZ} (\, \min(net_j^s \,, \, \mu_{B_j^s}(y_j^s) \,)) \tag{3}$$

The defuzzification operation for the output fuzzy set is determined heuristically by the designer according to the adopted forms of fuzzy sets. The genetic description of such a fuzzy-neural network is represented in the form of a chromosome structure with a stream of real values which correspond to the design parameters of all fuzzy sets included in the network. Such parameters are adjusted during the evolution process. Simulations of such a model in case of adopting gaussian functions as fuzzy sets, min. operation as T-norm and max. operation as S-norm, ensure its efficiency when compared with backpropagation neural networks [3], [19]-[20].

10.2 The Second Approach

This approach is derived directly from the first one by adopting only two types of fuzzy-neuron nodes, namely, AND and OR nodes. Such an approach is considered to be a logical form of the classical "*radial basis function*" (RBF) model. The proposed model is a 3-layer structure network: input, hidden and output layers. The hidden layer is the AND-layer and the output layer is the OR one. Links fuzzy sets for the hidden layer are selected to be gaussian while for the output layer, they are selected as a triangular shape. T-norm operation is set to be the algebraic product for the hidden layer while the S-norm operation is set to be the bounded-sum for the output layer. Nodes' fuzzy sets are reduced to be singleton with certain real weight values. Fig.2 illustrates the structure of the proposed "*Logical Radial Basis Function*" (LRBF) model. Similar to the first approach, genetic algorithms are used for adjusting the model parameters. Simulation experiments of the proposed LRBF ensure its efficiency when compared with the classical RBF networks [3], [21],[22].

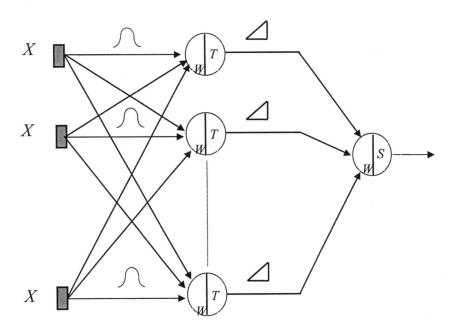

Fig. 2. The structure of the proposed logical RBF net. T is a T-norm operation, S is an S-norm operation and W is a real weighting value

11. Philosophy Of The Proposed IALS Models

The philosophy of the proposed IALS models is based on a clear view of the role of each intelligence technique in the hybrid model. The concepts of ANNs, FL and GAs are integrated together to obtain a new form of nonlinear mapping networks. In otherwords, each of the three intelligent techniques has its essential role in the proposed model. The following points illustrate the role of each intelligence technique [3]:

Fuzzy Logic: The fuzzy logic characteristics introduced to each node of the obtained network results in a fuzzy-neuron processing element which is equivalent to a fuzzy IF-THEN rule. Such a fuzzy-neuron node constructs a hypersurface in the pattern space rather than a hyperplane for the case of classical model of artificial neurons. This means higher classification and reasoning capabilities.

Neural Networks: Grouping such fuzzy-neuron processing elements in a multilayer network structure means higher nonlinear mapping capabilities and more transparent structure than the classical "multilayer perceptron" (MLP) nets. Also, such a network structure presents the model in the form of a multilevel fuzzy inference system which greatly reduces the required number of fuzzy rules to establish the nonlinear mapping.

Genetic Algorithms: Adopting GAs as the learning technique for the proposed hybrid models is essential. This makes the learning independent of the shapes of fuzzy sets, the types of T-norm or S-norm operations and the selected formula for the defuzzification operation. Moreover, there is no problem concerning continuous, or piecewise continuous parameters. However, it should be noted that genetic learning has a higher computational complexity than gradient-descent techniques.

From the above discussion, it is clear that the proposed models are strongly-coupled hybrid intelligent models in which the characteristics of the three intelligence techniques are merged together.

12. Conclusion

This paper introduces a generalized framework for developing hybrid intelligent systems based on integration of neural networks, fuzzy logic and genetic algorithms. The paper presents the concepts of intelligent artificial life system (IALS) as a generalized conceptual space which includes all possible integration schemes of such intelligence technologies. Concepts like order of intelligence, degree of intelligence and types of coupling are also illustrated. Based on such concepts, two approaches for hybrid, third order intelligent modelling were presented together with the philosophy of the hybrid integration scheme in the IALS space.

References

1. G. R. Madey, J. Weinroth and V. Shah, " Hybrid intelligent Systems: Tools for decision making in intelligent manufacturing", published in "Artificial Neural Networks for intelligent Manufacturing ", Edited By: C. H. Dagli, Chapman & Hall, London, 1994, Ch. 4, pp. 67-90.

2. D. Stacey, "Intelligent systems architecture: Design techniques", published in "Artificial Neural Networks for intelligent Manufacturing ", Edited By: C. H. Dagli, Chapman & Hall, London, 1994, Ch. 2, pp. 17-38.

3. H. A. Hefny, "A Novel Intelligent Modelling Approach Using Neural Networks, Fuzzy Logic and Genetic Algorithms", Ph.D. Thesis, Faculty of Eng., Cairo Univ., 1998.

4. C. L. Karr and J. G. Gentry, "Fuzzy Control of PH Using Genetic Algorithms", IEEE Trans. On Fuzzy Systems, vol. 1, No. 1, Feb. 1993.

5. D. Park and A. Kandel, "Genetic-Based New Fuzzy Reasoning Models with Application to Fuzzy Control", IEEE Trans. On Sys., Man, and Cybern., vol. 24, No. 1, Jan. 1994.

6. R. K. Belew , J. Mclnerney and N. N. Schraudoiph , "Evolving Networks: Using the Genetic Algorithm with Connectionist Learning", Proceedings of the Workshop of Artificial Life II, Addison-Wesley, 1992.

7. Y. Ichikawa and T. Sawa, "Neural Network Application for Direct Feedback Controllers", IEEE Trans. On Neural Networks, vol. 3, No.2, March 1992.

8. D. Whitley , S. Dominic, R. Das, and C. Anderson, "Genetic Reinforcement Learning for Neurocontrol Problems", Tech. Report. Dept. Computer Science, Colorado State Univ., 1992.

9. H. A. Elsimary , "A Novel Algorithm for Ensuring the Fault Tolerance of Artificial Neural Networks Using Genetic Algorithms", Doctoral Dissertation, Cairo Univ., 1993.

10. J. R. Jang , "ANFIS: Adaptive-Network-Based Fuzzy Inference System", IEEE Trans. On Sys., Man, and Cybern., vol. 23, No.3, May/Jun 1993.

11. C. T. Lin and C. S. G. Lee, "Reinforcement Structure / Parameter Learning for Neural Network-Based Fuzzy Logic Control Systems", IEEE Trans. On Fuzzy Systems, vol. 2, No.1, Feb. 1994.

12. E. Khan, "Neural Network Based Algorithms For Rule Evaluation & Defuzzification in Fuzzy Logic Design", International Neural Network Society Annual Meeting, Oregon Convention Center, Portland, Oregon, July 1993.

13. H. Ishibuchi , R. Fujioka, and H. Tanaka, "Neural Networks That Learn from Fuzzy If-Then Rules", IEEE Trans. On Fuzzy Systems., Vol. 1, No. 2, May 1993.

14. H. Narazaki and A. L. Ralescu, "An Improved Synthesis Method for Multilayered Neural Networks Using Qualitative Knowledge", IEEE Trans. On Fuzzy Systems, Vol.1, No. 2, May 1993.

15. M. M. Gupta, "On Fuzzy Neuron Models", Published in "Fuzzy Logic For Management Of Uncertainty", edited by Lotfi Zadeh, and Janusz Kacprzyk, John Wiley & Sons, Inc., 1992.

16. M. M. M.Chowdhury and Y. Li, "Messy Genetic Algorithm Based New Learning Method for Structurally Optimised NeuroFuzzy Controllers", Proceedings of The IEEE International Conference on Industrial Technology, Singapore, 1996, pp.274-278.

17. T. Fukuda and K. Shimojima, "Hirarchical Intelligent Robotic Systems: Adaptation, Learning and Evolution", Proceedings of International Conferenceon Computational Intelligence and Multimedia Applications (ICCIMA'97), Gold Coast, Australia, 10-12 Feb. 1997, pp. 1-5.

18. H. A. Hefny, A. H. Abdel Wahab and S. I. Shaheen , "Genetic-Based Fuzzy Neural Classifier ", In Proceedings of International Conference for Electronics, Circuits & Systems (ICECS), Cairo, Egypt, Dec. 15-18, 1997, pp.7-11.

19. H. A. Hefny, A. H. Abdel Wahab and S. I. Shaheen, "Genetic-Based Fuzzy Neural Network (GBFNN) A hybrid approach for approximation", In Proceedings of the International Conference on Computational Intelligence &Multimedia Applications (ICCIMA), GOLD COAST, AUSTRALIA, Feb. 10-12, 1997, pp. 122-125.

20. H. A. Hefny , A. H. Abdel Wahab , A. A. Bahnasawi and S. I. Shaheen , "Genetic-Based Fuzzy Neural Networks: A Third Order Intelligence Connectionest Model", Journal of Engineering and Applied Science, Faculty of Engineering, Cairo Univ., Vol. 45, No.1, Feb. 1998, pp. 87-101.

21. H. A. Hefny, A. H. Abdel Wahab , A. A. Bahnasawi and S. I. Shaheen , "Logical Radial Basis Function Networks (LRBF). A Hybrid Intelligent Approach For Function Approximation", In Proceedings of the Seventh Turkish Symposium on Artificial Intelligence and Neural Networks (TAINN'98), Bilkent Univ., Ankara, Turkey, June 24-26, 1998, pp.53-61.

22. H. A. Hefny, A. A. Bahnasawi, A. H. Abdel Wahab and S. I. Shaheen , "Logical Radial Basis Function Networks-A Hybrid Third Order Intelligent Approximator", Accepted For Publication in 'Advances in Engineering Software', August 10, 1998.

The Design of a Multi-tiered Bus Timetabling System

Hon Wai Chun[1] and Steve Ho Chuen Chan[2]

[1] City University of Hong Kong, Department of Electronic Engineering
Tat Chee Avenue, Kowloon, Hong Kong
eehwchun@cityu.edu.hk any
[2] Advanced Object Technologies Ltd, Unit 602A HK Industrial Technology Centre
72 Tat Chee Avenue, Kowloon, Hong Kong
steve@aotl.com

Abstract. This paper describes the design of the Bus Timetabling System (BTS) we have developed for one of the largest privately held bus companies in the world. This Bus Company operates close to 3,500 buses and employs over 7,500 bus drivers. The BTS generates a timetable for each bus route using "distributed constraint-based search" (DCBS). The DCBS algorithm combines techniques of constraint programming, heuristic search, with distributed scheduling. This allows multiple types of constraints to be considered, ensures that idle resources can be shared, and yet generates a timetable within reasonable time. The Bus Timetabling System also determines the bus captain duty assignments for the scheduled bus routes. This paper focuses on both the AI methodologies used and the design of the client-server multi-tiered architecture. We used a distributed object architecture to implement our distributed scheduling system.

1 Introduction

This paper describes the design of a practical application that combines state-of-the-art distributed AI methodologies with state-of-the-art distributed software architecture to perform bus timetabling. The Bus Timetabling System (BTS) uses a novel AI algorithm, which we call Distributed Constraint-based Search (DCBS) [3], that combines distributed [4, 10] and constraint-based [13, 14] scheduling with heuristic search. The DCBS algorithm automatically generates a bus timetable based on a set of service frequency requirements, duty assignment constraints, operational nature of the route, and the type of buses available. Even for a single bus route, the timetabling search space is quite large and highly non-linear; a traditional constraint-satisfaction problem (CSP) algorithm will not suffice [6, 9]. The DCBS algorithm solves this problem by distributing the scheduling responsible among a collection of scheduling agents.

The Bus Company we developed BTS for is quite unique in its operations. In other companies, bus timetabling and duty assignment are considered as independent scheduling tasks, or at least to have minimal dependencies. For our Bus Company, each bus is associated with a driver. If a driver takes a rest or meal break, the bus becomes idle. Therefore, duty assignment constraints, such as duty duration and meal requirements, must also be considered at the same time with the bus timetabling constraints such as service frequency requirements. BTS uses constraint-programming techniques to encode these different types of constraints.

To implement the distributed scheduling algorithm, we used distributed object computing (DOC) techniques. The architecture is basically a multi-tiered client-server architecture where the application server consists of a collection of distributed scheduling agents or processes. Each of these distributed scheduling process contains distributed objects, represented as DCOM components, that capture knowledge and constraints related to a bus route.

This paper first describes the distributed AI methodology used. It then describes the design of a distributed software architecture that implements this methodology.

2 The DCBS Algorithm

Figure 1 shows the overall structure of the DCBS bus-timetabling algorithm. The distributed scheduling system consists of a shared Timetabling Blackboard and a collection of Distributed Scheduling Agents. Each scheduling agent contains a DCBS algorithm to perform timetabling. There is one distributed scheduling agent per bus route to be scheduled.

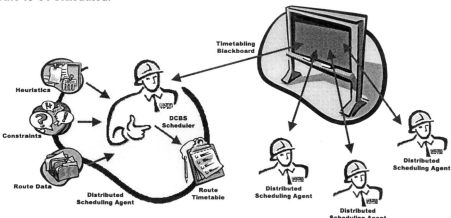

Fig. 1 The structure of the DCBS bus timetabling algorithm.

The timetabling blackboard is the main communication channel between all the distributed scheduling agents and contains service requests that cannot be fulfilled by a route's own set of buses. Distributed agents with idle resources will check the timetabling blackboard for additional work assignment.

Each distributed scheduling agent contains only data for the route it is responsible for scheduling. There are four types of route data. The first is general route parameters, such as the number of available buses, number of runs, route start and end times, terminuses to be served, etc. The second is a table of the journey times between terminuses of routes at different times of the day and day of the week. The third is the desired headway table for a route. The fourth is the operational parameters; such as the minimum rest/meal break duration, time periods when meal breaks should be taken, etc.

In addition, each distributed scheduling agent uses two types of scheduling knowledge to generate a route timetable – a set of timetabling constraints and a set of heuristics to help guide the scheduling search. Although each agent only generates a timetable for its own route, the timetabling blackboard allows buses and bus driver resources to be shared among different routes.

2.1 Bus Timetabling Constraints

BTS uses object-oriented constraint-programming techniques [8, 12] to encode the business logic or operational constraints. The following lists some of the key bus timetabling constraints that are considered by BTS:

- **Journey Time**. The first priority is to ensure that there is enough travelling time between terminuses while making all the necessary stops in between.

- **Layover Breaks**. Bus drivers are given a short *layover break* at the end of each journey; duration is roughly 10% of the journey time.

- **Meal Breaks**. Bus drivers must be assigned a *meal break*. For routes with uniform service frequencies, relief drivers are scheduled.

- **Service Frequencies**. The *service frequency*, or *headway*, defines how often buses leave each terminus at different times of the day.

- **Light Runs**. When there are not enough buses, additional buses may need to travel directly from the other terminus without passengers.

- **Inter-Workings**. When there are more buses than needed, these buses may be scheduled to service one or more other routes.

- **Home Depot**. Buses have assigned home depots where they are scheduled to return for storage overnight.

- **Supplementary Buses**. Sometimes *supplementary buses* will be used to service morning/afternoon peak hours. These drivers must work in *split-shift* duties.

2.2 Variables to be Scheduled

During the process of producing a bus timetable, the DCBS scheduling algorithm must schedule or assign values to many types of timetable unknowns. Some of these unknowns are encoded as constraint-satisfaction-problem (CSP) constrained variables.

- **Departure Time**. This depends on service frequency requirements. The exact service frequency will also depend on the bus's departure time.

- **Bus Sequence**. At any time, there may be several buses at a terminus. The scheduling algorithm must select a bus to depart next.

- **Journey Time**. The journey time will depend on the route, direction of travel and the time of the day.

- **Rest/Meal Break**. At the end of each journey, the scheduling algorithm must decide whether to assign a rest or meal break if parking is available.

- **Rest/Meal Break Duration**. The duration of the break should be assigned opportunistically to match headway requirements.

- **Light Runs**. The scheduling algorithm must decide when to schedule light runs or use supplementary buses.

- **Inter-Workings**. The scheduling algorithm must also decide when it is appropriate to free a bus for an inter-working run.

2.3 Scheduling Criteria

The DCBS timetabling algorithm was designed with several scheduling criteria and objectives in mind. These objectives are user adjustable such that each route may have a different set of scheduling objectives.

- **Minimise Total Buses Needed.** The scheduling algorithm should only use just enough buses to satisfy the service frequencies.

- **Minimise Supplementary Buses Needed**. The scheduling algorithm should minimise the use of supplementary buses if possible.

- **Minimise Fluctuation in Headway.** The time between each departure from a terminus must be as close to the frequency requirements as possible.

- **Optimise Resource Usage**. The scheduling algorithm should minimise idle time. Buses with long idle time are scheduled to service other routes.

In terms of AI methodologies, the DCBS algorithm we developed for bus timetabling uses distributed scheduling as a framework to reduce search space and yet provide a mean of co-operation between scheduling agents. It uses constraint programming to encode inter-related bus timetabling and duty assignment constraints, and heuristic search to improve overall timetabling performance. The following Section explains how the DCBS algorithm is actually implemented and the software technologies we used.

3 The Software Architecture

In terms of the implementation software architecture, the BTS has a multi-tiered client-server architecture. The system consists of three key software modules: the BTS Client Program, the BTS Scheduling Program, and the BTS Database Program.

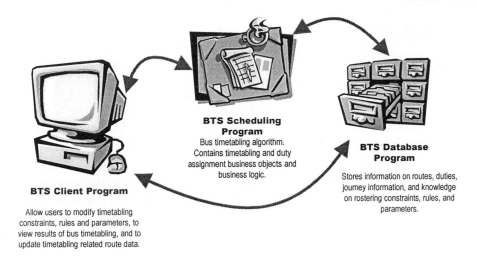

Fig. 2 The BTS multi-tiered software architecture.

- ■ **Client-Tier**. Each BTS workstation contains a copy of the "BTS Client Program." The BTS Client Program is built using Microsoft Visual Basic and contains user-friendly menus and screens that allow users to input and modify bus timetabling and duty assignment constraints, route information, headway tables, journey times, etc. Through the BTS Client Program, users view and edit BTS-generated bus timetables and duty assignments. The Client Program also allows the user to print operational and management reports.

- ■ **Server-Tier**. The "BTS Scheduling Program" is an application server built using distributed object technology and contains a collection of DCBS scheduling agents. Each scheduling agent contains all the necessary business objects and business logic needed to generate a bus timetable and duty assignment of a single route. The BTS Scheduling Program is installed only on the BTS server machine. In other words, the BTS server machine will be responsible for generating the bus timetables for all the remote BTS client workstations. The scheduling agents automatically load appropriate constraints, rules, and parameters from the BTS database. Once a bus timetable is generated, it is stored back into the BTS database for later retrieval, update, or modification.

- ■ **Database-Tier**. The third software component is the "BTS Database Program." The Database Program also resides on the BTS server machine and coordinates all data requests from all the remote BTS client machines. It supports a set of system administrative functions, such as database backup and restore, database maintenance, user security, etc.

3.1 Software Operation

Operating BTS is relatively straightforward. There are only three main steps:

- **Step 1.** The first step in generating a bus timetable and duty assignment for a particular bus route is to enter all information related to that route, such as headway frequencies, number of buses, terminuses, etc., into the BTS. In addition, the user can update any timetabling data that might have changed, such as journey times, operational requirements, etc. The user can also add, modify, or remove any constraints, rules, and parameters related to bus timetabling.

- **Step 2.** Once all necessary route and timetabling information have been entered and updated, the user can then generate a new bus timetable. Once a bus timetable has been generated, it is displayed to the user in the form of a user-friendly spreadsheet. The user can generate as many timetables for a route as needed with different sets of constraints and parameters (in "what-if" situations). The BTS user can interactively refine a bus timetable by making manual modifications and freezing portions of the timetable before requesting the BTS to produce a new version.

- **Step 3.** Once the bus timetable is finalised, the user can print operational and management reports and produce export files to transfer to the corporate mainframe machines.

4 The Distributed Environment

To encode the distributed AI methodologies in BTS, we decided to use distributed object computing (DOC) technologies. For distributed scheduling, each DCBS scheduling agent is encoded as a DCOM distributed object. C++ is used for implementation. To encode constraints, we used a commercial object-oriented constraint-programming class library [5, 11, 12]. This class library also provides hooks to encode heuristics that are used during constraint-based search. On top of the library is a scheduling framework [1, 2] that is designed to solve a general class of resource allocation problems. The DCBS timetabling blackboard is implemented as relational database tables.

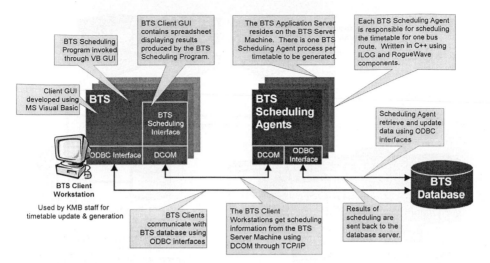

Fig. 3 Details of the BTS middleware infrastructure.

The client graphic user interface (GUI) is implemented using Microsoft Visual Basic. It communicates with BTS scheduling agents using DCOM and the BTS database using ODBC. Each request to generate a timetable from a client machine will fork a new scheduling agent process in the server machine. Each process will be responsible for the timetabling of one bus route. Both the client and application-server tier can communicate directly with the BTS database, which is implemented using the Microsoft SQL Server database.

5 The System Architecture

The following diagram illustrates the hardware architecture of the BTS. Only machines related to BTS are shown. The Bus Timetabling System operates within the Bus Company's corporate LAN. Timetables produced by the BTS are exported and transferred to corporate mainframe machines. The LAN connects all the remote BTS client workstations to the BTS server machine. BTS client workstations are installed at the corporate headquarters as well as several major bus depots. Each depot is responsible for producing timetables for its own bus routes.

The BTS server machine uses Microsoft NT Server as the operating system and executes a collection of BTS scheduling agent processes. This same machine contains the BTS SQL Server database.

The BTS client workstations use either Microsoft NT Workstation or Windows 98 as the operating system and execute the BTS Client Program. These machines can reside at any remote site and communicate with the BTS server machine through the Corporate LAN.

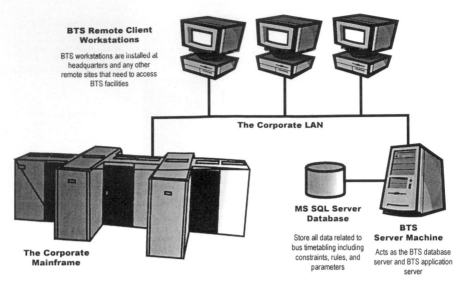

Fig. 4 The BTS system architecture.

6 Conclusion

This paper explained how we implemented distributed AI scheduling concepts using distributed object technologies to produce an advanced bus timetabling system. It outlined the methodologies and architectures used. Previously, the Bus Company used a semi-automatic program that requires many iterative tries before a satisfactory timetable can be produced. This is, of course, too time consuming and not efficient for their growing bus fleet. Our program generates a better schedule and in less time (within a minute). Previously, it took close to a day to process a single route.

Currently, we have finished Phase I of the project, which is a standalone version. The system is now undergoing extensive testing before integration with the Bus Company's LAN. Phase II of the project includes adding enhancements and additional functionality to BTS based on comment and feedback from the testing of the Phase I system. Full-scaled deployment is scheduled for late 1999.

Acknowledgement

The author would like to thank the Kowloon Motor Bus Company of Hong Kong for the cooperation received. Initial design and implementation were performed in part by Resource Technologies Limited.

References

1. H.W. Chun, K.H. Pang, and N. Lam, "Container Vessel Berth Allocation with ILOG SOLVER," The Second International ILOG SOLVER User Conference, Paris, July, 1996.
2. H.W. Chun, M.P. Ng, and N. Lam, "Rostering of Equipment Operators in a Container Yard," The Second International ILOG SOLVER User Conference, Paris, July, 1996.
3. H.W. Chun, "A Distributed Constraint-Based Search Architecture For Bus Timetabling and Duty Assignment," In the *Proceedings of the Joint 1997 Asia Pacific Software Engineering Conference and International Computer Science Conference*, Hong Kong, 1997.
4. E. Ephrati, "Divide and Conquer in Multi-agent Planning," In *Proceedings of AAAI-94*, 1994.
5. ILOG, *The ILOG Solver 3.2 Reference Manual*, 1997.
6. K. Komaya and T. Fukuda, "A Knowledge-based Approach for Railway Scheduling," *CAIA*, pp. 405-411, 1991.
7. V. Kumar, "Algorithms for Constraint Satisfaction Problems: A Survey," In *AI Magazine*, 13(1), pp.32-44, 1992.
8. C. Le Pape, "Using Object-Oriented Constraint Programming Tools to Implement Flexible "Easy-to-use" Scheduling Systems," In *Proceedings of the NSF Workshop on Intelligent, Dynamic Scheduling for Manufacturing*, Cocoa Beach, Florida, 1993.
9. H.-C. Lin and C.-C. Hsu, "An Interactive Train Scheduling Workbench Based On Artificial Intelligence," In *Proceedings of the 6th International Conference on Tools with Artificial Intelligence*, November, 1994.
10. D.E. Neiman, D.W. Hildum, V.R. Lesser, and T.W. Sandholm, "Exploiting Meta-Level Information in a Distributed Scheduling System," In *Proceedings of AAAI-94*, 1994.
11. J.-F. Puget, "A C++ Implementation of CLP," In *ILOG Solver Collected Papers*, ILOG SA, France, 1994.
12. J.-F. Puget, "Object-Oriented Constraint Programming for Transportation Problems," In *ILOG Solver Collected Papers*, ILOG SA, France, 1994.
13..G.L. Steele Jr., *The Definition and Implementation of a Computer Programming Language Based on Constraints*, Ph.D. Thesis, MIT, 1980.
14. P. Van Hentenryck, *Constraint Satisfaction in Logic Programming*, MIT Press, 1989.

Scheduling of a Production Unit via Critical Block Neighborhood Structures

Patrick Van Bael and Marcel Rijckaert

Chemical Engineering Department
De Croylaan 46, 3001 Heverlee (Belgium)
{patrick.vanbael, marcel.rijckaert}@cit.kuleuven.ac.be

Abstract. *This paper describes a scheduling problem and solution strategy of a chemical production unit. The scheduling problem consists of recipes to be optimally produced on shared resources where specific constraints have to be met. The solution strategy emphasis executable schedule generation and iterates towards an (near) optimal schedule construction. Therefore a combined algorithm of constraint propagation/satisfaction principles and an iterative improvement paradigm is used. The results show clearly that the automation has several benefits. The human scheduler's task is changed from schedule construction towards control of an automatic scheduler. The automatic scheduler handles more complex problems and produces substantial more effective schedules.*

1 Introduction

Many practical scheduling problems do not fit into the simplified scheduling models. The shop models like open or closed job shop models, flow shop models, one machine problems or multiple parallel machine problems have been described in literature and solved with general or even specific algorithms [2][3][5][6]. However, these simplified model representations can not always be successfully applied to the practical specific problem detail, which is common for industrial scheduling problems [12]. These problems usually consist of specific constraints, which have to be satisfied and can not always be modeled and solved through the use of one single global method [7][8]. Hence, one tries to solve large real-world scheduling problems by using modular algorithm structures or architectures. Two elements are important in modular environments:

☐ Hybridization: the use of different techniques combined in the same architecture to solve the same task;

□ Specialization: the combination of different modules using a similar technique for solving different tasks.

Modularity simplifies large complex tasks into a set of small, connected and manageable sub-modules, which can be solved separately whilst reducing complexity and development time.

Within a framework one can combine different (types of) scheduling techniques together with Artificial Intelligence techniques or even Operational Research techniques. The scheduling optimization problem of a chemical substance production unit can also be solved with the use of a more complex architecture.

In the following section the problem is described, the third section focuses on the algorithm architecture to solve the scheduling problem and the fourth section shows practical results accomplished with this algorithm architecture. The last section concludes with a discussion on these results.

2 Problem description

The chemical pilot production produces chemical products by demand. These demands or given by researchers and contain usually several rather similar products, i.e. orders with small variations in recipe structure. The scheduler tries to make daily schedules for the pilot production plant to accomplish these demands. Therefore the scheduler needs to satisfy the different constraints on the production unit, on the recipes and on the combined production load of the different recipes. Meanwhile he has to obtain a schedule, which agrees with the company policy and has to succeed in satisfying the different researchers needs, a fairly difficult and responsible task.

The chemical pilot unit is depicted in figure 1. The chemical products are made into one of the sixteen possible reactors, which are separated into two production areas. Each reactor can load raw materials stored in the storage tanks located nearby the respective production area. A shared loading system (belonging to half of the reactors) transfers the material from the raw material tanks into the reactors. Meanwhile a shared monitoring machine (belonging to all reactors) can take samples. The most important constraints on this production unit are capacity constraints and availability constraints. Only sixteen products can be made at the same time. Each production area can produce at most eight products with at most eight different ingredients to be added. The amount of ingredient product is limited to the maximal storage capacity of one tank and limited to the total storage capacity of the raw material storage tank area. The loading system is continuously available but a minimal setup time is needed for every load activity to fulfill. The monitoring system can only take samples every two minutes. The pilot production unit is only available for eight hours during a day.

A typical recipe of a chemical product is shown in figure 2. The basic recipe consists of two phases, which are not specifically serial. The first phase produces the chemical product while the second describes the monitoring procedure. The production phase consists of activities to add raw materials. Each activity has two time specifications. The first is the time period to wait after a chemical substance started production (K) and the second specifies the duration to add the specific component (t_1). At no time the relative time distances between two production actions can be altered, i.e. they are fixed. The monitoring phase defines the amount of samples to take from the chemical semi- or final product together with time indications to start each sample procedure. Each sample has a duration of two minutes (t_2). The time indications are also relative towards the beginning of the substance production (K). Every sample procedure can be moved backwards specified with its own value (t_3). All sample and production procedures can be moved all together within the buffer period (b) and is therefore limited. This because the base product initially added in the reactors has its own shelf life.

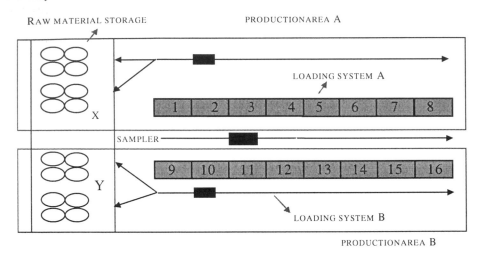

Fig. 1. Pilot production unit

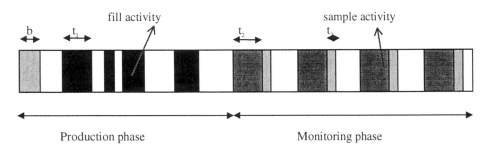

Fig. 2. Recipe: duration of an action (t_1: fill duration, t_2: sample duration, t_3: start slack time) and possible buffer period (b)

3 Algorithm description

The algorithm is in the first place developed to obtain feasible schedules, which is the main difficulty of the problem and in the second place to obtain an optimal feasible schedule. The problem will be simplified by giving a priori the orders to be scheduled for one day. The reason for this is that at the moment the priorities to select the orders for a given day are basically human priorities and the responsibility of a good selection is given to the human scheduler. Hence, the problem consists of scheduling the orders for a one-day schedule where the orders have a complete predefined recipe without starting values. Also a few preferences of the production units have to be taken into account. The final schedule contains data of which reactor is assigned to the orders and start times assigned of the different activities within the orders as well as data of the raw materials namely the amount and storage tank.

Since many constraints have to be unviolated and the priority is set to feasible schedules a two-module algorithm is build where the first module ensures a construction of a feasible schedule and the second optimizes the schedule. Figure 3 illustrates the principle and can be seen as a master-slave structure. The slave part is build with constraint satisfaction programming paradigms like constraint propagation/satisfaction. This technique ensures creation of feasible schedules. The master part guides the search towards optimal schedule generation. It uses an iterative improvement algorithm (simulated annealing) and guides the search towards the optimal schedule. An initial module ensures that a schedule can be generated without violating the capacity constraints of the production unit. This module searches for a possible split of orders over the two production areas without violating the raw material constraints. This module works as a preprocessing phase and is meant to discard overconstraint problems.

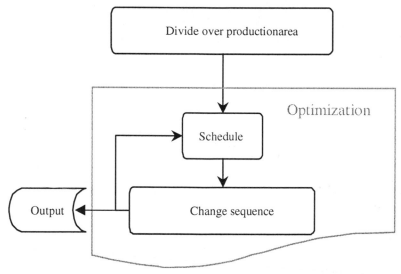

Fig. 3. General algorithm architecture

As can be seen in figure 3 the algorithm consists of three major modules: the schedule builder, the optimizer and the data-preprocessing module. Next these three modules will be explained in more detail.

The data-preprocessing module is meant to reduce the schedule problem size. Since a raw data set does not guarantee to contain a feasible solution, we ensure there is one by reducing the global optimization problem towards a simplified optimization problem with one criterion to optimize, i.e. the makespan or total duration of a one-day schedule. The global optimization problem optimizes the orders to be produced on the production unit by making a schedule with the smallest makespan that does not violate any of the constraints. But a simplified problem consists of a schedule with the smallest makespan where the orders are already assigned to the different resources. Moreover the simplified problem does not need the data of the raw materials any more. But this simplification is a problem of its own. The module uses a heuristic to assign the different orders to the available productionareas and takes into account the capacity constraints of the raw material storage tanks. The heuristic is a distance measure and separates the orders over the two productionareas. It selects first the two most distanced orders, i.e. the orders that are most different according to the raw materials needed. Next the other orders are added to the order with the shortest distance without violating the capacity constraints. If this gives an infeasible solution then the algorithm stops prematurely otherwise it goes on.

The schedule builder makes a feasible schedule with the orders. Therefor the orders or listed and one after the other assigned with the start times needed by the different activities it contains. The orders and the respective activities are assigned start times with the earliest possible start times available. This is done efficiently through constraint propagation techniques where a certain activity keeps only the available start times. Especially this constraint propagation/satisfaction ensures that no two activities can overlap or even fall within a minimum time-distance, as pre-specified, of each other.

The optimizer tries to reach the optimal schedule by iterative improvement. Therefore simulated annealing (SA) is used. A general description of this algorithm can be seen in figure 4 [1] and has already many times used with success in different domains [4][7][8].

INITIALISE $(i_{start}, c_0, L_0, k, i)$

repeat

for $l = 1$ to L_k do

begin

GENERATE $(j \in Si)$

if $f(j) \leq f(i)$ then $i = j$

else if $(\exp(\dfrac{f(i) - f(j)}{c})) > \text{random}[0,1])$ then $i = j$

end

$k = k+1$

CALC(L_k)

CALC(c_k);

until stop criterion

Fig. 4. Simulated annealing algorithm

Simulated Annealing is a near-optimal stochastic optimization algorithm and is based on a simulation of a natural optimization process, i.e. the annealing (cooling down) of a liquid metal to a solid metal with a minimal energy level. This physical process can be achieved by gradually lowering the temperature, giving atoms (of the metal) enough time to rearrange themselves into an equilibrium state at each temperature. Hence, annealing helps the material to avoid local minimal energy states and to find the global minimum energy state.

Simulated annealing as described in figure 4 is an analogy of the physical annealing process. It consists of a few general parameters like the initial temperature (c_0), the end temperature (c_t), the temperature decrement function ($CALC(c_k)$), the equilibrium definition (L_k) and a few special parameters like the neighborhood structure and the cost function (f) used. The most practical difficulty with this optimization algorithm is the neighborhood structure definition. Also the parameter value settings is difficult but usually gives good results after a few trial and error attempts. Usually a neighborhood structure is used where random swap and insert operations are performed. In this algorithm we use the critical neighborhood structure as first defined by Yamada and enhanced by Van Bael [9][10][11]. This neighborhood structure only swaps two activities within a critical block lying on a critical path. Figure 5 shows a critical path and critical blocks belonging to this path.

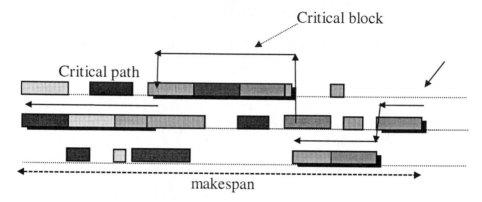

Fig. 5. A critical path and critical blocks. The neighborhood structure uses only activities from a critical block

The algorithm architecture can thus be seen as a specializing modular architecture. The difference between such a strategy and a single algorithm, which tries to optimize the problem with one cost function, is that no search time is wasted with the creation of unfeasible schedules. The sub-modules should thereby have a simpler task. In the past a lot of single algorithms have failed because it is very difficult to obtain a schedule which is feasible and near-optimal when one tries to optimize the whole constraint set at once.

4 Testcases

The algorithm architecture is written in Borland 4.5 C++ with a visual Basic 3.0 interface and runs on a pentium 133 Mhz. The algorithm is applied to a practical set of orders. Three cases show the improvement of the automated system. Figure 6 shows the data and results. Figure 7 shows a schedule from case 1 generated with the

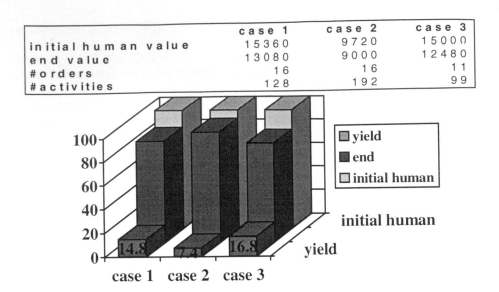

	case 1	case 2	case 3
initial human value	15360	9720	15000
end value	13080	9000	12480
#orders	16	16	11
#activities	128	192	99

Fig. 6. Data and results of three cases shown

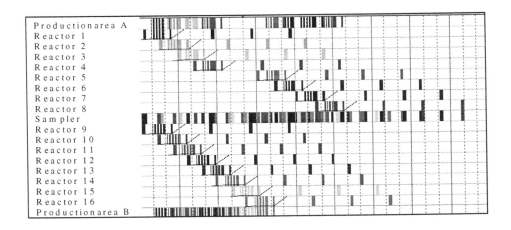

Fig. 7. Schedule of case 1 with a schedule evaluation of 15000

algorithm. One can see that the overall improvement ranges from about seven till seventeen percent. As is evident from figure 6, the more activities to schedule the lower the compaction can be. And from figure 7 we see that even if the schedule looks like no improvement can be made, the algorithm still can drop the makespan with over 5 percent.

The biggest improvement is naturally the automation of schedule generation. In the past a human scheduler was busy for one day to generate a schedule of equal quality as an initial solution where the algorithm architecture took no longer than 5 minutes on average to achieve the illustrated results. The improvements over a human scheduler are not only that the makespan is reduced, but also the way the scheduler works now. His schedule generation job has changed towards control of an optimizing system. This way he can spend more time generating complex recipes, thus optimizing the product quality.

5 Conclusion

In this paper, a modular algorithm architecture is developed which was used to solve the scheduling problem of a chemical pilot production unit. The algorithm is an iterative improvement/constraint satisfaction scheduling algorithm consisting of three modules. Each module solves a smaller sub-problem focusing on specific constraints and resulting in a near optimal feasible solution. With this algorithm preliminary results have shown that it is possible to solve a scheduling problem by focusing on specific difficulties of the practical scheduling problem. The algorithms' performance is an improvement over human schedule generation both in time and quality.

References

1. Aarts E., Korst J., Simulated Annealing and Boltzmann Machines. Wiley & Sons Inc. Chichester. 1989.
2. Baker K., Introduction to Sequencing and Scheduling, Wiley & Sons Inc. New York, 1974.
3. Ciriano T.A., Leachman R.C., Optimisation in Industry. Wiley & Sons Chichester. 1993.
4. Kalivas J., Adaption of Simulated Annealing to Chemical Optimization Problems, Elsevier Science, 1995.
5. Morton T., Pentico D., Heuristic Scheduling Systems. Wiley & Sons , Inc. New York, 1993.
6. Pinedo, M., Scheduling: theory, algorithms and systems, Prentice-Hall Englewood Cliffs (N.J.). 1995
7. Rayward-Smith, V., Applications of modern heuristic methods, Waller Henley on Thames. 1995
8. Rayward-Smith V. J., Osman I.H., Reeves C.R., Smith G.D., Modern heuristic search methods, Wiley & Sons Chichester. 1996.
9. Van Bael P., Solving Job Shop Problems with critical Block Neighboorhood Search, Proceedings of NAIC98, Amsterdam, 1998
10. Yamada, T., Rosen, B.E. and Nakano R., A simulated Annealing Approach to Job Shop Scheduling using Critical Block Transition Operation, In: Proc. Of IEEE ICNN (IEEE, Florida, 1994), pp 4687-4692
11. Yamada, T., Nakano, R., Job-Scheduling by Simulated Annealing Combined with Deterministic Local Search, In: META-HEURISTICS: Theory & Applications, edited by Osman, I. H., and Kelly, J. P., Kluwer Academic Publishers Boston, 1996
12. Zweben M., Fox M., Intelligent Scheduling, Morgan Kaufmann Publishers,Inc., 1994

An Algorithm for Station Shunting Scheduling Problems Combining Probabilistic Local Search and PERT

Norio TOMII, Li Jian ZHOU, Naoto FUKUMURA

Railway Technical Research Institute
2-8-38 Hikari-cho Kokubunji-shi TOKYO 185-8540 JAPAN
{tomii, zhou, fukumura}@rtri.or.jp

Abstract. Shunting scheduling problems at railway stations can be regarded as a sort of resource constrained project scheduling problem (RCPSP). Unlike the normal RCPSPs, however, shunting scheduling problems require that the number of work which consists of a project has to be dynamically changed in the process of solving the problem and some of the work has to be performed at a prescribed timing, for example. We propose an efficient algorithm for shunting scheduling problems combining probabilistic local search and PERT. Local search and PERT are combined so that the candidates for answers in the local search process are evaluated by PERT. This enables us to reduce the search space of the local search to a great extent and makes the algorithm work quite fast. We have confirmed the effectiveness of our algorithm through several experiments using practical train schedule data.

1 Introduction

Station shunting scheduling problems (sometimes just called, shunting scheduling problems in this paper) is to make shunting schedules of trains at railway stations. In other words, in a shunting schedule of a station, which side tracks and when they are assigned to each shunting work are prescribed. A shunting schedule is made at the same time when train schedules are renewed or when they are changed due to operation of seasonal trains, etc.

Shunting schedules play quite an important role in railways in connection with realizability of train schedule. That is, a train schedule prescribes arrival/departure times and tracks for each train, and it the shunting schedule that guarantees trains can appear exactly at the prescribed time at the prescribed track.

However, it is not always possible to make a shunting schedule under constraints prescribed by train schedules, because basically train schedules and shunting schedules are made independently. In such cases, it is desired to make a shunting schedule to satisfy the constraints of train schedule as much as possible.

Until now, shunting schedules have been made manually and no attempts for computerization are reported. To make a shunting schedule is quite a demanding work to require a lot of labor, time and experience. This is because various kinds of constraints exist to operate a lot of trains, in that trains must not conflict on a side track; the running time between a track and a side track must keep to a certain prescribed time; there must be a route for each shunting; a certain time interval has to be kept between trains which use a conflicting routes, etc.

We believe that a computer system that automatically makes shunting schedules will be quite beneficial to reduce the time and labor and obtain schedules of much higher quality. To this end, we need a scheduling algorithm which does not employ know-how or experience of human experts. It should not be dependent on specific conditions of each station, but applicable to various types of stations.

Shunting scheduling problems can be considered as a kind of resource constrained project scheduling problem (RCPSP) when an occupation of a track by a train is regarded as a *work*, and series of occupation of tracks or side tracks by trains as *project*. Tracks and side tracks are considered to be *resources*.

Recently, it attracts attention that the local search technique such as simulated annealing, tabu search and genetic local search are effective for scheduling problems. For example, it is reported that the local search technique is effective for job shop scheduling problems (JSSP)[1][2], which is a special case of RCPSP.

It is quite important when using the local search technique to examine how to define a notion of neighborhood and how to design a search algorithm in the neighborhood. As we will prove in Chapter 2, however, the conventional concept of neighborhood and searching algorithm cannot be used for shunting scheduling problems. This is because a shunting scheduling problem has such unique features that the search space is enormous because not only the orders of work but precise execution times of each work have to be decided and the problem structure itself has to be changed during the process of the problem solving.

In this paper, we propose a two-stage search algorithm to concur the problem that the search space is enormous. We divide the problem into two subproblems. One is a resource allocation subproblem and the other a shunting time decision subproblem. We first conduct local search in the search space where shunting times do not explicitly appear. Then, for the candidate solutions obtained in this process, we decide shunting times using PERT technique and then evaluate the candidates.

In order to deal with the latter feature, we introduce an idea to define a neighborhood space using transformation operators for a shunting scheduling network, which is a kind of PERT network expressing a shunting schedule.

Merits of our algorithm are as follows.
- We succeeded in reducing the search space of the algorithm to a great extent.
- We need not care about a process to adjust shunting times to make a candidate solution feasible because we can always get feasible solutions as a result of PERT calculation.

In chapter 2, we explain an outline of shunting scheduling problem together with its features from the viewpoint of scheduling problem. In chapter 3, we introduce an algorithm which combines probabilistic local search and PERT. In chapter 4, we verify the applicability of our algorithm to actual shunting scheduling problems.

2 Shunting scheduling problem at railway stations

2.1 An outline of shunting scheduling problem

Fig. 1 is an example of a shunting schedule at a station whose track layout is shown in Fig. 2. This diagram is called a shunting diagram illustrating a schedule of movements of trains at the station in course of time, which is depicted along the horizontal axis.

In Fig. 2, Tracks 1 to 3 are *tracks* (sometimes called passenger service tracks) where trains can directly arrive at and depart from. On the other hand, Track 5 and 6 are *side tracks* where trains are temporarily stored before they depart. Track 4 is called a *storage track*. As it doesn't have a platform on either side, it is not used for arrival or departure of trains, but used only for temporary storage of trains. As both side tracks and storage tracks are used for shunting, we call these tracks *side tracks* in this paper.

Movement of trains between a track and a side track and between two side tracks is called shunting. Shunting is necessary for the following three cases (1) a train which arrives at a track and is scheduled to depart from a different track. (2) a train which arrives and departs from the same track, but another train is planned to use the track between the arrival and the departure of the former train. (3) a train on a side track if other train uses it.

The third case happens when there are no other side tracks available for the latter train. We call such shunting *2-level shunting*.

The objective of shunting scheduling problem is to decide for trains which need shunting, side tracks to be assigned and shunting times.

Arrival and departure times of trains and tracks are prescribed by train schedules. We call these *planned arrival times*, *planned departure times*, *planned arrival tracks* and *planned departure tracks*, respectively. We also use the terms of *planned times* to group the former two notions and *planned tracks* for the latter two notions).

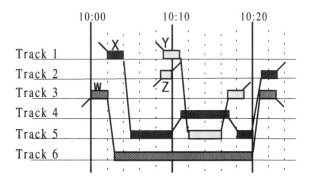

Fig. 1. Example of a shunting schedule

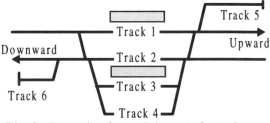

Fig. 2. Example of a track layout of a station

2.2 Constraints of shunting scheduling problems

The following constraints have to be considered in solving the shunting scheduling problem.

(1) Constraints concerning train schedules

A solution should not have inconsistency with planned times and tracks prescribed by the train schedule.

(2) Constraints caused by facility conditions

- Existence of routes: There must exist a physical route between the starting point of shunting and the destination point.
- Length of side tracks: Trains have to use a side track whose length is larger than their length.
- Conflict of side tracks: Only one train can be put on a side track at a time.

(3) Temporal constraints

- Minimum dwell times on tracks and side tracks have to be retained. These times are required to assure a time for passengers to get on or off and for inspection of inside the train.
- Shunting running time: We call the running time of trains between a track and a side track or between a side track and another side track the shunting running time. The shunting running times have to be as prescribed for particular starting and destination points. If the time is shorter, the shunting schedule is physically impossible to realize. The time shall not be longer, either, as the route is blocked for shunting during he running time and cannot be used by other trains.
- Crossover conflicts: At railway stations, two routes sometimes intersect each other. In Fig. 2, for example, the route from Track 1 to Upward direction and the route from Track 5 to Track 2 are intersecting. These two routes cannot be used simultaneously. If one of the two intersecting routes is used, trains which use the other route have to wait for a certain time, which is called the crossover conflict time, before it becomes open. This is called a crossover conflict constraint, a crucial condition that shall strictly be observed from the viewpoint of security.

The Crossover conflict constraint is quite important in making shunting schedules. If it is ignored, crossover conflicts might occur. This would quite probably cause trains to delay and make it impossible to arrive or depart on the scheduled time.

2.3 Criteria for evaluation of shunting scheduling

Among constraints described in the previous section, constraints caused by physical conditions, namely, constraints other than those imposed by train schedules and shunting running times are impossible to relax.

Constraints imposed by train schedules allow relaxation in the last resort. But of course, it is desired to satisfy planned times as much as possible. Although the shunting running times can be extended when necessary, it is not desirable to make it too long. To save energy, the frequency of shunting shall be limited to a minimum.

From the above discussions, we can conclude that the criteria for evaluations of shunting schedule are to keep the following minimum.
- Discrepancies from the planned departure and arrival times.
- Discrepancies from the planned shunting running times.
- Frequency of shunting operation.

2.4 Station shunting scheduling problem as a resource constrained project scheduling problem

When we compare the station shunting scheduling problem with RCPSPs commonly dealt with, the former has the following unique features.
(1) The number of works which consists of a project is not given and has to be dynamically changed during the process of problem solving. This is due to the fact that 2-level shuntings have to be considered.
(2) Not only orders of works but also their precise execution times have to be included in the solution. This is because we have to examine temporal constraints such as minimum dwell times, cross over conflict times, etc.
(3) In usual RCPSPs, constraints of interval times between works are not assumed. In contrast, the shunting scheduling problem has a strong constraint on the intervals between works. This is because shunting running times between side tracks shall definitely be fixed.
(4) The number of projects is quite large. It varies depending on the size of stations, but several hundred to a thousand trains are normally involved in a shunting schedule of one day.

3   Shunting scheduling algorithm combining probabilistic local search and PERT

3.1 Basic idea

Station shunting schedules which apparently do not satisfy some constraints are changed into feasible ones when the shunting times are appropriately adjusted. The station shunting schedule shown in Fig. 3 is infeasible, for example, because two trains are assigned to use the Sidetrack at the

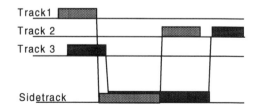

Fig. 3. An apparently infeasible shunting schedule

same time. By adjusting the shunting times, however, we can make it feasible.

This means that not only orders of works but also their execution times have to be determined to satisfy the given constraints in shunting scheduling problems. If we add the process of determination of execution times, however, the search space becomes too large to obtain an effective algorithm.

In this paper, we introduce a two-stage algorithm. We first produce a candidate solution in which assignment of side tracks are temporarily decided and execution times of shunting are not explicitly considered. Then, we try to adjust the execution time of each shunting so that all constraints are satisfied. We depict an outline of this algorithm in Procedure 1, and show the details in the following sections.

Step0 Initial solution: Produce an initial solution x by using heuristics. Let x be a temporary solution. Set $k=0$.

Step1 Local search: Search in the neighborhood of x and generate candidate solutions $x_1, x_2, ..., x_m$. Set $k=k+1$.

Step2 Evaluation: Evaluate $x_1, x_2, ..., x_m$. If a solution is found among them, output the solution. When a solution better than the temporary one is contained among them, let it be a new temporary solution.

Step3 Selection: Select one from $x_1, x_2, ..., x_m$ and let it be x. If $k > L$, then let the temporary solution be x. Go to **Step1**.

Procedure 1. Outline of the algorithm

3.2 Initial solution

By using heuristics, we decide assignment of a side track and an execution time for each shunting and produce an initial solution. Heuristics used here is just a simple one in that
- Decide an assignment of the side track depending on the pair of departing and arriving tracks.
- Decide shunting times using the minimal dwell times.
This means that we do not consider conflict of trains on side tracks at all.

3.3 Local search

Shunting scheduling network

We express a shunting schedule using a PERT network. We call the network *shunting scheduling network*. A shunting scheduling network is described as (N, A), where N is a set of nodes ; A is a set of arcs and $A \in N \times N$. A weight $w \geq 0$ is assigned to each arc. These definitions mean the following.
- A node corresponds to an event of arrival or departure of a train.
- An arc expresses the time base order of execution between nodes. The weight of an arc means the minimum time necessary between the occurrence of the two events on each side of the arc. We have five types of arcs as shown in Table 1.

A shunting scheduling network is constructed according to the shunting execution orders which are decided based on the execution times of shunting in the candidate solutions. In Fig. 4, we show the shunting scheduling network constructed from the shunting schedule in Fig. 1. (Characters in the nodes mean the combination of Train, Node number, Arrival or Departure. A means arrival and D means departure.)

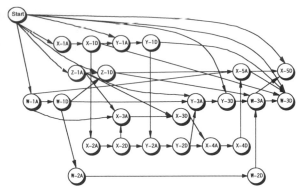

Fig. 4 An example of shunting scheduling network

Table 1. Arcs

Arc Type	Content
Operation	Between arrival and departure nodes of a train
Shunting	Between departure and arrival nodes of a shunting work
Track	Between departure and arrival nodes of consecutive trains
Crossover	Between nodes which use intersecting routes
Planned time	Between the Start node and nodes which have planned time

Calculation of shunting times

We calculate execution time of each node for the shunting scheduling network of the candidate solutions. The calculation method is almost the same as the conventional PERT method. However, the conventional method does not assure that shunting running times become equal to prescribed ones. By adding backtracking, the algorithm produces an answer which satisfies shunting running time constraints. Thus, calculation of shunting times for shunting scheduling network guarantees a feasible schedule which satisfies all physical constraints imposed by facility conditions and temporal conditions as far as possible.

1-operator neighborhood

When conducting the local search, it is quite important how to define a neighborhood space. For JSSP, several ideas are proposed[3][4][5][6]. For example, a neighborhood defined by an exchange of works existing in critical blocks, a neighborhood defined by a movement of works on critical blocks to the top or the bottom of the block have been proposed. In shunting scheduling problem, however, the problem structure itself has to be dynamically changed during the course of problem solving. Hence, the conventional ideas of neighborhood for JSSP do not work.

In this paper, we propose a neighborhood defined by transformation of shunting scheduling network. We define network transformation operators as shown in Table

2. Using these operators, we then define 1-operator neighborhood *Nb(P)* for a shunting scheduling network *P* as follows.

Nb(P)={shunting scheduling network obtained by applying a transformation operator to *P*}

Table 2. Network transformation operators

Operator	Content
Change of track	Change shunting work to use other side track
2-level shunting	Set 2-level shunting
Cancellation of 2-level shunting	Cancel 2-level shunting
Change of shunting order	Change shunting order
Crossover	Change direction of a crossover arc

3.3.4 Local search algorithm

We conduct local search in the neighborhood and get candidate solutions. We do not search the whole 1-operator neighborhood, but focus on a part where we can quite probably find a solution. Details are shown in Procedure 2.

Step1: Selection of a node: Select a node *x* at random from the nodes which do not satisfy planned time constraints or shunting running time constraints.
Step2: Transformation of shunting scheduling network:
(1) If *x* does not satisfy the planned schedule constraint, perform the following. Let *Y* be a set of nodes which satisfy the planned schedule constraint and *Cr(x)* be a set of paths from *x* to each *y∈ Y*, where *y* is a node which is first encountered along critical paths from *x* to the Start node. Apply each transformation operator in Table 3 to each arc on *Cr(x)* depending on the type of the arc, and get candidate solutions.
(2) If *x* does not satisfy the shunting running time constraint, apply Change of track operator to *x* and get a candidate solution.

Procedure 2. Local search algorithm

3.4 Evaluation

We evaluate the candidate solutions from the viewpoints of the number of nodes which do not satisfy the planned schedule constraint, the number of arcs which do not satisfy the constraint of shunting running times and the frequency of shunting works. Results of evaluation are used in the Selection step.

3.5 Selection

We select a candidate solution which will be

Table 3. Network transformation operators for each arc type

Arc type	Operators applied
Operation	Change of track, 2-level shunting
Shunting	Change of track
Track	Change of shunting order
Crossover	Crossover

handed to the next step from the set of candidate solutions found in the local search process. As for selection method, we use a roulette selection method, which is commonly used in Genetic Algorithm, based on the evaluation results of the candidates.

4. Results of experiments and considerations

4.1 Results of experiments

We performed experiments using actual train schedule data. The target station is a typical middle size station where three lines meet together, and it has 6 tracks and 2 side tracks. Its track layout

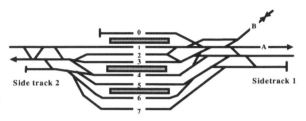

Fig. 5. Track layout of the station used for experiments

is shown in Fig. 5. Target time ranges for experiments are from 6 o'clock to 10 o'clock (Case 1) and from 16 o'clock to 18 o'clock (Case 2), both of which are the busiest hours in this station. The numbers of trains involved are 98 (Case 1) and 40 (Case 2) respectively.

The results of 10 experiments for each case are shown in Tables 4 and 5. Cand means the number of candidate solutions generated in the whole process. Step is a number of iteration of Procedure 1. Plan and Run mean the numbers of shunting works which do not satisfy planned schedule constraint and shunting running time constraint, respectively. Shunt means frequency of shunting works. We set L in Procedure 1 be 15. It took just a couple of minutes to get these solutions (Pentium 166MHz, 80MB memory).

Table 4. Results of Experiments (Case-1)

Exp	Cand	Step	Plan	Run	Shunt
1	40	9	0	0	28
2	21	7	0	0	28
3	21	7	0	0	28
4	73	17	0	0	28
5	26	8	0	0	28
6	26	8	0	0	28
7	31	9	0	0	28
8	42	9	0	0	28
9	90	17	0	0	28
10	33	9	0	0	28

Table 5. Results of Experiments (Case-2)

Exp	Cand	Step	Plan	Run	Shunt
1	50	6	0	0	18
2	11	2	0	0	18
3	27	6	0	0	18
4	49	6	0	0	18
5	37	9	0	0	18
6	12	2	0	0	18
7	11	2	0	0	18
8	36	8	0	0	18
9	27	6	0	0	18
10	12	2	0	0	18

4.2 Considerations

(1) From Tables 4 and 5, we can conclude that in each of 10 experiments, the algorithms succeeded in obtaining a practical solution.

(2) In experiments 4 and 9 of Case 1, a solution which satisfy all the constraints was not found in the first 15 steps and the algorithm returned to the

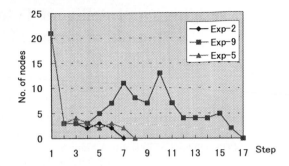

Fig. 7. Comparison of search processes

temporary solution and continued the search process. After that, a solution was found in 2 steps in both of those experiments. This shows that our strategy to return to the temporary solution successfully worked.

(3) We show in Fig. 7, the numbers of nodes which do not satisfy constraints in each candidate solution during execution of Experiment 2, 9 and 5 in Case 1. This graph proves that it is not a good idea to select the best candidate solution in each step and it is often effective to sometimes select a "bad" candidate solution. This proves rationale of using roulette selection method.

5. Conclusions

We introduced an algorithm combining probabilistic local search and PERT for station shunting scheduling problems. This algorithm is applicable to various kinds of track layouts and train schedules, since no knowledge or conditions specific to each station are employed. We have proved that the algorithm works quite fast and confirmed its effectiveness through experiments using actual train schedule data.

References

1. Yamada and Nakano: Job-Shop Scheduling by Genetic Local Search (in Japanese), *Transactions of Information Processing Society of Japan,* Vol.38,No.6 (1997).
2. Yamada and Nakano, Job-Shop scheduling by Simulated Annealing Combined with Deterministic Local Search, Kluwer Academic Publishers, MA, USA (1996).
3. Taillard, E.:Parallel Taboo Search Techniques for the Job-shop Scheduling Problem, *ORSA, J. on Comput.,* Vol.6, No.2 (1994).
4. Laarhoven, P.v., Aarts, E. and Lenstra, J.: Job Shop Scheduling by Simulated Annealing, *Oper. Res.,* Vol.40, No.1, (1992).
5. Dell'Amico, M. and Trubian, M.: Applying Tabu Search to the Job-shop Scheduling Problem, *Annals of OR,* Vol.41 (1993).
6. Brucker, P., Jurisch, B. and Sievers, B.: A Branch Bound Algorithm for the Job-Shop Scheduling Problem, *Discrete Applied Mathematics,* Vol.49 (1994).

Designing an Intelligent Tutoring System in the Domain of Formal Languages

Dusan Popovic[1] and Vladan Devedzic[2]

[1]Mihailo Pupin Institute, Volgina 15, 11000 Belgrade, Yugoslavia
dule@son.imp.bg.ac.yu
[2]FON - School of Business Administration, University of Belgrade, Jove Ilica 154, 11000 Belgrade, Yugoslavia
devedzic@galeb.etf.bg.ac.yu

The paper describes design of an intelligent tutoring system called FLUTE. The system has been recently developed in order to help students learn concepts from the domain of formal languages and automata. The basic idea of the FLUTE system is a systematic introduction of students into the system's domain, in accordance with both the logical structure of the domain and individual background knowledge and learning capabilities of each student. The main goal of the paper is to illustrate important design decisions and implementation details that are relevant even beyond the FLUTE system itself. The system is presented in the paper primarily from the pedagogical and design perspectives.

1 Introduction

This paper presents the design of the FLUTE (Formal Languages and aUTomata Education) intelligent tutoring system. This system is being developed to serve the purposes of teaching the formal languages and automata course at the Faculty of Electrical Engineering in Belgrade.

The basis for development of this system has been found in the experience of the authors of classical intelligent tutoring systems [1], [2], [3]. The WUSOR system functions on the principle of dynamic student modeling relying on constantly updating the temporary student model on the basis of the student's behavior during the education process. This principle can be applied to student modeling in the FLUTE system. Learning strategy is one of the essential components of tutoring systems. A learning strategy based on examples has been used in practice in several tutoring systems (*GUIDON,BIP,ELINT,LISPTUTOR*) [3], and the simplicity of its logical conception makes it applicable to the FLUTE system as well.

This paper is organized as follows. The next section presents the purpose and capabilities of the FLUTE system. It also notes what software and hardware platforms are used for system implementation. The third section presents the FLUTE system architecture and describes the roles of its components. A presentation of a session with FLUTE is given in the fourth section.

2 Purpose and Capabilities of the FLUTE System

One of the main problems in teaching formal languages and automata is abstractness of the topics treated. The effectiveness of the educational process depends predominantly on the teacher's ability and motivation to familiarize students with a topic using the simplest possible method. As the ultimate goal of each learning process is to apply the acquired knowledge, the teaching concept based on as many practical examples as possible appears to be most adequate solution. This is especially true for the field of formal languages and automata, where a large number of notions faced for the first time by a majority of students are mastered most adequately through the presentation of many appropriate examples. In general, such an approach ensures student's maximum motivation to master the treated topics in the best possible way and allows detecting problems that arise with respect to the correctness and application of the acquired knowledge.

The FLUTE tutoring system is intended to aid students both in mastering the basic notions from the subject of formal languages and automata and in subsequent more detailed studying of this field which represents a basic for further studies of the principles of programming languages and compilers. This tutoring system permits:

- a student to select an arbitrary class of grammar and test a series of examples of formal languages with respect to the membership of a given formal language in the selected grammar class; if the student hasn't completely mastered the basic notions, the system should allow him to select an appropriate example from the database of previously defined languages
- a student to test an automaton on the basis of the selected regular expression
- education to proceed in accordance with each student's knowledge level, which is achieved by incorporating precisely defined student models into the system
- each session to proceed according to a prespecified session control model
- constant monitoring of the student's knowledge level and responses according to the principles of intelligent tutoring and cooperative systems, which would provide considerably higher-level teaching in this field.

Intelligent tutoring systems have not been developed for the field of formal languages and automata so far; this adds to the importance of this project.

2.1 Hardware and Software Environment

One of the important components in software engineering, apart from a precisely defined structure of a software system (which facilitates system implementation and maintenance), is planning of the environment. This refers to both software environment (the underlying model of the system, programming language, and tools used for implementation) and the hardware platform (computer type, operating system). As for the software environment, the FLUTE system is based on the GET-BITS model of intelligent tutoring systems (see below), MS Visual C++ was selected as the implementation language, and interoperable software components [6] are used as principal building blocks in the system design and implementation. The system's prototype is developed on a PC/Windows 98 platform.

2.2 The GET-BITS model

The GET-BITS model of intelligent tutoring systems has been used in designing the FLUTE system [4], [5]. GET-BITS is a multilayered, object-oriented, domain-independent model of intelligent tutoring systems [4]. It is based on a number of design patterns and class hierarchies that have been discovered for teaching and learning processes. In support of the patterns, appropriate C++ class libraries have been developed. The class hierarchies start from the universal, abstract, and unifying concept of *knowledge element*. It reflects the possibility of representing all data, information and knowledge in an intelligent system in a unified way [5]. Many meaningful subclasses that are needed for building a wide range of intelligent tutoring systems are derived directly or indirectly from the knowledge element class (concepts like lessons, questions, tasks, answers, rules, plans, assessments, exercises, etc.). However, classes for knowledge representation are not the only tools needed to build an object-oriented intelligent tutoring system. Apart from knowledge of various kinds, in each such intelligent tutoring system there must also exist some control objects that functionally connect the system's modules, handle messages, control each session with the system, monitor student's reactions, etc. In other words, such objects provide control and handle dynamics of intelligent tutoring systems. GET-BITS also specifies classes of these control objects. For more detailed description of the model, see [4].

3 The System Architecture

FLUTE is a complex intelligent tutoring system containing expert knowledge from several domains. Each component is, depending on its function, in interaction with the remaining system components which results in expressed modularity of the entire system.

The architecture of the FLUTE system is depicted in Figure 1. The meanings of the tutoring system components shown in Figure 1 are described in the following subsections.

3.1 Expert module

The subject-matter to be mastered by the student is included in the expert module and is organized hierarchically by topics (themes), subtopics and lessons. All topics are displayed to the student in the main menu, but the selection of menu options is limited to those subtopics which match the student's current background knowledge level. This level is determined in accordance with the student model (see subsection 3.2.). Under each subtopic the student is supposed to learn an appropriate number of concepts and lessons in a sequence depending on the actual subtopic, i.e., lesson.

The expert module incorporates several important topics from the domain of formal languages and automata (elements of the of the set theory, the notion of formal languages, formal grammars, finite automata, Turing automata, relationships between formal grammars and automata, and decidability, solvability and complexity). This Section describes in detail the hierarchy of the knowledge (subtopics) covered by formal grammars and finite automata. Typical problems which the problem generator, as part of the FLUTE's pedagogical module, creates and presents to the student for knowledge testing purposes are described for each subtopic. Figure 2 shows the way topics and contents of the learning process are organized in the system.

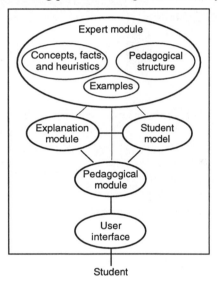

Fig. 1. The architecture of the FLUTE system

Grammars. Explanations of the notions used to define a grammar See also Figure 3). The student is supposed to learn the elements of the definitions of grammars: the sets of terminal nonterminal symbols, the set of derivation rules and the start symbol.

Determining a grammar class. Explanations of criteria used to determine the class of a given grammar (0-3). The examples included in the database of solved problems as well as the examples generated by the problem generator contain the grammar whose class the student is assumed to determine.

Exploring regular sentences. Determining whether or not a specified sentence belongs to a given grammar. Mastering this subtopic requires the knowledge of the preceding subtopics (the rules of defining a grammar and the rules of derivation). Solved and generated problems contain the grammar (defined in terms of the elements used to define a grammar) and the sentence whose membership of a given grammar is being determined.

Dead and unreachable symbols. Determining the set of dead and unreacheable symbols. Generated problems require the knowledge of the rules of derivation.

Uniqueness of a grammar. Determining the uniqueness of a given grammar. Solving the problems generated under this subtopic requires the knowledge of the rules of derivation. Generated problems contain the grammar defined in terms of the elements used to define the grammar whose uniqueness is being determined.

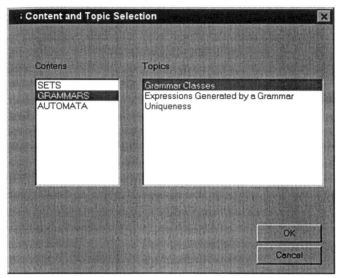

Fig. 2. Selection of contents and topics to learn about

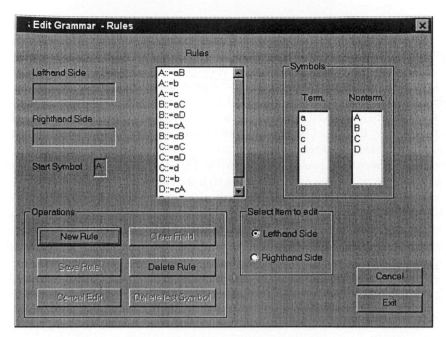

Fig. 3. Defining/editing a grammar

Finite automata. Explanations of the notions used to define finite automata and regular expressions. The problems generated at this level contain incomplete definitions of automata, and the student is expected to determine the missing defining element. The student is assumed to master the basic elements used to define automata (the set of input symbols, the set of automaton states, the function of state transition).

Sequence acceptance by the automaton. At this level the student is expected to be capable of determining, on the basis of a specified sequence and a state transition diagram, whether or not the specified sequence was accepted by the automaton.

Translating the rules of derivation into an automaton. Mastering the rules of translating a regular expression into an appropriate automaton. Problems contain the rules which the student is assumed to transform into a corresponding automaton.

Exploring determinism of automata. Determining whether or not a given automaton is deterministic. This level requires the knowledge of translating the rules of derivation into an appropriate automaton.

Automata conversion. Mastering the rules of translating a nondeterministic automaton into a deterministic one. This level requires knowing the rules of determining whether or not a given automaton is deterministic.

Minimizing an automaton. Minimizing a given automaton. The problems generated at this level contain a description of the automaton to be minimized.

Pushwodn automata. This subtopic refers to explanations of the notions used to define pushdown automata and is structured similarly to the more general subtopic 3.2.1. The problems generated at this level contain incomplete definitions of pushdown automata, and the student is expected to determine which element is missing from the definition.

3.2 Student model

Student records and models. Each tutoring session with the FLUTE system starts with student identification. Other system options cannot be used without passing this level. If the student is known, the system uses the student model stored in the user database in the form of a special object. If the student is a newcomer, the system uses an assumed student model specified in advance. Regardless of whether the user is known or new, the model is being updated during a session in accordance with the student's knowledge and behaviour. The model is saved at the end of a session.

The basic elements of student model. In the FLUTE system the student model is described in terms of a set of parameters which are calculated as simple functions of a number of quantities measured and updated automatically by the system during each session. These quantities include: the number and type of topics, subtopics and lessons passed by the student during preceding sessions, success (marks, grades) in mastering preceding lessons, the mean time needed by the student to answer a question stated by the system, the number of correct answers, the number of incorrect answers, the student's initiative during tutoring, the number of student's questions and help requests during a session, the level of knowledge and familiarity with previously mastered lessons after the expiry of a longer time period, learning continuity and frequency, the number of new examples entered into the database of solved problems by the student, etc. At the end of a session, the system presents to each student a report on his/her work. This report is created using the values of the parameters from the student model for that particular user. Privileged users are allowed to change student model elements.

3.3 Pedagogical module

The role of an educational system is to familiarize a student with a given topic as efficiently as possible taking into account the student's individual intellectual capabilities and background knowledge. One among the major tasks of such a system is to monitor the tutoring process itself. This is why great importance is attached to the pedagogical module in the FLUTE system. The main parts of FLUTE's pedagogical module are:

- a problem generator, intended for students knowledge testing
- pedagogical rules, on the basis of which the system determines the type of action to be taken at any moment
- The problem generator is described in Section 4. The elements of a pedagogical rule are as follows:
- the level (of knowledge) attained by a student at that moment, which follows from the student model and is directly associated with previously mastered lessons
- the number of trials a student is allowed to make to master a given level (notion, lesson), which depends on problem complexity and a student's current level.

FLUTE's pedagogical component determines the type of action to be taken using the results of communication with the knowledge base containing the student model and the knowledge base containing pedagogical rules. During problem generation, the pedagogical module communicates, if necessary, with the solved problems base.

3.4 Explanation module

FLUTE's explanation module is a user-oriented, rule-based system, using all the available information from the domain knowledge base (contents of lessons, objectives, topics, dependency graphs and examples), as well as from the student model, in order to answer the student's questions and provide desired explanations. Specifically, it is designed to, on request from the student:
- determine the contents of the answer/explanation;
- decide upon the explanation presentation style (level of details, hints, augmentation material to be included (e.g., illustrations, examples, pointers to related concepts and references), etc., depending on the question itself and on the relevant attributes of the student model (current knowledge, current progress, performance);
- select the knowledge model that provides the easiest explanation in cases when multiple knowledge models can be used (e.g., select what formalism to use in the explanation if something can be explained both by grammars and automata);
- compose the explanation and order its statements in a comprehensible way.

The questions the student can ask are selected from a context-sensitive menu (Why, How, What If, Related Topic, Illustration, and More Explanation).

3.5 User interface

When working with the FLUTE system, the user studies a particular lesson at every moment. In addition to the descriptions and explanations of the basic notions related to that lesson, at the level of a lesson the following options are available:
- reviewing the solved problems (from the corresponding database)
- entering a new problem by a student
- problem generation by the system
- textual help
- entering the generated problem into the problems-and-solutions database

As a student is permitted to enter and test a desired example, the system is provided with an error identification and correction mechanism. To prevent a chain of errors from occurring, each entry of an expression is immediately followed by checking for errors. Apart from checking the syntax and semantics of entered elements, at each level (in each lesson) the sequence of element entry is checked through interaction with the expert module. As an illustration, if a student wishes to test his own example from the domain of a grammar, he will not be allowed to enter the rules of derivation before entering the elements used to describe that grammar. Through such an user interface the FLUTE system yields a maximum efficiency in the tutoring process and avoids possible student frustration.

4 Problem Generation

In addition to a collection of solved problems (examples), the FLUTE system also offers the possibility of generating problems at each level and incorporating them into the problems-and-solutions database. It is possible to generate two basic types of problem:

- a problem involving an error to be identified by a student
- a problem requiring a student to solve it and enter the solution.

As far as a problem containing an error to be identified by a student is concerned, the system generates first a pseudorandom error code which carries the information about the character of the error and then the problem itself. For example, when a student is assumed to answer whether of not the definition of a grammar contains all the required elements, the system generates first the code of the error to appear in the problem (i.e. the code of the grammar defining element that will be omitted in the generated problem) and then the problem itself.

5 Session Contents

As far as a session is concerned, the designer of an intelligent tutoring system has to consider, among others, two important requirements:

- a session must proceed in a logically conceived way
- during a session, the system must be as cooperative as possible with respect to a student's previous knowledge level and pace of learning.

In educational systems without built-in intelligence the sessions were static in character, while the addition of intelligent components makes the tutoring systems very dynamic. How a session proceeds depends primarily on the student's previous knowledge, pace of learning, and application of knowledge.

One of the main tasks of the tutoring system is to identify a situation when a student is unable to master a particular level of knowledge. In such a case the system responds by:

- displaying already solved examples to allow the student to understand the difficulties that have come up in solving the problem
- generating simpler examples to aid the student in solving the problem
- if necessary, by returning the student to a preceding level

The FLUTE system's intelligent reaction is reflected in identifying situations when the student is unable to solve a particular problem on the basis of his theoretical knowledge. In such a case, the control structure (pedagogical module and student modeling module) assumes complete control over how the session will proceed further. A mechanism aiding the student to overcome a current problem is activated at that moment. The system is capable of generating tasks based on fundamental postulates, thus allowing the student to master the basics of the subject he is learning. If the student is unable to solve the generated tasks or a task he has tried to solve once, the system will return him to a preceding level. For example, if the student should determine whether a particular sentence belongs to given grammar, and if he fails in a few attempts, the system will generate tasks in which the student can practice derivation rules.

As the educational process may proceed in several stages, it is necessary to store data about a particular user relating to the student model. If that user is included in the user-database, the model from the user database will be referenced; if not, the system will start the creation of the explicit student model based on data given by the student himself.

At any moment during a session the user can obtain on-line information about a particular topic or concept from the domain of formal languages and automata. The system offers the user an aid in the form of:

- textual information about particular concepts
- solved examples relating to particular areas from the system's knowledge domain

The knowledge from the domain of formal languages and automata built in the system is organized hierarchically, which means that the educational process starts with basic examples and proceeds with more complex ones thus providing continuity of learning.

As this system substitutes a teacher, throughout a session it evaluates the student's knowledge, updates the student model and determines how the session will proceed further.

6 Conclusions

The FLUTE intelligent system is intended to offer the capabilities for not only acquiring elementary knowledge of the field of formal languages and automata but also upgrading this knowledge. The field is very important, because it provides a background for further studies of compilers.

The FLUTE system offers the following advantages:
- it provides an easy control of the tutoring process and easy verification of the knowledge he acquired
- the system behavior during a session varies in response to the student's learning pace, which provides maximum individualization of the educational process
- in addition to solved examples, the system is capable of generating new ones
- the number of tutoring systems for formal languages and automata is small, which adds to the importance of the FLUTE system
- its modular organization makes the system easy to expand
- the software and hardware used for the system implementation guarantee further development with minimum effort

One among the important properties of the FLUTE system is its expandability. Some of the aspects of future development of the system are as follows:
- adding new solved examples to the database
- incorporating a number of student modeling variants
- incorporating a module for the statistical evaluation of the system effectiveness
- expanding the knowledge base to support tutoring in the field of compilers

References

1. Vassileva, J.: An Architecture and Methodology for Creating a Domain-Independent, Plan-Based Intelligent Tutoring System. Edu. and Training Techn. Int. 27 (1990) 386-397
2. Anderson, J.R., Skwarecki, E.: The Automated Tutoring of Introductory Computer Programming. CACM, 29 (1986) 842-849
3. Wenger, E.: Artificial Intelligence and Tutoring Systems. M. Kaufmann, Los Altos (1987)
4. Devedzic, V., Jerinic, Lj.: Knowledge Representation for Intelligent Tutoring Systems: The GET-BITS Model. In: du Boulay, B., Mizoguchi, R. (eds.): Artificial Intelligence in Education", IOS Press, Amsterdam / OHM Ohmsha, Tokyo (1997) 63-70
5. Devedzic, V., Debenham, J.: An Intelligent Tutoring System for Teaching Formal Languages. In: Goettl, B.R., Halff, H.M., Redfield, C.L., Shute, V.J. (eds.): Lecture Notes in Computer Science, Vol. 1452. Springer-Verlag, New York (1998)
6. Szyperski, C.: Component Software: Beyond Object-Oriented Programming. ACM Press/Addison-Wesley, NY/Reading, MA (1998)

Leveraging a Task-Specific Approach for Intelligent Tutoring System Generation: Comparing the Generic Tasks and KADS Frameworks

Eman El-Sheikh and Jon Sticklen

Intelligent Systems Laboratory
Computer Science & Engineering Department, Michigan State University
3115 Engineering Building, East Lansing, MI 48824 USA
{elsheikh, sticklen}@cse.msu.edu

Abstract. There is a need for easier, more cost-effective means of developing intelligent tutoring systems (ITSs). A novel and advantageous solution to this problem is proposed here: the development of task-specific ITS shells that can *generate* tutoring systems for different domains within a given class of tasks. Task-specific authoring shells offer flexibility in generating ITSs for different domains, while still being powerful enough to build knowledgeable tutors. Two widely adopted task-specific methodologies, GTs and KADS, are examined and compared in the context of automatic generation of intelligent tutoring systems.

1 Introduction

In the last two decades, intelligent tutoring systems (ITSs) were proven to be highly effective as learning aides, and numerous ITS research and development efforts were initiated. However, few tutoring systems have made the transition to the commercial market so far. One of the main reasons for this failure to deliver is that the development of ITSs is difficult, time-consuming, and costly. Thus, there is a need for easier more cost-effective means of developing tutoring systems.

A novel solution to this problem is the development of a task-specific ITS shell that can automatically generate tutoring systems for different domains within a class of tasks. Task-specific authoring shells offer flexibility in generating ITSs for different domains, while still providing sufficient semantic richness to build effective tutors. The method proposed by our research follows the generic task (GT) approach for knowledge-based systems development. It involves the development of an ITS shell that can interact with any generic task-based expert system to produce a tutoring system for the domain knowledge represented in that system. Another task-specific approach is the Knowledge Acquisition and Documentation System (KADS). The focus of this report is two-fold: first, to investigate the suitability of the task-specific methodology for tutoring system generation, and second, to compare and contrast the two approaches, GTs and KADS, for this purpose.

2 The Need for Efficient ITS Development Techniques

2.1 What is an ITS?

Intelligent tutoring systems are computer-based instructional systems that attempt to determine information about a student's learning status, and use that information to dynamically adapt the instruction to fit the student's needs [10]. ITSs are also commonly referred to as *knowledge-based tutors*, because they are modular, and have separate knowledge bases for domain knowledge (which specify what to teach) and instructional strategies (which specify how to teach). The basic architecture of an ITS, shown in figure 1, consists of four main components: the student model, the pedagogical module, the expert model, and the communication module. ITS design is founded on two fundamental assumptions about learning:

- Individualized instruction by a competent tutor is far superior to classroom-style learning because both the content and style of instruction can be continuously adapted to best meet the needs of the learner.
- Students learn better in situations that closely approximate the actual situations in which they will use their knowledge, i.e., students learn best "by doing," learn via their mistakes, and by constructing knowledge in a very individualized way.

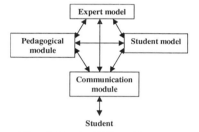

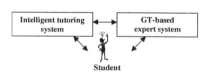

Fig. 1. Basic ITS architecture   **Fig. 2.** System-user interaction model

2.2 Learning Effectiveness of ITSs

For a long time, researchers have argued that individualized learning offers the most effective and efficient learning for most students. Intelligent tutoring systems epitomize this principle of individualized instruction. Recent studies have found that ITSs can be highly effective learning aides. Shute [7] evaluated several ITSs to judge how they live up to the two main promises of ITSs: (1) to provide more effective and efficient learning in relation to traditional instructional techniques, and (2) to reduce the range of learning outcome measures where a majority of individuals are elevated to high performance levels. Results of such studies show that tutoring systems do accelerate learning with no degradation in final outcome. For example, in one such study, students working with the Air Force electronics troubleshooting tutor for only 20 hours gained proficiency equivalent to that of trainees with 48 months of on-the-job experience. In another study, students using the LISP tutor completed programming exercises in 30% less time than those receiving traditional classroom instruction and scored 43% higher on the final exam.

2.3 Difficulty of Developing ITSs

Although ITSs are becoming more common and proving to be increasingly effective, a serious problem exists in the current methodology of developing intelligent tutoring systems. Each application is developed independently, usually from scratch, and is a time-consuming, laborious process. In one study of educators using basic tools to build an ITS, results indicated that each hour of instruction required approximately 100 person-hours of development time [6].

Another problem is that tutoring expertise is usually hard-coded into individual applications. There is very little reuse of tutoring components between applications because the field lacks a standard language for representing the knowledge, a standard interface to allow applications to access the knowledge and a set of tools to allow designers to manipulate the knowledge. In describing the state of building ITSs at the first international conference on intelligent tutoring systems, Clancey and Joerger [2] complained that "...the reality today is that the endeavor is one that only experienced programmers (or experts trained to be programmers) can accomplish. Indeed, research of the past decade has only further increased our standards of knowledge representation desirable for teaching, while the tools for constructing such programs lag far behind or are not generally available." Unfortunately, this problem remains unresolved.

Authoring tools for ITSs are not yet available, but authoring systems are commercially available for traditional computer-aided instruction and multimedia-based training. However, these systems lack the sophistication required to build "intelligent" tutors. Commercial authoring systems give instructional designers and domain experts tools to produce visually appealing and interactive screens, but do not provide a means of developing a rich and deep representation of the domain knowledge and pedagogy. Indeed, most commercial systems allow only a shallow representation of content.

2.4 ITS Shells as Cost-effective Solutions for Creating ITSs

ITS shells can speed up ITS development and reduce costs by providing essential infrastructure along with a conceptual framework within which to design and build educational systems. However, a commitment to an ITS shell's underlying philosophy entails other commitments that may or may not be suitable for application in a certain domain. Along one dimension, as systems become more general purpose, they become more complex, and generally, more difficult to use. Systems that are too specific will generally not allow enough flexibility, and thus provide little reuse. Along another dimension, shells can vary from being easy to learn and use, to offering complex and powerful ITS generation features. A system that is designed to be learned easily by novices may not provide the powerful features that experienced users need to efficiently produce tutoring systems. The tradeoffs revolve around the usability and flexibility of more general shells, versus the power that is derived from knowledge-rich shells based on explicit and detailed models. An "ideal" ITS shell should offer the following characteristics:
- Facilitate the development of pedagogically sound learning environments.

- Provide ease of use, so that authors can use it to develop tutoring systems with a minimal amount of training and computer programming skills.
- Support rapid prototyping of learning environments to allow authors to design, build, and test their systems quickly and efficiently.
- Allow the reuse of tutoring components to promote the development of flexible, cost-effective ITSs.

3 The Task-specific Paradigm for Knowledge-based Systems

Early Artificial Intelligence research followed general-purpose approaches, such as the use of rule-based systems. These first-generation systems had several problems, including rigid inferencing mechanisms and weak knowledge representation methods. Recognizing these problems, several schools of thought emerged forming what is currently known as second-generation knowledge-based systems [1, 5, 8, 11]. While these schools differ in detail, they share a common philosophy, each having deviated from the general tools approach by developing *task-specific* taxonomies of problem solving types or categories. The common assumption of these approaches is that human problem solving can be classified into categories of problems. Each category shares a common methodology, while the knowledge needed for solving individual problems differs from one problem to another.

3.1 General Description of the Task-specific Viewpoint

By grouping problems into task categories, task-specific approaches avoid the generality of first-generation techniques, while avoiding the pitfall of having to design a new technique for every problem. In other words, by carefully analyzing a class of problems, a framework can be formulated that can be applied for analyzing similar problems. Furthermore, task-specific architectures facilitate the knowledge acquisition process by focusing on high level descriptions of problem solving and not on low level implementation specifics. By implementing the results of the analysis in the form of an expert system development tool, the analysis and development of another expert system can be greatly simplified. The tools thus serve to guide the analysis of the problem.

While the different task-specific approaches share a common philosophy, they differ in detail. For example, the generic task approach is based on the hypothesis that knowledge should be represented differently according to its intended use [1]. While this mode of knowledge representation can lead to duplication in knowledge stored, it also leads to a more efficient mode of problem solving, since the knowledge in each instance is represented in a form suitable for its immediate use. On the other hand, other task-specific approaches such as the KADS approach [11] view knowledge as being independent from its intended use. Instead, knowledge is represented in an abstract neutral format that is expanded at run-time to generate the constructs needed. For an overview on other task-specific approaches, see [5, 8].

3.2 Description of the Generic Task Approach

The idea of generic tasks can be understood at one level as a semantically motivated approach for developing reusable software - in particular reusable shells for knowledge-based systems analysis and implementation [1]. Each GT is defined by a unique combination of: (1) a well-defined description of GT input and output form, (2) a description of the knowledge structure which must be followed, and (3) a description of the inference strategy utilized. To develop a system following this approach, a knowledge engineer first performs a task decomposition of the problem, which proceeds until a sub-task matches an individual generic task, or another method (e.g., a numerical simulator) is identified to perform the sub-task. The knowledge engineer then implements the identified instances of atomic GT building blocks using off-the-shelf GT shells by obtaining the appropriate domain knowledge to fill in the identified GT knowledge structure. Having a pre-enumerated set of generic tasks and corresponding knowledge engineering shells from which to choose guides the knowledge engineer during the analysis phase of system development.

Several of these atomic generic tasks are currently available, and are implemented in the Intelligent Systems Lab's toolkit. These include structured matching, hierarchical classification, and routine design.

3.3 Description of the KADS Approach

KADS is a methodology for knowledge-based systems development that originated in Europe [11]. Originally, the KADS (Knowledge Acquisition and Documentation System) project was initiated to provide support for analyzing and describing problem solving behavior in terms of generic inferences. Over time, the KADS view has evolved into a more comprehensive framework, called KADS-2, which has many layers spanning from analysis through strategies to implementation, and the acronym also came to stand for Knowledge Analysis and Design Support. The KADS methodology includes a life cycle model for task decomposition of the problem, a set of modeling languages and frameworks for describing the objects in the methodology, and a set of tools and techniques for data analysis, model construction, and knowledge representation.

In KADS, the development of a knowledge-based system is viewed as a modeling activity. The five basic principles underlying the KADS approach are: (1) the introduction of partial models as a means to cope with the complexity of the knowledge engineering process, (2) the KADS four-layer framework for modeling the required expertise, (3) the reusability of generic model components as templates supporting top-down knowledge acquisition, (4) the process of differentiating simple models into more complex ones, and (5) the importance of structure preserving transformation of models of expertise into design and implementation.

The most well-known ingredient of KADS is the four-layer model: a conceptual framework for describing the problem-solving expertise in a particular domain using several layers of abstraction. A major assumption is that generic parts of such a model of expertise can potentially be reused as templates in similar domains – these are the interpretation models. KADS distinguishes between a conceptual model of expertise independent of a particular implementation, and a design model specifying how an expertise model is operationalized with particular computational and representational techniques. This gives the knowledge engineer flexibility in specifying the required problem solving expertise, but also creates the additional task of building a system that correctly and efficiently executes the specification.

4 Leverage of a Task-specific Framework for ITS Generation

Task-specific authoring environments aim to provide an environment for developing ITSs for a class of tasks. They incorporate pre-defined notions of teaching strategies, system-learner interactions, and interface components that are intended to support a specific class of tasks rather than a single domain.

Task-specific authoring systems offer considerable flexibility, while maintaining rich semantics to build "knowledgeable" or "intelligent" tutors. They are generally easy to use, because they target a particular class of tasks, and thus can support a development environment that authors can use with minimal training. A task-specific ITS shell also supports rapid prototyping of tutoring systems since different knowledge bases can readily be plugged into the shell's domain-free expert model. In addition, they afford a high degree of reusability because they can be used to develop tutoring systems for a wide range of domains, within a class of tasks. Moreover, task-specific authoring environments are likely to be more pedagogically sound than other types of authoring environments because they try to utilize the most effective instructional and communication strategies for the class of tasks they address. Overall, the task-specific authoring approach fulfills many characteristics of an "ideal" ITS authoring shell.

5 ITS Generation Using the GT Approach

The GT development methodology offers considerable leverage in generating ITSs. The basic assumption of the GT approach is that there are basic "tasks" - problem solving strategies and corresponding knowledge representation templates - from which complex problem solving may be decomposed. ITSs can be automatically generated by developing a shell that can interact with any generic task-based expert system, and produce an intelligent tutoring system for the domain topic addressed by that system. An ITS shell for generic task-based expert systems can generate tutors for a wide range of domains, provided the domain applications can be represented as generic tasks. This approach facilitates the reuse of tutoring components for various domains, by plugging different domain knowledge bases into the shell.

The ITS architecture must be designed to understand and interact with any GT-type problem solver to produce a tutoring system for the domain addressed by the problem solver. The learner interacts with both the tutoring system (to receive instruction, feedback, and guidance), and the expert system (to solve problems and look at examples), as shown in figure 2. This interaction is the same for both GT-based and KADS-based expert systems. However, the way the interaction between the ITS shell and expert system is implemented would be different for each approach. Rather than re-implement the expert module for each domain, the ITS shell interfaces with a GT system to extract the necessary domain knowledge. This facilitates the reuse of the pedagogical module, student model, and user interface components for different domains. Linking the ITS's expert module to the problem solver deserves special consideration. Rather than encode domain knowledge explicitly, the expert module extracts and tries to make sense of the domain knowledge available in the expert system. Thus, the quality of the tutoring knowledge is affected by the knowledge representation used by the expert system. The GT methodology's strong commitment to both a semantically meaningful knowledge representation method and a structured inferencing strategy allows the extraction of well-defined tutoring knowledge. The expert module can extract three types of knowledge:

- decision-making knowledge (how the data relates to the knowledge)
- knowledge of the elements in the domain knowledge base
- knowledge of the problem solving strategy and control behavior

To make the knowledge available to the ITS shell, the expert system must understand its own knowledge representation. The structure of a generic task system must be extended to allow any such expert system to have a self-understanding of its knowledge structures and reasoning processes. A typical GT system is composed of agents, each of which has a specific goal, purpose, and plan of action. For the extraction of tutoring knowledge, each agent must be able to answer basic questions about itself using knowledge about its goal, actions, and context. The expert module of the tutoring shell uses these answers, along with an encoding of the expert system's structure to formulate domain knowledge as required by the ITS.

6 ITS Generation Using the KADS Approach

To examine how tutoring systems can be generated using the KADS framework, four main questions must be answered:

- How would ITS generation be approached from the KADS viewpoint?
- How would it be similar and different from ITS generation using the GT approach?
- What advantages, if any, would the KADS approach offer for ITS generation?
- What disadvantages, if any, would it have?

6.1 Extending KADS to Support ITS Generation

To support generation of intelligent tutoring systems, the KADS framework must be extended to allow two-way communication between the ITS shell and a KADS-based system. In one direction, the expert module of ITS shell must understand KADS task structures and interact with task models. In the other direction, KADS models and structures must support ITS functions and interactions by implementing tutor-related primitives. At the conceptual level, ITS generation can be done using the KADS methodology by: (1) extending the life cycle model to include the specification of tutoring functions and objectives at the analysis phase, and how these specifications could be transformed into a KBS that supports tutoring at the design phase; (2) extending the modeling framework to include concepts that describe tutoring requirements for use in conceptual, design and other models; and (3) augmenting the KADS tools with the techniques needed to build appropriate tutoring knowledge structures and primitives. The implementation of this support for ITS generation within KADS can be achieved by augmenting the reusable libraries of model structures with tutoring primitives. In other words, the interpretation models, which are reusable models of inference structures and task decompositions, need to be extended to answer requests from the ITS shell for tutoring functions.

The link between the ITS shell's expert model and the KADS executable model of expert problem solving is shown in figure 3. The expert model must understand the KADS framework for expertise. This is accomplished by encoding knowledge of the KADS expertise model within the ITS shell's expert model. This model contains the knowledge needed to understand and interact with the four KADS layers:

1. Strategic layer: The expert model must understand the strategies and plans of the expert system by using knowledge of the KADS strategic layer. This knowledge is used to guide the instructional processes of the ITS.
2. Task layer: Interaction between the ITS and the KADS model is needed at the task layer, so that the ITS can understand and make use of the expert system's tasks and goals. This can be done by creating goal structures for tutoring goals. Goals are specified as primitive, compound, or modality. Modality goals specify the communication between the user and ITS, and between the ITS and expert system.
3. Inference layer: At the inference level, the ITS needs to understand the inferencing primitives used by KADS conceptual models. This is achieved by encoding knowledge about the basic KADS knowledge sources and meta-classes.
4. Domain layer: In KADS, knowledge is stored in the domain layer as axiomatic structures. The expert model of the ITS shell stores knowledge about the representation format of these axiomatic structures to enable the shell to extract domain knowledge from the expert system.

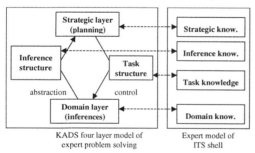

KADS four layer model of   Expert model of
expert problem solving   ITS shell

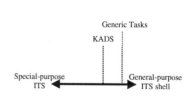

Fig. 3. ITS shell linkage to a KADS model   **Fig. 4.** ITS development spectrum

The lowest level in KADS that allows the achievement of reusability is the inference level. KADS defines interpretation models of inference structures that can be reused and combined to develop knowledge-based systems. To facilitate the construction of tutoring systems, the interaction between the ITS shell and the expert system must be done at the inference layer. In other words, the way the ITS shell interacts with a KADS-based system would be different for each inference model.

6.2   Comparison of KADS with the GT Approach

Overall, generating tutoring systems using the KADS approach is similar to the GT approach: both are done at the task level, whether by extending task structures or models. This similarity stems from their common philosophy, namely the classification of human problem solving into task-specific categories, and the reuse of such generic tasks for solving problems in different domains. The two approaches are also similar in that they both involve augmentation of the knowledge structures, task decompositions, and inference methods to support tutoring. Although similar in high-level methods, the two approaches differ in the specifics of how tutoring generation is achieved. In the KADS approach, the developer must specify the ITS extensions separately for the analysis space (conceptual model) and the design space (design model). Moreover, KADS stores domain knowledge in a task-neutral format that is independent of its intended use, unlike the GT approach that assumes that knowledge should be represented in a format suitable for the way in which it will be used.

An important distinction is that the inference primitives in KADS are more fine-grained than their GT counterparts, so the interface to the ITS shell must be specified at a lower level, equivalent to the level of internal operators in the GT methodology. The ITS framework can be specified in a more formal and concrete manner than with the GT approach, since the interactions between the ITS shell and expert system will be defined at a finer grain size. This facilitates the process of building the tutoring system architecture, since models of primitive inferences are available for the developer to use in defining the ITS-KADS interactions. However, the interaction between the ITS shell and a KADS-based system is complicated, since it must be done at the level of the inference layer. KADS supports the specification of tasks through the reuse of library-stored inference models. This implies that the ITS shell must support a model of interaction between the shell and the task for each inference model. In comparison, the GT approach specifies inferencing semantics for each class of tasks. This allows the ITS shell to represent a single task model for each task category.

Consider the scope of ITS development, as shown by the spectrum in figure 4. At one end of the spectrum, ITSs must be built from scratch for every domain, not allowing any reuse. At the other end, ITS shells are flexible enough to allow the reuse of domain knowledge and tutoring components for any domain. The GT approach falls more towards the general-purpose end of the spectrum than the KADS approach. For an ITS shell to interact with a GT-based system, it must understand and include support for each basic generic task. Such a shell can generate ITSs for any domain that can be represented as a generic task. The KADS approach falls closer to the middle of the spectrum. Although this approach does not require starting from scratch for each ITS to be built, it does require the implementation of a different ITS framework for each inference model defined in the KADS library. It is important to understand the impact of this distinction on the capability of each approach to generate ITSs. Both approaches require multiple frameworks within the ITS shell to model a knowledge-based system. The GT approach requires a framework for each base GT task and the KADS approach requires a framework for each inference model. In the GT approach, a single framework for each generic task is sufficient because a fixed set of semantics is enforced for each task, so all the associated inferencing strategies can be identified and encoded into that framework. However, in the KADS approach, generic inference models are reused to build task-based systems. These reusable models are at the inference level – one level below the task level. These models are at the level of the internal operators in the GT approach. Thus, in the KADS approach, the ITS shell must include many more finer-grained frameworks, one for each inference model, to support interaction with a task-based system.

One of the advantages of the KADS approach is its task-neutral knowledge representation format. Although such task-neutrality has not been well demonstrated in working KADS systems, the approach assumes that such task-neutral representation is possible. Domain knowledge is stored independently of its intended use and, thus, the ITS shell can share knowledge across different tasks. The advantage is the elimination of duplicate knowledge representations if the same domain is to be tutored from different task perspectives. However, this advantage could be a disadvantage since a task-neutral knowledge format does not provide a semantically rich representation for tutoring task-based domains. In other words, the quality of the tutoring knowledge would be affected by the knowledge representation used within the KADS approach. Another advantage of the KADS approach is that it provides a more formal design, development, and evaluation methodology for knowledge-based systems, which would allow the development of a more formal specification of the ITS framework.

ITS generation using the KADS approach has several disadvantages. For one, the migration from conceptual specification and design to implementation of an ITS is more challenging. The GT approach offers direct support for implementation in the form building blocks, which can be extended to include tutoring support. The implementation task is not as straightforward in the KADS framework since there is a gap between the conceptual modeling and design phases, and the implementation of a system that meets the required specifications. Another disadvantage alluded to above stems from the general task-neutral representation of domain knowledge adopted. The issue is how usable that knowledge is for tutoring from a task-specific viewpoint, and whether it needs to be represented in a more suitable format.

7   Conclusion

There is a need for easier, more cost-effective means of developing intelligent tutoring systems. This report suggests that a novel and advantageous solution to this problem is the development of a task-specific ITS shell that can generate tutoring systems for different domains within a class of tasks. Task-specific authoring shells offer flexibility in generating ITSs for different domains, while still being powerful enough to build knowledgeable tutors. We are currently engaged on a research path to develop the ability to automatically generate ITSs given a generic task expert system [3, 4]; the ITS generated covers the knowledge domain of the given expert system. We argue that a task-specific framework for ITS generation is highly leveraged, and this led us to pose the comparative question: which of the task-specific approaches will be of most utility in developing this ITS generation capability?

Two widely adopted task-specific methodologies, GTs and KADS, were selected for comparison. Although there are other candidates that might be compared, the

GT/KADS comparison proves illustrative. The analysis reveals that both frameworks can support the generation of ITSs for different domains. The salient result is that a GT framework is expected to be of greater leverage in generating ITSs because of the larger problem solving *granularity* of the GT approach compared to the KADS approach (see [9] for a complete treatment of this issue). In a sense, because of stronger semantic commitments taken following the GT approach, a system developer is strongly tied to a backing theory of problem solving. Yet, once implemented, the stronger commitments made in a GT-based system enable a *general* ITS overlay to be developed to generate ITSs on an automatic and largely knowledge-free basis.

The comparison here has been from a theoretical viewpoint. Our current research track includes the development of an ITS shell for GT-based expert systems. An experimental question we have posed, and will answer over the next several years is the following: to what extent is the hypothesized knowledge-free ITS generation within a GT framework possible? The most important question we hope to answer is the extent to which this generation is in fact knowledge-free.

References

1. Chandrasekaran, B. (1986). "Generic tasks in knowledge-based reasoning: high-level building blocks for expert system design." IEEE Expert 1(3): 23-30.
2. Clancey, W. and K. Joerger (1988). A Practical Authoring Shell for Apprenticeship Learning. Proc. of ITS'88: First International Conference on ITSs, Montreal, Canada.
3. El-Sheikh, E. and J. Sticklen (1998). A Framework for Developing Intelligent Tutoring Systems Incorporating Reusability. IEA-98-AIE: 11th International Conference on Industrial and Engineering Applications of Artificial Intelligence and Expert Systems, Benicassim, Castellon, Spain, Springer-Verlag (Lecture Notes in Artificial Intelligence, vol. 1415).
4. El-Sheikh, E. and J. Sticklen (1998). Using a Functional Approach to Model Learning Environments. Model Based Reasoning for Intelligent Education Environments, ECAI'98: European Conference on Artificial Intelligence, Brighton, UK.
5. McDermott, J. (1988). Preliminary Steps Towards a Taxonomy of Problem-Solving Methods. Automating Knowledge Acquisition for Expert Systems. S. Marcus. Boston, MA, Kluwer Academic Publishers.
6. Murray, T. (1998). "Authoring Knowledge-Based Tutors: Tools for Content, Instructional Strategy, Student Model, and Interface Design." The Journal of the Learning Sciences 7(1): 5-64.
7. Shute, V. J. and J. W. Regian (1990). Rose garden promises of intelligent tutoring systems: Blossom or thorn? Proceedings from the Space Operations, Applications and Research Symposium, Albuquerque, NM.
8. Steels, L. (1990). "Components of Expertise." AI Magazine 11(2): 28-49.
9. Sticklen, J., and Wallingford, E. (1992). On the Relationship between Knowledge-based Systems Theory and Application Programs: Leveraging Task Specific Approaches. First International Conference on Expert Systems and Development, Cairo, Egypt.
10. Wenger, E. (1987). Artificial Intelligence and Tutoring Systems: Computational and Cognitive Approaches to the Communication of Knowledge. Los Altos, CA, Morgan Kaufmann Publishers, Inc.
11. Wielinga, B., A. Schreiber, et al. (1992). "KADS: A Modeling Approach to Knowledge Engineering." Knowledge Acquisition 4(1): 5-54.

An Object-Oriented Robot Model and Its Integration into Flexible Manufacturing Systems

Christian Schäfer[1] and Omar López[1]

Dpto. de Automática (DISAM), Universidad Politécnica de Madrid,
Jose Gutierrez Abascal, 2, 28 006 Madrid, Spain
{chris, olopez}@disam.upm.es, schafer@ira.uka.de

Abstract. The paper describes an object-oriented, general model of a robot on one side and of a flexible manufacturing cell on the other side. The modeling aims at an improvement of information exchange between these commonly separated levels of the CIM hierarchy. Interactions between both levels are discussed with special emphasis on the integration of information about the physical properties of the robot into the flexible manufacturing cell. By this means a high adaptiveness on changing manufacturing situations is achieved. In addition the open architecture of the model facilitates the integration of future methods in manufacturing process planning and execution. The paper closes with two practical examples that illustrate the information exchange between the robot and the manufacturing cell level in concrete scenarios.

1 Introduction

A Flexible Manufacturing System (FMS) comprises highly automated, autonomous manufacturing resources (robots, buffers etc.) within a complex control architecture. Each resource –as stand-alone machine– has its own control; in an FMS environment the resource control is moreover linked to a higher control. Liu [7] defines an object-oriented framework that distinguishes and categorizes physical objects from control objects which are categorized separately. In Liu's model pairs of physical-control classes are defined. Traditional architectures of integration [9, 2] consider the need to integrate the resource's physical properties but little has been done to actually model this low-level information [8, 5] and to implement it into a higher control level. Moreover mechanical simulators such as ADAMS are generally not sufficiently open to be integrated into the manufacturing environment.

Figure 1 presents our modification of the CIM architecture provided by Siemens [9]. The left part of the figure shows the six levels into which a manufacturing company is divided. The central block defines the functions done at each level; in the same central block the arrows represent the direct information flows between the levels.

To focus the subject of this paper, we argue that manufacturing execution must be improved by integrating information generated from the lowest level of the enterprise where some physical parameters of the manufacturing resource

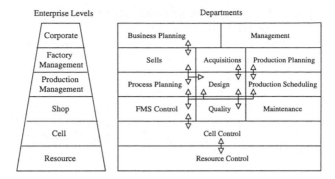

Fig. 1. FMS within CIM

directly affect the outcome of manufacturing execution. Manufacturing execution involves the three lower levels of the CIM hierarchy: Shop, cell and resource level. To integrate the physical parameters of the resources we decided to model an FMS as a composition of manufacturing resources (resources for short) associated with their control mechanisms. The model is developed using the object-oriented Unified Modeling Language UML [6]. In the model several *resource* classes are interrelated with *control* classes. The *control* classes provide methods to interface one level with the other.

2 Resource Control within a Flexible Manufacturing System

The FMS is governed from shop level. The shop level receives as input a production order and the process plan associated with that order. At this level some tasks must be controlled: Transport control, buffer control, storage control, tool control and process plan control. The shop controller decides in which cell a specific operation has to be executed.

Then the shop controller sends a command to the ASRS controller to put in place the ASRS and retrieve material or to storage a product. In the following the FMS controller informs the control of the chosen cell of the activities it has to perform. Conversely a cell controller informs the shop controller when a task is completed. The shop controller tells a cell buffer to stop a particular pallet. In the moment in which the target pallet arrives at the cell, the buffer tells the cell controller to move its robot manipulator. The cell controller sends a message to the manipulator controller. The manipulator controller is now activated to give the manipulator the coordinate information to go to. The manipulator's sensor tell the manipulator controller how the manipulator is performing the task. When the manipulator finishes handling the object, the manipulator controller sends an end_of_task message to the cell controller which in turn sends a message to the shop controller to be ready to receive new orders.

Much work has been done to model with object-oriented techniques the control architecture in a Flexible Manufacturing System. An extensive description of how different manufacturing resources interact with their own control and with the FMS control can be found in [10, 1, 3, 11]. There is much debate over the required number of distinct levels, however three natural levels are commonly identified. The *manufacturing resource* level, the *cell* level and the *shop* level as stated before. Each physical equipment with its own control defines the manufacturing resource level. The composition and layout of the manufacturing resources define the cell level. Processing and storage machines that share the services of material handling machines and buffer storage space together form a cell. The cell level has a centralized control in charge of defining the behavior of the resources. This behavior depends strongly on the physical properties of its resources and it is therefore at this point where the integration of information becomes essential to improve manufacturing performance.

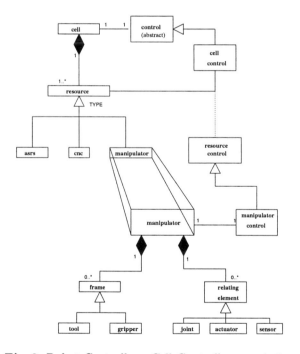

Fig. 2. Robot Controller - Cell Controller association

Figure 2 represents the proposed UML model. Although the shop level is omitted the model is coherent with the mentioned works. A Flexible Manufacturing System is composed of a number of cells, each composed of resources (robots, CNC machines, buffers, etc.). Conceptually and physically a control must exist over the hole system; this situation is modeled as an abstract class *control* with its sub class *cell control*. The class *control* manages coordination information for manufacturing execution.

For the manufacturing resource level, Smith [10] states that a resource has the ability to autonomously perform an action as indicated by the cell controller. The resource controller monitors these actions. Thus a relationship between the *resource* class and the *cell control* class exists. The association between these two classes gives birth to another association class, in this case the *resource control* class. The *resource control* class is in charge of commanding the resource according to the information received from its sensors.

Our model highlights the importance of considering the *resource* and the *resource control* as separate classes. In this way the possibility to capture a number of physical parameters is given. Taking into account that a manufacturing resource is composed of several objects, it is obvious that changing one of these objects results in a performance change of the whole resource. Finally the performance of the whole cell is touched.

In figure 2 a robot manipulator is modeled as a composition – in UML terminology – of different devices: sensors, grippers, joints, etc. . A more detailed description of these classes is presented in the next section.

A robot manipulator has attributes such as degree of freedom, robot type, speed range, force range, etc. which depend on the composition of the robot manipulator. The model makes a clear distinction between the robot manipulator as a mechanical artifact and its control as separate entities. The responsibility of the *manipulator control* class over the *manipulator* class is represented with a navigability from *manipulator control* to *manipulator*. From the cell level, a manipulator is an entity with capabilities to perform certain actions depending on the manipulator's characteristics. For example a cell controller must adapt (although not explicitly) the force exerted by the manipulator for pressing a cylinder into a workpiece according to design specifications and be confident the manipulator will maintain this desired feature. The manipulator's problem will be to 'tell' his components to do the job in the way the cell controller commands.

But changes are common and they do affect. When the cell controller decides to change operations (perhaps to complete a higher priority order decided at production scheduling) and to continue later with the initial operations, the parameters of the manipulator (the force exerted, the mass, etc.) may be modified. If the cell controller is not aware of these changes the process will be disadjusted. Integrating the physical properties inside the *manipulator control* class and defining methods to manipulate them in the *cell control* class is the best way to establish communication between one-another. Manufacturing execution is aware of modifications and performance is improved.

So having the *manipulator control* class as a sub class of the *resource control* class and in turn associating the *resource* class with the *cell control* class, the capability to integrate the physical properties of the resource, in our case the manipulator, is developed.

3   The model of the robot manipulator

The modeling of the robot manipulator is also based on a consequent object-oriented approach. This guarantees on the one hand a good integration into the manufacturing cell architecture as a *resource* through an easy-to-use, public interface. On the other hand the object-oriented approach facilitates the inclusion of new knowledge within the robot level. New knowledge can consist in information about constructive changes of the robot design as well as the application of new control strategies, numerical calculation methods etc. . The main advantages of the proposed robot model are that the management of the manufacturing cell has direct access to the real physical properties of the robot and its control. Thus production planning becomes more reliable and the management can effectively respond to modifications of the robot and/or its environment.

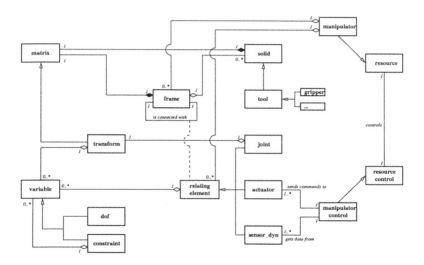

Fig. 3. Robot model in UML

Conceptually the robot manipulator is modeled separating as much as possible three key characteristics: The robot's mechanical properties, its control and the underlying mathematical tools. Figure 3 shows the developed class diagram for the robot manipulator using the Unified Modeling Language [6]. All class names are boldface in the figure and italic in the text. The classes are arranged from left to right in such a manner that their contents change from mathematical to physical signification. This division follows the general scientific principle that physical effects are by nature independent of the mathematical tools used to describe them.

The explanation of the model details starts on the outer right side of the figure 3. Here the model consists of the subclasses *resource* and *resource control* which are part of the resource level of the manufacturing cell. They provide for a standard integration and coordination of various machines inside the manufacturing cell (see figure 2). These classes and their links to the manufacturing *cell control* as well as the upper levels of the integrated manufacturing environment have already been dealt with in detail in the preceding section.

The physical characteristics and aspects of the robot are contained in the *manipulator* class. In this context the *manipulator* is defined as a stationary mechanism that positions and/or orients a tool to interact with the environment. A typical example of such a manipulator with its components is shown in figure 4. Various implementation attempts with different class schemas have lead to the conclusion that it is very convenient to divide the components of the manipulator into the two groups, *frames* and *relating elements*. These two classes can therefore be considered as the backbone of the robot model – they interconnect mathematical and physical properties.

A *frame* represents a coordinate system in which one or more rigid bodies are fixed. The rigid bodies carry masses and are referred to as *solids*. Generally but not necessarily for each link of the robot a single *frame* is chosen and the *solids* correspond to the material that forms the link. Inside each *frame* the *solids* are immovable and hence for each *frame* a constant mass and constant moments of inertia can be calculated. A special type of *solid* is a *tool* and in particular a *gripper*. Unlike the *solids*, the *tools* have the ability to interact with the robot environment e.g. to hold an object. In this case the mass and moments of inertia are recalculated for the *frame* to which the gripper belongs.

The second group of manipulator components are *relating elements*. They have in common, that they connect two (and only two) *frames*. Typical representatives are the sub-types *joints*, *actuators* and *sensors*. Note that the *relating elements* can connect arbitrary *frames* and not only *frames* that follow in the kinematic order (compare figure 4). Moreover one should be aware that in the proposed model the *relating elements* do not carry masses – the masses have to be included in the corresponding *frames*.

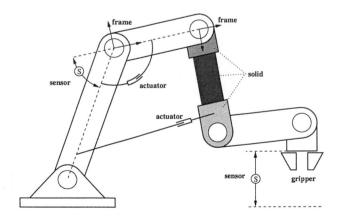

Fig. 4. Robot and its components

A *joint* defines the kinematic restrictions of one *frame* with respect to another. From a mathematical point of view this restriction is expressed by a *transform* matrix, that can include various *variables*. As an example may serve a rotary joint that is expressed by a Denavit-Hartenberg transformation matrix in which the angle of rotation is the only variable. Joints with more that one degree of freedom e.g. ball-socket joints are admitted too.

An *actuator* is a *relating element* that relates two *frames* by an external force or torque. The amount of force or torque is characterized by a *variable*. For robots usually rotary actuators that are aligned with the *joint* axes are applied. Nevertheless the model foresees linear actuation and even the possibility is given to move a link with several *actuators*.

Finally a *sensor* is defined as a manipulator element that measures relations between two *frames*. In the context of the proposed robot model the measured quantities are limited to dynamic quantities like distance, orientation, velocity, acceleration as well as force and torque. Commonly the *sensors* are aligned with the actuators or joints but independent sensors (e.g. that measure the distance between floor and gripper as seen in figure 4) are admitted too.

The modular design of the robot model allows to exchange or to add element like *solids*, *joints*, *actuators* or *sensors* in whatever way. Thus existing robot models can easily be adapted or new ones can be composed.

Besides the manipulator's mechanical components, an important part of the robot is its control. As shown in figure 3 the *manipulator control* on the one hand gets data from the *sensors* and on the other hand sends commands to the *actuators*. It governs the *variables* that these components include. The model supports low level control schemas like reactive control or tele-control as well as plan based control schemas that may require knowledge about the structure of the whole robot.

Now it remains to describe the mathematical part of the model which is found on the left side of figure 3. In the concept of the model the mathematical methods are treated apart from the the physical properties of the robot. As a consequence of this philosophy the mathematics become exchangeable within the model and the application of new mathematical methods is facilitated. A *matrix* library provides for most of the vector and matrix calculus, whereas the *variables* form a class apart. The *variables* are divided into two sub-groups: degrees-of-freedom (*dof*) and *constraints*. The *dof* are independent variables that can be set by the *manipulator control*. *Constraints* are either variables that can be calculated by use of the *dof* or conditions that have to be fulfilled like ranges of motion etc. . To illustrate the interplay between the mathematical part of the model take a look at the class *transform*. This class provides for a normalized transformation matrix in a *joint*. Depending on the degree of freedom of the joint it can include one or more *variables*. The number of parameters to derive the matrix depends on the applied method. So far the Denavit-Hartenberg method, Euler-angles and screw theory are implemented but the model structure is open to integrate table lookup or other methods to compute the transformation matrix.

4 Examples of information integration

The examples illustrate the integrating approach of the presented manufacturing cell architecture in two concrete scenarios. In both cases the relation between the detailed physical properties of the manufacturing resource and manufacturing execution is illustrated and immediate implications analyzed. The examples are simplified as much as possible to improve the understanding of the developed architecture. Nevertheless the architecture will hold for most complex scenarios with a high number of cell resources too.

The first example is about a single robot within a flexible manufacturing cell. The robot has n DOF and is non-redundant; it is composed only of rotational joints and all joints are directly actuated. Gravity, friction etc. are neglected to simplify the problem. The scenario foresees that the current task of the robot is to press cylinders into work pieces. The execution of the pressing operation requires a directed force $F_{\text{req}} = (F_{\text{req},x}, F_{\text{req},y}, F_{\text{req},z})^T$ (expressed in world space coordinates). Assume that one of the robot motors gets broken and is replaced by the maintenance with a motor of different torque.

On robot level this implies that the sequence of operations that are executed by the robot inside the manufacturing cell are checked. During the pressing operation the robot configuration is given by $\theta = (\theta_1, \theta_2, ..., \theta_n)^T$, where θ_i is the joint angle at the i-th joint. The relation between the required motor torques $\tau_{\text{req}} = (M_{\text{req},1}, M_{\text{req},2}, ..., M_{\text{req},n})^T$ and the required pressing force is given by

$$\tau_{\text{req}} = \mathbf{J}^T (F_{\text{req},x}, F_{\text{req},y}, F_{\text{req},z}, 0, 0, 0)^T \qquad (1)$$

where $\mathbf{J}$ is the Jacobian of the robot at the point θ. The replaced joint motor is now checked to assure if it can still provide for the necessary corresponding torque M_{req}. If the check results unsatisfactory the production operations have to be modified.

From a point of view of the manufacturing execution level the important fact is that the quality of the manufacturing process is highly correlated to the union force of the cylinder in the work piece. Manufacturing execution has been adjusted (although not necessarily on purpose) to the parameters of the previous robot motor and the process has been statistically controlled. But as has been demonstrated in the previous analysis the force exerted by the new robot motor is different which means the actual force exerted is different too. If the force of the new robot motor is beyond the limits, upper or lower, and this fact is not taken into account, the manufacturing process will no longer be controlled and as a consequence after manufacturing execution, the population percentage of finished products that will be out of limits will increase. The integration of information about the up-to-date physical properties of the robot is now essential to determine the source of the deviation.

In a second example the importance of information exchange between the robot level and the manufacturing cell level is outlined for cell redesign. Suppose a flexible manufacturing cell with a single robot and two buffers. Again the robot has only rotational joints and all joints are directly actuated. The manufacturing cell serves for a whole variety of different products. Two buffers supply the robot with the products and support the pieces while the robot is manipulating them. Note that all the work pieces are processed by the robot at the same positions in space which are determined by the position of the buffers. To improve the productivity of the flexible manufacturing cell, the robot is now equipped with more powerful motors.

At robot level this implies that the performance of the robot changes within its workspace. Regions in the workspace where the robot can exert high forces on the work piece have grown and might permit an optimization of the production process. A detailed analysis of the robot's workspace is carried out. Analysis criteria are not only traditional specifications like reachability but also manipulability [12, 4], precision etc. . All these criteria require a detailed information about all the physical properties of the robot. The resulting statements are not only about the robot as a whole (global) but the performance of the robot at each point inside its workspace is quantified. In this way, taking into account actual and estimated future manufacturing operations, the possible work points of the robot are determined and the buffers can be placed correspondingly.

At manufacturing execution level two implications are considered. The robot's movements were adjusted to guarantee that the robot reaches the same position to handle the pieces. The analysis tells us the robot workspace has increased and therefore the real positions the robot reaches are offset accordingly. Manufacturing execution must be stopped to reprogram the robot's positions. The second consequence is positive. As the space where the robot can exert high forces increased another buffer can be added to the original configuration of the cell having a bigger work area with the same (adjusted) quality parameters, reducing the time to finish more products. Again the importance to base the manufacturing cell level on physical properties of its resources (in this case the robot) is stressed.

5 Conclusions

One way to improve manufacturing execution as well as planning in flexible manufacturing systems is to integrate information of the lower levels of the CIM hierarchy. In particular the incorporation of knowledge about the physical properties of the manufacturing resources (e.g. robots) is stressed because the resources directly influence the outcome of the manufacturing execution. For this purpose an object-oriented architecture of the lower levels of the CIM hierarchy (shop, cell and resource level) has been developed and complemented by a detailed model for robot manipulators.

On the one hand the flexible manufacturing system is modeled as a composition of a number of cells, each composed of resources (robots, CNC machines, buffers etc.). Conceptually for each resource two classes are introduced: A resource class and its corresponding resource control class. On the other hand the robots are modeled based on a separation of three key characteristics: The robot's mechanical properties, its control and the underlying mathematical tools.

829

Within the mechanical part of the model it is furthermore necessary to distinguish kinematic passive elements (e.g. solids) from kinematic active elements (e.g. joints, actuators and sensors)

In the following the information flow inside the model architecture is analyzed and a high adaptiveness on changing manufacturing situations is shown. The principles of the presented approach are now exposed and further work will concentrate on the extension of the model, in particular on the modeling of other manufacturing resources.

Acknowledgements

Christian Schäfer likes to thank Prof. E. A. Puente and Prof. R. Dillmann for their organisational and scientific support, as well as Dr. F. Matía for suggestions concerning UML. Moreover Christian Schäfer acknowledges the financial contribution of the European Commission through an individual 'Marie Curie Research Training Grand'.

Omar Lopez thanks the economical support given by the Secretaria de Educacion Publica and the Universidad Autonoma del Estado de Hidalgo, Mexico. He also thanks Dr. Francisco Sastron at DISAM without whom this paper could not have been possible.

References

1. Q. Barakat and P. Banerjee. Application of object-oriented design concepts in CIM systems layout. *Int. J. Computer Integrated Manufacturing*, 8(5):315 – 326, 1995.
2. R. Bernhardt, R. Dillmann, K. Hörmann, and K. Tierney, editors. *Integration of Robots into CIM*. Chapman and Hall, London, 1992.
3. Sang K. Cha and Jang H. Park. An object-oriented model for FMS control. *Journal of intelligent manufacturing*, 7:387 – 391, 1996.
4. S.L. Chiu. Task compatibility of manipulator postures. *The International Journal of Robotics Research*, 7(5), 1988.
5. J.J. Craigh. *Introduction to Robotics*. Addison-Wesley Publishing Company, Massachusets, 1989.
6. Martin Fowler. *UML distilled. Applying the standard object modeling language*. Addison – Wesley, 1997.
7. Chih-Ming Liu. An object-oriented analysis and design method for shop floor control systems. *International Journal of Computer Integrated Manufacturing*, 11(5):379 – 400, 1998.
8. R. Paul. *Robot Manipulators*. MIT Press, Cambridge, 1982.
9. Siemens. *CIM: A management perspective*. Siemens Aktiengesellschaft, Berlin, Germany, 1990.
10. Jeffrey S. Smith and Sanjay B. Joshi. A shop floor controller class for computer integrated manufacturing. *Int. J. Computer Integrated Manufacturing*, 8(5):327 – 339, 1995.
11. Omar A. Suarex. Standard based framework for the development of manufacturing control systems. *International Journal of Computer Integrated Manufacturing*, 11(5):401 – 415, 1998.
12. T. Yoshikawa. Manipulability of robotic mechanisms. *The International Journal of Robotic Research*, 4:1004–1009, 1985.

The Economic Benefits of Internet-Based Business Operations in Manufacturing

Ergun Gide and Fawzy Soliman

School of Management,
University of Technology, Sydney, P.O. Box 123, Broadway, Sydney, NSW, 2007,
AUSTRALIA.
Ergun.Gide@uts.edu.au, Fawzy.Soliman@uts.edu.au

1 Introduction

Information Technology, and in particulay the Internet Technology, has been developing very rapidly and as such it is difficult to predict the level and extent of Internet usgae in business operations. However, expert predictions show that Internet-Based Business Operations already started to change the way of conducting business. The early the firms adopt the Internet-Based Business Operations, the more likely they will survive and compete with their rivals (Cronin, 1996a).

Business Operations over the Internet are very much in their infancy. They are rapidly becoming the new method to conduct business and to interact with customers, suppliers and partners. Electronic Business Operations cover many aspects of buying/selling relationships ans also many operations within manufacturing.

Electronic trading opportunities offer businesses the chance to compete at an international level. These Electronic Trading Opportunities, are being expanded to Web sites and many trading forums have emerged.

Many reliable surveys show that over the next ten years, the growth of Internet-Based E-Commerce will outstrip the growth of traditional commerce (Soliman and Gide, 1997). It is the commercialisation of the Internet that is leading the way to this remarkable growth in Electronic Activities. The Internet serves as a foundation for all of these new opportunities in commerce.

There has been a phenomenal growth in commercial presence on the Internet in recent times. In the last 2 years the commercial domain registrations of the entire Internet have grown to represent some 85% of all organisations. This effectively kills the myth that the Internet is an academic and research playground. Facts and figures from industry analysis show that (Gide and Soliman, 1997a):

1. Internet-Based E-Commerce is expected to reach $150 billion by the year 2000 and more than $1 trillion by the year 2010. Sales generated via the Web have

grown from $17.6 million in 1994 to nearly $400 million in 1995 (a growth rate of over 2100%). The number of sites using the Internet for product transactions has increased from 14% in 1995 to 34% in 1996 and to a projected increase of 44% in the next 3 years.

2. The Internet has reduced the number of letters, voice calls and faxes around the globe. Thirty per cent of Internet users in one survey stated that Internet usage had resulted in new business opportunities and 43% said that it has increased productivity (Soliman, 1998).

According to the Internet marketing research firm (ActivMedia, 1997), projections indicate that global Web sales through 2002 could total $1.5 trillion, or about 3% of combined Gross Domestic Product (GDP) for all countries worldwide. The study tracked eight Web business segments: manufacturing, computers and software, business and professional, consumer, travel, investment/finance, publishing, and real estate. In addition the market research firm Paul Kagan and Associates released 10-year revenue projections for the interactive media industry, showing that in the year 2007, the Internet-related income is expected to be $46 billion, having risen from a projected $11.1 billion for 1997. Electronic Commerce, revenue is expected to increase from $0.9 billion in 1997 to $11.7 billion over the next 10 years.

Books and computer hardware and software are the items most people purchase via the Web, according to data from the most recent study of Internet demographics by Nielsen Media Research and Industry Trade Association CommerceNet. The study shows that 5.6 million people have purchased books online, while 4.4 million people have purchased hardware, and 4 million people have purchased software via the Internet.

According to Nielsen Media Research and CommerceNet, 78 million people used the Web during the first six months of 1998, and 20 million of those users made purchases via the Web. The following are the highlights of shopping and purchasing activities from a recent study (Gide and Soliman, 1998):

- 48 million Web shoppers - increase of 37% from September 1997.
- 20 million Web purchasers - increase of 100% from September 1997.
- 71% of Web purchasers are men, 29% are women - unchanged from September 1997.
- Women represent 36% of all online book buyers and 12% of all online computer hardware buyers.
- Among persons age 16-24, the top items purchased on the Web are books, CDs/cassettes/videos, and clothing.
- Among persons 50 years and older, the top items purchased on the Web are books, software and computer hardware.
- Consumers under the age of 35 represent 65% of all persons buying clothing on the Web, and 64% of all persons buying CDs/cassettes/videos.

3 The Internet And Electronic Market

The phenomenal predictions of the size of the Internet market should be interpreted with some other factors in mind. There is scant evidence that the Internet is actually creating new sales. Certainly, the Internet is beginning to generate new sales channels, especially for products and services, which can be delivered digitally over the net. And there is no doubt that bank-assumed risk from credit card transactions through SET processes will accelerate traditional retail sales over the Internet. But these sales are still generally no more than sales substitution, or sales that would previously have been made by personal visits, mail order or the like.

Data on Web shopping and purchasing taken from a Nielsen Media Research (CommerceNet) study on Internet demographics released in August 1998 is shown in Table 1 below.

Table 1: Top Items Purchased on the Web, June, 1998 vs. September, 1997

Items Purchased	June, 1998 (million people)	September, 1997 (million people)
Books	5.6	2.3
Computer Hardware	4.4	2.0
Computer Software	4.0	2.8
Travel (airline tickets, hotel & car reservations)	2.8	1.2
Clothing	2.7	0.9

Some authorities are already claiming that the main benefit of the Internet to-date is better customer service. However, there is a wide debate about where future investments will be made, that is: a) to support business-to-business processes, or b) to back office processes. The technologies and products which will enable businesses to do business with each other over the Internet is generally agreed to be attracting between 5 and 8 times the near-term future investment.

4 Key Values Of Electronic Business Operations

There are various types of key measurements that must be tracked prior to embarking on a full implementation. Some of the important key elements to measure business value are:

➢ **Improving customer service:** Providing customers self-access to their accounts, transactions and orders, is a valuable service. The level of satisfaction for those customers interacting electronically will undoubtedly rise.
➢ **Reducing costs:** The most basic cost reductions could be related to publishing costs, which include the cost of production, printing and distribution. Furthermore, marketing and selling costs are also lower in an electronically enabled commerce environment.
➢ **Providing Business Intelligence:** In the Electronic Business Operations world, businesses need to know much more about their clients. Electronic commerce

❑ Consumers 35 years old and over represent 63% of all persons buying computer hardware on the Web, 59% of all persons buying software and 58% of all persons buying books.

Businesses are aggressively adopting inter-company trade over the Internet because they want to cut costs, reduce order-processing time, and improve information flow (Cronin, 1996b). For most firms, the rise in trade over the Internet also coincides with a marked decrease in telephone and facsimile use, allowing salespeople to concentrate on pro-actively managing customers' accounts rather than serving as information givers and order takers.

2 What Electronic-Business Operation Is?

Electronic Commerce, Electronic Trading and Electronic Business are often used interchangeably and many times there is a perception that these terms principally refer to the procurement cycle - the ordering and paying for goods or services either via electronic commerce technologies such as EDI or, more recently and growing in popularity, on-line Internet shopping.

Internet-Based E-Commerce is not an extension of EDI (Electronic Data Interchange) which has been primarily limited to computer-to-computer transactions, and has not been associated with major transformations of firms. Internet-Based E-Commerce is giving a new way to Electronic Business Operations, with different characteristics than traditional EDI and is an evolution from EDI (Soliman and Gide, 1997).

There is no exact definition of the Internet-Based Business Operations. Since, Internet commerce is still immature, so is the definition. However, one definition made by Kalakota (1996), as *"the process of converting digital inputs into value-added outputs"*. Basically, this process involves taking information as raw material and producing value added information-based products or services out of the original raw information (Soliman, 1998).

So, Electronic Business Operations refers to an on-line production process owned by intermediaries. Producers of information interact with services and other processed information, such as orders, payments or instructions. In reality, Internet Business Operation is about businesses and consumers adopting a new process or methodology in dealing with each other. These processes are in essence supported by electronic interactions that replace close physical presence requirements or other traditional means(Gide and Soliman, 1997b).

The Internet offers the greatest potential for Electronic Business Operations known to date. According to Steel (1996) "there are less than 100,000 EDI (Electronic Data Interchange) users world-wide after 40 years or so of endeavour".

makes it possible to market to specific individuals based on their patterns of (purchasing and browsing) behaviour. Hence they need to capture, and to analyse, as much information as possible about each individual purchase (or cancelled purchase) in order to build up customer's profiles. This is achieved in much the same way that neighbourhood stores once did, through personal acquaintance with the consumer and continuous contact. The use of this analysed data leads to what is being called "market response" systems or "adaptive marketing".

➢ **Process simplification:** Instead of using paper, using the World Wide Web (WWW) simplifies and speeds the approval process.

➢ **Generating new revenue:** The new Internet-Based Electronic Marketplace generates new revenue by selling new products and services specifically designed for the electronic marketplace. Existing product or services can also be sold on the Internet.

➢ **Taking faster decisions:** By receiving information about competition through an Intranet information retrieval database, it would be possible to develop a competitive strategy much faster than otherwise. The drivers for manufacturing are customer's needs and time. Time is a major source of competitive advantage and competitive pressures requiring production schedules to be shortened.

5 Current Challenges To Internet-Based Business Operations

There are two main drawbacks or challenges in using Internet-Based Electronic Business Operations, these are: security issues and payment tools. These two issues are receiving the highest priority and the best attention they deserve, both from vendors and users and implementers.

6 Security and Privacy Issues

While EDI users enjoy a high level of reliability and security, they are often restricted to the exchange of data with users of the same Value Added Network (VAN). For electronic commerce to really transform the way we do business a secure solution that works globally is required. To achieve this, a series of international standards needs to be agreed and vendors need to carry out a rigorous program of interoperability tests. Moreover, as trade moves beyond national boundaries, a common legal infrastructure must be agreed. For example, a contract that has been digitally signed in one country needs be recognised in other countries.

On the other hand, a recent Forrester Research report found that security has fallen from first place in 1995 to fifth place in 1996. This indicates that there is a growing confidence in solving the Internet security issues that have been very widely publicised. Even though security is a challenge it is not a barrier to Electronic Business Operations.

Security is fairly new to the Internet, so it has not matured yet. However, computer security professionals have known about the Internet security for years and are now improving it.

7 Payment Tools

There is confusion over the availability and choice of Internet payments tools. In addition, there are no interoperability standards to make one work with another. Over the past two years, new payment tools from small companies have emerged.

There are many traditional methods of payment available in the real world such as: Cash, Cheques, Credit Cards, Traveller's Cheques, Prepaid Cards, Debit Cards, Physical Tokens, Bank Notes, Secure Wire Transfers, Money Orders, Letters of Credit, etc. However, none of these mechanisms is directly transferable in an unmodified form to suit the Internet. This is because each method assumes a physical presence or that there is a delay incurred in the processing of funds so that fraud can be detected and stopped. Some of the new Electronic Business Operations payment tools that can be used in manufacturing and business operations are:

1. *Electronic Cash (Digital Cash)-* It is a token-based currency which translates into equivalent real currency units that are guaranteed by a bank. Usually, there is a trusted authority that allows the user to conduct and pay for transactions of this nature. This usually takes place after a pre-determined relationship has been established (eg DigiCash).
2. *Smart Cards-* Smart Cards can be used with or without a stored value. Usually, the user is able to pay with them without having to connect to a remote system. If they have a stored value which contains "real digital cash", they are known as "Cash Cards" because they replace carrying cash (eg Mondex).
3. *Electronic Cheques-* These are the equivalent of paper based cheques. They are initiated during an on-screen dialog which results in the payment transaction. Authentication and verification are usually performed instantaneously by using digital signatures and time-stamping controls during the transaction (eg CheckFree).
4. *Encrypted Credit Cards-* There are varying degrees of encryption implementations credit of credit cards over the Internet, with the SET (Secure Electronic Transactions) holding the most promise (eg CyberCash).

8 Types of Internet Business Operations

At present, there are 3 types of Internet Business Operations (shown in figure 1 below), these three types are:

➢ Business to Business
➢ Business to Consumer, and
➢ Business to Employee.

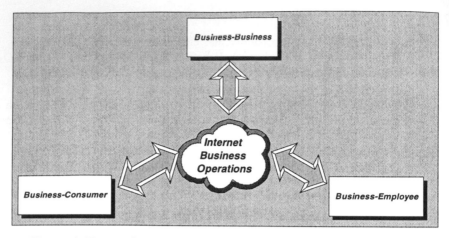

Figure 1: Types of Electronic Business Operations in the Virtual Market.

Business-to-Business Operation is complementary to EDI in that it is beginning to be used for non-production, non-replenishment applications. The widely used current terms used to describe the function of Electronic Business Operations are "Business to Business" and "Business to Consumer". The expression "business-to-business" is inexact and sometimes misleading. In Electronic Business Operations it is not always possible to tell who is accessing the automated point of sale/point of contact. It could be a retail consumer buying in wholesale quantities; it could be a business buying in retail quantities-and many other variants. Business-to-Business automated ordering processes are designed to empower business managers. The business server can only be accessed through the corporate Intranet, or an Extranet for "communities of interest".

The Business-to-Consumer Operation complements normal retail shopping, mail order and direct marketing. It can accommodate delivery of soft (digital) goods, such as published material, software, audio and video products.

Business-to-Employee Operation is beginning to develop a new market place. A checkpoint on an emerging application area. As with buying a T-shirt from the company shop; many companies now allow employee to buy using the corporate Intranet. A variant from an emulated business-to-consumer application is where employees may have purchases deducted from the payroll, or from allowances. Allowances or entitlements for clothes or equipment are often the norm in the armed services, police, fire services, airlines, banks, health services and so on.

9 Internet Benefits to Business Operations

To date the major benefits from the Internet include improved internal and external communications. The Web has specifically brought a new marketing medium and enhanced information resource. Innovative applications are starting to appear which allow for sales and database interrogation. Other benefits such as e-mail and file transfer functionality, Web utilisation gave many companies *'Internet presence'* and provided them with opportunities to develop and expand new services.

According to the Cisco Systems Inc. (a leading maker of Internet equipment), estimated more than $1 trillion to $2 trillion worth of goods and services will be sold on the Net by 2002. According to the Cisco, part of the reason for the low-ball estimates by market analysts is that many exaggerate the importance of business-to-business operations and underestimate the potential growth of the consumer market as the Net becomes more mainstream. On the other hand, Gartner, a leading industry consulting firm, has estimated business-to-business electronic operations will be 12 to 15 times larger than consumer markets for the next few years, with consumer sales only catching up with business markets midway into the next decade.

Businesses-to-business operations involve companies and their suppliers while consumer markets include home shopping, banking, health care and broadband--or high-power--communications to the home.

A model for realising the optimum level of benefits from using the Electronic Business Operations over the Internet is illustrated below in Figure 2 below.

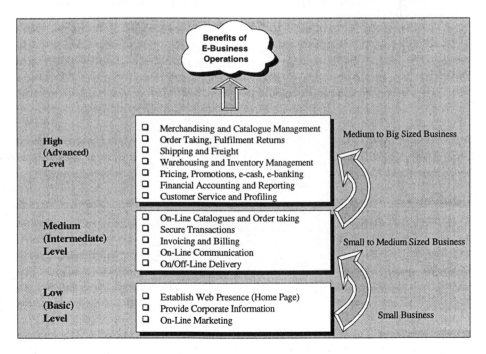

Figure 2: Model for Optimum Level of Adaptation and Benefits of Electronic Business Operations over the Internet.

In manufacturing, traditionally Design Engineering, Procurement and Production Departments communicate with each other using paper based methods. However the introduction of Internet–Based Electronic Business Operations and its superiority of over traditional EDI is adding new dimension to reducing the cost of manufacturing. So, in a typical manufacturing setting Design Engineering Department supply design drawings and specification to Procurement Department to procure material, commence production, and ultimately deliver goods to customers as per orders. In a general manufacturing setting, there are three types of flows: a) Material flow, b) Clerical flow and c) Information flow. Improvement in the movement of raw material, Work-In-Process and Finished Goods is likely to occur as a result of using the Internet-Based Electronic Business Operations. The main benefit to manufacturing from the Internet lies in the second and third types of flow.

The number of parts used in production could be in the order of thousands of items. These parts are usually purchased from suppliers on the basis of price, quality, and delivery on time and suppliers financial position and reputation in the industry.

Accordingly Material Procurement professionals must be equipped with timely and valuable information on parts and their suppliers. The Internet-Based Electronic Business Operations provide them with a fast and efficient way of obtaining comprehensive information of the market, feedback from the industry and the performance of suppliers.

The following figure (Figure 3) illustrates how clerical and production information can be efficiently and cost-effectively communicated throughout the supply chain using the Internet-Based Electronic Business Operations.

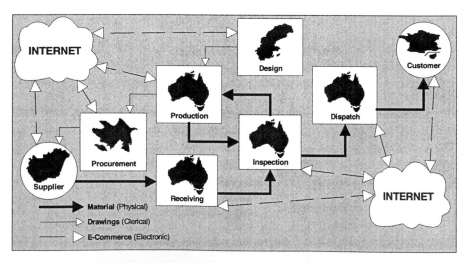

Figure 3: Supply Chain Communication in Manufacturing using Internet-Based Electronic Business Operations.

10 Conclusions

Business Operations over the Internet is very much in the early stages. Early indications are that the-Internet is a viable trading medium. The problems of cost, standards and a lack of interactivity will prohibit traditional batch-EDI scenarios. Success in Internet-Based Electronic Business Operations depends on how organisations strategically position their products and services through other Internet-based electronic communities and intermediaries, as well as on how they facilitate the interactions with their customers, suppliers, and partners. Even though Electronic Business Operations make sense *theoretically*, the reality is that it has to integrate with internal and external processes that are already in place. Sometimes, this integration is a challenge linked to a major re-engineering exercise accompanied by resistance to change. Moreover, since implementation of Electronic Business Operations is, in many cases, evolutionary, organisations need to react to change the business processes as demand increases.

Despite these benefits and success stories, a number of issues remain to be resolved such as security and privacy. Other issues regarding the growth of the Internet are the lack of: a public key infrastructure (particularly for international trade), governmental stance, access, reliability (service levels), integrated applications and understanding/awareness of the Internet-Based Electronic Business Operations capabilities, and finally the relative cost of required technologies.

References

[1]   ActiveMedia (1996), www.activemedia.com
[2]   Cisco Systems Inc. (1998), www.cisco.com
[3]   CommerceNet (1998), www.commerce.net
[4]   Cronin, M. J., (1996a), Global Advantage on the Internet, Van Nostrand Reinhold, USA
[5]   Cronin, M. J., (1996b), The Internet Strategy Handbook: Lessons from the New Frontier of Business, Harvard Business Press, USA.
[6]   Forrester Research's Business Trade & Technology Strategies Service, (http://www.internetnews.com/ec-news/cur/1997/07/3005-bb.html)
[7]   Gartner (1998), www.gartner.com.
[8]   Gide, E. and Soliman, F. (1998): "Framework for the Internet-Based E-Commerce in Manufacturing and business Operations", Leeds 15 – 16 October, pp 66-72.
[9]   Gide, E., and Soliman, F., (1997a): "Analysis of Conducting Business on the Internet," in the Proceedings of Inet-tr'97 Conference, Ankara, 21-23 November 1997.
[10] Gide, E., and Soliman, F., (1997b): "Key Drivers for Using the Internet in Australia," in the Proceedings of Inet-tr'97 Conference, Ankara, 21-23 November 1997.
[11] Kalakota, R., and Whinston, A. B., (1996), Frontiers of Electronic Commerce, Addison Wesley, USA.

[12] Kalakota, R. (1997) 'Debunking Myths About Internet Commerce', University of Rochester, USA.

[13] Nielsen Media Research (1998), www.nielsen.com

[14] Paul Kagan and Associates (1997), www.paulkagan.com

[15] Soliman, F., and Gide, E., (1997): "Impact of Internet-based E-Commerce on Manufacturing and Business Operations," in the Proceedings of Inet-tr'97 Conference, Ankara, 21-23 November.

[16] Soliman, F. (1998): "A Model for the Introduction of E-Business", Proceedings of the "NETIES'98: Networking for the Millennium", Leeds 15 – 16 October, pp 55-59.

[17] Steel, K (1996) University of Melbourne, Australia, Private communication.

CNC Manufacturing of Complex Surfaces Based on Solid Modeling

A. Elkeran[1], M.A. El-baz[2]

[1]Production Eng., Faculty of Eng., Mansoura University EGYPT
[2]Industrial Eng. And Systems, Faculty of Eng., Zagazig University EGYPT
elkeran@mum.mans.eun.eg

Abstract. The need for manufacturing moulds and stamping dies arises in industrial complex products. Most moulds and dies consist of a complicated curved surface that are difficult to produce. To represent these parts using CAD systems, both the wireframe and solid modeling are used. The wireframe model in most complicated shapes is not enough to recognize the machining parts. This paper presents an approach for the generation of NC machining instruction for complicated surfaces directly from a solid modeling. The work includes extracting data for the machined part using STL format, and constructs the topological surface for the various cutting planes. The work generates the tool path required for machining the part for both rough and fine operation. Roughing is done by pocketing procedures using zigzag cuts by flat end mill and considers the arbitrary shaped such as islands on the cutting planes. The finishing operation uses a small ball-end mill. The goal of the tool path generation is the reduction of the cutter location (CL) data file, and prevent the cutter gouging.

1. Introduction

Machining surfaces are an important feature in many engineering parts such as molds, turbine blade, automobile bodies, ... etc. Machining these objects on a numerically controlled (NC) machine is a challenging task to produce quality parts in a more effective, efficient and error free fashion. CAD/CAM can be considered as the discipline that uses computers in the design and manufacture of parts [1]. One of key feature in modern CAD/CAM system is the automatic generation of NC machining programs directly from a CAD model. Injection molds, stamping dies and other sculptured surface products are common place in modern design. To automate the generation of tool paths to machine these parts, the most fundamental information is the surface CAD model. To represent the required part using a CAD system both the wire frame and solid modeling are used. The generation of 3D CAD system uses solid modeling ensures the consistency and completeness of the part geometry. The solid modeling enables partly analysis, which make it possible to create molds and dies in more realistic way. The solid modeling presents an interactive graphical environment to construct the model of component in terms of standard procedures supported by the CAD system. The user generates the component model starting from a protrusion feature, then add or subtract other primitives. The modern 3D CAD system is able to export the part information date in terms of STL format [2] which contains the boundary representation structure of the surface in the form of small triangles. To construct the topological surface for the various cutting plans a Z-map method is used. Z-map is an efficient method to convert the STL data format to a topological of the surface. Based on the constructed surface, cutter path for both roughing and finishing operation is performed. To generate a tool path, gouging problem should be considered. Gouging frequently occurs in the region where the radius of curvature of the surface is small compared with the radius of the tool. To prevent gouging, it is essential to be able to detect all the interference regions on the part surface, and analyze the geometric relationships between the cutting tool and the part surface, assuming that the tool size has been decided. In roughing cuts, a large flat end mill is used to maximize the metal removal rate. In finishing operations a small ball end mill is used. The consideration of the tool path generation includes the generation of an offset surface from the surface model in both flat end mill and ball end mill, and use it to generate the cutter location (CL) date. In the following sections, the proposed methodology for each task will be introduced.

2. Solid modeling representation

The representation of three-dimensional (3D) CAD model forms the basic cornerstone of computer controlled manufacturing process. Initially computer were used to represent two-dimensional (2D) drawings with multiple views. These are very similar to hand drawings from which they were derived, but quickly. The potential to show three dimensional images such as isometric views prompted the development and generation of 3D wire frame representations. In such drawings, lines, arcs and circles were the prevalent primitives used. Hidden line algorithms were used to represent 3D geometries, but the CAD systems did not have any efficient way of displaying and storing complete 3D part information. When the capabilities of computers expanded, constructive solid geometry and boundary representation flourished. Constructive solid geometry (CSG) enables the application of boolean rules (union, intersection and subtraction) to geometric primitives such as a cube, a cylinder, a torus, a cone and a ball. The boundary representation (B-rep) represents a solid unambiguously by describing its surface in terms of vertices, edge and faces [3]. Various mathematical techniques have also evolved to represent free form curves and surfaces. Theses includes parametric splines, Bezier curves, and B-splines. All these formats enable the designer to represent a free-form curve using a parametric representation in each coordinate $(x, y,$ and $z)$ [4, 5]. The difference between the formats is the degree of the control of the curves. In parametric cubic curves for instance, if one control point is moves, the whole curve is affected. In Bezier and B-spline, depending on the degree of the curve, local control is achievable. Today's CAD systems typically use a combination of B-rep and CSG formats. The free-form representations vary, but the Non-Uniform Rational B-spline (NURBS) is one of the complex representation used in leading CAD software packages. It allows localized curvature modification and exact representation of primitives. Cell representation is sometimes termed as another solid modeling format which discretize surfaces into cells, typically triangles, squares or polygons. Since the CAD software is often the first step leading to a finite element analysis, cell representation is often available with the other representation formats. STL format described below is a form of a cell representation on the surface of a solid. STL or stereolithography interface specification is the most widely used of rapid prototyping processes. Two representations are commonly known, the ASCII representation and the binary representation. Both describe the coordinates of three points that form a triangle in space and its associated outpointing normal. The binary format results in much smaller file sizes, a typical Figure is 6:1 ratio, where as the ACSII format can be read and visually checked. A typical ASCII format is listed in Appendix A. The binary format used in this work due to the file size and speed of transfer is described such as follows:

The header record consists of 84 bytes, the first eighty are used for information about the file, author's name and other miscellaneous comments, the last four bytes represent the number of triangular facets. Next, for each facet, 50 bytes are used to represent the x, y, and z components of the normal to the facet, then x, y, and z coordinates of each vertex of the triangle. 4 bytes are used for each coordinate, resulting is 48 bytes per facet. The last two bytes are not used.

3. Z-map Generation

Once the geometry data of the product is translated from the solid model CAD system to STL format, the geometry data is converted into a Z-map data. The Z-map is an intermediate stage between the geometry and the tool cuter path [6, 7, 8]. The Z-map is constructed as a rectangular grid (mesh) in the xy-plane located at the highest z value of the geometry as illustrated in Fig. 1.

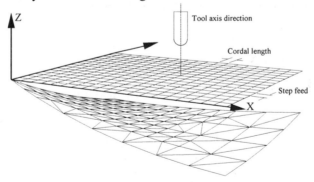

Fig. 1. The concept of Z-map

The grid snap is defined by the parameters entered for cordal length in x-axis, and step feed in y-axis . The smaller these values, the more accurate the determination of the topological surface, however also the more time needed for both calculation and milling. From each x, y node of Z-map, a vertical ray is projected onto the surface to determine the z value. The data are determined only for the rays, which located inside all the triangles of the surfaces. The smallest value for z coordinate represents a node on the machined surface. The check-up of the vertical ray is being inside a triangle plane or not can be carried out such as follows:

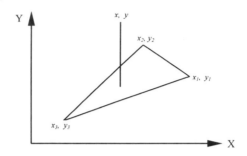

Fig. 2. Triangle cell representation

Fig. 2, illustrates a point (x, y) which represent a node of the Z-map and points (x_1, y_1), (x_2, y_2), and (x_3, y_3) represent the nodes of the triangle plane. The check carried out through the determination of the sign of the following equations [9]:

$$n_1 = - (y_2 - y_1)(x - x_1) + (x_2 - x_1)(y - y_1) \qquad (1)$$

$$n_2 = - (y_3 - y_2)(x - x_2) + (x_3 - x_2)(y - y_2) \qquad (2)$$

$$n_3 = - (y_1 - y_3)(x - x_3) + (x_1 - x_3)(y - y_3) \qquad (3)$$

If the sign of the three values of n_1, n_2, and n_3 are positive, the ray is being inside the triangle, otherwise, the ray is outside. This rule is valid only if the three points of the triangle are arranged in counter-clockwise direction.

4. Tool Path Generation

The machining of 3D-surfaces are commonly performed through the rough cutting and finishing operations. The rough cuts is used to remove enough material through several paths. Each path involves a pocketing operation in one cutting plane. The height of cutting plane depends on the maximum allowable cutter's depth of cut. The finishing operation is performed in only one path, because it is very time consuming. In the following section, the discussion of the tool path generation for roughing with flat end-mills and finishing with ball end mills is presented, then the generation of tool paths to prevent gouging is discussed.

4.1 Roughing with Flat End-Mills

Roughing cuts with flat end-mills are a series of pocketing on the cutting planes. A cutter path can be generated by moving the cutter over the entire feasible cutting region to remove the materials. Feasible cutting region is found through the continuos adjacent on the nodes of the Z-map. The depth of cut should not exceeds the permissible depths (z_{max}). Therefore, the cutting plane is determined according to the maximum permissible depth of cut. The cutting is performed through a zigzag movement for each plane, and the tool retracts to a safe plane if it encounter a protrusion edge. Then moves to the other edge of the protrusion to continue the cutting operation. To prevent gouging through the adjacent movement of the tool, the step movement of the cutter should not exceed the boundary of the surface.

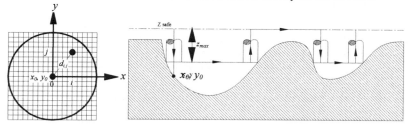

Fig. 3. Construction of cutter location for roughing with flat-end mill

As depicted in the Fig. 3, if (x_0, y_0) represents the present location of the flat end cutter of radius r, the gouge free height $(zc_{0,0})$ of the tool at this position is calculated as follows:

$zc_{0,0} = $ Min $(|z_{i,j}|)$ where
i,j includes all Z-map points within the cutter's projection area only ($d_{i,j} < r$)
if $zc_{0,0} >= z_{max}$ then $zc_{0,0} = z_{max}$ else $zc_{0,0} = z_{safe}$

4.2 Finishing with Ball End Mills

After the rough machining is done, most of the material has been removed from the machined part. Finishing operations are performed to remove material along the surface. Machining accuracy and machining time are the two major issues to be

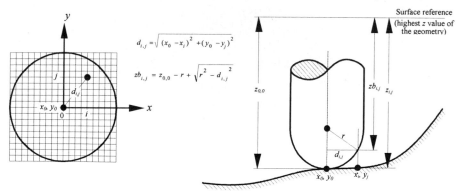

Fig. 4. Z-map representation of ball-end mill

considered in cutting tool selection and cutter path generation for finishing operations. In the ball end mill machining, the tool is bounded by a spherical face and one cylindrical face. The Z-map representation of ball end cutter bounded by the part surface is shown in Fig. 4 if only the ball part of the tool is considered.

As depicted in the Fig. 4, if (x_0, y_0) represents the present location of the ball end cutter of radius r, the gouge free height $(zc_{0,0})$ of the tool at this position is calculated as follows:

$h_{max} = \text{MAX} \left(|zb_{i,j}| - |z_{i,j}| \right)$, where $(h_{max}$ is the revise distance in the Z-direction)

if $h_{max} > 0$ *then* $zc_{0,0} = z_{0,0} - h_{max}$ else $zc_{0,0} = z_{0,0}$

i,j includes all Z-map points within the cutter's projection area only $(d_{i,j} < r)$

5. Example Part

The methodology proposed is now illustrated by solving a practical problen. The solid drawing is shown in Fig. 5 The proposed system takes the STL data information, and

Fig. 5. An example part to be machined

proceed the entire technique to generate the cutter paths for the machining of the complete surfaces. The complete NC-Code is directly sent to an NC controller to machine the desired part. Fig. 6 represents the cutter path for the finishing operation.

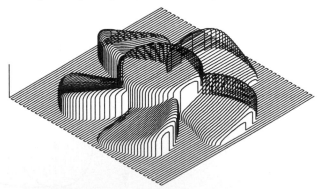

Fig. 6. Tool path for finishing with ball end mill

Cutter selected for roughing is a flat end-mill with a cutter radius of 10 mm, and the cutting planes for roughing are 5 mm. The cutter selected for the finishing is a ball-end-mill with cutter diameter of 6 mm. The grid snap for cordal length in x-axis is selected as 0.05 mm, and step feed in y-axis is 0.2 mm. The designed part is machined on a three axis BOXFORD vertical milling machine and the machined part is shown in Fig. 7.

Fig. 7. Machined part

6. Conclusion

This paper presents an integrated approach for the generation of CNC machining instruction for complicated surfaces directly from a solid model. The part geometry is represented using STL format. Z-map method is used for the physical representation of the surface to be machined. Cutter paths for both rough machining and fine operation are generated automatically. The goal of the tool paths generation is to prevent cutter gouging. To ensure the applicability of the generated NC program in practical machining process, the proposed system is tested through several machining parts and a good results are obtained.

References

1. Goover, M. P. and Zimmers., "CAD/CAM Computer-Aided Design and Manufacturing" Prentic-Hall, inc. New jersy, 1989.
2. Fadel, G. M. and Kirschman, C., "Accuracy Issues in CAD to RP Translations" http://www.mcb.co.uk/services/conferen/dec95/rapidpd/fadel/fadel.htm
3. D. Zhang and A. Bowyer, 'CSG Set-Theoretic Solid Modeling and NC Machining of Blend Surfaces', Proceedings of 2nd Annual ACM Conference on Computational Geometry, IBM, Yorktown Heights, June 1986.

4. Faux, I. D., and Pratt, M. j., "Computational Geometry for Design and Manufacture" Ellis Horwood, 1981.
5. M. E. Mortenson, "Geometric Modeling" , John Wiley & Sons., Inc., 1985
6. T. Saito and T. Takahashi, 'NC Machining with G-buffer Method,' Computer Graphics, vol. 25, no. 4, July 1991, pp 207-216.
7. R B Jerard,., Angleton, J.M., Drysdale, R.L. and P.Su, 'The Use of Surface Points Sets for Generation, Simulation, Verification and Automatic Correction of NC Machining Programs,' Proceedings of NSF Design and Manufacturing Systems Conf , Tempe, Az, Jan. 8 - 12, 1990, Society of Manufacturing Engineers, pp 143 -148.
8. Choi, B. K. "Surface Modeling for CAD/CAM" Elsevier Netherlands, 1991.
9. Jack E. Zecher "Computer Graphics for CAD/CAM Systems" Marcel Dekker, Inc. New York, 1994.

Appendix A

STL ASCII format example:
    facet normal 0.000000 e+00   0.000000 e+00   1.000000 e+00
        outer loop
            vertex   2.029000 e+00   1.628000 e+00 9.109999 e-0.1
            vertex   2.229000 e+00   1.628000 e+00 9.109999 e-0.1
            vertex   2.229000 e+00   1.672000 e+00 9.109999 e-0.1
        end loop
    end facet

Analysis of Manufacturing Lines Using a Phase Space Algorithm: Open Line Case

T. EL-FOULY - N. ZERHOUNI - M. FERNEY- A. EL MOUDNI

Laboratoire d'Automatique Mécatronique Productique et systèmique (L.A.M.P.S)
Espace Bartholdi Belfort Technopôle BP 525 90016 Belfort Cedex (France)
Tel : +(33) 3 84.58.23.97, Fax : +(33) 3 84.58.23.42
tarek.elfouly@lmp.enibe.fr

Abstract. This paper presents a study concerning the minimization of the evolution time of a class of manufacturing systems. The study passed by several steps beginning by modeling the manufacturing system using continuous Petri Nets. The second step uses the developed model to construct the different phases of evolution using the phase algorithm that will be defined later in the paper. The constructed phase schema helps in solving the proposed problem and helps at the same time in defining a control sequence that could be applied to the system. The variable parameter is the source speed or in other words the rate by which the pieces to be manufactured are supplied to the system. In this paper the class of an open manufacturing line will be presented. At the same time changing the source speed could have an important affect on the manufacturing line. We will study the effect of changing the speed of the source on the throughput of the system. One of the important advantages in using the developed algorithm is its ability to conserve the continuity of the system and also to show the dynamic behavior of the studied system.

1 Introduction

This paper presents an intermediate step towards a control theory that could be applied at first to manufacturing lines and that we aim to generalize it to be applied after that to production systems. At the beginning of this work the manufacturing line is modeled by a continuous Petri Net that will help in constructing the equations describing the evolution of the number of pieces resting in the stock. These equations are nonlinear equations, which made us think in cutting the evolution or the lifecycle of the manufacturing process into different phases, where a phase is a period of time where the dynamics of the system is constant without change. The phase variation is marked by the variation of the state equations of the system, which could be translated in other words as a change in the dynamics of the system. The effect of changing the initial state of the studied manufacturing line was also studied wit respect to changes in the machine speeds. After studying this effect the study took a direction towards fixing some parameters and varying another to reach a stationary or a desired state using the shortest path. This shortest path is dictated by an initial marking for the PN model and the initial transition speeds. Choosing the shortest path led to the study of the effect of changing the source speed to the level of the phase, which means selecting the most appropriate speed that minimizes the overall time when a desired state is defined. Choosing the most appropriate speed don't guarantee a good throughput, so a module was added to try to choose the source speed to compromise between the minimum evolution time and the throughput of the line. Constructing the different phases will lead to construct the phase space and by its turn will define the control series depending on the source speeds. In section (3) an introduction to Petri Nets is presented taking into consideration the concept of phases. In section (4), the algorithm that constructs the different phases is presented. In section (5) an example to show the utilization of this concept is presented followed by the general conclusion.

2. Petri Nets

The Petri Net is a graphical utility for describing the relation between events and conditions [7]. It permits to model the behavior taking into consideration the sequence of activities, the parallelism of activities in time, the synchronization of the activities, and the resource sharing between activities. A Petri Net is an oriented graph consisting of places [5], transitions and arcs. The arcs connect between a place and a transition or vice versa. These arcs indicate the direction of flow of marks. For a manufacturing system, the different places of the Petri net model the different buffers of the system while the transitions model the different machines [1].

A Petri Net is defined by 4 variables, 2 sets and 2 applications [5] PN = < P, T, Pre, Post>

Where:
$P = \{P_1, P_2, ... P_n\}$ is a set of places, $T = \{T_1, T_2, ... T_n\}$ is a set of Transitions.
Pre: $P \times T \rightarrow N$ $(P_i, T_j) \rightarrow$ Pre (P_i, T_j)=arc weight between P_i and T_j.
Post: $P \times T \rightarrow N$ $(P_i, T_j) \rightarrow$ Post (P_i, T_j)=arc weight between T_j and P_i.

Incidence Matrix: $W = $ [Post (P_i, T_j)]-[Pre (P_i, T_j)]

There are many types of Petri nets, each one models a particular class of systems. In which follows some types of Petri nets will be presented i.e. the timed Petri nets and the continuous Petri nets. There are two ways of assigning timing to the timed Petri nets, either by assigning the timing to the transitions or by assigning the timing to the places. The timed Petri nets model a system having the number of marks circulating in the system is not important. But if this number of marks is important (explodes) the continuous Petri nets are used in this case. A continuous Petri Net is a model where the marking in the places is real and the transitions are continuously fired [5]. There are two types of continuous Petri nets, the first is the constant speed continuous Petri net, which is characterized by having all of its transitions having a constant firing speed $V = 1/d$. The second is the variable speed continuous Petri net, which is characterized by having a variable firing speed, $V = (1/d)\mathbf{min}(1, m_1, \ldots, m_i)$ associated to its transitions.

An open manufacturing line is presented as shown in the following figure.

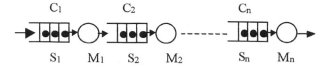

Fig. 1. Open Manufacturing Line

The manufacturing line presented in Fig.1 is a open one and it consists of n working benches, a source and an output stock. Each bench consists of a stock and a machine. The stocks are considered to have infinite capacities. The following Petri Net model could model this manufacturing line:

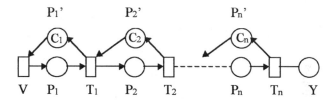

Fig. 2. Petri Net Model

The Petri net model presented in Fig 2. models the open manufacturing line of Fig 1. Where a machine M_j is modelled by a continuous transition T_j having the same speed U_j while a stock S_i is modelled by a continuous place Pi having the capacity C_i. An equation for the system's evolution is needed to be established. This equation must be function in the initial marking. Initially we consider all places are empty while the speed vector of the n transitions is given by:

$$U = [U_1, U_2, \ldots U_n].$$
$$M_0 = [0, 0, 0 \ldots 0].$$

The flow of pieces entering the system to be manufacturing is defined by the speed of the source machine given by the symbol V. The number of pieces produced by this system is given Y.

Equation (1) describes the evolution of the system as follows:

$$\dot{m}_1 = V - U_1 \min(1, m_1) \tag{1}$$
$$\dot{m}_2 = U_1 \min(1, m_1) - U_2 \min(1, m_2)$$

$$\bullet$$

$$\bullet$$

$$\dot{m}_n = U_{n-1} \min(1, m_{n-1}) - U_n \min(1, m_n)$$

And

$$\dot{Y} = U_n \min(1, m_n)$$

For the initial state we have all places initially empty without any marking we will have equation (2):

$$\dot{m}_1 = V - U_1 \cdot m_1 \tag{2}$$
$$\dot{m}_2 = U_1 \cdot m_1 - U_2 \cdot m_2 \qquad \text{If } M_0 = [0, 0, 0 \dots 0]$$
$$\vdots$$
$$\dot{m}_n = U_{n-1} \cdot m_{n-1} - U_n \cdot m_n$$

And :

$$\dot{Y} = U_n \cdot m_n$$

Or in matrix form:

$$\dot{M} = A \cdot M + B \cdot V \tag{3}$$

And the output:

$$\dot{Y} = C \cdot M + D \cdot V \tag{4}$$

Where:

$$A = \begin{pmatrix} -U_1 & 0 & 0 & \cdot & \cdot & 0 \\ U_1 & -U_2 & \cdot & \cdot & \cdot & 0 \\ 0 & U_2 & -U_3 & \cdot & \cdot & \cdot \\ \cdot & 0 & \cdot & \cdot & \cdot & \cdot \\ \cdot & \cdot & \cdot & \cdot & -U_{n-1} & 0 \\ 0 & 0 & \cdot & \cdot & U_{n-1} & -U_n \end{pmatrix}, B = \begin{pmatrix} 1 \\ 0 \\ 0 \\ \cdot \\ \cdot \\ 0 \end{pmatrix}, C = \begin{pmatrix} 0 \\ 0 \\ 0 \\ \cdot \\ \cdot \\ 1 \end{pmatrix}^T \tag{5}$$

, And $D = 0$

The analysis presented is particularly made to study the effect of changing V on the system's behavior. The simple analytical methods can't solve the previous system of equations because the matrix A could be a singular matrix depending on the simulated phase. The Maple V5 R5 software allows us to solve numerically such systems of equations based on the values of A, B,C and D. But the numerical solution of this system of equations is not the goal of this study because what we need is the analytical expressions describing the vector M(t). Using the expressions describing the different marking of the different places M(t) the different evolution times are calculated and then the smallest one is chosen. Choosing the minimum time of evolution for each phase allows us after that to construct the control sequence to be applied to the system to minimize the evolution time required to reach the desired state. A tool has been developed using the software tool Maple V5 R5 to calculate the different Evolution times. The tool gives after that the different phases of the system. In the next section an algorithm for constructing the phase schema and choosing the minimum evolution time is presented.

3. The Phase-Control Algorithm

The phase-control algorithm is the algorithm which helps us to construct the phase schema and at the same time to construct the control sequence to be applied to the system during the real evolution of the studied system. The inputs to the algorithm are the initial state and the control variable to be applied to the system. The initial state consists of the initial marking of the different places of the Petri net model, and the speed vector defining the speeds of the different transitions of the Petri Net model. The control variable in our case is the flow of pieces supplied to the system or in other words the speed of the source. The different steps of the algorithm could be summarized as follows:

❑ After providing the initial state and the control variable to the Petri Net model, the first phase could be constructed without any problems. This is due to the fact that the phase depends a lot on its initial state. For each studied phase there are 2 inputs and an output. The 2 inputs are the initial marking of the different places of the Petri Net model for this phase which is by its turn the final marking for the previous phase. The Second input is the optimal evolution time for the previous phase. The output from a phase is the equations defining the evolution of the marking of the places of the Petri Net model.

❑ The second step in this algorithm is providing the outputs of the first phase to the controller. The controller is the module that will calculate the optimal evolution time for the first phase and at the same time it will calculate the corresponding source speed. This is done as shown in the previous algorithm, where the equations describing the evolution of the marking during the studied phase is applied to the controller. The desired state to be reached is checked if it belongs to this phase, if it belongs the algorithm stops and the system has reached the desired state, if not the algorithm continues with the next step. The equation describing the fact of choosing the best time is as follows:

$$\underset{V\in[V_{min},V_{max}]}{Min}\left(M(t_1,V)\right) \Rightarrow t_{1opt}, V_{1opt}, M(t_{1opt}, V_{1opt})$$

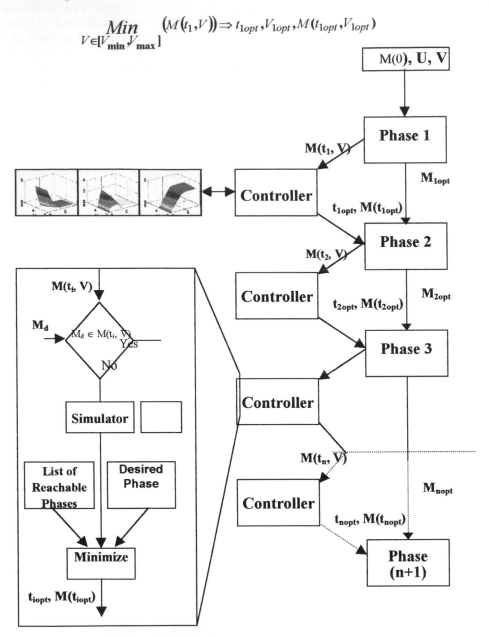

Fig. 3. The Phase-Control Algorithm

❏ The Controller performs a three-dimensional simulation to determine the place that will be responsible for the next phase variation. There are many reasons to have a phase variation, for example if the marking of a place reaches the unity value or if the marking in a place reaches its maximum capacity. This phase is checked after that with the list of reachable phases by the studied system and also with the desired phase which contains the desired state. Using these simulations the minimum evolution time and the corresponding marking of the places are calculated.

❏ The output of the controller is after that applied to the next phase. This algorithm continues until a stationary state is reached or the desired state is reached in the case of searching for a one.

4. Illustrative Example

The system presented in this example is an open manufacturing line consisting of 3 working benches each consists of a machine and an accompanying stock. This line is presented as in Fig. 4 :

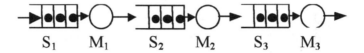

$$S_1 \quad M_1 \quad S_2 \quad M_2 \quad S_3 \quad M_3$$

Fig. 4. Open Manufacturing Line with 3 working benches

This closed manufacturing line is modelled by the Petri net model of Fig.5 where a stock is modelled by a continuous place and a machine is modelled by a continuous transition.

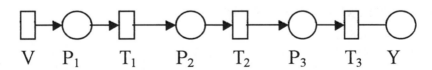

$$V \quad P_1 \quad T_1 \quad P_2 \quad T_2 \quad P_3 \quad T_3 \quad Y$$

Fig. 5. Corresponding Petri Net model

The equations that describes the marking evolution in the different places of the Petri Net model :

$$\dot{m}_1(t) = V - U_1 \cdot \min(1, m_3(t)) \tag{6}$$
$$\dot{m}_2(t) = U_1 \cdot \min(1, m_1(t)) - U_2 \cdot \min(1, m_2(t))$$
$$\dot{m}_3(t) = U_2 \cdot \min(1, m_2(t)) - U_3 \cdot \min(1, m_3(t))$$

And :

$$\dot{Y}(t) = U_3 \cdot \min(1, m_3(t))$$

The speed vector is defined by $U = (4 \quad 3 \quad 2)$

The initial marking vector is defined by $M_0 = \begin{pmatrix} 0 \\ 0 \\ 0 \end{pmatrix}$

And $Y(0) = 0$.

And since we work with limited capacity stocks, then the vector presenting the maximum capacity of the corresponding places is given by:

Capacity $= (10, 3, 2.3)$

Using the defined initial values, the equations that defines the marking evolution in the different places of the Petri net could be given by:

$$\dot{m}_1(t) = V - 4 \cdot \min(1, m_1(t)) \tag{7}$$
$$\dot{m}_2(t) = 4 \cdot \min(1, m_1(t)) - 3 \cdot \min(1, m_2(t))$$
$$\dot{m}_3(t) = 3 \cdot \min(1, m_2(t)) - 2 \cdot \min(1, m_3(t))$$

And :

$$\dot{Y}(t) = 2 \cdot \min(1, m_3(t))$$

The speed V is of course constant during the phase but the equations describing the evolution of the marking during the phase will be defined as function in V so as to be applied to the algorithm that will calculate the appropriate value of V to be applied to this phase. Then the equations describing the marking in the different places of the PN model during the first phase could be given by :

$$m_1(t) = V * \left(\frac{1}{4} - \frac{1}{4} e^{-4t} \right) \qquad m_2(t) = 4 \cdot V \cdot \left(\frac{1}{12} - \frac{1}{3} e^{-3t} + \frac{1}{4} e^{-4t} \right)$$

$$m_3(t) = 12 \cdot V \cdot \left(\frac{1}{24} - \frac{1}{8} e^{-4t} + \frac{1}{3} e^{-3t} - \frac{1}{4} e^{-2t} \right)$$

And the output

$$Y(t) = 24 * V * \left(\frac{1}{24} t - \frac{13}{288} + \frac{1}{32} e^{-4t} - \frac{1}{9} e^{-3t} + \frac{1}{8} e^{-2t} \right)$$

The equations previously shown are those during the first phase. These equations will be solved to minimize the time t, then the obtained results are supplied as the input to the second phase. This process continues until the algorithm finds a stationary repetitive state or the desired state.

A phase variation could be obtained when the marking in one of the places reaches the unity value which marks a variation in the dynamics of the system or when one of the places reaches its maximum capacity which will be followed by stopping the machine(transition) supplying this place with the pieces.

The overall curve describing the relation between the marking in the different places and the evolution time is as follows:

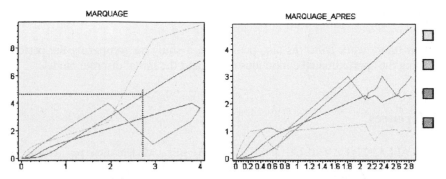

Fig. 6. Evolution without/with applying the algorithm

The results presented in the previous curves shows the effect of applying the algorithm to the overall performance of the system. The first remark is obtained by comparing the level of pieces in the first stock with the original one. The level or number of pieces ranges to a certain limit between 2 boundaries while in the original case the number of pieces could arrive to infinity. In the original case we need to use an infinite capacity stock or the source of pieces to the system must be stopped which is not logical. But in our case the number of pieces is limited which is considered as a great advantage. The second remark is the throughput of the system which could be seen higher than the original case without applying the algorithm and which could be better also by adjusting the initial conditions and the maximum and minimum allowable speed for the source.

5. Conclusion

In this paper we presented a modeling method using continuous Petri Nets and the basic concept of hybrid automata. This method is based on a new concept of cutting the evolution of the system on different phases. Each phase is characterized by different dynamics. But at the same time this method saves the continuity of the system and it also saves the dynamics of the system during a certain phase and this could be recognized in the equations describing the marking of the different places of the PN model.

Choosing the minimum time of simulation could be considered as a big step towards an optimal control that could minimize the time necessary for reaching a desired state.

The constructed phase schema helps in future simulation or running of the studied system. The fact of having the phase schema helps in determining if a desired state is reachable or no.

We could also conclude from the obtained results that in the worst cas the throughput of the system won't be affected, and at the same time the level of pieces in the different stocks is guranteed to rest less than a certain limit or the stocks must have an infinite capacity.

Our future work concerns also performing a study on comparing the performance of using the approximated continuous model and the initial discrete model.

References

1 H. ALLA, J.B.CAVAILLE, J. LE BAIL, G. BEL, Modélisation par réseaux de Petri Hybrides et évaluation de Performances de systèmes de production par lots, (Rapport Intermédiaire, 1992)

2 M. ALLAM, H. ALLA, Modeling Production Systems by Hybrid Automata and Hybrid Petri Nets, Control of industrial systems, IFAC-IFIP-IMACS ,1 ,1997 ,463-468.

3 R. ALUR, C. COURCOUBETIS, T. A. HENZINGER, and P. H. HO, Hybrid automata: an algorithmic approach to the specification and analysis of hybrid systems, In R. L. Grossman, A. Nerode, A. P. Ravn, and H. Rischel, editors, Workshop on theory of Hybrid systems, Lecture Notes in Computer Science 736, 1993, 209-229.

4 R. ALUR, C. COURCOUBETIS, N. HALBWACHS, T.A. HENZINGER, P.-H. HO, X. NICOLLIN, A. OLIVERO, J. SIFAKIS, S. YOVINE, The algorithmic Analysis of Hybrid systems, Theoretical Computer Science 138, 1995, 3-34.

5 R. DAVID, H. ALLA, Du Grafcet aux Réseaux de Petri. Hermès:Paris.

6 T. EL-FOULY, N. ZERHOUNI, M. FERNEY, A. EL MOUDNI, Modeling and analysis of manufacturing systems using a hybrid approach, $3^{\text{ème}}$ conference Internationale sur l'Automatisation des Processus Mixtes, ADPM'98 Reims, 1998, P 79 – 85.

7 J. PETERSON, Petri Net Theory and the Modeling of Systems, Prentice-Hall, 1981.

8 J. A. STIVER, P. J. ANTSAKLIS, M. D. LEMMON, A logical DES approach to the design of hybrid systems, Technical Report of the ISIS Group (Interdisciplinary Studies of Intelligent Systems) ISIS-94-011, University of Notre Dame, October 1994.

A Formal Approach to Lingware Development

Bilel Gargouri, Mohamed Jmaiel, and Abdelmajid Ben Hamadou

Laboratoire LARIS FSEG-SFAX B.P. 1088 - 3018 TUNISIA
Bilel.Gargouri@fsegs.rnu.tn

Abstract. This paper has two purposes. First, it presents a formal approach for the specification and development of lingware. This approach is based on the integration of the main existent formalisms for describing linguistic knowledge (i.e., Formal Grammars, Unification Grammars, HPSG, etc .) on the one hand, and the integration of data and processing on the other one. In this way, a specification of an application treating natural language includes all related aspects (linguistic and processing) in a unified framework. This approach promotes the reuse of proved correct specifications (linguistic knowledge specifications, modules and subsystems).

Second, it proposes an environment for the formal specification and development of lingware, based on the presented approach.

1 Introduction

Since several years, many efforts have been multiplied to develop software in the area of Natural Language processing (NLP). Despite these efforts, several difficulties persist at the development level as well as at the maintenance one. These difficulties result from the complexity of this area, notably the dichotomy data-processing, the diversity of data to represent (especially linguistic data), the multitude of formalisms proposed for their representation where each formalism presents a certain rigor in the description of one kind of knowledge, the variety of processing and the strong interactions between data and processing, etc.

The investigation of some development approaches applied to applications related to the NLP domain at all levels (i.e., lexical, morphological, syntactic, semantics and pragmatic) [8, 15] allowed us to observe an almost total abssence of using methodologies that integrate all phases of the software life cycle. Especially, at the first development level, we observed a near absence of the formal specification phase.

On the other hand, we observed a lack of formal validation of the approaches used in the development process and consequently a lack of guarantee on the performances of the obtained results. Similarly, we noticed the no recourse to rigorous integration methods to solve the problem of the dichotomy data-processing.

The use of formal tool was restricted, in most cases, to the description of the language (i.e., grammars) and to reduced part of the processing aspects. Generally, only algorithmic language have been used for this reason. Few are those who used a high level formal specification language [12, 16].

Starting from the results of an experience that we have carried out on a real application of NLP which consists to realise its complete and verified specification (w.r.t. the initial informal one), we have been able to highlight the need to apply formal methods in the lingware development. Moreover, we have proposed to generalize the application of formal methods and we gave some methodological criteria for the choice of the appropriate formal method [10].

In this paper, we present a formal approach for the specification and the development of lingware. This approach is based on the integration of the main existent formalisms for the description of linguistic knowledge (i.e., Formal Grammars, Unification Grammars, etc .) on the one hand, and the integration of data and processing on the other hand. In continuation, we will give an idea on the experience that we have carried out. Then we will present foundations and contributions of our approach as well as a comparison between the proposed approach and those of the main existent environments of lingware development. Finally, we will present the environment based on our formal approach.

2 The use of formal method in NLP

2.1 Presentation of the experience

To measure the impact of using formal method in the NLP context, we have carried out a complete and proved correct specification of the CORTEXA system (Correction ORthographique des TEXtes Arabes) [2]. This experimentation has been accomplished with the VDM formal method.

Besides the availability of the documentation concerning the conceptual data and the code, the choice of the system CORTEXA is also motivated by the diversity of approaches used to represent knowledge and processing.

The application of VDM for the specification of CORTEXA is motivated by the fact that this method is based on predicates that give a high expressive power. In addition, the formal method VDM, which is based on simple and rich notation, allows to specify both linguistic and processing aspects, and provides a development methodology based on formal refinements and transformations. Moreover, VDM has made its evidences in the development of several systems of information such as real time system [1].

Starting from the informal requirements specification, we developed the abstract specification of the CORTEXA system (also called implicit specification) that includes, among others, formal specifications of its functions, its actions and its correction rules. Then, this specification has been formally proved correct and complete. Finally, we generated a design specification (also called direct or explicit specification) from the abstract specification by applying VDM refinement and transformation rules. This design specification is easily transformed into code to reach the realization phase.

2.2 The interests of formal method in NLP

Although this experimentation was limited in time (it has been carried out on one year approximately) and in its context (it has been interested to one system and not to several), it allowed us to appreciate the request to formal methods in the development process of applications related to the NLP [9]. Moreover, it enabled us to extract some global advantages dedicated to the area of the NLP that consolidate the advantages presented above in the framework of general software development. These specific advantages can be summarised and argued in the following points.

First, at the level of requirement specification, applications of the NLP are generally very ambitious. But it is widely known that the existent models and tools to represent linguistic knowledge are very limited. However, the use of formal tool in the first stages of development (i.e., analysis) allows to detect the limits of the system to develop, especially, at the level of linguistic aspects. Consequently, this enables us to start the design stage on a real version of the system and to anticipate the possibilities of extension and reuse.

Furthermore, the complexity of NLP, the linguistic data diversity and the strong interactions that exist between data and processing make the design step very difficult and may lead to incoherence problems. The use of formal methods at the design level allows first, to solve the dichotomy data-processing problem by the integration or by the control of coherence (i.e., by formal proofs). Then, to put in obviousness, by successive refinements of interesting reusable abstractions such that modules or subsystems that should be available in a library [5]. These abstractions correspond, for example, to standard modules of the NLP at the phonetic, morphological or syntactic level.

Moreover, the use of a unified notation gives the possibility to integrate in the same application a variety of linguistic knowledge specified with different formalisms. This will allow to have a best coherence in the final specification to produce.

Finally, one can add that formal methods allow to verify the property of multi-language of an application by data and processing integration in the same specification. Indeed, the use of formal proofs on a such specification enables to test the possibility of applying some routines on different linguistic data.

3   Proposed approach

The solution that we propose for the specification and the development of lingware is based on an approach that profits from formal method advantages (i.e., VDM, B, etc .). This methods have made their evidences in the specification and the development of software in general. The idea is, therefore, to use these methods in the natural language context while providing solutions for the problem of the dichotomy data- processing and the integration of the main existent formalisms of linguistic knowledge description.

The expressive and unified notation (i.e., the specification language), associated to the formal method, can play the role of a pivot language for linguistic knowledge descriptions. In addition, we can use the same language to specify, in a unified framework, both data and processing. Consequently, we can apply the development process associated to the retained formal method.

The major problem, that we have to deal with, consists of how to represent various linguistic knowledge descriptions, initially done with different formalisms, within a unique notation. The investigation of the formal descriptions of the main existent formalisms [13, 15], led us to bring out their representations within a unique formal language (i.e., VDM-SL) while ensuring the equivalence between the original and the resulted descriptions.

Therefore, we propose, for simplicity and convenient reasons, to acquire linguistic knowledge descriptions with the initial formalisms then to realize an automatic passage to the pivot representation by using appropriate rules as shown by Figure 1. This facilitates the user task, since he doesn't need to deal with the representation of the different linguistic descriptions in the pivot language.

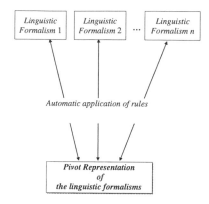

Fig. 1. Integration of the main existent formalisms for linguistic knowledge description

Consequently, the data and processing integration within a unique specification, as indicated in Figure 2, can be realized once these two types of specifications were presented by the same notation. This enables to solve the dichotomy data-processing problem.

We note that the pivot representation does not replace, at least at this level, the use of initial formalisms for the linguistic knowledge description which remain more convenient to manipulate. The use of the pivot representation remains transparent to the user.

The integration of data and processing within a unified formal specification allows to apply the development process associated with the retained formal method, which is based on formal refinements and transformations of the requirements (abstract) specification. The development process seems to be independent of the natural language context. However, the use of some theorems related to this context is envisaged. This enables us to bring new profits through the use of this approach.

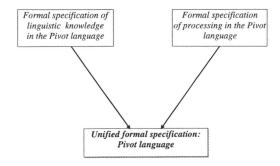

Fig. 2. Integration step of data and processing within a unified specification

4 Development environment using the proposed formal approach

4.1 Architecture of the environment

The figure below, presents the architecture of the lingware development environment based on our formal approach. It shows, notably, the acquisition of processing specification and the linguistic knowledge descriptions, the integration of the main existent formalisms, the integration of data and processing. In addition, it shows the application of the refinement and correctness proof process, and presents the library of specifications and the different interfaces of the environment.

In this environment, each partial description of linguistic knowledge (that concerns only some aspects of natural language), initially informal, is acquired within the appropriate formalism. A set of interfaces is available, in which each interface is concerned with one formalism. After the input of a partial formal description, this will be automatically transformed, applying a set of rules, into a pivot representation. Linguistic knowledge representations, in pivot language, will be stored in the specification library.

On the other hand, an interface is available to the acquisition of the processing specification in the pivot language. The obtained formal specification will be stored (after validation) in the library, in order to integrate it with data descriptions in a unified specification. The processing specification should refer to linguistic data to use.

Thereafter, refinement and validation process will be applied on the obtained unified specification. Finally, the validated and complete specification will be stored in the library.

4.2 Linguistic knowledge specification

Several researches on the linguistic knowledge representation have proven the power and the rigor of the used formalisms such that GU, GF and HPSG in

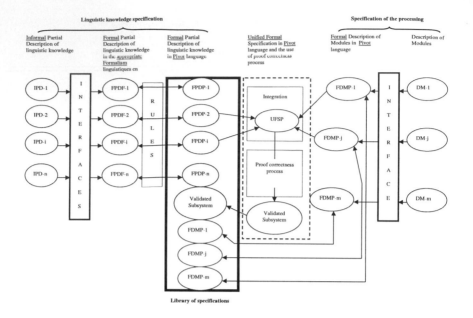

Fig. 3. Architecture of the environment of lingware development

the description of natural language knowledge [8]. This makes the use of these formalisms indispensable.

Our approach takes into account this fact. It maintains, consequently, the use of the main existent formalisms for linguistic knowledge description. This offers, in addition to rigor, a certain facility in the description, the acquisition and the exploitation of these knowledge. Indeed, all descriptions of the natural language knowledge will be given in the appropriate formalism via a specific interfaces. These descriptions would be then transformed automatically (by applying appropriate rules) in the pivot language. These formal descriptions can be , thereafter, integrated in the lingware specification and/or stored in the specification library. The acquisition of new linguistic knowledge takes into account the content of the specification library.

4.3 Processing specification

The processing specification will be presented directly in the pivot language. This specification, which contains the specification of processing (subsystems, modules, functions, etc.) and other kind of data description (not necessary linguistic data), refers also to the linguistic knowledge to use.

An appropriate interface allows the acquisition and the validation (notably syntactic verification) of the processing specification. Once this specification is validated, it will be stored in the library and one can proceed to its integration with the specification of the necessary linguistic knowledge which should be already available in the library. The new processing specification is based, among others, on the reuse of the content of the library.

4.4 Integration of data-processing

The dichotomy data-processing represents an important problem of the NLP. The integration of these two components within a unified specification allows to solve this problem and to apply, consequently, all relative refinements and correctness proof process of the formal method. This permits to profit, in general, from the formal method advantages.

The integration is guaranteed thanks to the use of a unified specification language. This expressive pivot language (i.e., VDM-SL), enables the representation of both linguistic knowledge and processing. However, establishing a rigorous and compatible integration method, that can provide better profits, is envisaged.

4.5 The library of specifications

The role of the specification library is to promote the reuse. This library contains three kinds of specifications, namely, the linguistic knowledge, the processing, and some modules and validated subsystems specifications (data and processing).

The design of this library depends, among others, on the pivot language and the integration method to use. Consequently, this design will be realized later.

5 Representation of the main formalisms in VDM

In order to illustrate our approach, we give, in this section, a formal specification in VDM-SL of two types of grammars : formal grammars and unification grammars. These grammars, play a very important role in the description of linguistic knowledge.

Indeed, formal grammars have been the firsts proposed in this area. They were the origin of the majority formalisms. Similarly, unification grammars have made their evidences in the linguistic knowledge description. Starting from these grammars, several other formalisms have been proposed (i.e., GPSG, HPSG, etc.).

5.1 Formal grammar representation

In this section, we propose the representation of formal grammar [3] in the VDM-SL, precisely a free context grammar.

- Set of non-terminal characters : $NTERM = \{A, B, ..., Z\}$
- Set of terminal characters : $TERM = \{a, b, ..., z\}$

– A context free grammar in VDM-SL is a Formal-G2 type :

$$Formal\text{-}G2 \quad :: \; T : \textbf{set of } (\textbf{seq of } TERM)$$
$$N : \textbf{set of } (\textbf{seq of } NTERM)$$
$$S : \textbf{seq of } NTERM$$
$$P : \textbf{set of } (N \longrightarrow \textbf{set of } (\textbf{seq of } (T \cup N)))$$

$$\textbf{\textit{where inv-Formal-G2}}() \triangleq ((T \neq \emptyset) \wedge (N \neq \emptyset) \wedge (T \cup N = \emptyset) \wedge$$
$$(S \in N) \wedge (P \neq \emptyset) \wedge (S \neq []))$$

5.2 Representation of the unification grammars

To represent the unification grammars in VDM-SL, our framework was essentially based on a formal description of this formalism [11]. In this description, an unification grammar is defined as a 5-uplet :

$$G = (S, \, (\Sigma, r), \, T, \, P, \, Z)$$

where

S : Set of sorts, finite and not empty

(Σ, r) : Set of function related to S composed from the Σ letter functions and the $r :\longrightarrow S^* \times S$ function that gives a$row(u, \, s)$ for each f symbol in Σ The u represents the number of repetition of f in S^* and s represents the type of the function f

T : A finite set of terminal symbols

P : A finite set of rules such : $P = \{A \mid A$ is a term on Σ and a is a sequence such that $a \in (T \cup T_\Sigma)^*\}$ where T_Σ is the set of ground terms generated from Σ

Z : The start symbol which have ϵ repetition in $(\Sigma, \, r)$

VDM-SL representation

– Set of types : S : **set of** (**seq of** (A)) where $A \stackrel{\text{def}}{=} \{a, b, ..., z\}$
– The set of characters function :

$$fct \; :: \; \textbf{set of } (\textbf{seq of } (A))$$
$$\textbf{\textit{where inv-fct}}() \triangleq fct \cap Sorts = \emptyset$$

– The function $r : Rank :: \textbf{map } fct \textbf{ to}(\textbf{seq of } Sorts \times Sorts)$
– The signature Σ :

$$sign \; :: \quad S \; : \; Sorts$$
$$Sig \; : \; fct$$
$$r \; : \; Rank$$

$$\textbf{\textit{where inv-sign}}() \triangleq (\textbf{Dom}(r) = sig) \wedge$$
$$rng(r) \subseteq (\textbf{seq of } Sorts \times Sorts)$$

– Set of terminal : $Terminal :: \textbf{set of } (\textbf{seq of } (A))$

– The start symbol:

$$Start \ :: \ S \ : \ Sorts$$
$$\textbf{where inv-Start()} \ \triangleq \ (\textbf{Rank}(Start) = (\epsilon, \ Start))$$

– Set of production rules associated to the signature Σ :

$$Rule \ :: \quad PG \ : \ A$$
$$PD \ : \ B$$
$$\textbf{where inv-Rule()} \ \triangleq \ \forall \ r \ \in Rule.(\textbf{mk-Rule}(a, \ b) \ a \ \in \ T_\Sigma,$$
$$b \in \ \textbf{seq of} \ (T \ \cup \ T_\Sigma)$$

– A unification grammar in VDM-SL corresponds to the type GU:

$$GU \ :: \quad \Sigma : sign$$
$$T : Terminal$$
$$Z : Start$$
$$P : Rule$$

$$\textbf{where inv-GU}() \ \triangleq \ ((T \ \neq \emptyset) \ \wedge \ (Z \ \in Sorts) \ \wedge \ (r(Z) \in S) \ \wedge$$
$$(\forall \ r \ \in P, V = \textbf{mf-P}(a, \ b)). \ a \ \in \ T_\Sigma,$$
$$b \in \ \textbf{seq of} \ (T \ \cup \ T_\Sigma))$$

6 Comparison between the proposed formal approach and the main development environments in NLP

Several environments and platforms of lingware development already exist such that EAGLES [7], ALEP [14],etc. These environments manipulate, in general, a particular family of formalisms for linguistic knowledge description (i.e., formalisms based on the unification) what limits the possibility to consider others types of knowledge.

Furthermore, some of these environments allow the integration of the data and the processing, in general using the object oriented approach, but they don't perform a formal correctness proofs. The effort is rather put on the reuse.

These environments propose, generally, their proper languages for the linguistic knowledge description and processing. Some of these environments use specific concepts (i.e., task, action, object and resolution for ALEP) which requires a special training to their users.

Contrarily to approaches used in these environments, our formal approach finds its originality in the next points :

– The use of a formal method that covers all the life cycle of the lingware to develop;
– The pivot representation for the majority of description formalisms of linguistic knowledge while keeping, for simlicity, the latter as interfaces with users;
– The integration of data-processing in a unified framework;
– The reuse of provably (w.r.t. initial informal specification) correct and complete specifications.

7 Conclusion

Besides the consolidation of formal method advantages in the development process of general software, the approach presented in this paper, offers some specific advantages to the natural language area. These specific advantages concern the linguistic level (i.e., linguistic knowledge representation) as well as the lingware life cycle level.

This approach promotes the contribution of several experts in different formalisms. Similarly, it allows separated interventions of experts in domains of natural language and software engineering without being expert in all formalisms, profiting thus from the dichotomy data-processing .

Currently, our works are concentrate especially on the formal validation of the different representations in VDM, the design of the specification library and the automatic passage to the code parting from the validated specification.

References

1. Barroca L.M. and McDermid J.A.: Formal methods: use and relevance for the development of safety-critical systems. The Computer Journal,(1992) vol.35, N.6
2. Ben Hamadou A.: Vérification et correction automatiques par analyse affixale des textes écrits en langage naturel : le cas de l'arabe non voyellé. Thése Es-Sciences en Informatique, Faculté des Sciences de Tunis, (1993)
3. Chomsky Naom.: Structures syntaxiques, (1959), Le Seuil,Paris.
4. Cliff B. Jones: Systematic software development using VDM. Prentice Hall. International (1986)
5. Darricau Myriam, Hadj Mabrouk Habib, Ganascia Jean-Gabriel: Une approche pour la réutilisation des spécifications de logiciels. Génie logiciel, (1997) N.45, pp.21-27
6. Dawes John: The VDM-SL reference guide. Pitman publishing, (1991)
7. Erbach, J. Dorre, S. Manandhar, and H. Uszkoreit: A report on the draft EAGLES encoding standard for HPSG. Actes de TALN'96, (1996), Marseille, France
8. Fuchs Cathrine: Linguistique et Traitements Automatiques des Langues. Hachette (1993)
9. Gargouri B., Jmaiel M. and Ben Hamadou A.: Intérêts des Méthodes Formelles en Génie Linguistique. TALN'98, (1998) 10-12 Juin, Paris, FRANCE
10. Gargouri B., Jmaiel M. and Ben Hamadou A.: Vers l'utilisation des méthodes formelles pour le développement de linguiciels. COLING-ACL'98, (1998) 10-14 Août 1998, Montréal,Québec,CANADA
11. Haas Andrew: A Parsing Algorithm for Unification Grammar. Computational Linguistics, (1989) Volume 15, Number 4
12. Jensen K., Heidorn G. E., Richardson S. D.: Natural Language Processing :The PLNLP Aproach. Kulwer academic publishers, (1993)
13. Miller Philip et Torris Thérèse: Formalismes syntaxiques pour le traitement automatique du langage naturel, Hermes Paris (1990)
14. Myelemans Paul: ALEP-Arriving at the next platform. ELSNews,(1994), 3(2):4-5
15. Sabah Gérard: L'intelligence artificielle et le langage : volume 1et 2, Editions Hermés, (1989)
16. Zajac Rémi: SCSL : a linguistic specification language for MT. COLING'86, (1986) Bonn, 25-29 August

A Formal Knowledge Level Process Model of Requirements Engineering

Daniela E. Herlea[1], Catholijn M. Jonker[2], Jan Treur[2], Niek J.E. Wijngaards[1,2]

[1] Software Engineering Research Network, University of Calgary,
2500 University Drive NW, Calgary, Alberta T2N 1N4, Canada
Email: danah@cpsc.ucalgary.ca

[2] Department of Artificial Intelligence, Vrije Universiteit Amsterdam,
De Boelelaan 1081a, 1081 HV, Amsterdam, The Netherlands
URL:http://www.cs.vu.nl/~{jonker,treur, niek}. Email: {jonker,treur, niek}@cs.vu.nl

Abstract. In current literature few detailed process models for Requirements Engineering are presented: usually high-level activities are distinguished, without a more precise specification of each activity. In this paper the process of Requirements Engineering has been analyzed using knowledge-level modelling techniques, resulting in a well-specified compositional process model for the Requirements Engineering task.

1 Introduction

Requirements Engineering (RE) addresses the development and validation of methods for eliciting, representing, analyzing, and confirming system requirements and with methods for transforming requirements into specifications for design and implementation. A requirements engineering process is characterised as a structured set of activities that are followed to create and maintain a systems requirements document [4], [8], [9], [10] . To obtain insight in this process, a description of the activities is needed, the inputs and outputs to/from each activity are to described, and tools are needed to support the requirements engineering process.

No standard and generally agreed requirements engineering process exists. In [8], [10] the following activities are expected to be core activities in the process:

- *Requirements elicitation*, through which the requirements are discovered by consulting the stakeholders of the system to be developed.
- *Requirements analysis* and negotiation, through which requirements are analyzed in detail for conflict, ambiguities and inconsistencies. The stakeholders agree on a set of system requirements as well.
- *Requirements validation*, through which the requirements are checked for consistency and completeness.
- *Requirements documentation*, through which the requirements are maintenained.

In [5] also the activity *modelling* is distinguished. In [9] the main activities *elicitation*, *specification* and *validation* are distinguished. Other approaches in the literature distinguish other activities, for example, *requirements determination* [12]. These activities overlap with some of the activities mentioned above.

Various knowledge modelling methods and tools have been developed [3] and applied to complex tasks and domains. The application of a knowledge modelling method to the domain of Requirements Engineering in this paper has resulted in a compositional process model of the task of Requirements Engineering. In the literature, software environments supporting Requirements Engineering are described, but no knowledge level model is specified in detail.

requirements engineering

  1   elicitation

        1.1 problem analysis

        1.2 elicitation of requirements and scenarios

        1.3 acquisition of domain ontology and knowledge

  2   manipulation of requirements and scenarios

        2.1 manipulation of requirements

               2.1.1 detection of ambiguous and non-fully supported requirements

               2.1.2 detection of inconsistent requirements

               2.1.3 reformulation of requirements

                      2.1.3.1 reformulation into informal requirements

                      2.1.3.2 reformulation into semi-formal requirements

                      2.1.3.3 reformulation into formal requirements.

               2.1.4 validation of requirements

               2.1.5 identification of clusters of requirements

        2.2 manipulation of scenarios

               2.2.1 detection of ambiguous and non-fully supported scenarios

               2.2.2 detection of inconsistent scenarios

               2.2.3 reformulation of scenarios

                      2.2.3.1 reformulation into informal scenarios

                      2.2.3.2 reformulation into semi-formal scenarios

                      2.2.3.3 reformulation into formal scenarios

               2.2.4 validation of scenarios

               2.2.5 identification of clusters of scenarios

        2.3 identification of relationships between requirements and scenarios

  3   maintenance of requirements and scenarios specification

        3.1 maintenance of requirements and scenarios documents

        3.2 maintenance of traceability links

Fig. 1. Overview of the process abstraction levels

In the approach presented in this paper requirements and scenarios are considered equally important. Requirements describe (e.g., functional and behavioural) properties of the system to be built, while scenarios describe use-cases of interactions between a user and the system; e.g., [6], [11]. Both requirements and scenarios can be expressed in varying degrees of formality: from informal, to semi-formal (structured natural language description), to formal (using temporal logic).

The compositional knowledge modelling method DESIRE (see [2] for the underlying principles, and [1] for a detailed case description) has been applied to obtain the formal process model the task of Requirements Engineering. The compositional process model constructed for the Requirements Engineering task is described in Sections 2 to 5. A discussion is presented in Section 6.

2 Process Composition within Requirements Engineering

An overview of the different processes and their abstraction levels within the process Requirements Engineering is shown in Fig. 1. For each of the processes a composition relation has been specified which defines how it is composed of the processes at the next lower level of process abstraction. Note that this specification only specifies the process abstraction relations between the processes and neither the manner in which the processes interact, nor the order in which they are performed. The latter aspects are part of the process composition specifications which are discussed in Sections 2.1 and further. A specification of a process composition relation consists of a specification of the information links (static perspective) and a task control specification (dynamic perspective) which specifies when which processes and information links are performing their jobs (see [2]).

2.1   Process Composition of Requirements Engineering

The process composition of requirements engineering is described following the different levels of process abstraction depicted in Fig. 1. The composition relation (static perspective) for the first two levels of process abstraction is shown in Fig. 2: the process requirements engineering is composed of the processes elicitation, manipulation of requirements and scenarios, and maintenance of requirements and scenarios specification.

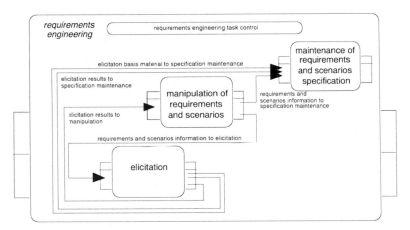

Fig. 2. Process composition of requirements engineering: information links

Within the component requirements engineering a number of information links are distinguished. The names of these information links reflect which information can be exchanged through the information link between the two processes.

The process elicitation provides initial problem descriptions, requirements and scenarios elicited from stakeholders, as well as domain ontologies and knowledge acquired in the domain. The process manipulation of requirements and scenarios manipulates requirements and scenarios to resolve ambiguous requirements and scenarios, non-supported (by stakeholders) requirements, and inconsistent requirements and scenarios. This process reformulates requirements from informal requirements and scenarios, to more structured semi-formal requirements and scenarios, and (possibly) finally to formal requirements and scenarios. It also provides relationships among and between requirements and scenarios. The process maintenance of requirements and scenarios specification maintains the documents in which the information requirements and scenarios are described, including information on traceability.

Each of the processes depicted in Fig. 2 can be characterized in terms of their interfaces (input and output information types), as shown in Table 1.

process	input information type	output information type
elicitation	• requirements and scenarios information	• elicitation results • elicitation basic material
manipulation of requirements and scenarios	• elicitation results	• requirements and scenarios information
maintenance of requirements and scenarios specification	• elicitation results • requirements and scenarios information • elicitation basic material	• elicitation results • requirements and scenarios information • elicitation basic material

Table 1. Interface information types of direct sub-processes of requirements engineering

The dynamic perspective on the composition relation specifies control over the sub-components and information links within the component requirements engineering, as depicted in Fig. 2. Task control within requirements engineering specifies a flexible type of control: during performance of each process it can be decided to suspend the process for a while to do other processes in the mean time, and resume the original process later.

2.2 Knowledge Composition of Requirements Engineering

The information types described in the interfaces of the component requirements engineering and its direct sub-components are briefly described in this section. All of these information types specify statements *about* requirements and/or scenarios. In turn a requirement is a statement that some behavioural property is required, expressed by the object-level information types in Fig. 3. To be able to express, for example, that a requirement is ambiguous, or that a scenario has been elicited, or that a requirement is a refinement of another requirement, requirements and scenarios expressed as statements on the object level, are terms at the meta-level.

The information types specified in the interfaces of the component requirements engineering and its direct sub-components all refer to the information type requirements meta-descriptions.

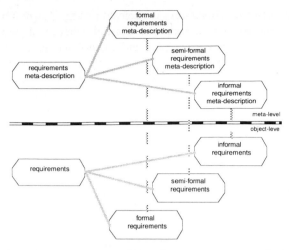

Fig. 3. Information types and meta-levels related to meta-description of requirements

The information types for scenarios are similar to the information types for requirements. The information type requirements and scenarios information is based on three information types: requirements information, scenarios information, and relations between requirements and scenarios. In turn, the information type requirements information is based on three information types: current requirements, clusters of requirements, and relations among requirements. The information type scenarios information is based on three similar information types: current scenarios, clusters of scenarios, and relations among scenarios.

3 Composition of Elicitation

The first two levels of process abstraction for elicitation are shown in Fig. 4. The processes problem analysis, acquisition of domain ontology and knowledge, and elicitation of requirements and scenarios are distinguished within the process elicitation.

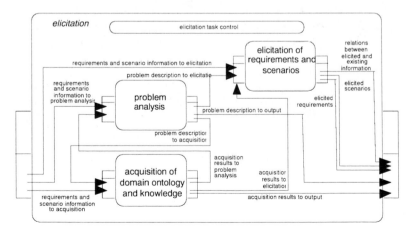

Fig. 4. Process composition relation of elicitation : information links

874

The three sub-processes of elicitation, as depicted in Fig. 4, are closely intertwined. The process problem analysis extracts the (initial) perceived problem from the stakeholders. It can also determine that requirements and scenarios are needed for another level of process abstraction. The process acquisition of domain ontology and knowledge acquires from stakeholders ontologies and knowledge of the domain, possibly related to existing requirements and scenarios. The process elicitation of requirements and scenarios elicits requirements and scenarios from stakeholders on the basis of identified problems, existing requirements and scenarios. Each of the processes depicted in Fig. 4 can be characterized in terms of their interface information types, as shown in Table 2.

process	input information type	output information type
acquisition of domain ontology and knowledge	• requirements and scenarios information • problem description	• acquisition results
problem analysis	• requirements and scenarios information • acquisition results	• problem description
elicitation of requirements and scenarios	• requirements and scenarios information • acquisition results • problem description	• elicited requirements • elicited scenarios • relations between elicited and existing information

Table 2. Input and output information types of the direct sub-processes of the process elicitation.

The dynamic perspective on the process composition within elicitation, task control, specifies flexible control, similar to the control one process abstraction higher.

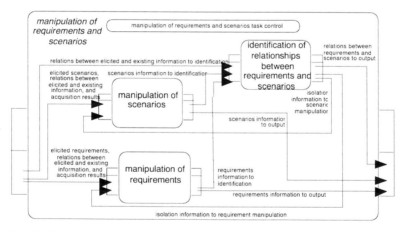

Fig. 5. Process composition of manipulation of requirements and scenarios: information links.

4 Composition of Manipulation of Requirements and Scenarios

The process composition relation between the first two levels of process abstraction for manipulation of requirements and scenarios are shown in Fig. 5. The processes manipulation of requirements, manipulation of scenarios, and identification of relationships between requirements and scenarios are distinguished within the process manipulation of requirements and scenarios.

The process manipulation of requirements is responsible for removing ambiguities, resolving non-fully supported requirements (by stakeholders), and resolving inconsistencies, while striving for progressive formalisation of requirements. This process also produces the relationships among requirements. The process manipulation of scenarios is similar to the process manipulation of requirements. The process identification of relationships between requirements and scenarios establishes which requirements are related to which scenarios, and vice versa. Each of the processes depicted in Fig. 5 can be characterized in terms of their interface information types, as shown in Table 3.

process	input information type	output information type
manipulation of requirements	• elicited requirements • relations between elicited and existing information • acquisition results • isolation information	• requirements information
manipulation of scenarios	• elicited scenarios • relations between elicited and existing information • acquisition results • isolation information	• scenarios information
identification of relationships between requirements and scenarios	• requirements information • scenarios information	• relations between requirements and scenarios • isolation information

Table 3. Interface information types of the processes within manipulation of requirements and scenarios.

Also in this case the dynamic perspective on the composition relation specifies flexible control over the sub-components of the component manipulation of requirements and scenarios.

5 Composition of Manipulation of Requirements

The first level of process abstraction within manipulation of requirements is shown in Fig. 6. The processes reformulation of requirements, validation of requirements, detection of ambiguous and non-fully supported requirements, detection of inconsistent requirements, and identification of functional clusters of requirements are distinguished within the process manipulation of requirements.

The process detection of ambiguous and non-fully supported requirements analyses the requirements for ambiguities and the extent of non-supportedness of requirements (by stakeholders). The process detection of inconsistent requirements analyses the requirements for inconsistencies among requirements. The process reformulation of requirements plays an important role within manipulation of requirements: problematic requirements are reformulated into (less problematic) requirements by adding more and more structure to requirements: from informal to semi-formal to formal. The process validation of requirements has interaction with stakeholders to establish the supportedness of a requirement in relation to a stakeholder, and whether pro and con arguments exist for a requirement. The process identification of clusters of requirements identifies clusters of requirements on the basis of clustering criteria.

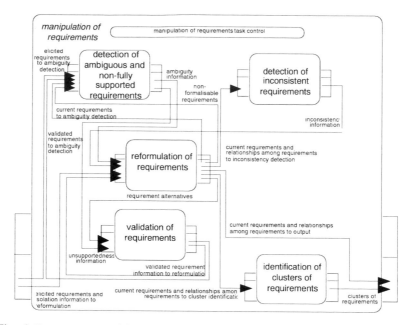

Fig. 6. Process composition of manipulation of requirements: information links.

As before, each of the processes depicted in Fig. 6 can be characterized in terms of their interface information types, as shown in Table 4. Also in this case, task control specifies flexible control. The process manipulation of scenarios has a structure similar to manipulation of requirements.

6 Discussion

The compositional knowledge modelling method DESIRE has been applied to the task of Requirements Engineering. The resulting compositional process model has been presented in some detail in this paper. The process model has been constructed on the basis of studies of available literature, and the application of requirements engineering techniques to a real-life case study: the design of a Personal Internet Assistant [7].

process	input information type	output information type
detection of ambiguous and non-fully supported requirements	• elicited requirements • current requirements • validated requirements information • non-formalisable requirements	• ambiguity information • unsupportedness information
detection of inconsistent requirements	• current requirements • relations among requirements	• inconsistency information
reformulation of requirements	• elicited requirements • ambiguity information • inconsistency information • validated requirements information • isolated scenarios	• current requirements • relations among requirements • requirement alternatives • non-formalisable requirements
validation of requirements	• requirement alternatives • unsupportedness information	• validated requirements information
identification of clusters of requirements	• current requirements • relations among requirements	• clusters of requirements

Table 4. Interface information types of processes within manipulation of requirements.

The processes have been described at different levels of process abstraction, with descriptions of their interfaces, a static composition relation (possibilities for information exchange) and a dynamic composition relation ('control flow'). The static composition relation does not prescribe a particular task control through the process composition. The task control is formulated in terms of conditions which trigger particular activities. The task control specification reflects the amount of flexibility and iterative nature of sub-processes of the requirements engineering process.

The compositional process model presented in this paper has been formally specified and provides more details and structure for the requirements engineering process than process models described in the literature on requirements engineering. For example, in [8], [10] the following activities are considered core activities in the requirements engineering process: 'requirements elicitation', 'requirements analysis and negotiation', 'requirements documentation', and 'requirements validation'. The first three of these core activities form the top level composition of the process model introduced in this paper. In contrast to the references mentioned, in the model introduced here a detailed specialisation of these three main processes is added. In the process model introduced the fourth main activity, 'requirements validation' is considered an integrated part of the manipulation processes both for requirements and scenarios, and is modelled within these processes: detection of inconsistent requirements, detection of inconsistent scenarios, validation of requirements, validation of scenarios.

To further investigate the applicability of this compositional process model, additional requirements engineering experiments will be conducted. The formally specified compositional process model for the task of requirements engineering can be employed in the design of automated tools for requirements engineering (e.g., [5]), supporting the activities of (human) requirement engineers on the basis of an agreed shared model of the requirements engineering task.

References

1. Brazier F M T, Dunin-Keplicz B M, Jennings N R, and Treur J., Formal Specification of Multi-Agent Systems: a Real World Case In: Lesser V (ed.) *Proceedings First International Conference on Multi-Agent Systems ICMAS'95* (1995) pp. 25-32 MIT Press. Extended version in: Huhns M and Singh M (eds.) *International Journal of Cooperative Information Systems IJCIS* Vol. 6 No 1 (1997) pp. 67-94.

2. Brazier, F.M.T., Jonker, C.M., and Treur, J., Principles of Compositional Multiagent System Development. In: J. Cuena (ed.), *Proceedings of the 15th IFIP World Computer Congress, WCC'98, Conference on Information Technology and Knowledge Systems, IT&KNOWS'98*, 1998, pp. 347-360.

3. Brazier, F.M.T., and Wijngaards, N.J.E., An instrument for a purpose driven comparison of modelling frameworks. In: Plaza, E., and Benjamins, R. (eds.). *Proceedings of the 10th European Workshop on Knowledge Acquisition, Modelling, and Management (EKAW'97).* Sant Feliu de Guixols, Catalania, Lecture Notes in Artificial Intelligence, vol. 1319, Springer Verlag, 1997, pp. 323-328.

4. Davis, A. M., Software requirements: Objects, Functions, and States, Prentice Hall, New Jersey, 1993.

5. Dubois, E., Du Bois, P., and Zeippen, J.M., A Formal Requirements Engineering Method for Real-Time, Concvurrent, and Distributed Systems. In: *Proc. of the Real-Time Systems Conference*, RTS'95, 1995.

6. Erdmann, M. and Studer, R., Use-Cases and Scenarios for Developing Knowledge-based Systems. In: *Proc. of the 15th IFIP World Computer Congress, WCC'98, Conference on Information Technologies and Knowledge Systems, IT&KNOWS'98* (J. Cuena, ed.), 1998, pp. 259-272.

7. Herlea, D., Jonker, C.M., Treur, J. and Wijngaards, N.J.E., *A Case Study in Requirements Engineering: a Personal Internet Agent.* Technical Report, Vrije Universiteit Amsterdam, Department of Artificial Intelligence, 1999. URL: http://www.cs.vu.nl/~treur/pareqdoc.html

8. Kontonya, G., and Sommerville, I., *Requirements Engineering: Processes and Techniques.* John Wiley and Sons, New York, 1998.

9. Loucipoulos, P. and Karakostas, V., *System Requirements Engineering.* McGraw-Hill, London, 1995.

10. Sommerville, I., and Sawyer P., *Requirements Engineering: a good practice guide.* John Wiley & Sons, Chicester, England, 1997.

11. Weidenhaupt, K., Pohl, M., Jarke, M. and Haumer, P., Scenarios in system development: current practice, in *IEEE Software*, pp. 34-45, March/April, 1998.

12. Yadav, S. et al., Comparison of analysis techniques for information requirements determination. Communications of the ACM, vol. 27 (2), 1988.

Pieces of Mind: Component-Based Software Development of Artificial Intelligence

Jovan Cakic[1], Vladan Devedzic[2]

[1]Visegradska 6/2, 11000 Belgrade, Yugoslavia
cakic@EUnet.yu
[2]Department of Information Systems, FON
School of Business Administration, University of Belgrade
POB 52, Jove Ilica 154, 11000 Belgrade, Yugoslavia
devedzic@galeb.etf.bg.ac.yu

Abstract. New phase in Object-Orientated methodology evolution, known as component-based development, has finally opened process of industrial software production in past few years. Potential benefits for Artificial Intelligence are still waiting to be discovered, but it is clear that component-based development could bring AI closer to the software industry, by bringing easy and ready-to-use solutions. This paper will try to point in potential benefits that component-based development could introduce into the world of AI.

Introduction

If you would like to include certain AI feature into a commercial, most likely C++, application, you would face the fact that development of such functionality is extremely painful and long lasting task. If knowledge and inference were not suppose to be an essential part of your application functionality, you would be probable give-up. An idea of interface enhancement, which would give your application certain intelligent behavior is difficult to achieve because there is no appropriate tool on the market letting you both:
- develop such functionality, and
- integrate it into your software

In past years, AI/ES have been developed as stand-alone systems, but the software industry puts emphasis on integration with new and existing applications, usually implemented in C++. AI has made a significant advancement, but "the rapid growth path of AI/ES was truncated because many current expert systems require special software and hardware"[David Hue]. The bridge between theory of AI and popular programming languages has to be made for thew benefits of both AI and software industry.

AI captured into the C++ class structure can bring to commercial programming a new way of embedding intelligence into commercial software and a chance for AI to come out of laboratories throughout the big door. Component-based development could do even more.

Benefits of C++ object-oriented development are well noun, and this paper will not try to make a new analysis of this problem. It is a basic assumption that the need for C++ object-oriented development in the domain of AI exists. The question is: What next?

Strategic question: Stay with class-based development or go for software components?

In the reality of software development today, established by market, such as:
- rapid changes of development methodology and implementation technology
- imperative on development productivity and software efficiency

there is no time and place for halfway solutions. On the other hand, on the user side, there is a similar frustration with:
- massive offer and rapid changes
- need for efficiency and lack of time to learn how-to-use

In short, developer needs quick and flexible solutions easy to implement and user needs software powerful and still easy to use. In that kind of game, AI must embrace the need for change through acceptance of:
- modern object-oriented programming languages and development tools
- evolution of object-oriented methodology

Evolution of object-orientation over the past decade

Object-oriented software development achieved widespread commercial acceptance in the late 1980s. The hallmark of object-orientation at that time was the use of classes, which allowed developers to model state and behaviour as a single unit of abstraction. This bundling of state and behaviour helped enforce modularity through the use of encapsulation. In classic object-orientation, objects belonged to classes and clients manipulated objects via class-based references. One characteristic that dominated the first wave of object-oriented development was the use of implementation inheritance.

To use implementation inheritance properly, a nontrivial amount of internal knowledge is required to maintain the integrity of the underlying base class. This amount of detailed knowledge exceeds the level needed to simply be a client of the base class. For this reason, implementation inheritance is often viewed as white-box reuse.

New approach to object-orientation is to only inherit type signatures, not implementation code. This is the fundamental principle behind interface-based (or component-based) development, which can be viewed as the second wave of object-orientation. Interface-based programming is a refinement of classical object-orientation, built on the principle of separation of interface from implementation. In interface-based development, interfaces and implementations are two distinct concepts. Interfaces model the abstract requests that can be made of an object ("what"). Implementations model concrete instantiable types that can support one or more interfaces ("how"). An interesting aspect of second-wave systems is that the implementation is viewed as a black box—that is, all implementation details are considered opaque to the clients of an object. [David Chappell]

Stay with class-based development or go for software components? This paper will try to find an answer to that question. However, the goal of this paper is not to make a detailed comparative analysis of class-based and component-based development, but only to underline few, for AI most interesting, characteristics of both approaches.

Class-based development

Language Dependence. All class libraries are language dependent.

Platform Dependence. To accomplish platform independence with class-based development it is necessary to publish class library in a form of source code. Generally, this is not a very good strategic move.

White-Box Reuse. It is necessary to expose a nontrivial amount of internal knowledge, though the source code or documentation, in order to allow developer to use implementation inheritance properly in the design process.

Half-product. Class library is a programming language extension. This is only a potential functionality waiting to be discovered and used. Developer needs to understand basic domain concepts (AI in our case), read the documentation (must be good) and try to get as much as possible during the implementation stage. Implementation results depend on developer's programming skills and quality of library code and associated documentation and Help.

Quality Standards. It is very hard to establish quality standard for the written code.

Source Code Exposure. A class library can be available it two forms:

* *Source code.* This solution offers a possibility for platform and compiler independence, code customization, redesign and debugging. However, selling the code is strategically wrong decision.
* *Precompiled module* (DLL file). This is platform and compiler dependent development. Constant maintenance and upgrading is necessary, as well as quality Help and documentation.

As a conclusion, class libraries are only the first but very important step in the in-house software developing process. The next step should be component-based development of the final product.

Component-based development

Language Independence. Language independence is built-in feature of component-based development model. The programming language used for implementation is unimportant for component functionality and collaboration with other programs and components.

Platform Independence (portability). Platform independence is supposed to be built-in feature of component-based development model. This is not entirely true for all component models, but it is excepted to be completely achieved in the near future.

Black-Box Reuse. All implementation details are considered opaque to the clients of an object.

Final Product. The software components are final products, with completely developed functionality.

Software Component Quality and Testing Standards. The reliability of the application as a whole is separated from reliability of its components. When is constructed from reliable components, the reliability of the application is evidently enhanced.

Additional benefits come from:

Development Specialization. Software developers will became specialist component engineers for a particular domain, in our case AI. For the users of software components this means easy and fast integration of AI into commercial applications, without need for profound understanding of this specific domain.

Decoupling development of separate components. Component-based development model strongly supports software modularity and independent component development.

Built-in Distributed architecture (local/remote transparency). Responsibilities for building distributed architecture are taken away from application developers as built-in feature of component-based development model. Transparency of the network communication allows to the client of an object to be completely unaware of their geographic location.

Transformation and Maintainability (Plug-and-Play concept). Easy assembling, reconfiguration and swapping components form available component libraries offers ease of build, redesign, upgrade and repair.

Ongoing Research Framework

The idea of software components with AI is not a new one. The market already offers a several number of ActiveX controls in the domain of AI (fuzzy logic, genetic programming, expert systems...). The objective of our research is to make a systematic approach to the construction of a software component library, which will cover a wide area of intelligent systems.

Research foundation – OBOA class library

The ongoing research is suppose to be next step in the developing process of an object-oriented model of intelligent systems, called OBOA (OBject-Oriented Abstraction) [7]. The model represents a unified abstraction of different concepts, tools and techniques used for representing and making knowledge in intelligent systems.

The OBOA class library has been developed in five levels of abstraction, using object-oriented methodology and model of orthogonal software architecture. The ultimate practical goal of developing the OBOA model has been to use it as the basis for building a set of foundation class libraries for development of intelligent systems and their blocks and components.

Table 1. Levels of abstraction in OBOA class library

Level	Objective	Domain-dependence	Semantics
1	Integration	Domain – dependent	Multiple system integration and co-ordination.
2	System	Domain – dependent	Complex concepts, tasks, problems and tools.
3	Blocks	Domain – independent	Representational units and functional blocks (knowledge and data structures and steps and activities to solve common classes of problems.
4	Units	Domain – independent	Knowledge representation techniques and reasoning mechanisms.
5	Primitives	Domain – independent	Basic data structures and operations. Domain – independent level

Research Framework

The research, which will include design and implementation of several software components with AI, will be performed in the following adopted framework:
- The OBOA class library
- Component-based development in Component Object Model from Microsoft
- UML and RationalRose as CASE toolMicrosoft Visual C++ and Active Template Library as the implementation tool

The choice of the Component Object Model and ActiveX technology from Microsoft could be put in question by some authorities in the field, but the author opinion is that the chosen model offers, at the moment, the best:
- development tools and documentation support
- performance, flexibility and ease of utilization
- commercial perspective
- comparing to the alternative solutions.

Pieces of mind: A vision for the future

The level of integration of AI into commercial software through usage of software components will generally depend on requirements, but also on feasibility. Two possible scenarios are to be distinct:

- complete integration into software during the development stage, as a task for the future
- integration into finished software

AI components integration during the development stage

Complete and proper integration of AI into commercial software can be done only as an integral part of overall application development process. By proper infrastructure implementation and use of software components, the ultimate aim to accomplish is adaptable level of AI integration during the entire life cycle of the application.

In classic, "non-intelligent" applications, user must, on his own:
- make problem (goal) formulation
- set a resolution strategy, by taking in consideration application potentials, his domain knowledge and some given requirements
- solve the problem using the application

Instead of previous scenario, AI integration into commercial software should provide:
- expert assistance during the problem formulation
- autonomous choosing of resolution strategy by taking into account current application potentials (in fact, self-potentials) and available domain knowledge
- resolution strategy execution and refining, error and performance control, results analysis and visualization

At the final stage, AI package will completely take over application control from user, and the application will become truly intelligent. The role of the user will be to set the goals, and the application will do the rest. To accomplish such role, application should become:

Aware of self-potentials. Because of massive offer and constant change, it is the fact that users of today don't have the time neither the patience to explore all functional possibilities offered by commercial applications. Therefore, users often absorb only the basic application's functionality. To ease and optimize the usage, certain software developers are introducing the automation for several predefined scenarios of software utilization (for example Wizard concept introduced by Microsoft). This automation is not only macro-like sequence of commands – a certain level of domain knowledge is also applied. The reason for this automation is a reasonable assumption that particular goals will be attained better if the application takes over the control from the user and engages itself in an optimal way. The next step in the right direction will be a potential for general and adaptable automation, based on user-defined goals and application functional capabilities. If we are developing application to be open for constant evolution, this application must be aware of its current functionality.

885

Open to constant functional evolution. This is a new approach comparing to the traditional application incremental, release-by-release, improvement process. The application should have an infrastructure, which will allow plug-and-play usage of software components in the way similar to the hardware components (approach well known not only in the world of computers). That is an adaptable architecture, built:

- from the choice of software components on the market
- independently from original application developer
- according to the user's needs

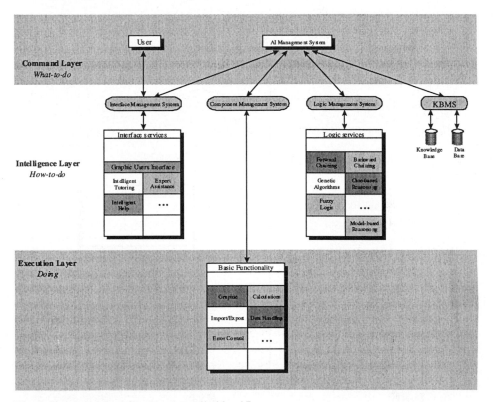

Fig. 1. Commercial application controlled by AI

To realize such an idea in the software world, it is necessary to adopt fundamental industrial approach – product build of components. This approach will lead to the software wide open to the AI integration. A concept of a possible architecture is shown in Fig. 1.

For such an application it is expected three levels of functionality:

Command layer. This should be the top layer in the application functional hierarchy. The primal task should be to take over the application control from user. This should increase application productivity and improve the quality of obtained solutions.

Intelligence layer. Components on this level should take care of:

- software controls integration and management
- execution of given tasks with optimal software resources engagement

Execution layer. This is the level of basic application functionality, as we know from software of today.

Geographic location of software components is completely insignificant, which is one of the key features of component-based development.

In the first development stage of proposed architecture, components from lower part of Intelligence layer will be implemented. This include components for:

- Logic services
- Interface services

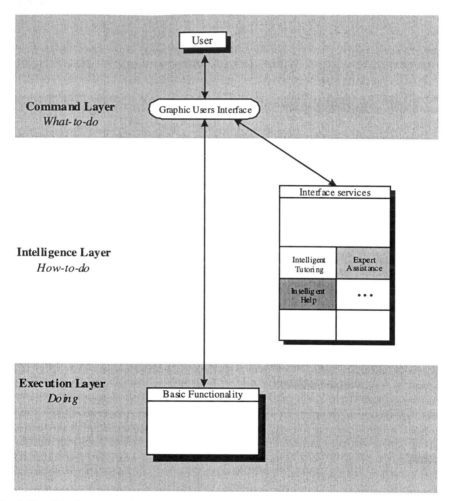

Fig. 2. Possible AI support for existing commercial application

The next development stage is implementation of components for integration, management and control from upper part of Intelligence layer. This include components for:

- Interface Management System
- Component Management System
- Logic Management System
- Knowledge Base Management System (KBMS)

The final stage is implementations of Command Layer components – AI Management System as a top level commands and control mechanism.

AI components integration into finished software

The AI integration into existing software can be accomplished via application Graphic Users Interface. A possible case of this partial integration is shown in Fig. 2.

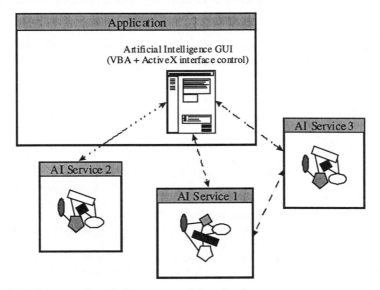

Fig. 3. Possible AI support for existing commercial application

Concrete example of an ActiveX control integration into an application which supports Visual Basic for Applications (all MS Office applications, AutoCAD, CorelDraw in the near future, ...) is shown in Fig. 3.

Interface Services set of controls is implemented as ActiveX Automation Servers and they offer different types of AI functionality. ActiveX control inserted into the application GUI represents predefined interface for that AI support. Also is possible to make inserted ActiveX control invisible and let the user create his own interface for AI support. In that case, ActiveX control will play the role of API to the available AI services, provided by Interface Services set of ActiveX controls.

This is the way to implement different AI features, like:

- Intelligent catalogs
- Intelligent Help system
- all kinds of Expert Systems from the current application domain

Conclusion

It is necessary for all researchers and developers in the domain of AI to accept the technological reality of industrial software development of today if they want to enter and stay on the software market. Actually, this includes:
- modern object-oriented programming languages and development tools
- evolution of object-oriented methodology
- need for the shortest and the fastest path from an idea to its realization

Quest for the best solutions is always an imperative for academic research programs. In the reality of software engineering best solutions are too expensive – the focus is on optimal solutions. For the moment, C++ and component-based development with ActiveX technology are solutions are the choice software industry can afford.
Software components and component-based development will probably not become a magic tool for solving all kinds of problems, but they will be certainly move AI few steps into the right direction. I hope that the ongoing research of AI component-based development will approve this approach.

References

1. Dr. Dobbs Journal and Microsoft Corporation, *The Component Object Model: Technical Overview*, Dr. Dobbs Journal (1994)
2. David Hu, *C/C++ for Expert Systems*, MIS: Press (1989)
3. Kim W. Tracy and Peter Bouthoorn, *Object-oriented Artificial Intelligence Using C++*, Computer Science Press (1996)
4. Don Box, *Essential COM*, Addison-Wesley Pub Co (1998)
5. David Chappell, How Microsoft Transaction Server Changes the COM Programming Model, Microsoft System Journal, January (1998)
6. V Devedzic, *Organization and Management of Knowledge Bases – An Object-Oriented Approach*, Proceedings of The Third World Congress on Expert Systems, vol.II, Seoul, Korea (1996) 1263-1270
7. D Radovic, V Devedzic, *Object-Oriented Knowledge Bases: Contents and Organization*, Proceedings of The IASTED International Conference on Modeling, Simulation and Optimization, CD edition, Gold Coast, Australia (1996)
8. ActiveX Programming with Visual C++, QUE (1997)
9. Special Edition Using ActiveX with Visual C++, QUE (1997)
10. Special Edition Using MFC and ATL, QUE (1997)

CNC Automatic Programming System for Wire-EDM

T. T. EL-Midany[1], A. Elkeran[1], H. T. RADWAN[1]

[1] Faculty of Eng. B.O.Box #2 Mansoura University 35516, EGYPT
Email: elkeran@mum.mans.eun.eg

Abstract. This paper introduces a developed CAM system for wire-EDM to automatically select the sequencing of the wire path considering fixation stability. The system creates technological aspects based on geometry features. Many of the wire cutting techniques such as trim cut, correction, taper angle, threading & breaking points, and the standard radius are automatically added. For small area of polygons No-Core are automatically selected. Many files such as NC file, technology data file, position file, and job file are automatically generated to control the cutting process. Also, the system simulates the cutting process for verifying check, minimizes errors, and to save machine time.

1. Introduction

Electrical discharge machining, the process normally referred to as EDM, came into industrial use shortly after World War II [1]. Its initially applications were in "tap-busting," the electrical erosion of broken taps in parts and die sections too valuable to discard. It was soon discovered, however, that the process of electrical erosion could be controlled to machine cavities and holes. After that, the wire EDM was used to execute the through cutting of EDM [2].

Wire cutting process is widely used for making stamping dies, tools, templates, extrusion dies, and progressive dies. It is also used for prototype production of parts to be made later by die stamping or CNC milling [3-5].

A computer support as an aid to part programming was not required during the early period of NC use in the wire-EDM. The parts to be machined were of two dimension configurations requiring simple mathematical calculation. With the increased use of NC systems and growth in complexity of production, the manual programming became a tedious work, time consuming. And Manual method depends to a very great extended on the experience of the operator, which leads to many human errors. And so, the part programmer was no longer able to calculate efficiently the required tool path. Therefore, the use of CAM systems as an aid to part programming became a necessity.

Many trails are achieved to develop the wire-EDM programming methods. Most current wire-EDM CAM systems neglected some of the wire-EDM technique [6-9]. For example, the sequence of the wire path, the generation of the standard radius, the fixation stability, correction (wire offset), conical cutting, trim cut, No-Core cut, calculating the threading & breaking point of the wire, etc.

The development of a wire-EDM CAM system to overcome the previous limitations, was the aim of this work. The results of the proposed system show a great saving in time and the user effort. Precision & monitoring for all steps of the cutting process are illustrated.

2. The Proposed System User Interface

When developing CAM software, one must consider how the interaction between the user and the program will take place. The portion of the software that provides this link is referred to as the user interface. User interfaces have become so important to most users that it is often primary factor in determining which software package a user ends up using. The proposed system user interface, was designed by the Microsoft Visual Basic ver. 5 to satisfy the following factors:
1. Control the flow of data through the program by making use of available input devices.
2. Detect the errors and recovery as well as the ability to cancel any operation midway through the process.
3. Providing graphical feedback to user inputs (such as blinking selected entities), and preventing unpleasant such as program "crashes."

3. The Proposed System Functions

The proposed system was designed to satisfy the following factors:
1. Sequencing the wire cutting motion in a way to minimize the machining cutting time.
2. Considering the fixation stability to avoid the drop of the work piece.
3. Analyzing the part-geometry and adding special features to the contour to improve the quality of the cutting process.
4. Optimizing the wire cutting parameters for individual part-elements according to the part-features and technology data.
5. Graphically simulating the cutting process of the spark erosion to verify the cutting sequence.
6. Generating NC part-programs directly from CAD geometry database.

4. The Proposed System Structure

The structure of the proposed programming system is illustrated in Fig. 1.

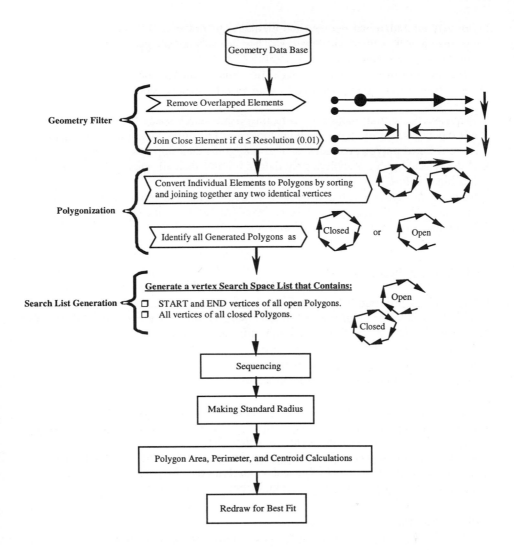

Fig. 2. Geometry Processor Flow Chart

4.1 Geometry Processor

The geometry processor Fig. 2 analyses the extracted part-geometrical elements, and decides an optimized cutting sequence. The criteria for an optimized sequence are the fixation stability for the remaining sheet and the cutting time. During this step, part geometrical elements are processed according to the following steps:

1. Filter part-geometrical elements by removing the overlapped elements and joining together any close vertexes if the distance between them less than kerf width.

2. Convert the individual geometrical elements to polygons by sorting and joining together any two identical vertexes. Then identifying all generated polygons as closed or open polygons.
3. Generate a vertex searching space list that contains start and end vertices coordinates of all open polygons as well as all vertexes coordinates of all closed polygons.
4. Sequence the cutting motion by selecting from search space list the polygon that has the closest vertex to the previous tool (wire) position, taking into consideration if the selected polygon includes another polygons, then skip this selection.
5. If the selected polygon includes another polygons, then all inner polygons must be machined before the outer polygon.
6. Remove the selected polygon from search space list and repeating the step no. 4 until the end of all nesting parts. Insert standard radius for the sharp corners.
7. Calculate polygon area, perimeter, and centroid.

4.2 Technology Processor

The technology processor Fig. 3 analyses the data stored in the wire cutting libraries, which is available for modifying from the "Wire Option" dialogue box, and decides the optimum cutting conditions. During this step, cutting conditions [10-12] are processed according to the following steps:
1. Search for the polygon area, if the polygon area ≤ 3 mm, the technology processor automatically adds the codes for No-Core cutting in the Numerical Control file (*.cnc).
2. Add trim cut and the correction in the technology and NC files.
3. Define the threading and breaking points of the wire in the NC file, and add the taper angle in the NC file and technology data file.
4. Make the multiple cuts and employing the dimensions (mm/inch) in the job file.
5. Add the taper angle in the NC file and technology data file.
6. Store all cutting files, NC file, position, technology file, and job file.

4.2.1 Position File
Position file contains the command for positioning the machine head for x-y and z-axis without wire and cutting.

4.2.2 Numerical Control File
This file contains the elements of the drawing in the ISO code format. The basic structure of this file contains lines $\Rightarrow$ G01, curves CCW$\Rightarrow$ G03, curves in CW$\Rightarrow$ G02, stop $\Rightarrow$ M00, etc.

4.2.3 Technology Data File
The cutting libraries use this file to define the technology functions.

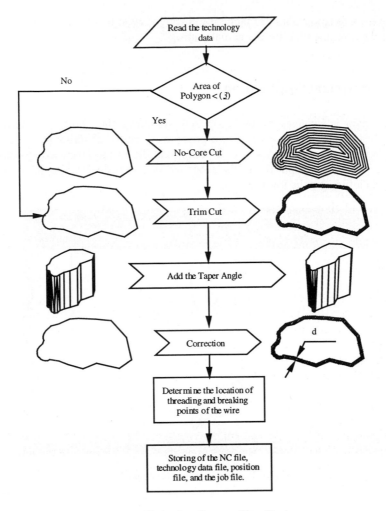

Fig. 3. Technology Processor Flow Chart

4.2.4 Job File
The job file is used to call the three files, position, NC, and technology file.

5. Simulation Check

The Computer simulation is the dynamic use of an electronic model that represents a real or proposed system. Simulation aims to make the best use of resources (machine, material, and men: called entities in simulation jargon) by helping the user to decide what is needed, and where the best position to locate it. The proposed system uses the simulation check to save hours of machining through few minutes of verifying check, and to save the machine time for the actual cutting.

6. Proposed System Application

One of the main advantages of the proposed system is the ability of machining compound dies in one path, instead of two programs (one for blanking punch and the other for piercing die). Figure 5 shows a compound die in the proposed system and the wire path. The main output file of the proposed system is the NC file, position file, technology data file, and the job file.

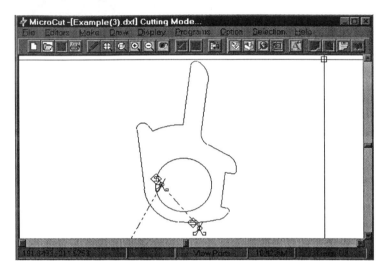

Fig. 5. Proposed system (example1)

6.1 Sample List of NC file (example1.cnc)

```
$Proposed system Ver 1.01
cod(iso) ldr(6)
lit(%)
N001 G01 X+056678 Y+104772 M22
:00002 M63
N003 D01 P01 S95 T95 G43
N004 G03 X+045895 I+022948 G44
N005 G01 X-005000 G40
N006 G03 X-045353 Y+004961 I-022948 S01 T01 G44
N007 M00
N008 G03 X-000543 Y-004961 I+022405 J-004961
N009 G01 X+005000 G40
N010 G45
N011 M62
```

```
N012 G01 X+030703 Y-037406 M22
:00013 M63
N014 D02 P02 S95 T95 G43
N015 G03 X+016257 Y+002572 I-005515 J+087525 G44
N016 G01 X-000314 Y+004990 G40
N017 G03 X+008303 Y+006480 I-003266 J+012743 S02 T02
G44
N018 G03 X-001751 Y+003432 I-002057 J+001113
N019 G01 X-004649 Y+000612
...
N043 M02
```

6.2 Sample List of Technology data file (example1.ted)

```
% I 13 12 11 00 00 00 14 (AEGDIR_3)
T00 15 05 19 002 02 02 02 002 002 02 02
S00 05 07 03 03 17 30 03 02 02 02
T01 20 30 28 002 02 05 02 002 002 02 02
S01 05 07 03 14 17 30 03 02 02 06
T02 20 30 28 002 02 05 02 002 002 02 02
S02 05 07 03 14 17 30 03 02 02 06
P01 0
P02 0
D01-228
D02+228
M02
```

6.3 Sample List of position file (example1.pof)

```
cod( POSITION File )
lit(!)
Pos 01
PaX+056678
 PaY+105348
 Pos 02
 PaX+030703
 PaY-037406
 !
$$    Position Without Wire   = 168.0207
```

6.4 Sample List of job file (example1.job)

```
Dimensions COMPLETE mm
POF C:\Program Files\Microsoft Visual
Basic\pof\Example(1).pof
TEC C:\Program Files\Microsoft Visual
Basic\ted\Example(1).ted
CNC sav RETRY FROM : 9 TIMES WITH T00 TECH SKIP STP IF
NOT THREAD STOP
```

```
CNC  C:\Program Files\Microsoft Visual
Basic\CNC\Example(1).cnc   EXEC STP EXEC WIE EXEC T01
EXEC S01 FROM 001 TO 999
FIN
```

7. Conclusion

The application affirms that the time saved through the proposed CAM system starting from the initial calculation to the generation of the NC file is about 50 % or more compared with EasyCut.

By using the proposed CAM system, sequencing the wire cutting motion in a way to minimize the machining cutting time, can be accurately and simply performed. Suitable wire cutting techniques are selected based on the geometry specifications. For example, No-Core cut was selected if the contour area was < 3 mm, as stored in the ISO code cutting library. The type of polygon (Punch/Die) is identified by an automatic method based on the fixation stability and the geometry specification. The location for the threading and the breaking point of the wire is selected by an automatic method to attain the optimum location.

Adding the Standard Radius for the sharp corners of the polygon to avoid the wire undercutting, taper angle, trim cut, and correction are achieved by an automatic method.

References

1. Herbert, Y. W., "Manufacturing Process", PP 303-323 Printice-Hall, Inc. Englewood cliffs, New Jersey, 1997.
2. Jain, Kalpak S., "Manufacturing Engineering Technology", 2nd Ed., Addison Wesley Publishing Co. Inc., 1991.
3. John, E. Fuller, "Electrical Discharge Machining", Vol. #16, Metal Handbook, 9th Ed., 1988.
4. Tomlin D. S., "Fundamentals of the Past, Present, and Future of Wire Technology for EDM", Part 2, Vol. 10, PP 8-11, 1988.
5 Houman, "What is EDM Power supplies and machines," Society of Manufacturing Engineers," 1983
6. Beltrami, I. and etal, "Simplified Post Processor for Wire EDM", Vol. #58 No. 4, PP 385-389, Journal of Materials Processing Technology, AGIE Ltd., Losone, Switzerland, April 15th, 1996
7. AGIE Training Center, "EasyCut Getting Start", Issue 3.1, Pub. by ESC Ltd., England, March, 1991.
8. Noaker, P., "EDM gets with the Program", Vol. #110, No. 6, PP 55-57, Manufacturing Engineering, June 1993.
9. Zecher E., G., " Computer Graphics for CAD/CAM systems", PP 139-146, Marcel Dekker, Inc., New York, Pasel, Hong Kong, 1991.
10.Drozda T. J, and Wick C., "Tool and Manufacturing Engineering Handbook", Vol. 1, Machining, 4th Ed., Society of manufacturing engineering, PP 14-59, 1993
11.Clauser H., "Ultrasonically Forming Intricate Multiple Graphite Electrodes, EDM Dig., Vol. 9, PP 12-15, 1987.
12 .Rhoades L. J., "Removing the EDM Recast," Society of Manufacturing Engineers, 1987

Authors Index

Springer
and the
environment

At Springer we firmly believe that an international science publisher has a special obligation to the environment, and our corporate policies consistently reflect this conviction.

We also expect our business partners – paper mills, printers, packaging manufacturers, etc. – to commit themselves to using materials and production processes that do not harm the environment. The paper in this book is made from low- or no-chlorine pulp and is acid free, in conformance with international standards for paper permanency.

Lecture Notes in Artificial Intelligence (LNAI)

Lecture Notes in Computer Science